Automotive Chassis

Brakes, Suspension, and Steering

Automotive Chassis

Brakes, Suspension, and Steering

TIM GILLES

Santa Barbara City College
Santa Barbara, California

THOMSON
DELMAR LEARNING

Australia Canada Mexico Singapore Spain United Kingdom United States

Automotive Chassis: Brakes, Suspension, and Steering
Tim Gilles

Vice President, Technology and Trades SBU:
Alar Elken

Editorial Director:
Sandy Clark

Sr. Acquisitions Editor:
David Boelio

Developmental Editor:
Matthew Thouin

Marketing Director:
David Garza

Channel Manager:
William Lawrensen

Marketing Coordinator:
Mark Pierro

Production Director:
Mary Ellen Black

Production Editor:
Barbara L. Diaz

Art/Design Specialist:
Cheri Plasse

Technology Project Manager:
Kevin Smith

Technology Project Specialist:
Linda Verde

Editorial Assistant:
Andrea Domkowski

COPYRIGHT 2005 by Thomson Delmar Learning, a division of Thomson Learning, Inc. Thomson, the Star logo, and Delmar Learning are trademarks used herein under license.

Printed in the United States of America
1 2 3 4 5 XX 06 05 04

For more information contact Thomson Delmar Learning Executive Woods
5 Maxwell Drive, PO Box 8007, Clifton Park, NY 12065-8007
Or find us on the World Wide Web at
www.delmarlearning.com

ALL RIGHTS RESERVED. No part of this work covered by the copyright hereon may be reproduced in any form or by any means—graphic, electronic, or mechanical, including photocopying, recording, taping, Web distribution, or information storage and retrieval systems-without the written permission of the publisher.

For permission to use material from the text or product, contact us by
Tel.   (800) 730-2214
Fax   (800) 730-2215
www.thomsonrights.com

Library of Congress Cataloging-in-Publication Data:
Gilles, Tim
  Automotive chassis : brakes, suspension, and steering / Tim Gilles.
   p. cm.
  Includes index.
   ISBN: 1-4018-5630-6
  1. Automobiles—Chassis.   I. Title.
  TL255.G55 2005
  629.2′4′0288—dc22
                        2004062041

NOTICE TO THE READER

Publisher does not warrant or guarantee any of the products described herein or perform any independent analysis in connection with any of the product information contained herein. Publisher does not assume, and expressly disclaims, any obligation to obtain and include information other than that provided to it by the manufacturer.

The reader is expressly warned to consider and adopt all safety precautions that might be indicated by the activities herein and to avoid all potential hazards. By following the instructions contained herein, the reader willingly assumes all risks in connection with such instructions.

The publisher makes no representation or warranties of any kind, including but not limited to, the warranties of fitness for particular purpose or merchantability, nor are any such representations implied with respect to the material set forth herein, and the publisher takes no responsibility with respect to such material. The publisher shall not be liable for any special, consequential, or exemplary damages resulting, in whole or part, from the readers' use of, or reliance upon, this material.

TABLE OF CONTENTS

PREFACE xvii
ABOUT THE AUTHOR xviii
DEDICATION xviii
SUPPLEMENTS xix
ACKNOWLEDGMENTS xx

SECTION ONE
INTRODUCTION AND SAFETY

■ CHAPTER 1
Introduction To Chassis Systems and Repairs 3

Objectives 3
Introduction 3
Brake System 3
    Drum and Disc Brakes 4
    Antilock Brakes 5
Suspension and Steering Systems 6
    Suspension System 6
    Steering System 6
    Vehicle Frames 6
    Springs 7
    Suspension Linkage 7
Shock Absorbers 8
    Additional Suspension and Steering Parts 9
Electronic Suspension Systems 9
Service Information 9
    Shop Manuals 9
    Identifying the Vehicle 10
    Manufacturer Service Manuals 10
    Generic Service Manuals 12
    Lubrication Service Manual 13
    Microfiche 14
    Computerized Service Information 14
    Parts and Labor Estimating Manuals ... 15
    Technical Service Bulletins 15
    Hotline Services 15
    Internet Service Information 15
    Hollander Interchange Manuals 15
    Trade Magazines 15

Customer Relations 15
    Telephone Service 16
Service Records 16
    Computer Records 17
Keep the Car Clean 17
    Fender Covers 17
Linen Service 18
Wholesale and Retail Distribution of Auto Parts ... 18
    Jobbers 18
    Retail Chain Stores 20
    Dealership Parts Departments 20
    Service Stations 20
Parts Pricing 20
    Ordering Replacement Parts—The Aftermarket and Original Equipment 21
Measuring 21
    Measuring a Brake Drum 21
Measuring with a Micrometer 23
    Reading a Micrometer 23
    Metric Micrometers 24
    Combination Digital Mikes 25
    Electronic Caliper 25
Checking Rotor Runout with a Dial Indicator 26
Review Questions *26*
Practice Test Questions *26*
Micrometer Practice *28*

■ CHAPTER 2
Getting Started in the Shop: Lifting a Vehicle/Shop Equipment and Safety 29

Objectives 29
Introduction 29
Hydraulic Equipment 29
    Hydraulic Jacks 30
    Vehicle Support Stands 30
Vehicle Lifts 31
    Lift Types 31
    Frame-Contact Lifts 31
    Special Auxiliary Adapters/Extenders ... 33
    Wheel-Contact Lifts 33
        Wheel-Free Jacks 34

TABLE OF CONTENTS

In-Ground Lifts..................................... 34
    Frame-Engaging Lifts 34
    Axle-Engaging Lifts......................... 34
    Semihydraulic and Fully Hydraulic Lifts 35
    In-Ground Lift Maintenance 35
    Surface Mount Lifts 36
        Surface Mount Lift Maintenance 36
Lifting a Vehicle Safely 37
    Center of Gravity 37
    Safe Lift Operation Procedures 37
Air Compressors 38
    Air Lines 38
    Air Compressor Maintenance................. 38
Shop Equipment and Safety 39
    General Personal Safety 40
    Protecting Your Eyes...................... 40
    Back Protection Safety 41
    Ear Protection............................. 41
    Clothing and Hair 41
    Shoes and Boots 42
    Hand Protection 42
Fire Safety 42
    Fire Extinguishers 42
    Flammable Materials 43
        Fuel Fires............................ 43
        Gasoline Safety........................ 44
    Electrical Fires and Battery Explosions........ 44
Breathing Safety 45
Electric Shock 45
    Electrical Safety 45
Machinery and Tool Safety........................... 46
    Hydraulic Presses 46
        Press Safety 46
    Drills..................................... 46
        Drilling Safety Precautions............. 46
    Grinder 47
        Grinder Safety Precautions............. 47
        Wire Wheel Safety Precautions 48
    Compressed Air Tool Safety Precautions 48
        Impact Wrench Safety 49
        Air Chisel Safety 49
        Die Grinder/Air Drill Safety 50
    Hand Tool Safety 50
Welding Equipment 51
General Safety Around Automobiles 52
    Before a Test Drive 52
    Working Around Belts 52
Shop Habits....................................... 52
Dealing with Toxic Materials and Chemicals........... 53
    Hazards in Dealing with Brake Dust 53
    Controlling Brake Dust..................... 53
    HEPA Vacuum........................... 53
    Brake Washer 54
    Asbestos Disposal Regulations 54
    Protecting Your Skin 54
    Dermatitis............................... 54
    Barrier Creams 55
    Material Safety Data Sheets 55
Solvents Common to the Automotive Industry ... 55
    Hazards Related to Solvents 55
    Solvent Safety 56
Review Questions 57
Practice Test Questions 57
Shop Safety Test 58

SECTION TWO
BRAKES

CHAPTER 3
Brakes Fundamentals 61

Objectives 61
Introduction 61
Brake System Operation 61
Types of Brakes 63
    Drum Brakes 63
    Disc Brakes 63
    Hybrid Brake System 63
    Power Brakes........................... 63
    Introduction to Antilock Brake Operation..... 63
        Pedal Feel 64
Weight Transfer................................... 64
    Function of Brakes in Stopping 64
    Coefficient of Friction 65
Brake Fade....................................... 66
    Types of Brake Fade 66
    Preventing Brake Fade 66
Federal Motor Vehicle Safety Standards 66
Braking Requirements............................. 67
Brake Linings 67
    Friction Material Design................... 67
    Variations in Brake Linings 67
    Bonded or Riveted Linings................. 67
Hydraulic Brake Fluid.............................. 68
    Synthetic Brake Fluid..................... 68
    Mineral Oil Brake Fluid 69
Complete Brake Job 69
    Ethics in Brake Work 70
    Chassis Systems and the Lemon Law......... 70

Review Questions 71
Practice Test Questions 71

■ CHAPTER 4
Master Cylinder and Hydraulics 73

Objectives 73
Introduction 73
Hydraulic Brake System Operation. 73
Pascal's Law. 73
Master Cylinder Operation. 74
   Low Pedal Master Cylinder Action. 75
   Tandem Master Cylinder 77
   Master Cylinder Check Valve. 78
   Master Cylinder Reservoir 78
   Master Cylinder Location. 79
Diagonally Split Hydraulic System. 79
Quick Take-Up Master Cylinder 80
   Quick Take-Up Operation. 81
Metering and Proportioning Valves 82
   Metering Valve 82
   Proportioning Valve 83
      Height-Sensing Proportioning Valves 84
   Pressure Differential Switch/Brake
   Warning Light 84
   Master Cylinder Fluid Level Switch 85
   Combination Valve 85
ABS Brake Pressure Control 86
Brake System Electrical Switches and
Warning Lights. 86
   Stoplight Switches 86
   Parking Brake Switch 86
   Brake Pad Wear Indicators 86
Brake Hose and Tubing. 86
   Brake Hose 87
      Diagnosing Brake Problems by Pinching
      the Hose 88
   Tubing. 88
      Steel Tubing 88
      Tubing Fittings 88
Flared Connections. 89
   High-Pressure Compression Fittings 90
Review Questions 91
Practice Test Questions 92

■ CHAPTER 5
Master Cylinder and Hydraulic System Service 94

Objectives 94
Introduction 94
Hydraulic System Service 94
   Check Brake Pedal Travel and Feel. 94
   Fluid Check. 95
   Fluid Leaks 95
Master Cylinder Inspection 96
   Check the Vent Port 96
   Using the Vent Port to Check for Air in
   the System 97
   Plugged Reservoir Vent. 97
   Master Cylinder Testing 97
   Master Cylinder Removal. 97
   Master Cylinder Disassembly 97
   Reservoir Installation 98
   Residual Check Valve Service 99
Bench Bleeding a Master Cylinder 99
   Bench Bleeding Using a Syringe. 100
   Bench Bleeding Quick Take-up Master
   Cylinders. 100
   Bench Bleeding with a Reverse Fluid
   Injector 100
   Reinstalling the Master Cylinder 101
   Adjusting Brake Pedal Free Travel 101
Brake Fluid Service 101
   Adding Brake Fluid 101
   Fluid Contamination 101
   Hydraulic System Flushing. 102
   Routine Brake Fluid Replacement 102
   Fluid Change Interval. 102
Brake Fluid Testing 102
   Moisture Content Testers 103
   Refractometer Testing. 103
   Voltmeter Testing 103
   Fluid Test Strips. 103
   Selecting the Correct Brake Fluid 104
Bleeding Brakes. 104
   Brake Bleeding Sequence 104
   Manual Brake Bleeding. 105
   Manual Bleeding with a Hose 106
   Pressure Brake Bleeding 106
      Pressure Bleeder Adapters. 107
   Vacuum Brake Bleeding 108
   Fluid Injection 108
   Gravity Bleeding. 109
   Bleeding Antilock Brake Systems (ABS) 109
   Brake Bleeding Problems 109
Pressure Testing. 110
   Proportioning Valve Diagnosis. 110
Brake Warning Lamp Diagnosis 111
   Hydraulic Safety Switch Service 111

TABLE OF CONTENTS

Master Cylinder Fluid Level Switch Service... 111
Stoplight Switch Service... 111
Brake Hose Service... 111
   Restricted Brake Hoses... 111
   Brake Hose Installation... 112
   Copper Washers... 112
Brake Line/Tubing Service... 112
   Cutting Tubing... 113
   Bending Tubing... 114
   Flaring the Ends of Tubing... 115
   SAE Flare... 115
   Installing Flared Tubing... 116
   ISO Flaring... 116
   Union Repairs... 117
Review Questions... *117*
Practice Test Questions... *118*

■ CHAPTER 6
Drum Brake Fundamentals and Service... 120

Objectives... 120
Introduction... 120
Drum Brake Parts... 120
   Bonded and Riveted Linings... 121
   Drum Brake Lining Terms... 121
   Anchor and Wheel Cylinder... 121
   Wheel Cylinder Parts... 121
   Wheel Cylinder Housing... 122
   Drum Brake Springs... 123
Variations in Drum Brake Design... 124
   Dual-Servo Drum Brake... 124
   Leading-Trailing Drum Brake... 125
Drum Brake Adjustment... 125
   Dual-Servo Self-Adjustment... 125
   Leading-Trailing Self-Adjustment... 127
   Driver Habits and Self-Adjusters... 128
Drum Brake Service... 128
   Disassembly and Inspection... 128
   Inspect Axle and Wheel Seals... 128
   Removing a Brake Drum... 128
   Brake Cleaning... 130
   Wheel Cylinder Inspection... 131
   Rebuild or Replace?... 131
   Brake Lining Inspection... 131
   Brake Shoe Removal... 132
   Holddown Springs... 132
   Rebuilding Wheel Cylinders... 133
   Honing Wheel Cylinders... 133
   Measuring a Wheel Cylinder after Honing... 134

Reassembling a Wheel Cylinder... 135
Removing Wheel Cylinders... 135
Inspecting and Cleaning Brake Parts... 135
   Self-Adjuster Service... 136
   Parking Brake Lever Installation... 136
   Drum Brake Hardware Kit... 136
   Checking Return Spring Condition... 136
   Results of Weak Return Springs... 137
Ordering Replacement Parts... 137
   Check Shoe Arc... 138
Brake Reassembly... 138
   Keep Brakes Clean during Reassembly... 138
   Replacing Drum Brake Shoes... 138
   Installing the Self-Adjuster... 139
   Initial Clearance Adjustment... 139
   Manual Brake Clearance Adjustment... 139
   Self-Adjuster Clearance Adjustment with Drum Installed... 140
Review Questions... *140*
Practice Test Questions... *141*

■ CHAPTER 7
Disc Brake Fundamentals and Service... 142

Objectives... 142
Introduction... 142
Disc Brake Operation... 142
   Disc Brakes Advantages... 142
   Disadvantages of Disc Brakes... 144
   Disc Brake Pads... 144
   Brake Pad Wear Indicators... 144
Fixed and Floating Disc Calipers... 144
   Floating Calipers... 144
   Floating Caliper Design Variations... 146
      Pin Slider Caliper... 146
      Center Abutment Caliper... 146
      Swing Caliper... 146
   Caliper Pistons... 147
Four-Wheel Disc Brakes... 147
Disc Brake Service... 147
   Road Test... 147
   Chassis Problems... 148
   Disc Brake Inspection... 148
   Friction Material Inspection... 148
   Rotor Inspection... 149
   Checking Rotor Runout with a Dial Indicator... 149
   Should Rotors Always Be Machined?... 150
   Inspecting the Caliper... 150

Replacing Disc Linings 150
Rear Disc Brake Cautions 152
Disc Caliper Rebuilding 153
Caliper Disassembly 153
Rebuilt Calipers 154
Cleaning and Inspecting Caliper Parts. 154
Inspecting the Pistons 154
Caliper Reassembly. 155
Installing Disc Pads in the Caliper 156
Noise Prevention. 158
Mounting Hardware and Lubricants 158
Lubricants and Shims 159
Rear Disc Pad Installation. 159
Caliper Installation. 160
Disc Brake Problem Diagnosis 160
Glazed Linings 160
Rotor Problems 160
Pedal Pulsation 161
Brake Pull 161
Uneven Wear 161
Test-Drive after a Brake Job. 161
Review Questions *162*
Practice Test Questions *162*

■ CHAPTER 8
Friction Materials, Drums, and Rotors 164

Objectives 164
Introduction 164
Drum and Disk Brake Lining Materials 164
Semimetallic Linings 165
Nonasbestos Organic Linings. 165
Fully Metallic Linings 165
Ceramic Linings 165
Carbon Fiber Linings 165
Noise Control, Performance, and Wear 166
Lining Selection 166
Evaluating Friction Materials 166
SAE Brake Tests 166
Aftermarket Friction Material Certifications . . 166
D3EA Certification 167
Brake Manufacturers Council
Certification (Beep) 167
Lining Edge Codes 167
Breaking in New Linings 168
Brake Drums 168
Brake Rotors 168
Directional Rotors. 168
Ceramic Brake Rotors 169

Composite Rotors 169
Grooved and Drilled Rotors 170
Drum and Rotor Service 170
Hard Spots 170
Drum Wear Inspection 170
Drum Wear Limits 171
Disc Rotor Inspection 171
Rotor Thickness and Runout Measurements . . 171
Cause and Effect of Rotor Runout and
Variations in Thickness 172
Machining Drums and Rotors 172
Should Drums and Rotors Always Be
Machined? 173
Drums and Rotor Lathes. 173
Multiuse Brake Lathe 173
Mounting Rotors and Drums 174
Silencing Band 175
Tool Bits 176
Check and Adjust Arbor Speed. 176
Scratch Cut 176
Precut the Inner and Outer Ridges. 177
Machining Brake Drums. 177
Position the Tool Holder 178
Turning a Drum 178
Rotor Machining. 179
Rotor Surface Finish 180
Wash the Machined Surfaces 180
Checking Lathe Accuracy. 181
On-Vehicle Rotor Machining 181
Hub and Lug Removal 182
Review Questions *183*
Practice Test Questions *184*

■ CHAPTER 9
Other Brake Systems: Parking Brake and Power Brake 185

Objectives 185
Introduction 185
Parking Brake Fundamentals 185
Parking Brake Lever 185
Parking Brake Cables and Equalizer 185
Parking Brake Warning Lamp. 187
Types of Parking Brakes 188
Drum Brake Parking Brake 188
Disc Brake Parking Brakes. 188
Auxiliary Drum Parking Brake 188
Integral Disc Brake Parking Brake 189
Screw and Nut Parking Brake Caliper. 189
Ball and Ramp Parking Brake Caliper. 190
Auxiliary Transmission Parking Brake 190

Parking Brake Service 190
   Servicing the Parking Brake Cable 191
   Parking Brake Cable Adjustment 191
   Disc Brake Caliper Self-Adjustment Reset 193
Parking Brake Warning Light Service 193
   Accidents Caused by Incorrect Parking
   Brake Adjustment 194
Power Brake Operation 194
   Types of Power Brakes 194
Vacuum Brake Boosters.................... 195
   Brake Booster Check Valve 195
   Charcoal Filter 196
   Vacuum Booster Operation 196
   Power Booster Air Filter 198
   Types of Vacuum-Suspended Brake Boosters .. 198
      Pedal Feel 198
      Reaction Disc Booster................ 198
      Plate and Lever Booster 199
      Tandem Brake Booster 199
   Power Booster Mounting Location 200
   Function of Vacuum in the Brake Booster 200
   Auxiliary Vacuum Reservoir............... 202
   Auxiliary Vacuum Pump 203
Power Brake Service 203
   Vacuum Booster Operation Test........... 203
   Vacuum Booster Leak 204
   Brake Drag 204
   Vacuum Supply Checks 204
   Checking Vacuum 205
   Leaking Power Booster Front Seal 206
   Replacing a Power Brake Booster 206
   Booster Pushrod Clearance Adjustment..... 207
Hydraulic Power Brake Boosters.............. 207
Mechanical Hydraulic Power Brakes 207
   Hydro-Boost Inspection 208
   Hydro-Boost Repair and Replacement 208
   Hydro-Boost Bleeding................... 209
Electric Hydraulic Power Brakes 209
   Powermaster Service................... 210
   Replacing a Powermaster Booster......... 211
Review Questions 211
Practice Test Questions 211

■ CHAPTER 10
Antilock Brakes, Traction, and Stability Control 213
Objectives 213
Introduction 213
Antilock Brakes 214
   Pedal Feel 214

Antilock Brake System Components 214
   Electronic Control Unit 215
   Wheel Speed Sensors 215
      Wheel Speed Sensor Installations 216
   Hydraulic Control Valve Assembly 217
Types of Antilock Brake Systems 217
   Integral ABS 218
      Pumps and Motors 218
      Reservoir 219
      Pump Control Switches 219
   Nonintegral ABS 219
   Two-Wheel ABS 219
   Four-Wheel ABS 220
Antilock Brake System Operation............. 220
   Single Channel Antilock Brake Operation.... 220
   Three- and Four-Channel ABS Operation 221
   Electromagnetic Antilock Brakes 222
   ABS System Safeguards................. 222
   Brake Performance during an ABS Stop 223
   Traction Control....................... 223
   Electronic Stability Control 224
Antilock Brake System Service 225
   Red Brake Warning Light 225
   Amber Brake Warning Light.............. 226
   Diagnosing ABS Problems 226
ABS Brake Fluid Service.................... 227
   Inspecting Brake Fluid Level 227
   Depressurizing ABS Pressure............. 227
   Flushing and Bleeding ABS............... 227
   Wheel Speed Sensor Service 228
      Testing a Sensor 228
      Metal Shavings.................... 229
      Replacing a Wheel Speed Sensor 230
      Sensor Air Gap 230
      Tone Ring Inspection 230
   Additional Precautions with ABS 230
   Integral ABS Service 231
   Rear-Wheel Antilock Brake Service 231
   Electromagnetic ABS Service 231
   Common ABS Electrical Problems 231
Review Questions 232
Practice Test Questions 232

SECTION THREE
TIRES AND WHEELS

■ CHAPTER 11
Bearings, Seals, and Greases 235
Objectives 235
Introduction 235

Plain Bearings 235
Frictionless Bearings 235
    Direction of Bearing Load 236
Ball Bearings 236
Roller Bearings 237
    Tapered Roller Bearings 238
    Needle Bearings 238
Wheel Bearings 238
    Drive Axle Bearings 239
    Full-Floating Axles 239
    Front-Wheel Drive Bearings 239
Greases 240
    Grease Performance Classification 240
    Extreme Pressure Lubricants. 240
    Chassis Lubricants 241
    Wheel Bearing Grease 241
    Universal Joint Grease 241
    Multipurpose Grease 241
    Solid Lubricant Greases 241
Wheel Bearing Seals 241
    Other Seal Designs 242
Seal Tolerance 243
Wheel Bearing Service 243
Repacking Wheel Bearings 243
    Remove the Seal 244
Bearing Inspection and Diagnosis 245
    Adding Grease to Bearings 246
    Grease the Inside of the Hub 246
    Inspect the Spindle 247
Repacking Disc Brake Wheel Bearings 248
Wheel Bearing Adjustment 248
Selecting and Installing a Cotter Pin 249
Diagnosing Wheel Bearing Noise 249
Replacing Bearing Races 250
Servicing Front-Wheel Drive Bearings 250
    Front Bearing Inspection 251
    Front Bearing Replacement 251
        Bearing Replacement—Bolt-on Type 252
        Bearing Replacement—Press-on Type 252
    Removing a Press-Fit Bearing 253
    Remove the Hub from the Bearing 253
    Remove the Bearing from the Knuckle 254
    Install the New Bearing in the Knuckle 255
    Install the Hub 255
    Removing and Replacing a Bearing Using
    Pullers 256
Reassembly of Axle Stub Shaft to Bearing 256
Review Questions *257*
Practice Test Questions *257*

■ CHAPTER 12
Tire and Wheel Fundamentals 259
Objectives 259
Introduction 259
Tire Construction 259
Tubeless Tires 260
Traction 260
Tire Tread 260
    Tread Pattern Designs 262
    Tire Tread Material 262
        Rubber 263
    Tire Cord 263
Tire Ply Design 263
Tire Sidewall Markings 265
    Tire Size 265
    Speed Rating 267
        New High Speed Ratings 270
    Load Rating 270
Gross Vehicle Weight 270
    DOT Codes 270
All-Season Tires 271
    Snow Tires 271
    Snow Chains 272
    Run-Flat Tires 272
    Compact Spare Tire 273
    Retreads 273
    Low Pressure Warning 273
    Direct and Indirect Pressure Monitors 273
Tire Quality Grading 274
    Tread Wear 275
    Traction Grade 275
    Temperature Grade 275
Changing Tire Size 275
    Tire Size Equivalents 275
    Overall Tire Diameter 276
Wheels 277
    Custom Wheels 277
    Wheel Offset 277
Lug Studs 279
    Lug Nuts 279
Tire Valve Stems 279
Review Questions *280*
Practice Test Questions *280*

■ CHAPTER 13
Tire and Wheel Service 281
Objectives 281
Introduction 281
Tire Inflation 281

TABLE OF CONTENTS

Checking Air Pressure..................281
Checking and Adjusting Tire Pressure 282
Tire Wear.........................284
Sidewall Checks 285
Tire Rotation......................286
Removing and Tightening Lug Nuts 287
Repairing Wheel Studs 290
Removing and Mounting Tires on Rims 290
    Deflate the Tire..................290
    Tire Changers 291
    Unseating the Beads 291
    Removing the Tire from the Wheel 292
Inspecting the Tire and Wheel 292
    Check the Bead Seat 292
    Valve Stem Service 293
    Rubber Lube 293
    Directional Tires 294
Install the Tire......................294
    Inflating the Tire 294
    Seating the Beads 294
    Mounting High Performance Tires........297
    Bead Roller and Tulip Clamp Tire Changer ...297
    Install the Valve Core...............298
Tire and Wheel Runout 298
    Checking Wheel Runout 299
Tire Repair........................300
    Inspecting the Tire 300
Repairing a Tire.....................300
    Preparing a Tire for Repair 301
    Plugging the Hole.................302
        Ream the Hole 302
        Install the Plug................302
        Buff the Inner Liner 303
    Patching The Tire 303
        Install the Patch 303
    Combination Plug-Patches............304
    Liquid Puncture Sealants 304
Tire and Wheel Balance 304
    Wheel Weights 305
Types of Wheel Balance 305
    Static Balance 305
    On-the-Car Balancers 307
    Couple Imbalance 307
    Dynamic Balance 307
        Computer Balancers.............308
    Centering a Wheel on the Balancer........309
        Preparation for Mounting the Wheel 309
    Mounting Hub-Centric Wheels 309
    Mounting Lug-Centric Wheels...........309
    Program the Wheel Balancer 310

    Balance the Wheel and Tire 310
    Match Mounting 311
Force Variation 311
Mounting the Wheel on the Car 312
Tire Chain Service..................312
Review Questions *313*
Practice Test Questions *314*

SECTION FOUR
SUSPENSION AND STEERING

■ CHAPTER 14
Suspension Fundamentals 317

Objectives 317
Introduction 317
Sprung and Unsprung Weight 317
Frame and Suspension Designs 318
    Body-over-Frame 318
    Unibody 318
    Crash Protection................320
    Other Frame Types 320
Springs 321
    Coil Springs 321
    Spring Rate 322
    Wheel Rate 322
        Linear-Rate Coil Springs.........323
        Variable-Rate Coil Springs 323
    Torsion Bar 323
    Leaf Spring 324
    Air Springs 325
Suspension Construction 325
    Independent and Solid Axle Suspensions 325
    Control Arms 326
    Bushings....................326
    Ball Joints 327
Suspension Types 327
    SLA Double Wishbone Suspension 327
    MacPherson Strut Suspensions..........329
        Modified Strut 329
    High-Performance Suspensions 329
        Multilink Suspensions 329
Shock Absorbers 331
    Hydraulic Shock Absorber Operation........331
        Compression and Rebound Resistance ...332
        Bump Stops and Limiters.........334
        Aeration of Fluid 334
    Gas Shocks 334
    Rear Shock Mounts...............335
Air Shocks/Leveling Devices.............335

Other Front End Parts 336
    Stabilizer Bar 336
    Spindle and Ball Joints 337
Suspension Leveling Systems 337
    Automatic Suspension Leveling 338
    Electronically Controlled Shock Absorbers . . . 339
        Magneto-Rheological Fluid Shock
        Absorber 340
    Active Suspensions 340
Review Questions 341
Practice Test Questions 342

■ CHAPTER 15
Suspension System Service 343
Objectives 343
Introduction 343
Diagnosing Suspension System Problems 343
Shock Absorber Service 343
    Testing a Shock 344
    Shock Mounts 346
    Bleeding/Purging Shocks 346
    Air Shocks 346
    Gas Shocks 346
MacPherson Strut Service 347
    Removing the Strut 347
    Replacing a Strut Cartridge 349
    Adding Oil to the Strut Body 349
    Inspecting the Upper Strut Bearing 350
    Installing the Coil Spring 350
    Reinstalling the Strut Assembly 350
Bushing Service 350
    Control Arm Bushings 350
    Strut Rod Bushing Service 352
    Stabilizer Bar Service 352
Spindle Service 352
Ball Joint Service 352
    Measuring Ball Joint Wear 353
        Wear Indicator Ball Joints 354
    Separating Ball Joint Tapers 354
    Replacing the Ball Joint 355
Coil Spring Service 356
    Adjusting Spring Height 356
    Coil Spring Replacement 356
    SLA Coil Spring Replacement 357
    Torsion Bar Removal and Replacement 359
    Leaf Spring Service 360
Wheel Alignment 360
Electronic Suspension Service 360
    Electronic-Controlled Shock Absorbers 361
Review Questions 362
Practice Test Questions 362

■ CHAPTER 16
Steering Fundamentals 364
Objectives 364
Introduction 364
Steering Gears 364
    Recirculating Ball and Nut Steering Gear 365
    Rack-and-Pinion Steering 365
Steering Linkage 366
    Parallelogram Steering Linkage 367
        Ball Sockets 367
        Tie-rods 367
        Steering Arm 368
    Rack-and-Pinion Steering Linkage 369
    Steering Damper 369
Steering Column 369
    Tilt and Collapsible Steering Columns 369
Power Steering 370
    Power Steering Pump 370
        Power Steering Pump Operation 371
        Steering Power Consumption 371
    Power Steering Hose 371
    Drive Belts 372
        Belt Material 372
        V Belts 372
        Belt Cords 372
        V-Ribbed Belts 373
        Serpentine Belt Drive 373
Types of Power Steering 374
    Electronically Controlled Variable Effort
    Power Steering 375
Four-Wheel Steering 377
    Why Four-Wheel Steering? 377
Types of Four-Wheel Steering 378
    Hydraulic Four-Wheel Steering 378
    Electric Four-Wheel Steering 378
Electronically Controlled Electric Steering 380
Review Questions 381
Practice Test Questions 381

■ CHAPTER 17
Steering Service 382
Objectives 382
Introduction 382
Fluid Level Checks 382
    Type of Fluid 382
Diagnosing Steering Problems 382

- Noise 383
- Hard Steering 383
- Tire Wear 383
- Steering Part Inspection 383
- Steering Linkage Inspection 383
  - Steering Gear Looseness 383
  - Parallelogram Inspection 383
    - Idler Arm Inspection 383
    - Pitman Arm and Center Link Inspection 383
    - Tie-Rod Inspection 384
  - Rack-and-Pinion Steering Linkage Inspection 384
- Steering Linkage Repairs 384
  - Idler Arm Replacement 385
  - Pitman Arm Replacement 385
  - Tie-Rod End Replacement 385
  - Rack-and-Pinion Tie-Rods 386
- Steering Wheel, Column, and Air Bag Service 387
  - Air Bags 387
    - Air Bag Clock Spring 388
  - Steering Wheel Service 389
  - Steering Column Service 389
    - Flexible Joint Replacement 390
- Steering Gear Service 390
  - Recirculating Ball and Nut Steering Gear Service 390
  - Manual Rack-and-Pinion Steering Service 391
    - Rack-and-Pinion Steering Looseness 391
- Power Steering System Service 391
  - Reservoir Service 391
  - Power Steering System Flushing 391
    - Power Steering Flush Methods 392
  - Bleeding the System of Air 392
  - Power Steering Pump Replacement 393
  - Repairing Power Steering Pump Oil Leaks 393
  - Pulley and Seal Removal and Replacement 393
  - Power Steering Pressure Diagnosis 394
  - Power Steering Pump Service 395
    - Flow Control Valve Service 395
    - Pump Disassembly 395
    - Pulley Shaft Seal Service 395
  - Drive Belt Service 397
    - Belt Inspection and Adjustment 397
    - V Belt Inspection 397
    - Belt Alignment 398
    - Replacing Belts 398
    - V-Ribbed Belt Replacement 399
    - Belt Tension 400
    - Tightening a Belt 401
    - V Belt Tension 401
    - V-Ribbed Belt Tension 401
  - Power Steering Hoses 401
    - Power Steering Hose Service 402
    - Refilling the Power Steering System 402
  - Power Steering Gear Service 402
    - Replacing Rack-and-Pinion Units 403
    - Variable Power Steering Service 404
    - Electronically Controlled Power Steering System Service 404
- *Review Questions* *404*
- *Practice Test Questions* *405*

CHAPTER 18
Wheel Alignment Fundamentals 406

- Objectives 406
- Introduction 406
- Toe 406
- Camber 407
- Caster 409
  - Caster and Tire Wear 410
- Steering Axis Inclination 410
  - How SAI Works 411
  - Scrub Radius 411
    - Positive and Negative Scrub Radius 411
    - Incorrect Scrub Radius 412
- Turning Radius 412
- Tracking 413
- Setback 413
- Special Handling Characteristics 413
  - Slip Angle 413
  - Understeer and Oversteer 414
- *Review Questions* *414*
- *Practice Test Questions* *415*

CHAPTER 19
Wheel Alignment Service 416

- Objectives 416
- Introduction 416
- Prealignment Inspection 416
- Tire Wear Inspection 416
  - Tire Wear from Camber 417
  - Tire Wear from Toe 417
  - Other Toe Wear Factors 417
- Ride Height Check 418
- Toe Change 418
- Torque Steer 419

Suspension and Steering Looseness 420
Test Drive 420
    Tire Checks 420
    Power Steering Checks 420
    Inspection Checklist 421
Wheel Alignment Procedures 422
Measuring Alignment 422
    Measuring Camber 423
    Measuring Caster 423
        Road Crown and Pull 423
    Adjusting Caster and Camber 424
        Plan Ahead 424
    Measuring Steering Axis Inclination 426
        Measuring Included Angle 428
    Measuring Toe 428
        Methods of Calculating Toe 429
        Toe in FWD and RWD Vehicles 429
    Adjusting Toe 429
    Centering the Steering Wheel 431
        Toe Change Check 431
    Measuring Turning Radius 433
General Wheel Alignment Rules 433
    Caster and Camber 433
    Toe 434
Four-Wheel Alignment 434
Performing a Four-Wheel Alignment 436
    Compensating the Alignment Heads 436
    Measuring Caster and Camber 436
    Adjusting Rear-Wheel Alignment 437
        Adjusting Rear Camber 437
        Adjusting Rear Toe 437
    More Wheel Alignment Rules 440
Review Questions *440*
Practice Test Questions *441*

SECTION FIVE
AXLES AND JOINTS

■ CHAPTER 20
Front-Wheel Drive CV Joint Fundamentals and Service 443

Objectives 443
Introduction 443
Front Drive Axles 444
Axle Shaft Parts 444
    CV Joints 444
        Plunge CV Joints 445
        Fixed CV Joints 445

CV Joint Construction 445
    Fixed Joints 445
    Types of Plunge Joints 446
Axle Shaft Design 446
    CV Joint Boots 447
Front-Wheel Drive Service and Repair 447
CV Joint Boot Service 447
    Boot Kits 447
    Axle Inspection and Diagnosis 447
CV Joint Diagnosis 448
    Fixed Joint Problems 448
    Plunge Joint Problems 448
Axle Shaft Removal 448
    Loosen the Axle Nut 448
    Remove the Stub Shaft from the Hub 449
    Remove the Axle 450
    Boot Removal 451
        Clips and Snaprings 451
CV Joint Replacement 452
Fixed Joint Disassembly and Inspection 453
    Install the Boot 454
CV Joint Boot Clamps 454
    Installing the Small-End Clamp 455
Servicing an Inner Tripod Joint 456
Double Offset Plunge Joints 456
Cross Groove Joint Service 456
Rebuilt Half Shafts 457
Installing the Axle 457
Review Questions *458*
Practice Test Questions *458*

■ CHAPTER 21
Rear Wheel Drive Shafts, Universal Joints, and Axles 459

Objectives 459
Introduction 459
Drive Shaft (RWD) 460
    Slip Yoke 460
    Universal Joints 461
    Two-Piece Drive Shaft 462
Drive Shaft Angle 462
Constant Velocity Joints 463
Drive Shaft Diagnosis 465
    Drive Shaft Balance 466
    Drive Shaft Inspection 466
Universal Joint Diagnosis 466
Drive Shaft Service 466
Universal Joint Disassembly 467
Universal Joint Reassembly 468

Drive Shaft Installation... 470
Two-Piece Drive Shaft Service... 470
Drive Axles and Bearings... 470
   Semi-Floating Axle Bearing Types... 470
      Bearing-Retained Axle... 470
      C-Lock Axle... 470
Independent Rear Suspension... 471
Gear Oils... 471
   API Gear Oil Ratings... 472
   Limited Slip Gear Oils... 472
   Lubricant Leaks... 473
Axle Bearing Diagnosis... 473
   Listening for Noises... 473
Axle Bearing Service... 473
   Removing a Bearing-Retained Axle... 473
   Removing a C-Lock Axle... 474
   Axle Seal Service... 474
Axle Bearing Replacement... 475
   Pressed Fit Bearing Replacement... 475
Axle Bearing Installation... 476
Reinstall the Axle... 476
Full-Floating Axle Service... 477
Differential Pinion Seal Replacement... 478
Differential Bearing Problems... 479
Review Questions... *479*
Practice Test Questions... *480*

■ CHAPTER 22
Drive Line Vibration Service... **481**

Objectives... 481
Introduction... 481

Vibration Analysis... 481
Types of Vibrations... 482
Vibration Test Instruments... 483
Vibration and Frequency Testing... 483
   Road Test... 484
   Testing in the Service Bay... 484
   Drive Shaft Runout... 484
   Repairing or Replacing a Drive Shaft... 485
Other Causes of Vibration... 485
Drive Shaft Balance... 485
   Checking Drive Shaft Balance... 486
Checking Drive Shaft Angle... 487
Review Questions... *488*
Practice Test Questions... *488*

APPENDIX A: Practice ASE Certification Exam for Brakes (A5)... 491

APPENDIX B: Practice ASE Certification Exam for Suspension and Steering (A4)... 497

APPENDIX C: Brakes NATEF Task Correlation Grid for Brakes... 504

APPENDIX D: NATEF Task Correlation Grid for Suspension and Steering... 508

APPENDIX E: DOT Tire Codes... 512

BILINGUAL GLOSSARY... 519

INDEX... 539

PREFACE

Within a very short time after beginning my first automotive chassis teaching assignment over 30 years ago, I realized there was a need for a better book for my class. I began to collect materials in preparation for writing a book. I also taught engine rebuilding and the need in that area was even more pronounced, so that's where I started my writing career—but I never forgot my chassis project. I continued to do research for my classes and collect information because I wanted to write this book. Due to the successes of repeated editions of my engines book another, very large, project came my way. That text, *Automotive Service: Inspection, Maintenance, and Repair,* is now in its second edition and has also been a successful project.

I've had offers to write other books, but like my mom and dad always said, "Do what you do well." When advising students in my college's automotive program, I always add to that, "Do what you do well—and like." To me, the chassis/undercar area is an especially fun area to teach. I'm really pleased to have the opportunity to finally put together the first of what I hope will be many editions of this book.

ABOUT THIS BOOK

Automotive Chassis provides the reader with the comprehensive knowledge needed to service, diagnose, and repair automotive brake, suspension, and steering systems. Taking a generic rather than product-specific approach, the text provides all of the need-to-know information in an easy-to-understand format. Appropriate for entry-level as well as more experienced technicians, this text provides opportunities for the learner to develop critical diagnostic and problem-solving skills.

Organization of the Text

This text follows a logical organization, from which a student can begin working in the shop right away:

- Section One offers an overview and introduction to the text, covering brakes, suspension and steering, parts and service information, and measuring. Also included is information on shop equipment, safety, and hazardous materials.
- Section Two deals with the brake system, including fundamentals, master cylinder and hydraulics, drum brakes, disc brakes, friction materials, parking and power brakes, and antilock brakes.
- Section Three discusses wheel bearings, greases and seals, and tires and wheels.
- Section Four deals with frame and suspension systems, shock absorbers, steering systems, and wheel alignment.
- Section Five includes content on front axle and CV joints and rear-wheel drive shafts, universal joints, and axles. Drive line vibration is covered in this section as well. Although much of this material is not a part of the ASE task list for the areas of suspension and steering and brakes, it is included in this text because these repairs are commonly performed by shops specializing in the undercar service area.

Features of the Text

Learning the theory, diagnosis, and repair procedures for today's complex chassis systems can be challenging. To guide learners through this material, we have included a series of features that will ease the teaching and learning processes.

Objectives. Each chapter begins with a list of objectives that state the expected learning outcomes that should result by completing a thorough study of the chapter contents.

Shop Tips. Found throughout the chapters, these tips cover those tasks commonly performed by experienced technicians.

Cautions. Cautions are urgent warnings that personal injury or property damage could occur if careful preventive steps are not taken.

Case Histories. These true stories describe automotive-related situations encountered by the author. They provide the reader with insight into the critical thinking skills necessary to diagnose automotive chassis problems.

Safety Notes. Since safety is a major concern in any automotive shop, this text contains numerous Safety Note boxes to focus the reader's attention on important safety information.

Notes. Throughout the text, Notes are included to highlight important topics.

Vintage Chassis. Boxed information on vintage brakes and steering/suspension systems (and related parts) puts today's newer technologies in historical perspective and offers the reader insights into the development of the automotive undercar.

Review Questions. At the end of each chapter there are questions of varying types (true/false, multiple choice, short answer, and sentence completion) that provide an opportunity for reinforcement and review of key concepts presented in the chapter.

About the Author

Tim Gilles has authored and coauthored several textbooks. He has been an automotive teacher since 1973 and is currently a professor in the Automotive Technology Department at Santa Barbara City College in Santa Barbara, California. He holds the industry certifications of ASE Master Engine Machinist and ASE Master Automotive Technician. He has a master of arts degree in Occupational Education from Chicago State University and a bachelor of arts degree from Long Beach State University in Long Beach, California.

Tim has been active in professional associations for many years, as president and board member of the California Automotive Teachers (CAT) and as a board member and election committee chair of the North American Council of Automotive Teachers (NACAT). He is a frequent seminar presenter at association conferences. He is a long-time member of the California Community College Chancellor's Trade and Industry Advisory Committee, is active in several industry associations, including AERA, STS, ARC, and IATN, and has served several terms as a board member of the Santa Barbara Chapter of the Automotive Service Council (ASC).

Dedication

The completion of this book was made possible with the help of a great many individuals. *Automotive Chassis* is dedicated to them and especially to my parents for their inspiration, and to my wife, Joy, and children, Jody and Terri.

SUPPLEMENTS

An instructor's CD is available and includes answers to the chapter-end review questions and ASE practice tests, PowerPoint presentations with images for every chapter of the book, a customizable computerized test bank, and electronic job sheets that may be used for shop activities.

Thomson Delmar Learning also offers a number of excellent automotive video and CD-ROM products. For additional information, please visit our Web site at www.autoed.com.

ACKNOWLEDGMENTS

The author would like to extend special thanks to Thomson Delmar Learning Developmental Editor Matt Thouin for the special effort he extended to this project and for his encouragement and extraordinary attention to detail. Special thanks also go to the Delmar production team, including Rachel Baker, Barbara Diaz, and Cheri Plasse for their exceptional effort and dedication to bringing this new text to publication. Additional thanks are due to Joy Gilles, who organized and cataloged the entire art manuscript, and to Santa Barbara City College automotive teacher Bob Stockero. Bob very carefully reviewed the entire manuscript, presented many excellent suggestions, and critiqued the changes that resulted. Others who provided valuable technical support include: Tom Birch, author and Yuba College professor emeritus; Jay Buckley of Bendix Brakes; Bob Freudenberger contributed to some of the Vintage Chassis features; Richard Meyer of Tokico shocks; and Bill VandeWater of Bridgestone/Firestone.

The author and publisher would like to offer special thanks to the following reviewers for their comments, criticisms, and suggestions on this first edition text:

Darrell Bush
Metropolitan Community College
Omaha, NE

John Eichelberger
St. Philip's College
San Antonio, TX

George Hritz
College of Marin
Kentfield, CA

Phil Krolick
Linn-Benton Community College
Albany, OR

Dan Perrin
Trident Technical College
Charleston, SC

Mario Schwarz
Santa Fe Community College
Gainesville, FL

Raymond K. Scow, Sr.
Truckee Meadows Community College
Reno, NV

Bob Stockero
Santa Barbara City College
Santa Barbara, CA

SECTION One

Introduction and Safety

■ **CHAPTER 1**
Introduction to Chassis Systems and Repair

■ **CHAPTER 2**
Getting Started in the Shop: Lifting a Vehicle/Shop Equipment and Safety

CHAPTER 1

Introduction to Chassis Systems and Repair

■ OBJECTIVES

Objectives: Upon completion of this chapter, you should be able to:
- ✔ Understand the major parts of the chassis system.
- ✔ Locate service information in books and electronic form.
- ✔ Explain the aspects of good customer relations.
- ✔ Understand wholesale parts distribution and be able to order the correct parts for a repair.
- ✔ Demonstrate ability to accurately measure with a micrometer.

■ KEY TERMS

adaptive suspension system
aftermarket
automotive parts wholesaler
booster
chassis
cost leaders
drum mike
flat-rate manual
generic manuals
jobber
list price
lubrication service manual
microfiche
net price
original equipment manufacturer (OEM)
passive suspension system
remove and replace (R&R)
repair order (RO)
runout
service record
steering system
stock
subframe
suspension linkage
suspension system
technical service bulletins (TSBs)
tires, batteries, and accessories (TBA)
variably priced items
vehicle identification number (VIN)
warehouse distributor (WD)
work order (WO)

■ INTRODUCTION

Chassis, a vehicle specialty area commonly called *undercar repair,* is the topic of this book. The vehicle chassis includes the suspension and steering systems, frame, tires, brakes, axles, and wheels. This first chapter introduces you to basic information about overall chassis design, the realities of ordering parts, repair orders, service information, and measuring. Measuring is presented here because it is a very important skill you will need in the area of chassis repair. It is also something you can learn to do while you are preparing to work on actual vehicles.

■ BRAKE SYSTEM

Brakes are the part of the chassis system responsible for stopping the moving vehicle (**Figure 1.1**). Brake systems perform "work" by using *friction* to convert the mechanical energy of motion, called *kinetic energy,* into heat energy. A rotating part—a drum or rotor—interacts with nonrotating friction materials called brake linings or pads. When you step on the brake pedal, pressurized fluid forces the friction materials against the surfaces of drums or rotors located behind the wheels. Tubing, which runs the length of the vehicle frame, is connected to rubber hoses at both front wheels and above the rear axle (**Figure 1.2**). These flexible connections from the tubing to the hydraulic brake system parts allow the wheels to move up and down on their springs without damaging the tubing.

Stepping on the brake pedal pushes a piston in a *master cylinder* that moves fluid through the brake tubing to hydraulic cylinders at each wheel (**Figure 1.3**). Hydraulic pressure pushes on pistons at the wheel, forcing the friction surfaces against the drums or rotors to stop the car.

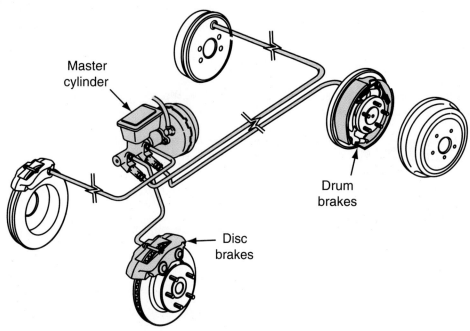

Figure 1.1 A disc and drum combination brake system.

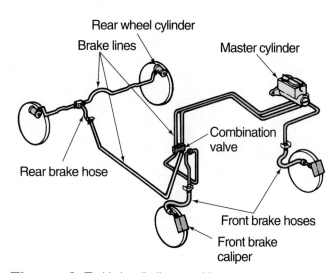

Figure 1.2 Hydraulic lines and hoses connect the master cylinder to the wheel brake units.

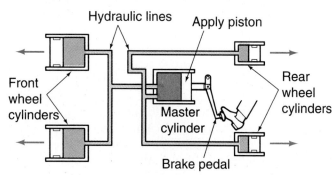

Figure 1.3 A piston in a master cylinder moves fluid through brake tubing to hydraulic cylinders at each wheel.

Drum and Disc Brakes

There are two main types of brakes, *drum* and *disc*. Disc brakes are usually found in the front, with drum brakes in the rear (**Figure 1.4**). Drum brakes provide good initial stopping ability as well as a mechanism for an effective and inexpensive mechanical parking brake. Several different drum brake designs are covered in Section Two. In the past, almost all vehicles were equipped with drum brakes. Today, drum brakes are found in some rear brake applications, with disc brakes used mostly in the front.

Disc brakes have a rotor and caliper similar to those used on bicycle brakes (**Figure 1.5**). Since the mid-1960s, many North American cars have been equipped with front disc brakes. Due to their design, a means of assisting brake pedal pressure is required with most disc brakes. The most effective and economi-

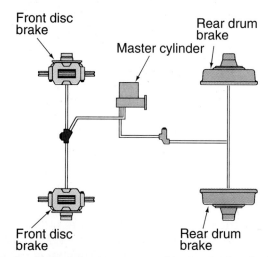

Figure 1.4 A brake system with disc brakes in the front and drum brakes in the rear.

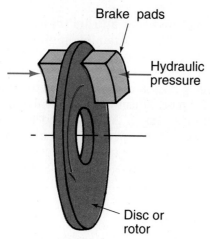

Figure 1.5 Disc brake systems have a rotor and caliper with brake pads, similar to those used on bicycle brakes.

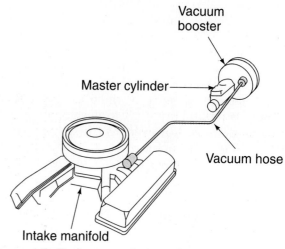

Figure 1.6 A vacuum booster is installed behind the master cylinder.

cal ways to boost brake pressure is to use a vacuum-assisted brake device called a **booster** installed behind the master cylinder (**Figure 1.6**). Vacuum is supplied by the engine.

Antilock Brakes

Most newer cars are equipped with computerized *antilock brake systems (ABS)* to prevent the wheels from locking up, especially during braking on ice, water, or snow. Electronic sensors at each wheel monitor the speed of wheel rotation. If a wheel locks up, the ABS pulses the pressure to that wheel at up to fifteen times per second. During normal operation, the ABS acts like conventional brakes; it only functions differently when a wheel locks up.

Vintage Chassis

The first cars used "spoon" brakes like those found on horse carriages. A spoon brake is a block of wood that is levered against a wheel. The next brake design to come along was the contracting brake, in which a steel strap wrapped around the outside of the wheel hub is tightened by the driver. **Figure 1.7** shows an early external band brake applied by hydraulic pressure.

In 1903, a Frenchman named Louis Renault dramatically improved stopping power with the internal expanding drum brake. This brake design has curved friction material–lined "shoes" that are pushed against the inside of a drum. Internal expanding drum brakes are still in wide use today, but a brake drum is like a cast iron cooking pot, and getting rid of heat has always been a big problem with this brake design. The development of disc brakes helped solve that problem. A disc brake uses friction elements that "pinch" a flat rotor which is exposed to outside air. One of the earliest disc brake versions appeared on the front wheels of an electric car designed by Elmer Ambrose Sperry in 1898. Then came the 1949 Crosley with its "spot" brakes (round pads). Front wheel disc brakes became popular on domestic cars in the 1960s.

Another early challenge was how to transmit the signal to stop from a driver's foot to the brake. Early

Figure 1.7 An early external band brake. *(Courtesy of Tim Gilles)*

on, the pedal replaced the hand lever, but mechanical means of applying the brakes (cams, cables, and levers) remained for some time. However, these systems were dangerous and could cause skidding, because they were impossible to equalize perfectly and required constant adjustment. The idea to use an

(continued)

Vintage Chassis (continued)

enclosed liquid to do the job was proposed back in 1897. The concept was that hydraulic systems would solve the problem because they provide equal pressure throughout the area in which the brake fluid is enclosed. It took many years to develop dependable hydraulic designs, however. The first American car with hydraulically actuated brakes was the 1921 Dusenberg, followed by the 1924 Chrysler. In 1967, a federal law required that all cars sold in the United States must have two separate hydraulic circuits.

ABS is the hot topic today. ABS pulses the brakes on hard stops or slippery surfaces, preventing the wheels from locking up and causing a skid. Patent applications were made for mechanical versions of ABS in the mid-1920s. Electronic systems were offered for a while in the early 1970s. Neither of these systems was dependable or affordable enough to be acceptable. Today, most vehicles come with ABS as standard equipment.

■ SUSPENSION AND STEERING SYSTEMS

The suspension and steering systems support the weight of the vehicle and allow it to be steered. **Figure 1.8** shows suspension and steering systems mounted on a subframe. These systems are discussed in Section Four.

Suspension System

The **suspension system** supports the vehicle, cushioning the ride while holding the tire and wheel correctly positioned in relation to the road. Suspension system parts include the springs, shock absorbers, control arms, ball joints, steering knuckle, and spindle or axle.

Steering System

The **steering system** works with the suspension system. It allows the driver to steer the car while providing a comfortable amount of steering effort. Steering system parts include the steering gear, the steering linkage, the steering wheel, and the steering column. There are two styles of steering. One design, called *parallelogram steering,* has a gear box and parallelogram linkage attached to a pair of tie-rods (**Figure 1.9**). The other, called *rack-and-pinion steering,* is a simple long rack with straight tie-rods extending from its ends (**Figure 1.10**).

Vehicle Frames

Various types of frame designs are used in cars and trucks. Cars are designed to be as lightweight as possible to improve fuel economy. Many newer cars have front-wheel drive. Almost all older, heavier cars had rear-wheel drive with long, heavy frames running the complete length of the car from the front bumper to the back bumper.

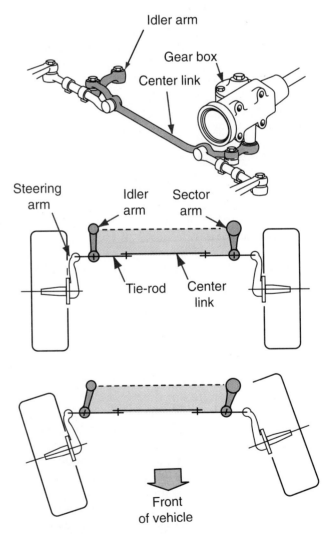

Figure 1.9 A steering gear box and parallelogram linkage.

Figure 1.8 Suspension and steering systems.

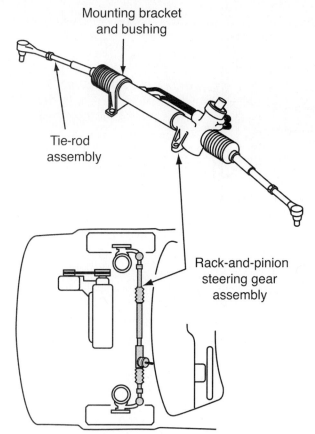

Figure 1.10 Rack-and-pinion steering with linkage.

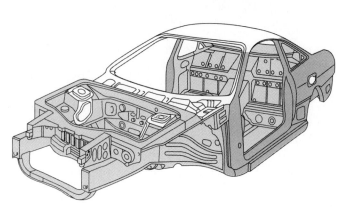

Figure 1.11 Unitized body construction.

Cars with front-wheel drive have a **subframe** in the front that includes the engine, transaxle, and steering and suspension systems (see **Figure 1.8**). These cars, as well as those with modern rear-wheel drive, rarely have a frame that runs the entire length of the car. Instead, they have a unitized body and frame with a sheet metal floor pan and small sections of frame at the front and rear (**Figure 1.11**). More information on frame designs is provided in Chapter 14.

Springs

Springs support the load of the car, absorbing the up and down motion of the wheels that would normally be transmitted directly to the frame and the car body. There are three types of springs: the coil spring, the torsion bar, and the leaf spring. The ones shown in **Figure 1.8** are coil springs.

Suspension Linkage

Suspension linkage provides a means of attaching the tires and wheels to the rest of the chassis. The suspension system allows tires and wheels to move independently of the chassis and can include a mounting point for springs. Suspension linkages are often called *control arms*. They are connected to the chassis using plastic or rubber bushings.

When a wheel moves up as the spring compresses, this is called *compression* or *jounce;* the opposite wheel movement is called *rebound*. During compression or rebound, the control arm moves up or down, twisting the rubber bushing and allowing parts to pivot without any metal-to-metal contact.

Automotive front suspensions are usually of two main designs: the *short/long arm suspension (SLA)* and the *MacPherson strut* (**Figure 1.12**). The SLA suspension uses two control arms of unequal length so that when the spring is compressed, the wheel will tilt *inward* at the top. This allows the track width of the front tires to remain constant (**Figure 1.13**). If the control arms were the same length, the tire would slide from side to side as it went over bumps.

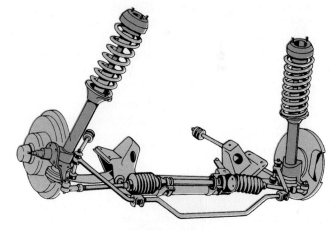

MacPherson strut

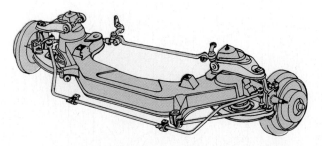

Short/long arm

Figure 1.12 Two popular suspension designs. *(Courtesy of Federal-Mogul Corporation)*

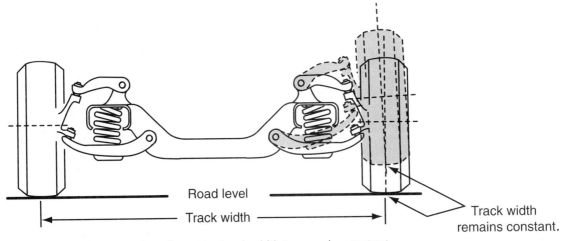

Figure 1.13 The SLA suspension allows the track width to remain constant.

Rear suspensions can be independent, but usually they have rigid axles. Some use leaf springs, but most have coil springs.

■ SHOCK ABSORBERS

A vehicle's tires absorb road shock (**Figure 1.14**), and the springs do as well. Additionally, vehicles have four shock absorbers, commonly called *shocks,* one at each wheel. Shock absorbers do not actually absorb shock; their function is to dampen spring oscillations by converting the energy from spring movement into heat energy. Also called *spring dampers,* shock absorbers are designed to resist, or *damp out,* excess and unwanted motion in the suspension. When a spring is compressed, it absorbs energy. During spring rebound, this energy is released. The first rebound is followed by several more compression–rebound cycles (**Figure 1.15**). Properly functioning shock absorbers are critical in maintaining proper tire-to-road contact.

To review, shock absorbers convert the kinetic energy from the excess motion of the spring into heat energy as they damp out the spring oscillations. Without this damping, the vehicle would bounce in an unsafe and uncomfortable manner (**Figure 1.16**).

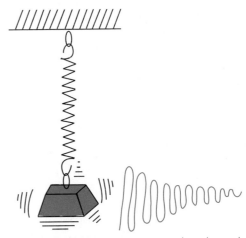

Figure 1.15 When a compressed spring rebounds, it begins to oscillate.

Many smaller cars use the MacPherson strut design. It incorporates the shock absorber into the front suspension (**Figure 1.17**), using only a single control arm on the bottom.

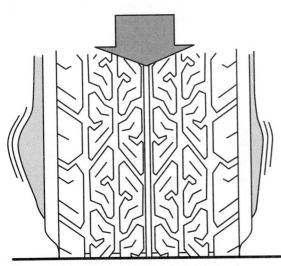

Figure 1.14 The tires absorb road shock.

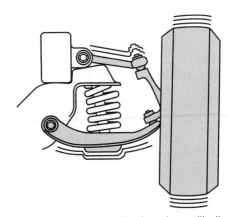

Figure 1.16 A worn shock absorber will allow the wheel to hop.

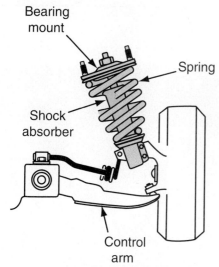

Figure 1.17 A MacPherson strut assembly.

Additional Suspension and Steering Parts

Other parts are attached to the suspension to help control a vehicle's ride. Parts like stabilizers and strut rods are insulated from front suspension parts and the frame with rubber bushings (**Figure 1.18**).

The spindle support arm, also called the *steering knuckle*, includes the axle on which the wheel bearing is mounted. Ball joints (see **Figure 1.17**) connect the control arm to the spindle. A ball joint allows motion in two directions, moving with the same up-and-down motion as the bushings on the other end of the control arm. A ball joint also allows the spindle to pivot for steering.

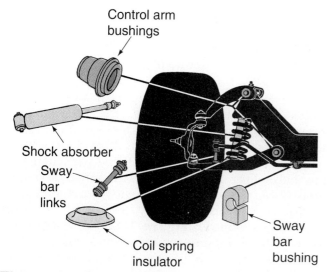

Figure 1.18 Parts are insulated from each other with rubber bushings. *(Courtesy of Federal-Mogul Corporation)*

The MacPherson strut suspension system uses only one ball joint because this system has only one control arm. A pivot bearing at the top of the strut allows it to rotate for steering.

ELECTRONIC SUSPENSION SYSTEMS

Normal suspension systems are called **passive suspension systems.** Passive systems have either a firm ride or soft ride. Their height varies according to mechanical forces on the suspension, and they do not adjust to these changes. Some of today's vehicles have automatic systems, called **adaptive suspension systems.** Electronically controlled suspension systems are used by many manufacturers of luxury vehicles. These systems keep the vehicle at the same height when weight is added to different parts of the car. Some of the advanced systems vary the damping capability of the shock absorbers as well. Most of the systems have air or conventional springs.

More information on the design and service of chassis parts is found in Section Four of this text.

SERVICE INFORMATION

In 1965, a mechanic needed to read about 5,000 pages of material in order to be very knowledgeable about automobiles. According to CARQUEST Corporation, the amount of published information needed for automobile repair was less than a quarter of a million pages in 1974. In the year 2000, it was estimated that a technician needed access to approximately 1.6 billion pages of information. Today, a technician needs to have access to even more information. In fact, the amount of information required for vehicle servicing is doubling every six months.

Repair manuals in 1965 were relatively small. Service information for only one of today's makes of vehicles usually consists of more than one manual or information source stored digitally and accessed by computer.

There are many procedures for technicians to learn in order to successfully service and repair automobiles. This part of the chapter deals with how to locate service information. Several types of service information are available, including traditional service manuals, computer libraries, and microfiche.

Shop Manuals

Shop manuals, also called repair manuals, are books that give written instructions for repairs. Illustrations provide helpful hints. This section describes the various types of shop manuals.

Before working on any make of vehicle for the first time, it is a good idea to consult a repair manual to find instructions, as well as precautions, that can help you accomplish your task successfully. Repair manuals are written for experienced technicians and make the

Vintage Chassis

Some early automobiles could reach surprisingly high speeds. From a safety standpoint the suspension, steering, and braking systems used in these vehicles were very primitive. Designs carried over from horse and buggy days. Direct linkage from a tiller to the drag link could result in a broken arm if you hit a bump (**Figure 1.19**). It was not until the 1901 Packard that a steering wheel was used on a car.

Various types of steering mechanisms were designed to reduce kickback and improve mechanical advantage. The best idea was the recirculating ball steering gear with migrating ball bearings on the steering shaft to reduce the high friction between the internal gears of the earlier gear boxes. There were early attempts at power steering, but quality systems did not appear until after World War II. The war effort resulted in much research on hydraulics for aircraft servos and tank controls. In 1951, Chrysler applied this new technology to the production of the first automotive power steering system.

The rack-and-pinion steering gear found on many vehicles today uses a round gear on the steering wheel shaft. It meshes with a long, flat gear cut into the top of a long shaft that connects between the front wheels. Rack-and-pinion steering is not a new idea. In fact, it was used on huge steam tractors in the mid-nineteenth century. The 1886 Daimler used a curved-rack version to control its center-pivoted axle.

Figure 1.19 An early Ford car that uses a tiller instead of a steering wheel. *(Courtesy of Tim Gilles)*

The planetary steering used on Ford's Model T was based on the same principle.

Rack-and-pinion steering was abandoned on early vehicles due to severe kickback and its limited mechanical advantage. It was not until the 1950s that European engineers rediscovered its potential and began to improve its design to eliminate those problems. In 1971, Ford followed the lead of its European division by producing the Pinto, the first volume domestic car to come with rack-and-pinion steering.

---

assumption that certain terms are understood and that commonly practiced procedures will be followed.

Keep in mind that it is important to keep parts organized when you are disassembling a component or system. Be sure to label parts for correct reassembly. As you progress in your automobile repair skills, you will see why it is important to be methodical. When in doubt, always check the repair manual. Even with the help of a good manual, it is very difficult to put back together something that you did not take apart yourself.

Identifying the Vehicle

Although you may know the year and model of a vehicle, sometimes you may need to know more about the specific vehicle. Each vehicle comes with a **vehicle identification number (VIN).** Passenger vehicles manufactured for sale in the United States have the VIN located on a plate on top of the dashboard, just under the windshield, on the driver's side of the vehicle (**Figure 1.20**). Since 1981, each VIN has been made up of seventeen numbers and letters. Each letter or digit stands for something. For instance, on domestic cars the engine code is the eighth character and the year of the vehicle is the tenth character (**Figure 1.21**). Check the manufacturer's service manual for the meaning of each character for that make of car.

Manufacturer Service Manuals

Manufacturers develop their own service manuals to be used by technicians in their dealerships (**Figure 1.22**). Each manual covers one year and model of vehicle. Dealer technicians have access to a full set of service manuals that list every service operation in detail. Manufacturers' manuals provide the most detailed service information for one particular year and make of car. Because of this, they are costly. Independent repair shops do not usually own manufacturers' service manuals unless they specialize in the repair of particular vehicles. **Figure 1.23** shows a table of contents from a manufacturer's service manual. One section of the car is covered in each section. For instance, brakes are covered in Section 5 and engines are covered in Section 6.

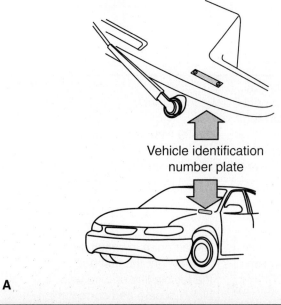

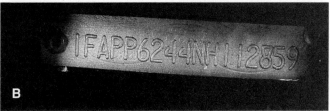

Figure 1.20 (A) The vehicle identification number (VIN) is viewed through the windshield. (B) Photo of a VIN plate. *(B, Courtesy of Tim Gilles)*

Figure 1.22 A manufacturer's service manual.

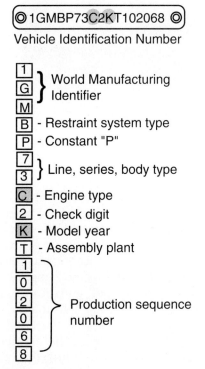

Figure 1.21 Each digit in the VIN stands for something.

TABLE OF CONTENTS	SECTION NUMBER
GENERAL INFOR. AND LUBE	
General Information	0A
Maintenance and Lubrication	0B
HEATING AND AIR CONDITIONING	
Heating and Ventilation (Non-A/C)	1A
Air Conditioning System	1B
V-5 A/C Compressor Overhaul	1D3
STEERING, SUSPENSION, TIRES, AND WHEELS	
Diagnosis	3
Wheel Alignment	3A
Power Steering Gear & Pump	3B1
Front Suspension	3C
Rear Suspension	3D
Tires and Wheels	3E
Steering Column, On-Vehicle Service	3F
Steering Column—Std., Unit Repair	3F1
Steering Column—Tilt, Unit Repair	3F2
DRIVE AXLES	
Drive Axles	4D
BRAKES	
General Information—Diagnosis and On-car Service	5
Compact Master Cylinder	5A1
Disk Brake Caliper	5B2
Drum Brake—Anchor Plate	5C2
Power Brake Booster Assembly	5D2
ENGINES	
General Information	6
2.0 Liter L-4 Engine	6A1
3.1 Liter V6 Engine	6A3
Cooling System	6B
Fuel System	6C
Engine Electrical—General	6D
Battery	6D1
Cranking System	6D2
Charging System	6D3
Ignition System	6D4
Engine Wiring	6D5
Driveability and Emissions—General	6E
Driveability and Emissions—TBI	6E2
Driveability and Emissions—PFI	6E3
Exhaust System	6F

Figure 1.23 The table of contents from a typical manufacturer's service manual.

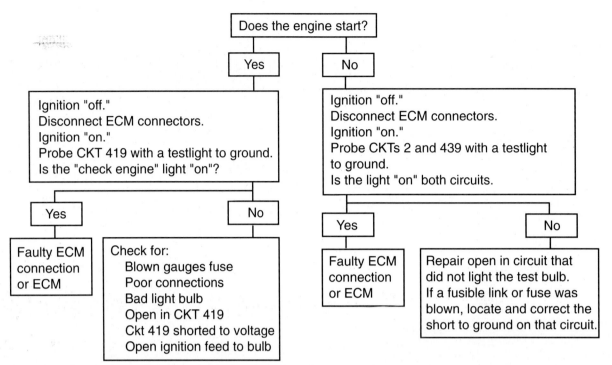

Figure 1.24 A typical step-by-step troubleshooting chart from a service manual.

Three types of information are contained in shop manuals:

1. Diagnostic or troubleshooting information (**Figure 1.24**)
2. Step-by-step repair information
3. Specification charts (**Figure 1.25**)

TORQUE SPECIFICATIONS	
Fan assembly motor-to-fan	3.3 Nm (29 lb. in.)
Fan-to-radiator support bolt	9 Nm (80 lb.in.)
Hose clamps	
Heater hose	1.7 Nm (15 lb. in.)
Radiator hose	3.4 Nm (30 lb. in.)
Lower air deflector to impact bar	25 Nm (19 lb. ft.)
Throttle body inlet pipe bolt (3.1L)	25 Nm (19 lb. ft.)
Transmission oil cooler fitting at radiator	27 Nm (20 lb. ft.)
Transmission oil cooler pipe connections	20 Nm (15 lb. ft.)
Radiator outlet pipe to block (2.0L)	27 Nm (20 lb. ft.)
Radiator-to-radiator support bolt	10 Nm (90 lb. in.)
Thermostat housing bolt	27 Nm (20 lb. ft.)
Coolant pump-to-block bolt	
2.0L	25 Nm (19 lb. ft.)
3.1L	24 Nm (18 lb. ft.)
Coolant pump-to-front cover bolt	
3.1L	10 Nm (90 lb. in.)
Coolant pump pulley-to-pump bolt	
2.0L	24 Nm (17 lb. ft.)
3.1L	21 Nm (15 lb. ft.)
Surge tank bolt	4 Nm (35 lb. in.)
Surge tank pipe to block bolt	27 Nm (20 lb. ft.)
Temperature sending or gauge unit	

Figure 1.25 A specification chart from a service manual.

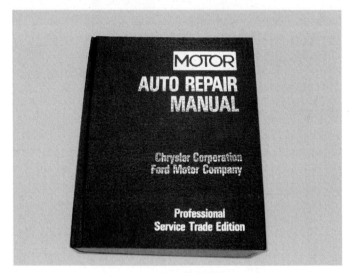

Figure 1.26 A generic service manual. *(Courtesy of Tim Gilles)*

Generic Service Manuals

Several companies produce **generic manuals** that deal with all makes of foreign or domestic cars (**Figure 1.26**). There are usually two types of generic manuals: those that deal with mechanical repairs to such things as brakes, suspension systems, and engines; and those devoted to the areas of fuel, emission, ignition, and air-conditioning systems. Separate manuals are available for light trucks.

Comprehensive generic repair manuals include sections for each make of car arranged in alphabetical order by maker. An index of cars is located at the front of the manual (**Figure 1.27**). On the first page of each

Chapter 1 Introduction to Chassis Systems and Repair 13

section is a service operations index that can be used to locate a particular repair procedure. Many manuals have separate sections for repairs of components such as carburetors, starters, alternators, transmissions, or differentials. A few pages of each section are devoted to specifications for such things as tune-ups, electrical adjustments, wheel alignment, and tightening specifications (**Figure 1.28**).

Lubrication Service Manual

The manual used frequently during lubrication/safety checks is a **lubrication service manual** (**Figure 1.29**). This manual has all of the most-used specifications. It is arranged in three basic sections:

1. Lubrication and maintenance information
2. Capacities
3. Underhood service information

The lubrication and maintenance section includes lube and maintenance instructions, lubrication diagrams and specifications, vehicle lift points, and preventative maintenance intervals. The capacities section includes cooling and air-conditioning system capacities, cooling system air bleed locations, wheel and tire specifications, and wheel lug torque specifications. The underhood service information section includes underhood service instructions, and specifications and diagrams for engine tune-up; mechanical, electrical, and fuel systems; and belt tensions.

GENERAL INFORMATION
Acura
Audi
BMW
Chrysler Mitsubishi
Daihatsu
Ford Motor Company
General Motors
Geo
Honda
Hyundai
Infinity
Isuzu
Jaguar
Lexus
Mazda
Mercedes-Benz
Nissan
Saab
Subaru
Suzuki
Toyota
Volkswagen
Volvo

Figure 1.27 In the front of a repair manual is an index of cars alphabetized by manufacturer.

Figure 1.28 Some of the manual pages are devoted to specifications for different systems. *(Courtesy of Mitchell 1)*

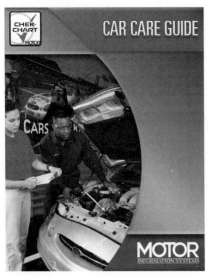

Figure 1.29 A lubrication service manual. *(Courtesy of Chek-Chart Publications)*

Microfiche

Some service manuals have been published on **microfiche,** small plastic film cards that are magnified by a microfiche reader (**Figure 1.30**). The advantage of microfiche is that it takes up very little space. Many microfiche readers have a copying feature so that the technicians can take service information to the repair area. Some parts departments also use microfiche. In recent years, most companies that used microfiche have switched to computers.

Figure 1.30 Microfiche saves space by storing information on a small card that is magnified by a microfiche reader.

Computerized Service Information

One of the most popular sources of service information is the computer information library (**Figure 1.31**). Due to the tremendous amount of information required to service new vehicles, electronic media has become commonplace in the service industry. The systems are easy to use and information is easy to access. The technician uses the computer's keyboard, a mouse, or a light pen to make selections from a menu on the screen (**Figure 1.32**). When a technician locates the information in the computer, it can be read from the screen or printed on a printer so it can be taken to the service bay or workbench. Computer information libraries also provide labor time estimates.

Figure 1.31 A computerized information library using compact disks capable of storing a vast quantity of information. *(Courtesy of Tim Gilles)*

Figure 1.32 Information selections are made from a menu on the screen. *(Courtesy of Mitchell1)*

Parts and Labor Estimating Manuals

When parts fail they are replaced with either new or rebuilt parts. When the technician replaces a part, this is known as a **remove and replace (R&R)** job. The estimated time it should take a professional technician to perform an R&R job is listed in the **flat-rate manual**, also called a *parts and time guide* (**Figure 1.33**). The flat-rate manual also lists the estimated cost of parts. An estimator or a business owner will use this type of manual to provide as accurate service estimates to customers as possible. As noted, computer information libraries also provide labor time estimates.

Technical Service Bulletins

Manufacturers and industry professional associations publish **technical service bulletins (TSBs)** for subscribers and members. Sometimes members report on experiences discovered through trial and error in these bulletins. At other times, information is contributed by manufacturers. Manufacturers make extensive use of TSBs to inform technicians of important service changes to their vehicles.

Hotline Services

Hotline services provide answers to service questions by telephone. Manufacturers provide telephone help for the technicians in their dealerships. There also are subscription services for independent repair shops to get information by phone.

Some manufacturers also have modem systems that can transmit computer information from a vehicle to another location via phone lines. After the car's diagnostic link is connected to the modem, the technician in the service bay runs a test sequence on the automobile. The system downloads the latest repair information on that particular model of car. If that information does not repair the problem, a technical specialist at the manufacturer's location will review the data and propose a repair.

Internet Service Information

Technicians make extensive use of the Internet. Tire, brake, and suspension part manufacturers have very detailed Web pages with much helpful information for technicians.

The International Automotive Technicians Network (IATN), made up of many thousands of technicians, is one example of how technicians share diagnostic tips for solving tough problems. The URL is http://www.members.iatn.net.

Hollander Interchange Manuals

Sometimes a part is not available from a dealer. Salvage yards use an interchange manual that lists parts that interchange from one vehicle to another (**Figure 1.34**). For instance, front spindles and brake calipers from one model of Chevrolet might fit an Oldsmobile of a different year. Interchange manuals are updated with new material each year. Manuals are available for both domestic and imported vehicles.

Trade Magazines

There are many professional magazines targeted to service technicians. Examples include *Motor, Motor Service, Brake and Front End, Motor Age,* and *Import Service*. In addition to articles on new developments, opinions, and procedures, these magazines usually include information from the latest TSBs. Many of the magazines include a mail-in reader service postcard. Readers can circle numbers on the card that relate to items about which they would like to receive more information.

■ CUSTOMER RELATIONS

Automotive repair shops are in business to make a profit. If a shop does not make a profit, it goes out of business. Good customer relations are the key to a

Figure 1.33 A flat-rate manual.

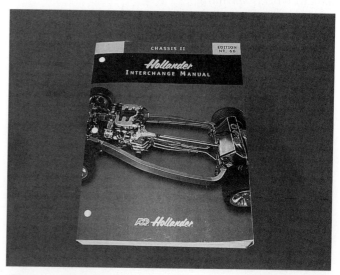

Figure 1.34 A Hollander Interchange Manual. *(Courtesy of Tim Gilles)*

successful business. It is important to make a good first impression with customers so that they are confident in the shop. In shops other than those of large dealerships, most technicians will need to communicate directly with the customer at least some of the time.

Dealerships have professional salespersons who greet customers and write up repair orders. These people are called *service writers* or *service advisors*. When a customer has a simple concern about a car's performance, an untrained service writer will not be a problem. However, one of the complaints often heard about large dealerships is that customers have to take their cars back again in order to have the original problem corrected. In fact, in several surveys, the number of complaints incorrectly diagnosed and not repaired on the initial visit is nearly one-half. Often the problem can be traced to faulty diagnosis by an undertrained service writer.

Service writing in smaller shops is usually done by the owner, who is also a technician. Since the customer gets to speak directly with a technician about a car's problem, the result is often a satisfied customer after only one trip to the shop. Part of the solution to the problem for larger shops is to hire service writers who understand the operation of the various systems of the car.

On the other hand, product service training for new model vehicles is more readily available to and is used more often by dealership personnel. This, and the ease of warranty service, often means that the dealership is the best choice for new vehicle repairs.

Telephone Service

Although the owner or manager of a business often answers the telephone, technicians in small shops usually share telephone responsibilities. Phone calls should be answered promptly and courteously, with technicians stating the name of the business and the person who is speaking.

Most often there is a phone extension in the repair area. Although it is sometimes an annoyance to stop a job in order to answer a telephone call, phone calls are an important source of business. The telephone often presents customers with their first impression of the business. When a customer's first impression is that the business is professionally managed, things will go easier from that point on.

SERVICE RECORDS

A service record is written for every car entering a repair shop. A multiple-copy, numbered **service record, repair order (RO),** or **work order (WO)** is used for legal, tax, and general recordkeeping purposes. One of the RO copies is given to the technician. It lists the necessary repairs and is used to make notations of completed repairs and items that require attention. One of the copies of the RO is for the customer. It includes the cost estimate. The remaining copy is for the shop's files.

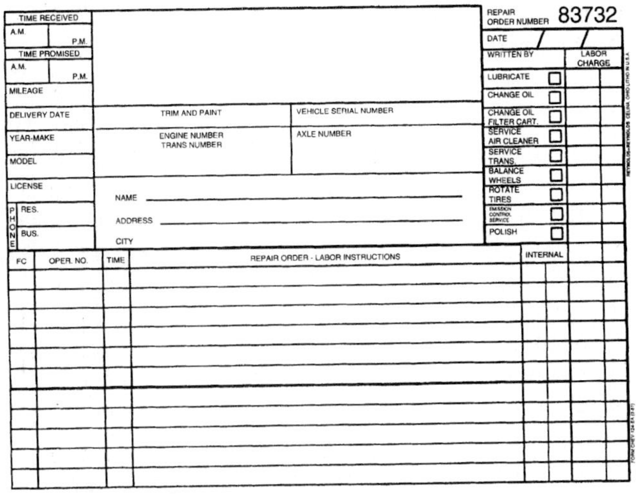
TIME RECEIVED A.M. P.M.
TIME PROMISED A.M. P.M.
MILEAGE
DELIVERY DATE
YEAR-MAKE
MODEL
LICENSE
PHONE RES.
BUS.
TRIM AND PAINT
VEHICLE SERIAL NUMBER
ENGINE NUMBER
TRANS NUMBER
AXLE NUMBER
NAME
ADDRESS
CITY
REPAIR ORDER NUMBER 83732
DATE
WRITTEN BY
LABOR CHARGE
LUBRICATE
CHANGE OIL
CHANGE OIL FILTER CART.
SERVICE AIR CLEANER
SERVICE TRANS.
BALANCE WHEELS
ROTATE TIRES
EMISSION CONTROL SERVICE
POLISH
FC
OPER. NO.
TIME
REPAIR ORDER - LABOR INSTRUCTIONS
INTERNAL

Figure 1.35 A repair order.

The repair order (**Figure 1.35**) is important for several reasons:

- It fully identifies the customer and the vehicle.
- It gives the technician an idea of why the car is in the shop for repairs.
- It specifies the shop's hourly labor rate.
- It gives the customer an estimate of the cost of the repair.
- It gives the time the vehicle will be ready for the customer.
- It contains the customer's signature, which gives approval for the repair and affirms that the customer will pay for the shop's services as stated.
- It is a legal document.

Repair orders are numbered for future reference. Also entered on the repair order are the:

- Odometer reading of the car's mileage
- Make of the vehicle (Ford)
- Model of the vehicle (Mustang)
- Model year of the vehicle (2004)
- Vehicle license number (1VAL239)

Information that will identify customers includes their:

- Name
- Mailing address
- Home phone number
- Business phone number
- Signature

Below the area that contains customer information is room for labor instructions. This area is filled in with the customer's complaint, possible cause(s), and repairs to be performed. The customer should be carefully questioned about the symptoms. Customers often have a preconceived and incorrect notion of what is wrong. Examples of questions that might be asked if the complaint is that a vehicle stalls include:

- Does it pull to the left all the time when driving?
- Does it pull to the left only during a stop?
- Does it pull only for an instant, then go straight?
- Have you checked your tire pressures?
- Do you hear any unusual noises when stopping or going over bumps?
- When was the last time you had the car worked on?
- What type of work was done?

There is a structured method of questioning that can be followed when asking questions of the customer.

- First, ask for a general description of the problem.
- Ask whether the problem happens in the front, in the back, or under the hood.
- Identify symptoms: Do you hear, feel, or smell something?

With the information listed on the RO, technical skills can be used to identify the problem.

Computer Records

Most successful repair shops use personal computers for such things as keeping records, maintaining a running inventory and ordering parts, and tracking employee productivity (**Figure 1.36**). Items stored in the computer's memory can include personal notes regarding particular customers and what occurred during their most recent visit to the shop. A computer locates a customer's records using either the vehicle's license number, the owner's name, or the owner's address. The license number is the most popular means of access.

Even when records are kept on a computer, most shops also maintain *hard copy* records. Although some shops have a computer terminal in the service bay, this is uncommon. It is important that a shop keep written records of all repairs and recommendations. Litigation (legal cases) often results in court losses due to a lack of adequate records. A complete RO includes a written estimate of the cost of the repair and a record of phone conversations with a customer when an earlier approved estimate required updating.

■ KEEP THE CAR CLEAN

Be sure that your hands, shoes, and clothing are clean before getting into a customer's car. Work shoes often have grease on their soles. A dirty steering wheel or carpet will guarantee an unhappy customer, no matter how good a repair job was done. Many shops use carpet mats made of paper to help keep a customer's carpet clean (**Figure 1.37**). Some shops also use seat covers.

Fender Covers

Fender covers are used whenever underhood work is done (**Figure 1.38**). Greasy hands or brake fluid can ruin the finish on a vehicle. Fender covers also protect against scratches from such things as belt buckles. Cars cost many thousands of dollars. When working on a car, be sure to respect it as much as the owner does.

Figure 1.37 A carpet pad protects the car's carpet from grease. *(Courtesy of Tim Gilles)*

Figure 1.36 Most shops use a personal computer for shop management activities.

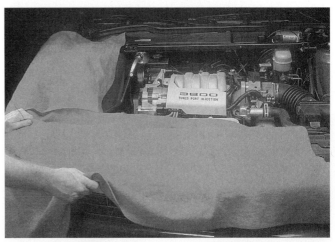

Figure 1.38 Fender covers are used when working under the hood.

Figure 1.39 Shops have linen service for uniforms and shop towels. *(Courtesy of Mission Linen supply)*

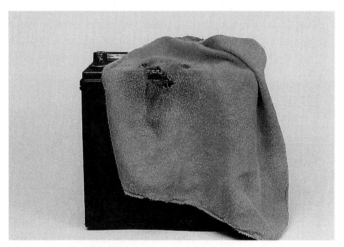

Figure 1.40 This shop towel was damaged by battery acid. *(Courtesy of Tim Gilles)*

Figure 1.41 Coveralls and a shop coat. *(Courtesy of Tim Gilles)*

■ LINEN SERVICE

Most shops use a weekly linen service for shop towels and uniforms (**Figure 1.39**). The shop is billed for the cost of any shop towels that have signs of acid on them. Shop towels are usually dyed red. When they are exposed to battery acid or other acids, they turn blue. This alerts the linen service to the probability that the rags will disintegrate in the laundry (**Figure 1.40**).

Sometimes shops own uniforms; at other times they rent from the linen service. Some shops get shirt and pant service, and others get shop coat or coverall service (**Figure 1.41**). Shop clothing is often made of materials that are resistant to battery acid.

■ WHOLESALE AND RETAIL DISTRIBUTION OF AUTO PARTS

The purpose of this section is to familiarize you with some of the terms used in the automotive service business. Approximately one in every six jobs is related to the automobile. The automotive **aftermarket** is supplied from the manufacturer to the customer through a large parts and service distribution system (**Figure 1.42**). Both the **original equipment manufacturers (OEMs)** and a large number of independent parts manufacturers sell parts to over 1,000 **warehouse distributors (WDs)** throughout the continent. Warehouse distributors are large distribution centers that sell to **automotive parts wholesalers,** known in the industry as **jobbers.**

Jobbers

Jobbers sell parts, accessories, and tools to several different markets (**Figure 1.43**). More than half of a typical jobber's customers are independent repair shops and service establishments. Fleet companies, farmers, trade and industrial accounts, and, occasionally, new car dealerships are some of the other customers a jobber might have. The *do-it-yourself (DIY)* clients of a typical parts business account for over a quarter of a jobber's customers. **Figure 1.44** shows the products most often purchased by walk-in customers.

Chapter 1 Introduction to Chassis Systems and Repair 19

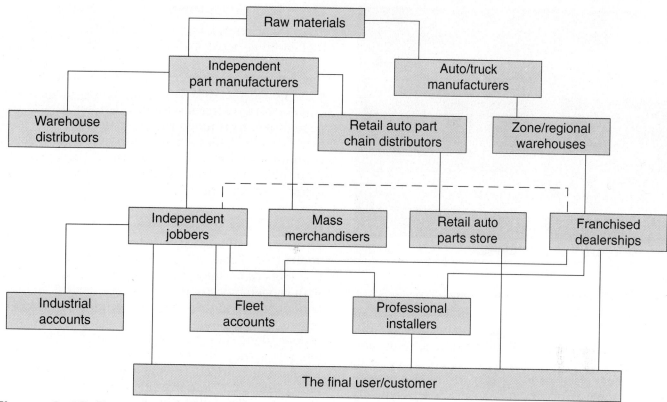

Figure 1.42 The auto parts supply network.

Figure 1.43 A wholesale and retail parts store. (Courtesy of Tim Gilles)

Services provided to customers are often a prime element in gaining repeat business. Prompt delivery of phoned-in parts orders is provided to regular wholesale customers. Often a parts jobber will have a machine shop as well (**Figure 1.45**). Having a machine shop enables a jobber to perform such tasks as drum and rotor machining, engine machining, flywheel surfacing, power steering pressure hose construction and repair, and hydraulic press work.

CLASS CODE	NUMBER OF PARTS	PROBABILITY OF SALE	NUMBER OF PARTS LIKELY TO SELL	NUMBER OF PARTS LIKELY NOT TO SELL
1	1,800	99.6%	1,793	7
2	700	97.0%	679	21
3	1,300	91.1%	1,195	105
4	3,400	77.8%	2,645	755
5	6,000	52.8%	3,170	2,830
6	7,000	31.3%	2,190	4,810
7	8,000	16.1%	1,290	6,870
8	50,000	4.9%	2,450	47,550

Figure 1.44 Probability of sale by part class.

Figure 1.45 A parts store with a machine shop. (Courtesy of Tim Gilles)

Figure 1.46 Mass merchandisers often have large service facilities. *(Courtesy of Tim Gilles)*

Retail Chain Stores

Many parts retailers are managed by national or regional retail chains. *Chain stores* usually deal in faster moving items such as alternators, starters, tools, and car accessories. They often do not have as large an inventory of diverse parts as a jobber store. Some retail stores are owned by one or several owners but buy everything through the distribution system of the main company. Prices, especially on sale items, are sometimes lower than at retail stores.

Mass merchandisers, such as large department store chains, sell fast-moving, popular items such as oil and filters, windshield wipers, and tune-up items to the DIY market. They often have a large service facility for doing minor repairs on automobiles (**Figure 1.46**).

Dealership Parts Departments

Auto dealers have their own parts departments. A good dealership parts department will maintain a large, organized inventory of original parts. Dealer parts departments usually have a separate counter for the technicians who work in the shop. Some dealerships have phone connections or an intercom in the service bay so that technicians can call in parts orders before walking to the parts counter. Some dealerships even have personnel who bring the cars to the service bay for the technician. This is one way that dealers can help technicians be more productive in a busy shop.

The operation of a dealership's parts business is often determined by its owner. Some dealers are very aggressive in pricing and services offered to the independent market, and others are not. Discounts are offered in varying degrees to wholesale accounts. It is not uncommon for dealerships to deliver parts to independent repair shops within a 50-mile radius in some markets. Although they are competitors, independent parts establishments and dealers often have a give-and-take arrangement, supplying each other with parts when they are needed in a hurry.

Service Stations

A service business is very concerned with the cost of its high-volume parts. A shop that sells five oil filters a day might be able to purchase them in quantity at half the cost and keep them *in stock* (on the premises). The same holds true with **tires, batteries, and accessories (TBA).** Service stations often buy TBA at very good discounts through the oil company whose products they sell. A higher volume shop will often have a computerized inventory from which parts are automatically ordered by computer when the supply falls below a certain level.

■ PARTS PRICING

Prices are usually based on what the jobber pays. The ability to find good prices on parts is an important part of a jobber's business. Jobbers are sometimes affiliated with cooperative groups. This gives them the ability to band together so that they can secure better prices by buying in huge quantities. They still continue to be independent owners, however, and may buy from whomever they choose.

Parts are priced for the customer according to either list, net, or variable pricing (**Figure 1.47**). The **list price,** also called the *retail price,* is the suggested price of an item and is what the car repair customer pays. In reality, very few stores charge list price to the DIY customer who walks into a parts store. Most parts are discounted, although in varying amounts. The **net price** is the price that the wholesale auto repair customer pays.

Some items like oil, oil filters, coolant, and spark plugs are priced by most establishments with little or no markup, because most customers are aware of their prices. These parts, called **cost leaders** or *loss leaders,* are considered to be part of the cost of promoting a business (**Figure 1.48**). **Variably priced items** are those that are slow movers and not readily available to

Part Number	List	User Net	Dealer	Part Number	List	User Net	Dealer
5188	3.08	2.62	2.05	5282	20.94	17.80	13.94
5189	6.55	5.57	4.36	5283	22.40	19.04	14.92
5193B	13.48	11.46	8.98	5284	13.69	11.63	9.12
5195C	14.81	12.59	9.87	5287A	14.02	11.92	9.34
5197D	13.60	11.56	9.06	5288C	17.66	15.01	11.78
5197D	17.99	15.29	11.98	5289B	32.96	28.01	21.95
5198D	16.32	13.87	10.87	5290A	13.80	11.73	9.19
5200C	12.67	10.77	8.45	5291A	13.28	11.29	8.85
5201B	16.57	14.08	11.03	5292B	28.81	24.50	19.19
5202B	16.55	14.06	11.02	52393B	16.41	13.95	10.93
5204	18.73	15.92	12.47	5294	12.36	10.51	8.23
5208A	13.66	11.62	9.10	5296A	15.93	13.54	10.62
5210C	15.11	12.84	10.06	5297	10.94	9.30	7.29
5211E	20.19	17.16	13.45	5298	2.70	2.30	1.80

Figure 1.47 An example of list, net, and dealer prices.

Chapter 1 Introduction to Chassis Systems and Repair

Figure 1.48 Oil, oil filters, coolant, and spark plugs are cost leaders, or loss leaders.

customers. These items can be marked up to a higher amount to help compensate for cost leaders.

When a service or repair business (installer) buys auto parts from a jobber, the price the installer is charged sometimes depends on the volume of business that the installer does with that parts establishment. More volume on a monthly basis can mean a higher percentage of discount to the shop. The part is marked up to the retail price for the customer who brings in a car for repair. This system can make a difference in the profitability of a service business.

Ordering Replacement Parts—The Aftermarket and Original Equipment

Factory replacement parts are categorized as OEM. Stock means the replacement part is the same as the original manufacturer part. *Aftermarket* is a broad term that refers to parts that are sold by the non-OEM market. Many OEM parts are manufactured by the same manufacturers as aftermarket parts.

■ MEASURING

Measuring is very important in brake and suspension work. Before brake parts can be replaced, you must document wear or runout of brake rotors and/or drums. State consumer protection agencies regularly run sting operations to check for fraudulent repairs. These repairs mostly relate to the unnecessary replacement of parts. A shop can incur fines and lose its license in the event of an infraction. You can see why your employer will insist that you do an accurate assessment of these items on each vehicle.

The first chapters in this book deal with brakes. Learning the correct way to take measurements now will make it possible for you to perform accurate and professional brake inspections, and you will have time to practice for more complex brake work. The following discussion will familiarize you with the design and use of the special micrometers and dial indicators used in brake and suspension work. Information on the symptoms and causes of brake rotor and drum wear are covered in detail in Chapter 8.

Measuring a Brake Drum

Automotive shops that do brake work own a special tool for measuring brake drums. This device is actually a dial indicator, but in the automotive repair industry it is referred to as a *brake micrometer* (**Figure 1.49**), commonly called a drum mike. Its center shaft has graduated indentations that adjust to fit drums of different diameters. The shaft is marked with size designations and reads like a ruler. The reading on the dial indicator is added to the graduated shaft setting. For instance, the shaft setting in **Figure 1.50** is 11 inches. The dial indicator reading of 0.375 inch is added to this for a total of 11.375 inches. Brake micrometers are also available with a center shaft that reads in meters (**Figure 1.51**) for working on imported vehicles.

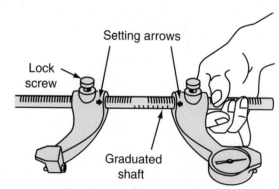

Figure 1.49 A brake drum mike.

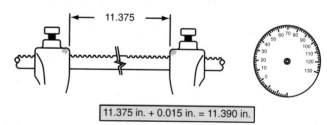

Figure 1.50 Reading a drum mike. Add the dial indicator reading to the setting on the graduated shaft.

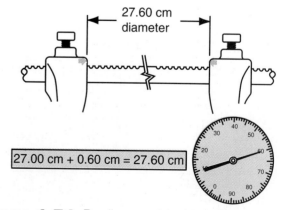

Figure 1.51 Reading a metric drum mike.

22 SECTION 1 INTRODUCTION AND SAFETY

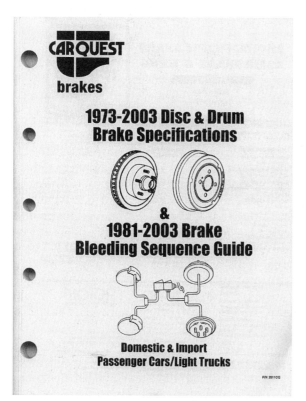

Figure 1.52 A brake dimension booklet. *(Courtesy of Car Quest)*

1.53). The original size dimension is called the *nominal diameter*. The maximum diameter is cast into the outside of the brake drum (**Figure 1.54**). If the specification found cast into the drum is metric, you can determine the drum's original size equivalent in inches using the specification manual.

There are two movable parts on a drum mike, one with an anvil and one with a dial. A small arrow on the inside edge of each movable part must be aligned with the correct line on the ruler on the center shaft (see **Figure 1.50**). If the nominal diameter is 10 inches, adjust the *anvil* side of the drum mike to align with the corresponding number on the graduated shaft and firmly set the lock screw. Next, set the *dial* side to 10

Figure 1.54 The maximum diameter is cast into a brake drum. *(Courtesy of Tim Gilles)*

Brake dimension booklets (**Figure 1.52**) are available from all of the major brake part manufacturers, usually at no cost. Dimension booklets will list inch specifications for metric brake drum sizes (**Figure**

BRAKE SPECIFICATIONS GUIDE

MANUFACTURER			ROTOR					DRUM				CALIPER		WHEEL	
				Thickness					Diameter						
YEAR MODEL	CALIPER ILLUS.	Maximum Parallel Variation	Runout Limit	Nominal Thickness	Minimum Machining	Discard At Or Under	Finish Microinch	Runout Limit	Nominal Diameter	Maximum Machining	Discard At Or Over	Caliper to Bracket	Bracket or Caliper to Knuckle	Nut Torque (Ft.-Lbs.)	Bearing Adj.
DAIMLERCHRYSLER Continued															
2003-00 Neon..........11		.0005	.003³	.861-.871	.803	⁹	15-80	.006	7.875	7.904	7.921	—	16	100	—
Rear disc		.0005	.003³	.344-.364	.285		15-80	—	—	—	—	—	16	100	—
2003-98 Concorde, Intrepid, LHS, 300M..........10		.0005	.003³	1.019-1.029	.960	⁹	15-80	—	—	—	—	—	16	100	—
Rear disc		.0005	.003³	.458-.478	.409		15-80	—	—	—	—	—	16	100	—
2002-97 Prowler		.0005	.003	.939-.949	—	.881	15-80	—	—	—	—	20	—	100	—
Rear Disc		.0004	.002	.793-.782		.736	16-91	—	—	—	—	—	—	100	—
2002-01 Viper		.0005	.003	1.244-1.264	—	1.197	15-80	—	—	—	—	24	80	—	—
		.0005	.003	.871-.850		.803	15-80	—	—	—	—	24	38	80-100	—
2000-92 Viper		.0005	.003	1.244-1.264	—	1.197	15-80	—	—	—	—	24	85	—	—
		.0005	.003	.871-.850		.803	15-80	—	—	—	—	24	85	80-100	—
2000-96 Caravan, Voyager, Town & Country..........11		.0005	.003³	.939-.949	.881	⁹	15-80	.006	9.842	9.874	9.904	—	30	100	(A)
Rear disc: 2000-97		.0005	.003³	.482-.502	.443		15-80	—	—	—	—	—	16	100	—
1996		.0005	.003³	.468	.449	.409	15-80	—	—	—	—	—	16	100	—
2000-95 Breeze, Cirrus, Stratus, Sebring Convertible..........11		.0005	.003³	.900-.911	.843	⁹	15-80	.006	7.875	7.904	7.921	—	16	100	—
Rear disc		.0005	.003³	.350-.360	.285	⁹	15-80	—	—	—	—	—	16	100	—
1999-95 Neon..........11		.0005	.003³	.782-7.92	.724	⁹	15-80	.006	7.875	7.904	7.921	—	16	100	—
Rear disc		.0005	.003³	.344-.364	.285	⁹	15-80	—	—	—	—	—	16	100	—
1997-94 LHS, New Yorker..........10		.0005	.003³	.940-.950	.912	.882	15-80	.006	8.661	8.691	8.705	—	16	100	—
Rear disc		.0005	.003³	.458-.472	.439	.409	15-80	—	—	—	—	—	16	100	—
1997-93 Concorde, Intrepid..........10		.0005	.003³	.940-.950	.912	.882	15-80	.006	8.661	8.691	8.705	—	16	100	—
Rear disc		.0005	.003³	.458-.472	.439	.409	15-80	—	—	—	—	—	16	100	—
1995-91 Caravan, Voyager, Town & Country..........7		.0005	.003³	.944	.911	.881	15-80	.006	9.00	9.060	9.090	30	160	100	(A)
AWD..........7		.0005	.003³	.944	.911	.861	15-80	.006	11.00	11.060	11.090	30	160	100	(A)
All other FWD..........6, 7, 8		.0005	.004³	.935	.912	.882	15-80	.006	7.875	7.904	7.921	25-35	130-190	100	(A)
w/8.66" drum		—	—	—	—	—	—	.006	8.661	8.691	8.705	25-35	130-190	100	(A)
Rear disc: TC		.0005	.005	.345	.321	.291	15-80	—	—	—	—	15-18	80	100	—
Others: Solid rotor		.0005	.005	.472	.439	.409	15-80	—	—	—	—	17	—	100	(A)
Vented rotor		.0005	.003	.866	.827	.797	15-80	—	—	—	—	17	—	100	(A)
1992-91 Monaco		.001	.003	.945	NS	.886	NS	NS	8.858	NS	8.917	15-22	70	95	—
Rear disc		.001	.003	.394	.374	.345	NS	—	—	—	—	—	21⁶	95	—
1990 Monaco		.001	.003	.945	NS	.886	NS	NS	8.858	NS	8.917	15-22	70	63	—
Rear disc		.001	.003	.394	.374	.345	NS	—	—	—	—	—	21⁶	63	—
1990-89 Omni, Horizon, Turismo..........6		.0005	.004³	.498	.461	.431	15-80	.006	7.875	7.904	7.921	18-26⁶	130-190	95	(A)
Caravan, Voyager, Town & Country..........7		.0005	.004³	.861-.870	.833	.803	15-80	.006	9.00	9.060	9.090	25-35	130-190	95	(A)

Figure 1.53 Metric and inch specifications from a brake dimension booklet. *(Courtesy of Car Quest)*

Chapter 1 Introduction to Chassis Systems and Repair 23

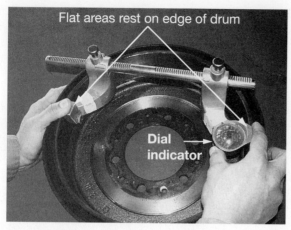

Figure 1.55 The flat areas on the drum mike rest on the edges of the drum. *(Courtesy of Tim Gilles)*

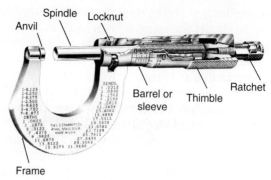

Figure 1.56 A cutaway view of an outside micrometer. *(Courtesy of the L. S. Starrett Co.)*

inches and set the lock screw. When the setup is complete, take a reading on the drum.

The anvil and dial ends have small flat areas that fit on the top of the brake drum (**Figure 1.55**). Hold the anvil in place as you rock the dial end until it centers and you get the highest reading on the dial. Add the dial reading to the nominal diameter size to which you adjusted the gauge. For instance, if the gauge says 0.027 inch and the nominal size is 10 inches, the reading is 10.027 inches. Compare this with the *discard diameter* dimension. Following remachining, the diameter will need to be at least 0.030 inch less than that amount.

NOTE: *To avoid confusion, brake dimension booklets usually lists "machine to" dimensions as well as discard diameters.*

Make at least four readings at 45-degree intervals around the drum to check for *out-of-round wear*. If the drum appears unworn but is 0.006 inch or more out of round, machining of the drum will be required.

There are other tools, such as inside micrometers and special calipers with pointed ends, that can be used to measure deep grooves in the drum prior to machining, but these tools are not common in shops. Determining the depth of a groove is most easily done after the drum is mounted on a brake lathe (see Chapter 8).

■ MEASURING WITH A MICROMETER

Rotor parallelism and thickness are measured with a micrometer. Micrometers have several advantages over other types of measuring instruments. They are clear and easy to read. They measure consistently and accurately, and they have a built-in adjustment to compensate for wear. A cutaway view of an outside micrometer is shown in **Figure 1.56**.

Some micrometers have extra features such as a ratchet, which ensures that spindle pressure is always consistent. Another handy feature is a locknut, which can hold a reading so it cannot be changed accidentally. Some micrometers have a decimal equivalent chart stamped on the frame.

A special micrometer for disc brake rotor measurement has a point on one of its measuring surfaces called an anvil (**Figure 1.57**). This feature allows the depth of a groove in the rotor surface to be measured.

Reading a Micrometer

The micrometer, or mike, operates on a simple principle (**Figure 1.58**). The spindle (**Figure 1.58a**) has 40

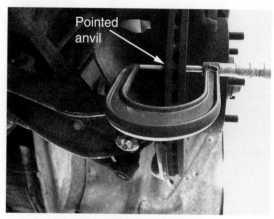

Figure 1.57 A special micrometer for measuring disc brake rotors. *(Courtesy of Tim Gilles)*

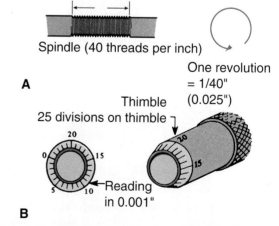

Figure 1.58 Relationships between (A) micrometer spindle pitch (40 threads to an inch); spindle/thimble movement (1/40 in., or 0.025 in., per revolution); and (B) graduations on the thimble of an inch-standard micrometer.

threads per inch, so one revolution of the thimble will advance or retract the spindle 1/40th inch, or 0.025 inch, (**Figure 1.58b**). On the sleeve, or barrel, each line represents 0.025 inch, and a new line is uncovered every time the thimble is turned one complete revolution. Measurement amounts are shown in **Figure 1.59**.

As each line on the thimble passes the zero line on the hub, the mike opens another 0.001 inch. To read the exact number of thousandths, note the number on the thimble that lines up with the zero line. In **Figure 1.60**, this amount is 0.008 inch. If you add 0.350 to 0.008 you get 0.358 inch, which is the actual measurement in this reading.

NOTE: *The measurement reflects the actual size of the part.*

In **Figure 1.61**, the reading is 0.358 inch. The reading is obtained by adding the barrel measurement

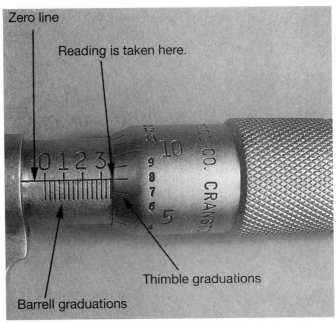

Figure 1.60 Reading a micrometer. *(Courtesy of Tim Gilles)*

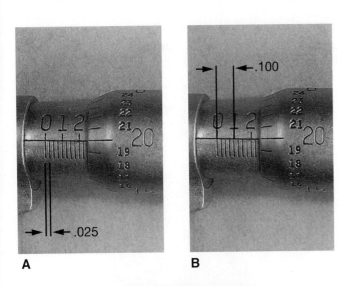

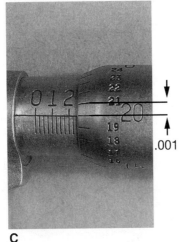

Figure 1.59 Micrometer measurements. (A) Each line on the barrel is 0.025 in. (one revolution of the thimble). (B) Each number on the barrel is 0.100 in. (four revolutions of the thimble). (C) Readings on the thimble are 0.001 in. (1/25 of a revolution of the thimble). *(Courtesy of Tim Gilles)*

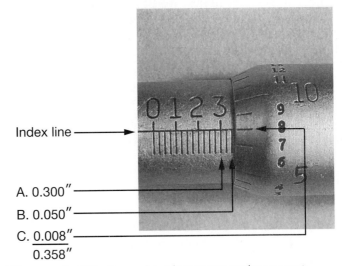

Figure 1.61 To read a micrometer to the nearest 0.001 in., add A, B, and C together. *(Courtesy of Tim Gilles)*

to the thimble measurement. The reading is rounded off to the nearest 1/100 inch (0.001 inch).

Metric Micrometers

Except for the graduations on the hub and thimble, metric micrometers look just the same as English mikes. One turn of the thimble turns the spindle 0.5 mm as opposed to 0.025 inch. The hub is graduated in millimeters and half millimeters (**Figure 1.62**). The lines below the index line are the half millimeters (0.5 mm). Two revolutions of the thimble equal 1.0 mm on the hub. The thimble usually has 50 divisions. Every fifth line has a number from 0 to 45. Each graduation on the thimble is equal to 0.01 mm (1/100 millimeter).

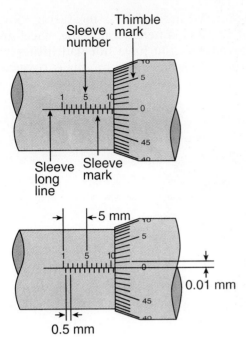

Figure 1.62 Metric micrometer markings and graduations.

The metric micrometer is read in the same manner as its English counterpart. Just add the millimeter and half millimeter readings on the barrel to ¹⁄₁₀₀ (0.01) millimeters on the thimble.

Combination Digital Mikes

Combination micrometers give both metric and English readings. Two versions are available. One of them gives inch readings in the conventional manner, whereas the other reads the metric system. LCD readings taken from the *digital readout window* (**Figure 1.63**) are accurate to 0.0001 and are convertible between inch and metric.

Electronic Caliper

An *electronic caliper* (**Figure 1.64**) works the same way as an ordinary caliper but its measurement is shown on a liquid crystal display (**LCD**) to an accuracy of

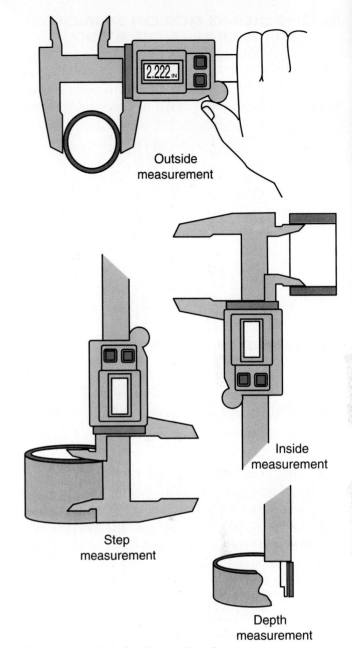

Figure 1.64 An electronic caliper.

Figure 1.63 A digital micrometer.

0.0005 inch. The digital reading can be changed between the English and metric systems with the touch of a switch. This is a very handy tool but tend to be more costly.

REMEMBER: *Later chapters in this text will present more information on disc and drum service and machining. The intent of this exercise is for you to learn how to use the measuring instruments accurately prior to performing brake repairs.*

CHECKING ROTOR RUNOUT WITH A DIAL INDICATOR

When a rotor is warped from heat or incorrect machining, this is called **runout**. Rotor runout causes the rotor to swing from side to side as it rotates (**Figure 1.65**). A dial indicator is used to check rotor runout.

Although a dial indicator is a measuring tool, its use will not be covered here because the wheel must first be removed and you have not had lifting or safety instruction. Checking rotor runout is covered in the section on disc brake service in Chapter 7.

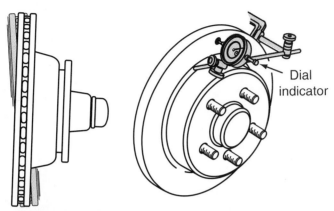

Figure 1.65 Disc brake rotor runout is checked with a dial indicator.

REVIEW QUESTIONS

1. Two main types of brakes are _____ and _____.
2. What gives a brake booster its power?
3. If a wheel locks up, what does an ABS system do with pressure to the wheel?
4. During normal stopping, does the ABS system function?
5. In what year did federal law require all cars sold in the United States to have two separate hydraulic circuits?
6. Which steering system has no linkage, only two tie-rods extending from its ends?
7. List four types of vehicle springs.
8. List two main designs of automotive front suspensions.
9. When the wheel moves up as the spring compresses, this is called _____. Opposite wheel movement is called _____.
10. Name two parts (not shock absorbers) that absorb road shock.
11. What is the name of the front suspension design that incorporates the shock absorber into the front suspension?
12. How many numbers and letters are there in the VIN used by manufacturers to identify a particular vehicle?
13. What are three types of information found in lubrication service manuals?
14. What is the name of the manual that is used to determine the length of time to complete a job?
15. What is the name of the special tool for measuring brake drums that is actually a dial indicator?

PRACTICE TEST QUESTIONS

1. Step-by-step repair information is included in the owner's manual provided with a new vehicle. True or False?
2. Technician A says that a flat-rate manual gives the amount of time it should take to complete a repair job. Technician B says that a flat-rate technician will be paid for the amount of time it actually takes to complete the job. Who is right?
   A. Technician A   B. Technician B
   C. Both A and B   D. Neither A nor B
3. Technician A says that one revolution of the micrometer thimble equals 1/40 inch. Technician

B says one revolution of the micrometer thimble equals 0.025 inch. Who is right?
- **A.** Technician A
- **B.** Technician B
- **C.** Both A and B
- **D.** Neither A nor B

4. Which of the following steering systems has two tie-rods?
   - **A.** Gear box with parallelogram linkage
   - **B.** Rack-and-pinion steering
   - **C.** Both A and B
   - **D.** Neither A nor B

5. Technician A say a stock part is OEM. Technician B says *aftermarket* refers to parts sold by the non-OEM market. Who is right?
   - **A.** Technician A
   - **B.** Technician B
   - **C.** Both A and B
   - **D.** Neither A nor B

6. Drum brakes are more likely to require a power booster than disc brakes. True or False?

7. Technician A says the MacPherson strut suspension system uses only one ball joint. Technician B says the SLA suspension uses two control arms of equal length. Who is right?
   - **A.** Technician A
   - **B.** Technician B
   - **C.** Both A and B
   - **D.** Neither A nor B

8. Technician A says most older cars have front-wheel drive. Technician B says most newer cars have frames running the complete length of the car. Who is right?
   - **A.** Technician A
   - **B.** Technician B
   - **C.** Both A and B
   - **D.** Neither A nor B

9. Shock absorbers dampen spring oscillations by converting the energy from spring movement into heat. True or False?

10. Technician A says if the drum appears unworn but is 0.006 inch or more out-of-round, machining of the drum will be required. Technician B says rotor parallelism and thickness are measured with a micrometer. Who is right?
    - **A.** Technician A
    - **B.** Technician B
    - **C.** Both A and B
    - **D.** Neither A nor B

MICROMETER PRACTICE

Enter the micrometer readings below. *(Courtesy of Tim Gilles)*

1.

2.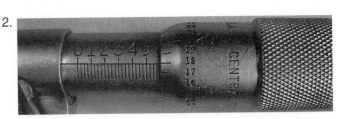

3.

4.

5.

6.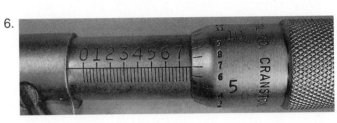

7.

8.

CHAPTER 2

Getting Started in the Shop: Lifting a Vehicle/Shop Equipment and Safety

■ OBJECTIVES:

Upon completion of this chapter, you should be able to:

- ✓ Describe the different types of lifting equipment.
- ✓ Demonstrate safe practices in lifting a vehicle with jacks and lifts.
- ✓ Understand the types and operation of air compressors.
- ✓ Use shop tools and equipment safely.
- ✓ Explain fire prevention and hazards.
- ✓ Identify hazards associated with cleaning solvents.
- ✓ Explain the contents of a material safety data sheet for a specific chemical.
- ✓ Complete a shop safety test.

■ KEY TERMS

arc welding
barrier cream
center of gravity
Class A fire
Class B fire
Class C fire
Class D fire
frame-contact lift
high-efficiency particulate air filter (HEPA filter)
hydraulic
oxyacetylene welding
puddle
wheel-contact (drive-on) lift

■ INTRODUCTION

This chapter introduces you to the things you need to know as you begin working in an automotive shop. Chapter content includes information on lifting equipment, safely raising a vehicle, and safety and shop practices, including dealing with chemicals.

Major pieces of equipment shared among all of a company's employees are typically owned by the employer. A well-equipped general shop will own most of the equipment required to assist the technician.

Hydraulic Equipment

Hydraulic machinery used in repair shops includes jacks, lifts, and presses. They all operate in the same manner, using hydraulic pressure to increase force. It is possible to increase pressure by hydraulic means when a small piston is used to move a larger one. **Figure 2.1** shows hydraulic fluid pressure of 5 pounds per square inch (psi) acting on a 2-square-inch area to produce a force of 10 pounds. When that 5 psi is applied to a 4-square-inch piston, the force that results is doubled to 20 pounds. The amount of movement required of

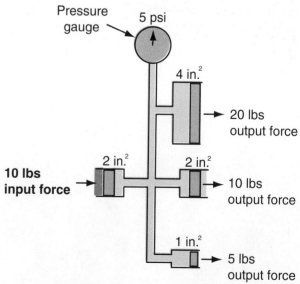

Figure 2.1 Hydraulic pressure is increased when a small piston is used to move a larger one.

SECTION 1 INTRODUCTION AND SAFETY

Figure 2.2 A bottle jack.

the jack or press handle (input piston) will be more, but less effort will be needed.

If you understood the previous paragraph, you will know that a 2-ton jack will require more frequent handle movement than a 1.5-ton jack. If you have ever used a 10-ton bottle jack (**Figure 2.2**), you will remember that the handle must be pumped many times to move the fluid necessary to cause the jack ram to move a small distance. Sometimes a more powerful jack is needed.

Hydraulic Jacks

The hydraulic floor jack, called a *service jack* (**Figure 2.3**), is used to raise and lower vehicles. The jack is positioned under an area of the vehicle frame or at one of the correct lift points shown in a service manual (**Figure 2.4**). Most of today's vehicles do not have full frames and may be damaged if lifted improperly.

Do *not* jack under the vehicle floor pan or beneath front or rear linkages that can be bent. When lifting a rear-wheel drive vehicle from the rear, you can usually lift it from the center (**Figure 2.5**). Use caution when

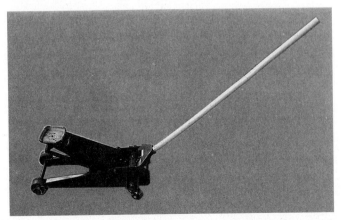

Figure 2.3 A hydraulic floor service jack. *(Courtesy of Tim Gilles)*

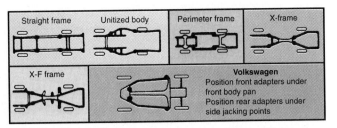

Figure 2.4 Lift points for different vehicle frame types. *(Courtesy of Automotive Lift Institute)*

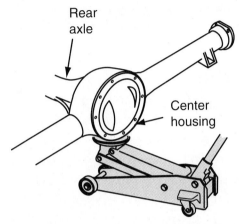

Figure 2.5 The rear of a vehicle can usually be lifted at the center.

 Position the jack so its wheels can roll as the vehicle is lifted. Otherwise, the lifting plate may slip on the frame, or the jack may tip over.

 An apprentice was preparing to do a brake job on a vehicle. He raised it using a floor jack and completed the work. The vehicle was test-driven, and the customer took it home. The next day the customer returned, complaining that his passenger seat was stuck all the way forward and would not adjust. Upon further investigation, the shop foreman discovered that the floor was pushed up under the seat, binding it up and preventing it from moving. What do you think caused this and how would you go about correcting the problem? Hint: Preventing this problem by using the correct lift points would have been much easier than correcting it.

lifting newer front-wheel drive vehicles because trailing link suspension parts can be easily damaged.

Vehicle Support Stands

Hydraulic floor jacks (service jacks) are used extensively in automotive repair. Cars have fallen and crushed people who used jacks that failed. It is foolish to crawl under a vehicle that is not resting *solidly* on vehicle support stands, commonly called *jack stands*.

There are several different types of vehicle support stands. A popular professional type is shown in **Figure 2.6.** Always use support stands in pairs (**Figure 2.7**), positioned in the recommended locations on the

> **CASE HISTORY**
> *An apprentice was working on a vehicle that was supported only by a hydraulic service jack. The master technician he was working with told him to get a pair of jack stands to support the vehicle. While he was getting the support stands, the service jack failed. Had he been under the car, he would have been crushed.*

> **CASE HISTORY**
> *A California auto shop student was killed when the truck he was working on fell off its support stands while he was adjusting the clutch. The car was not supported at a solid location on the frame.*

> **SHOP TIP**
> Place a piece of thick plywood or a steel plate under support stands that are used on asphalt.

frame. Be sure to use them only on a level concrete surface. On a hot day, the legs of a support stand can dig into asphalt and allow the vehicle to fall.

■ VEHICLE LIFTS

Most shops have hydraulic or electric lifts to raise vehicles high in the air for convenience during undercar repairs. Automotive lifts are often called *hoists* or *racks*. The two main categories are the *in-ground lift* and the *above-ground*, or surface mount, lift. When the lifting mechanism of a lift is located below the floor, the lift is an in-ground lift.

Lift Types

Within the two categories of lifts—surface mount and in-ground—there are several styles, but two main types:

1. the **frame-contact lift** (**Figure 2.8**).
2. the **wheel-contact (drive-on) lift** (**Figure 2.9**).

There are also in-ground lifts called axle-engaging lifts, and other lifts called pad lifts (covered later). An advantage to frame-contact, axle-engaging, or pad lifts is that they allow the vehicle's wheels to hang free, which makes it easier to perform tire, brake, and suspension work. These lifts usually provide better access to the underside of the vehicle.

Frame-Contact Lifts

Frame-contact lifts have adapters at the end of adjustable lift arms. The lift adapters come into contact with the vehicle frame at the manufacturer's specified lift points, which are at the rocker panels or on a section of the frame (**Figure 2.10**). Be sure to check the vehicle service manual before trying to lift a car. When a car is lifted improperly, body, suspension, or

Figure 2.6 Jack stands.

Figure 2.7 Always use vehicle support stands in pairs.

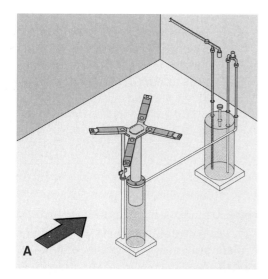

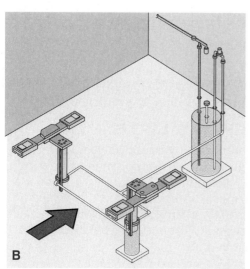

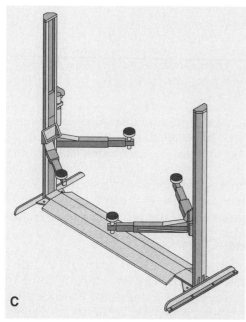

Figure 2.8 (A) A single-post frame-contact lift. (B) A two-post frame-contact lift. (C) A surface mount frame-contact lift. *(Courtesy of Automotive Lift Institute)*

Figure 2.9 A wheel-contact lift. *(Courtesy of Dover)*

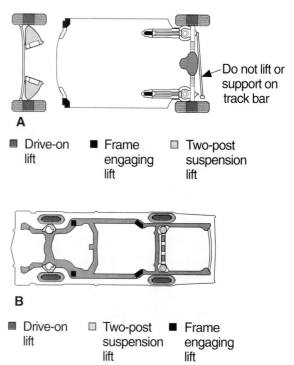

Figure 2.10 Lift points for various types of lifts. (A) Unitized body. (B) Body and frame.

steering components can sometimes be bent. The windshield might even pop out or the vehicle can fall.

In 1992, the Society of Automotive Engineers adopted a standard (SAE J2184) for vehicle lift points. Some manufacturers label their vehicles with decals depicting the recommended lift points. These are located inside the passenger side front door. Permanent markings on the underside of the chassis also

 Adapters must be placed at the manufacturer's recommended lifting points and they must be set to raise the vehicle on a level plane or the vehicle will be unstable.

Figure 2.11 Adapters flip up to accommodate different frame heights. *(Courtesy of Automotive Lift Institute)*

Figure 2.12 Some adapters have a screw adjustment. *(Courtesy of Automotive Lift Institute)*

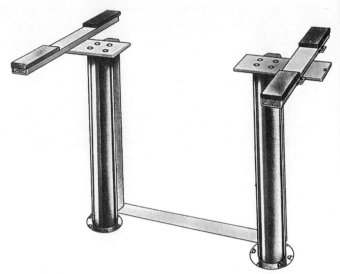

Figure 2.13 A rocker panel, or pad, lift. *(Courtesy of Automotive Lift Institute)*

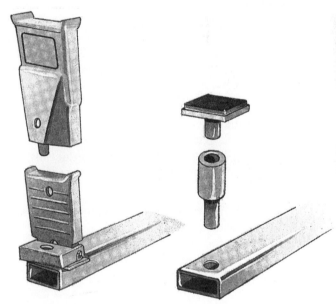

Figure 2.14 Extenders can be used with pickup trucks and vans. *(Courtesy of Automotive Lift Institute)*

mark the lift points. These markings can include a hole, a boss, or a ¾-inch triangular depression. Lift points can also be found in the service literature.

Some adapters flip up. Sometimes called *foot pads*, these adapters adjust to several positions to accommodate different heights between the lift points in the front and rear of the vehicle and to provide clearance between the rocker panel and the lift arm (**Figure 2.11**). Some foot pads have screw threads that allow them to be positioned higher or lower (**Figure 2.12**). Some adapters have rubber pads. Be sure that they are in good condition and that they are not covered with oil or grease that could make them slippery.

Passenger cars with unibody construction are lifted at points on the rocker panel, just under the bottom outside edge of the car body. There are special rocker panel lifts, called pad lifts (**Figure 2.13**), that can be used for these cars and cars with perimeter frames. Do not use pad lifts for trucks.

Special Auxiliary Adapters/Extenders

Sometimes extenders are used on the adapters (**Figure 2.14**). If the lift points on the vehicle are undercoated, a special adapter might be needed when using a lift that has steel adapters. Damaging the undercoating can void the vehicle's rust protection warranty. Some sport utility vehicles, light trucks, and vans require special adapters to provide clearance between the lift arm and the rocker panel. Most lift manufacturers make these available. Do not use makeshift extenders.

Wheel-Contact Lifts

A wheel-contact lift is one in which all four of the vehicle's wheels are supported (see **Figure 2.9**). Some of these drive-on lifts are of the in-ground type; others are of the four-post, surface mount type. The four-post lift has the advantage of easier installation and removal. Four-post wheel-engaging lifts are especially popular in service, engine and transmission repair, wheel alignment, and muffler shops.

When driving a car onto one of these lifts be sure that the tires are an equal distance from the edges of the ramps. Wheel-contact lifts have secondary stops

SAFETY NOTE: After the vehicle is raised, the jack is lowered onto a mechanical safety latch.

(wheel chocks) for roll-off protection at the front and rear (**Figure 2.15**). After positioning the vehicle on the lift, always use manual wheel chocks to prevent the car from rolling.

NOTE: *GM light trucks with four-wheel steering have a 5-inch wider wheelbase in the rear. These vehicles sometimes will not fit on a wheel-contact lift.*

Wheel-Free Jacks. On a wheel-contact lift, the wheels are supported unless a wheel-free jack is used (see **Figure 2.15**). These jacks, which are air-powered or hydraulic, are used to raise either end of the car for wheel-free work.

When using one of these jacks, keep your hands clear and be sure to extend each of the lift arms an equal amount to avoid uneven loading. Be sure that a wheel-free jack is lowered all the way before driving onto or off the lift.

Some advantages of wheel-contact lifts are:

- The vehicle may be raised safely, even without the engine in it. Trying to lift a vehicle without the engine can be dangerous on a frame-contact lift, because the unbalanced vehicle might tip off the lift.
- The center of the underside of the vehicle is accessible on the twin-post hydraulic and four-post types. This is important for exhaust system and transmission/drive line work.

- For quick, easy service the vehicle requires no setup of lift adapters.
- Wheels are in contact with the wheel ramps. This is necessary for wheel alignment work.

IN-GROUND LIFTS

Most in-ground lifts used for raising automobiles have either one or two pistons, often called *posts*. Two-post lifts have the pistons located in one of two positions:

1. One in front of the other
2. Side by side

Newer in-ground systems must be enclosed with double walls to prevent oil from them leaking into the ground.

Frame-Engaging Lifts

In-ground frame-contact lifts are either *single-post* or *two-post* style (see **Figure 2.8A and B**). The two-post style lift is either *drive-through* (**Figure 2.16**) or *drive-over* (**Figure 2.17**). Drive-through lifts are more open in the center to allow easier access to the underside of the vehicle. The vehicle is driven between the lift arms.

The lift arms on a drive-over lift are located closer together so that vehicles can be driven over the arms without bumping into them. As vehicles have been downsized, drive-through lifts have become more popular. Using a drive-over lift to work on the underside of a small car is very difficult because access is extremely limited. Two-post lifts provide better access to the underside of vehicles than single-post hoists.

Above-ground lifts also have the advantage of being relatively portable, so they can be more easily transported if the shop has to move. They also do not present a groundwater pollution problem in the event of a hydraulic fluid leak the way an in-ground hoist does.

Axle-Engaging Lifts

An axle-engaging lift, also called a *suspension contact* lift, has at least two posts (pistons) positioned front to rear (**Figure 2.18**). The front post is movable so that it can accommodate vehicles with wheelbases of different lengths.

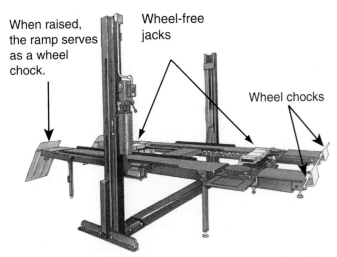

Figure 2.15 This lift has safety stops at each end. *(Courtesy of Automotive Lift Institute)*

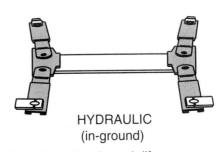

HYDRAULIC (in-ground)

Figure 2.16 A drive-through lift.

Chapter 2 Getting Started in the Shop: Lifting a Vehicle/Shop Equipment and Safety

vehicle by the wheels; this option works for front-wheel drive vehicles.

Semihydraulic and Fully Hydraulic Lifts

In-ground lifts that are powered by compressed air are either *semihydraulic* or *fully hydraulic*. Other in-ground lifts use an electric motor–driven pump to pressurize the hydraulic oil. Air pressure is not necessary in this system. *Semihydraulic lifts* have a self-contained air/oil reservoir. *Fully hydraulic lifts* have a separate air/oil reservoir. When compressed air is put into the tank containing the hydraulic oil, it pressurizes the oil to lift the vehicle.

There are separate oil and air controls on fully hydraulic lifts (**Figure 2.19**). Semihydraulic lifts use only one control. The air control is opened first. When the air has had a chance to pressurize the air/oil tank, the second control handle is moved to allow the pressurized oil to lift the car.

When the shop's air compressor has not been on long enough to build adequate air pressure or when a compressor is too small or the shop has been using too much air, the lift will work very slowly.

In-Ground Lift Maintenance

An oil leak can cause an in-ground lift to vibrate while lifting a vehicle or prevent the lift from rising to its full height. If you hear air escaping or see or suspect an oil leak, stop using the lift immediately and release the air pressure.

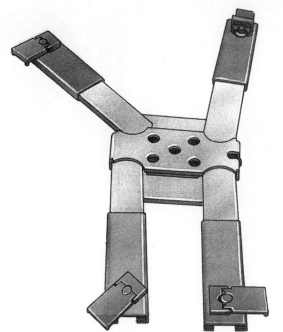

Figure 2.17 A driveover lift. *(Courtesy of Automotive Lift Institute)*

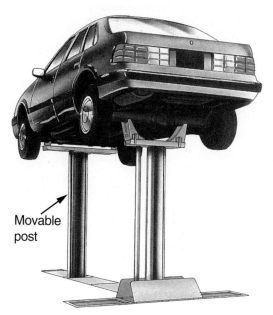

Figure 2.18 An axle-engaging or suspension contact lift. *(Courtesy of Automotive Lift Institute)*

The front post of the lift has adjustable arms that are positioned just under the lower control arm, as far out toward the wheel as possible. The rear post engages the rear axle housing on rear-wheel drive vehicles. Some front-wheel drive cars have a *support rail* between the rear axles that allows the use of this type of lift, but this type of lift is not recommended for a majority of front-wheel drive cars. Other lifts of this style provide the option of lifting the rear of the

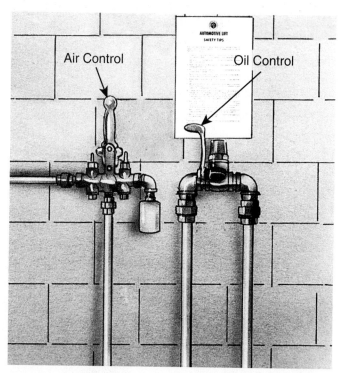

Figure 2.19 Fully hydraulic hoists have two controls. *(Courtesy of Automotive Lift Institute)*

Lifts are usually repaired only by qualified service personnel. Sometimes an experienced technician or business owner will perform maintenance to lifts. However, always follow the lift manufacturer's maintenance instructions.

NOTE: *Be sure that all of the air pressure is exhausted from the lift before attempting to check or add oil.*

Remove the fill plug carefully by hand. Do not use an impact wrench. There could still be air pressure in the air/oil tank. Use only the specified oil.

Surface Mount Lifts

Surface mount lifts have been installed in shops since 1970. They are mounted completely above the floor (see **Figures 2.8c and 2.9**). An electric motor operates either a screw drive (**Figure 2.20**) or a hydraulic pump and cylinders.

Ease of installation is one advantage of a surface mount lift. It can be removed and reinstalled in another location if a business owner is renting a shop and has to move. A surface mount lift is also easier to maintain or repair because it requires no excavation. In addition, surface mount lifts are also less likely to leak oil into the ground, which is an important environmental concern.

The most popular surface mount lift is a two-post, drive-through, frame-engaging type. It has two lifting carriages that each support two swing arms. The carriages are synchronized so that they go up and down together. A steel chain, cables, synchronized motors, or hydraulic circuits are used to keep the two carriages moving together. **Figure 2.21** shows a chain synchronizer.

The synchronizing parts and the driving system are attached to the posts either overhead or across the floor. When they are across the floor, sometimes a wide groove is cut in the concrete. Parts that connect the two posts are submerged in the groove so that the

Figure 2.21 This lift has a chain-type synchronizer. *(Courtesy of Automotive Lift Institute)*

floor remains unobstructed. This is especially convenient when a transmission jack is used.

Surface Mount Lift Maintenance. To bolt a surface mount lift, the floor must have at least 5 inches of solid concrete. The bolts that hold the lift to the concrete floor should be inspected periodically for tightness (**Figure 2.22**). They can and do vibrate loose. If cracks develop in the concrete, use of the lift should be discontinued until it can be checked by a professional. Always follow the lift manufacturer's maintenance instructions.

Figure 2.20 A screw drive hoist. *(Courtesy of Automotive Lift Institute)*

Figure 2.22 Periodically check the bolts of a surface mounted lift for tightness. *(Courtesy of Automotive Lift Institute)*

LIFTING A VEHICLE SAFELY

Lifts have an excellent safety record but unfortunately vehicles occasionally fall off them. Such accidents do not occur often, but when they do they can be serious and sometimes life-threatening. When a vehicle comes down, the accident is usually due to carelessness, misuse, or neglected maintenance. Training in the use of the lift is mandatory before attempting to lift any vehicle.

The American National Standards Institute (ANSI) and Automotive Lift Institute (ALI) have set the American National Standard (ALOIM-1994) for automotive lifts. This standard provides safety requirements for operation, inspection, and maintenance of lifts. It requires annual inspection of automotive lifts by a qualified lift inspector. The annual inspection is designed to keep automotive lifts in good operating condition.

Center of Gravity

Find the **center of gravity** (CG) of the vehicle and position it over the posts of the lift. The center of gravity is the point between the front and the rear wheels where the weight will be distributed evenly. Different positions are used for front-wheel drive (FWD) and rear-wheel drive (RWD) cars (**Figure 2.23**).
According to ALI:

- On RWD cars the center of gravity is usually below the driver's seat.
- On FWD cars, the center of gravity is usually slightly in front of the driver's seat, beneath the steering wheel.

NOTE: *Some two-post above-ground hoists have offset arms that allow the doors to be opened, but others do not. Do not position the vehicle to the front or rear of the posts just so that the door can be opened easier. Positioning the center of gravity is critical.*

On one- or two-post lifts, position the vehicle's center of gravity over the posts. On four-post models, position the car equally between the front and rear.

When lifting a vehicle, be sure that the lift or jack contacts the frame at the recommended lift point (see **Figure 2.4**). To be sure that the vehicle is firmly supported, raise it about 6 inches off the floor and then push on its bumpers.

You are lifting a vehicle that weighs several thousand pounds. The vehicle must be correctly spotted on the lift so that the center of gravity is correct.

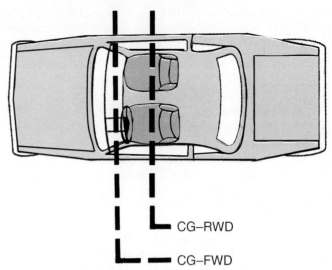

Figure 2.23 Center of gravity (CG) positions for front-wheel drive (FWD) and rear wheel (RWD) vehicles. *(Courtesy of Automotive Lift Institute)*

When lifting a vehicle on a frame-contact hoist, place the lift adapters at the specified locations and raise the vehicle about 6 inches off the floor. Then, shake it to see that it is firmly placed.

Safe Lift Operation Procedures

1. If there are any problems with the lift, do not use it. See your supervisor immediately. Do not take chances.
2. When lifting, first raise the vehicle until its wheels are about 6 inches off the ground. Then, jounce the vehicle and double-check the contact between the adapters and the frame to be sure the vehicle is safely engaged.
3. Be certain that all four lift pads are contacting their lift points and bearing a load. It is not unusual for three lift arms to be touching the car with the fourth one free to move. If a lift arm can be moved after the car is in the air, the car is unevenly loaded. Lower the car and reposition the arm.
4. If a lift arm is positioned improperly, lower the vehicle slowly to the ground and reposition the arm.

NOTE: *Some lifts have swing arm restraints that hold the unloaded arm in position to prevent accidental movement. However, these restraints are not designed to prevent the car from falling if it is not properly positioned.*

5. When performing vehicle repairs on a vehicle on a frame-contact lift, do not use a large

(continued)

(continued)

prybar or do anything else that might knock the vehicle off the adapters. When tight bolts are encountered, it is best to use an air impact wrench on them.
6. Be sure that the lift contact points on the vehicle are in good condition and that there is no oil or grease on them.
7. Some lifts have different length arms in the front than they do in the rear. These are called asymmetrical arms. Be sure to consult the manufacturer's instructions before using this type of lift.
8. Newer lifts have a safety locking device that holds the post should a hydraulic failure occur (**Figure 2.24**). Be sure it is engaged. If the lift is not raised high enough for it to engage or if the lift is not equipped with a safety device, use four high-reach supplementary stands (tall jack stands).

NOTE: *The locking device must be disengaged before the lift can be lowered.*

9. The lift area should be clean. There should be no grease or oil on the floor. Hoses, extension cords, and tools should be placed back where they belong.
10. Insurance companies usually prohibit customers from being in the lift area. Do not allow customers to drive their cars onto any lift.
11. Be sure that the lift has adequate capacity to lift the weight of the vehicle. If the vehicle contains any loads—inside the passenger compartment, in the trunk, or in the bed of a pickup—the center of gravity will be affected and the vehicle will be unsafe to lift.
12. Be sure that the lift is all of the way down before attempting to drive a car onto or off it.
13. Before lowering a vehicle, be sure to alert anyone nearby. Check again to be certain that no tools or equipment are below the car. All of the car's doors should be closed.

AIR COMPRESSORS

Compressed air is used to power tools in the shop in much the same way that electricity is used. One of the advantages of using compressed air is that there is no danger of electrical shock, especially in auto body work where water is sometimes used when sanding. Compressed air can be used to blow off parts, to apply paint with a spray gun, to power hand tools, or to lift vehicles with an air jack.

Air compressors resemble small engines (**Figure 2.25**). They have one or more pistons and one-way

Figure 2.24 A locking device is engaged when the lift is at its highest travel. *(Courtesy of Automotive Lift Institute)*

check valves. When a piston moves down, a check valve allows air to be drawn into the compressor's cylinder. When the piston moves back up, another check valve directs the air into a holding tank, called a *receiver tank* (**Figure 2.26**).

Air Lines

Air delivery hoses and lines must be large enough so that sufficient air gets from the compressor to the tool. When lines are too small, the pressure must be raised to compensate. Raising the pressure heats the air, which is hard on tools and causes the compressor to work harder. When a line is longer than 150 feet in length, a minimum pipe size of 1 inch is recommended.

All air lines must be at least as big as the outlet on the compressor, and there can be no small diameter sections of line between the compressor and the tool. A small line acts as an orifice, dropping pressure at the tool. This happens when someone installs a *hose coupling* of the wrong size. A 5/16-inch hose with a 1/4-inch coupling will permit less air flow after the air passes the restriction caused by the smaller coupling. This also happens when someone mistakenly installs an air regulator that is too small; for instance, a 3/8-inch regulator in a 1/2-inch air line. **Figure 2.27** shows a typical air tool setup.

Air Compressor Maintenance

Water is produced as outside air is condensed by the compressor into the receiver tank. The tank requires

Figure 2.25 An air compressor resembles a small gasoline engine. *(Courtesy of Campbell Hausfeld Inc.)*

Figure 2.26 The receiver tank acts as a reservoir for the air. *(Courtesy of Campbell Hausfeld Inc.)*

periodic *bleeding,* or *blow down,* to drain it of water. Moisture in the air supply line is reduced if the receiver tank is drained daily.

A valve or faucet is usually located at the bottom of the tank. Some compressed air systems have water filters (traps) at the outlet of the compressor to remove the moisture in the air. Some shops have oilers in the line to provide lubrication for air tools. Other air systems have special dryers to provide high-quality air for spray painting.

The oil level in the compressor should be checked regularly, especially if it appears that there is oil leakage. The oil should be changed every three months.

The compressor needs to breathe clean air, much the same as an internal combustion engine. The air filter must be cleaned regularly. If it becomes restricted, the compressor will overheat.

Check the drive belts regularly for signs of wear or looseness. A loose belt will slip and wear the pulley.

■ SHOP EQUIPMENT AND SAFETY

Personal safety is the highest priority of any organization or endeavor. All automotive areas of specialization have safety implications, but the chassis systems—brakes and suspension—are among the most important when it comes to safety. Customers literally put their lives, and the lives of their friends and families, into your hands when you perform repairs to their brakes or suspension systems.

At first glance, you would not suppose that working on cars could be very dangerous, other than the obvious things like having a vehicle fall on you. Accidents are always unplanned and often the result of carelessness or ignorance.

Employers are responsible for providing safety training for all their employees and for providing a safe working environment. According to federal law, each shop must have a safety training program. One shop owner in Los Angeles was prosecuted in the death of an employee for not providing proper safety training.

Employees or students are obviously responsible for their personal safety and the safety of others. Accidents that occur in an automotive shop often happen because safety considerations are not as obvious when repairing automobiles as they are in such trades as roofing or carpentry. Accidents are often caused by carelessness resulting from a lack of experience or knowledge, or from being in a hurry and taking short cuts.

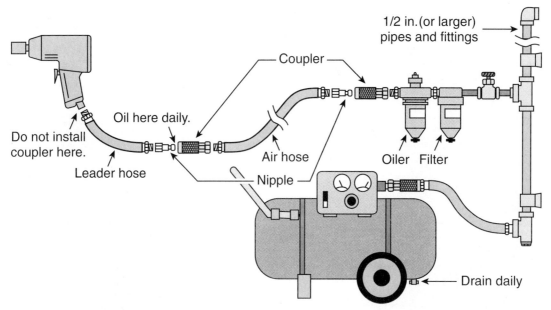

Figure 2.27 A typical setup for use of an air tool.

Injury accidents are often caused by someone other than the person injured. The person who caused the accident may feel guilty about the harm they have caused someone else. In the event of an accident, be sure to inform your instructor or supervisor, who will know what procedures to follow.

Injured persons often suffer from shock and should not be left unattended. When an injury does not appear to be serious enough to call an ambulance, a companion should go with the injured party to seek professional help.

Every shop should have someone trained to handle emergencies. The American Red Cross offers thorough first-aid training.

General Personal Safety

A first-aid kit (**Figure 2.28**) contains items for treating some of the small cuts and abrasions that often happen. Fires and accidents involving such things as lifts and battery chargers happen occasionally in automotive shops. But the most common injuries are mostly preventable injuries involving the back or eyes.

Personal protective equipment (PPE) including gloves, goggles, and respirators can prevent injuries (**Figure 2.29**).

Protecting Your Eyes

Eye injuries are a common occurrence in automotive shops. Continual use of safety glasses or goggles is recommended. Prescription safety glasses are an advantage because the user always wears them.

Wearing eye protection will prevent most eye injuries. When using machinery, eye protection is mandatory. Several types of eye protection are shown in **Figure 2.29**.

Figure 2.28 A first-aid kit.

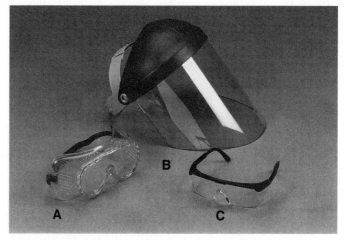

Figure 2.29 Eye protection: (A) Goggles. (B) A face shield. (C) Safety glasses.

Safety goggles or a face shield should be worn when using shop tools and equipment. When parts are pressed or pounded on, they can *explode*. Rotating tools can throw pieces of metal or grit, causing eye injuries. Face shields (see **Figure 2.29B**) are convenient because they can stay with the piece of equipment and are easily adjusted to fit your head.

To prevent eye injuries, eye protection must be worn during the following times:

- Whenever working around moving parts and machinery (grinding, cutting, drilling, washing, etc.)
- When blowing off parts with compressed air
- When working under a vehicle
- When working on air conditioning
- When flame cutting or welding

If your eye is accidentally contaminated with a dangerous liquid, flush it thoroughly in an eyewash fountain (**Figure 2.30**) or other water source.

Back Protection Safety

Tires and suspension parts can be relatively heavy. Following safe lifting procedures will prevent most back injuries (**Figure 2.31**). Follow these safety precautions when lifting:

CASE HISTORY *A technician lifted a flat tire out of the trunk of a car. As he reached forward and tugged the tire up and out of the trunk, he felt a small pop in his lower back. The result was a herniated disk in his lower back, an injury that will affect him for the rest of his life.*

1. Be sure to get help when moving heavy items. The normal tendency is to say that items are not that heavy because you dislike asking somebody for help. If something is in an awkward position for lifting, leverage and the position your back is in can make it easier for an injury to occur.
2. If an item is too heavy to lift, use the appropriate equipment.
3. Before moving a heavy item, plan the route that the item will be carried and how it is to be set back down when you get there.
4. Lift slowly.
5. Do not jerk or twist your back. Shift your feet instead.
6. Bend your knees and *lift with your legs, not your back!* Also, keep your lower back straight when lifting. Think about thrusting your stomach out.

Ear Protection

Hearing damage happens due to exposure to loud noises over a period of time. Therefore, ear protection should be worn when loud air tools are used.

NOTE: *Once it has been damaged, your hearing will not recover.*

Clothing and Hair

Clothing (shirttails) or hair that hangs out can get caught in moving machinery or under a creeper. Keep

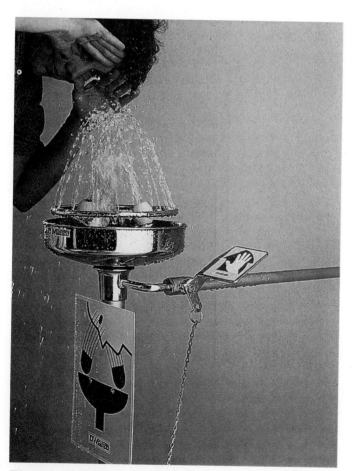

Figure 2.30 An eyewash fountain. *(Courtesy of Western Emergency Equipment)*

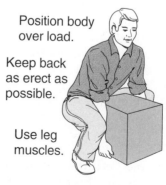
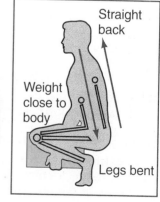

Figure 2.31 Lifting precautions.

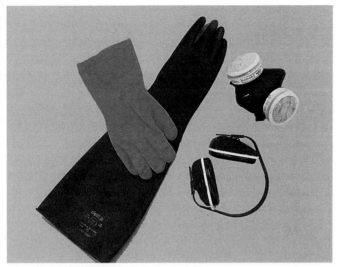

Figure 2.32 Personal protective equipment (PPE) includes gloves, a respirator, and ear protection. *(Courtesy of Tim Gilles)*

Figure 2.33 Fire extinguishers.

long hair tied back or under a cap. Shirttails should be tucked in or shop clothing can be worn over the shirt.

Shoes and Boots

Leather shoes or boots offer much better protection than tennis shoes or sandals. Soles are available that help prevent slips and resist damage from petroleum products. Boots and shoes that have the toe reinforced with steel inserts are widely available. Remember, because steel toed boots are safety equipment, they are deductible from your income tax.

Hand Protection

Cuts and scrapes are common when repairing automobiles. Be sure to keep your fingers away from moving machinery and hot surfaces. Some solvents used to clean parts can penetrate your skin. Use the correct type of glove when handling chemicals (**Figure 2.32**). Some inexpensive gloves can allow chemicals to pass through them. Chemicals and gloves are discussed in more detail later in the chapter.

■ FIRE SAFETY

Two major items should be considered when dealing with fires. Common sense is important. If a fire jeopardizes your personal safety, leave the area immediately and call for help. If you can safely remove the source of fuel or heat to a fire, do so. This might include shutting off fuel to a fuel fire or disconnecting the electrical source from an electrical fire.

Fire Extinguishers

A fire extinguisher is a portable tank that contains water or foam, a chemical, or a gas (**Figure 2.33**). There are four kinds of fires, each calling for a different type of fire extinguisher (**Figure 2.34**).

 SAFETY NOTE Be sure to have a large enough fire extinguisher on hand. An engine compartment fire often requires a larger extinguisher, depending on how far the fire has progressed.

1. A **Class A fire** is one that can be put out with water. Such things as paper and wood fuel these kinds of fires.
2. A **Class B fire** is one in which there are flammable liquids such as grease, oil, gasoline, or paint.
3. A **Class C fire** is electrical.
4. A **Class D fire** is fueled by a flammable metal such as magnesium or potassium.

Either a carbon dioxide (CO_2) or dry chemical fire extinguisher can be used on *Class B and C fires.*

A gauge on the top of a fire extinguisher tells whether it is fully charged or if the charge pressure has leaked off. Fire extinguishers in business establishments are periodically inspected by the local fire department or a fire extinguisher service company, but be sure to check the gauge on all shop fire extinguishers on a regular basis.

A popular fire extinguisher that people keep in their vehicles or boats is a 2A-10BC. The 2A means that the extinguisher can put out a 2-cubic-foot fire of a material like paper. The 10B means that it can put out a 10-square-foot fire in liquid. The C means that it can extinguish electrical fires.

For auto repair shops, the size should be at least a 20B.

Locate and check the type of fire extinguisher(s) in your shop. They should be located in a place *other* than where a fire would most likely start. For instance, do not locate a fire extinguisher right over the welding

	Class of Fire	Typical Fuel Involved	Type of Extinguisher
Class **A** Fires (green)	**For Ordinary Combustibles** Put out a Class A fire by lowering its temperature or by coating the burning combustibles.	Wood Paper Cloth Rubber Plastics Rubbish Upholstery	Water*[1] Foam* Multipurpose dry chemical[4]
Class **B** Fires (red)	**For Flammable Liquids** Put out a Class B fire by smothering it. Use an extinguisher that gives a blanketing flame-interrupting effect; cover whole flaming liquid surface.	Gasoline Oil Grease Paint Lighter fluid	Foam* Carbon dioxide[5] Halogenated agent[6] Standard dry chemical[2] Purple K dry chemical[3] Multipurpose dry chemical[4]
Class **C** Fires (blue)	**For Electrical Equipment** Put out a Class C fire by shutting off power as quickly as possible and by always using a nonconducting extinguishing agent to prevent electric shock.	Motors Appliances Wiring Fuse boxes Switchboards	Carbon dioxide[5] Halogenated agent[6] Standard dry chemical[2] Purple K dry chemical[3] Multipurpose dry chemical[4]
Class **D** Fires (yellow)	**For Combustible Metals** Put out a Class D fire of metal chips, turnings, or shavings by smothering or coating with a specially designed extinguishing agent.	Aluminum Magnesium Potassium Sodium Titanium Zirconium	Dry powder extinguishers and agents only

*Cartridge-operated water, foam, and soda-acid types of extinguishers are no longer manufactured. These extinguishers should be removed from service when they become due for their next hydrostatic pressure test.

Notes:
(1) Freeze in low temperatures unless treated with antifreeze solution, usually weighs over 20 pounds, and is heavier than any other extinguisher mentioned.
(2) Also called ordinary or regular dry chemical. (sodium bicarbonate)
(3) Has the greatest initial fire-stopping power of the extinguishers mentioned for class B fires. Be sure to clean residue immediately after using the extinguisher so sprayed surfaces will not be damaged. (potassium bicarbonate)
(4) The only extinguishers that fight A, B, and C class fires. However, they should not be used on fires in liquified fat or oil of appreciable depth. Be sure to clean residue immediately after using the extinguisher so sprayed surfaces will not be damaged. (ammonium phosphates)
(5) Use with caution in unventilated, confined spaces.
(6) May cause injury to the operator if the extinguishing agent (a gas) or the gases produced when the agent is applied to a fire is inhaled.

Figure 2.34 A guide to fire extinguisher selection.

bench or next to the solvent tank. If a fire began there, you would not be able to get to the extinguisher. Locating fire extinguishers on both sides of a common work area is a good idea.

Flammable Materials

Absorbent materials such as *grease sweep* and rags soaked in oil or gasoline should be stored in covered metal containers (**Figure 2.35**). Keeping oily materials separated from air prevents them from self-igniting, a process called *spontaneous combustion*. Flammable materials should be stored in a flammable storage cabinet (**Figure 2.36**). Grease sweep is discussed later in this chapter.

Fuel Fires. Gasoline is a major cause of automotive fires. Liquid gasoline is not what catches on fire. Rather, it is the *vapors* that are so dangerous. Gasoline vapors are heavier than air, so they can collect in low places in the shop. They can be ignited by a spark from

Figure 2.35 Keep combustibles in safety containers.

Cold fuel spray can cause a bulb to shatter, and the hot filament can start a very dangerous fuel fire.

Figure 2.36 An approved flammable storage cabinet. *(Courtesy of Justrite Manufacturing)*

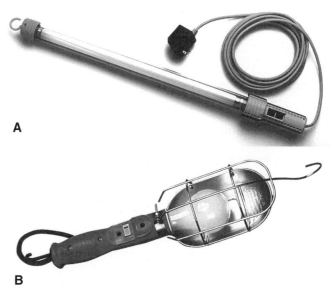

Figure 2.37 Two kinds of shop lights: (A) A fluorescent tube safety light. (B) The light shown here should be used with a shatterproof bulb. *(Courtesy of Woods Industries, Inc.)*

a light switch, a motor, electrical wires that have been accidentally crossed, or a dropped shop light.

There are two kinds of portable electric lights used in automotive shops (**Figure 2.37**). One popular kind has a fluorescent bulb that is enclosed in a plastic tube. The other uses an incandescent lamp, or light bulb. A fire hazard exists when using an incandescent lamp, especially around high-pressure fuel systems. You can also be burned on the lamp's metal shield.

Gasoline. The following procedures/precautions should be taken with gasoline:

1. Gasoline should be stored only in an approved safety container (see **Figure 2.35**), never in a glass jar.
2. Never use gasoline to clean floors or parts. Parts cleaning solvent (*Stoddard solvent* is the industry term) has a higher flash point than gasoline. A flammable liquid's flash point is the temperature at which the vapors will ignite when brought into contact with an open flame.
3. Careless cigarette smoking or failing to immediately and thoroughly clean gasoline spills can create a dangerous situation. People get used to working around gasoline and then begin to ignore how hazardous it can be.
4. Do not attempt to siphon gasoline with your mouth. Accidentally breathing gasoline into the lungs can be fatal.

A fire will go out if its source of oxygen is removed or if the temperature is lowered below the fuel's flash point. Firefighters can use a fog of water to cool a fuel fire and cut off its oxygen, but shop personnel should be knowledgeable about fire extinguishers and how to use them.

NOTE: *Do not attempt to douse a fuel fire with water. Water will cause the fire to spread because fuel is lighter than water and floats on top of it.*

Electrical Fires and Battery Explosions

If there is an electrical fire, the vehicle's *battery must be disconnected* as fast as possible so the fire can be put out. Electrical fires are prevented by disconnecting the battery before working on a vehicle's electrical system or around electrical components, such as the starter or alternator (**Figure 2.38**).

Batteries give off explosive hydrogen when they charge. Sparks must be avoided around batteries. Unbolt the ground cable first. This prevents the possibility of a spark occurring if a wrench accidentally completes a circuit between a "hot" cable and ground.

NOTE: *The ground cable is the one that is bolted to the engine block. Do not assume that ground is the negative cable. On some older vehicles, the positive cable is ground. Also, do not assume that a red cable is positive and a black cable is negative.*

Figure 2.39 shows a battery that exploded when a technician disconnected a battery charger while it was still charging. Luckily, he was wearing glasses, although he complained of ringing ears for a couple of weeks and his clothes were ruined by the spray of acid.

Figure 2.38 Remove the battery ground cable first.

Figure 2.40 Use an exhaust ventilation system when working around a running engine. (Courtesy of Tim Gilles)

Figure 2.39 This battery exploded when a technician disconnected the battery charger while it was still charging. (Courtesy of Tim Gilles)

■ BREATHING SAFETY

There are many potential breathing hazards encountered in brake and suspension work, including paints, cleaning chemicals, grinding or cutting dust, asbestos, and vehicle exhaust. Some of these hazards, such as vapors from cleaning chemicals, are only breathed by accident during service work. Others (dust and fibers) can be present in the air and must be systematically guarded against. Some materials are only dangerous after long-term exposure.

Sometimes you will need to work around a running engine. Be sure the parking brake is firmly applied and try to stand to the side of the vehicle. Exhaust gas can contain large amounts of carbon monoxide. Protect yourself and others from exhaust gas by using an exhaust evacuation system (**Figure 2.40**).

Asbestos hazards are covered later in this chapter.

■ ELECTRIC SHOCK

Twelve-volt direct current (DC) electrical systems, like the ones used in automobiles, do not cause serious electrical shock's unless the engine has a high discharge-type ignition. Shop equipment, however, is powered by either 110V or 220V alternating current (AC), which can be very dangerous.

Electrical Safety. The following procedures/precautions should be taken with electricity:

- Be sure that a tool is turned off before plugging it in so that a spark does not jump from the outlet to the plug.
- Three-wire electrical tools are the best choice for commercial work. The extra terminal is for ground (**Figure 2.41**). If you use a homeowner-type tool with a two-wire plug it should be double insulated.

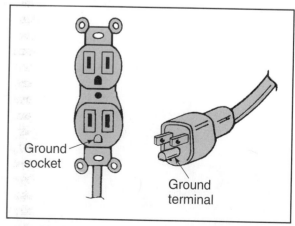

Figure 2.41 The third wire terminal is for ground.

MACHINERY AND TOOL SAFETY

There are many types of machinery used in automotive shops. Common sense will prevent most of the injuries and accidents associated with machinery:

- Do not talk to someone who is operating a machine.
- Do not talk to someone when you are operating a machine.

Hydraulic Presses

There are many sizes of presses. Some are mechanical, such as the arbor press (**Figure 2.42**), and some are hydraulic. The largest press (**Figure 2.43**) is the most versatile, having both fast and slow pumping speeds and the ability to separate axle bearings, which sometimes require 25 tons of pressure.

Many special-use press fixtures are available. A *bearing separator plate* is often used for press work. Be sure to support it where the bolt holds the two halves of the tool together (**Figure 2.44**). If the separator is installed in the press 90 degrees to the correct position, the bolts will be bent and the tool can be damaged.

Press Safety. A hydraulic press can be very dangerous because parts being separated are under enormous pressure—so much, in fact, that they can explode. Common automotive presses develop hydraulic force in the range of 20 to 50 tons (40,000–100,000 pounds). Bearings have been known to explode like a bomb, injuring the user with shrapnel. Be sure to use common sense:

- Use applicable safety guards.
- Wear face protection.
- When a press has fast and slow speeds, use the fast speed to press parts whenever possible. This position does not provide full hydraulic pressure.
- Use extreme caution as the pressure applied to the part becomes higher.

Drills

Drills are used often in vehicle repair. When a part is off a vehicle, a stationary drill press is used (**Figure 2.45**). Other times a portable drill motor is best. There are a variety of sizes of handheld drill motors (**Figure 2.46**).

You will need to consider the speed of a drill press or portable motor depending on the type of metal and the size of hole being drilled. A cutting lubricant is used to drill all metals except cast iron or aluminum, which can be drilled dry or cooled with kerosene.

The part that grips the drill bit is called the *spindle chuck,* or chuck (**Figure 2.47**). A drill bit is tightened between the jaws of the chuck with a *chuck key.* Be sure to remove the chuck key before operating the drill.

A drill motor is classified according to the largest drill bit that its chuck can accommodate. The following are approximate speed ranges for different sized drill motors:

- ¼-inch drill motor—1,500–2,000 rpm (revolutions per minute)
- ⅜-inch drill motor—1,000 rpm
- ½-inch drill motor—500 rpm

Notice that the larger drill bits turn at slower speeds.

Some drills are variable speed; others are also reversible. Drills usually have a button located next to the trigger that can be used to hold the drill in the On position. To shut the drill off, just press the trigger again.

Drilling Safety Precautions. Drill bits can break and chips of metal can fly so *always* wear eye protection.

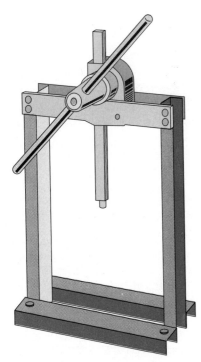

Figure 2.42 An arbor press.

Figure 2.43 A large, 50-ton press. *(Courtesy of Snap-on Tools)*

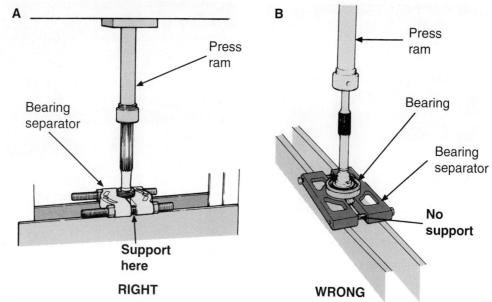

Figure 2.44 Support the bearing separator under the bolts so the tool is not damaged.

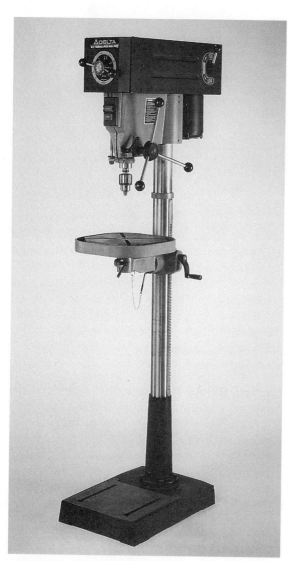

Figure 2.45 A drill press. (Courtesy of Delta International Machinery Corp.)

When drilling, *release pressure occasionally* to allow chips to break off before they become too long and dangerous. Be especially careful when a drill bit starts to break its way through the bottom of the piece being drilled. This is when the drill bit is likely to catch.

Grinder

Almost every shop has a grinder (**Figure 2.48**). Grinders are either bench or pedestal mounted (floor type). A typical grinder is equipped with a grinding wheel on one side and a wire brush for cleaning parts on the other side. Different size motors are available. The motor should be powerful enough to maintain speed under pressure. The grinding wheel is kept square by dressing it with the tool shown in **Figure 2.49**.

A grinder is often used to sharpen tools. A water pot is usually attached to its front because when metal is ground it must be constantly quenched. If the metal is allowed to become too hot, either of two things can occur:

1. Metal that is heated and then quenched will be made more brittle.
2. Metal that is heated and allowed to cool slowly will be softened.

Grinder Safety Precautions. Serious accidents can occur if a grinder is not used properly. Use the following safety precautions:

1. *Stand to the side when starting the motor.* The grinding wheel is more likely to explode during startup because of the inertia of the wheel.
2. *Wear face protection.*
3. Keep the face of the grinding wheel clean and true (see **Figure 2.49**).

48 SECTION 1 INTRODUCTION AND SAFETY

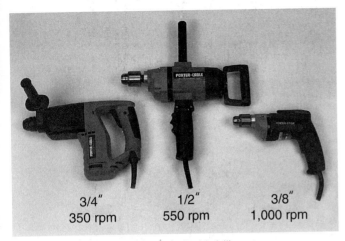

Figure 2.46 Various handheld drill motors.

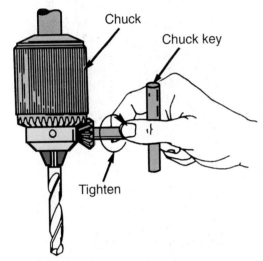

Figure 2.47 The drill bit is tightened into the chuck with a chuck key.

Figure 2.48 A bench grinder with a wire wheel. *(Courtesy of Tim Gilles)*

Figure 2.49 Dressing a grinding wheel. *(Courtesy of Tim Gilles)*

Figure 2.50 Position the tool rest as close to the grinding wheel as possible. *(Courtesy of Tim Gilles)*

4. *Position the tool rest as close to the wheel as possible* so that nothing can get trapped between the wheel and the tool rest (**Figure 2.50**).
5. *Do not grind on the side of the grinding wheel. This can cause the wheel to explode.*

Wire Wheel Safety Precautions. Grinders in automotive shops are usually equipped with a wire wheel on one side. The wire wheel is used to remove carbon, paint, or rust from parts and to de-burr freshly machined parts. Wear leather gloves if they are available, and do not push too hard against the wheel. Finger injuries can result if your hand slips. Vise grip pliers can be used to hold smaller parts.

Compressed Air Tool Safety Precautions

Many common tools used by technicians are air powered (**Figure 2.51**). Compressed air is very useful to a

Chapter 2 Getting Started in the Shop: Lifting a Vehicle/Shop Equipment and Safety 49

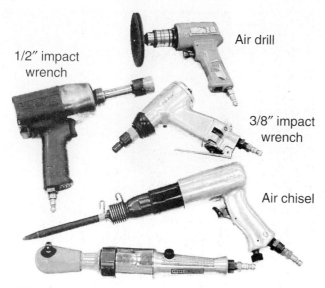

Figure 2.51 Miscellaneous air tools. (Courtesy of Tim Gilles)

Figure 2.52 A safety blowgun (top) and a rubber-tipped, high-pressure blowgun (bottom).

Figure 2.53 Regular and impact sockets. (Courtesy of Tim Gilles)

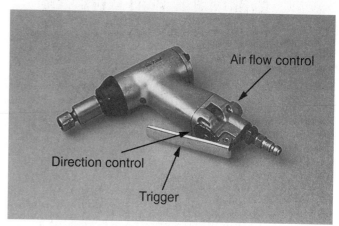

Figure 2.54 A ⅜" impact wrench with a universal impact socket. (Courtesy of Tim Gilles)

technician but can be dangerous when used improperly. Horseplay has no place in a shop. A blast of air can result in a broken eardrum. Blowing compressed air into a body orifice can result in death. The following are some safety tips to follow when using compressed air:

- Always wear eye protection when blowing off parts. Debris can blow into eyes.
- Do not blow air against your skin; the high-pressure compressed air used in auto repair shops can penetrate skin.
- Compressed air is used to power chassis grease guns. Pressurized grease can penetrate skin.
- Hold onto the air hose when uncoupling an air line so it does not fly through the air. When possible, bleed off the air from an air line before uncoupling an air hose.
- Rubber-tipped blowguns are not safety regulated like other blowguns, which are approved by the federal Occupational Safety and Health Administration (OSHA). When held against a part, full shop air pressure is available at the tip. **Figure 2.52** compares safety and rubber-tipped blowguns.

Impact Wrench Safety. Air impact wrenches are used extensively in auto repair. They save time and make work easier when used with knowledge and common sense. The following are some safety tips when using impact wrenches. Use approved *impact sockets*; ordinary chrome sockets can break (**Figure 2.53**). Be sure that the socket is secured to the air socket. The clip at the end of the tool's square drive can become worn so that it no longer holds the socket. Do not turn on the impact wrench unless the socket is installed on a nut or bolt, especially when using a wobble socket (**Figure 2.54**). The socket can fly off the impact wrench, possibly injuring someone or damaging a vehicle.

When the impact wrench fails to loosen a fastener, use a large breaker bar. This is also true when using a ratchet, which can be broken by excessive force (**Figure 2.55**).

Air Chisel Safety. Air chisels are used for such things as cutting metal and replacing front suspension parts. An injury can occur if an air chisel is not used properly. Before pulling the trigger, be sure to have the tool

SECTION 1 INTRODUCTION AND SAFETY

Figure 2.55 (A) Internal parts of a ratchet head make up a ratchet repair kit. (B) A breaker bar prevents the need for a ratchet repair kit. *(Courtesy of Tim Gilles)*

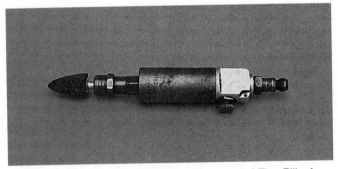

Figure 2.56 A die grinder. *(Courtesy of Tim Gilles)*

bit against the workpiece. Otherwise, the tool might fly out of the air chisel.

Die Grinder/Air Drill Safety. Air drills and die grinders are used with small grinding wheels, burrs, and wire brushes. A die grinder (**Figure 2.56**) is similar to an air drill but turns at a much higher speed. There are several safety precautions to follow when using these tools:

- These tools are very loud so be sure to use ear protection.
- Die grinders turn at speeds of up to 20,000 rpm. When using a grinding wheel or wire wheel at high speeds, be sure that the wheel is rated for such use (**Figure 2.57**).

Figure 2.57 Die grinders must be used with wheels with a high speed rating. *(Courtesy of Tim Gilles)*

Hand Tool Safety

Cuts and scrapes are common when working on cars. Experienced technicians cut themselves less often than beginners. This is because they use hand tools properly. Here are some general safety procedures to follow when using hand tools:

1. Before using a tool, inspect it to see that it is not damaged or cracked.
2. Maintain tools in safe working condition.
3. Do not use worn or broken tools.
4. When using screwdrivers, do not hold small components in your hand in case the blade slips.
5. Do not put sharp tools in your pocket.
6. Be sure to always use eye protection when using hammers and chisels.
7. Mushroomed chisels and punches should be reground before use (**Figure 2.58**).

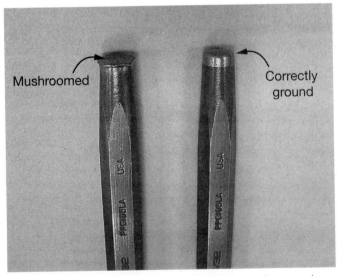

Figure 2.58 Regrind a punch with a mushroomed head. *(Courtesy of Tim Gilles)*

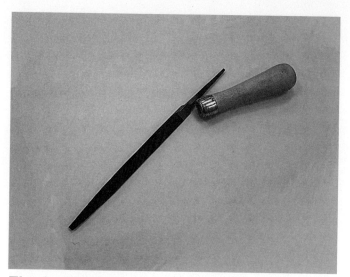

Figure 2.59 Install a handle on the tang of a file. (Courtesy of Tim Gilles)

8. Be sure that a file has a handle installed on its tang before using it (**Figure 2.59**).

■ WELDING EQUIPMENT

There are two main types of welding equipment, electric and gas. Both types melt metal into a **puddle**, which is moved by the operator to complete the weld. This is a mechanical skill and requires some practice. Electric welding is called **arc welding**, or *stick* welding. Gas welding is called **oxyacetylene welding**, or *torch* welding.

Oxyacetylene welding has an advantage over arc welding in that it is more versatile. Gas welding equipment can be used to cut and heat metals. A torch is often used to heat rusty or seized parts, such as exhaust components or stuck brake bleeder screws, for easier removal. Press-fit parts can be heated for easier installation. The obvious disadvantage is that there is an open flame, which can result in burns and start fires.

Arc welding is inexpensive and fast but usually requires a 220-volt electrical hookup, which some shops do not have. A ground (negative) clamp is hooked to the piece to be welded. A welding rod attached to the positive clamp is "struck" against the workpiece in a motion similar to striking a match. The electric arc melts the metal when electrons flow across the arc between the positive and negative electrodes.

Oxyacetylene welding uses two compressed gases, oxygen and acetylene, stored in separate steel cylinders (**Figure 2.60**). Hoses connect the cylinders to the torch where the gases are mixed. When the gases are mixed in the correct proportion and ignited, intense heat is created at the tip of the torch. The heat is intense enough to melt steel for welding or cutting.

Each cylinder is equipped with a regulator that controls the amount of pressure of the gas coming out of the cylinder. The regulator has two gauges: one shows the pressure inside of the cylinder and the other shows the pressure going to the torch.
The following are welding safety tips:

■ Be certain to mark hot metals with chalk so some someone is not accidentally burned.

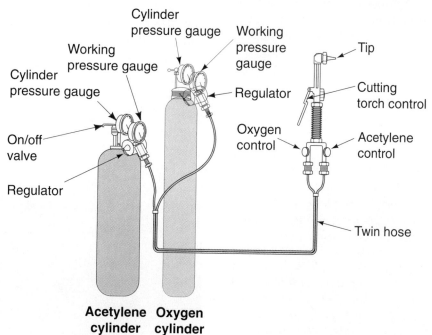

Figure 2.60 Oxyacetylene gas welding and cutting equipment.

- Arc welding produces *ultraviolet rays,* which can cause cataracts. A helmet with the appropriate shade of lens must be used to prevent damage to eyes. Also, light reflected from arc welding can flash burn unprotected eyes. This is why a protective screen is often used when welding.
- Welding can result in splatter of molten metal. Be sure to wear appropriate protective gear, including leather gloves; protective clothing; and face protection.
- Bystanders must have appropriate protective gear, as well.
- Cutting with a torch can result in a great deal of flying molten metal. Be sure that flammable materials are not in the area and that molten metal does not burn through the hoses.

GENERAL SAFETY AROUND AUTOMOBILES

Do not leave things on the floor of the shop that people can trip on. Hydraulic service jacks (floor jacks) should be left with the handle up, rather than hanging down low where someone could trip on them. Floor creepers should be placed vertically against a wall or workbench when not in use (**Figure 2.61**).

Before a Test Drive

Before driving a customer's car, remember to check the operation of the brakes and look at the condition of the tires. Do not test-drive a car with obvious safety hazards until they are corrected. It makes no sense to test-drive a car with dangerous brakes.

Working Around Belts

Part of chassis repair includes the power steering pump, which is belt-driven. Belts can be dangerous. One of the most common injuries among farmers is lost fingers because farm machinery has many belts. Fingers are often severed when they are caught between a belt and pulley.

Be sure that a helper understands what you are asking him or her to do. Assuming that they under-

Figure 2.61 Stand a floor creeper on end when not in use. *(Courtesy of Tim Gilles)*

stand can result in an accident like the one described in the Case History.

Before attempting a belt adjustment, be sure that the keys are out of the ignition. If someone cranks the engine over, your fingers can be cut off.

SHOP HABITS

Automotive shops can be kept cleaner when technicians get into the habit of using shop towels when working. Greasy, oily tools and hands should be wiped clean to prevent the mess from being spread around the rest of the shop.

Common sense around the shop dictates that spills be cleaned up with an absorbent material, such as rice hull ash or kitty litter grease sweep (**Figure 2.62**). Grease sweep is swept up and reused until it becomes too wet. It is better to use it when slightly wet because you can avoid the dust that results from new grease sweep.

A technician was installing a distributor and was adjusting the timing on an engine. He turned the engine to the correct position by grasping the fan belt and turning it. He asked a helper to turn the key on because he wanted to make the distributor trigger a spark. The helper turned the key all the way on and turned the engine over. The mechanic still had his fingers between the pulley and belt. When the crankshaft started to turn, he instinctively pulled his hands back. He was lucky not to lose any fingers but lost all of the fingernails on his right hand.

Figure 2.62 Grease sweep. *(Courtesy of Tim Gilles)*

One of the results of environmental regulation is that grease sweep is classified as hazardous if it is used to soak up used motor oil. Waste disposal companies provide a service in which superabsorbent cloths are used to soak up spills. The disposal companies then collect the cloths for proper treatment.

NOTE: *Because saturated grease sweep can be flammable, it must be stored in a metal can with a lid.*

Bioremedial oil absorbent products are newer materials that are being used instead of grease sweep. These products have microbes that "eat" oil or fuel, converting it to harmless CO_2 and water. Concrete floors cleaned with this material are left clean and slip resistant. The biggest advantage to this method is that it reduces or eliminates the need for hazardous disposal.

There are also nontoxic water-based degreasers. Be aware of regulations regarding sewage treatment. Do not allow the runoff from cleaning operations to enter a sanitary or storm drain.

■ DEALING WITH TOXIC MATERIALS AND CHEMICALS

There are many hazardous materials common to the automotive workplace. Handling these materials is regulated by various government agencies.

Hazardous materials can be absorbed through the skin, ingested by eating or drinking, or they can enter your bloodstream through the lungs by breathing fumes.

Hazards in Dealing with Asbestos

In the past, asbestos was the primary material in brake linings. It was still used into the early 1990s but was removed from friction materials because it was a health hazard. The U.S. and Canadian governments banned asbestos in new and replacement linings made in the two countries. Imported brake linings were not banned, however. Although the chance of encountering asbestos in brake materials has become quite small over time, asbestos might still be a part of the dust you are exposed to during brake work.

Some materials can pass right through a respirator. Asbestos is a good example. In fact, the respirator shown in **Figure 2.63** will probably not completely protect your lungs from asbestos fibers. Asbestos is a long crooked fiber, very small in diameter. It is so small a special **high-efficiency particulate air filter,** called a **HEPA filter,** is required for breathing protection.

Your lungs are designed to be able to rid themselves of inhaled impurities. Asbestos, however, is a sticky fiber that is absorbed into the lungs and irritates tissue. Asbestos is known to cause damage to lungs and has been linked to a specific cancer of the lung cavity. It causes asbestosis, an irreversible disease where breathing becomes more and more difficult.

Figure 2.63 This respirator will not provide complete protection from asbestos. *(Courtesy of Tim Gilles)*

NOTE: *Combining lung irritants like asbestos with cigarette smoking creates an especially dangerous situation.*

Modern brake friction materials have evolved to the point that they handle heat as well as or better than asbestos. Today's cars use smaller disc brakes that must be able to absorb more heat and run at higher temperatures. When brake linings overheat, they become glazed and no longer function effectively. Asbestos pads become glazed at lower temperatures than modern friction materials. Asbestos linings are also prone to faster wear at the higher temperatures in the modern braking system. Chapter 8 discusses brake friction materials in detail.

Controlling Brake Dust

Whether or not the brakes are made of asbestos, brake dust can include copper or iron particles that can also be hazardous to breathe. Be certain to observe proper dust control measures. Federal safety standards limit a worker's allowable exposure to brake dust to no more than 0.1 fiber per cubic centimeter (cc) of air during an 8-hour workday. Brake dust equipment includes special vacuum cleaners, brake washers, respirators, and dust masks. Dry brushing is expressly prohibited during brake cleaning.

HEPA Vacuum

One asbestos cleaning method found more often in the earlier years of dust control uses a special apparatus called a *HEPA vacuum* to draw off the dust and capture the particles in a special filter before they can enter the air (**Figure 2.64**). A regular vacuum is insufficient. Asbestos will pass right through it.

A HEPA vacuum completely encloses the brake drum or disc assembly. Gloves and a window allow the technician to clean the brake assembly. After using the vacuum to suck up loose material, other

54 SECTION 1 INTRODUCTION AND SAFETY

Figure 2.64 A HEPA vacuum can vacuum asbestos without letting it escape into the shop environment.

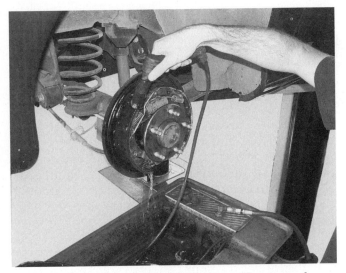

Figure 2.65 A brake parts washer. *(Courtesy of Tim Gilles)*

particles are dislodged using compressed air and are sucked into the vacuum.

NOTE: *Be sure to inspect the vacuum enclosure to see that it is tight against the brake backing plate before beginning work.*

Brake Washer

A more popular method of cleaning brakes is to wash the brake assembly at each wheel with a special brake washer using water and an organic solvent, or wetting agent, prior to disassembly of the parts (**Figure 2.65**). The technician should attempt to force the wetting agent to flow through the back edges of the brake drum into the brake assembly *before* the drum is removed. When a drum is removed first, a quantity of brake dust can escape into the surrounding air. A catch basin is positioned under the brake assembly. After the drum is removed, the brake parts are thoroughly wetted before you start to clean.

Asbestos Disposal Regulations

The Environmental Protection Agency (EPA) regulates the handling of asbestos waste. In most communities, it is acceptable to bury asbestos vacuum bags and waste materials in the local garbage dump or landfill provided they are sealed in a way that they cannot leak. A vacuum filter that is full must be wetted with a mist of water before removing it from the vacuum. Filters are sealed and labeled before disposal. When using a brake washer for asbestos protection, if the system has a filter it must be disposed of in the same way as a HEPA filter.

An alternative method for shops that do fewer than five pairs of brakes in one week uses a low-pressure water system, or fine mist spray bottle, to be used to wet the parts before beginning work. All parts are then wiped clean with a cloth that must be disposed of in an approved, properly labeled container or laundered in a way that prevents the release of asbestos fibers.

Protecting Your Skin

Your skin protects your body. However, the molecules of some toxic liquids are smaller than the membrane of human skin that allows sweat to escape. These liquids can penetrate the skin and enter your bloodstream. The use of protective gloves has become common in the automotive industry. Be certain the gloves you use provide the necessary amount of protection. The material of some gloves allows the same toxins that can penetrate your skin to penetrate the gloves. Obviously, if you have a cut or scrape, materials can then enter your bloodstream.

Dermatitis

Automotive technicians often receive chronic exposure to cleaning materials. All organic solvents can cause skin problems. Repeated contact causes the skin's protective fats and oils to dissolve. When the skin's protective barrier is removed, it becomes dry and cracked, leaving it vulnerable to absorption of toxic materials.

> **SHOP TIP** Blue nitrile gloves will protect you from many chemicals that common latex cannot protect against. Many technicians routinely wear blue nitrile gloves on their hands at all times. Nitrile gloves are not damaged by most solvents. They also provide some measure of protection from skin damage and penetration of the skin by solvents.

Using a petroleum-based solvent to clean parts by hand without wearing gloves is common. Years of this type of abuse can result in a painful condition called *dermatitis,* in which the skin on the hands cracks, and sometimes bleeds and peels. Dermatitis can also be caused by use of the wrong soap for the dirt being removed. Special hand lotions and creams are designed to treat dermatitis. Be sure to wash your hands thoroughly before applying a lotion or cream. If your hands are burned or swollen from solvent contact, seek medical attention.

NOTE:

- *Do not put hand creams on unclean skin as treatment. Hand cream can seal in the solvent irritant, causing even more damage.*
- *Be sure the gloves you use do not allow penetration by the solvent being used. Neoprene, an oil-resistant artificial rubber commonly used for hoses and gloves, protects against some types of solvents but not others. Latex, a material commonly used in making the surgical gloves popular with many mechanics, will also allow penetration by some chemicals.*

Barrier Creams

To aid in cleaning hands, **barrier creams** are often rubbed on the skin prior to working in a greasy environment. Although these creams are quite effective in making hand cleaning easier, they do not protect the skin against penetration by solvents and are not a substitute for gloves.

Material Safety Data Sheets

When a shop uses a hazardous material, the employer must supply employees with a material safety data sheet (MSDS) as shown in **Figure 2.66**. An MSDS provides details about the chemical's composition, lists possible health and safety problems, and gives precautions for its safe use. The supplier of a chemical is required by law to provide an MSDS upon request.

In addition to your local parts supplier, the Internet is a convenient source for accessing material safety data sheets. Repair shops collect the sheets for the materials they use and put them in a binder located where employees have ready access to them.

An MSDS must include any hazardous materials that are contained in a product. It lists the percentage of the product that is made up of each of the hazardous ingredients. For instance, an aerosol brake cleaner might be made up of 55 percent perchloroethylene, 25 percent methylene chloride, and so on. The MSDS lists exposure limits, which estimate the danger of short- and long-term exposure to the material. Flammable limits for a material are also listed on an MSDS. A material's *lower explosive limit* represents the lowest vapor concentration at which it will ignite.

```
HEXANE
==========================================================
MSDS Safety Information
==========================================================
Ingredients
==========================================================
Name: HEXANE (N_HEXANE)
% Wt: >97
OSHA PEL: 500 PPM
ACGIH TLV: 50 PPM
EPA Rpt Qty: 1 LB
DOT Rpt Qty: 1 LB
==========================================================
Health Hazards Data
==========================================================
LD50 LC50 Mixture: LD50:(ORAL,RAT) 28.7 KG/MG
Route Of Entry Inds _ Inhalation: YES
Skin: YES
Ingestion: YES
Carcinogenicity Inds _ NTP: NO
IARC: NO
OSHA: NO
Effects of Exposure: ACUTE:INHALATION AND INGESTION ARE HARMFUL AND MAY BE FATAL.
INHALATION AND INGESTION MAY CAUSE HEADACHE, NAUSEA, VOMITING, DIZZINESS, IRRITATION
OF RESPIRATORY TRACT, GASTROINTESTINAL IRRITATION AND UNCONSCIOUSNESS. CONTACT
W/SKIN AND EYES MAY CAUSE IRRITATION. PROLONGED SKIN MAY RESULT IN DERMATITIS (EFTS
OF OVEREXP)
Signs And Symptions Of Overexposure: HLTH HAZ:CHRONIC:MAY INCLUDE CENTRAL
NERVOUS SYSTEM DEPRESSION.
Medical Cond Aggravated By Exposure: NONE IDENTIFIED.
First Aid: CALL A PHYSICIAN. INGEST:DO NOT INDUCE VOMITING. INHAL:REMOVE TO FRESH AIR. IF
NOT BREATHING, GIVE ARTIFICIAL RESPIRATION. IF BREATHING IS DIFFICULT, GIVE OXYGEN.
EYES:IMMED FLUSH W/PLENTY OF WATER FOR AT LEAST 15 MINS. SKIN:IMMED FLUSH W/PLENTY
OF WATER FOR AT LEAST 15 MINS WHILE REMOVING CONTAMD CLTHG & SHOES. WASH CLOTHING
BEFORE REUSE.
==========================================================
Handling and Disposal
==========================================================
Spill Release Procedures: WEAR NIOSH/MSHA SCBA & FULL PROT CLTHG. SHUT OFF
IGNIT SOURCES:NO FLAMES, SMKNG/FLAMES IN AREA. STOP LEAK IF YOU CAN DO SO W/OUT
HARM. USE WATER SPRAY TO REDUCE VAPS. TAKE UP W/SAND OR OTHER NON-COMBUST MATL &
PLACE INTO CNTNR FOR LATER (SU PDAT)
Neutralizing Agent: NONE SPECIFIED BY MANUFACTURER.
Waste Disposal Methods: DISPOSE IN ACCORDANCE WITH ALL APPLICABLE FEDERAL, STATE AND
LOCAL ENVIRONMENTAL REGULATIONS. EPA HAZARDOUS WASTE NUMBER:D001 (IGNITABLE
WASTE).
Handling And Storage Precautions: BOND AND GROUND CONTAINERS WHEN TRANSFERRING LIQUID.
KEEP CONTAINER TIGHTLY CLOSED.
Other Precautions: USE GENERAL OR LOCAL EXHAUST VENTILATION TO MEET
TLV REQUIREMENTS. STORAGE COLOR CODE RED (FLAMMABLE).
==========================================================
Fire and Explosion Hazard Information
==========================================================
Flash Point Method: CC
Flash Point Text: 9F,_23C
Lower Limits: 1.2%
Upper Limits: 77.7%
Extinguishing Media: USE ALCOHOL FOAM, DRY CHEMICAL OR CARBON DIOXIDE. (WATER MAY BE
INEFFECTIVE).
Fire Fighting Procedures: USE NIOSH/MSHA APPROVED SCBA & FULL PROTECTIVE
  EQUIPMENT (FP N).
Unusual Fire/Explosion Hazard: VAP MAY FORM ALONG SURFS TO DIST IGNIT SOURCES & FLASH
BACK. CONT W/STRONG OXIDIZERS MAY CAUSE FIRE. TOX GASES PRDCED MAY INCL:CARBON
MONOXIDE, CARBON DIOXIDE.
==========================================================
```

Figure 2.66 A material safety data sheet (MSDS).

SOLVENTS COMMON TO THE AUTOMOTIVE INDUSTRY

Solvents are liquid organic chemical compounds used to dissolve solid substances. They are very popular in the automotive industry for cleaning parts because they work so well and can make jobs go much faster. Solvents are mostly made from petroleum or synthetic materials, and some can dissolve plastics and resins like those used in gaskets and sealers. They also evaporate quickly, leaving relatively little residue. A high evaporation rate is important for applications like gasket cements and paints, which must dry quickly.

Hazards Related to Solvents

All solvents are toxic to a certain extent, however, and must be used with care. Used correctly, solvents can be very handy tools. Always read the MSDS for any solvent you plan to use.

There are several chemical classes of solvents. Most are flammable and some can explode. Some can react with other substances or heat to create a different hazardous material. For example, chlorinated solvents are not flammable but when exposed to flame can become phosgene gas. Phosgene gas is the nerve gas that killed so many soldiers in World War I.

Symptoms of solvent exposure that show up suddenly are called acute. Acute symptoms include rashes and burns, nausea, or headaches. Solvents can cause irritation and damage to your skin, eyes, and respiratory tract, including the nose and throat. Continued irritation can result in chronic bronchitis and other lung diseases. These are called *local effects* because the harm is caused at its original point of contact.

Read the label on the container of any cleaning material you use. Some commonly used automotive solvents, like methyl ethyl ketone (MEK), can be absorbed through the skin. When a chemical enters the body in this way, it can travel to other organs such as the liver, kidneys, muscles, heart, or brain. This is called *systemic exposure*. The liver converts solvents to less toxic substances and the kidneys filter the results of that conversion and prepare them to leave the body. The liver and kidneys can be damaged by the substances they are attempting to eliminate.

Health effects from solvent exposure can result from repeated exposure over time, or from a single exposure. *Latent effects* can take weeks or even years to show up and are often irreversible; they include cancers, heart disease, blood disease, and an inability to bear healthy children. Symptoms of chronic exposure to dangerous chemicals have been documented over time. Heart damage was linked to solvents during a 5-year period in which more than 100 people died from sniffing solvents. Benzene, heptane, trichloroethylene, and methylene chloride are some of the chemicals that can cause arrhythmias. *Arrhythmia* is a term that applies to abnormal patterns in a heart's pumping cycle.

Some solvents can cause neurological symptoms. Central nervous system involvement can result in narcosis (unconsciousness), dizziness, headaches, fatigue, and nausea. At high concentrations, symptoms can mimic drunkenness, followed by death.

The most common solvents used in the automotive industry include hexane, mineral spirits, perchloroethylene, xylene, toluene, heptane, and petroleum napthas. In the MSDS example given earlier, the solvent perchloroethylene, or perc (also called tetrachloroethylene), was listed as the major ingredient in one popular brake cleaner. Perc and toluene have been implicated as spontaneous abortion risks in fathers who have been exposed in the workplace. It is important to read the label on the container before allowing a solvent to repeatedly come in contact with your skin.

Hexane is found in gasoline, rubber cements, and many spray cleaners (**Figure 2.67**). Check the label on the container and read the MSDS. Hexane can affect the central and peripheral nervous systems. Peripheral nervous system involvement causes a slowing of the speed of nerve impulses from your spine to your arms and legs accompanied by numbness, weakness, or paralysis. This is called peripheral neuropathy. Symptoms can be similar to multiple sclerosis.

Figure 2.67 Check the labels on the chemicals in your shop. *(Courtesy of Tim Gilles)*

Another disorder of the peripheral nervous system is polyneuropathy. Symptoms include muscle spasms, leg pain and weakness, and tingling in the arms. Solvents implicated as possible causes include benzene, n-hexane, n-heptane, and toluene.

Heptane is a central nervous system depressant. Dizziness and loss of coordination can result from short-term exposure (2,000 parts per million for 4 minutes). Long-term exposure can produce minimal peripheral nerve damage.

Sometimes two or more substances react with one another and can become more hazardous. This is called synergism. One example is the higher instance of health problems among asbestos workers who also smoke. In another example, when methyl ethyl ketone (MEK) is combined with n-hexane severe nerve damage can result. Carbon tetrachloride causes serious liver damage, especially when combined with alcohol.

Some solvents are known to cause cancer in humans and animals. Benzene, found in gasoline, can cause leukemia. Experts suspect that chlorinated solvents (those whose names include "chloride" or "chloro") might be carcinogens, that is, they might cause cancer.

Solvent Safety

Many automotive solvents have been reformulated in response to health findings. If you are working repeatedly with the same chemical, it is especially prudent to read the label and familiarize yourself with the MSDS. Also, minimize your exposure as much as possible by wearing proper protective gloves and respirators.

REVIEW QUESTIONS

1. The two main categories of lifts are the in-ground lift and the _____ lift.
2. What kind of jack is used to raise the wheels off a wheel-contact lift?
3. What kind of lift has two controls, one for air and one for hydraulic oil pressure?
4. What kind of lift would you want to install if you were renting a building and had a short-term lease, in-ground or surface mount?
5. Two ways that a surface mount lift is driven are with a hydraulic pump and cylinders _____ and drive.
6. If one of the lift arms can be moved after the vehicle is in the air, what would you do?
7. When would a two-stage air compressor be used?
8. If an air compressor runs for 6 out of every 10 minutes, what is its duty cycle?
9. The useful capacity of an air compressor is measured in standard _____ per minute.
10. List the approximate speeds of the following drill motors:

    ¼" _____ rpm (revolutions per minute)

    ⅜" _____ rpm

    ½" _____ rpm
11. When the impact wrench fails to loosen a fastener, would you use a large breaker bar or ratchet?
12. Electric welding is called _____ welding. Gas welding is called _____ welding.
13. Asbestos fiber is so small that a special filter, called a _____ filter, is required for adequate breathing protection.
14. Which gloves will protect you better from many chemicals, blue nitrile or latex?
15. When a chemical enters the body through the skin and travels to other organs, this is called a _____ exposure.

PRACTICE TEST QUESTIONS

1. Technician A says a 2-ton jack will require more frequent handle movement than a 10-ton jack. True or False?
2. Fluid pressure of 10 psi acting on a 2-square-inch area will produce a force of _____ pounds.
   - A. 2.5
   - B. 10
   - C. 20
   - D. 5
3. Two technicians are discussing the use of a bearing separator in automotive press work. Technician A says to support it where the bolt holds the two halves of the tool together. Technician B says if it is rotated 180 degrees to the correct position, its bolts can be bent. Who is right?
   - A. Technician A
   - B. Technician B
   - C. Both A and B
   - D. Neither A nor B
4. Technician A says unsafe practices around a 12-volt automotive electrical system can cause electrocution. Technician B says 110-volt AC current is not dangerous, but 220-volt AC is. Who is right?
   - A. Technician A
   - B. Technician B
   - C. Both A and B
   - D. Neither A nor B
5. Technician A says that if a car is lifted improperly, parts of the suspension or steering can be bent. Technician B says that unibody cars can be lifted from the edge of the car body at the rocker panel. Who is right?
   - A. Technician A
   - B. Technician B
   - C. Both A and B
   - D. Neither A nor B
6. Technician A says that on rear-wheel drive cars, the center of gravity is usually just below the driver's seat. Technician B says that on front-wheel drive cars, the center of gravity is usually just below the driver's seat. Who is right?
   - A. Technician A
   - B. Technician B
   - C. Both A and B
   - D. Neither A nor B
7. Technician A says to position a car on a two-post surface mount hoist with equal length adapter arms so that the driver's side door can be opened all the way. Technician B says to raise the car until its wheels are about 6 feet off the ground and shake it to see if it is safely mounted. Who is right?
   - A. Technician A
   - B. Technician B
   - C. Both A and B
   - D. Neither A nor B

8. Technician A says that using a ⅜-inch regulator in the middle of a ½-inch air line will result in less airflow through the lines. Technician B says that water must be periodically bled from an air compressor. Who is right?
   A. Technician A   B. Technician B
   C. Both A and B   D. Neither A nor B

9. Technician A says a single-speed drill motor with a ½-inch chuck will turn at a faster speed than one with a ⅜-inch chuck. Technician B says a faster speed is used with smaller drill bits. Who is right?
   A. Technician A   B. Technician B
   C. Both A and B   D. Neither A nor B

10. Technician A says that metal that is heated and then quenched will be made more brittle. Technician B says that metal that is heated and allowed to cool slowly will be softened. Who is right?
    A. Technician A   B. Technician B
    C. Both A and B   D. Neither A nor B

SHOP SAFETY TEST

Select your safety test answers from the following list:

Air	Flash	Impact	Vapor
B	Fuel	Instructor/Supervisor	Vehicle support
Back	Gasoline	Legs	or jack
Belt	Grinders	Manual	Water
Blowgun	Ground	Personal protective equipment	Wheel
Data Sheet	Ground	Re-ground	Wrench
Drill	Handle	Side	
Electrical	Handle	Six	
End	HEPA	Skin	
Eye	Hydraulic jack	Smaller	
Eyes	or service jack	Spontaneous	
False	Hydraulic	Talk or speak	

1. If you are injured while working in the shop, you should inform your _____ immediately.

2. The most common injuries in an automotive shop involve the _____ and _____.

3. Lift with your _____, not your back.

4. If a _____ is burning out of control, leave the area immediately and call for help.

5. A flammable liquid fire is called a Class _____ fire.

6. Keep oily materials separated from air to prevent _____ combustion.

7. What form of gasoline is the most dangerous—liquid or vapor? _____

8. The temperature at which a flammable liquid's vapors will ignite when brought into contact with an open flame is called the _____ point.

9. Disconnect the battery _____ cable before working on a vehicle electric system.

10. Before an _____ fire can be extinguished, the battery must be disconnected.

11. Do not attempt to put out a gasoline fire with _____.

12. Never use _____ to clean floors or parts.

13. When a shop uses a hazardous material, the employer must supply employees with a Material Safety _____.

14. A _____ vacuum is a special vacuum used for asbestos and other small particles.

15. Mushroomed punches or chisels should be _____ before use.

16. Always wear _____ protection when using hammers, chisels, pullers, batteries, air-conditioning machinery, compressed air, and in other hazardous situations.

17. Always use a _____ on the tang of a file.

18. Compressed _____ can penetrate skin and must be used cautiously.

19. Do not _____ to someone who is operating a machine.

20. A _____ bit has a tendency to grab the work when it just begins to break through the metal.

21. Stand to the _____ when starting the grinder.

22. Position the tool rest on a grinder as close to the _____ as possible, so that nothing can get trapped between the wheel and the tool rest.

23. A rubber-tipped _____ works at full shop air pressure.

24. When using an air impact wrench, be sure to use an extra thick socket called an _____ socket.

25. When an air impact _____ is used with a wobble socket, the socket can fly off if the tool is started when the socket is not on the bolt head.

26. When using an _____ chisel, be sure to have the tool bit against the workpiece before you squeeze the trigger.

27. Die _____ turn at speeds of up to 20,000 rpm. Be sure to use wheels that are rated for that high a speed.

28. Because they can develop force of from 20 to 40 tons, _____ presses can cause parts to explode.

29. Three-wire electrical tools are the best choice for commercial work. The extra terminal is for _____.

30. Locate the correct lift points in the service _____ before using a hydraulic service jack or lift.

31. When lifting a vehicle on a frame-contact lift, place the lift adapters at the specified locations and raise the vehicle about _____, then shake it to see that it is firmly placed.

32. Use a _____ _____ to raise and lower a vehicle only.

33. Do not work under a jacked vehicle unless it is resting *solidly* on _____ stands.

34. Hydraulic service jacks (floor jacks) should be left with the _____ up, rather than hanging down low where someone could trip on them.

35. The keys should be out of the ignition any time a _____ is being inspected or adjusted.

36. When a floor creeper is not in use, it should be stood on _____ so it is not accidentally stepped on.

37. Molecules of some toxic liquids _____ than the skin membrane can penetrate the skin and enter the bloodstream.

38. Do not treat solvent-irritated _____ with hand cream. It can seal in the solvent irritant, causing even more damage.

39. Gloves, goggles, and respirators are called _____ _____ _____, or PPE.

40. Brake dust that does not contain asbestos is not hazardous to breathe. True _____ False _____

SECTION TWO

Brakes

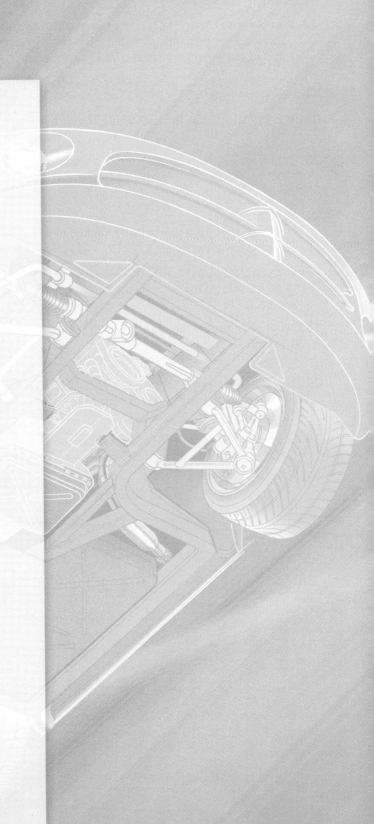

■ **CHAPTER 3**
Brakes Fundamentals

■ **CHAPTER 4**
Master Cylinder and Hydraulics

■ **CHAPTER 5**
Master Cylinder and Hydraulic System Service

■ **CHAPTER 6**
Drum Brake Fundamentals and Service

■ **CHAPTER 7**
Disc Brake Fundamentals and Service

■ **CHAPTER 8**
Friction Materials, Drums, and Rotors

■ **CHAPTER 9**
Other Brake Systems: Parking Brake and Power Brake

■ **CHAPTER 10**
Antilock Brakes, Traction, and Stability Control

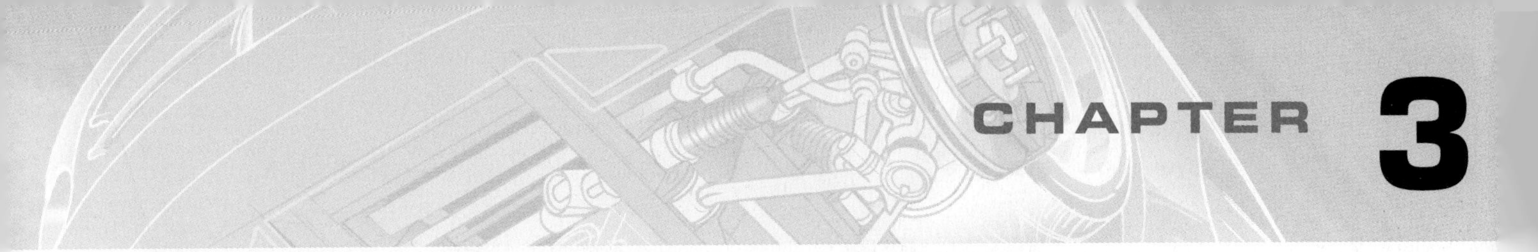

CHAPTER 3

Brakes Fundamentals

■ OBJECTIVES

Upon completion of this chapter, you should be able to:

✔ Describe the operation of a brake system.
✔ Explain the basic principles of braking, including friction, pressure, and heat dissipation.
✔ Explain how the laws of physics work to stop a vehicle through a brake system.
✔ Understand Federal Motor Vehicle Safety Standards (FMVSS).
✔ Describe the effects of surface area and friction material on the stopping ability of a brake system.
✔ Understand the effects of coefficient of friction.
✔ Understand the different DOT brake fluid specifications.

■ KEY TERMS

antilock brake system (ABS)
brake fade
brake linings
brake shoes
coefficient of friction
disc brakes
drum brakes
force
friction
gas fade
kinetic energy
lining fade
master cylinder
mechanical fade
parking brake
self-energizing
service brake
servo-action

■ INTRODUCTION

Automotive brakes are made up of two systems. The primary system, which stops the car, is called the **service brake** system. The secondary system, which consists of the **parking brake**, usually uses service brake parts. This chapter presents an overview of the operation of the braking system, including operation of drum and disc brakes, power brakes, antilock brakes, the physics of braking, brake fade, Federal Motor Vehicle Safety Standards, and hydraulic brake fluid.

■ BRAKE SYSTEM OPERATION

Automotive service brakes stop the car through the interaction of a rotating part—a drum or rotor—with a nonrotating part—friction materials called brake linings. In a modern hydraulic brake system, pressurized fluid is used to transfer motion and amplify force on friction materials applied against the surfaces of drums or rotors at the wheels. Depressing the brake pedal moves a piston in a **master cylinder** (**Figure 3.1**). The piston in the master cylinder moves a small amount, forcing fluid through the brake tubing to hydraulic cylinders at each wheel. Disc brake caliper pistons move a few thousandths of an inch and drum brake cylinder pistons move a fraction of an inch to force the friction surfaces against the drums or rotors. This results in a resistance to the flow of the brake fluid and pressure builds to amplify the pedal force exerted by the driver's foot.

The pressure acts on hydraulic cylinders at each wheel to produce **force** (**Figure 3.2**). Hydraulic brake tubing runs the length of the vehicle frame. Rubber hoses to both front wheels and above the rear axle provide a flexible connection from tubing to brake system components. Typical vehicles have three hoses, although cars with four-wheel independent suspensions have four hoses.

The brake pedal mechanically increases the leverage by about 4 to 1. Chapter 4 includes information about Pascal's Law of Pressure and describes in more detail how force is additionally multiplied by the brake hydraulic system.

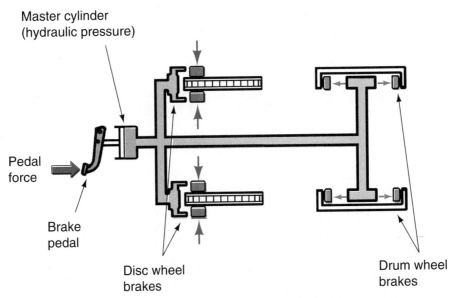

Figure 3.1 Depressing the brake pedal moves a piston in a master cylinder.

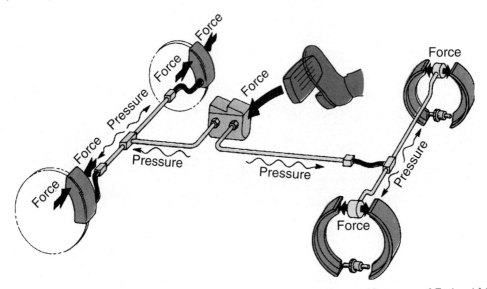

Figure 3.2 Pressure acts on hydraulic cylinders at each wheel to produce force. *(Courtesy of Federal Mogul Corporation)*

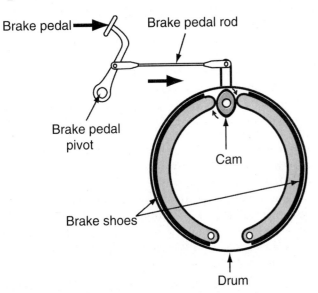

Figure 3.3 A mechanical drum brake.

Vintage Chassis

Early vehicles did not use fluid and hydraulic pressure to apply brakes. They used mechanically operated internal expanding drum brakes, which had levers and linkages (**Figure 3.3**). Luxury cars in the early 1920s were the first to use hydraulically applied brakes. In the mid-1920s some moderately priced cars began to use hydraulically operated brakes. Mechanical brakes were still used on Ford vehicles until 1938.

■ TYPES OF BRAKES

There are two main types of brakes: **drum brakes** and **disc brakes.** In the past, almost all vehicles were equipped with drum brakes on all four wheels. Today they are only found in some rear brake applications. Late-model vehicles use disc brakes in the front, with drum or disc brakes in the rear.

Drum Brakes

Drum brakes provide good initial stopping ability and also provide a means for a good, inexpensive, mechanical parking brake. In drum brakes, metal drums are bolted to the wheels. The linings and braking components are mounted on a fixed backing plate (**Figure 3.4**). Several different drum brake designs are covered in Chapter 6.

During a stop, the front shoe on a drum brake wedges into the brake drum. This motion is called **self-energizing,** or **servo-action** (see Chapter 6). Springs force the shoes back toward the wheel cylinder after braking.

Drum brakes have different kinds of self-adjusting mechanisms. With most designs, when the linings have worn a certain amount, self-adjustment takes place whether traveling forward or in reverse. Some drum brakes adjust when the parking brake is applied.

Disc Brakes

Disc brakes were developed during World War II for use in aircraft. They have a rotor and caliper similar to those used on a bicycle (**Figure 3.5**). Some American cars had disc brakes in the early 1960s and front disc brakes have been standard equipment since the early 1970s. A power assist is usually required with disc brakes.

Front disc brakes and rear disc brakes without an integral emergency brake are *self-adjusting.* More fluid is required as the caliper adjusts due to disc brake pad wear, so some master cylinders have reservoir chambers of different sizes. In this case, the larger one is for the disc brakes and the smaller one is for the drum brakes.

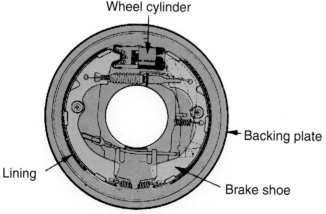

Figure 3.4 Drum brakes are mounted on a backing plate.

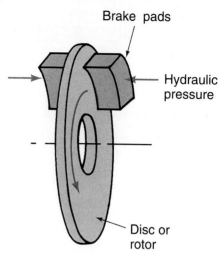

Figure 3.5 Disc brake systems have a rotor and caliper with brake pads, similar to those used on a bicycle.

Hybrid Brake System

Four-wheel drum brakes were standard equipment until the early 1970s. In the mid- to late 1960s, some cars began to use the hybrid system. A hybrid brake system combines disc brakes on the front and drum brakes on the rear. Federal brake performance standards issued in 1967 and the inherent superior qualities of disc brakes virtually eliminated the use of drum brakes for front wheels on cars built since the early 1970s. Drum brakes were still used on the rear.

Luxury and performance cars usually have four-wheel disc brake systems. Many light trucks and SUVs (sport utility vehicles) come equipped with four-wheel disc brakes as well.

Power Brakes

Disc brakes do not have the self-energizing action of dual servo drum brakes. A disc brake has the ability to apply more stopping energy at the wheels, but it also requires more application force. There are several ways to increase the force applied by the brake pedal. The most obvious way is to step harder on the brake pedal. But this is not always practical due to variations in the size and weight of the driver. For modern brake systems, the most effective and economical means of boosting brake pressure is to use a vacuum-assist brake device called a *booster,* installed behind the master cylinder (**Figure 3.6**).

Introduction to Antilock Brake Operation

The ability of the brakes to do their job is limited by the grip of the tires on the road surface. Even with the best quality tires, if a car skids there will be a loss of stopping ability and control. If the driver could release pressure on the brake pedal just before the wheel locked up, the skid could be avoided.

Most newer vehicles are equipped with computerized **antilock brake systems (ABS)** to keep the

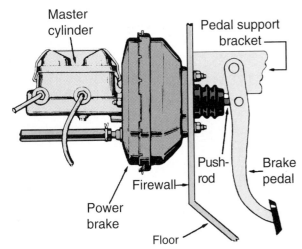

Figure 3.6 A typical power brake combined with a master cylinder.

wheels from locking up. ABS systems were first used on B-47 jet bombers in 1947 for better stopping on slick runways. Today, most large passenger and military planes have ABS. In the late 1970s, many European luxury cars imported to the United States came equipped with ABS. Antilock brakes were first used on some American cars in 1985, when Ford introduced the Teves system. Other domestic manufacturers soon had ABS systems.

During normal operation ABS systems act like conventional brakes. They only function differently when a wheel locks up. ABS systems have electronic sensors at each wheel. Each sensor monitors the speed of the rotation of the wheel by measuring how fast a toothed ring moves past the sensor (**Figure 3.7**). If a wheel locks up, the ABS system pulsates (modulates) the pressure to that wheel in much the same manner as a race car driver pumps the brakes during a turn to avoid a skid. ABS systems can pulse much faster than a human, however. They can pulsate the pressure to the brake system up to 15 times per second.

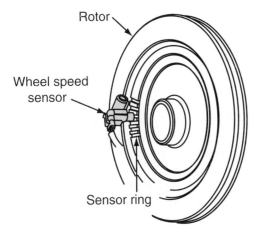

Figure 3.7 An ABS sensor and sensor ring mounted inside the rotor.

NOTE: *At stock car races, a pulsating chirp can be heard coming from ABS-equipped cars as the tires lock and unlock during turns.*

Pedal Feel. An ABS-equipped car has a different pedal feel than a car with a conventional braking system. When the ABS is activated, the driver feels a small bump followed by rapid pulsation in the brake pedal. Pedal pulsation is noticed in some systems more than in others. When the road is bumpy, when there is loose gravel, or when the road is slick, less pedal force is needed to activate the ABS system. During normal stops, there is no pedal pulsation.

■ WEIGHT TRANSFER

During a stop, weight shifts forward, causing the rear wheels to lift and the front wheels to dive. Much of the weight of the vehicle shifts forward onto the front brakes (**Figure 3.8**) so rear brakes typically have a longer service life than front brakes. A vehicle's weight distribution is determined to a great extent by the positioning of the heavier powertrain components—the engine and the transmission. In a rear-wheel drive car 60–70 percent of the weight shifts to the front during a stop. In a front-wheel drive car, over 80 percent of the weight shifts forward (**Figure 3.9**).

Function of Brakes in Stopping

The more a car weighs and the faster it is traveling, the greater its braking requirements will be. The amount of kinetic energy to be changed into heat during a stop can amount to several times the power developed by the engine. The temperature in the brake linings can

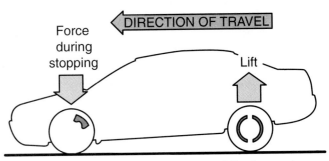

Figure 3.8 During a stop, weight shifts forward, increasing demands on front brakes.

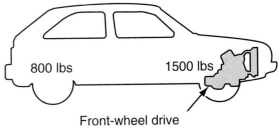

Figure 3.9 Front-wheel drive weight distribution shifts to the front.

approach 600°F. A car with a 300-horsepower engine that accelerates from 0 to 60 miles per hour (mph) in 7.5 seconds will typically require a brake system that can go from 60 to a complete stop in less than one-fifth that time. This is the equivalent of about 1,500 horsepower. A smaller car with 300 horsepower does not require brakes as large as a heavier car of the same power.

As speed doubles from 30 to 60 mph, the required stopping distance increases by nearly four times. At very high speeds, a speed increase of 5 mph can increase stopping distance by about 75 feet.

Energy is the ability to do work. Different forms of energy include mechanical, heat, electrical, and chemical energy. Energy cannot be consumed; it can only be changed to another type of energy. Brake systems perform "work" by using **friction** to convert the mechanical energy of motion, called **kinetic energy**, into heat energy (**Figure 3.10**). When a car is traveling at highway speed, a large amount of kinetic energy is stored in the speed (velocity) and weight (mass) of the vehicle (**Figure 3.11**). This is called *inertia* and it wants to remain in motion. When you apply the brakes to stop the car, friction is used to overcome inertia and change kinetic energy to heat energy.

Kinetic energy is at work when the car is moved forward or decelerated. The amount of kinetic energy changes with the weight of the car (its mass) and how fast its speed is changing. If speed is equal, doubling the weight of the car doubles the kinetic energy (**Figure 3.12**). But doubling the speed results in an increase in kinetic energy of four times (**Figure 3.13**).

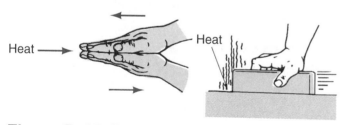

Figure 3.10 Friction changes kinetic energy into heat (thermal) energy.

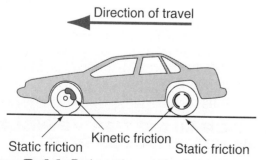

Figure 3.11 Both static and kinetic friction are at work during braking.

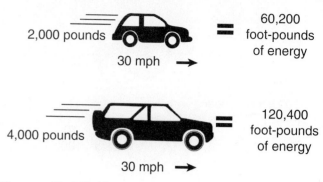

Figure 3.12 Kinetic energy increases proportionally with vehicle weight.

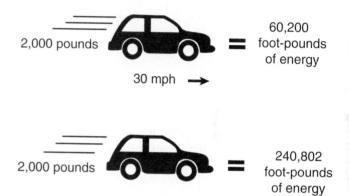

Figure 3.13 Kinetic energy increases exponentially with vehicle speed.

This is why really fast cars have really big brake rotors. The amount of frictional surface results in more stopping power.

When you want to stop faster, you push harder on the brake pedal. Several factors dictate how hard you need to push on the pedal. These factors include the diameter of drums and rotors, the surface area of the pads and rotors or drum and shoes, the friction coefficient of the linings, the amount of force created by the hydraulic system, the mechanical leverage in the brake pedal and linkage, and the extra pedal pressure created by the brake power booster. These all can increase friction, resulting in more stopping power and heat.

Coefficient of Friction

The amount of dry friction varies depending on the force applied, the material that the friction surfaces are made of, and the finish (roughness) of the friction surface. For example, worn, polished cobblestones or marble on a street in a really old town produce less friction than asphalt.

Friction is the force that resists movement between any two contacting surfaces. The ratio of the force holding two surfaces in contact to the force required to slide one over the other is known as the **coefficient of friction.** If it takes 70 pounds of force to drag one

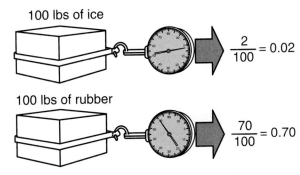

Figure 3.14 Coefficient of friction.

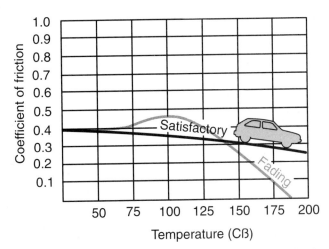

Figure 3.15 Coefficient of friction drops off as brakes fade.

piece of material across another, the coefficient of friction is 70 divided by 100, or .70 (**Figure 3.14**). Coefficient of friction varies with the temperature of the surfaces in contact, the rubbing speed, and the condition of the surfaces.

■ BRAKE FADE

During prolonged periods of heavy braking, brake rotors and drums can absorb heat faster than it can be transferred to the surrounding air. Removing heat from brakes is called *heat dissipation*. During a hard stop, temperatures can increase about 50°F per second. Yet, dissipating that second's worth of temperature increase might take 20 seconds.

Excessive temperatures can be caused by hauling heavy loads, stopping from high speeds, or continuous downhill braking on mountain roads. Brake coefficient of friction drops with excessive temperatures. This is called **brake fade**. In addition to the danger presented by the decrease in stopping ability, excessive heat can damage brake parts. Heat will cause the surface of linings to glaze and lose their ability to create friction. Glazed linings also tend to squeak or squeal. Rotors can warp, causing the brake pedal to pulsate during stopping. Drums can become out-of-round when the parking brake is set while the drums are still hot.

Drum brakes do not dissipate heat as well as disc brakes because drum brake linings are enclosed within the brake drum so heat is trapped. Disc brakes reject heat more readily because both sides of a brake disc are exposed to air nearly all of the time. The only part not exposed is the area where the caliper covers the rotor.

Types of Brake Fade

There are three types of brake fade: lining fade, mechanical fade, and gas fade. **Lining fade** can occur in disc or drum brakes. As the temperature of the lining increases beyond a certain point, its coefficient of friction drops off, requiring more pedal effort to stop the car. As the chart in **Figure 3.15** shows, the friction coefficient of typical brake linings rises slightly as the temperature rises. With a further temperature increase, friction coefficient drops rapidly. As fade becomes worse with the addition of more heat, stopping ability will drop off radically and may disappear altogether.

Mechanical fade occurs only in drum brakes when the drum expands as its temperature rises. This causes the pedal to move closer to the floor before the brakes apply. Pumping the pedal will result in a return to normal pedal height, until the next stop. The Chapter 4 discussion of master cylinder operation explains why this happens.

Gas fade is very rare. During extreme overheating of the brake system the pad's organic binding agent can release a thin layer of hot gas. This causes the pad to hydroplane, reducing contact between the pad and the metallic friction surface. Larger friction surfaces are more prone to trapping gas. To prevent this, slots can be cut in the pads or rotors to help cool and route away the gases.

Preventing Brake Fade

Brake parts are designed to fit the vehicle. High brake temperatures can be partially prevented with the installation of larger linings, rotors, and drums. Front brakes must be able to absorb more heat than rear brakes, so their linings have more surface area. Heavier vehicles also require wider linings with more surface area to carry off the increased heat generated when stopping a greater weight.

Fins can be cast into the outside of brake drums to increase the amount of surface area available for heat loss. Larger cars and light trucks use ventilated rotors and some rotors are made of more exotic materials (see Chapter 9). Semimetallic and carbon fiber lining materials resist fade, but they must rise further in temperature before reaching their maximum friction coefficient.

■ FEDERAL MOTOR VEHICLE SAFETY STANDARDS

Brake systems are regulated in the United States under the Federal Motor Vehicle Safety Standards (FMVSS) established by the U.S. Department of Transportation. Part 571 governing automotive brake standards was first issued in March 1967 and has been amended many times since then. There are many automotive sections to the FMVSS standards; the following sections apply to brakes:

- FMVSS 105/135   Hydraulic Brake Systems
- FMVSS 106   Brake Hoses
- FMVSS 116   Brake Fluids

The FMVSS established minimum braking requirements. Standard 105 regulates parking brakes, instrument panel warning lamps, automatic clearance adjustment, and the reservoir (including labeling). FMVSS105 was the original standard for 1999 and earlier vehicles. Cars and light trucks produced since 2000 must meet the tougher FMVSS 135 standard.

Besides hydraulic brake systems, other automobile regulations include those for lamps, fluids, lug nuts, discs and hub caps, seat belts, and tire air pressure monitoring.

■ BRAKING REQUIREMENTS

Brake systems are designed to fit the type of vehicle. High-performance cars require greater stopping power. Also, trucks and pickups intermittently have substantially increased braking requirements. Lightweight, economy cars do not share the same braking needs, so their parts can be considerably smaller.

Safety legislation has produced stringent standards that manufacturers must meet for automotive brakes. These standards have resulted in much improved braking ability for mass-produced vehicles. Federal standards define the minimum level at which brakes must perform. Requirements are specifically defined for vehicles of every size and weight. The stringent brake performance standard includes eighteen different stages. Stopping ability is not allowed to become less as brakes become worn. Brake fade and recovery are part of the test sequence.

The federal standards require that a crippled vehicle with only half of its hydraulic system remaining still come to a complete stop within about 2 to 3 times its normal stopping distance. Other federal tests include stops with the ABS and power brakes disabled.

■ BRAKE LININGS

The friction materials used in cars and trucks are called **brake linings** (**Figure 3.16**). On drum brakes, linings are commonly called **brake shoes.** On disc brakes, they are called *pads*.

Friction Material Design

When selecting a suitable brake lining, the reputation of the manufacturer is an important consideration. Brake manufacturers rely on repeat customers and

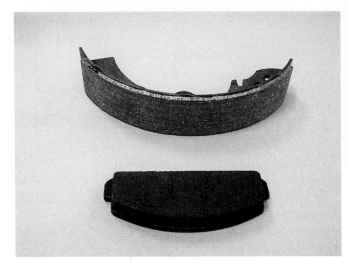

Figure 3.16 Drum brake shoe and a disc brake pad. *(Courtesy of Tim Gilles)*

make substantial expenditures in research and development in order to engineer competitive products.

There are many considerations when designing a brake lining. Linings must be able to resist fade caused by wetness or an increase in temperature. When fade does occur, the lining must be able to quickly recover. Federal standards dictate that friction materials perform adequately. But a most important customer consideration is that brakes be quiet. Additionally, a lining must not wear out too quickly and it must not cause excess wear to the surface of rotors or drums.

Variations in Brake Linings

Lining materials vary according to the application and the friction characteristics desired. A friction material that is hard will have a longer life span, but it will be more likely to make noise. Hard linings will not stop as easily because they have a lower coefficient of friction. A softer lining might stop better but will probably wear faster and fade at lower temperatures.

Some brakes stop well when they are cold but do not work well when hot. Others work very well when hot but will not stop the car well until they have had a chance to heat up after a few stops (**Figure 3.17**). Both of the previous circumstances can be very dangerous. The best policy is to use only the type of lining material that came on the vehicle as original equipment and to use premium quality materials. Other than risking a customer's safety, there could be a major liability for a repair shop if an accident results from the use of inferior linings.

The types and advantages of various friction materials are covered in greater detail in Chapter 8.

Bonded or Riveted Linings

Linings are either *bonded* (glued) or *riveted* to the disc backing or drum brake shoe. **Figure 3.18** compares bonded and riveted disc brake pads. Some newer pads

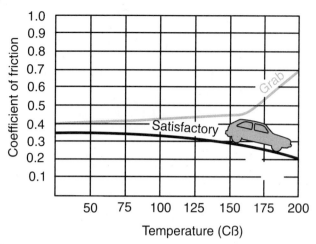

Figure 3.17 Linings of various materials perform differently with changes in temperature.

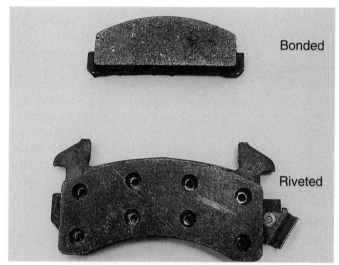

Figure 3.18 Bonded and riveted brake pads. *(Courtesy of Tim Gilles)*

Figure 3.19 An integrally molded pad has holes either full or partially full of lining material. *(Courtesy of Tim Gilles)*

are *integrally molded,* with the metal disc backing plate molded to the friction material as an integral unit. From the back of the pad, you can observe holes that are either full or partially full of lining material (**Figure 3.19**).

■ HYDRAULIC BRAKE FLUID

Hydraulic brake fluid meets a standard set by the Society of Automotive Engineers (SAE)—Standard J1703— as well as FMVSS 116. Three brake fluid number classifications have been assigned by the Department of Transportation (DOT): DOT 3, 4, and 5. The DOT number is always listed on the fluid container (**Figure 3.20**). DOT 5 is synthetic brake fluid. DOT 3 and DOT 4 are fluids made from a *polyglycol* base, similar to that of engine coolant.

Glycol based fluids are *hygroscopic,* which means that they absorb water. Being hygroscopic is an advantage for brake fluid because it can absorb moisture that enters the system. This prevents the formation of water drops that could boil or freeze. Dispersing the moisture throughout the fluid also prevents localized corrosion that could cause holes to form in brake lines.

NOTE: *Brake fluid can absorb a large enough quantity of water to be ruined in as little as an hour if left uncovered. According to the EIS Brake Company, a typical vehicle's brake fluid that has been changed in the last 18 months will have accumulated 2–3 percent water. A water content of 3 percent lowers the boiling point of DOT 3 brake fluid by 25 percent. DOT 4 fluid absorbs moisture at a slower rate but is more affected by moisture contamination. With a 3 percent water concentration, its boiling point can be as much as 50 percent lower. When there is a choice between using fluid from a large, partly empty container or two unopened smaller containers, the two smaller containers are the best choice.*

Because brakes produce heat, brake fluid becomes hot. Brake fluid is designed to have a high boiling temperature. Water has a far lower boiling point (212°F)

Figure 3.20 The DOT number is always on the label of a brake fluid container.

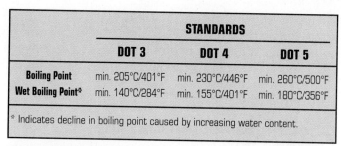

Figure 3.21 DOT specifications for both dry and wet boiling points.

than brake fluid (over 400°F). If brake fluid absorbs enough water, the fluid will boil in the lines. This will result in a loss of braking efficiency, a brake pedal that feels "spongy" during application, and even a total loss of braking ability.

DOT specifications list both *dry* and *wet boiling points* (**Figure 3.21**). The dry specification is for new fluid and the wet specification is for fluid that has absorbed 3 percent water. Typically a brake fluid reaches its wet boiling point within about 2 years. DOT 3 fluid has a dry boiling point of 401°F. and a wet boiling point of 284°F. DOT 4 fluid has a higher boiling point, 446°F dry and 401°F wet.

Fluid that meets DOT 3 or DOT 4 specifications can be used in drum and disc systems. DOT 4 is not a better fluid than DOT 3 simply because it has a higher number. Most domestic manufacturers specify DOT 3. Some DOT 3 fluids are heavy-duty and have a boiling point as high as DOT 4. Under perfect conditions, in which no moisture is allowed to contact the fluid, DOT 3 is said to have a longer life than DOT 4.

The correct fluid to use in a particular vehicle will be listed on the reservoir. It is important not to fill the reservoir all the way to the top. Space must be left to allow for fluid expansion as the brakes heat up. The maximum fill level is about ¼ inch from the top of the reservoir chamber. Be careful not to allow brake fluid to contact the car's finish because it can damage paint. Rinse thoroughly with soap and water in case of a spill. Wiping spilled brake fluid from a fender using a shop towel is more likely to damage the paint.

Glycol-based fluids are *not generic* (not all the same). They are a mixture of various substances, with up to ten different ingredients blended in the fluid. Some fluids perform better than others. All brake fluids are made of four components:

1. A lubricant to keep parts sliding freely
2. A solvent diluent, which determines the fluid's viscosity and boiling point
3. A modifier-coupler, which changes the amount of swelling of exposed rubber parts
4. Inhibitors to prevent corrosion and oxidation

The lubricant that makes up about 20–40 percent of a brake fluid's content is a synthetic polyglycol, such as polyethylene or polypropylene. Most of the content of the fluid (50–80 percent) is the solvent-diluent glycol ether. In some fluids, ethylene glycol (the same compound as automotive coolant) is also used for this purpose because it controls rubber swelling. A wide range of chemicals make up the small remainder of the fluid (0.5 to 3 percent).

NOTE: *Brake fluid prices often vary with the price of engine coolant due to the amount of glycol used.*

Synthetic Brake Fluid

Synthetic silicone-based brake fluid was developed in the early 1970s, requiring a new DOT category—DOT 5. The DOT 5 standard requires a minimum dry boiling point of 500°F and a wet boiling point of 356°F. The wet boiling point does not come into play, however, with silicone fluid because it is *not* hygroscopic. Because it does not absorb moisture, it is called a *lifetime fluid*.

Silicone fluid has a "spongy" pedal feel because it is more compressible than ordinary brake fluid. The pedal will also have more travel before the brakes are fully applied. This is because silicone fluid contains about three times as much dissolved air as glycol-based fluid. Glycol has 5 percent dissolved air and silicone has 15 percent.

NOTE: *Silicone fluid tends to develop air bubbles when it is cycled rapidly so manufacturers recommend it not be used in ABS applications. Check the manufacturer's recommendation for the correct fluid. ABS systems usually call for DOT 3 fluid. Currently, there are no manufacturers that recommend DOT 5 for their cars.*

A newer brake fluid first used by manufacturers in 1999 is called DOT 5.1. However, it is glycol- not silicon-based.

Mineral Oil Brake Fluid

Mineral oil brake fluid, called hydraulic system mineral oil (HSMO), is green rather than the light gold of ordinary glycol-based brake fluid. It is a good lubricant, it is not hygroscopic, and it has a high boiling point. But HSMO will cause rubber parts in conventional brake systems to swell. It is not compatible with ordinary brake systems and is not covered by a DOT classification. Some European manufacturers (Rolls Royce, Citroen, and Audi) have used it in their brake systems, although no manufacturers use it today.

■ COMPLETE BRAKE JOB

Figure 3.22 is an inspection checklist that shows what one manufacturer recommends for a complete brake job. A complete brake job includes all of the internal hydraulic parts of the wheel cylinders and disc calipers as well as new brake fluid throughout. New hardware and springs are often included. Drums or rotors are

BRAKE SYSTEM COMPONENTS	REQUIRED SERVICE	ESTIMATE PARTS	ESTIMATE LABOR
Disc pads*	☐ Replace front ☐ Replace rear		
Disc caliper*	☐ Recondition ☐ Replace		
Disc hardware*	☐ Replace		
Disc rotor*	☐ Resurface ☐ Replace		
Grease seals*	☐ Replace		
Front wheel bearings*	☐ Repack ☐ Replace		
Brake shoes*	☐ Replace front ☐ Replace rear		
Brake drums*	☐ Resurface ☐ Replace		
Wheel cylinders*	☐ Recondition ☐ Replace		
Brake hardware*	☐ Replace		
Parking brakes	☐ Adjust/Lube ☐ Replace		
Power brake booster	☐ Service ☐ Replace		
Master cylinder	☐ Recondition ☐ Replace		
Brake fluid*	☐ Flush system ☐ Add new fluid		
Lines, Hoses, Combination valve	☐ Replace		
Stop light	☐ Replace lamp ☐ Replace switch		
ABS Diagnosis	☐ Requires service		
Other			

* Total brake service includes reconditioning or replacement of these items

Figure 3.22 Items in a complete brake job.

often machined using a lathe (see Chapter 8). A front axle brake job on a rear-wheel drive car usually includes a front-wheel *bearing repack* (see Chapter 11). A master cylinder and new hoses are usually *not* included in a brake job.

Ethics in Brake Work

A common practice in recent years was for large merchandisers to advertise incomplete brake jobs at very low prices. Once the car was raised on the lift, the cost of the job often rose rapidly as needed items were sold to the customer. When the job was finished, the price was much higher than it would have been if the customer had patronized his or her regular repair shop. Consumer affairs agencies in some states have determined this practice to be unfair advertising. When an incomplete brake job is advertised in those states, a disclaimer must be included that says something like, "This job may cost substantially more when other necessary parts are added."

Chassis Systems and the Lemon Law

Manufacturers offer specific warranties on their new vehicles. A typical warranty is 3 years or 36,000 miles, whichever comes first. Some cars (mostly luxury) have longer warranties. Brakes are not usually subject to the same warranty period. After a period of 1 year or 12,000 miles without problems, brakes are considered to be a maintenance item.

NOTE: *A technician working in a new car dealership must be aware that repairs to brakes within the first year can sometimes result in future warranty problems for the manufacturer.*

The *lemon law* is a well-known term for the federal Magnuson-Moss Warranty Act, which established standards regarding consumer product warranties. Under the lemon law, a car buyer has the right to expect that a vehicle will perform according to promises made by the manufacturer. It is expected that the buyer will test the vehicle prior to purchase so any preexisting, or inherent, operating conditions can be verified. The vehicle is considered to be exempt from warranty complaints if it is "operating as designed." Consider the following two situations:

1. Customer 1 buys a car without taking it on a test drive. He later complains that the power steering makes an objectionable noise when turning.
2. Customer 2 buys a car that she has thoroughly test-driven. She finds it to be exactly what she's looking for and there are no problems. Five months later, the car has developed an annoying brake squeak. She returns to the dealership and has it repaired. The problem returns and she has it repaired once again. This happens repeatedly and the dealer cannot fix the problem.

Customer 1 has no grounds for complaint. This make of vehicle had noisy steering when it was new; therefore the use, value, or safety of the vehicle is not compromised. Customer 2 is involved in a cloudy situation. According to the warranty act, the vehicle has been in for the same repair at least four times during the first 18 months. Is this sufficient cause for the manufacturer to buy the vehicle back as a "lemon"? Probably not, but performing a "repair" and charging it back to the manufacturer under warranty could cause that to happen.

REVIEW QUESTIONS

1. How many brake hoses do most cars use to make a flexible connection from steel tubing to brake system components?

2. By approximately how much does the brake pedal mechanically increase leverage?

3. List the term that describes the action of the leading shoe on a drum brake as it wedges into the brake drum.

4. A power assist is usually required with which brakes—disc or drum?

5. A brake system that combines disc brakes on the front and drum brakes on the rear is called a _____ system.

6. Where is a typical power brake booster located?

7. Which brake linings usually have a longer service life, front or rear?

8. If it takes 60 pounds of force to drag one piece of material across another, what is the coefficient of friction?

9. Which brake design dissipates heat better, drum or disc?

10. On drum brakes, linings are commonly called _____. On disc brakes, they are called _____.

11. Hard linings do not stop as easily because they have which coefficient of friction, higher or lower?

12. A softer lining will fade at which temperatures, higher or lower?

13. List the terms that describe the two ways that linings are attached to the disc backing or drum brake shoe.

14. Which of the DOT brake fluid classifications are made from a polyglycol base similar to engine coolant?

15. The dry specification is for new fluid and the wet specification is for fluid that has absorbed _____ percent water.

PRACTICE TEST QUESTIONS

1. Which of the following is/are true about drum brake self-adjustment?
    A. Some drum brakes self-adjust when braking in reverse
    B. Some drum brakes self-adjust when the parking brake is applied
    C. Both A and B
    D. Neither A nor B

2. Technician A says drum brakes require more fluid as they self-adjust to compensate for lining wear. Technician B says when a master cylinder reservoir has chambers of different sizes, the larger one is for drum brakes and the smaller one is for disc brakes. Who is right?
    A. Technician A B. Technician B
    C. Both A and B D. Neither A nor B

3. Which of the following is/are true about brake fade?
    A. When disc brakes fade, the pedal moves closer to the floor before the brakes apply
    B. Pumping the pedal when drum brakes have faded will result in a return to normal pedal height, until the next stop
    C. Both A and B
    D. Neither A nor B

4. Technician A says harder friction materials wear out faster. Technician B says harder friction materials have more of a tendency to make noise. Who is right?
    A. Technician A B. Technician B
    C. Both A and B D. Neither A nor B

5. Technician A says that the parking brake system is usually made up of service brake system parts. Technician B says that integral disc parking brakes are better than drum parking brakes. Who is right?
    A. Technician A B. Technician B
    C. Both A and B D. Neither A nor B

6. All of the following use a self-adjustment mechanism *except*:
    A. Front drum brakes
    B. Rear disc brakes with an integral emergency brake
    C. Rear drum brakes
    D. Front disc brakes

7. Technician A says that some heavy-duty DOT 3 brake fluids have a boiling point as high as DOT 4. Technician B says it is important that the master cylinder reservoir not be filled completely to the top. Who is right?
   A. Technician A   B. Technician B
   C. Both A and B   D. Neither A nor B

8. Which of the following is/are true about glazed brake linings?
   A. They tend to squeak or squeal
   B. They tend to grab harder during a stop
   C. Both A and B
   D. Neither A nor B

9. Technician A says if brake fluid is not changed often enough, the fluid can boil in the lines. Technician B says water in the brake fluid can cause a spongy brake pedal. Who is right?
   A. Technician A   B. Technician B
   C. Both A and B   D. Neither A nor B

10. Which of the following are true about glycol-based brake fluids?
    A. DOT 5.1 brake fluid is glycol-based
    B. Glycol-based fluids are hygroscopic
    C. DOT 3 and DOT 4 are glycol-based
    D. All of the above

CHAPTER 4

Master Cylinder and Hydraulics

■ OBJECTIVES

Upon completion of this chapter, you should be able to:
- ✔ Understand Pascal's law and how it applies to hydraulic brakes.
- ✔ Explain how force and pressure work within a hydraulic system.
- ✔ Explain the operation of a single-piston master cylinder.
- ✔ Describe the parts and operation of a tandem master cylinder.
- ✔ Explain the differences in operation between longitudinal and diagonally split hydraulic systems.
- ✔ Explain the operation of a quick take-up master cylinder.
- ✔ Describe the different types of hoses, tubing, and connections.

■ KEY TERMS

banjo fittings
combination valve
diagonal split
flare nut
flared line
longitudinal split
low pedal
metering valves
pressure chamber
primary cup
proportioning valves
replenishing port
residual check valve
residual pressure
secondary cup
self-adjusters
tandem master cylinders
vent port

■ INTRODUCTION

Early brakes had mechanical linkages. Because they were only on the rear wheels, adjustment was not as critical a factor. Mechanical linkage on the front wheels would have been a difficult option because steering the wheels would cause the brakes to apply. The development of hydraulic brakes was the answer to this problem. Pressure in an enclosed hydraulic system is always equal, so the tendency of uneven adjustment to cause brake pull is reduced.

Vintage Chassis

Hydraulic brakes were invented by Malcolm Loughead in his California shop in 1918. He and his brother later changed the spelling of their name to Lockheed and started the Lockheed Aircraft Company.

■ HYDRAULIC BRAKE SYSTEM OPERATION

Pressurizing liquid to transfer motion or to multiply and apply force is called *hydraulics*. When the brake pedal is depressed, it moves a piston in the master cylinder (see Chapter 3, Figure 3.1). This pushes fluid under pressure through brake lines and hoses to each wheel, where the hydraulic pressure acts on pistons to produce force (**Figure 4.1**).

■ PASCAL'S LAW

Pascal's law is fundamental to the development of your automotive diagnostic ability. It applies to such items as the hydraulic brake system, suspension and steering, the lubrication system, engine bearing clearances, the pressurized cooling system, engine compression, and air conditioning. Laws of hydraulic pressure can even be applied to electrical fundamentals. One of the most important applications of Pascal's law is in the hydraulic brake system.

Pascal's law states that "pressure in an enclosed system is equal and undiminished in all directions."

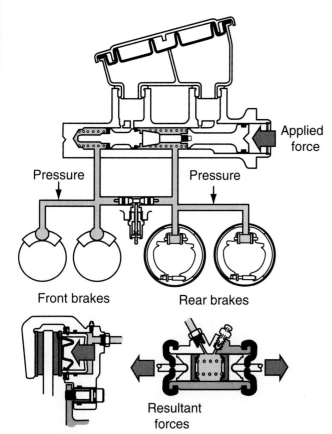

Figure 4.1 Pressure acts on hydraulic cylinders at the wheel to produce force.

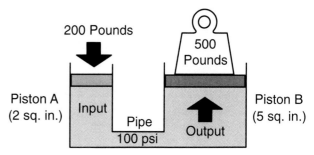

Figure 4.2 The force applied when the master cylinder operates is increased with a larger diameter wheel cylinder.

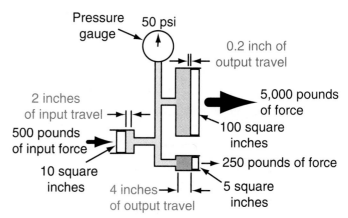

Figure 4.3 Application of hydraulic formulas. Notice that as output force increases, output travel (motion) decreases.

Enclosed means that the fluid is not moving. Hydraulic results can be determined using the following formulas:

Pressure = force ÷ area
Force = pressure × area

The following facts apply to the hydraulic braking system:

- The force applied to the brake linings when the master cylinder operates is increased with a larger diameter wheel cylinder.
- When larger wheel cylinders are used, the distance that the pedal has to travel before building up pressure increases.
- To increase the pressure coming from the master cylinder, the diameter of its bore must be *smaller*.

In the example shown in **Figure 4.2**, the input piston is smaller than the output piston. Using the formula *pressure equals force divided by area,* you can determine that with 200 pounds of force applied to a 2-square-inch piston the resulting pressure is 100 pounds per square inch (psi).

$$\frac{200 \text{ pounds}}{2 \text{ square inches}} = 100 \text{ psi}$$

When the 100 psi developed by Piston A is applied to the 5-square-inch piston on the right-hand side of the sketch, the resulting force is 500 pounds. The correct formula for determining this is *force equals pressure times area*:

100 psi × 5 square inches

In **Figure 4.3** you can see the effects of pressure on pistons of two different sizes. Notice that the smaller piston travels much further in response to movement of the input piston than the larger piston.

MASTER CYLINDER OPERATION

When a driver depresses the brake pedal, a linkage applies force to the back of the master cylinder. The master cylinder pressurizes brake fluid and routes it to the wheel cylinders and calipers in the vehicle's brakes. Most master cylinders have two chambers, but we will cover the operation of a simple, single-piston master cylinder first. Service technicians who do not understand the operation of the hydraulic system often perform brake repairs simply by replacing parts. This is unfortunate because it sometimes results in repairs that are not necessary while searching for the solution to a problem. To be able to efficiently diagnose brake system problems, you need to thoroughly understand the operation of the master cylinder and the complete hydraulic system.

Connected to the foot pedal, the master cylinder supplies hydraulic pressure to operate the wheel cylinders during braking (**Figure 4.4**). **Figure 4.5** shows a simple single-piston master cylinder with one seal

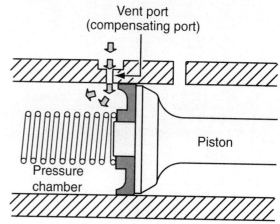

Figure 4.6 Fluid from the reservoir fills the master cylinder bore through the compensating port.

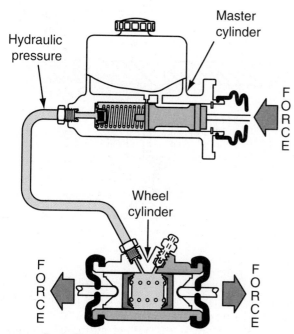

Figure 4.4 The master cylinder, which is connected to the foot pedal, supplies hydraulic pressure to operate the wheel cylinders during braking.

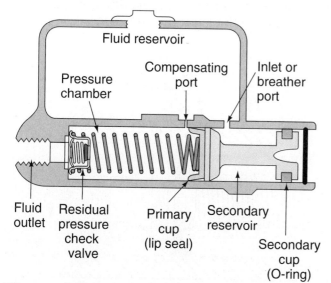

Figure 4.5 A single-piston master cylinder.

called the **primary cup**. The primary cup compresses fluid when the pedal is depressed. Another seal, called the **secondary cup**, keeps fluid from leaking out the back of the master cylinder bore. This seal does not seal against pressure.

NOTE: *A master cylinder has two ports for each piston. These ports have had many names over the years. SAE standard J1153 names the front port the **vent port** and names the rear port the **replenishing port**. The front port is also commonly called the compensating port, whereas the rear port is also called the inlet port.*

Fluid from the reservoir fills the master cylinder bore through the vent (compensating) port (**Figure 4.6**). When the pedal is applied, the primary cup moves forward in the master cylinder bore and passes the vent port. With the vent port isolated, fluid movement is restricted. As the pedal is forced further down, fluid pressure builds in the hydraulic system.

In the simple drum brake system shown in **Figure 4.7**, the wheel cylinder has a pair of rubber cup lip seals. The lips of the cup seals face inward against the pressurized fluid created by the master cylinder. The lip on the master cylinder primary cup also faces toward the fluid. Notice the direction of the primary cup lip seal in the master cylinder. When a seal is installed in this direction, it *seals*. A seal installed backward will leak.

Low Pedal Master Cylinder Action

As brake linings wear, clearance between the friction surfaces increases. Disc brakes and most drum brakes are self-adjusting. When there are no **self-adjusters** or if a self-adjusting mechanism fails to work, a drum brake will develop extra clearance. This clearance causes the brake pedal to move closer to the floor before the brakes apply, which is called **low pedal**.

When excessive wear of the friction materials occurs, the pedal might be able to travel all the way to

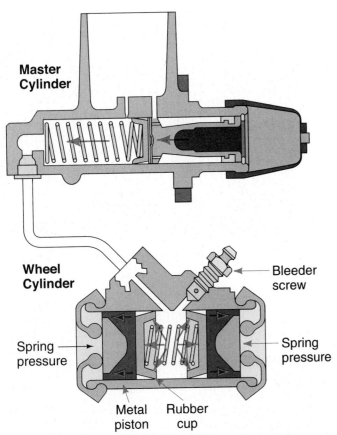

Figure 4.7 The lips of the rubber cups must face toward the brake fluid.

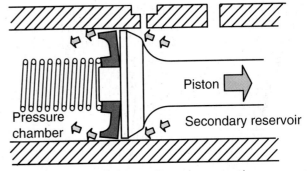

Figure 4.8 Fluid moves from the secondary reservoir to the pressure chamber when the brakes are adjusting.

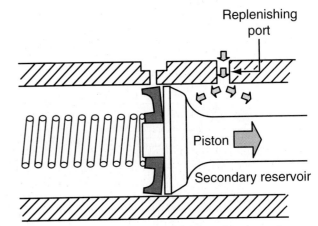

Figure 4.9 Fluid is replenished to the secondary area through the inlet port.

the floor. When this situation develops, a second application of the pedal should result in a higher pedal that will stop the car. What causes this to happen?

When the brakes are adjusted correctly, only a small volume of fluid in the master cylinder is moved to operate the brakes. When a low pedal is released and then quickly reapplied, fluid is momentarily trapped in the cylinders at the wheels. The fluid cannot return quickly to the master cylinder because the inside diameter of the steel brake supply tubing is very small. The small size of the tubing restricts the quick movement of the brake fluid back to the master cylinder. Large diameter tubing would allow the fluid to move more quickly, but this would be undesirable.

When the pedal is released, the primary cup moves back quickly in the master cylinder bore. An area in front of the primary cup, called the **pressure chamber** (see **Figure 4.6**), is momentarily empty of fluid and at a lower pressure (less than atmospheric pressure). Stored just behind the primary cup is a reservoir of fluid at atmospheric pressure. The sealing lips on the primary cup lip face away from the fluid pressure, so fluid can leak past the cup from the back to the front. This refills the temporarily empty pressure chamber in the master cylinder bore (**Figure 4.8**). The secondary reservoir is refilled through the replenishing (inlet) port (**Figure 4.9**).

When the pedal is quickly reapplied for the second time, there will now be enough fluid in the system to cause the linings to contact the drum and pressure will build on the fluid in the wheel cylinders. When the pedal is released once again, the brake shoe return springs force the excess fluid *slowly* back through the brake line from the wheel cylinder. The excess fluid bleeds off to the reservoir through the now-uncovered vent (compensating) port (**Figure 4.10**).

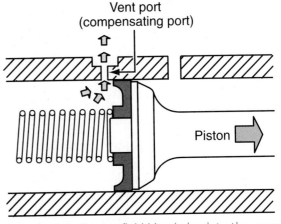

Figure 4.10 Excess fluid bleeds back to the reservoir through the compensating port.

Until the brakes are correctly adjusted, brake pedal height will increase on the second pedal application. Low pedal height can also be caused by an internal leak in one part of the master cylinder, but, in this case, a second application of the brake pedal will not increase pedal height.

Tandem Master Cylinder

Older cars were equipped with a simple single piston master cylinder like the one in the previous discussion. These cars were more dangerous than cars with modern brake systems. When a car with a single-piston master cylinder had a hydraulic system failure, it would experience total brake system failure.

Modern cars have master cylinders with two separate hydraulic systems. The first dual chamber master cylinder was used on a 1960 American Motors Rambler and several other cars used them in 1962. Since 1967, these **tandem master cylinders** have been mandated by law (FMVSS 105) on all cars.

During the following discussion of tandem master cylinder operation, remember Pascal's law: *Pressure is equal everywhere in an enclosed system.*

A tandem master cylinder has one cylinder bore with two separate pistons and chambers (**Figure 4.11**). During normal stopping, the primary cup on the rear (primary) piston pushes fluid to the brakes it serves. It also pushes fluid forward against a rearward-facing lip seal on the front (secondary) piston. When the friction

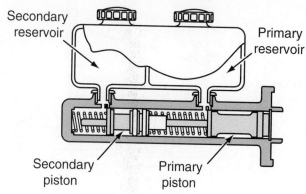

Figure 4.11 A tandem master cylinder.

materials begin to apply force at the wheels, pressure begins to build in both the front and rear systems at the same time.

When one-half of this system fails, pedal height will be lower, but the remaining half of the hydraulic system should have enough braking capacity to stop the car. During a failure in the part of the brakes served by the rear piston, a stub on the front of that piston bottoms out against the back of the front piston (**Figure 4.12**). Thus, the primary piston is applied mechanically instead of hydraulically.

When the hydraulic system served by the front part of the master cylinder experiences a failure, it cannot build up pressure. A stub on the end of the piston

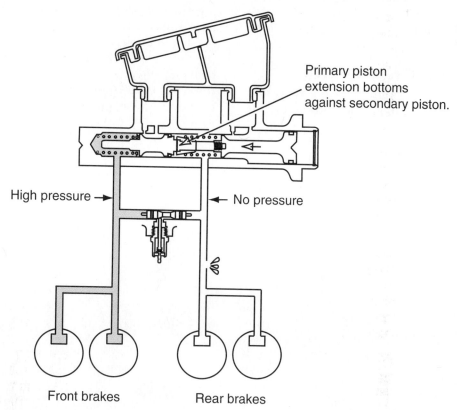

Figure 4.12 The rear piston bottoms against the front piston when the rear half of the hydraulic system fails.

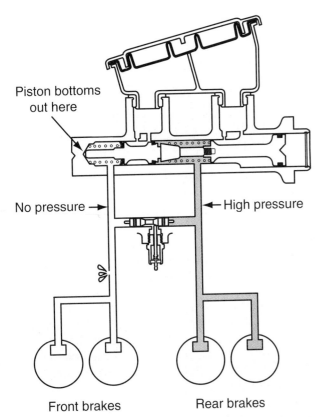

Figure 4.13 The front piston bottoms against the front of the bore, allowing pressure to build in the rear half of the system.

bottoms out at the end of the bore (**Figure 4.13**). This allows the rear piston to build up pressure as it forces fluid against the rearward-facing seal on the back of the front piston.

Master Cylinder Check Valve

Older drum brake master cylinders have a **residual check valve** at the fluid outlet. The purpose of this check valve is to keep a small amount of **residual pressure** (6–25 psi) in the brake system when the brakes are not applied. This helps keep the wheel cylinder rubber cup sealed against the wall of the wheel cylinder to prevent air from entering.

Disc brake systems do *not* have a check valve. If they did, the pressure would overcome the return action of the disc brake seal, causing the brakes to drag and wear. (This topic is covered later in this chapter.) At about the same time that drum brakes disappeared from the fronts of vehicles, residual check valves also disappeared from master cylinders serving rear drum brakes. A residual check valve would be a problem in split diagonal hybrid systems, and *cup expanders* on the ends of the center spring in the wheel cylinder make the valve unnecessary (**Figure 4.14**).

The residual check valve on a tandem master cylinder is a *flapper valve*. It is also called a *duck bill valve*. A flapper valve allows fluid to flow through its center in only one direction. It blocks fluid attempting to flow in the opposite direction. **Figure 4.15** shows the operation of a flapper-type residual check valve.

Master Cylinder Reservoir

Most of today's master cylinders are aluminum with a plastic reservoir sealed to the cylinder with rubber grommets (**Figure 4.16**). They are called *composite* master cylinders because they are made of two materials. Aluminum is lighter than cast iron and less expensive to manufacture. Plastic reservoirs are transparent to allow the fluid level to be observed without removing the reservoir cover, which could lead to dirt or moisture contamination.

The master cylinder fluid level fluctuates through the vent port as the brakes are used because the fluid heats up or cools down (see **Figure 4.10**). Therefore, the cover to the master cylinder must include a feature to prevent a vacuum lock as the expanded brake fluid

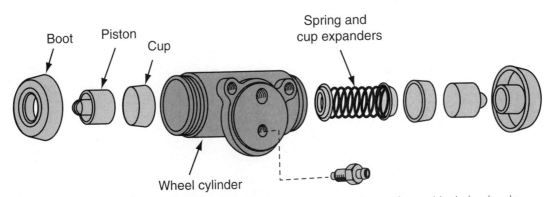

Figure 4.14 Cup expanders on the ends of the center spring take the place of a residual check valve.

Chapter 4 Master Cylinder and Hydraulics 79

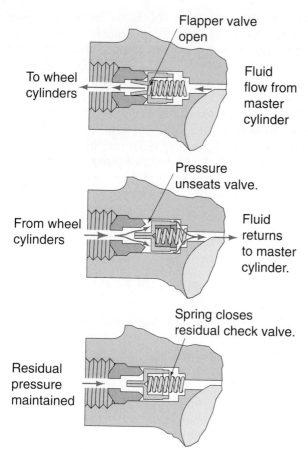

Figure 4.15 Operation of a master cylinder residual check valve.

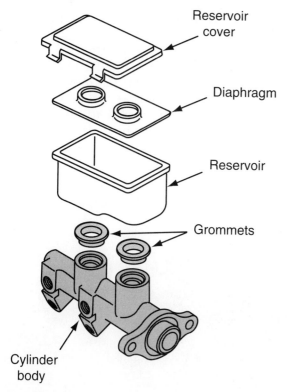

Figure 4.16 A master cylinder with a removable plastic reservoir.

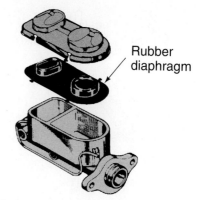

Figure 4.17 A reservoir cover with a rubber diaphragm.

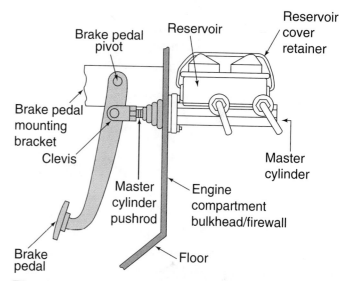

Figure 4.18 A master cylinder with a suspended pedal is mounted on the bulkhead.

cools down. If a vacuum lock were to occur, air would be drawn into the brake fluid past the secondary cup at the rear of the master cylinder.

Master cylinder reservoirs are prevented from vacuum locking in one of two ways. There is either a flexible rubber diaphragm in the cover (**Figure 4.17**) or screw cap, or there is a plastic float. These devices allow atmospheric pressure to act on the fluid in the reservoir without the fluid becoming contaminated with moisture from the outside air.

Master Cylinder Location

Master cylinders on almost all cars and light trucks since the 1950s have been mounted on the bulkhead, which is also called the firewall (**Figure 4.18**). Older vehicles had the master cylinder mounted under the floorboard (**Figure 4.19**).

■ DIAGONALLY SPLIT HYDRAULIC SYSTEM

Tandem systems are either **longitudinal split** (front/rear) or **diagonal split** (**Figure 4.20**). A longitudinal split system operates the front and rear brakes as

80 SECTION 2 BRAKES

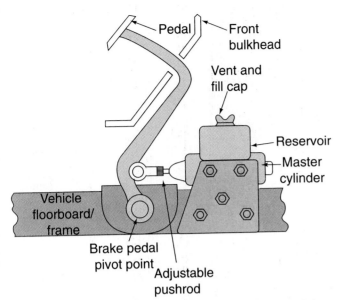

Figure 4.19 Some vintage cars and trucks had the master cylinder located under the floorboard.

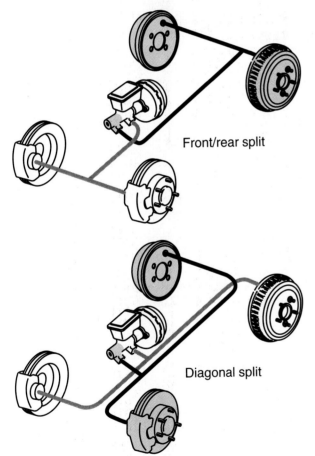

Figure 4.20 Tandem brake cylinders are hydraulically split in different ways.

separate hydraulic systems. Most rear-wheel drive vehicles use this system. The diagonal split system operates the brakes on opposite corners of the vehicle. This system is used on most front-wheel drive vehicles, which can experience weight transfer of about 80 percent at the front during a hard stop. If the front brakes were longitudinally split and the front hydraulic system failed, the 20 percent of braking done by the rear brakes would contribute very little to stopping the vehicle.

The solution to the weight shift problem is to split the hydraulic system diagonally, rather than front to rear. With a split diagonal system 50 percent of the braking will still be available in the event of a hydraulic failure in one half of the system. The circuit that includes the left front and right rear brakes comprises one-half of the brake hydraulic system. The other circuit includes the right front and left rear brakes.

Front suspension geometry helps negate the brakes' tendency to pull to one side or the other in the event of a partial hydraulic system failure. Chrysler was the first manufacturer to mass-produce a diagonally split system in 1978.

■ QUICK TAKE-UP MASTER CYLINDER

To increase fuel economy, some disc brake calipers are designed to cause less drag when the brakes are not applied. They have specially designed piston O-ring seals and grooves that cause the piston and pads to retract further than normal from the rotor. Retracting the piston further means that more fluid is needed in order to take up the clearance before pressure can build when the pedal is applied.

One master cylinder design moves a larger amount of fluid when the pedal is first applied. These master cylinders are called *quick take-up, step-bore, fast-fill,* or *dual-diameter bore* master cylinders. The bore has two different diameters (**Figure 4.21**). The rear part of the primary piston is larger in diameter than the front part (**Figure 4.22**). A larger bore in the master cylinder would normally result in very low braking force at the wheel cylinders. Remember, smaller master cylinder bores create more braking force but move less fluid.

The larger part of the bore, called the low-pressure chamber, allows its piston to quickly move a large volume of fluid when the pedal is first applied.

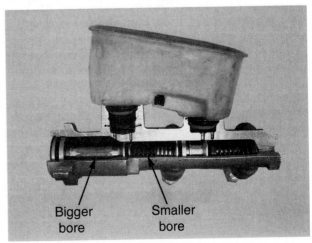

Figure 4.21 A quick take-up master cylinder has a bulge in the rear of the casting. *(Courtesy of Tim Gilles)*

Quick Take-Up Master Cylinder

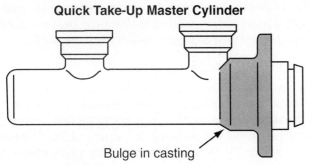

Figure 4.22 A section at the rear of a quick take-up master cylinder is larger than the rest of the bore.

When the friction materials contact the braking surfaces, the smaller part of the bore comes into play to give a large pressure boost during stopping. The result is less pedal travel.

Quick Take-Up Operation

The rear seal on the large part of the primary piston is a cup-type seal. This seal only pumps fluid at low pressure to move the pistons until the pads contact the rotors. The area of the piston between the large, low-pressure secondary cup and the smaller, high-pressure primary cup is called the *low-pressure section*.

The rear piston cup moves about twice the volume of fluid as that moved by the front piston cup. Remember how a simple master cylinder works when a brake adjustment is needed: fluid can flow past a lip seal in one direction but not the other. As the brakes are applied, the rear cup pushes a high volume of fluid past the back side of the lip of the front cup into the smaller high-pressure chamber (**Figure 4.23**). The large low-pressure cup continues to force extra fluid around the lip of the smaller seal, pushing the secondary piston forward as well. This extra fluid movement can only continue until all of the clearances between the linings and metal friction surfaces are taken up and back pressure in the system increases.

A spring-loaded quick take-up valve (also called a fast-fill valve) is located in the bottom of the rear reservoir. It replaces the vent and replenishing ports of an ordinary master cylinder. During the beginning of pedal application its check valve remains closed. This maintains enough pressure on the fluid in the larger low-pressure chamber at the rear of the master cylinder's primary piston to fill the smaller high-pressure chamber in front of it. The result is that pressure begins to build more quickly.

As long as the quick take-up valve remains closed, fluid bypasses the smaller primary piston cup and moves the secondary piston forward. When pressure in the low-pressure valley area of the piston rises to 70–100 psi, the check-ball in the quick take-up valve opens (**Figure 4.24**). This relieves pressure by allowing unnecessary extra fluid to return to the reservoir. If this pressure were allowed to build, it would result in a very hard brake pedal as the large low pressure cup remained responsible for creating force.

REMEMBER: *The rear piston is larger than the front piston. The pressure in the chamber between the pistons produces a force back toward the brake pedal that reduces braking force (**Figure 4.25**).*

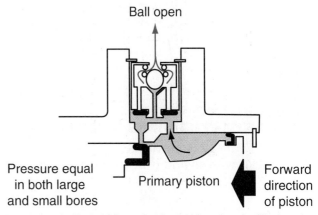

Figure 4.24 When the front chamber is filled and pressure builds in the rear chamber the valve opens, allowing pressure created by further fluid movement in the low-pressure chamber to be diverted to the reservoir.

Figure 4.23 Fluid in the low-pressure chamber is pumped past the cup seal into the high-pressure port during a stop.

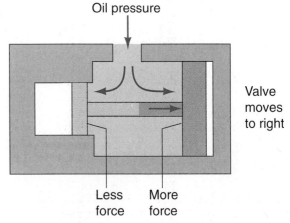

Figure 4.25 Fluid pressure will move the piston in the direction of the larger area.

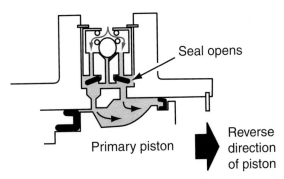

Figure 4.26 When the pedal is released, pressure in the low-pressure chamber drops quickly. The valve opens, allowing it to refill.

After the quick take-up valve opens, the smaller bore primary cup takes over and provides the pressure. From this point on, the quick take-up master cylinder works in the same way as an ordinary master cylinder. When the pedal is released, pressure in the large bore quickly drops very low and atmospheric pressure on the brake fluid in the reservoir opens the seal in the quick take-up valve to refill the large chamber with fluid for the next stop (**Figure 4.26**).

There are variations in system design when the hydraulic system is diagonally split or if the car has rear-wheel disc brakes. Some master cylinders have two quick take-up valves, one for the front and one for the rear.

METERING AND PROPORTIONING VALVES

Additional hydraulic control valves were required on older vehicles with hybrid (disc/drum) brake systems. Called **metering valves** and **proportioning valves**, they were needed to balance braking force between the front and rear brakes (**Figure 4.27**).

Metering Valve

The metering valve (**Figure 4.28**) is used on front disc brakes when a car has rear drum brakes. Its purpose is to prevent the front brakes from applying until the rear drum brake shoes overcome return spring pressure and contact the drums. The metering valve does not operate until system pressure reaches about 10 psi. Keeping the valve open below this pressure allows the fluid to expand and contract normally when the brakes are not applied. Once pressure builds to 10 psi, the metering valve shuts off pressure to the front brakes until the rear brake hydraulic system has developed pressure of 75 to 300 psi. The amount of pressure depends on the vehicle. When the metering valve operates the stem can be seen moving out, away from the body of the valve. The movement of the valve can be felt very slightly at the brake pedal during very light pedal application.

Preventing the front brakes from applying until the rear brakes start to build pressure serves two purposes:

1. It keeps front pads from doing too much light braking, preventing them from wearing out too soon.

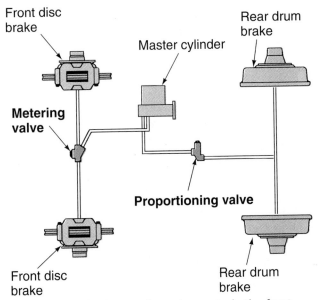

Figure 4.27 A metering valve controls the front brakes and a proportioning valve controls the rear brakes.

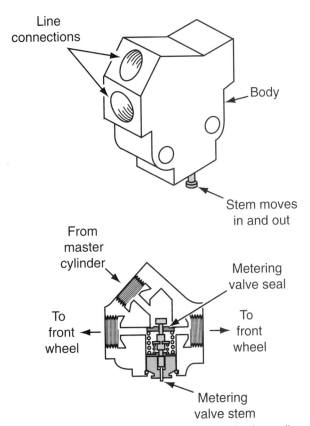

Figure 4.28 A metering valve keeps the front disc brakes from operating until rear drum linings are in contact with the brake drum.

2. It helps prevent dangerous skids that can occur on slick surfaces if the front brakes were to apply before the rear brakes.

Without a metering valve, the rear brake cylinders would have to overcome the drum brake return spring pressure before the rear brakes could begin to operate. Front disc brakes do not have return springs, so they would be able to start to apply before the rear brakes, even though both sides of the system had equal pressures. With four-wheel disc brakes, the metering valve is unnecessary.

Front-wheel drive vehicles with split diagonal systems do not use a metering valve. With FWD, the vast majority of the braking is done by the front brakes, so you want them to apply as soon as possible to overcome the torque of the front driving wheels. Because the weight is forward on these cars, the problem of front wheel lockup is greatly reduced and the metering valve is unnecessary. This is fortunate because a separate metering valve would need to be installed for each front wheel in a split diagonal system.

Proportioning Valve

The ability of the brakes to do their job is limited by the grip of the tires to the road surface. During hard stopping, a vehicle's weight shifts forward, away from the rear wheels. The proportioning valve (**Figure 4.29**) was introduced in the late 1960s to help prevent the rear wheels from locking up during a panic stop or when brakes are applied hard.

Disc brakes require greater force to apply than drum brakes. Dual servo drum brakes are self-energized (see Chapter 3), which causes the problem of rear wheel lockup on some trucks and cars. Drum and disc hybrid systems are not the only ones to use a proportioning valve. Four-wheel disc equipped vehicles also use them.

During a low-pressure stop, the proportioning valve does nothing. However, when pressure reaches a predetermined level, the proportioning valve prevents a further increase in pressure. The pressure at which this occurs is called the *split point* because greater pressure is now applied to the front wheels, while the pressure going to the rear wheels remains the same (**Figure 4.30**). As master cylinder pressure

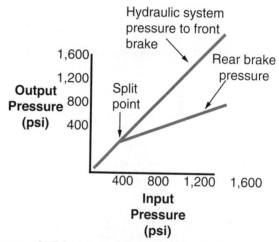

Figure 4.30 The split point happens when pressure reaches a predetermined level and the proportioning valve prevents a further increase in pressure.

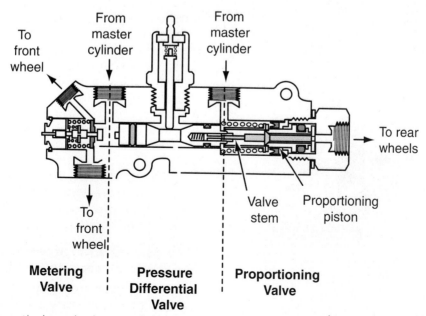

Figure 4.29 A proportioning valve is part of this combustion valve.

continues to increase, the proportioning valve once again opens. It will cycle open and closed, allowing pressure to the rear brakes to increase but not as much as pressure to the front brakes. This difference in percentage of pressure between the front and rear brakes is called *slope*. When the brakes are released, a spring opens the valve and fluid returns to the master cylinder.

Early systems had a single proportioning valve in the line to the rear brakes. With split diagonal systems, two proportioning valves are needed. A dual valve can be located near the master cylinder (**Figure 4.31**), or the valves can be built into the master cylinder (**Figure 4.32**) or threaded into the master cylinder outlet ports (**Figure 4.33**). Some late model antilock brake systems use dynamic proportioning, which eliminates the traditional proportioning valve (see Chapter 10).

Height-Sensing Proportioning Valves. Some proportioning valves are load sensitive. As chassis-to-ground height changes, they can either bypass or

Figure 4.33 These proportioning valves are installed at the master cylinder outlets. *(Courtesy of Tim Gilles)*

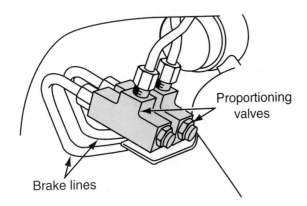

Figure 4.31 Dual proportioning valves.

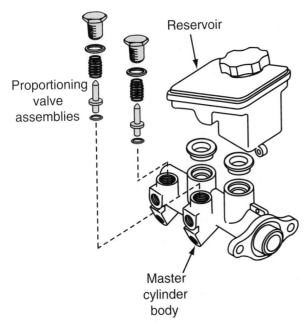

Figure 4.32 These proportioning valves are built into the master cylinder.

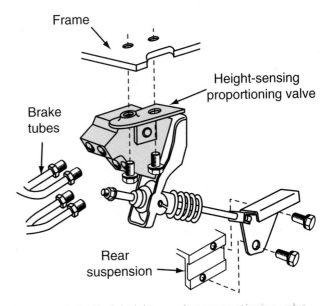

Figure 4.34 A height-sensing proportioning valve installed between the rear suspension and frame.

function normally to control stopping pressure to the rear brakes according to how much weight is on the rear of the vehicle (**Figure 4.34**). When the vehicle does not have a load, the proportioning valve functions normally, reducing pressure to the rear brakes after the split point. When the vehicle is loaded in the rear, the arm on the valve moves to allow full hydraulic pressure to reach the rear brakes.

Height-sensing proportioning valves were originally used on pickup trucks, but now they are often found on front-wheel drive passenger cars as well. This is because a four-door FWD car loaded with a family and luggage can be prone to having a considerable change in rear braking requirements.

Pressure Differential Switch/Brake Warning Light

Tandem systems are equipped with a brake warning light that alerts the driver when half of the hydraulic system has failed. When pressure on one side of the system drops, a piston inside the *hydraulic safety switch* (**Figure 4.35**) moves off-center. This completes an electrical circuit that illuminates a warning lamp on the driver's compartment instrument panel (**Figure 4.36**). On some cars, the same lamp illuminates when the parking brake is applied.

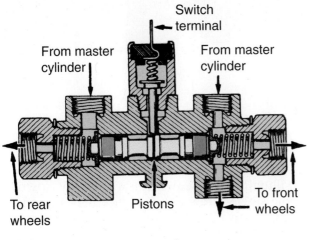

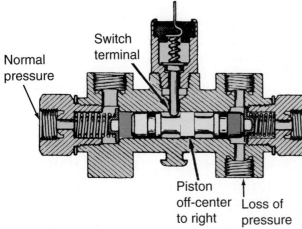

Figure 4.35 During a hydraulic system failure, the piston in the hydraulic safety switch moves off-center to complete an electrical circuit to the dash light.

Figure 4.36 A brake warning lamp. *(Courtesy of Tim Gilles)*

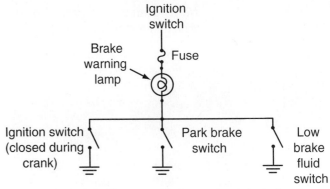

Figure 4.37 This warning lamp circuit includes the parking brake and low brake fluid switches.

Many cars have a low brake fluid level warning system instead of a hydraulic safety switch. This system also illuminates the brake warning lamp. Some vehicles have wear sensors in the disc brake pads that also turn on the same lamp when the pads wear beyond their limit. **Figure 4.37** shows the wiring circuit for the complete warning lamp circuit.

Master Cylinder Fluid Level Switch

Some master cylinders have a fluid level switch located in the master cylinder reservoir. There are several designs. A typical fluid level switch has a float (**Figure 4.38**) with contacts above it. When the float level drops beyond the allowable limit, the contacts close and complete the circuit to the warning light. This system can replace the pressure differential switch because a leak in the system will be evident by a drop in master cylinder fluid level.

Combination Valve

Since the 1970s, metering and proportioning valves have usually been combined with a hydraulic safety switch in one valve called a **combination valve**

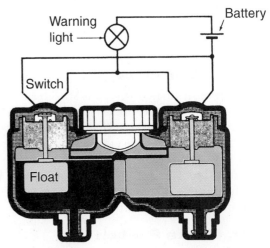

Figure 4.38 A float switch is used in place of a hydraulic safety switch.

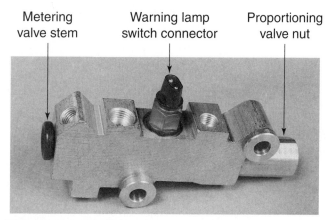

Figure 4.39 A three-function combination valve with a metering and proportioning valve and a safety switch.

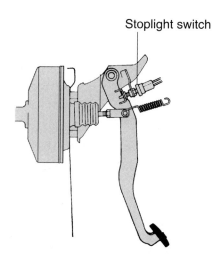

Figure 4.40 A stoplight switch.

(**Figure 4.39**). The three-function combination valve is the most common. Other combinations include a metering valve coupled with a proportioning valve or a safety switch with a metering valve.

■ ABS BRAKE PRESSURE CONTROL

Most newer cars are equipped with antilock brakes (ABS) as standard equipment. Sensors at the wheels provide input to a computer that monitors for unacceptable differences in speed between the wheels and modulates the brakes to compensate. ABS computer control is more accurate than earlier mechanical devices like the metering and proportioning valves. With ABS, these valves are no longer necessary. Earlier, mechanical proportioning valves were called *fixed* proportioning valves because the pressures at which they would activate were always constant. *Dynamic* proportioning is the term used in newer ABS systems that modulate rear brake pressures based on the speed of the wheels at the front and rear axles.

■ BRAKE SYSTEM ELECTRICAL SWITCHES AND WARNING LIGHTS

Several electrical switches and warning lights can be part of the brake system.

Stoplight Switches

The stoplights are turned on by a stoplight switch, usually mounted on the brake pedal arm (**Figure 4.40**). When the pedal is depressed, electrical contacts in the switch complete the circuit to the lights. In some older cars, stoplight switches are operated by hydraulic pressure. Some stoplight switches have extra contacts for cruise control, the torque convertor clutch, or antilock brake control.

Parking Brake Switch

When the parking brake switch, located on the parking brake lever or pedal, is moved slightly, a warning lamp illuminates. The switch is a normally closed single-pole single-throw (SPST) device that grounds the warning lamp circuit to illuminate the instrument panel lamp.

Brake Pad Wear Indicators

Some vehicles use electric brake pad wear indicators that illuminate a light in the instrument panel when brake pads wear to a certain point (see Chapter 7). Some systems use a small spring-loaded switch. Others use a pellet that is cast in the brake pad. The pellet wears with the pad. When it reaches its wear limit, the warning lamp on the instrument panel illuminates.

■ BRAKE HOSE AND TUBING

Steel hydraulic brake tubing runs the length of the vehicle frame. Rubber hoses on both front wheels and above the rear axle provide a flexible connection from the steel tubing to brake system components (**Figure 4.41**).

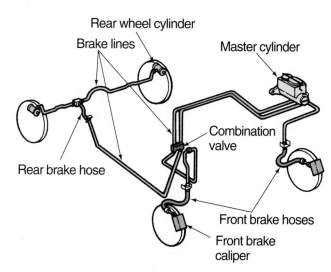

Figure 4.41 Hydraulic lines and hoses connect the master cylinder to the wheel brake units.

Brake Hose

Flexible rubber connections are required because the front wheels pivot during steering and the rear axle moves up and down as the springs deflect. When the vehicle has independent rear suspension, each rear wheel is served by its own flexible hose.

NOTE: *During hard stops, hoses must carry very high pressure. Because brake hoses are a safety item, Federal Motor Vehicle Safety Standard (FMVSS) 106 requires that hoses be able to withstand pressures of at least 4,000 psi for 2 minutes without bursting. Following that, the test raises the pressure 25,000 psi per minute until the hose bursts. This final point is the burst pressure for the hose. The federal safety standard requires that the outside of a brake hose be labeled with either HR, which means that the hose expands normally, or HL, which says that the hose has a lower amount of expansion.*

Brake hoses come in different lengths and have fittings swaged or crimped on the ends at very high pressure (**Figure 4.42**). The hose is made of rubber reinforced with woven fabric, usually rayon. The purpose of the fabric is to prevent the hose from expanding under pressure. The hose has an inner layer of rubber that comes into contact with the brake fluid (**Figure 4.43**). Most brake hoses have two *plies* that make up their middle fabric layers.

The *jacket* of the hose is the outside layer that protects the plies. The outside of a hose is subjected to damage from ozone in the atmosphere, which can cause the rubber to crack. The hose shown in **Figure 4.44** is a hose above the differential in a rear-wheel drive car. The hose has cracks in its exterior.

NOTE: *Water molecules are smaller than brake fluid molecules and over a long period of time, water vapor from the outside migrates to the brake fluid. This can happen even though brake fluid does not leak through the woven fibers and outer rubber layer of a brake hose.*

FMVSS 106 also calls for hoses to have either raised ribs or two 1/16-inch stripes painted on them. During installation stripes can alert the technician to whether the hose is twisted too much. Excessive twist can cause internal damage. At least one end of the hose has a swivel end to prevent twisting. When both ends have

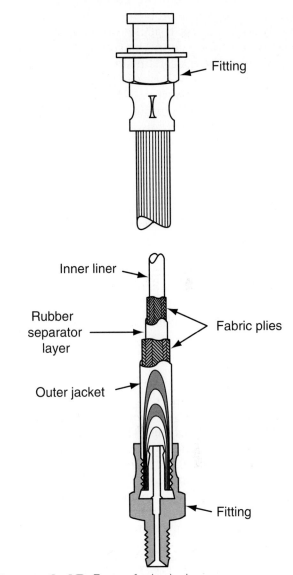

Figure 4.43 Parts of a brake hose.

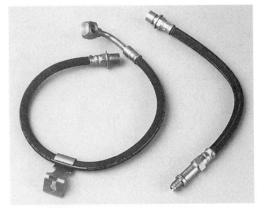

Figure 4.42 Brake hoses have fittings crimped on their ends.

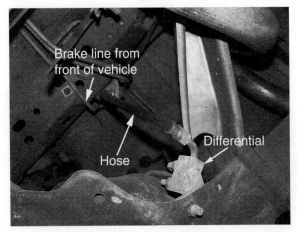

Figure 4.44 The rubber outer layer of this hose is cracked. *(Courtesy of Tim Gilles)*

been tightened, the hose is fixed securely and neither end can move.

Diagnosing Brake Problems by Pinching the Hose. A tool called a line clamp is used by technicians when troubleshooting brake problems. When using this tool, there is always a risk that the hose can be damaged. However, the line clamp has smooth jaws and only compresses the hose a certain amount, so it is less likely to damage a hose. Never use vise grips or any other type of pliers. Be aware that if you damage a hose, you are responsible for replacing it, unless it was already rotten or had a dried-out jacket. Also, some steel braided hoses used in the hot rod aftermarket use plastic liners. These can easily be ruined if they are pinched.

Tubing

Tubing found on automobiles, often called *line,* does not have threads at its ends like pipe. The size of tubing is determined by its outside diameter (OD). Pipe and hose are sized by their inside diameter (ID). Compare three pieces of ½-inch diameter pipe: one plastic, another steel, and the other copper. All of them have the same inside diameter but are very different in appearance (**Figure 4.45**).

Steel Tubing. Steel tubing, galvanized to prevent rust, is used for brake lines. It is *double-walled* so that it can bend easily. Copper-plated sheet steel is used to make the tubing. One type of tubing, called *seamless tubing,* is made by rolling the steel sheet until it has two layers and one seam, providing a double wall. Then, the copper is fused to this seamed tube using a *furnace brazing* process. When the copper melts, fusing to the tubing, the seam disappears.

Another type of brake tubing is made in two different layers, with the seams separated by at least 120 degrees. Furnace brazing is used to bind the copper into the seams of this tubing as well.

NOTE: *SAE Standard J1047 specifies that an 18-inch length of tubing must be able to withstand an internal pressure of 8,000 psi.*

Most brake tubing used on automobiles has an outside diameter of 3/16-inch. This is the size used for disc brakes. Drum brake systems sometimes use ¼-inch OD tubing. Tubing is also available in metric sizes, specified by SAE standard J1290. Common metric sizes include 4.75 mm, 6 mm, and 8 mm. An SAE standard for brake tubing dictates that it must be able to be bent in a full circle around a mandrel five times its diameter and not kink or be damaged in any way.

Sometimes there is a steel coil around the outside of a brake tube (**Figure 4.46**). The coil, called "armor," protects the line from damage where it might accidentally rub against something. The coil also allows the tubing to be bent without kinking.

When tubing is subjected to vibration or flexing, it is usually coiled (**Figure 4.47**). This is especially common on lines below the brake master cylinder. Because it is mounted on the car body and the brake lines are secured to the frame, flexing must occur there.

Tubing Fittings. There are many types of fittings used to join tubing to components. When using fittings be sure that they are the correct ones for the application. Flare fittings, compression fittings, and pipe fittings all have different types of threads and seats within them.

> **SAFETY NOTE**
>
> Brake tubing is made of steel. Copper tubing should *not* be used for brake or power steering lines. It is *single-walled* and does not have sufficient burst strength. It also work hardens from vibration, which can cause it to crack. Brakes and power steering systems develop very high pressures, sometimes in excess of 1,000 psi.

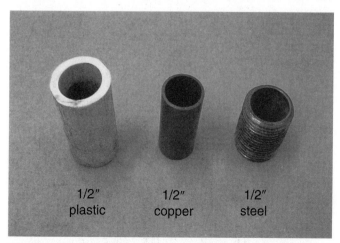

Figure 4.45 Compare the inside and outside diameters of ½-inch plastic, copper, and steel pipe. *(Courtesy of Tim Gilles)*

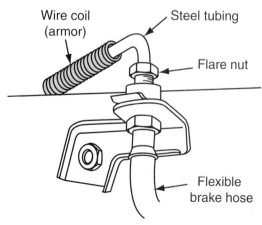

Figure 4.46 Armor protects the brake tubing from abrasion.

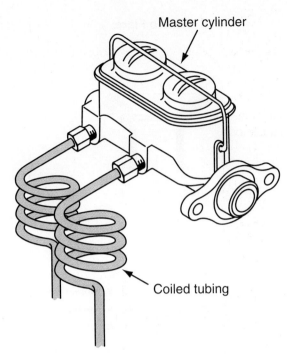

Figure 4.47 Tubing is coiled to prevent damage from vibration.

 When two parts with different types of threads are accidentally joined, damaged threads and a leak can result.

NOTE: *Threads can be either male or female. Female threads are those that are internal. Male threads are external.*

Sometimes manufacturers use connectors with oversized threads to prevent assembly errors. When replacing or adapting a part, such connectors can sometimes cause problems. *Step-up* or *step-down* adapters (**Figure 4.48**) are available for making these connections.

Connectors (**Figure 4.49**) are sometimes used between tubing brake parts. They can also serve as adapters between different types of fittings, such as when a pipe thread and flare fitting are joined.

Figure 4.48 Adapter fittings allow different sizes and types of fittings to be connected together. *(Courtesy of Tim Gilles)*

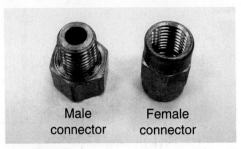

Figure 4.49 Connectors are used to join tubing to parts. *(Courtesy of Tim Gilles)*

Figure 4.50 Elbows are used for making sharp bends.

An *elbow* is sometimes used when a sharp turn is made. Bending the tubing takes too much space to do. Elbows can be either 30 degrees, 45 degrees, or 90 degrees (**Figure 4.50**).

■ FLARED CONNECTIONS

Flare fittings are used when two steel brake lines are connected. A flare is suited for high-pressure applications and *must* be used for such things as brakes or power steering. There are two kinds of flares used in automobiles, the *SAE type 45-degree double flare* or the *International Standards Organization (ISO) flare*, also called a *bubble flare* (**Figure 4.51**).

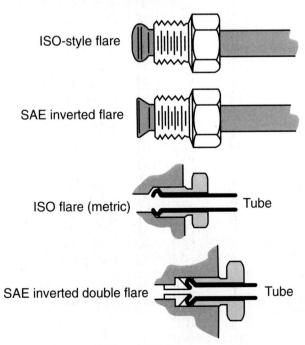

Figure 4.51 SAE and ISO flares

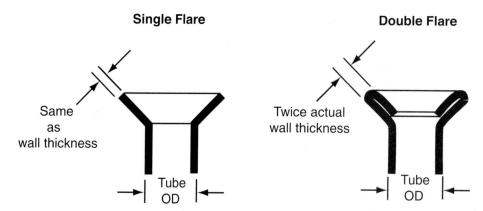

Figure 4.52 Single and double flares. Double flaring is used for brake tubing.

The end of an SAE **flared line** is tapered outward. The flare on the end of the line is wedged between the flare fitting and the **flare nut** on the line.

NOTE: Long and short flare nut designs are available. The long nut is used when a part is subject to vibrations because it supports the tubing farther away from the connection. Precut lengths of flared steel tubing are usually equipped with one long flare nut and one short one.

SAE automotive fittings are flared at 45 degrees. There are also 37-degree SAE flares. Be sure you are using SAE 45-degree fittings. The male nut will have a 45-degree angle and the female fitting will have a 42-degree angle. This intentional mismatch creates an interference fit for positive sealing.

Brake lines are always double flared (**Figure 4.52**) because single-flared lines tend to split the tubing at the seam. A 45-degrees SAE-type double flare is usually used with an *inverted flare* nut (**Figure 4.53**). The inverted flare nut is more common on automobiles. The standard flare is found in such applications as household natural gas lines.

ISO flares have been used on automobiles since the early 1980s. They used to be common only on imported vehicles but are now common on domestic vehicles as well. A bubble or ridge is formed in the line a short way back from its end. A single flaring operation can be done without danger of splitting the line, (as with a single 45-degree flare).

Like the SAE flare, the ISO flare uses an interference fit between the male and female parts. The outer surface angle is about 32.5 degrees, the flare seat is 30 degrees, and the angle at the end of the flare nut is 35 degrees. These fittings are popular with manufacturers because they seal well and simplify machining. An added advantage is that no seat is necessary as with the SAE flare (see **Figure 4.53**).

High Pressure Compression Fittings

Tapered pipe threads are not usually found in brake systems. This is because the female part of the fitting

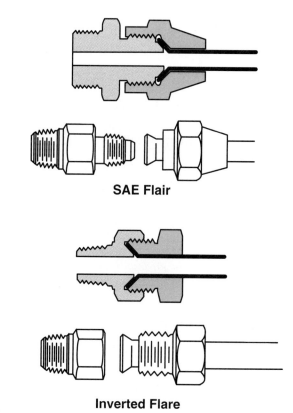

Figure 4.53 Compare inverted and SAE flares. *(Courtesy of Plews/Edelmann Division of Gates)*

SAFETY NOTE: Low-pressure applications such as vacuum lines often use flareless compression fittings with a brass sleeve that pinches tightly into the wall of the tubing to provide the seal (**Figure 4.54**). This type of compression fitting should not be used on high-pressure applications such as brakes or power steering systems.

can split when overtightened due to the taper of the thread. High-pressure male compression fittings are used instead, usually with soft copper compression washers that compress when the fittings are tightened.

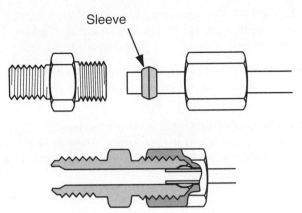

Figure 4.54 A compression fitting like the one shown here should *not* be used to repair a brake line.

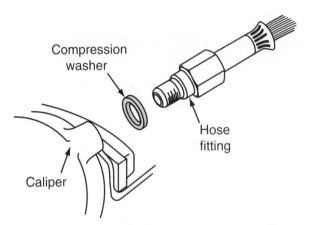

Figure 4.55 A brake line uses a copper sealing washer at one end.

SHOP TIP Once a compression fitting with a copper washer is tightened, the hose is no longer free to rotate. Therefore, the compression fitting must be installed before the other end of the hose is installed.

through a hollow bolt. A copper gasket is used to seal both sides of the fitting.

NOTE: *The copper gasket, which is often reusable, is easily lost during a brake service. The best practice is to keep a supply of these gaskets on hand and replace them any time a brake hose is removed.*

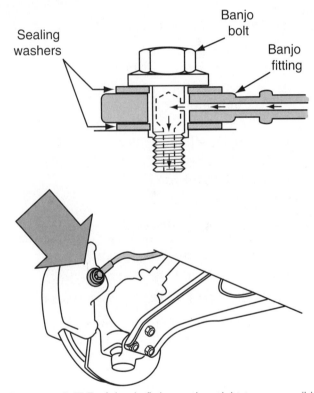

Figure 4.56 A banjo fitting makes tight turns possible.

Figure 4.55 shows a typical high-pressure compression fitting at the end of a brake hose. At other times, the end of the male thread has a tapered seat that seals at the bottom of a female fitting.

Brake tubing always has male fittings on both ends. Brake hose fittings have either female threads, male threads, or banjo fittings. **Banjo fittings**, named because they resemble a banjo, allow brake fluid to make a 90-degree turn in a tight space (**Figure 4.56**). Brake fluid goes into the banjo fitting and exits

REVIEW QUESTIONS

1. Pressurizing liquid to transfer motion or multiply and apply force is called _____.
2. Pressurized fluid acts on hydraulic cylinder pistons to produce _____.
3. Which wheel cylinder—larger or smaller—requires more pedal movement before pressure buildup can occur?
4. The front master cylinder port is called the _____ port.
5. The rear master cylinder port is called the _____ port.
6. Since the year _____, tandem master cylinders have been mandated by law (FMVSS 105) on all cars.

7. Cup expanders on the ends of the center spring in the wheel cylinder make a _____ pressure check valve unnecessary.

8. As brake fluid heats up or cools down, master cylinder fluid level fluctuates through the _____ port.

9. If a front-wheel drive car had a failure in its master cylinders front hydraulic system, which brake hydraulic system would provide very little stopping power—longitudinally or diagonally split?

10. With a quick take-up master cylinder, does the smaller or larger part of the bore take over to boost pressure after the friction materials contact the braking surfaces?

11. Which valve prevents rear drum brake wheels from locking up when brakes are applied hard, a metering valve or a proportioning valve?

12. Many newer cars have a low brake fluid level warning system instead of a _____ _____ switch.

13. The size of tubing is determined by its inside/outside (**choose one**) diameter. Pipe and hose are sized by their inside/outside (**choose one**) diameter.

14. A _____ fitting is used when two steel brake lines are connected.

15. Which flare nut is used when a part is subject to vibration, short or long?

PRACTICE TEST QUESTIONS

1. Which of the following is/are true about hydraulic pressure in the brake system?
   A. A larger diameter wheel cylinder increases the amount of force applied from the master cylinder.
   B. A master cylinder with a smaller diameter piston increases pressure to the wheels.
   C. Both A and B
   D. Neither A nor B

2. Technician A says wheel cylinder cup seal lips face outward. Technician B says a seal installed backwards will leak. Who is right?
   A. Technician A   B. Technician B
   C. Both A and B   D. Neither A nor B

3. A drum brake car has a solid, low brake pedal that rises to normal following a quick, second application of the pedal. Technician A says the brakes probably need an adjustment. Technician B says there is an internal leak in one half of the master cylinder's hydraulic system. Who is right?
   A. Technician A   B. Technician B
   C. Both A and B   D. Neither A nor B

4. Technician A says a residual check valve is not used in hybrid split diagonal systems. Technician B says a residual check valve cannot be used in disc brake hydraulic systems. Who is right?
   A. Technician A   B. Technician B
   C. Both A and B   D. Neither A nor B

5. Which is/are true about disc brake calipers?
   A. Some disc calipers are designed to cause less drag when the pedal is released.
   B. Some disc calipers have O-ring seals and grooves designed to retract the pistons and pads farther from the rotor than normal.
   C. Both A and B
   D. Neither A nor B

6. Two technicians are discussing quick take-up master cylinders. Technician A says the bore in a quick take-up master cylinder has two different diameters. Technician B says the front part of the primary piston is larger in diameter than the rear part. Who is right?
   A. Technician A   B. Technician B
   C. Both A and B   D. Neither A nor B

7. Which of the following is true about ABS-equipped vehicles?
   A. They must have a metering valve.
   B. They must have a proportioning valve.
   C. Both A and B
   D. Neither A nor B

8. Technician A says the metering valve helps prevent rear drum brake linings from wearing out too soon. Technician B says front-wheel drive vehicles with split diagonal systems do not use a metering valve. Who is right?
   A. Technician A   B. Technician B
   C. Both A and B   D. Neither A nor B

9. Which of the following is true about flared connections?
   A. A male flare nut with a 45-degree angle is coupled to a female fitting with a flatter angle to improve sealing.
   B. Bubble flares (ISO flares) are used only on imported vehicles.
   C. Both A and B
   D. Neither A nor B

10. Technician A says a hydraulic system failure with a single-piston master cylinder will result in a total brake system failure. Technician B says when one-half of a tandem brake hydraulic system fails, pedal height will be lower. Who is right?
    A. Technician A   B. Technician B
    C. Both A and B   D. Neither A nor B

CHAPTER 5

Master Cylinder and Hydraulic System Service

■ OBJECTIVES:
Upon completion of this chapter, you should be able to:
- ✔ Diagnose brake hydraulic system problems.
- ✔ Troubleshoot master cylinder problems.
- ✔ Service master cylinders.
- ✔ Describe master cylinder bleeding procedures.
- ✔ Explain various methods of brake fluid testing.
- ✔ Demonstrate all of the various methods of brake fluid flushing and bleeding.
- ✔ Diagnose brake switch and warning lamp problems.
- ✔ Service brake hoses and tubing.

■ KEY TERMS
flaring tool
reverse fluid injection (RFI)
vacuum bleeding

■ INTRODUCTION

Brake system hydraulic parts are usually serviced as part of a complete brake job. This chapter discusses servicing the master cylinder, brake lines, and hoses, and brake fluid bleeding. Repairs to drum brake wheel cylinders and disc brake calipers are covered in Chapters 6 and 7.

Rubber seals within hydraulic system components are affected by heat. During manufacturing, rubber is cured by a baking process called vulcanizing. In the brake system, further heating of the rubber causes it to continue to cure. Over time, the rubber becomes less pliable and can crack.

Sometimes a repair shop will inspect hydraulic components and only repair an item that is faulty. Most shops perform hydraulic system service during a brake job. One practice is to remove master cylinders, wheel cylinders, and disc calipers and replace them with new or remanufactured components. Another practice is to disassemble, clean, and inspect components before reassembling them with new internal parts that come in rebuilding kits.

The metal bores of hydraulic cylinders oxidize as they are exposed to moisture. Brake fluid's hygroscopic nature causes it to attract and absorb water. Cylinders are made of iron and aluminum, both of which can be damaged by moisture. Some cylinders have both aluminum and iron parts. These two dissimilar metals can create voltage like a battery, resulting in damaging electrolysis in which metal leaves one part and is deposited somewhere else.

■ HYDRAULIC SYSTEM SERVICE

A thorough inspection of the braking system should be performed before attempting any repairs. Start with the pedal and master cylinder.

Check Brake Pedal Travel and Feel

Apply the foot brakes and check the travel of the brake pedal. *Pedal reserve* is the distance that pedal is above the floor after the brakes begin to stop the car (**Figure 5.1**). Some manufacturers specify a pedal reserve travel measurement. Low pedal reserve can indicate a leak in one half of the system. When the pedal goes all the way to the floor, this indicates a serious problem involving failure in both halves of the brake system (see Chapter 4).

NOTE: *Most of the time you will notice a slight increase in pedal height, even if the brakes are fully adjusted.*

The pedal should feel firm against your foot. If the pedal feels soft and "spongy" there is probably air in the system that can be removed by bleeding the brakes.

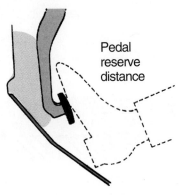

Figure 5.1 Check brake pedal travel.

> **SHOP TIP** Brake adjustment can be checked by pumping the pedal rapidly. If the height of a firm pedal rises significantly on the second application, the brakes need adjustment.

NOTE: *Cars with ABS brakes will have a feeling of pedal pulsation during hard stops. Some of these systems will also experience a rise in pedal height during hard stops.*

Fluid Check

Next check the fluid level. External leaks are usually visible at a brake backing plate, hoses, or the back of the master cylinder. Even when no fluid is leaking out, there could still be an internal leak. A leaking master cylinder is one example of when seals bypass fluid internally. If the pedal slips slowly toward the floor under very light pressure, the master cylinder is probably leaking internally.

Remove the cover(s) (**Figure 5.2**) and check the level of the fluid in both halves of the master cylin-

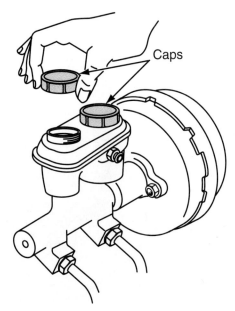

Figure 5.2 A master cylinder with two reservoir caps.

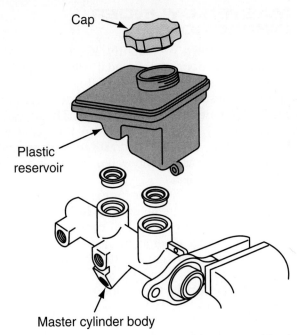

Figure 5.3 A master cylinder with a removable plastic reservoir.

der reservoir. Reservoirs can have one or two caps. **Figure 5.3** shows a plastic reservoir with a single cap. Plastic reservoirs are transparent to permit observation of the fluid level without removing the reservoir cover. Always remove the reservoir cap or cover, however, to verify fluid level. Stains from dirty fluid can give a false impression of fluid level.

NOTE: *Do not fill the reservoir all the way to the top. Space must be left to allow for fluid expansion as the brakes heat up. Expanding brake fluid can begin to apply the brakes as pressure builds in the reservoir. The maximum fill level is about ¼ inch from the top of the reservoir chamber.*

Fluid Leaks

When the master cylinder fluid level is low (and brake pads are not worn), look for an external leak elsewhere in the brake system. If the seal at the rear of the master cylinder bore (secondary cup) leaks, fluid can escape. When a wheel cylinder is leaking, the inside of a tire

> **SHOP TIP** To tell the difference between an axle leak (gear oil) and brake fluid leak, try washing off the fluid with water. Brake fluid is water-soluble, but gear oil is not. Also, brake fluid causes paint damage. Bubbled paint on the brake backing plate sometimes indicates a brake fluid leak.

will often be streaked with fluid and dirt. Occasionally the outside of the tire may even show signs of fluid leakage. A wet rear tire can be due to a leaking axle seal.

MASTER CYLINDER INSPECTION

Hold the pedal down very lightly. It should not fade away toward the floor. If the pedal drifts toward the floor and there is no visible sign of fluid leakage anywhere in the system, the master cylinder is leaking internally past one of its primary cups. This is called *bypassing*. Usually, pushing harder on the pedal will mask the problem, causing the brakes to work properly as the seal distorts against the wall of the cylinder and cuts off the internal leak.

Check the master cylinder for external leaks. On a master cylinder with a metal cover (**Figure 5.4**), leaks can occur when a bent metal bail causes the cover to loosen. Another place where a master cylinder can lose fluid is at the secondary cup on the rear piston. If the vehicle does not have a power booster, fluid will be visible on the carpet or on the inside of the firewall or bulkhead. When there is a power booster, fluid will leak between the front of the power booster and the master cylinder. A small amount of seepage is acceptable.

When disc brake linings wear, fluid level drops because the pistons move out in their bores to compensate for the lining wear. If one of the master cylinder chambers is larger and its fluid level is lower than the other one, the disc linings served by that chamber are probably worn.

NOTE: *When a secondary piston seal is defective, fluid can be pumped from one of the fluid chambers into the other. Look for this condition when one chamber is low and the other is overflowing.*

Check the Vent Port

Fluid movement past the vent port (often called the compensating port) should always be visible when the pedal is applied (**Figure 5.5**). The vent port is a very small opening. Brake fluid absorbs heat from the hydraulic cylinders at the wheels, causing it to expand. If the vent port is restricted, the brakes will not release all the way. This produces a pedal that is high and very hard. When the driver tries to accelerate the vehicle again after coming to a stop, the brakes continue to hold. This is called *brake drag*.

Check to see that brake pedal travel is adjusted correctly, with a small amount of free play at the brake pedal (**Figure 5.6**). A lack of pedal free play can cause the primary cup to remain ahead of the vent port even after the brakes are released, which produces the same symptoms as a plugged vent port.

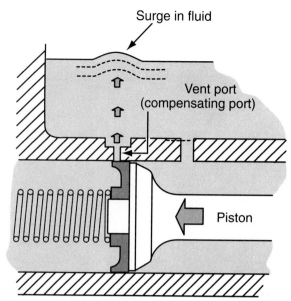

Figure 5.5 Fluid movement past the vent should always be visible when the pedal is applied.

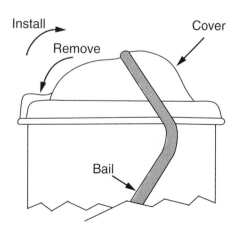

Figure 5.4 This bail can only be removed in one direction.

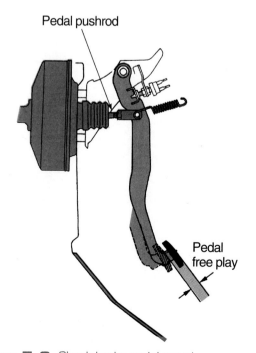

Figure 5.6 Check brake pedal travel.

Using the Vent Port to Check for Air in the System

To check for air in the hydraulic system, first check the fluid level in the master cylinder and add fluid if it is low. Then follow these procedures:

1. Replace the cover loosely on the reservoir.
2. Have a helper apply the brakes rapidly ten to twenty times, holding pressure on the pedal after the last push.
3. Remove the cover from the master cylinder and carefully watch the movement of fluid while the pedal is released.
4. A large spurt from the vent (compensating) port of either reservoir chamber indicates air in the system (**Figure 5.7**).
5. Note which chamber has the high fluid spurt. This is the side of the system that has the trapped air and will require bleeding.

Air trapped in the system is compressed during the repeated pedal pumping. The squirt results when the pedal is released, uncovering the vent port.

NOTE: *Be sure fender covers are installed on the fender. Splashed brake fluid can ruin paint. Wear eye protection.*

Plugged Reservoir Vent

When a master cylinder reservoir cover vent is obstructed air can be drawn in at the back of the master cylinder. This aerates the fluid and makes it compressible, resulting in a spongy pedal instead of a firm one. When the vent is part of the steel cover of a master cylinder, it is easily checked for an obstruction when checking the fluid level.

Master Cylinder Testing

Sometimes a master cylinder can be difficult to remove, so you might want to test it in place. First, bench bleed it in place. This procedure is covered later in this chapter. Plug the master cylinder outlet ports with plugs from the brake bleeding kit. Then, apply the brake pedal. With no air in the fluid, it should move very little. If it moves more than expected, once again carefully bleed the master cylinder. If pedal movement is still more than expected, the master cylinder is leaking internally.

Master Cylinder Removal

When removing the master cylinder from a vehicle, be sure to use fender covers to protect the paint from accidental spills. First, loosen and remove the two flared metal lines using a flare nut wrench (**Figure 5.8**).

If there is no power booster, remove the bolts or nuts holding the master cylinder against the bulkhead and disconnect the pedal pushrod from under the dash. When there is a power booster, unbolt the nuts that hold the master cylinder in place (**Figure 5.9**).

Remove the master cylinder and dump the brake fluid in a suitable container for disposal. A bulb syringe (which looks like a turkey baster) can be used to remove most of the fluid from the reservoir prior to master cylinder removal.

Master Cylinder Disassembly

When a master cylinder is defective, disassemble it and use your knowledge of master cylinder operation to

> **SHOP TIP** Current EPA regulations allow brake fluid to be disposed of in engine oil. Local regulations differ. Check the environmental regulations in your area before disposing of brake fluid.

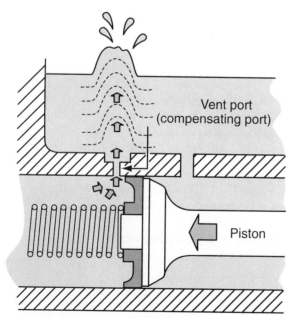

Figure 5.7 A heavy surge from the vent (compensating) port indicates air in the system.

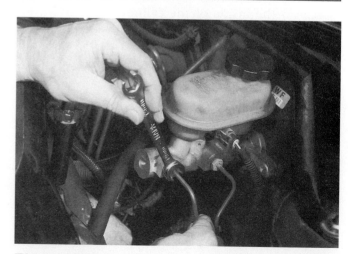

Figure 5.8 Remove the two flared metal lines using a flare nut wrench.

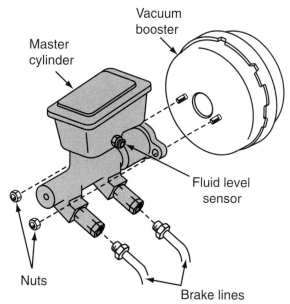

Figure 5.9 Remove the master cylinder.

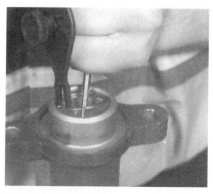

Figure 5.11 Hold the piston against piston return spring pressure as you remove the circlip.

locate the source of the problem. New or rebuilt master cylinders are available. When the cylinder bore is not damaged by corrosion, a rebuild kit may be used to restore the master cylinder to good operating condition (**Figure 5.10**). Liability issues and the high cost of parts and labor have made the practice of master cylinder rebuilding less popular than it was in the past. Most shops now replace master cylinders with new or factory rebuilt units.

To disassemble a master cylinder, push the piston deeper into the bore using a large Phillips head screwdriver or metal rod. While holding the piston in against piston return spring pressure, remove the circlip at the outside of the bore (**Figure 5.11**). Remove the piston from the bore. Sometimes there is a stop screw in the bottom of the center of the bore that must be removed before the front piston can be removed (**Figure 5.12**). Push the piston deeper into the bore while you remove the stop screw.

Clean the inside of the cylinder bore and use a light to inspect it. If the bore is corroded or pitted, the

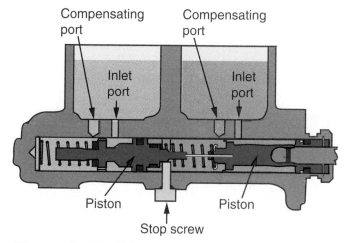

Figure 5.12 This master cylinder has a stop screw at the front of the rear master cylinder piston.

master cylinder must be replaced. If the bore is in good condition, it can be reassembled using new parts from the rebuild kit.

Cast iron master cylinder bores are reconditioned using a drill motor and a hone (see Chapter 6). Hone stones are abrasive. They cannot be used on aluminum cylinders, which have larger pores than cast iron. During manufacture, aluminum cylinders are anodized or rolled to close up the larger pores of the aluminum. If the bore is honed, the pores are reopened, which can cause newly installed rubber sealing lips in the rebuild kit to wear excessively.

NOTE: *Quick take-up master cylinder service is the same as with other master cylinders.*

Reservoir Installation

If the master cylinder reservoir is plastic (**Figure 5.13**), remove it, clean it with hot water, and dry it. Lubricate new grommets with brake fluid or silicone lubricant before reinstalling the reservoir on the body of the master cylinder (**Figure 5.14**). Be careful not to install the reservoir backwards.

Figure 5.10 A typical master cylinder rebuild kit.

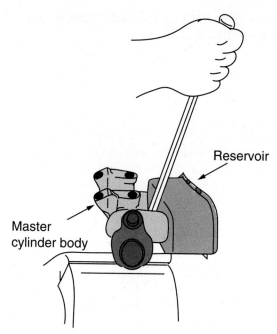

Figure 5.13 Removing a plastic reservoir.

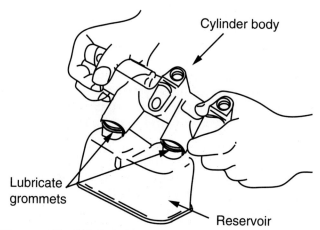

Figure 5.14 Lubricate new grommets with brake fluid or silicone lubricant.

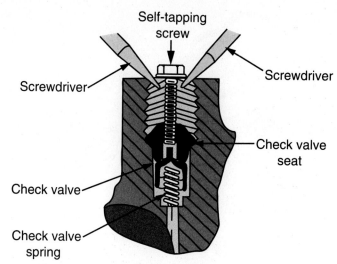

Figure 5.15 Removing a residual pressure check valve. (Courtesy of Dana Corporation/Brake Parts)

Residual Check Valve Service

Some master cylinders used with drum brakes use a residual pressure check valve. In tandem cast iron master cylinders the residual check valve is located behind a pressed-fit brass insert in the fluid outlet, under the fluid line's flare nut. To remove the check valve, install a self-tapping screw in the hole in the middle of the insert. Then pry it out with a pair of screwdrivers (**Figure 5.15**).

NOTE: *Rapid lining wear will occur if a check valve is used in a disc brake system.*

A restricted check valve can cause excessive pedal travel before the brakes apply. To test the operation of the check valve, apply and release the brakes. Then, open a wheel cylinder bleed screw. A brief spurt of fluid is normal as the residual pressure is released.

■ BENCH BLEEDING A MASTER CYLINDER

Some master cylinders, especially those mounted at an angle, are susceptible to trapping air in high places in brake fluid cavities. Quick take-up cylinders are especially prone to this. Some master cylinders are very difficult to bleed after they are installed on the vehicle. To prevent this, bench bleed all master cylinders before installation by filling them with fluid and bleeding any air from their bore.

Figure 5.16 shows a master cylinder in a vise with two tubes returning fluid to the reservoir. *Slowly pumping the apply rod forces air and fluid out of the piston chambers.* New, airless fluid enters the master cylinder through the reservoir. You can also do this by holding your fingers over the fluid outlets while releasing the pedal rod to prevent air from being sucked back into the piston chamber. Position a drain pan under the master cylinder to collect any fluid that

---

CASE HISTORY

An automotive student replaced the factory-installed front drum brakes on a 1968 Camaro with disc brakes. He used the original master cylinder. About a year later, his girlfriend enrolled in the same brake class. She brought her boyfriend's car in for installation of new disc brake linings (for the third time since the disc brakes were installed). From what she had learned in class, she had a hunch that the residual pressure check valve might still be in the master cylinder. She removed it and solved her boyfriend's problem.

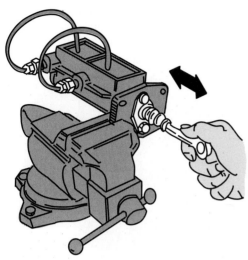

Figure 5.16 Bench bleeding a master cylinder. (Courtesy of Federal Mogul Corporation)

escapes with air when the pedal rod is applied. Repeat the process until no more air is visible in the fluid coming from the master cylinder.

Bench Bleeding Using a Syringe

One bench bleeding process uses a rubber-tipped syringe to pull fluid from one master cylinder outlet port while the other port remains plugged (**Figure 5.17**):

1. Carefully secure the master cylinder in a vise.
2. Remove the plug from one of the master cylinder outlet ports and draw brake fluid from the master cylinder into the body of the syringe until it is half full.
3. Hold the syringe with its rubber-tip end facing up while you depress the plunger to expel any air in the fluid.
4. Inject the fluid back into the master cylinder fluid outlet. If air bubbles are visible in the fluid reservoir as the fluid is reinjected, repeat the process until there are no bubbles.
5. Repeat the process for the other chamber until all air is bled from the master cylinder.

Bench Bleeding Quick Take-up Master Cylinders

Quick take-up master cylinders have a stepped bore. The rear piston is larger than the front one. When the pedal is first depressed, the large rear piston pushes lots of fluid under low pressure until all pad-to-rotor clearance is removed. This occurs with less pedal travel compared to a straight bore master cylinder.

To identify whether you have a quick take-up cylinder with a stepped bore, feel the bottom of the master cylinder. The casting will be larger near the rear.

A good way to bench bleed quick take-up master cylinders is by using the syringe method outlined earlier. You can also bleed the cylinder without a syringe by using a large Phillips screwdriver to slowly depress the rear piston three-quarters of the way into the bore without bottoming out the piston.

Bench Bleeding with a Reverse Fluid Injector

Reverse fluid injection is a popular newer method of master cylinder bleeding (**Figure 5.18**). The procedure is as follows:

1. Install a plastic plug in the master cylinder's primary (rear) outlet port.
2. Inject brake fluid into the outlet port supplied by the secondary piston using slow, steady strokes. When no air bubbles can be seen in the fluid in the front reservoir, stop injecting fluid.
3. Install a plug in the secondary outlet.
4. Remove the plug from the primary outlet and repeat the process with the secondary chamber.

After both chambers are bled, install plugs in both outlets and push on the pushrod. If there is no air in the master cylinder bore, resistance should be firm and the piston should not move inward.

When a master cylinder has four outlet ports, inject fluid into the bottom port until fluid drips from the top port. Then plug the bottom port and inject fluid into the top one until air bubbles are no longer visible in the reservoir.

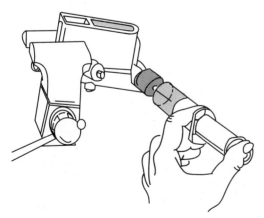

Figure 5.17 Using a rubber-tipped syringe to pull fluid from one master cylinder outlet port.

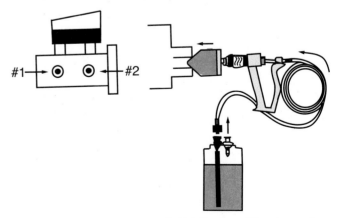

Figure 5.18 Reverse fluid injection. (Courtesy of Phoenix Systems, L.L.C.)

The injector can also be used to vacuum bleed the master cylinder.

With quick take-up master cylinders, push the master cylinder piston into the bore about ½ inch until the primary seals are positioned past the vent ports. This ensures that the low pressure area of the bore is filled with fluid during bleeding.

Reinstalling the Master Cylinder

When bench bleeding is completed, the master cylinder is ready to be reinstalled. Typically, two nuts hold it against the power booster. When installing a master cylinder, install the flared lines before tightening the mounting nuts. This allows the master cylinder to be wiggled while starting the threads into one another.

To avoid cross-threading when reinstalling flared lines, turn the flare fitting counterclockwise first to help seat the threads. Then, carefully turn the fitting clockwise to tighten it. Jiggle the metal tubing while turning the fitting as you turn it all the way against its seat, using only your fingers. Then, use a flare nut wrench to tighten it securely. If there are brass fittings or proportioning valves on the master cylinder outlet, be sure to use another wrench to hold them from turning while you firmly tighten the flare fitting.

Adjusting Brake Pedal Free Travel

Pedal free travel does not change during normal service. However, it is always checked when the master cylinder is replaced. Clearance between the brake pushrod and the back of the primary master cylinder piston or power booster should be less than ⅛ inch. Any clearance is multiplied by the leverage ratio of the brake pedal and linkage. For instance, with a pedal leverage ratio of 4:1, ⅛ inch of clearance at the pushrod translates to ½ inch free play at the pedal (4 × ⅛ inch). With 1/16 inch of free play, the brake pedal will have ¼ inch of free play (**Figure 5.19**).

■ BRAKE FLUID SERVICE

Some technicians make the mistake of removing the wheels before inspecting the fluid in the master cylinder. This can create a problem, especially if you discover fluid contamination that could prevent you from reassembling the brakes for safety and liability reasons. Remove the master cylinder reservoir cover and inspect the brake fluid (**Figure 5.20**). It should appear relatively clean and transparent.

Adding Brake Fluid

The reservoir should have a sufficient amount of fluid in reserve to allow the front brake pads to wear completely out. Some repair shops have a policy forbidding the practice of adding, or topping off, brake fluid.

NOTE: *Glycol brake fluid can ruin a vehicle's paint. Be careful when adding or removing brake fluid from a master cylinder.*

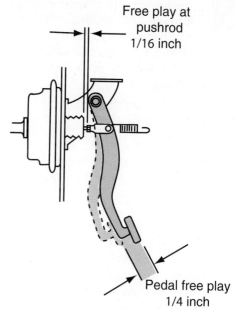

Figure 5.19 With 1/16 inch of pushrod free play, the brake pedal will have ¼ inch of free play.

Figure 5.20 Remove the master cylinder reservoir cover and inspect the brake fluid.

Fluid Contamination

If the reservoir cover is a steel one with a gasket, check the gasket to see that it is not torn or swollen (**Figure 5.21**). A swollen gasket can be a sign of contaminated brake fluid.

Figure 5.21 Check the gasket to see that it is not torn or swollen.

CAUTION: A brake fluid reservoir that is completely empty indicates a serious problem that should be addressed before the vehicle is driven.

NOTE: *Avoid accidental contamination of the brake hydraulic system with petroleum products. Sometimes rubber parts are used in the brake system, and rubber is not resistant to petroleum damage.*

Late-model brake seals are mostly resistant to petroleum damage, but dust boots are still made of natural rubber. As an experiment, put a rubber dust boot and wheel cylinder cup into a jar of engine oil. Allow them to remain there overnight.

Rubber parts that contact oil will swell. When this happens, the following steps are necessary:

1. Completely disassemble the entire hydraulic system.
2. Thoroughly flush the system with alcohol.
3. Install new rubber parts.

Hydraulic System Flushing

Occasionally a brake hydraulic system is accidentally contaminated with a petroleum-based product and must be flushed. Alcohol is used to clean the hydraulic system because it removes petroleum without leaving a residue. Alcohol also absorbs any water that might be present and carries it out during the flush. Following a system flush, use traditional bleeding methods as fluid is restored to the hydraulic system.

Routine Brake Fluid Replacement

When water is absorbed by brake fluid, the inside of the brake system corrodes. To avoid moisture contamination, the brake system is flushed whenever a brake job is performed, and sometimes more often. Manufacturers are now recommending more frequent fluid flushes because most cars are equipped with ABS brakes, which have very small passages that are more sensitive to contamination than older brake systems.

Most of today's brake systems are "bimetal." This means system parts are made of more than one type of metal. Common hydraulic system materials include aluminum, iron, steel, and copper. Master and wheel cylinder housings and pistons are often made of aluminum. Housings are also made of cast iron and pistons are made of steel. Steel hydraulic tubing is lined with copper.

You probably remember from science lessons that two dissimilar metals in an electrolyte solution can produce a voltage. New brake fluid includes additives to prevent it from becoming an electrolyte, but these additives become depleted. Moisture accumulation in the hygroscopic brake fluid also causes battery (galvanic) action. The corrosion that results plugs valves in the ABS and causes pitting in the wheel cylinders and master cylinder.

One of the reasons glycol is the material of choice for brake fluids is that it has an affinity for water. Water that is absorbed by the fluid will not be able to act adversely on one particular place where it might settle. In climates of high humidity (lots of moisture in the air), brake fluid has been shown to absorb an average of 3 percent of its volume in water over a period of a year and a half. If the fluid is not changed, moisture content can reach 8 percent or more in another year. Once the fluid reaches its saturation point, it cannot absorb any more water.

Remember from Chapter 3 that DOT 3 fluid has a minimum dry boiling point of 401°F and its saturated boiling point must be above 284°F. The following are DOT 3 boiling points with different volumes of moisture accumulation. They will show you the importance of maintaining moisture-free brake fluid from a safety standpoint.

Percent Water Volume	Boiling Point
1 percent	369°F
2 percent	320°F
3 percent	293°F

NOTE: *DOT 4 fluid absorbs fluid more slowly, but its boiling point drops off faster. It loses half of its boiling point when saturated by 3 percent.*

Fluid Change Interval

During average driving, brake linings usually wear out fast enough for brake fluid to remain in good condition between brake jobs. A general rule of thumb is to change fluid every two years or 30,000 miles, but always check the manufacturer's recommendations. ABS systems sometimes call for more frequent fluid changes.

Service bulletins for Ford, Mercedes, BMW, Volvo, and others call for a brake fluid change every twenty-four months or sooner. Some manufacturers do not list a fluid change requirement, but others specify intervals as low as 18,000–24,000 miles.

■ BRAKE FLUID TESTING

Some brake manufacturers estimate that half of all operating vehicles have never had a brake fluid change. Several methods of testing brake fluid are used in the industry.

Chapter 5 Master Cylinder and Hydraulic System Service **103**

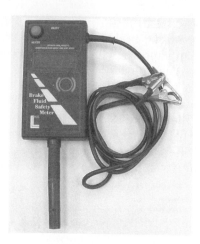

Figure 5.22 Moisture content testers. *(Courtesy of Tim Gilles)*

Moisture Content Testers

Testers are available that check hydraulic brake fluid's moisture content. **Figure 5.22** shows two of them. Over time, moisture gets into the system through air that permeates rubber brake hoses. Newer hoses have EPDM (ethylene propylene diene monomer) inner liners that prevent this. Moisture can also enter the hydraulic system past rubber seals that have hardened with age. An aging seal might prevent brake fluid from leaking out but still permit smaller molecules of air and water to seep in.

Refractometer Testing

Refractometer testing is another method of evaluating brake fluid (**Figure 5.23**). A drop of fluid is placed on the face of the tester and the result is read by viewing a gauge inside the instrument. Refractometers are designed for testing the specific gravity of specific fluids, including brake fluids, coolant, and battery electrolyte.

Voltmeter Testing

A voltmeter can be used to measure the conductivity of the brake fluid in the master cylinder housing. One

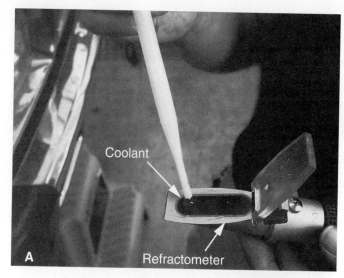

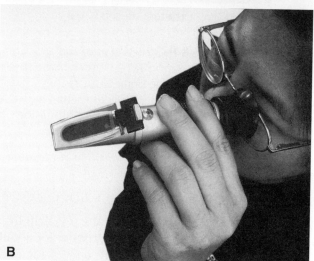

Figure 5.23 Using a refractometer. (A) Remove a coolant sample from the radiator and place a drop of coolant on the viewing surface. (B) Read the freeze protection through the eye piece. *(Courtesy of Tim Gilles)*

probe is put into the fluid and the other is connected to ground. A reading of 0.3 DC volt is the maximum.

Fluid Test Strips

As brake fluid wears out, it becomes thicker. Test strips can be used to check brake fluid for wear. These test strips are called FASCAR™ strips (**Figure 5.24**). FASCAR

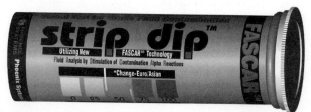

Figure 5.24 FASCAR test strip. *(Courtesy of Phoenix Systems, L.L.C.)*

Figure 5.25 Immerse test strip in brake fluid. (Courtesy of Phoenix Systems, L.L.C.)

stands for *fluid analysis by stimulation of contamination*. Test strips indicate the amount of breakdown in the anticorrosive additives in the brake fluid as they change in color from white to purple. The color change occurs with an increase in the concentration of alpha ions in the brake fluid.

NOTE: *Test strips do not give an indication of the amount of moisture in the brake fluid.*

To test the fluid with a test strip, follow these steps:

1. Immerse the reaction zone of the test strip in the brake fluid in the master cylinder for about a second (**Figure 5.25**).
2. Shake any excess fluid from the test strip.
3. Wait 30–90 seconds.
4. Compare the test strip's reaction zone with the FASCAR rating scale.
5. If the rating is 75 or higher the system should be flushed and the brake fluid replaced.

Selecting the Correct Brake Fluid

There are different types of brake fluids (see Chapter 3). Always use high-quality brake fluid of the type specified by the manufacturer.

No manufacturers currently recommend DOT 5 silicone brake fluid for their cars. If you encounter an older car with DOT 5 fluid, do not mix it with a glycol-based fluid (DOT 3 or DOT 4). Rubber brake seals are made of EPDM or SBR rubber. SBR seals can swell when exposed to silicone, causing them to soften or to leak.

NOTE: *Silicone brake fluid looks different than glycol-based fluid. Because dye is added to the clear silicone base, it is purple in color.*

■ BLEEDING BRAKES

To operate correctly, the hydraulic system must be able to push a solid column of fluid. Air or moisture in the hydraulic system can result in a soft, "spongy" pedal. Air is compressible; brake fluid is not. Water in the brake fluid can vaporize when it becomes hot, adding gas pockets to the fluid.

When brakes are bled, air and fluid are removed from the hydraulic system through a bleed screw (**Figure 5.26**). Bleed screws are located at the highest point on the back of the wheel cylinder or caliper (**Figure 5.27**). When the bleed screw is loosened, a hollow opening is uncovered (**Figure 5.28**) and fluid escapes through the center of the screw.

NOTE: *The master cylinder can be bled independently from the rest of the hydraulic system. (That procedure was covered in the section on master cylinder service earlier in this chapter.)*

Brake Bleeding Sequence

Manufacturers specify different bleeding sequences, depending on the design of the brake system. Correct

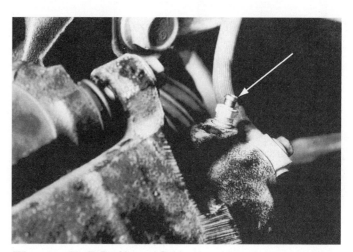

Figure 5.26 Air and fluid are removed through the bleed screw.

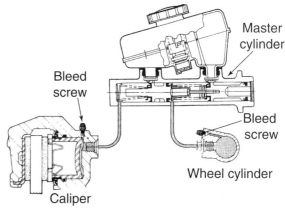

Figure 5.27 Bleed screws are located at the highest point on the back of the wheel cylinder or caliper.

Chapter 5 Master Cylinder and Hydraulic System Service

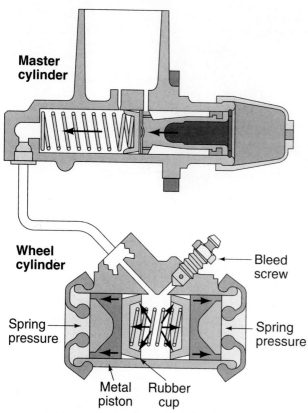

Figure 5.28 Loosening the bleed screw uncovers a passage for fluid flow.

> **SHOP TIP**
> - Sometimes a bleed screw is rusted, making it impossible to remove. A trick that works well for loosening stuck threads is to heat the screw and quench it with a wax stick. Then quickly remove the screw.
> - To prevent breaking a tight bleed screw, insert the solid end of a drill bit into the bleed hole and attempt to turn it with a six point wrench. The drill should fit snugly into the hole in the bleed screw.

brake bleeding procedures are very important. Check the service literature for the appropriate procedure. **Figure 5.29** is a shop library page showing how specific some of the procedures can be.

Most front-wheel drive vehicles have diagonally split hydraulic systems. These require a special bleeding sequence, described in the service literature. If the brakes are not bled in the correct sequence, a solid pedal might never be achieved. Some import calipers have two or three bleed screws per caliper. Follow the manufacturer's specified procedures for bleeding these calipers.

NOTE: *Brake manufacturers report that perfectly good power brake boosters are often returned under warranty because of incorrect brake bleeding procedures.*

Brakes can be bled in several ways including manually, with a pressure bleeder, by vacuum, by reverse injection, or by gravity. The different procedures are covered below.

Manual Brake Bleeding

Bleeding brakes manually is best done by two people. During this service, be sure that the pedal is depressed and released slowly. Here is a typical manual bleeding procedure:

1. Loosen the bleed screw approximately one turn.
2. Ask an assistant to push the brake pedal to the floor.

IMPORT CAR AND LIGHT TRUCK BRAKE BLEEDING SEQUENCE			
Application	Year	Sequence	Remarks
Nissan			
Altima with ABS	1993–02	RR, LF, LR, RF	Disconnect actuator electrical connectors; bleed front and rear ABS actuators after bleeding brake lines.
Altima and Quest (without ABS)	1993–02	RR, LF, LR, RF	On models equipped with Load-Sensing Proportioning Valve (LSPV), bleed LSPV before bleeding brake lines.
Axxess	1990	LR, RF, RR, LF	Check fluid level frequently during bleeding sequence.
Maxima	1995–02	RR, LF, LR, RF	Turn ignition off; disconnect negative battery cable. Bleed wheel in sequence.
Maxima	1986–02	LR, RF, RR, LF	On models with ABS, turn ignition off and disconnect ABS actuator electrical connectors before bleeding.
Maxima	1984–85	RR, LR, RF, LF	Check fluid level frequently during bleeding sequence.
Pathfinder and Pickup	1980–02	LR, RR, LF, RF	Bleed Load Sensing Proportioning Valve (LSPV) before bleeding wheel cylinders. On models with rear anti-lock brakes, bleed ABS actuator after bleeding brake lines.
Pathfinder	1996	RR, LF, LR, RF	On 4-wheel ABS models, disconnect negative battery cable before bleeding brakes in sequence.
Pickup	1978–79	RF, LF, LR, RR	Bleed master cylinder lines and front combination valve first. Bleed center and rear combination valve last.

Figure 5.29 Brake bleeding procedures for one manufacturer's vehicles.

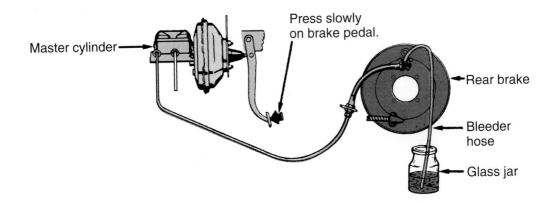

Figure 5.30 Manual brake bleeding procedure.

3. Close the bleed screw.
4. Release the brake pedal.
5. Repeat the process until the fluid coming out of the bleed screw is free of air.
6. Check the master cylinder fluid level often and refill it before the fluid level drops too low. This prevents air from entering the hydraulic system.

NOTE: *When manually bleeding brakes, pumping the brake pedal is not recommended because it tends to aerate the fluid.*

One problem with manual brake bleeding is that the master cylinder pistons are moved deeper into their bores than they usually travel. If the master cylinder is dirty, the seals will move past the area of the bores normally contacted by the seals, which is constantly wiped clean by the seals. When the seals are pushed across the dirty area beyond the normal seal wipe area during manual bleeding, the seals can be damaged.

Manual Bleeding with a Hose

Another manual bleeding method uses a hose installed from the open bleed screw into a container with brake fluid in it (**Figure 5.30**). The pedal is depressed and released several times. When no more bubbles show in the jar, bleeding is complete.

This method can be done without a helper, but you will not be able to observe the bubbles in the container while you are applying the brake pedal. However, if you continually refill the master cylinder reservoir and use a sufficient amount of brake fluid, this should not be a problem.

Pressure Brake Bleeding

Pressure bleeding is another method of bleeding brakes. One advantage of this method over manual bleeding is that the master cylinder seal is not dragged through the dirt in the bottom of the master cylinder bore.

A pressure bleeder is a canister with two chambers separated by a rubber diaphragm. There is an air chamber and a fluid chamber (**Figure 5.31**). The air chamber is filled with compressed air, which pushes the diaphragm against the fluid. Compressed air contains

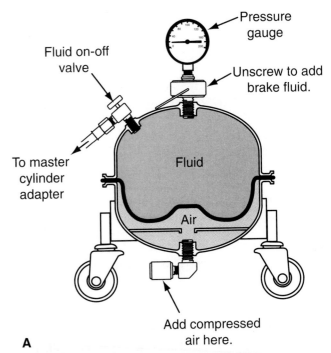

Figure 5.31 (A) A pressure brake bleeder. (B) A typical pressure bleeder and adapters.

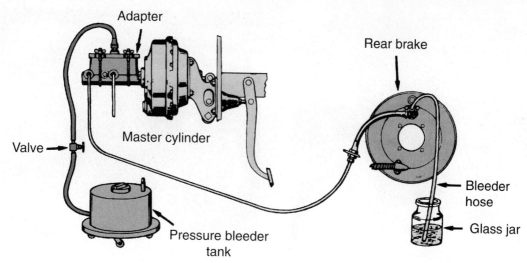

Figure 5.32 Pressure bleeder setup.

> **SHOP TIP** It is best if your helper does not depress the pedal further than three-quarters of full travel. Have your coworker position his or her left foot under the pedal to prevent excess pedal travel. If your helper releases the pedal before you have tightened the bleed screw completely, air will be drawn back into the hydraulic system as the piston in the master cylinder retracts.

substantial moisture that would be absorbed by the brake fluid if the diaphragm were not there to keep the air and fluid separated.

Prior to bleeding the brakes, the master cylinder is filled to within ½ inch of its top. Pressurized brake fluid is pumped through a hose to an adapter on the top of the master cylinder (**Figure 5.32**). When the adapter has been tightened in place, the hose from the bleeder is connected to it using a quick coupler. A valve is opened, pressurizing the entire hydraulic system. A bleed screw at the wheel is opened and fluid and air are pumped out. When the fluid is clean and free of air, the brake at that wheel is bled and the bleed screw is retightened.

Pressure Bleeder Adapters. With the correct adapters, pressure bleeding is the fastest way to bleed brakes. But adapters also have drawbacks. There are many master cylinders and a wide range of adapters are needed to be able to service them all. For a shop that works on one or two makes of vehicles, this will not be an issue. A shop that works on many types of vehicles will require more adapters. However, this might not be a problem for a large shop that specializes in brake work.

The following are some other pressure bleeding tips:

- Air in the pressure chamber should be 15–20 psi.
- On diagonal split systems, the master cylinder must be bled first. With a hose over the bleed screw leading into a clear container, open the bleed screw. Allow fluid to escape until it is completely clean and has no air bubbles.
- When bleeding brakes with a pressure bleeder, the metering valve must be disabled using a special tool. Otherwise, the metering valve will prevent fluid from reaching the wheel cylinders (**Figure 5.33**). This is not a problem when *manually* bleeding brakes using the brake pedal because the metering valve can be opened with pedal pressure.

> **SHOP TIP** The adapter must be fastened correctly to the master cylinder reservoir or fluid will leak out as soon as the valve on the adapter is opened. If the vehicle is raised in the air at this point in the process, a substantial amount of fluid could leak out before the valve is shut off. The vehicle will need to be lowered so the adapter can be remounted correctly, so it is best to check for leaks before raising the vehicle.

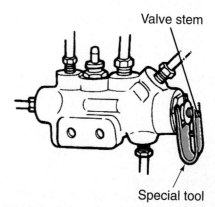

Figure 5.33 Installing this tool on the metering valve allows fluid to reach the wheel cylinders when pressure bleeding.

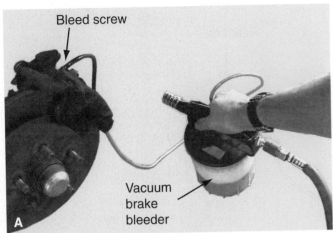
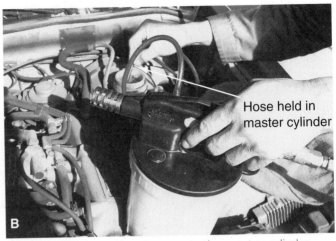

Figure 5.34 (A) A vacuum brake bleeder. (B) The vacuum brake bleeder can be used to empty the master cylinder reservoir. (Courtesy of Tim Gilles)

Do not forget to remove the metering valve tool following bleeding.

- When you are finished using a pressure bleeder, its valve must be shut off and it must be bled of air. Otherwise, a slow leak in the valve, the hose, or the quick coupling connection can result in the bleeder pumping out all of its fluid on to the shop floor.

Vacuum Brake Bleeding

Vacuum bleeding, also known as Vacula™ or *suction bleeding*, is very popular in the brake service industry. This bleeding method does not require adapters for the master cylinder. Setup is fast and the shop floor remains clean because fluid is collected into a small handheld vacuum container as bleeding is completed.

A vacuum bleeder pulls fluid from the master cylinder through the bleed screw by suction. A hose from the bleeder is connected to the open bleed screw. The vacuum bleeder is connected to a compressed air supply hose (**Figure 5.34**). Compressed air creates a vacuum as it moves past an opening in the brake bleeder's empty fluid container.

> **SHOP TIP** Some technicians like to use the Vacula during *break bleeding* to empty the *master cylinder* reservoir because it is fast and clean (**Figure 5.34b**).

During bleeding, air can be observed in the clear fluid hose. It is normal to see small air bubbles as air leaks past the threads of the bleed screw (**Figure 5.35**). The air will not cause harm because it does not enter the hydraulic system, but instead goes into the container with the spent fluid.

> **SHOP TIP** Using silicone grease on the bleeder threads can minimize air leaks through the bleed screw.

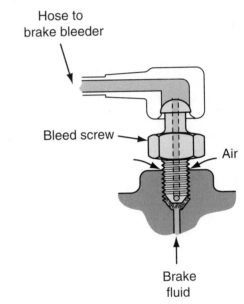

Figure 5.35 Air drawn past bleeder screw threads may create foam in the fluid.

Fluid Injection

The newest brake bleeding method is **RFI—reverse fluid injection**, or Phoenix Injection™ (**Figure 5.36**). Fluid is injected through the bleed screws from the underside of the vehicle (**Figure 5.37**). The master cylinder reservoir is emptied at the start of the procedure. The fluid injector is also capable of vacuum and can be used to pump fluid from the reservoir into a container.

Air tends to rise to higher places in the hydraulic system, where it often becomes trapped. Other bleeding methods are sometimes not successful in removing trapped air bubbles. With reverse fluid injection, the upward flow of fluid finds trapped air bubbles and carries them out of the master cylinder into the reservoir. The reservoir refills during the injection process.

Chapter 5 Master Cylinder and Hydraulic System Service

Figure 5.36 A Phoenix injector. (Courtesy of Phoenix Systems, L.L.C.)

Figure 5.37 Reverse fluid injection. (Courtesy of Phoenix Systems, L.L.C.)

Reverse fluid injection has become popular with ABS systems, which often have small fluid passageways. Dirt tends to settle out of the brake fluid at the bottom of the master cylinder reservoir. With other bleeding methods this dirt can be forced into the ABS, where it might plug a small passage. Injecting fluid from the bottom of the brake system to the top eliminates this potential problem.

Gravity Bleeding

Gravity bleeding is a simple and safe, but time-consuming, way of bleeding brake and clutch systems. To gravity bleed a system, open the bleed screws and allow the system to drain. Be sure to keep the master cylinder fluid level above the bottom of the reservoir.

Bleeding Antilock Brake Systems (ABS)

Some ABS systems require a scan tool to cycle the ABS during bleeding. Check the manufacturer's bleeding procedures before bleeding brakes. The gravity method will work on many, but not all, of these vehicles.

Brake Bleeding Problems

Several problems can occur during brake bleeding. One problem is trapped air, which results in a chronic low, spongy pedal. Pockets of air form in high spots in the hydraulic system, such as the bends of brake lines and hoses (**Figure 5.38**). Some master cylinders are mounted on a slant, which also results in trapped air. With the nose of the master cylinder higher, air cannot escape from the vent port and the pressure chamber remains filled with air. This problem is solved by removing the master cylinder and bench bleeding it.

Figure 5.39 shows typical bleed screw locations for drum and disc brakes. Notice the bleed screw in **Figure 5.40** is at the top of the caliper to allow air to escape during bleeding.

> **SHOP TIP** In most cases, gravity bleeding will work on ABS systems that require special procedures.

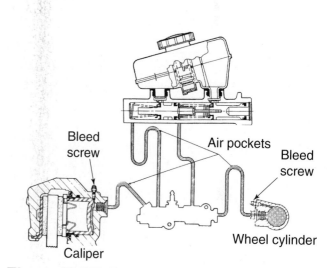

Figure 5.38 Bends in brake hoses and tubing can trap air and make bleeding more difficult.

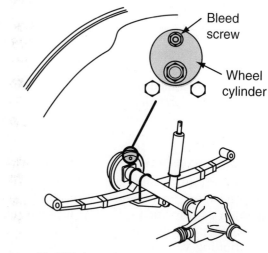

Figure 5.39 Location of a drum brake bleed screw.

SECTION 2 BRAKES

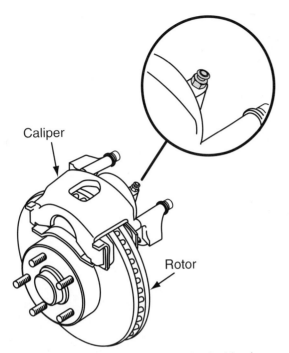

Figure 5.40 Location of a disc brake bleed screw.

> **SHOP TIP** Some technicians lift one end of the vehicle to level the master cylinder level during brake bleeding.

NOTE: *Disc brake calipers are different on the left and right sides of the vehicle. When a caliper is installed on the wrong side of the car, the bleed screw will be on the bottom. Purging air from the caliper will be difficult, if not impossible.*

Sometimes a brake pedal is consistently low and soft, even after thorough and correct brake bleeding practices. Air bubbles can become trapped within calipers. Air trapped between the piston and bore by the piston seals can be difficult to remove. Sometimes trapped air can be released by tapping lightly on the caliper with a hammer during brake bleeding (**Figure 5.41**).

■ PRESSURE TESTING

Sometimes it is necessary to verify pressures at the wheels in order to diagnose a brake system problem. A pressure gauge is installed in place of the bleed screw (**Figure 5.42**). The pressure test kit contains various adapters to fit in place of the many different bleed screw sizes and shapes.

Proportioning Valve Diagnosis

When a proportioning valve malfunctions problems sometimes develop. Front brakes typically wear out twice as often as rear brakes. If they are wearing out while rear brakes are showing very little wear, suspect the proportioning valve as a possible cause. The customer will probably complain of poor stopping, as well.

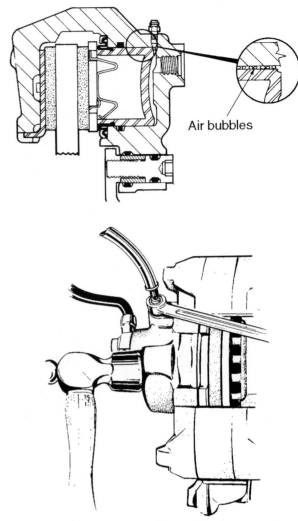

Figure 5.41 Tap the caliper with a hammer to release trapped air.

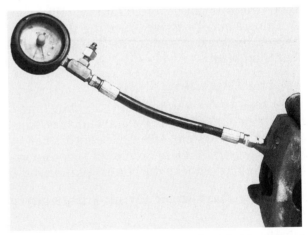

Figure 5.42 A pressure gauge installed in place of the bleed screw. *(Courtesy of Tim Gilles)*

Possible causes of proportioning valve problems include:

■ A height-sensing proportioning valve could be adjusted improperly.

- Aftermarket springs or a lift kit could be keeping the vehicle too high in the rear to allow the valve to bypass as it should when the vehicle is heavily loaded.

■ BRAKE WARNING LAMP DIAGNOSIS

Hydraulic brake systems have either a low brake fluid switch or a pressure differential sensing device called a *hydraulic safety switch*. If either of these devices triggers a signal saying that brake fluid is low or there is low hydraulic pressure in one-half of the system, a *red* brake light on the instrument panel should illuminate. When the key is on, the light will also come on whenever the parking brake is on (even a little bit). On some cars, a bulb check feature illuminates the light during engine cranking.

The amber light is assigned to the ABS system. See Chapter 10 for more information on amber light diagnosis. The red light should illuminate when the engine cranks and the amber ABS light should illuminate some time during key-on, engine crank, or engine start. Both lights should come on and go back off before driving begins. If a light does not come on or if it does not go off *after* coming on, there is a problem in the brake system or the electrical circuit for the lamp.

NOTE: *If a stoplight is out or if the two lights are of different intensity, the ABS can be disabled.*

Hydraulic Safety Switch Service

Hydraulic safety switches are found in vehicles that do not have fluid level sensing systems. Most of these valves are self-centering and spring-loaded. When there is a drop in system pressure, the dash light *only* comes on during a stop and does *not* remain on.

Most safety switches share the same bulb that indicates when the parking brake is applied (**see Figure 4.35**). To test the bulb in these systems, simply turn on the key with the parking brake applied. The bulb should illuminate.

Master Cylinder Fluid Level Switch Service

Fluid level switches are found on vehicles that do not have a hydraulic safety switch. To test the lamp on the instrument panel turn the key to the on position. It should light for a few seconds. The float in the master cylinder reservoir can be checked to see that it moves freely. It should turn the lamp on and off as it goes from full length (fluid empty) to its depressed position.

Stoplight Switch Service

To check the operation of the stoplight switch, apply the brakes and verify that the stoplights come on. If not, check to see if the ignition key must be turned to the on position first. Then, check the fuse. There are two wires to the stoplight switch. Sometimes there are more than two wires when a vehicle has cruise control. Remove the connector from the switch and use a jumper wire to connect them together. If the stoplights work with the wires connected together, the stoplight switch is faulty.

On North American vehicles, the switch is normally open and power is often supplied to the switch through the turn signal switch. Some stoplight switches have extra contacts for cruise control, the torque convertor clutch, or antilock brake control. Use a wiring diagram to determine how the circuit operates.

■ BRAKE HOSE SERVICE

Hoses are normally firm yet flexible. Inspect them for physical damage, such as cracking, swelling, softening from exposure to oil, bubbles in the rubber outer layer, or chafing caused when the hose rubs against a tire (**Figure 5.43**).

Restricted Brake Hoses

Brake hoses sometimes experience internal failure, especially when a careless technician allows a disc brake caliper to hang from a hose during service to brakes, suspension, or wheel bearings.

> **Vintage Chassis**
>
> In some early systems from the 1960s, the bulb indicating a hydraulic system failure would illuminate and remain on until it was recentered by a technician following a repair. Centering the valve can be done without special tools by bleeding one side of the system and then slowly bleeding the other side to position the valve. But if you work on many classic cars from the late 1960s, there is a special centering tool that can be installed in place of the electrical switch during a brake bleed to prevent the valve from moving off center.

> **SHOP TIP:** When a warning lamp is illuminated, disconnecting the switches one at a time until the light goes out will usually identify the cause of the problem.

> **SHOP TIP:** According to the SAE term list, a brake on-off switch is called a *BOO switch*.

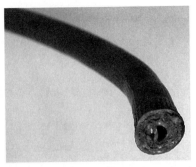

Figure 5.44 This hose collapsed on the inside. (Courtesy of Tim Gilles)

> **SHOP TIP** With the brakes released, when a bleed screw is opened and fluid spurts out, trapped brake fluid from a collapsed hose is indicated.

could cause delayed engagement of the brake or act as a one-way valve, preventing fluid pressure from releasing the brake after application.

When the brakes do not fully release, excessive brake pad wear results. A partially plugged brake hose will cause the car to pull at the beginning of a stop. When the pressures equalize, the pull goes away.

Another, though somewhat unusual, cause of a restricted brake line is when the circular metal bracket that positions the hose rusts internally. This constricts the hose, resulting in a restriction of fluid flow.

Brake Hose Installation

Hoses must be installed so they do not rub on other parts. Ribs on the exterior rubber of the hose help protect against occasional rubbing, such as when a wheel is fully extended during rebound from taking a speed bump at too high a speed. Some technicians replace brake hoses in pairs.

Replacing a hose requires the removal of a retaining clip (**Figure 5.45**).

Copper Washers

Copper washers that seal one end of the hose to the caliper or wheel cylinder are typically reused. However, repeated use of a copper washer will deform it. Replace it if it is deformed or appears to be excessively compressed. Some hoses use banjo fittings, which have two copper washers (**Figure 5.46**). Be aware that if one is lost, the fitting will leak.

BRAKE LINE/TUBING SERVICE

Brake lines, also called tubing, are routed to the wheel cylinders along the car frame, suspension, and rear axle (**Figure 5.47**). Metal clips, sometimes with rubber insulation, hold the tubing tight to prevent vibration. Brake lines are usually made of double-walled

Figure 5.43 (A) This brake hose has a bubble. (B and C) These hoses were damaged when they rubbed against a tire. (Courtesy of Tim Gilles)

Another cause of internal failure results from clamping the hoses with locking pliers. There are special tools designed to clamp brake hoses without damaging them, but many technicians avoid clamping hoses altogether.

A hose can appear fine from the outside, but its inside fine may be collapsed (**Figure 5.44**). A collapsed hose will trap fluid inside the front brake and cause excessive wear or a brake that pulls to one side. Internal damage is difficult to diagnose. Depending on which way the internal rubber flap is torn, the hose

Chapter 5 Master Cylinder and Hydraulic System Service   **113**

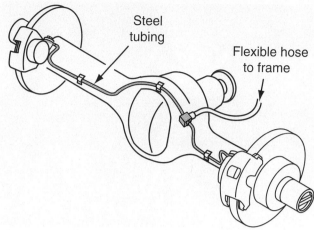

Figure 5.47 Brake lines are routed to the wheel cylinders along the car frame, suspension, and rear axle.

Figure 5.48 Tubing is available in precut lengths. (Courtesy of Tim Gilles)

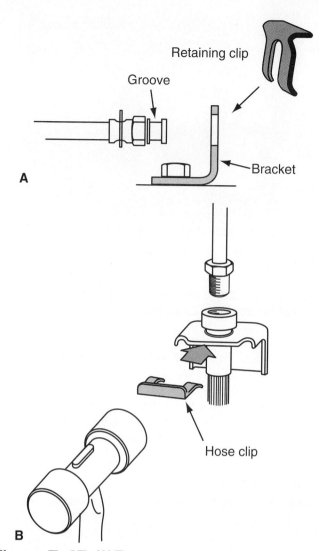

Figure 5.45 (A) The retaining clip fits in a groove at the end of the hose fitting. (B) Typical brake hose mounting and line connection.

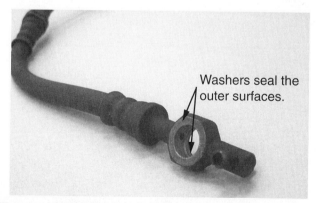

Figure 5.46 A banjo fitting is used with copper washers. (Courtesy of Tim Gilles)

steel, coated with a material that prevents rust, and flared at the ends.

Tubing diameter ranges from $1/8$ inch to $3/8$ inch. It comes in several precut lengths with flared ends and fittings installed (**Figure 5.48**), or in 25 foot rolls. Precut brake tubing usually has a short flare nut at one end and a long flare nut at the other.

When replacing a brake line, be careful to copy the original line as closely as possible and use all of the retaining clips that were installed at the factory. If you are replacing a piece of tubing that is not straight, measure it using a piece of string. This will help to determine the correct length, including bends. Sometimes a loop or bend can be put into a precut line to shorten it to the desired length without having to cut it and reflare one of its ends.

When unrolling bulk tubing, be careful not to kink it. Unroll it in the same direction that it was rolled. When fittings are to be installed on the tubing, add an additional $1/8$ inch at each end to allow for forming of the flare ($1/4$ inch total).

When loosening a flare fitting, a tubing wrench or flare nut wrench is used. Flared lines are held against a seat in the fitting. *Always* use two wrenches (**Figure 5.49**). The second wrench holds the female part of the fitting. The female fitting will tend to rotate when the male flare fitting is turned with a wrench. If this is allowed to happen, the line will become kinked (**Figure 5.50**).

Damaged steel brake lines can be cut and repaired or new lines can be fabricated using a flaring tool. Use only seamless steel tubing. Do not replace steel tubing with copper.

Cutting Tubing

When cutting tubing to length, be sure to cut it so that it is square on the end. It is best to use a *tubing cutter* (**Figure 5.51**). A hacksaw will leave a rough edge that

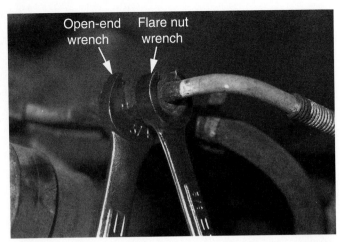

Figure 5.49 A flare nut wrench is used with an open end wrench to loosen a fuel line.

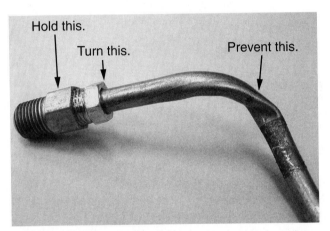

Figure 5.50 This fuel line was damaged when the apprentice tried to loosen the flare fitting without using two wrenches. *(Courtesy of Tim Gilles)*

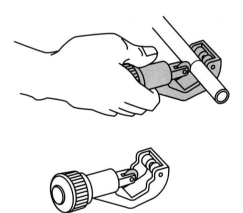

Figure 5.51 Using a tubing cutter to cut tubing.

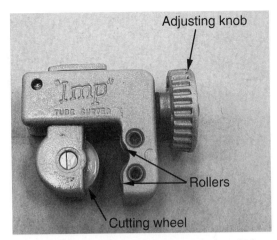

Figure 5.52 A small tubing cutter is handy for working in tight spaces on the vehicle. *(Courtesy of Tim Gilles)*

tighten it or the cutter will cut through too soon and the tubing could be damaged.

Following cutting, a burr usually remains on the end of the tubing. This should be removed with the reamer blade that most tubing cutters are equipped with. Be sure to remove any chips from the end of the tubing after completing the cut.

Some tubing cutters are very small. In tight quarters they can be used for repairing damaged tubing on the vehicle. The tubing cutter shown in **Figure 5.52** is especially handy for this task.

Bending Tubing

When bending tubing, remember that too sharp a bend will result in a kink or restriction. Tubing can be bent with a *tubing bender* (**Figure 5.53**). Slowly form it

> **SHOP TIP** If you do not have a tubing bender, tubing can be formed by bending it over a large piece of pipe. A water pump V belt pulley held in a vise also works well.

Figure 5.53 A tubing bender. *(Courtesy of Tim Gilles)*

might not be square and metallic debris can be left in the tubing. When the end is dressed off with a file, metal chips can get into the end of the tube. If these are not thoroughly cleaned out, serious damage to a component can result.

The tubing cutter is first tightened against, then rolled around the tubing. The handle is tightened to advance the cutter as the tubing is cut. Do not overusing the tubing that is being replaced as a guide when possible. Be careful not to bend the tubing too sharply.

It is better to install fittings and flare both ends before bending tubing. Otherwise, if the bend is too close to the flare, the flaring tool will not be able to clamp to the tubing. It is best if the bend is not too close to the flare. Leave at least a couple of inches when possible.

A *bending spring* (**Figure 5.54**) can be installed over the tubing. It will help keep the tubing from becoming kinked during the bend. When using a bending spring, one end of the line must not have its fitting installed yet. This is so that the spring can be installed and removed from the line.

Steel lines should not have long, straight runs. They are difficult to remove and replace and they can fatigue at the connections. When a short section of tubing is replaced, at least one end should have a bend so that the connection can flex (**Figure 5.55**).

Long runs should be supported with clamps. Heavy connections such as distribution blocks must be securely mounted. The ends of the tubing should align with the fitting. Be sure that the threads can be easily turned all the way into the fitting.

Flaring the Ends of Tubing

A **flaring tool** is used to flare the ends of the tubing with either an SAE double flare or an ISO (metric double) flare (see Chapter 4). Different tools are required

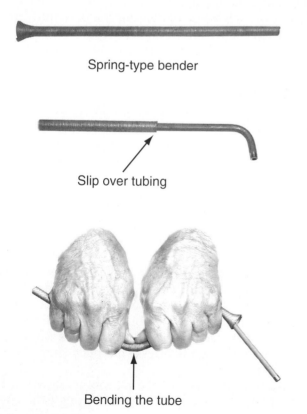

Figure 5.54 A bending spring helps prevent the tubing from becoming kinked.

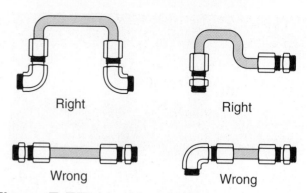

Figure 5.55 A bend in a rigid length of tubing can prevent vibration damage.

to perform these jobs. Be sure to use the same kind of flare that was used on the original line. The different types of flares are not interchangeable.

SAE Flare

An SAE flare can be either a single or double flare. Single flares are not used on small automotive tubing because they will split the tubing (**Figure 5.56**).

SAE double flaring is a two-step process. The brake line is clamped in a special tool while its end is formed. The tube is then folded over itself to complete the double flare. Use the following steps:

1. First, slip the fitting onto the line.
2. Select the correct size hole in the flaring tool bar.
3. Clamp the line in the flaring tool bar. It should extend out of the bar by the width of the flaring tool adapter (**Figure 5.57**).
4. Use a threaded flaring cone and clamp to form the end of the tubing (**Figure 5.58**).
5. **Figure 5.59** shows the appearance of the end of the tubing following the forming operation.
6. Insert the adapter in the end of the tubing and tighten down on it with the threaded flaring tool until it bottoms out (**Figure 5.60A**).

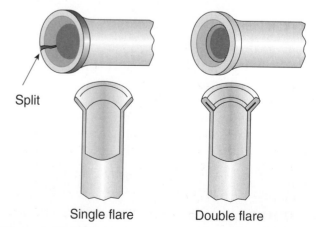

Figure 5.56 A single flare will result in a split in the tubing.

116 SECTION 2 BRAKES

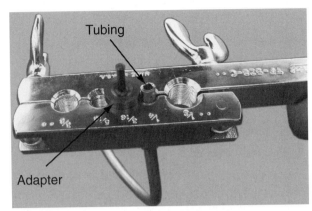

Figure 5.57 Clamp the tubing in the flaring bar with it protruding about the thickness of the adapter. *(Courtesy of Tim Gilles)*

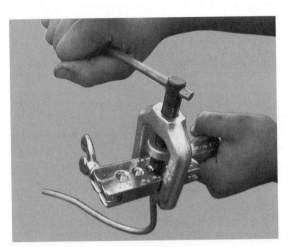

Figure 5.58 A threaded flaring cone and clamp are used to form the end of the tubing. *(Courtesy of Tim Gilles)*

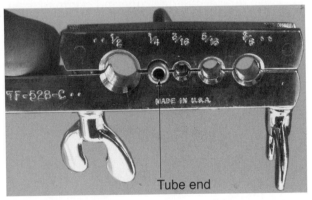

Figure 5.59 The end of the tube following the forming operation. *(Courtesy of Tim Gilles)*

7. Remove the adapter and tighten the flaring tool against the line again to complete the flare (**Figure 5.60B**).

If the flare is not formed properly, you will have to cut the end off the line and form another flare.

REMEMBER: *Always put the fitting on the line before flaring it. Also, leave enough space between a bend and the flared fitting so that the fitting can slide.*

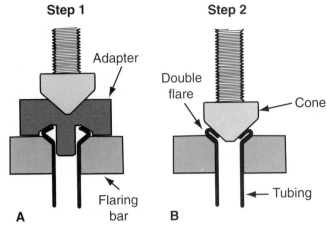

Figure 5.60 (A) The adapter is forced into the end of the tube. (B) The flaring cone completes the double flare.

> **SHOP TIP** Thread sealers or Teflon tape are *not* necessary when using flare fittings. A flare fitting seals internally and the threads should not be exposed to liquids. If the threads get wet, the flare is not a quality joint.

Installing Flared Tubing

When you install a length of tubing, leave the first flare nut loose after turning it by hand into the threads in the fitting. After starting the fitting on the other end of the tubing, tighten them both. Hold the fitting with a wrench while tightening the flare nut. Do not overtighten the fittings. After the flare is brought into contact with the fitting, turn it an additional one-sixth turn only.

ISO Flaring

ISO flares, commonly called bubble flares, have been found on automobiles since the early 1980's. They used to be common only on imported vehicles but are now common on domestic vehicles, as well. A bubble or ridge is formed in the tubing a short way back from its end (**Figure 5.61**). A single flaring operation can be done without danger of splitting the tubing (like with a single 45E flare). Two different types of flaring tools are available for creating bubble flares. Both of them grip the outside of the tubing while a pressure screw is tightened against a mandrel to form the bubble on the tubing. **Figure 5.62** shows one type in use. To achieve the correct shape to the bubble, the tubing is clamped even with the top of the clamping fixture. **Figure 5.63** shows the tubing before and after it has been shaped by the mandrel.

Another type of ISO flaring tool is shown in **Figure 5.64.** The tubing is inserted through the end of the clamping nut. When the clamping nut is tightened into threads in the flaring tool body, a collet is

Chapter 5 Master Cylinder and Hydraulic System Service

Figure 5.61 The end of an ISO flare is formed like a bubble. *(Courtesy of Tim Gilles)*

Figure 5.62 A bubble flaring tool in use. The mandrel beneath the pressure screw forms the end of the tubing. *(Courtesy of Tim Gilles)*

Figure 5.63 (A) Tubing clamped in the flaring bar prior to forming. (B) The appearance of the tubing after flaring. *(Courtesy of Tim Gilles)*

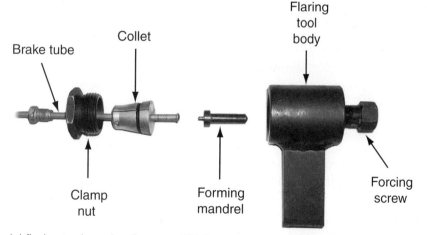

Figure 5.64 A special flaring tool used to form an ISO flare.

compressed to grip the outside of the tubing. The tubing must be positioned inside the tool mandrel at the correct height, using the gauge end of the flaring tool. First assemble the two halves of the tool mandrel and clamp them in a vise. Then, reverse the flaring tool/gauge and thread it into the tool mandrel to form the bubble flare.

Union Repairs

Sometimes a union is used to repair a damaged line. Unions that require the ends of the line to be flared first are the only kind that can be used to join brake lines. When there is enough straight line remaining behind a flare, a section of the line slightly less than the length of the fitting is cut from the old line.

REVIEW QUESTIONS

1. When brake cylinders with both aluminum and iron parts create a voltage, _____ causes metal corrosion.

2. The maximum fill level of a master cylinder is about _____ from the top of the reservoir chamber.

3. When a primary/secondary [choose one] cup seal leaks, fluid can leak out of the master cylinder.

4. To tell the difference between gear oil leaking from an axle seal and brake fluid leaking from a wheel cylinder, what would you do?

5. When a rear-wheel drive car's master cylinder has two chambers of different sizes, which brakes are served by the larger of the two?

6. If one master cylinder chamber is low on fluid and the other is overflowing, a _____ piston seal might be defective.

7. Fluid movement should always be visible through the _____ port when the pedal is first applied.

8. Quick take-up master cylinders are especially prone to trapping _____ in high places in brake fluid cavities.

9. Clearance between the brake pushrod and the back of the primary master cylinder piston or power booster should be less than _____.

10. List three ways of testing brake fluid.

11. Which bleeding tool requires adapters for the master cylinder?

12. What is the name of the brake bleeding method that pushes fluid from the bottom, carrying fluid and trapped air bubbles to the master cylinder reservoir?

13. What is the name of the bleeding method in which you open the bleed screws and allow the system to drain?

14. If front brakes are wearing out while rear brakes are showing very little wear, the _____ valve could be the cause.

15. Are thread sealers or Teflon tape necessary when using flare fittings?

PRACTICE TEST QUESTIONS

1. Technician A says a low pedal can indicate a leak in one-half of the brake hydraulic system. Technician B says if the pedal slips slowly toward the floor under very light pressure, a wheel cylinder is probably leaking internally. Who is right?

    A. Technician A   B. Technician B
    C. Both A and B   D. Neither A nor B

2. Technician A says if the height of a firm pedal rises significantly when it is pumped twice in rapid succession, the brakes need to be bled. Technician B says if the pedal feels soft and "spongy," there is probably air trapped in the brake fluid. Who is right?

    A. Technician A   B. Technician B
    C. Both A and B   D. Neither A nor B

3. The inside of a tire and wheel has wet, dirty streaks and the paint on the backing plate is bubbled. Technician A says this is most likely due to a leaking axle seal. Technician B says this could be due to a leaking wheel cylinder. Who is right?

    A. Technician A   B. Technician B
    C. Both A and B   D. Neither A nor B

4. When brake linings wear, which of the following is/are true?

    A. Fluid level will appear low in disc brake systems
    B. Fluid level will appear low in drum brake systems
    C. Both a and b
    D. Fluid level remains the same

5. A tandem master cylinder has a lower fluid level in one of its chambers. Technician A says the disc pads are probably worn if this is a diagonally split hydraulic system. Technician B says the disc pads are probably worn if this is a longitudinally split hydraulic system. Who is right?

    A. Technician A   B. Technician B
    C. Both A and B   D. Neither A nor B

6. Technician A says an aluminum wheel cylinder is honed lightly prior to installing new rubber cup seals. Technician B says a cast iron master cylinder is honed lightly prior to installing new rubber cup seals. Who is right?
   A. Technician A   B. Technician B
   C. Both A and B   D. Neither A nor B

7. Mineral oil has accidentally been added to the master cylinder reservoir. Which of the following is true?
   A. The entire hydraulic system must be disassembled
   B. The hydraulic system must be thoroughly flushed with alcohol
   C. All rubber parts must be replaced
   D. All of the above
   E. None of the above

8. Technician A says test strips can be used to indicate the amount of moisture in the brake fluid. Technician B says when test strips change from white to purple, the brake fluid is worn out. Who is right?
   A. Technician A   B. Technician B
   C. Both A and B   D. Neither A nor B

9. A brake hose has collapsed internally. Technician A says this can cause excessive wear to a brake pad on one side of the caliper. Technician B says this can cause a brake pull to one side at the beginning of a stop, then go away. Who is right?
   A. Technician A   B. Technician B
   C. Both A and B   D. Neither A nor B

10. Technician A says to pump the brake pedal several times with bleeding brakes. Technician B says a restricted vent port will cause brake drag. Who is right?
    A. Technician A   B. Technician B
    C. Both A and B   D. Neither A nor B

CHAPTER 6

Drum Brake Fundamentals and Service

■ OBJECTIVES:
Upon completion of this chapter, you should be able to:
- ✔ Understand drum brake parts and operation.
- ✔ Explain variations in drum brake design.
- ✔ Describe the operation of the different types of self-adjusting drum brake mechanisms.
- ✔ Perform a drum brake inspection.
- ✔ Demonstrate drum brake disassembly.
- ✔ Understand service and repair procedures to wheel cylinders.
- ✔ Describe drum brake reassembly and adjustment procedures.

■ KEY TERMS
anodized aluminum
wheel cylinder kit

■ INTRODUCTION

Until the mid-1960s, almost all vehicles were equipped with drum brakes. Today, drum brakes are found in some rear brake applications, combined with disc brakes in the front. Disc/drum systems are called hybrid brakes. Advantages of drum brakes are that they provide good initial stopping ability and also a means for an effective and inexpensive mechanical parking brake.

■ DRUM BRAKE PARTS

The brake *shoe* is made of stamped steel with a friction lining surface attached to it (**Figure 6.1**). The metal *rim* of a brake shoe is slightly narrower than the depth of the drum. To provide reinforcement, a *web* is welded to the shoe rim. There is a seat for the shoe *anchor* at one end, and a notch where force is applied by a pin from the wheel cylinder piston. The outside edges of the shoe are crimped in several places. When the brakes are applied and released, the crimped areas along the inside edge of the shoe slide against corresponding raised platform areas on the brake's *backing plate* (**Figure 6.2**). Springs, parking brake parts, and the self-adjuster are also attached to the web of the shoe.

Some brake shoes are made of cast aluminum because they are lighter, resulting in less unsprung weight. Aluminum is not as strong as steel, especially when heated, but in performance applications where cost is not an objective, minimizing the weight of the wheel assembly is highly desirable.

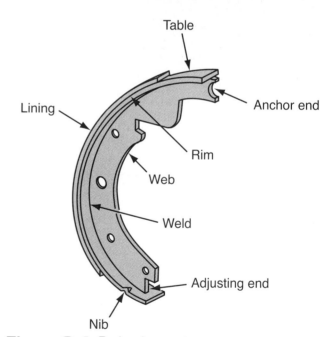

Figure 6.1 Brake shoe parts.

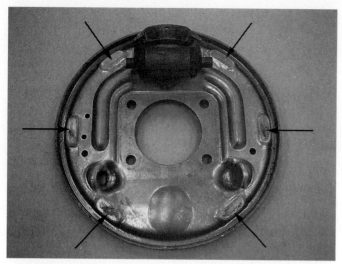

Figure 6.2 A brake backing plate and wheel cylinder. The brake shoes slide against the raised platforms. (Courtesy of Tim Gilles)

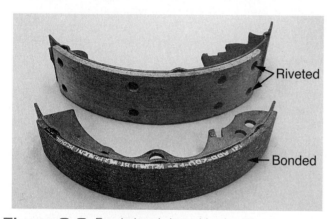

Figure 6.3 Bonded and riveted brake shoes. (Courtesy of Tim Gilles)

Bonded and Riveted Linings

Linings are either bonded or riveted to the shoe rim (**Figure 6.3**). Resin adheres friction material to a bonded lining. Bonded linings run hotter than riveted linings because the resin binder insulates the lining from its steel backing, preventing heat transfer.

Friction material on a riveted lining is held to the shoe by brass or aluminum rivets. Riveted linings are usually quieter than bonded linings because bonded linings are more rigid. Rigidity leads to a higher likelihood of noise-causing vibration. Chapter 8 has more information on lining friction materials.

Drum Brake Lining Terms

Braking force is applied by the wheel cylinder to the *toe* end of the brake lining. The opposite end of the shoe is called the *heel*. The *primary lining* faces the front of the vehicle. The rear facing lining is the *secondary lining*. To allow the primary and secondary linings to wear evenly, linings sometimes vary in length and can be made of different materials (**Figure 6.4**). The different drum brake designs are covered later in the chapter.

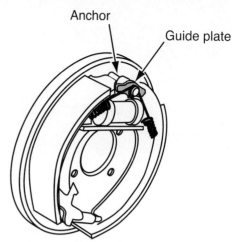

Figure 6.4 A guide plate fits against the outer ledge of the anchor to center the shoe at the correct depth within the drum.

Anchor and Wheel Cylinder

During a stop, the wheel cylinder applies force to the brake shoe. The anchor must be large in diameter so it can oppose the torque put on the shoe as it slows the rotating brake drum. A typical anchor has an outer ledge that a guide plate fits against. The guide plate centers the shoe at the correct depth within the drum.

Wheel Cylinder Parts

Most drum brake wheel cylinders have two rubber sealing cups that apply force to metal pistons in each end of the metal cylinder bore (**Figure 6.5**). Pistons are made of iron, anodized aluminum, or plastic. The brake shoe has either a raised piston contact section (**Figure 6.6A**) or a separate pushrod that connects it to a cavity in the outer end of the piston (**Figure 6.6B**).

Opposing rubber cup seals press against the inside faces of the pistons, which are flat. The lips on the cup seals face toward the center of the body of the wheel

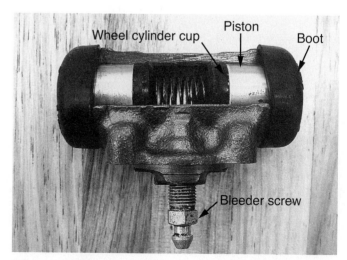

Figure 6.5 The position of parts in a wheel cylinder. (Courtesy of Tim Gilles)

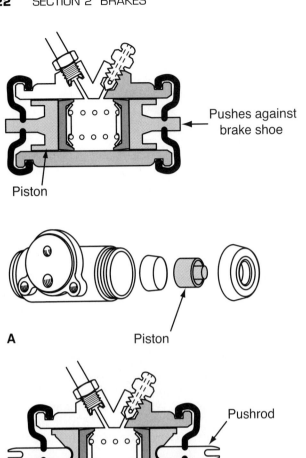

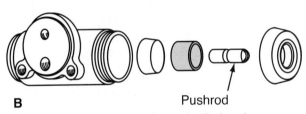

Figure 6.6 Two types of wheel cylinder piston designs. A pushrod connects the brake shoe to outer end of piston in view B.

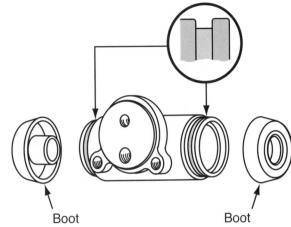

Figure 6.7 Dust boots fit into grooves in the wheel cylinder body.

cylinder, so pressure can force them to seal against the outside of the cylinder bore during a stop.

Occasionally, wheel cylinder pistons have grooves where a sealing O-ring is installed, similar to grooves and seals in a typical master cylinder piston.

The wheel cylinder is sealed from contaminants at its ends by rubber dust boots to prevent rust or electrolysis. Dust boots typically attach to a groove in the outside of the cylinder (**Figure 6.7**). The center of the dust boot has a small hole that surrounds the pushrod or piston boss. A few wheel cylinders (GM in particular) have internal sealing dust boots, held into a groove inside the end of the bore by a metal ring.

Older hydraulic systems used residual pressure check valves to maintain a constant hydraulic seal between the lips of the seals and the walls of the cylinder, even when the pedal was not applied. Modern wheel cylinders use a spring and cup expanders to separate the rubber cups when there is no pressure in the cylinder.

Wheel Cylinder Housing

Wheel cylinders are made of either aluminum or cast iron. A wheel cylinder is typically held to the backing

---

Vintage Chassis

Until the mid-1950s, some vehicles had adjustable anchors. If you ever work on a vintage car, check to see if it has an adjustable anchor. It is easy to adjust. Simply loosen its nut on the back side of the backing plate and adjust the brake shoes tight against the drum. Then tighten the anchor and finish the brake adjustment.

Vintage Chassis

Other wheel cylinder designs no longer in common use in modern cars include single-piston cylinders and step bore cylinders. One single-piston design has the wheel cylinder slide on the backing plate to apply one lining as the piston applies pressure, much like a floating disc brake caliper. Other times, single cylinders can be used, one for each shoe.

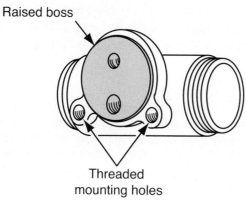

Figure 6.8 A large metal boss is cast into the back of the cylinder.

plate with two small screws, but this presents a problem. During stopping, a good deal of torque is applied against the wheel cylinder. To prevent this torque from being absorbed by the wheel cylinder's small mounting screws, a large metal boss is cast into the back of the cylinder (**Figure 6.8**). The wheel cylinder boss fits snugly into a hole in the backing plate, preventing the wheel cylinder from pulling loose.

Drum Brake Springs

Drum brakes use several types of springs that serve different purposes:

- *Return springs* or *retracting springs* push the wheel cylinder pistons back into the cylinder bore after the brakes are released (**Figure 6.9**).
- A *holddown spring* keeps the crimps on the rim of the brake shoe held against their corresponding raised bearing areas on the backing plate. The typical holddown spring is a coil with a pin (sometimes called a *nail*) that extends through a hole in the backing plate (**Figure 6.10**). A similar holddown spring that uses a nail is U-shaped (**Figure 6.11**). Some holddown springs are of the beehive type (**Figure 6.12**).

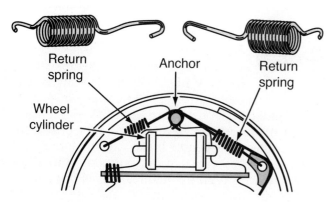

Figure 6.9 Return springs.

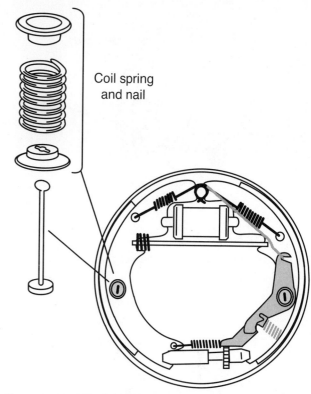

Figure 6.10 A holddown spring keeps the shoe against the backing plate. This one is a coil-and-nail type.

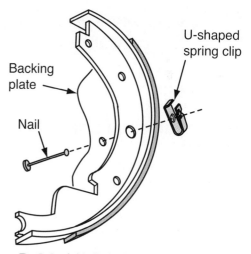

Figure 6.11 A U-clip holddown spring also includes a "nail."

- Another spring, on the self-adjuster, pulls against the self-adjusting lever during an adjustment and holds it in the correct position when it is inactive. Self-adjuster operation is covered later in this chapter.
- The parking brake strut is loose when the parking brake is not applied. A spring prevents the strut from rattling (**Figure 6.13**).

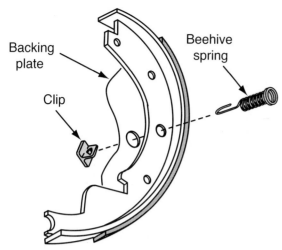

Figure 6.12 A beehive-type holddown spring and clip.

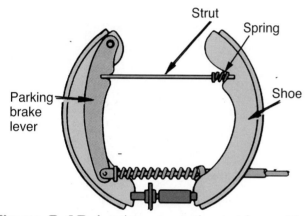

Figure 6.13 A spring prevents the strut from rattling.

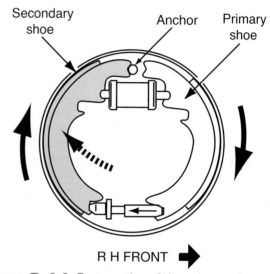

Figure 6.14 Servo action of the primary shoe against the secondary shoe.

Figure 6.15 A dual-servo brake's front lining has a shorter length of friction material than its rear lining. *(Courtesy of Tim Gilles)*

and applies additional pressure against the secondary (rear) shoe (**Figure 6.14**). When stopping in reverse, the secondary shoe is energized and pushes on the primary shoe.

The primary shoe acting on the secondary shoe causes the rear shoe to do a higher percentage of the braking. If you compare the linings on primary and secondary shoes, you will find the front linings have a shorter length of friction material than the rear (**Figure 6.15**). This is so they will wear at similar rates. Sometimes different friction materials are used on primary and secondary linings as well.

■ VARIATIONS IN DRUM BRAKE DESIGN

Several different drum brake designs have been used on motor vehicles during their long history, with two of them still used extensively today. The choice of design depends on the application. Some brakes work better with the heavier loads of rear-wheel drive vehicles. Others are used primarily on front-wheel drive cars, which have less weight in the back of the car. The lack of load on the rear wheels creates more of a tendency for rear-wheel lockup during a stop.

Dual-Servo Drum Brake

During stopping, a drum brake's leading shoe digs into the brake drum. This design is self-energizing. Originally called Bendix brakes, this brake design is also called *dual-servo* or *duo-servo* because *both* brake shoes are self-energizing. They use servo action in which a small force is applied to make a larger force. With no anchor at the bottom between the front and rear shoes, the primary (front) shoe floats during braking

Vintage Chassis

The inventor of the Bendix starter drive, Vincent Bendix, bought the patent rights to Perrot brakes and started the Bendix Brake Company in 1923. He concentrated his efforts on developing better brakes for automobiles. The Bendix brake, the first four-wheel automotive brake system, was introduced in 1928, when vehicles typically had rear-wheel drive, with brakes only on the rear wheels.

Self-energized brakes have the toe end of the primary shoe at the top. The toe end of the secondary shoe is at the bottom because that is where force is initially applied to the shoe.

Leading-Trailing Drum Brake

Some vehicles use a *leading-trailing* brake (**Figure 6.16**), also called a *simplex brake*. A leading-trailing brake is a *non-servo* brake with an anchor at the bottom end of each shoe. The leading (front) shoe is self-energized, while the trailing shoe is not. Because the front shoe is self-energized, it wears more than the rear. Leading-trailing brakes have the toe located at the wheel cylinder end of the shoe, with the heel at the anchor end.

Weight distribution in front-wheel drive cars can be 80 percent or more toward the front. Leading-trailing brakes are used because they are less effective in stopping than servo-type brakes. Front-wheel drive cars with split diagonal hydraulic systems (see Chapter 4) and light-duty trucks without ABS use leading-trailing rear brakes because they minimize the tendency for the relatively underweighted rear wheels to lockup. On older vehicles without antilock brakes, this usually eliminates the need for a proportioning valve.

Another brake design, called the *duplex brake,* has two leading shoes and uses a single-piston wheel cylinder at each leading end of its shoe. Some older domestic and imported vehicles had single-piston wheel cylinders at *opposite* ends of the brake shoe. In this design, called a *leading-leading brake,* each shoe is energized during forward stopping, but in reverse there is no energization on either shoe.

■ DRUM BRAKE ADJUSTMENT

Disc brakes automatically adjust themselves, but drum brakes require periodic adjustment. As drum brake linings wear, increased clearance between the lining and the drum results in more brake pedal travel before pressure can build and the brakes start to apply. The disc brakes in the front and the drum brakes in the rear are connected hydraulically in the master cylinder. The tandem master cylinder design prevents the front brakes from applying before the rear brakes contact the brake drum. When drum brakes need to be adjusted pedal travel drops for the entire system.

If drum brake clearance becomes excessive, or when a hydraulic failure occurs in the half of the master cylinder that serves the rear brakes of a front/rear split hydraulic system, the rear brake master cylinder piston must bottom out against a stop before the disc brakes can apply (see Chapter 4).

A typical drum brake adjuster has a threaded shaft attached to an integral starwheel. When it is turned, extending the thread, excess clearance is taken up. Drum brakes have had self-adjusting mechanisms since the early 1960s. During self-adjustment, the integral starwheel rotates the threaded shaft, taking up excess clearance (**Figure 6.17**). **Figure 6.18** compares typical self-adjusters for leading-trailing and duo-servo brakes. Leading-trailing brake starwheel adjusters are part of the parking brake strut. Dual-servo brakes have a star adjuster that fits between the bottoms of the floating brake shoes.

Dual-Servo Self-Adjustment

Most dual-servo brakes use a cable-type self-adjuster. Many General Motors cars use a lever-type adjuster.

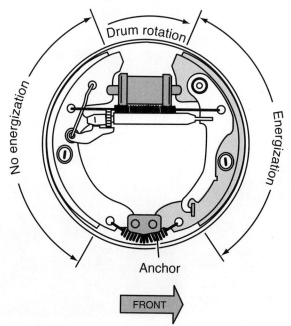

Figure 6.16 In a leading-trailing brake, or simplex, only the front lining is energized.

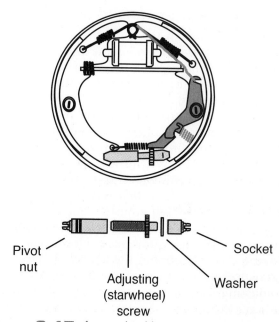

Figure 6.17 A starwheel is turned to adjust for brake clearance.

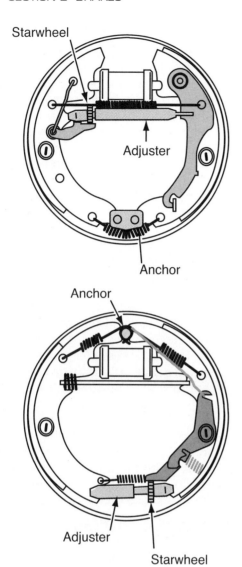

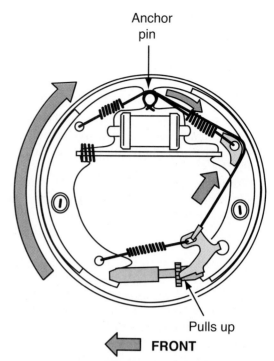

Figure 6.19 The top of the shoe moves away from the anchor when there is clearance between the shoe and the drum.

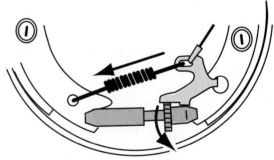

Figure 6.20 When the brake is released the spring pulls the adjusting lever, turning the starwheel.

Figure 6.18 Typical starwheel adjuster locations for leading-trailing and dual-servo brakes.

Figure 6.21 This adjusting lever pulls up against the starwheel to adjust the brakes. *(Courtesy of Tim Gilles)*

Both adjusters work in the same manner. Dual-servo self-adjusters operate only when the brakes are applied during a stop while backing up. Wear to the friction material causes excessive lining-to-drum clearance. When this happens, the top of the rear brake shoe moves further away from the anchor pin while stopping. In one common cable-type self-adjusting system, the adjusting cable lifts the adjusting lever against its spring (**Figure 6.19**). When the brake is released, the spring pulls the adjusting lever back down. This turns the starwheel to adjust the brakes (**Figure 6.20**).

Many dual-servo self-adjusters have the adjuster lever located above the starwheel teeth, but some self-adjusters have the lever positioned below the starwheel (**Figure 6.21**). Positioning the adjusting lever below the starwheel results in a more positive adjustment.

Adjusting lever and starwheel parts have left- or right-hand threads that are not interchangeable from side to side. They are labeled left (L) or right (R) to prevent incorrect installation (**Figure 6.22**). If you install

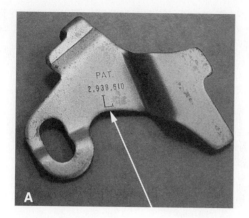

B

Figure 6.22 (A) Self-adjuster levers are not interchangeable from side to side. (B) The end of the adjusting screw has a left (L) or right (R) designation. *(Courtesy of Tim Gilles)*

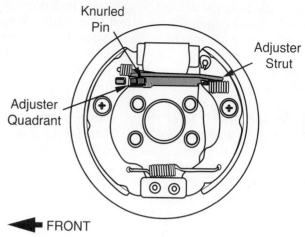

Figure 6.23 An expanding strut self-adjuster is part of the parking brake strut assembly.

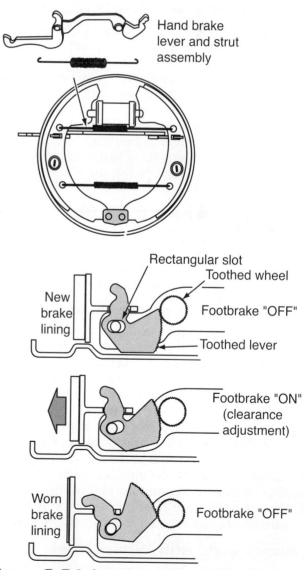

Figure 6.24 A rotating ratchet repositions its teeth in response to increasing clearance.

a left-side starwheel on the right side of the vehicle, self-adjustment will cause more brake clearance instead of less.

Leading-Trailing Self-Adjustment

There are several leading-trailing brake self-adjuster designs that operate as part of the parking brake strut assembly which fits between the two brake shoes. The most common type is called an *incremental adjuster* because during a self-adjustment its starwheel moves only one tooth at a time. The starwheel is part of the parking brake strut and operates in either forward or reverse, whenever too much clearance develops between the lining and drum.

An *expanding strut* self-adjuster also works off the parking brake rod and strut assembly (**Figure 6.23**). If there is too much lining-to-drum clearance when the parking brake is applied, the strut and telescoping rod assembly expands to readjust the clearance when the parking brake is released.

Another parking brake strut adjuster has a rotating ratchet that repositions in response to increasing clearance (**Figure 6.24**). Teeth on a movable lever are in mesh with teeth on a fixed wheel. Shoe return springs hold the outer edge of the toothed lever against a rectangular slot in the web of the brake shoe. The clearance between the other edge of the toothed lever and the slot in the brake shoe web determines the amount of clearance between the brake lining and the drum.

During a stop, hydraulic pressure forces the linings against the drum. After the clearance between the leading shoe and toothed lever is taken up, further movement of the shoe disengages the teeth on the lever and wheel. As the lever moves further it pivots, repositioning itself on the teeth of the wheel.

When linings wear excessively and movement of the lever runs out of teeth, the lever bottoms out against the shoe. From this point, no further adjustment can take place. Brake pedal and parking brake travel will increase with further lining wear, warning the driver of the need for service.

Other self-adjusters include ones that use the movement of a cam to adjust clearance.

Driver Habits and Self-Adjusters

Too tight an adjustment is not a problem with self-adjusters, but underadjustment sometimes occurs. This is usually due to a rusted adjusting mechanism. Sometimes this occurs because the driver does not use his or her brake properly. If a driver has a habit of backing up without using the service brakes and shifting from reverse to drive, the brakes will not self-adjust and threads in the self-adjuster can freeze up. Similarly, if a driver does not use the parking brake, relying instead on the transmission park selector, the brakes will not self-adjust.

■ DRUM BRAKE SERVICE

The previous information in this chapter has been provided to familiarize you with the construction and operation of drum brakes. The remainder of this chapter deals with drum brake inspection and repair procedures.

Disassembly and Inspection

After a brake drum is removed, several things must be inspected prior to completing your repair. Wear is a normal occurrence, but you should be on the lookout for *unusual* wear so the condition that caused it to occur will not reoccur after your parts replacement. During disassembly, look closely for misalignment problems and wear to the backing plate and brake drum. If everything looks good, a standard brake job should provide a reasonable and economical result for the customer. If other problems are apparent, probably because the customer continued to drive the car after the linings were worn to the metal backing, repair costs are likely to increase substantially (**Figure 6.25**).

Inspect Axle and Wheel Seals

Before removing the brake drum, look under the vehicle at the brake backing plates for signs of fluid leakage

Figure 6.25 (A) This brake lining has worn totally away and the drum is worn very thin. (B) The customer continued to drive his pickup until the shoe wore all the way through this drum, a very dangerous situation. *(Courtesy of Tim Gilles)*

at the axles or grease leaking from the wheel bearings. A leaking drum brake wheel cylinder will also leave the backing plate wet.

Removing a Brake Drum

The wheel and drum must be removed in order to inspect drum brake linings. First, mark the drums so they can be reinstalled in the same position (for instance, LR for left rear). Some drums are held against the hub by retaining screws (**Figure 6.26**). Others are held in place with a Tinnerman nut, also called a Pal nut (**Figure 6.27**).

Some cars have hubs with removable wheel bearings that can be repacked with grease during this service (see Chapter 11). These are found on older cars with front-wheel drum brakes and on the rear-wheel drum brakes of many front-wheel drive cars. The center hole on the drums of this design pilots off the bearing hub. This is also true of rear-wheel drive cars, where the drum

Chapter 6 Drum Brake Fundamentals and Service   129

Figure 6.26 Some drums have retaining screws. *(Courtesy of Tim Gilles)*

Figure 6.27 Some drums are retained with a Tinnerman nut. *(Courtesy of Tim Gilles)*

> **SHOP TIP** Brake fluid is soluble in water. Axle lubricant is not. If you are uncertain whether a lubricant leak is brake fluid or oil, try washing the backing plate off with water. If it beads up, it is not brake fluid.

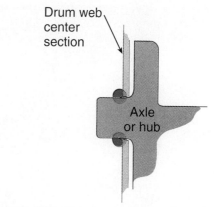

Figure 6.28 The drum rusts to the axle flange.

Figure 6.29 Turning screws into threaded holes in the drum forces the drum off of the axle flange. *(Courtesy of Tim Gilles)*

centers off a machined area on the axle flange. Removal of these drums is often made difficult due to rusting that occurs between the center hole of the drum and the pilot flange (**Figure 6.28**).

There are two common methods used to remove frozen brake drums. First, spray the flange area with penetrating oil. Many rear drums on front-wheel drive cars and imported light trucks have two threaded holes. Screws are tightened into these holes (**Figure 6.29**). The end of the screw pushes against the flange to force the drum off. Tighten the screws a half turn at a time, first on one side and then on the other.

When the drum has no threaded holes, it will usually loosen following a couple of sharp raps with a hammer between the wheel studs (**Figure 6.30**). Be careful not to damage a wheel stud. Pounding here is relatively safe because the drum is supported by the axle flange, which is very sturdy. However, **Figure 6.31** shows a drum that was ruined by someone who did not understand the relationship between the axle and drum.

NOTE: *Pounding on the rear lip of the drum will probably damage the drum.*

Some rear drum assemblies on front-wheel drive cars have sealed permanent wheel bearings that are tight on the axle. These are removed using a slide hammer (**Figure 6.32**). Be sure to consult vehicle service instructions before using a slide hammer.

When a car owner continues to drive a car after the linings have worn down to the shoe, the shoe will wear a channel in the drum. Then, the self-adjusters operate, adjusting the shoe into the worn out area of

Figure 6.30 To break loose rust between the axle flange and drum, give a sharp, firm rap with a hammer above the area supported by the axle flange. *(Courtesy of Tim Gilles)*

Figure 6.31 A drum ruined by someone who did not understand the relationship between the axle and drum. *(Courtesy of Tim Gilles)*

Figure 6.32 Removing a pressed-fit hub and drum assembly with a slide hammer. *(Courtesy of Tim Gilles)*

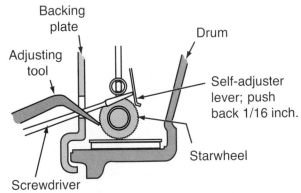

Figure 6.33 Move the self-adjusting lever out of the way so you can turn the starwheel.

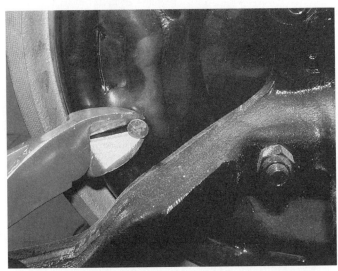

Figure 6.34 Cutting holddown pins to allow removal of a severely worn brake drum. *(Courtesy of Tim Gilles)*

the drum. If this has happened, the drum cannot be removed, unless the brake adjustment is backed off. The self-adjuster lever must be moved out of the way while performing the adjustment (**Figure 6.33**).

The heads of the shoe holddown pins (nails) are visible against the backing plate when viewed from the back side. When shoes are worn into the drum, cutting the heads off the holddown pins can allow faster removal of the drum (**Figure 6.34**).

Brake Cleaning

Before inspecting or disassembling brakes, clean the entire assembly. Brake dust is dangerous to breathe. It can be controlled with a high-efficiency particulate arresting (HEPA) vacuum or a low-pressure wet brake

> **SHOP TIP** Brake hardware kits usually have new holddown pins, but be certain of this before damaging the old ones.

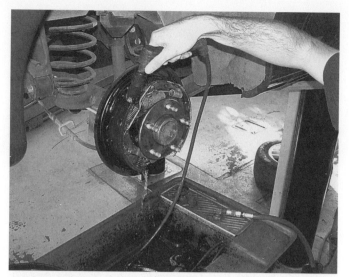

Figure 6.35 A brakes parts washer. *(Courtesy of Tim Gilles)*

washer (**Figure 6.35**). Low-pressure wet brake washers are the most popular way of cleaning brake assemblies. They are fast, convenient, and do a good job of cleaning the brakes in preparation for service. There are companies that provide regular service for these cleaners.

Wheel Cylinder Inspection

Carefully lift the edges of the rubber dust boots away from both sides of each wheel cylinder to check for leakage and rust (**Figure 6.36**). A small amount of brake fluid dampness is normal and provides lubrication of the pistons and seals. More important, fluid

 Brake dust is hazardous to breathe. Repair shops are required by law to have a brake parts washer or a HEPA vacuum when servicing brakes.

Figure 6.36 Inspect under the wheel cylinder boot for leakage and rust. *(Courtesy of Tim Gilles)*

should not drip from the wheel cylinder when the dust boot is pulled back to inspect it. Also, a cast iron wheel cylinder bore should not show rust and an aluminum piston should not show oxidation.

NOTE: *Rust (oxidation of iron or steel) is a serious problem. About one-seventh of the iron produced each year by industry is used to replace iron parts that have rusted. The exact process that occurs during rusting is not known. What is known is that oxygen and water are necessary for rusting to take place.*

Rebuild or Replace?

If there is any sign of contamination, the wheel cylinder should be replaced. When new linings are installed, the pistons and seals will be repositioned slightly, which can cause them to begin to leak. Unlike disc caliper pistons, which move very little during braking, drum brake wheel cylinder seals have a larger wiping distance during brake application.

Another consideration when deciding whether to service drum brake cylinders is how long the brakes have been in service without repair. Rear drum brakes last longer than front disc brakes and it is not uncommon for a car to have been driven for several years before needing the rear brakes replaced. Disassembly of the cylinder and replacement of the rubber parts is often the most prudent service choice for the customer. If the wheel cylinder begins to leak and the fluid contaminates the linings, they will need to be replaced even if they have substantial friction material remaining.

NOTE: *When a set of linings has become contaminated with axle lubricant or grease, replacement is recommended. Following thorough cleaning, the outside surface of the lining will appear to be clean, but lubricants or chemicals can still be absorbed into the lining. When the lining becomes excessively hot, the chemicals and lubricants can rise to the surface, resulting in erratic braking.*

Brake Lining Inspection

Inspect the thickness of the brake linings. **Figure 6.37** shows the relationship between the lining (friction material) and the brake shoe. Riveted linings should have at least 1/32 inch of friction material remaining above the rivet heads. **Figure 6.38** shows a riveted brake lining that is worn all the way to the rivet heads.

Some vehicles have a small hole in the brake backing plate through which the thickness of the linings

SHOP TIP Factory minimum specifications could be less, but a good rule of thumb for advising customers when to replace brake linings is that a bonded lining's friction material should be at least the same thickness as its metal backing.

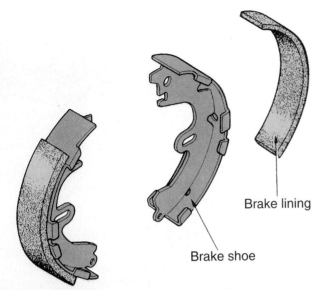

Figure 6.37 The brake lining and shoe.

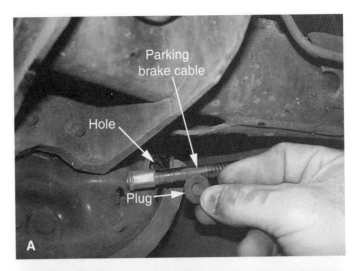

Figure 6.38 These riveted linings are worn almost to the rivet heads. *(Courtesy of Tim Gilles)*

Figure 6.39 (A) Remove the protective plug. (B) Shine a light through the hole to observe lining thickness remaining. *(Courtesy of Tim Gilles)*

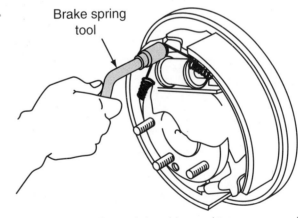

Figure 6.40 A special tool is used to remove and replace brake springs.

can be verified without removing the drum. A protective plug is removed to uncover the hole. Shine a light into the hole and observe the lining thickness (**Figure 6.39**).

Brake Shoe Removal

Removal of dual-servo brake linings is done with special tools. **Figure 6.40** shows a typical tool used to remove and reinstall brake return springs. Brake springs in leading-trailing brakes can be removed with special tools (**Figure 6.41**). Often, they are removed using pliers.

Holddown Springs

Holddown springs pull the brake shoes against the brake backing plate (see Figures 6.10–6.12). Coil-type holddown springs are removed with pliers or a special tool. **Figure 6.42** shows a tool used to remove beehive-type holddown springs. A Phillips screwdriver of the right size can also be used, but it is more difficult to use. Coil spring holddown springs can be removed using pliers, but the tool shown in **Figure 6.43** is much easier to use.

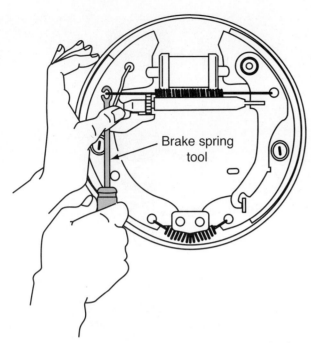

Figure 6.41 A special tool used to remove brake springs in leading-trailing brakes

Figure 6.43 (A) Coil-type holddown springs are removed with pliers or a special tool. (B) Parts of a coil-type holddown spring assembly. *(Courtesy of Tim Gilles)*

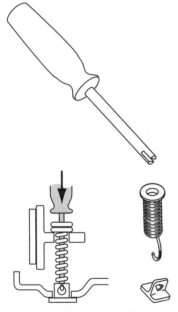

Figure 6.42 A special tool used to remove a beehive-style holddown spring.

SHOP TIP Before disassembly, be sure to carefully inspect the linings being replaced. Sometimes there are differences in materials or lengths of the primary and secondary linings. Until you have done several brake jobs on the same make of car, leave one side assembled while you complete repairs to the other side. This will leave you a "mirror image" to consult as you reassemble the brakes.

Rebuilding Wheel Cylinders

Many shops find it more cost-effective to replace wheel cylinders rather than rebuild them. Other shops rebuild wheel cylinders that are made of cast iron and are not corroded. Many can be rebuilt while still bolted to the backing plate. After removing the brake linings, remove the dust covers from the wheel cylinders. At one side of the wheel cylinder, push in on the piston to force the parts out the other side (**Figure 6.44**).

NOTE: *The wheel cylinder bolts may need to be removed on some vehicles in order to allow the pistons to be removed.*

Honing Wheel Cylinders

Aluminum cylinders cannot be honed, but iron cylinders are honed and cleaned prior to reassembly. Two kinds of hones are available. One type has either two or three stones and the other is a flex hone similar to those used in engine rebuilding (**Figure 6.45**). A two-stone

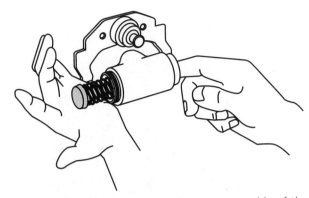

Figure 6.44 Push on the piston on one side of the cylinder to force the parts from the cylinder.

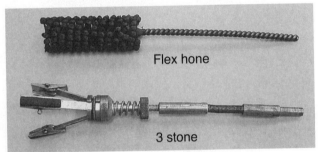

Figure 6.45 Two types of brake cylinder hones. (Courtesy of Tim Gilles)

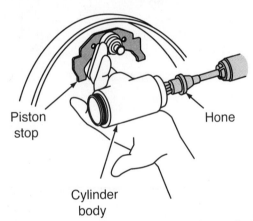

Figure 6.46 With piston stops, a wheel cylinder has to be unbolted from the backing plate so the pistons can be removed before honing.

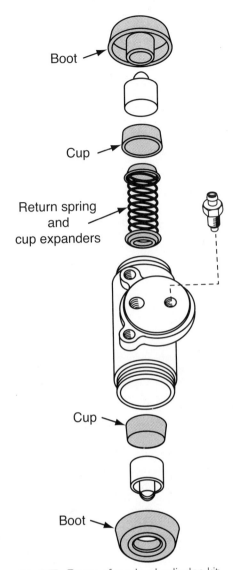

Figure 6.47 Parts of a wheel cylinder kit.

hone can get into smaller bores. It is used for rebuilding master cylinders and smaller import car cylinders.

NOTE: *Hones with stones are damaged if they are removed from the cylinder bore while still spinning.*

After the wheel cylinder is lightly honed to clean it up (**Figure 6.46**) and the pistons are cleaned, new rubber parts are installed from a **wheel cylinder kit** (**Figure 6.47**).

Measuring a Wheel Cylinder after Honing

When the size of the wheel cylinder has been increased excessively during honing, the cylinder must be replaced. Most automotive and light truck wheel cylinders range between ¾ inch and 1³⁄₁₆ inches in diameter. For wheel cylinders in this size range, if the piston to bore clearance exceeds 0.005 inch the cylinder is too large. A cylinder that is too large can allow the rubber sealing lips to be forced into the space between the bore and its piston. This can result in a cup that sticks in its bore or damage to the lips of the seal.

The bore can be measured with a telescoping gauge and micrometer before comparing it to the size of the piston. Another method of measurement is to put a thin feeler gauge in the bore and attempt to slide the piston alongside it. If maximum clearance is 0.005 inch, use a 0.006 feeler gauge (**Figure 6.48**).

NOTE: Anodized aluminum *cylinders, found on many late-model cars, should* not *be honed because their hard surface layer will be removed. They will corrode and can damage new seals. If in doubt as to whether a wheel cylinder is cast iron or aluminum, check with a magnet. A magnet will not be attracted to an aluminum cylinder.*

Chapter 6 Drum Brake Fundamentals and Service 135

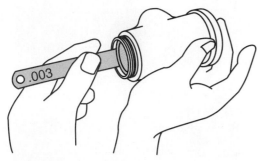

Figure 6.48 Use a thin feeler gauge to measure clearance between the piston and bore.

Clean an aluminum cylinder with alcohol or hot water and a nylon or soft bristle brush or an old tooth brush. If an aluminum cylinder's bore has pit marks it must be replaced rather than rebuilt.

Reassembling a Wheel Cylinder

During reassembly, lubricate parts thoroughly with brake fluid. Reassemble the wheel cylinder as shown in Figure 6.5. The lips on the wheel cylinder cups must face toward the fluid. A seal lip installed backwards will result in fluid leaking from the wheel cylinder when the brakes are applied (**Figure 6.49**).

NOTE: *Apprentices often learn which direction a seal is supposed to face when they complete their first brake job on their own. If you install the seals in the wrong direction, brake fluid will leak very quickly from the wheel cylinders the first time the brakes are applied. Be sure to check for puddles beneath the rear wheels before returning the car to the customer.*

After the wheel cylinder is assembled, use a clamp to hold the assembly together while installing the linings (**Figure 6.50**).

NOTE: *Use of a wheel cylinder clamp is only advisable when the wheel cylinder is new or has been disassembled and cleaned.*

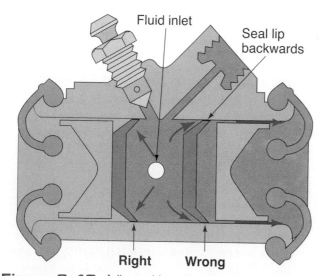

Figure 6.49 A lip seal installed backwards will leak.

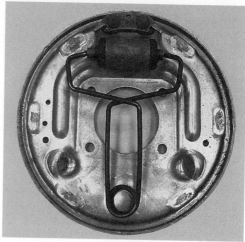

Figure 6.50 A brake cylinder clamp. (Courtesy of Tim Gilles)

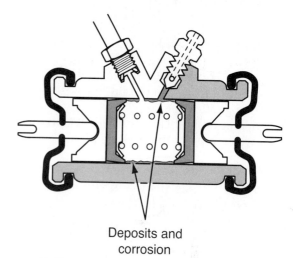

Figure 6.51 The area of the wheel cylinder between the cups accumulates dirt and becomes corroded during the service life of the cylinder.

The area of the wheel cylinder between the cups accumulates dirt and becomes corroded during the service life of the cylinder (**Figure 6.51**). If a wheel cylinder cup seal is pushed into the dirty area beyond its normal range of travel, a leak is almost certain to develop.

Removing Wheel Cylinders

When a wheel cylinder bore is pitted or corroded, it must be replaced. Use a flare nut wrench to remove the brake tubing fitting on the back. The wheel cylinder is usually held to the backing plate by two screws. Sometimes a cylinder is fastened to the backing plate with a clip.

■ INSPECTING AND CLEANING BRAKE PARTS

Prior to reassembling the brakes, clean and inspect the rest of the parts. When a serviceable wheel bearing is part of the drum hub assembly, clean and repack the

Figure 6.52 Lightly lubricate the pads on the backing plates. Do not apply too much lubricant like this student did. *(Courtesy of Tim Gilles)*

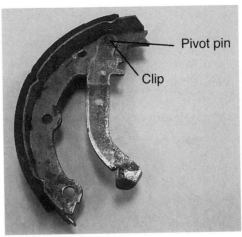

Figure 6.53 The parking brake actuating lever is attached to the secondary lining by a pivot pin and clip. *(Courtesy of Tim Gilles)*

Figure 6.54 Remove the old park brake lever pin from the brake shoe and reuse it. This one has a reusable E-clip. *(Courtesy of Tim Gilles)*

bearings with grease and replace the seal as described in Chapter 11.

Clean the backing plates and look for grooves in the shoe platforms where the linings slide. If the platforms are in good shape, apply a very small amount of high-temperature lubricant to them (**Figure 6.52**). If they are worn, replace the backing plate.

Self-Adjuster Service

Self-adjusters become rusty and should be disassembled so their threads can be cleaned.

NOTE: *Be careful not to lose the thin washer if it is included in the assembly.*

Following cleaning, brake assembly lubricant is applied to self-adjuster threads. Thread the parts together and operate the self-adjusters to see that they move freely. If one self-adjuster is rusted or dirty and is not working, brake pedal height can drop in just a few thousand miles.

Check the parts of the actuating mechanism to be sure there is no unusual wear. Sometimes cables become frayed or stretched or their ends become damaged.

Parking Brake Lever Installation

The parking brake actuating lever is attached to the secondary lining by a pivot pin and clip (**Figure 6.53**). It is usually much easier to attach the lever to the secondary lining before installing the lining on the backing plate. The pivot pin fits tightly in a hole in the lining. The pin that the lever pivots on does not usually come with the new linings and must be removed from the old ones (**Figure 6.54**). Sometimes the pin does not fit into the hole in the replacement lining because paint used during the rebuilding process has become lodged in the hole. Running a drill bit of the correct size through the hole will solve the problem. Be sure not to drill the hole too large or the lining will be ruined. Some pivot levers are held onto the pin with a reuseable E-clip. Others use replaceable clips that are installed with pliers.

Drum Brake Hardware Kit

Many shops routinely replace all brake springs with every brake job. A drum brake hardware kit includes return springs for the primary and secondary shoes, holddown springs and pins (sometimes called nails), and a connecting spring. The kit (**Figure 6.55**) includes items for both drum brakes on an axle.

Checking Return Spring Condition

If springs are to be reused, visually check their condition to be sure they are not loose or damaged (**Figure 6.56**). The paint on the springs should be in good con-

> **SHOP TIP** Removing the parking brake cable from the parking brake actuating lever can sometimes be awkward. Most technicians leave the parking brake cable attached to the parking brake actuating lever and simply install it to the pivot pin on the new lining.

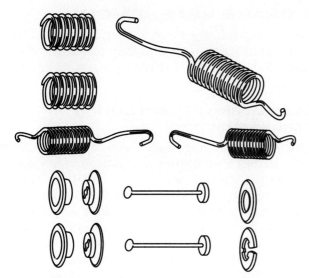

Figure 6.55 Drum brake hardware for one wheel.

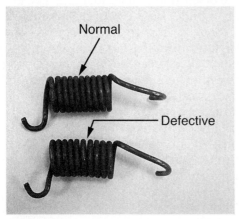

Figure 6.56 Inspect springs for damage. *(Courtesy of Tim Gilles)*

dition, indicating that they have not been overheated. When a return spring has been removed from the brake assembly, see if a piece of paper can be inserted between its coils. If so, replace the spring. Another test is to hold the spring up to a light. If light is visible between the coils of the spring, replace it.

Results of Weak Return Springs

A weak return spring can allow the primary shoe to apply with more pressure than the engineers allowed for. This can result in a brake that applies earlier than the one on the opposite side of the vehicle, resulting in a quick pull of the steering wheel to one side. A weak spring can also cause a shoe to release more slowly than normal, causing a drag.

REMEMBER: *Pascal's Law says pressure within the hydraulic system is equal when there is no movement of the fluid. But force at the wheels depends on the application of that pressure. If the return spring on one side of the vehicle is weaker than that on the other side, the amount of application force could be slightly different. If the vehicle has power steering, this might not be evident to the driver.*

SHOP TIP If a spring is dropped on a hard surface, it should not ring or bounce.

ORDERING REPLACEMENT PARTS

Consult the appropriate service manual before working on a vehicle. Major brake part suppliers provide brake service manuals to users of their products.

Brake linings are replaced in pairs. Parts are purchased in *axle sets*. An axle set includes all four shoes for wheels at either end of the vehicle. Be sure to carefully compare your new parts to the old ones. Manufacturers often use the same parts on different vehicle models, but sometimes there are small differences in the parts. A wheel cylinder or a shoe anchor might be slightly different, or a shoe might be slightly wider than one found on another model of the same manufacturer's vehicle.

CASE HISTORY *A technician removed brakes from a Nissan Pathfinder that had been seriously damaged when the customer continued to drive the vehicle after the friction material had worn to the shoe. Besides the brake shoes, other parts were damaged (**Figure 6.57**). The shop attempted to locate parts at the dealer. They were told that parts had to be shipped from Japan and were expensive. The customer located a salvage yard that had a Nissan pickup. The parts that he brought to the shop looked the same, but in reality they were slightly different. After installation, the brake drum would not fit. The vehicle took up shop space for two weeks until the correct part could be shipped to the dealer.*

Figure 6.57 These brake parts were badly damaged when a customer continued to drive the car well after the brakes were worn out. *(Courtesy of Tim Gilles)*

Professional parts suppliers are most often reliable when they supply you with replacement parts. When in doubt, they will sometimes deliver two parts for you to carefully compare with the old ones.

One question the parts supplier might ask is, "What size is the engine?" The parts person is following a procedure to determine the actual part you need. A ¾ ton Suburban 2500 with a 454 cubic inch 7.4L engine might need a larger set of brakes than the same vehicle with a 5.7L (350-cubic-inch) engine. The parts supplier's time is valuable. Before you make your call for parts, measure the width of the linings and the diameter of the brake drum.

Check Shoe Arc

Before installing the brake shoes, hold them against the inside of the drum and rock them slightly from heel to toe. There should be a very small amount of clearance (approximately 0.005 inch) at the ends (**Figure 6.58**).

The important thing for you to check is that no clearance exists between the center of the shoe and the brake drum. This can cause noise and can make a car pull to one side.

Vintage Chassis

On vintage cars, when drum brakes were used on all four wheels, shoes were custom fitted to fit the drum. This was called "shoe arcing." Today's brake shoes are prearced, which will allow them to wear into the drum.

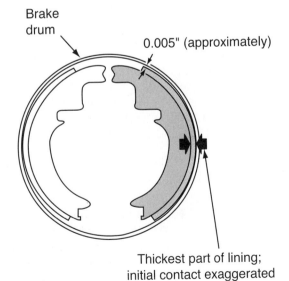

Figure 6.58 The ends of new brake linings have a small amount of clearance.

■ BRAKE REASSEMBLY

When reassembling drum brakes, carefully inspect replacement parts to be sure they match those being replaced. During the assembly process be careful to keep parts clean.

Keep Brakes Clean during Reassembly

Sometimes technicians are careless about keeping brake parts clean during installation. They simply clean the assembled parts after installation. This is a poor shop practice. Other than the fact that some grease might still remain in the relatively porous surface of the fresh lining, problems can occur with the cleaning material being trapped in the surface, too. Denatured alcohol and brake cleaner are two popular materials for cleaning brake parts. But these products should be used only while parts are disassembled.

In the event that you accidentally get some contaminants on brake surfaces during assembly and need to clean them, be sure to allow the shoes to dry thoroughly before installing the brake drums. Denatured alcohol or brake cleaner normally evaporate at a very rapid rate, but these materials can pose a problem on new brakes because they will not be able to evaporate if the drum is installed immediately after cleaning. During the test drive, vapors that come off the linings could cause the linings to "hydroplane" on the surface of the drum, reducing braking performance and resulting in a spongy pedal.

Replacing Drum Brake Shoes

When reinstalling brake shoes on the backing plates, it is sometimes easier to first assemble the bottom spring to the brake linings. Then the linings can be folded into each other, making installation easier (**Figure 6.59**).

In dual-servo brakes, the secondary lining is always longer. In leading-trailing brakes, the leading shoe is occasionally longer than the trailing shoe. Some linings will be glued to the shoe rim in an off-center position for engineering reasons (**Figure 6.60**).

Be sure to carefully inspect the shoes as you remove them so they can be reassembled in the correct position. Unfortunately, sometimes a car will have had a

> **SHOP TIP** Masking tape is available in wide rolls. It takes very little time to apply a layer of masking tape to the friction surfaces of drum brake shoes prior to assembling them to the backing plate. This makes cleanup very quick and easy. Manufacturers would probably do this for you, but masking tape adhesive must be removed within a short time after it is installed or removal is very difficult.

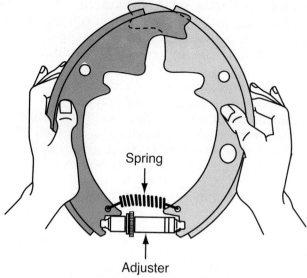

Figure 6.59 The linings can be folded into each other to make installation easier.

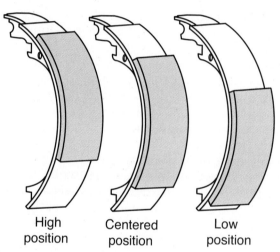

Figure 6.60 Friction linings bonded in different positions on the shoe.

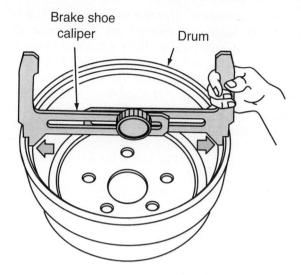

A Setting tool to drum

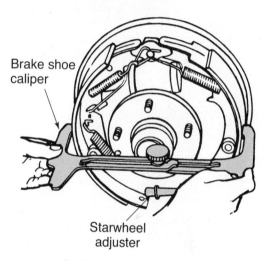

B Setting brake shoes to tool

Figure 6.61 A brake adjusting gauge. (A) Adjust the tool to fit the drum. (B) Adjusting the brake shoes to fit the tool will provide a good initial drum-to-lining clearance.

previous brake repair in which the parts were reassembled incorrectly. A clue will be that one brake lining is severely worn while the other has very little wear.

Installing the Self-Adjuster

Clean, lubricate, and return self-adjusters to the smallest adjustment position prior to installing them.

NOTE: *Self-adjusters must be reinstalled on the correct side of the car because they have either right- or left-hand threads and cannot be interchanged.*

Initial Clearance Adjustment

After brake linings are installed on the backing plates, an initial clearance adjustment is made before the drums are installed. A brake adjusting gauge is adjusted to the size of the drum (**Figure 6.61**). Then, the starwheel of the brake adjuster is turned until the shoes expand to the size of the adjusting gauge. This allows for about 0.01 inch of clearance between the lining and drum. If additional adjustment is necessary after the brake job is completed, drive the car slowly in reverse while applying the brakes repeatedly until the brake pedal height no longer increases significantly on the second application of the pedal.

Manual Brake Clearance Adjustment

Unless you are working on a vintage car, rear drum brakes are adjusted by an automatic mechanism. However, technicians are quite often called upon to service classic cars. You will not want to appear ignorant to your customer, so it is best to know how to adjust brakes. Newer self-adjusting brakes also sometimes require initial adjustment, so an understanding of brake adjustment will help you with that service as well.

Older car brakes required manual adjustment as their linings became thinner with wear. Some of the self-adjusting methods used on cars were described earlier in this chapter. Drum brakes typically have an adjuster that features a threaded stud, called a starwheel, and a long nut. The starwheel is part of the threaded stud (see **Figure 6.17**).

Early brakes were adjusted with a tool called a *brake spoon* (**Figure 6.62A**) through an access hole in the backing plate (**Figure 6.62B**) or the front of the drum. The brake spoon fit into the teeth in the starwheel. With the car raised, a mechanic would rotate the tire by hand while prying against the starwheel with the brake spoon until the linings locked the drum and the wheel could no longer turn. From that point, the starwheel was backed off (unwound) until the wheel spun freely once again. This would typically be about seven or eight teeth, although cars like the Volkswagen with a duplex brake (two wheel cylinders and two leading shoes) might require backing off only two or three teeth before the wheel turned free.

The most common non-servo brake adjuster was an eccentric cam. Another type of adjuster used a tapered wedge between the heels of the shoes.

Figure 6.62 (A) A brake spoon used for adjusting drum brakes. (B) Inserting a brake spoon through the adjuster access hole in the backing plate. *(Courtesy of Tim Gilles)*

Self-Adjuster Clearance Adjustment with Drum Installed

To adjust lining-to-drum clearance after the drum is installed, you will need to access the starwheel through a hole in the backing plate or front of the drum as is done with manual adjusters. While turning the wheel by hand, use a brake spoon to rotate the starwheel until the wheel will no longer turn.

When tightening the starwheel to remove clearance, the self-adjuster does not have to be held out of the way. The starwheel will move easily in that direction. To loosen the adjustment in order to provide the desired clearance, the self-adjusting mechanism is held out away from the starwheel (see **Figure 6.32**). After the starwheel has been tightened, loosen it about five to ten teeth until the wheel turns freely once again. Older cars and trucks without self-adjusters use the same procedure, but the self adjusting mechanism does not have to be held out of the way during loosening.

REVIEW QUESTIONS

1. Edges of the shoe are crimped in several places that rub against corresponding raised platform areas on the brake _____ _____.

2. Which lining type runs hotter, bonded or riveted?

3. Which lining type is usually quieter, bonded or riveted?

4. Force is applied by the wheel cylinder to which end of the brake lining, the heel or the toe?

5. Do the lips on the cup seals face toward or away from the center of the body of the wheel cylinder?

6. When a drum brake's leading shoe digs into the brake drum during a stop, this is called _____.

7. Bendix brakes are also called the _____ _____ brakes because *both* brake shoes are self-energizing.

8. When the front shoe floats at the bottom during braking, applying additional pressure against the rear shoe, this is called _____ action.

9. The disc brakes in the front and the drum brakes in the rear are operated hydraulically by the _____ cylinder.

10. Which drum brake parts are labeled left or right to ensure that they are installed on the correct side of the vehicle?

11. Leading-trailing brakes have several self-adjuster designs that operate as part of the _____ brake strut assembly.

12. The rear brake drums on many front-wheel drive cars and imported trucks have two threaded holes. What are they for?

13. List two methods of cleaning brake dust before disassembling brakes.

14. Name two materials that might not be able to evaporate if the drum is installed immediately after cleaning.

15. A brake adjusting gauge adjusted to the size of the drum allows for about _____ of clearance between the lining and drum.

PRACTICE TEST QUESTIONS

1. Technician A says the leading-trailing brake is used on most front-wheel drive cars because this brake design has less tendency to lock the rear wheels during a stop. Technician B says leading-trailing brakes use servo action. Who is right?
   - A. Technician A
   - B. Technician B
   - C. Both A and B
   - D. Neither A nor B

2. Technician A says Bendix brakes have a longer primary lining than secondary lining. Technician B says sometimes primary and secondary linings are made of different friction materials. Who is right?
   - A. Technician A
   - B. Technician B
   - C. Both A and B
   - D. Neither A nor B

3. Technician A says dual-servo self-adjusters operate only when the brakes are applied during a stop while backing up. Technician B says if you install a left-side starwheel on the right side of the vehicle, self-adjustment will cause more brake clearance instead of less. Who is right?
   - A. Technician A
   - B. Technician B
   - C. Both A and B
   - D. Neither A nor B

4. Technician A says the self-adjuster in some dual-servo brakes is part of the parking brake strut and operates when stopping in either forward or reverse. Technician B says some leading-trailing brakes self-adjust when the parking brake is released. Who is right?
   - A. Technician A
   - B. Technician B
   - C. Both A and B
   - D. Neither A nor B

5. A car has self-adjusting brakes with a low brake pedal that rises to normal on the second application. Technician A says this is usually due to a rusted self-adjusting mechanism. Technician B says this can happen because the driver does not use his or her brake properly. Who is right?
   - A. Technician A
   - B. Technician B
   - C. Both A and B
   - D. Neither A nor B

6. Technician A says a small amount of brake fluid dampness under a wheel cylinder boot is normal. Technician B says a small quantity of fluid will normally drip from the wheel cylinder when the dust boot is pulled back to inspect it. Who is right?
   - A. Technician A
   - B. Technician B
   - C. Both A and B
   - D. Neither A nor B

7. Technician A says linings that have become contaminated with axle lubricant can usually be cleaned successfully with brake cleaner. Technician B says linings that have been cleaned will likely cause braking problems when hot. Who is right?
   - A. Technician A
   - B. Technician B
   - C. Both A and B
   - D. Neither A nor B

8. Technician A says aluminum cylinders cannot be honed. Technician B says a three-stone hone can go into smaller bore master cylinders and imported wheel cylinders. Who is right?
   - A. Technician A
   - B. Technician B
   - C. Both A and B
   - D. Neither A nor B

9. A vehicle has a weak brake shoe return spring. Technician A says this can allow that brake to apply faster than the one on the other side of the vehicle. Technician B says this can resulting in a quick pull of the steering wheel to one side. Who is right?
   - A. Technician A
   - B. Technician B
   - C. Both A and B
   - D. Neither A nor B

10. Technician A says a single rusted self-adjuster can result in a drop in brake pedal height. Technician B says when drum brakes need to be adjusted in a tandem master cylinder hydraulic system, pedal travel drops for the entire system. Who is right?
    - A. Technician A
    - B. Technician B
    - C. Both A and B
    - D. Neither A nor B

CHAPTER 7

Disc Brake Fundamentals and Service

■ OBJECTIVES:

Upon completion of this chapter, you should be able to:

✔ Describe the operation of disc brakes.
✔ Explain the advantages and disadvantages of disc brakes.
✔ Describe the different disc caliper designs.
✔ Diagnose disc brake problems.
✔ Disassemble, rebuild, and reassemble disc brake calipers.
✔ Explain how to prevent disc brake noise.

■ KEY TERMS

fixed caliper
floating caliper
rotor

■ INTRODUCTION

American cars have had front disc brakes since the early 1970s. A few cars had them beginning in the early 1960s. Modern passenger cars and light trucks have disc brakes in the front and many have them on all four wheels.

■ DISC BRAKE OPERATION

Disc brakes have a **rotor** (disc) and a caliper, like a bicycle brake (**Figure 7.1**). **Figure 7.2** shows the parts of a typical disc brake system.

A rubber, square-cut O-ring seals the disc brake apply piston to its bore. Unlike drum brakes, disc brakes do not require a return spring. When the brakes are applied, the O-ring distorts (**Figure 7.3**). When the brakes are released, the seal retracts to its original position, pulling the piston back and allowing the linings to release the rotor. The lining, or pad, glides on the surface of the rotor, wiping off contaminants.

Disc Brake Advantages

One advantage of disc brakes is that they self-adjust. As linings wear, the piston slides forward in its bore to take up the slack (**Figure 7.4**). As this occurs, the caliper requires more fluid. On a master cylinder with reservoirs of different sizes, the larger one is for the disc brakes and the smaller one is for the drum brakes.

When a rear disc brake includes a mechanism for the parking brake, there will be an automatic adjuster

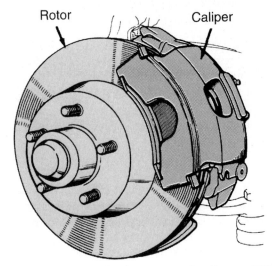

Figure 7.1 A disk brake rotor and caliper resembles a bicycle brake.

(**Figure 7.5**). Operating the parking brake causes the piston to thread outward to adjust the clearance. Operation and service of the parking brake are covered in Chapter 9.

Besides their self-adjusting capability, disc caliper brakes have several other advantages over drum brakes. Disc brakes are light in weight, are relatively inexpensive, and dissipate heat better. A rotor will not distort when subjected to brake lining force like a brake drum

142

Chapter 7 Disc Brake Fundamentals and Service

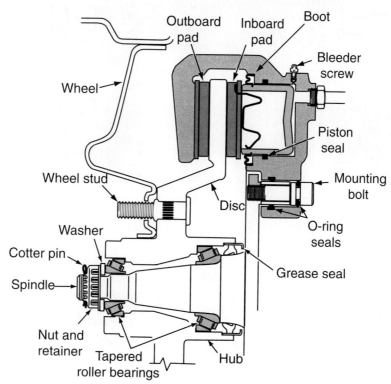

Figure 7.2 A typical disc brake system shown here with the wheel bearing assembly.

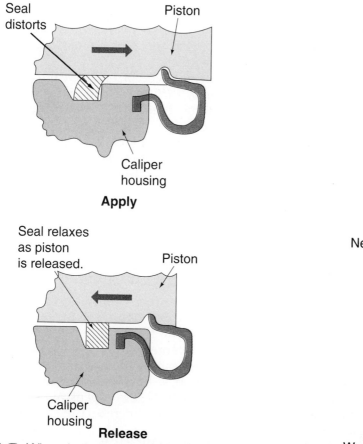

Figure 7.3 When the brakes are applied, the seal distorts. When the brakes are released, the seal retracts, pulling back the piston.

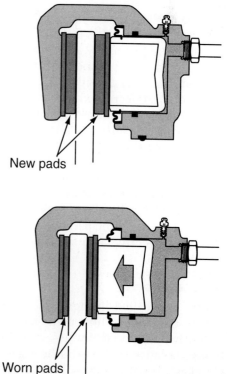

Figure 7.4 As the disc linings wear, the piston moves out in its bore to self adjust.

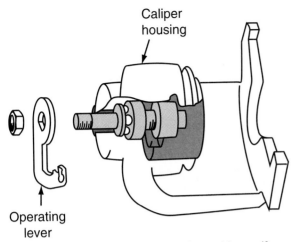

Figure 7.5 A rear disc brake caliper with a self-adjuster operated by the parking brake.

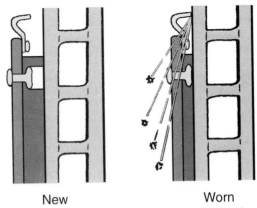

Figure 7.6 When the lining is worn too much, the metal warning sensor contacts the brake rotor.

does. When the friction material on a disc pad wears, it falls off the rotor as it turns. Drum brakes trap these contaminants between the shoe and drum. This causes even more wear and can cause braking efficiency to be affected.

When drum brakes were used for front brakes, it was not unusual for brakes to cause a pull to one side or the other. The pull could result from leverage differences due to different sized drum diameters on the same axle, unequal lining wear, or differences in the application of friction because of the fit of the shoe to the drum arc. Disc brakes are far less likely to cause pull because the brakes on each side of the vehicle give the same braking effort.

Disadvantages of Disc Brakes

One disadvantage to disc brakes is that the large rotor surface is more prone to amplify any noise that results from vibration. Harder linings and thinner rotors magnify the problem. Another disadvantage of disc brakes is that they are not as effective when used as a parking brake. Drum brakes work better because they self-energize, but disc brake pads have a more difficult job trying to hold on to the slick surfaces of the rotor. Imagine trying to grip a piece of vertical glass between two other pieces of glass. The gripping power of the pads is effective enough when they are forced against the rotor by the hydraulic system, but when using only hand or foot effort on the parking brake, the force applied is not as effective as a self-energized drum parking brake.

Disc Brake Pads

Brake pads are also called linings. Disc brake lining materials can be the same as those used on drum brakes. Front-wheel drive cars typically use semimetallic or ceramic linings because of the extra heat generated in the front brakes. Sometimes new linings are slightly tapered because the trailing edge of the lining runs hotter, which causes it to wear more. Also, the leading and trailing edges of the pads are sometimes tapered to reduce vibration, making them quieter. Chapter 8 discusses lining materials in detail.

Brake Pad Wear Indicators

Different methods are used to warn of excessive disc brake pad wear. Some disc brake linings include an *audible wear sensor* (**Figure 7.6**), a small metal tab attached to one of the pads. The tab rubs against the rotor when the lining wears thin. The noise that results alerts the driver that brake work is needed before serious damage occurs to the rotor. Audible sensors are the most common type of wear sensors. The noise that they make is often heard when the brakes are *not* applied, when the sensing tab drags and vibrates against the rotor.

A different type of brake wear sensor is called a *visual sensor*. Each brake pad has an electrically controlled sensor (**Figure 7.7**). When the pads are excessively worn, this sensor triggers a warning light on the dashboard to illuminate.

A small number of cars use a *tactile sensor*. When the friction material becomes too thin, a high spot on the rotor contacts the metal backing on the pad. This completes the warning system circuit, causing the brake pedal to pulsate and alerting the driver to seek service.

■ FIXED AND FLOATING DISC CALIPERS

A disc brake caliper is either a **fixed caliper** or a **floating caliper** (**Figure 7.8**). Early American disc brakes used the fixed caliper design. Fixed calipers have one or more pistons on each side of the caliper. With a fixed caliper there is more piston area for the fluid to act on, so a power brake is often not necessary. Fixed calipers are more expensive to produce and are no longer used on American cars.

Floating Calipers

Floating calipers, also called *sliding calipers,* are a simpler design than a fixed caliper. They have fewer parts

Chapter 7 Disc Brake Fundamentals and Service **145**

and are easier to service, but they usually require a power assist to achieve normal pedal pressure.

A floating caliper has one or two pistons located on only one side of the caliper. Typical floating calipers have only one piston, although some use two. The caliper slides as the piston moves out of its bore (**Figure 7.9**). When all of its sliding surfaces are in good operating condition, hydraulic action applied to the pad on one side of the caliper results in an equal force on the other side's pad.

In a properly operating single-piston floating caliper fluid pressure is applied to the back of the caliper piston, which forces the inside pad against the surface of the rotor. Newton's first law states, "For every action

Figure 7.7 (A) An electronic disc pad wear sensor. (B) Removing the sensor from the pad. *(Courtesy of Tim Gilles)*

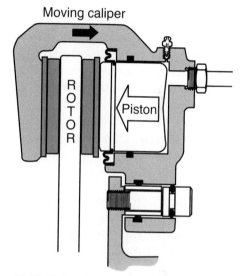

Figure 7.9 A floating caliper moves as the brakes are applied.

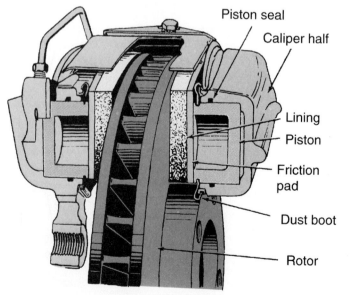

Fixed Caliper

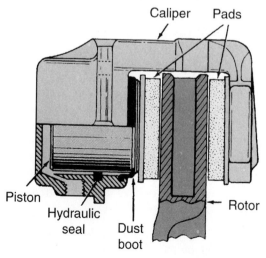

Floating Caliper

Figure 7.8 A fixed and a floating caliper.

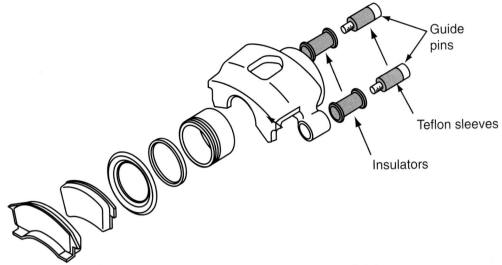

Figure 7.10 A pin slider caliper slides on guide pins that allow it to twist slightly.

there is an equal and opposite reaction." The inside pad "reacts" against the caliper, causing it to slide until the outboard pad contacts the rotor with an equal amount of force.

Floating Caliper Design Variations

Different floating caliper designs include the pin slider, the center abutment, and the swing caliper.

Pin Slider Caliper. The majority of cars use a pin slider caliper. A pin slider caliper slides on guide pins during and after brake application (**Figure 7.10**). The guide pins allow the caliper to twist slightly. This allows for minor changes in pad alignment but still allows the caliper to return to its original position after the brakes are released.

Repeated sliding would result in excessive wear to the guide pins or the holes in the caliper. Instead bushings or O-rings (also called insulators or sleeves), made of rubber or teflon, prevent metal to metal contact between the guide pins and the caliper (**Figure 7.11**).

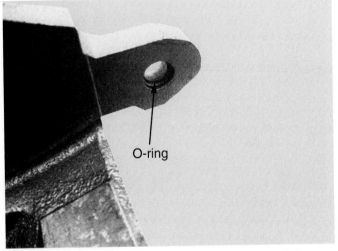

Figure 7.11 A rubber insulating bushing or O-ring prevents metal-to-metal contact between the guide pins and the caliper. (Courtesy of Tim Gilles)

Center Abutment Caliper. Some American vehicles, mostly older ones, use a sliding caliper design called a *center abutment caliper*. It has two wedged sliding surfaces called ways (**Figure 7.12**). The caliper slides between the ways. A small amount of clearance between the abutment and ways is normal because it allows the caliper to slide. Excessive clearance would result in noise as the caliper rocks against the ways when the brakes are applied.

NOTE: *Ways is a term from the machine tool industry describing the guides that a machine tool slides against when it moves on its mount.*

Swing Caliper. Some import cars use a caliper called a *swing caliper*. The caliper is attached to the spindle

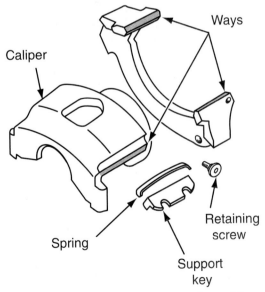

Figure 7.12 A center abutment caliper slides between two wedged sliding surfaces called ways.

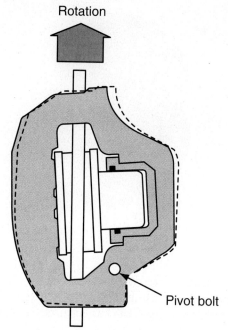

Figure 7.13 A swing or pivoting caliper. Note that the caliper pivots to apply the outboard pad. Also note the normal taper of the lining.

support arm by a pivot pin. As the brakes are applied and released, the caliper pivots back and forth on the pin. Unlike the other floating caliper designs, the piston does not remain parallel to the rotor so the linings on these cars are wedge shaped (**Figure 7.13**).

Caliper Pistons

Caliper pistons are hollow, shaped like a cup. They are installed with the open side against the back of the friction pad. This provides very little surface area for heat to be able to transfer back into the brake fluid. Most pistons are made of steel, and some are chrome-plated. Fiberglass-reinforced phenolic resin pistons are also used in some calipers. Plastic pistons like these are light in weight, do not conduct heat readily, and do not corrode like metal ones do.

■ FOUR-WHEEL DISC BRAKES

Disc brakes are used either on front brakes only, or on all four wheels. Four-wheel disc brakes have been used on most luxury and performance vehicles since the end of the 1970s. European cars have used four-wheel disc brakes for many years. They are more costly than the disc/drum hybrid system. Their first use on an American car was in 1965 on the Corvette.

Rear disc brake systems can have either fixed or floating calipers. They are typically the same as front calipers, although the diameter of the rear pistons will be smaller and they have some sort of arrangement for a parking brake (see Chapter 9).

■ DISC BRAKE SERVICE

The following section of this chapter deals with the diagnosis and repair of disc brake systems. **Figure 7.14** shows several possible brake system complaints.

Road Test

Prior to repair of a problem, a road test is done to verify the complaint. Before a road test, check the following items:

- Be sure there is fluid in both reservoir sections of the master cylinder.
- Check to see that pedal height is normal. If not, do not road-test the car.
- Check tire pressures to see that they are equal and inflated to manufacturer's specification.
- Verify that the tires are the specified size and that none of them is smaller or larger than the rest.

During the road test, drive the car on a deserted, flat, straight section of road. As you drive, notice whether the vehicle pulls to one side or the other. Check also to see if there is a pull during braking. If there is, swap the front tires. Sometimes a pull caused by a tire will only occur when braking. If the pull changes to the other direction, the tires are at fault.

When a car pulls to one side during braking, the other side's brakes might be malfunctioning. Look for a lack of wear on the brake pads on the side that is not

Brake System Complaints

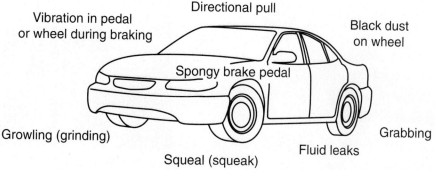

Figure 7.14 Possible brake system complaints.

working correctly. Remove the caliper and check for a sticking piston or sticking caliper slide pins.

When a brake pulls to one side when the car is not braking, look for a collapsed brake hose or pinched steel brake line. Cracking open the bleed screw on the side with the restriction will sometimes result in a squirt of fluid as the built up pressure is released.

During the test drive, look for pedal pulsation, a wheel grabbing or locking up, noises, or other unusual conditions.

Chassis Problems

Problems related to the chassis can cause symptoms felt in the brakes, such as brake pull. Chassis parts that can be at fault include loose wheel bearings, worn ball joints or bushings, and worn steering linkage parts.

Disc Brake Inspection

Brakes are more easily inspected with the car raised on a lift so that some visible checks can be made without too much difficulty.

During the inspection, list all of your findings and recommendations on the repair order. After completing the brake inspection, inform the customer of his or her repair options. If the customer has not already authorized repairs, they should be contacted by phone as soon as possible, especially if the shop's appointment schedule allows an option for performing the repairs while the car is already in the shop.

Friction Material Inspection

Disc brakes are inspected to see that a sufficient amount of friction material remains on the pads (**Figure 7.15**). When the friction material wears completely away, rotor damage will result. Brake pads are typically replaced before the thickness of the remaining friction material on the steel backing plate reaches $1/8$ to $3/16$ inch. Some shops recommend pad replacement when the remaining pad material is the same thickness as the steel backing.

Some customers use their brakes well beyond the point where metal is wearing on metal. This most often

Figure 7.16 These rotors are worn so badly that one of them actually separated into two pieces. (Courtesy of Tim Gilles)

results in much more costly repairs due to damage to the rotors, and sometimes the calipers (**Figure 7.16**).

On most vehicles, you can remove a front wheel and visually inspect the pad thickness without removing the brake caliper. Sometimes the thickness of the friction material is visible through a hole in the caliper (**Figure 7.17**). On other calipers a flashlight can be used to look down the side of the caliper (**Figure 7.18**). Check to see that the brake pad backing is thinner than the lining material, just as you do with drum brakes.

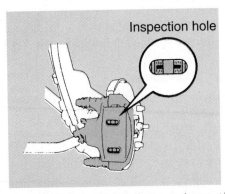

Figure 7.17 Some calipers have an inspection hole to see if the lining is worn excessively.

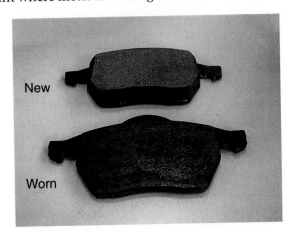

Figure 7.15 New and worn disc pads. (Courtesy of Tim Gilles)

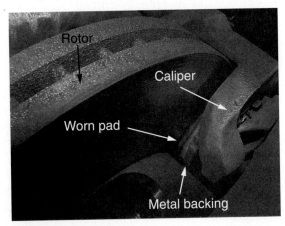

Figure 7.18 Inspecting disc lining thickness. (Courtesy of Tim Gilles)

Chapter 7 Disc Brake Fundamentals and Service **149**

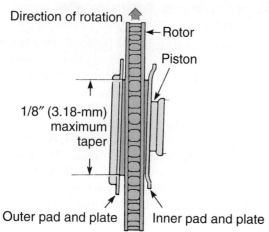

Figure 7.19 Tapered wear of less than ⅛ inch is considered normal.

CAUTION The rotor holds heat for some time. You could be burned if you touch a hot rotor on a car that has recently been driven.

rection will be required. If you cannot see visual runout, check for runout with a dial indicator.

Checking Rotor Runout with a Dial Indicator

Rotor runout causes the rotor to swing from side to side as it rotates (**Figure 7.21A**). Ideally, runout should be less than 0.003 inch although some manufacturers

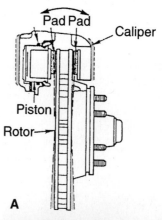

Figure 7.21 (A) Runout causes the rotor to swing from side to side as it rotates. (B) Be certain that the mating surfaces of the rotor and hub are clean. (B, Courtesy of Tim Gilles)

Disc pads occasionally wear unevenly. If one lining is worn more than the other on a floating caliper disc brake, a caliper slide is probably rusted or dirty. This prevents the caliper from floating freely.

Some tapered wear on disc pads results from the tendency of the caliper to twist during braking. This is considered normal as long as the taper is less than ⅛ inch (**Figure 7.19**).

When a symptom such as noise or pull is present, the caliper must be removed for a more thorough visual inspection of the pad. Sometimes disc pads must be replaced because they have worn unevenly or are heat damaged and/or cracked (**Figure 7.20**). The rotor will often suffer damage as a result, so inspect it carefully.

Rotor Inspection

Inspect the rotor to see that it does not have visual wear. Scoring and unusual wear will be visible as the rotor is turned. Feel it with your hand.

While rotating the rotor, rest a screwdriver or pencil on the caliper mount and hold it near the rotor as a reference while watching for visible runout. Runout that can be seen visually is already excessive and cor-

Figure 7.20 Pad damaged by excessive heat. (Courtesy of Tim Gilles)

allow as much as 0.008 inch. Some late-model cars specify as little as 0.0005 inch (½ thousandth) of runout.

NOTE: *Runout can be a result of hub distortion caused when lug nuts are tightened with an impact wrench. It can also be caused by dirt or rust on the mating surfaces of the rotor and hub. Be certain these areas are clean (**Figure 7.21B**).*

A runout measurement is taken off the outboard surface of the disc brake rotor using a dial indicator (**Figure 7.22**). Attach the indicator to a rigid part. If the rotor has a serviceable wheel bearing, temporarily remove end play at the adjusting nut. You do not need to tighten the bearing. Simply remove any looseness to ensure the accuracy of this test. A roller on the end of the dial indicator provides for a more stable measurement as the rotor is turned against it (**Figure 7.23**).

Figure 7.22 A dial indicator is mounted on a rigid part to measure rotor runout. (Courtesy of Tim Gilles)

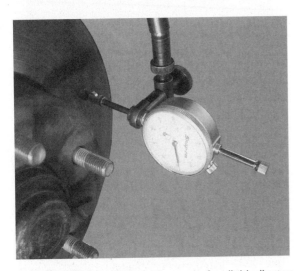

Figure 7.23 A roller on the end of a dial indicator reduces friction and provides a more accurate measurement. (Courtesy of Tim Gilles)

Should Rotors Always Be Machined?

A rotor in good condition will probably provide a better surface than a freshly machined rotor. A used rotor surface on properly operating brakes will be smooth and have some friction material embedded within the pores of the metal. A used rotor can provide better stopping performance than a new one.

When a used rotor is not machined, it is important that the corresponding rotor on the other side of the vehicle not be machined. Otherwise a brake pull can result. Rotors are *always* machined in pairs. If there is any doubt as to the quality of a rotor's surface finish and straightness, remachine the rotor. See Chapter 8 for more information on rotor machining.

Inspecting the Caliper

Carefully examine the caliper for signs of problems. Check the caliper piston dust boots to see if they are torn. Damaged dust boots will allow in moisture that will ruin the caliper. There should be no sign of fluid leakage from the caliper piston seals.

Replacing Disc Linings

Disc linings are usually easy to replace. Sometimes the linings can even be replaced without removing the caliper from its mount. This type of caliper will have an access hole in the top through which the pads can be removed and installed (**Figure 7.24**).

Most caliper designs require removal of the caliper to remove the pads. **Figure 7.25** shows typical pin slider and center abutment caliper removal. Some center abutment calipers use a tapered pin between the ways. **Figure 7.26** shows removal of this caliper.

Some calipers are easily removed by loosening two bolts on the back of the spindle support. Sometimes

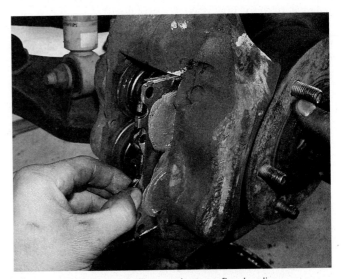

Figure 7.24 Installing pads on a fixed caliper with access through the top of the caliper. (Courtesy of Tim Gilles)

Chapter 7 Disc Brake Fundamentals and Service **151**

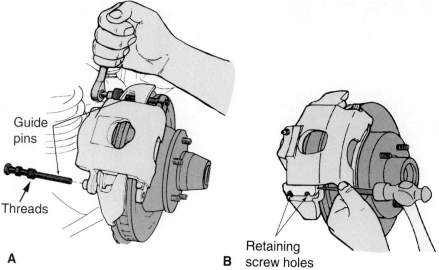

Figure 7.25 Removing a caliper: (A) pin slider; (B) center abutment. *(Courtesy of Federal Mogul Corporation)*

Figure 7.26 This caliper uses wedges between the ways on the caliper and its support. View A shows a special tool for removing the caliper slide. Be sure all of the parts are clean and free of rust. *(Courtesy of Tim Gilles)*

there are other bolts as well. Removing these bolts when doing a simple pad replacement or rotor turning will result in needless disassembly of the caliper (**Figure 7.27**).

When the caliper is unbolted, use wire or a hook to hold it so its weight is not allowed to hang on the hose (**Figure 7.28**). The hose can be damaged internally (see Chapter 5).

Before replacing linings, spin the rotor and inspect it for roughness on the front and back that could result from a worn-out brake pad. Be sure to check both sides of the rotor. Sometimes one pad wears out but the other does not.

Prior to installing replacement pads in a floating caliper check the condition of its slides. Move the caliper back and forth to verify that it can slide freely on its pins or ways.

When replacing linings on floating calipers, use a C-clamp or large pliers to move the piston back in its

Figure 7.27 When removing a caliper be sure to remove the correct bolts. *(Courtesy of Tim Gilles)*

Figure 7.28 Do not allow the caliper to hang from the brake hose. *(Courtesy of Tim Gilles)*

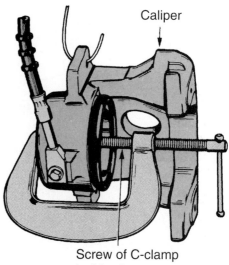

Figure 7.30 If the piston was not pushed into the bottom of a single-piston floating caliper's bore prior to caliper removal, a C-clamp can be used to do it now.

bore (**Figure 7.29**). This allows the caliper to be removed more easily. It also allows the new linings, which are considerably thicker than the worn ones, to fit during reinstallation of the caliper.

NOTE: *Before attempting to push the piston back in its bore, open the bleed screw on the back of the caliper. Then, move the piston all of the way back into its bore. Tighten the bleed screw immediately so it is not accidentally left loose. If left loose, the master cylinder will empty of fluid.*

Opening the bleed screw before retracting the piston is very important. Rust and sediment result as moisture accumulates in the brake fluid. Disc brake calipers and wheel cylinders are the lowest points in the hydraulic system and tend to be the dirtiest areas. When performing a simple brake pad replacement, some unknowing technicians will bottom out the pistons in the caliper bores without first opening a bleeder screw. This forces the sediment from the low

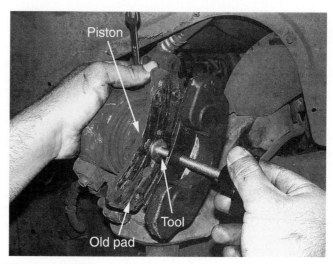

Figure 7.31 A piston compressor used with the old pad to compress the piston. *(Courtesy of Tim Gilles)*

parts of the system back into the ABS and master cylinder reservoir.

When the caliper has been removed from the vehicle, a floating caliper piston can be pushed into its bore as shown in **Figure 7.30**. Attempting to compress pistons in a fixed caliper or a two-piston floating caliper presents a problem. Pushing in on one piston results in the other piston coming out of its bore. All pistons must be compressed at the same time. A tool for this purpose is shown in **Figure 7.31**. The old pad is used to assist in this operation. Another tool for fixed calipers and multiple piston calipers is shown in **Figure 7.32**.

Rear Disc Brake Cautions

Servicing and repairing rear-wheel disc brakes calls for a few special considerations. As with any repair that is

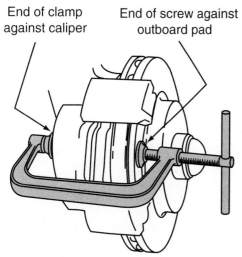

Figure 7.29 Use a C-clamp or a large pair of pliers to move the piston back in its bore.

Figure 7.32 This tool spreads the pads to compress pistons in fixed calipers and two-piston floating calipers. (Courtesy of Tim Gilles)

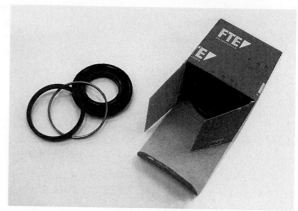

Figure 7.33 A caliper rubber parts kit. (Courtesy of Tim Gilles)

new to you, be certain to consult service literature before beginning your repair.

Rear-wheel disc brakes have a built-in parking brake consisting of either a miniature drum brake located inside of the disc rotor or an integral mechanism that clamps the disc brake pads to the rotor (see Chapter 9). The latter design will have either a screw or a cam attached to a lever on the back of the caliper. If there is a lever, do not try to force the piston into the bore. Check the service information for instructions before proceeding.

When the parking brake has not been used regularly, the self-adjusting mechanism can sometimes rust, freezing up the caliper. These calipers are complex and there is a better than average chance that parts will not be reusable or readily available at a reasonable cost. Therefore, many shops install rebuilt units instead of servicing them themselves. If you use a rebuilt caliper be sure to keep all of the parts on the old caliper so it will be acceptable as a core return.

The drum-in-hat rear disc parking brake design (with a miniature drum brake) should last the life of a vehicle. The thickness of the parking brake lining need only be sufficient to hold the car. It need not dissipate heat or resist wear because it is used solely to hold the vehicle when parked. It only suffers wear or damage if the vehicle is driven when the parking brake is partly applied.

Disc Caliper Rebuilding

A caliper rubber parts kit (**Figure 7.33**) contains a new boot, a piston seal, and sometimes rubber O-rings (for sliding caliper bolts). A square-cut piston seal is held in position by a groove in the bore (see **Figure 7.2**). A rubber boot keeps contaminants away from the sealing area.

Caliper Disassembly

To disassemble a caliper using compressed air and a rubber-tipped blowgun, position a piece of wood or a folded shop towel between the piston and caliper (**Figure 7.34**). Some technicians like to place a shop towel over the piston to prevent brake fluid from spraying onto them.

NOTE: *A typical blowgun for general shop use is OSHA-approved and will only provide 35 psi of air pressure. A rubber-tipped blowgun will provide full shop air to the caliper. Chapter 2 describes the two types of blowguns.*

CAUTION Before applying air pressure, be sure that your fingers are out of the way. Apply pressure in short bursts to gently push the piston from the bore. Full shop air pressure from a rubber-tipped blowgun can pop the piston out with dangerous force. Pressure behind the piston will become equal to shop air pressure and output force will be multiplied by the area of the piston (remember Pascal's law).

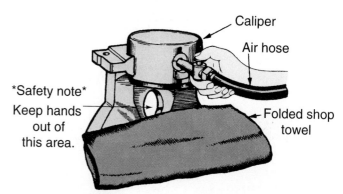

Figure 7.34 Use compressed air and a rubber-tipped blowgun to remove a caliper piston.

SHOP TIP Occasionally, shop air pressure will not be enough to move a piston that is stuck in a caliper. Typical shop air pressure usually ranges between about 100–130 psi, but the master cylinder is capable of producing several hundred pounds of pressure. Reconnect its hose and use the high pressure of the brake hydraulic system to free the stuck piston.

If a caliper has more than one piston, position a piece of wood against the loose piston(s) to prevent it from coming all the way out of its bore before the stuck one does. If one piston is out of its bore, the air pressure needed to push the stuck piston from its bore will not be able to build but will leak out through the other piston's open bore (Pascal's law, once again).

There are special tools available for removing stuck pistons manually (**Figure 7.35**). When a piston is stuck in a caliper bore, there is a good chance its surface finish is corroded. In this case, many shops will elect to purchase a rebuilt caliper rather than attempt to rebuild it.

Rebuilt Calipers

Many shops find it more cost-effective to install rebuilt calipers rather than rebuild calipers themselves. Loaded and unloaded calipers are available from re-

SHOP TIP
- When a piston is stuck in a caliper, purchasing a loaded caliper will probably be your best alternative.
- Compare the friction codes on the boxes to make sure the pads are the same for the left and right calipers.

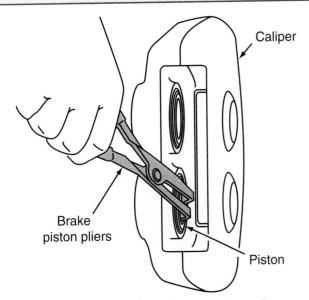

Figure 7.35 A special tool for removing caliper pistons.

Figure 7.36 A loaded caliper. *(Courtesy of Tim Gilles)*

manufacturers. Unloaded calipers are also called *bare calipers*. The loaded caliper comes assembled with new friction pads, hardware, and shims (**Figure 7.36**).

Cleaning and Inspecting Caliper Parts

Remove the piston and dust boot from the caliper. Some dust boots need to be pried from the caliper (**Figure 7.37**) and others come off easily. Remove the old piston seal from the bore (**Figure 7.38**). Thoroughly clean the bore. **Remember:** The bore is not a sealing surface so it does not need to be honed like a drum brake wheel cylinder bore. In drum brake cylinders, the seal rides on the surface of the cylinder bore. With disc brakes, the caliper pistons are the sealing surface rather than the bore in the caliper. The caliper bore does need to be clean, however, and there are special honing tools available to use for this purpose.

Inspecting the Pistons

Inspect the pistons for scratches, rust, and corrosion. The outer surface of the piston slides against the

Figure 7.37 Removing the dust boot from the caliper.

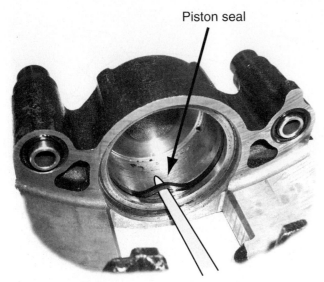

Figure 7.38 Removing the old seal from the caliper bore.

square O-ring seal in the caliper bore. If there are flaws in the surface finish of the piston (**Figure 7.39**), it must be replaced. According to brake manufacturers, chips or flaws are allowed near the top of the boot groove if they are less than ½ inch in length and are no deeper than the bottom of the boot groove.

Varnish can be cleaned from the piston using spray brake parts cleaner.

> **SHOP TIP** If you use emery cloth to clean a piston, its polished sealing surface will be ruined. Stubborn varnish buildup can be cleaned with crocus cloth. Crocus cloth, unlike abrasive emery cloth, is a very fine polishing cloth that does not leave scratches or grit.

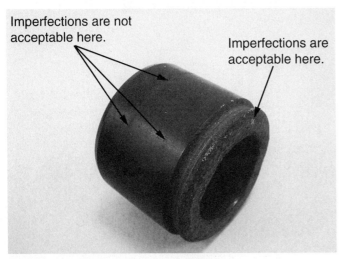

Figure 7.39 This phenolic caliper piston has imperfections in its sealing surface and must be replaced. *(Courtesy of Tim Gilles)*

Caliper Reassembly

There are different procedures for caliper reassembly. **Figure 7.40** shows typical parts for floating caliper assembly. Commercial brake assembly lubricants make reassembly of disc brake calipers easier, but brake fluid can also be used. Apply a liberal amount of lubricant to the piston seal, install it in the channel in the bore, and push on it to seat it (**Figure 7.41**).

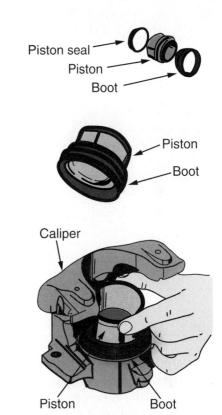

Figure 7.40 Caliper reassembly.

Figure 7.41 Lubricate the new seal and install it in the caliper.

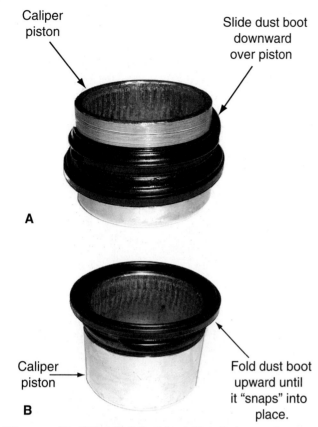

Figure 7.42 (A) A dust boot installed on a piston. (B) Fold the boot back to prepare the piston for installation.

Sometimes the outer lip on the dust boot is installed in its place on the piston before pushing the piston into the bore. The boot is installed on the piston until it seats in the groove in the piston. Then it is pulled down, turning it inside out (**Figure 7.42A**). The seal is pulled back over itself into a position ready for installation into the bore (**Figure 7.42B**).

Boot installation procedures vary. On some calipers, the seal is installed into position on the caliper prior to installing the piston in its bore. Hold the piston above the bore as you fit the outside of the dust boot to the caliper (**Figure 7.43**). Then push the piston into the bore.

Push the piston into the bore, carefully and gently working it back and forth as it slides past the seal until it reaches the bottom of the bore. This is done to make enough room for the pads when the caliper is reinstalled.

SHOP TIP When reassembling some large bore single-piston calipers, it is helpful to apply short bursts of air into the caliper's fluid inlet to help square up the piston and inflate the dust boot while working the piston down into the bore.

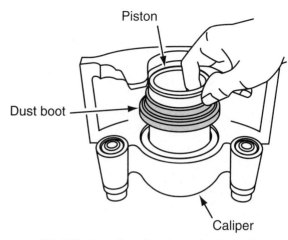

Figure 7.43 Installing the piston in the caliper.

NOTE:

- *If force is required, something is wrong and you will ruin the seal if you continue trying to install the piston.*
- *Once a caliper has been disassembled, a new seal must be used. The old seal will usually expand, making reassembly impossible.*

Some caliper boots are installed to the caliper after the piston is installed in the bore. After the piston is installed, silicone grease is applied to the area where the caliper boot will seat prior to installing the boot (**Figure 7.44**). The seal is pushed into the bore and the boot is put into position (**Figure 7.45**). The correct sized tool is selected from the tool kit and used to pound the boot into place (**Figure 7.46**).

Installing Disc Pads in the Caliper

Most fixed calipers use interchangeable inner and outer pads. When working with fixed calipers, pads

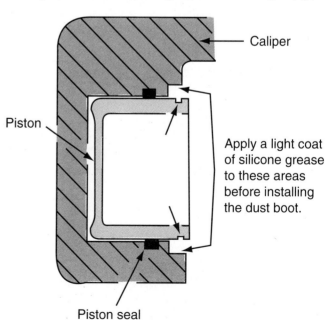

Figure 7.44 Lubricate these points prior to installing the dust boot in the caliper.

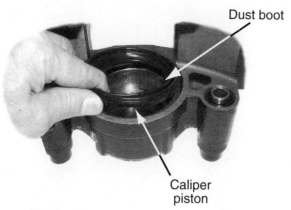

Figure 7.45 Push the piston into the bore and install the dust boot into the caliper.

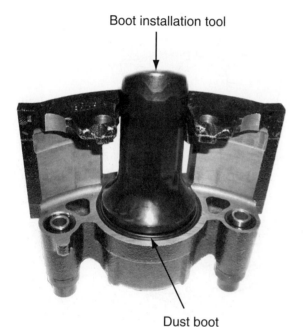

Figure 7.46 A special tool is used to pound the boot into place.

Figure 7.47 Two types of anti-rattle clips. *(Courtesy of Tim Gilles)*

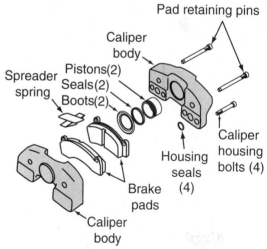

Figure 7.48 Parts of a two-piston fixed caliper. Do not separate the caliper halves unless a caliper rebuild kit is used.

can usually be slid in from the top after caliper installation. They are kept in place by one or two steel pins. Each pin is held in place by a retaining clip or small cotter pin.

When the brakes are released, the pads are free to move. An anti-rattle bracket or clip holds tension against the pads and pins to prevent vibration and noise (**Figure 7.47**). In some calipers, the inside pad floats and the outside pad is fastened tightly to the caliper.

When replacing pads on fixed calipers, remember not to disassemble the two caliper halves. There are rubber seals between the caliper halves that will need to be replaced if it is disassembled. **Figure 7.48** shows the parts of a two-piston fixed caliper.

Inner and outer pads on floating calipers are usually different. Sometimes pads are designed to be used only on one side of the vehicle. Be sure to install them on the correct side. When a pad is installed backwards and the brake wear sensor is at the leading edge rather than the trailing edge, the sensor might not make noise when the pads are excessively worn.

NOISE PREVENTION

Brake noise is a common complaint from customers. Vibration is the leading cause of disc brake noise. During brake application, the lining sticks to the rotor momentarily before slipping and then sticking again. This causes a high-frequency vibration that results in annoying squeaks and squeals.

Noise during a stop can include squeaks, metal-to-metal sounds from excessive wear, rattles due to loose parts, and rubbing from a distorted backing plate. A very common cause of brake noise occurs when the metal brake pad back vibrates against the metal caliper piston. Harder lining materials have a tendency to make noise, especially when cold, but they last longer and provide better hot stopping than softer linings. Many newer vehicles come from the factory with semimetallic linings that are more prone to low-frequency vibration and noise. Some consider a small amount of noise from these linings to be normal.

It is important that disc linings be firmly attached to the apply piston or caliper to avoid vibration. Some floating calipers have tabs on the pads that hold them tightly to the caliper. These are adjusted to fit, using either a hammer (**Figure 7.49**) or pliers (**Figure 7.50**).

Mounting Hardware and Lubricants

Most pads have small parts, often called anti-rattle clips, that position the pad and keep it from moving in the caliper (**Figure 7.51**). Some are attached to the back of the pad and others work off guide pins. This hardware is serviced when brake pads are replaced. Sometimes replacement pads come with new hardware, especially with original equipment parts. Some

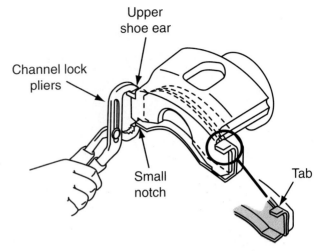

Figure 7.50 Crimping the outer shoe ears tightly to the caliper.

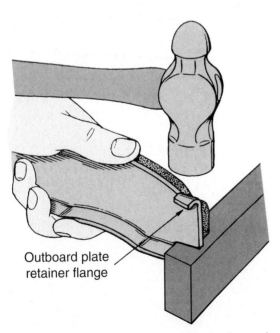

Figure 7.49 Using a hammer to fit the disc pad to the caliper.

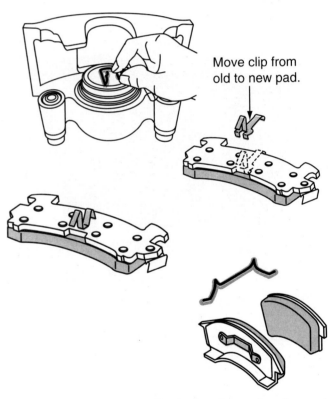

Figure 7.51 Various disc brake clips and springs.

aftermarket brake manufacturers do not provide new hardware with their brakes. The old hardware is reused or must be separately ordered if damaged or broken. Damaged or missing anti-rattle hardware can cause a brake to squeak or drag.

Aftermarket glue/insulator materials can be spread on the metal back of the pad before it is installed in the caliper (**Figure 7.52**). After it gets hot and vulcanizes the parts to one another, it helps prevent vibration and dampens noise. With high heat, these materials can sometimes be short-lived, however. Some newer

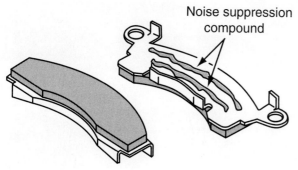

Figure 7.52 Noise suppression compound spread on the back of the pad.

calipers come with dampening materials sandwiched between stainless steel shims.

Lubricants and Shims

Lubricants and/or shims (especially with imports) are often provided by the manufacturer (**Figure 7.53**). Some shims have a rubber insulation coating. Others use thick grease or anti-seize compound applied to the mating sides of the steel brake pad backing and shim. This dampens vibration that could result in brake squeal. Careful installation of the shims and lubricant is important. Be sure to coat both sides of the shim with lubricant and follow specific instructions that are supplied with them.

Metal-to-metal contacts must be lubricated to prevent noise. Lubricant is always applied to caliper sliding surfaces. Various lubricants include synthetic brake grease, silicone, anti-seize compound, or molybdenum-disulfide (called "moly lube"). A satisfactory brake lubricant must be suitable for use in a high-temperature environment.

NOTE: *Do not use wheel bearing grease on brake parts.*

Be careful not to allow lubricant to contaminate the friction surface of the pad. Prior to applying lubricant, carefully clean the contact surfaces of the ways on center abutment calipers with fine sandpaper. These surfaces are designed to be separated by a very small clearance.

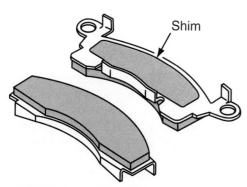

Figure 7.53 Insulating shims and/or lubricants are used to dampen vibration and prevent noise.

Rear Disc Pad Installation

When installing pads on rear disc brakes with an emergency brake that is not a separate drum within the rotor, the piston will have notches that must be aligned with the pegs protruding from the back of the pad (**Figure 7.54**). These keep the piston from rotating when the parking brake is applied. If only the pads are being replaced and the caliper is not being rebuilt, the piston must be screwed back into the caliper before the pads can fit. You cannot force them using a C-clamp as with front brakes. **Figure 7.55** shows a tool that is used to turn the piston into the caliper.

Figure 7.54 (A) Note the pin on the back of a pad used with a threaded caliper piston. (B) A threaded-type rear disc brake piston. *(Courtesy of Tim Gilles)*

Figure 7.55 A tool used to retract the piston. *(Courtesy of Tim Gilles)*

Figure 7.56 Torque caliper mounting bolts to specifications.

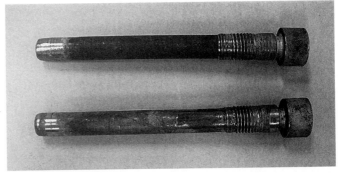

Figure 7.57 Worn caliper guide pins. *(Courtesy of Tim Gilles)*

Caliper Installation

Install the caliper and verify that it moves freely. Use a torque wrench to tighten the bolts that hold it to the steering knuckle (**Figure 7.56**). Most caliper bolts are internal hex, Allen head, or Torx®. Use the appropriate adapter with the torque wrench. Torque specifications are found in vehicle brake specifications booklets (see Chapter 1). Brake calipers vibrate and are prone to coming loose if not tightened properly.

Floating calipers must be able to slide freely on their mounts. A pin slider floating caliper will have some type of bushing to insulate the guide pins that the caliper slides on. These O-ring bushings along with the other rubber parts are usually included in caliper kits and are important for noise reduction. When they require lubrication, use the correct lubricant supplied with the brake kit. Do *not* use a petroleum-based lubricant that might damage a rubber bushing.

Be aware of the design and operation of the caliper. Check all sliding surfaces to see that they are not rusted and are in good condition. Inspect sliding caliper guide pins. Worn guide pins (**Figure 7.57**) can cause a brake to drag or squeak after it is released. If guide pins are not clean and free of rust, they must be replaced.

 **CAUTION** Immediately after installing both of the calipers, be sure to apply the brake pedal. Only hand pressure is necessary, so this can be done through the open driver's door while the car is on the lift. Sometimes more than one application of the pedal is required to push the linings against the rotor. If the car is backed off the lift before checking the brake pedal, an accident can occur.

Bushings and sleeves often become so gummed up that the caliper cannot slide easily. This can cause one lining to wear more than the other. Another possible result is a brake that will not release all the way after a stop. This produces an intermittent annoying squeal that goes away whenever the brakes are applied.

■ DISC BRAKE PROBLEM DIAGNOSIS

Several complaints can result from disc brake problems, including noise and braking and handling problems. Brake noise is the most common complaint. The following are some causes and solutions to disc brake problems.

Glazed Linings

Brake linings are not completely cured until a vehicle has been driven several hundred miles. Overheating linings when they are new damages the resins in the lining. The resin softens and solidifies on the surface of the linings, resulting in a glaze that causes brake squeal. Caution new owners not to use the brakes hard for the first 100 miles of city driving. Newer linings are more thoroughly cured than older ones were.

Glazed linings are *not* the primary cause of a squeal. Brake squeal is caused by vibration. The rotor and the linings both have a smooth and polished texture. Sanding a glazed lining might make the noise go away for a short time because the texture of the pad has been temporarily changed, but the noise will return once the pad has worn to its normal texture once again. If brake pads are glazed, they must be replaced.

Rotor Problems

Rotor wear problems or incorrect machining can cause several different complaints which are covered here. Hard spots on rotors can cause squeaks. Hard spots occur when the rotor becomes overheated. Hard spots and other material specific to disc rotors are covered in Chapter 8.

A very common cause of brake noise is incomplete brake machining. When a rotor is machined on a brake lathe, small metal chips are left on the brake rotor sur-

CASE HISTORY

A customer bought a used half-ton pickup truck with oversized tires. He brought the truck into a repair shop, complaining of very high brake pedal effort during stopping. The shop tested his power brake unit and felt that it was good. The customer insisted that the truck occasionally stopped normally so the shop replaced the brake booster and the problem persisted. The shop installed a smaller set of wheels and tires from another vehicle, had the customer test-drive the vehicle, and the problem was solved.

face. After a rotor is machined, it is sanded to produce a nondirectional finish (see Chapter 8). Following rotor sanding, it is very important to wash the rotor before installing it on the vehicle. Washing the rotor removes metal dust. If the dust is allowed to remain on the rotor surface, it will become embedded in the brake pads. The result is annoying brake noise for the rest of the life of the pads.

Pedal Pulsation

During your study of drum brakes, you learned that pedal pulsation can result from an out-of-round brake drum. Pedal pulsation can also result when disc brake rotors have been overheated or machined incorrectly.

Rotor runout and thickness variation are often the result of incorrect torquing of the wheel lug nuts. During a stop, the brake pedal will pulsate or vibrate if the piston moves in and out of the caliper. Chapter 8 describes how rotor runout results in wear that causes variations in rotor thickness. The piston moves in and out as the disc pads follow these thickness variations.

In extreme cases of rotor runout or when a wheel bearing adjustment is loose, the rotor can knock the pad against the piston while the brakes are released. This pushes the piston deeper into its bore. The pedal will have to travel farther and may go all the way to the floor the next time it is stepped on.

When floating caliper guides are sticky, runout of the rotor can also cause a grabbing brake.

Brake Pull

A disc brake does not have return springs like a drum brake. Following a stop, the distorted O-ring returns to its original shape, returning the lighter weight piston to its original position in the much heavier caliper.

A caliper can place thousands of pounds of force on the pads and rotor. Even when there is a sticky piston or a dirty or rusty sliding surface, the caliper will probably generate enough force to apply the brake. The piston, however, will not be able to release sufficiently to prevent the pads from dragging on the rotor.

Another cause of brake pull is when a vehicle has tires of unequal sizes, especially on the front. During braking, a car will pull to the side with a smaller diameter tire because the smaller diameter has a greater leverage exerted on it.

One cause of hard braking effort is the installation of oversized tires on a pickup truck. The leverage against the brakes makes the power brake and master cylinder ineffective for the new braking requirements.

Uneven Wear

A technician will sometimes find brake pads that are unevenly worn. A floating caliper frozen on its guide pins or sliding ways can cause this. Uneven wear can also occur when a fixed caliper has one or more frozen or lazy, sticky pistons. The following are uneven pad wear symptoms and results:

- When the outboard pad is excessively worn, look for a hardware problem as the cause.
- When the inboard pad is excessively worn, look for a sticking piston as the cause.

When brake pads on one side of the vehicle wear differently than those on the other side, look for an internally collapsed brake hose as a possible cause. To test for this, raise the wheels off the ground and have an assistant apply the brakes a few times. Right away after the brakes are released, try to turn the wheel on the side with the excessive lining wear. If the wheel is hard to turn, quickly loosen the caliper bleed screw and see if the wheel turns easily. Also watch to see if fluid squirts from the bleed screw. A disc brake system would not have residual pressure after brake release, so pressure must be trapped in the caliper due to a blockage somewhere between the brake caliper and master cylinder. This test only works if there is a severe blockage and might not work with a partly blocked hose or a hose with a loose flap that acts like a one-way check valve (see Chapter 5).

■ TEST-DRIVE AFTER A BRAKE JOB

It is very important to test-drive a vehicle that has had brake service before returning it to the customer. You should break in the linings in the correct manner and ensure that the vehicle is performing as desired. Never allow the customer to perform the test drive.

Check for correct pedal height. This is verified by applying the brake pedal a second time in rapid succession while watching for a rise in pedal height. If the pedal height rises significantly, the rear brakes require a more accurate adjustment. When rear brake adjustment is off by a large amount this can result in only the front brakes working, combined with lower than normal brake pedal height.

Seat the linings in the manner prescribed by the manufacturer. A typical break-in procedure involves braking the vehicle gently from 30 to 15 mph, allowing the brakes to cool for a couple of minutes after a few stops. Then, continue for a total of ten to fifteen stops. This will allow the pads to wear in and conform to the rotor surfaces. Breaking in the linings in the correct manner will also help avoid brake noise. It allows any uncured resins in the pad to finish curing without overheating, which would produce a glazed lining that squeaks or squeals. Advise the customer to avoid hard stops for the first week or so.

Before you return cars to customers, educate them regarding the type of brake adjustment they have. Depending on the type of brakes they have, remind them that they need to either apply the emergency brake often or to apply the service brakes when operating in reverse to keep the emergency brake hardware in operating condition.

REVIEW QUESTIONS

1. On a master cylinder with reservoirs of different sizes, which one is the larger one for, the disc or drum brakes?
2. _____ sensors are the most common type of wear sensors.
3. Two designs of disc brake calipers are the fixed or the _____ caliper.
4. Which caliper design usually requires a power assist if normal pedal pressure is to be achieved?
5. The _____ _____ is the most common type of disc brake caliper.
6. Which caliper slides on two wedged sliding surfaces called ways?
7. Caliper pistons are installed with which side against the back of the friction pad—the open or closed side?
8. When there is a pull during braking, swap the front tires. If the pull changes to the other direction, which are at fault—the tires or the brakes?
9. When there is a symptom present, such as noise or pull, the _____ must be removed for a more thorough visual inspection of the pad.
10. Before attempting to push the piston back in its bore, open the _____ _____ on the back of the caliper.
11. A _____ caliper comes assembled with new friction pads, hardware, and shims.
12. _____ is the leading cause of disc brake noise.
13. If caliper _____ pins are rusted, they must be replaced.
14. _____ a glazed lining might make the noise go away for a short time because the texture of the pad has been temporarily changed, but the noise will return once the pad has worn down to its normal texture once again.
15. Following rotor machining and sanding, it is very important to _____ the rotor to remove metal dust that will become embedded in the brake pads.

PRACTICE TEST QUESTIONS

1. Technician A says disc brakes do not require a return spring. Technician B says the piston seal pulls the piston back to release the rotor. Who is right?
   A. Technician A   B. Technician B
   C. Both A and B   D. Neither A nor B

2. Technician A says front disc brakes require no self-adjusters. Technician B says hard braking effort can be caused by the installation of oversized tires. Who is right?
   A. Technician A   B. Technician B
   C. Both A and B   D. Neither A nor B

3. Technician A says disc brakes are far less likely to cause pull because the brakes on each side of the vehicle give the same braking effort. Technician B says disc brakes tend to be quieter than drum brakes. Who is right?
   A. Technician A   B. Technician B
   C. Both A and B   D. Neither A nor B

4. Technician A says when a car pulls to one side during braking, the other side's brakes might be malfunctioning. Technician B says brake pads on the side that is not working correctly are often worn less. Who is right?
   A. Technician A
   B. Technician B
   C. Both A and B
   D. Neither A nor B

5. Technician A says sometimes a pull caused by a tire will only occur when braking. Technician B says worn suspension bushings can cause a brake pull. Who is right?
   A. Technician A
   B. Technician B
   C. Both A and B
   D. Neither A nor B

6. Technician A says a freshly machined rotor will provide a better surface than a used rotor in good condition. Technician B says a used rotor can provide better stopping performance than a new one. Who is right?
   A. Technician A
   B. Technician B
   C. Both A and B
   D. Neither A nor B

7. Technician A says that if a rotor on one side of the car is machined, the other side must be machined. Technician B says if only one rotor is machined, a brake pull can result. Who is right?
   A. Technician A
   B. Technician B
   C. Both A and B
   D. Neither A nor B

8. Technician A says if you use crocus cloth to clean a caliper piston, its polished sealing surface will be ruined. Technician B says the caliper bore must have a finely honed finish for the new rubber parts to seal against. Who is right?
   A. Technician A
   B. Technician B
   C. Both A and B
   D. Neither A nor B

9. Technician A says inner and outer pads on floating calipers are usually identical. Technician B says sometimes pads are designed to be used on one side of the vehicle only. Who is right?
   A. Technician A
   B. Technician B
   C. Both A and B
   D. Neither A nor B

10. Front disc brakes have an intermittent annoying squeal that goes away whenever the brakes are applied. Technician A says to look for one lining worn more than the other. Technician B says the brake might not be releasing all the way. Who is right?
    A. Technician A
    B. Technician B
    C. Both A and B
    D. Neither A nor B

CHAPTER 8

Friction Materials, Drums, and Rotors

■ OBJECTIVES

Upon completion of this chapter, you should be able to:
- ✔ Explain the differences between the various types of brake friction materials.
- ✔ Understand federal original equipment (OE) and aftermarket brake standards.
- ✔ Interpret SAE brake lining edge codes.
- ✔ Explain differences in brake drums and discs.
- ✔ Diagnose drum and disc wear.
- ✔ Understand brake lathe principles.

■ KEY TERMS

finish cut
rough cut
scratch cut

■ INTRODUCTION

Service to drums and rotors was covered in the two previous chapters. This chapter deals with friction materials and the design, inspection, and machining of drums and rotors. Today's friction materials are quite sophisticated. In recent years, they have become vehicle-specific rather than "one size fits all."

Disc brakes are installed on almost all front wheels. Brake drums are found on the rear wheels of many cars, SUVs, and light trucks. They are also prevalent on all four wheels of most older vehicles. Rotors are found not only on front wheels but often on all four wheels.

■ DRUM AND DISC BRAKE LINING MATERIALS

Drum brake linings are commonly called *shoes* and disc brake linings are called *pads*. During the following discussion of friction materials, shoes and pads will be referred to as *linings*. Designations for types of brake linings include asbestos, nonmetallic organic, semimetallic, metallic, ceramic, and carbon. The basic friction material of a brake lining is the powder or woven fiber mat. Friction modifiers and fillers are held into the friction material by phenolic resins and binders.

Older brake lining materials were classified as organic or inorganic. Asbestos linings were called *organic* and semimetallic linings were called *inorganic*.

Vintage Chassis

Brakes in the earliest cars were the same as those used on the horse and buggy. Friction materials included wood, leather, and camel hair, materials which could start to burn when heated sufficiently. In the early 1900s braking was greatly improved with the development of a new friction material that combined asbestos and copper wire mesh. This new material did not burn and braking was greatly improved. The company that developed this new material was called Raymond, which later became the Raybestos Brake Corporation.

Asbestos has not been used for original equipment (OE) brakes since the early 1990s, although it is still available as an imported aftermarket lining material.

Dictionary definitions of *organic* include "chemical compounds containing carbon," or "something derived from a living organism." The "organic" designation does not seem to be an accurate one for brake linings, but it is commonly used. Newer OE friction material classifications include *semimetallic, nonasbestos organic (NAO),* and *ceramic.*

Semimetallic Linings

Semimetallic linings, original equipment on some new front-wheel drive cars, include a mixture of sponge iron and steel wool fibers to add strength and temperature resistance. Semimetallic linings are very good for quickly removing and absorbing heat from the rotor or drum. The hotter they get, the better they work and the heat transfer does *not* affect the service life of the lining. Traditional semimetallic linings contain 60 percent steel by weight. For noise reduction, low-metallic linings with less than 30 percent steel have become popular, especially in European cars.

The following are some disadvantages of semimetallic linings:

- As semimetallics wear, a dark brown dust develops on the front wheels. The dust is a combination of graphite from the lining compound and oxidized metal (rust) (**Figure 8.1**).
- They have a tendency to make noise.
- They tend to wear rotors more quickly.

Nonasbestos Organic Linings

Nonasbestos organic (NAO) linings are made without iron or steel. Most NAO linings use aramid fibers or fiberglass. They have many more ingredients than semimetallic linings so their engineering development is more difficult. NAO linings are most often found on low-temperature disc pads and drum brake linings.

Fully Metallic Linings

Fully metallic linings have been used for many years on vehicles that must operate under very heavy-duty conditions and in auto racing. These linings have the best fade resistance of any lining, but they require higher pedal pressure and cause more drum and rotor wear than other types of linings. They are effective for extremely high-heat applications but work *very* poorly when cold.

Metallic linings are fused together using a process called *sintering*. Using heat and pressure, this process forms powdered metal without melting it. Bonded metallic linings can de-bond from the shoe backing under the intense heat generated during severe brake loads. To prevent this, metallic drum brake linings are often cut into three segments. Holes are drilled in them and countersunk to accept the installation of brass screws.

Ceramic Linings

Ceramic linings were first installed on some new vehicles in 1985. Today, about half of all vehicles have ceramic linings as original equipment and aftermarket ceramic linings are also available. Rather than steel wool, these linings use ceramic and copper fibers to control heat. The ceramic material dampens some of the noise inherent in semimetallic linings. Any vibrations that remain are at a frequency above the range of human hearing. Besides reducing noise, an added benefit is that the brake dust created during lining wear is light in color and does not rust like steel.

Ceramic linings have more stable friction characteristics than semimetallics. Their coefficient of friction does not drop as fast as they become hot. Ceramic lining compounds are more complex, however. Some ceramic compounds have twenty different ingredients. Testing and development are complicated, which in turn, makes ceramic linings more costly to produce. Sometimes other types of linings work as well or better.

Carbon Fiber Linings

Carbon fiber linings, or carbon fiber reinforced carbon (CFRC), are premium linings used on police cars, race cars, and aircraft. They are semimetallic composites consisting of a carbon mix reinforced with embedded high-density hollow carbon fibers instead of powder. The pads are integrally molded to the metal back to reduce noise (**Figure 8.2**).

> **SHOP TIP** When installing metallic linings on a motor home or light truck to increase fade resistance, install them on one axle or the other but not on all four wheels. This will permit a safer amount of braking when the brakes are cold.

Figure 8.1 Dark brown dust on the wheel is a result of semimetallic pad wear. *(Courtesy of Tim Gilles)*

Figure 8.2 When pads are integrally molded, the lining material is visible through holes on the back side of the lining. *(Courtesy of Tim Gilles)*

Carbon fiber linings are more expensive than other linings, but their coefficient of friction remains constant as temperature changes. Unlike full metallics, they do not suffer from a nearly nonexistent coefficient of friction when cold. They are also quiet and have a low rate of wear, sometimes twice that of normal linings.

Noise Control, Performance, and Wear

Brakes are designed to be *application-specific*. This means they are designed to fit a particular vehicle. There are many compounds used in friction materials so linings can be very different from one vehicle to the next.

Changing the ingredients used in making the pad can help suppress noise. NAO, low-metallic, and ceramic pads have less tendency to produce noise. Ordinary semimetallic linings are prone to produce noise as the steel material in the pad rubs against the iron rotor. Brass, copper, or ceramic fibers used instead of steel or iron help quiet the linings without affecting the life or performance of the lining.

When softer linings are used for noise reduction, a shorter service life can be expected. Also, replacement brakes for vehicles equipped with ABS must be very close to the friction coefficient of the OE lining or the antilock brake system calibration might not be accurate.

Due to weight transfer, front linings are expected to wear out about twice as often as rear linings. If front brakes are being replaced for the second time but the rear brakes are still in a like-new condition, they might not be working correctly. Check for correct rear pressure that could be blocked by a load-sensitive proportioning valve (see Chapter 5) or brake linings that are the wrong friction material for the application.

Lining Selection

Most brake producers provide linings of several levels of quality and performance. Premium linings are original equipment. They offer good quality stopping power, low noise and dust, and are relatively inexpensive.

Premium brake pads are acceptable for everyday use but not for off-road competition or racing. The demands of repetitive, high-speed stopping found in these situations are beyond the capability of premium pads. Upgraded linings, called "performance street," are more performance-oriented. Owners of pickup trucks and commuter SUVs occasionally used off-road or for trailer towing sometimes choose these brake linings. They have higher pedal sensitivity, stopping power, and resistance to brake fade.

Compromises with stock pads include faster wear to the pad or rotor, increased noise, and a moderate increase in brake dust. Brake dust is released as the friction material in a standard brake lining "carbonizes" during normal braking temperatures. Some performance brake linings give off significantly less dust in street driving because they resist carbonizing until heated to over 1,000°F.

High-performance/track pads have the best stopping power and fade resistance, even after repeated high-speed stops. They are metallic and must be hot before they can provide maximum stopping power. These linings are good at the race track but cool down too fast between stops to be used on the street. Other shortcomings of these pads include significant increases to pedal pressure requirements, rotor and lining, brake dust, and noise.

■ EVALUATING FRICTION MATERIALS

Friction materials that appear to be the same can perform very differently. The following section lists ways that friction materials are evaluated.

SAE Brake Tests

The Society of Automotive Engineers (SAE) has developed tests for the evaluation of friction materials so manufacturers can compare their lining performance to the FMVSS 135 minimum performance standards for new brakes. SAE J1652 was developed in 1994 to test front disc brakes. SAE J2430 came out in 1999 for evaluating the combined performance of front and rear brakes.

Aftermarket Friction Material Certifications

FMVSS 135 applies only to new car production. Industry-based certification programs have been developed to assure customers that aftermarket linings have performance characteristics equal to or better than the new car original equipment manufacturer (OEM) standards set by FMVSS 135. One part of the FMVSS 135 standard maintains the same stopping distance requirements with 25 percent less pedal effort. To meet this requirement, aftermarket manufacturers have had to change their friction materials.

If older aftermarket linings are used on cars manufactured under FMVSS 135 standards, the car will not stop as fast as it did with the OEM linings. According to a National Highway Traffic and Safety Administration (NHTSA) study in 1990, the average aftermarket friction material was significantly different from OEM materials. The result was an increase in stopping distance of 10 percent or more.

Prior to the development of these evaluation programs, brake material was assumed to meet safety standards. Unfortunately, there were no federal standards for aftermarket brakes and inferior materials were sometimes supplied to unsuspecting customers. To ensure industry credibility, aftermarket manufacturers needed a means of certifying that their friction materials were consistent with the federal safety standards required of original equipment.

The following are two examples of industry programs developed to give technicians and consumers a means of evaluating aftermarket friction materials. These dynamometer testing programs replaced an older road test, where road conditions and the actions of the driver could create variations in results. A dynamometer is to a car what a treadmill is to a human. With a vehicle in a special dynamometer test bay, the wheels can be spun and braking tested under controlled conditions.

D3EA Certification. The D3EA—dual dynamometer differential effectiveness analysis—certification (**Figure 8.3**) was developed and marketed in the late 1990s. The D3EA uses front and rear brake dyno testing verified by an independent test lab to evaluate how appropriate friction materials operate in different applications. The procedure determines whether the replacement friction materials are equal to or better than OEM standards.

Brake Manufacturers Council Certification. The Brake Manufacturers Council (BMC) program, called BEEP for brake effectiveness evaluation procedure, uses a single-end brake dynamometer to test brakes against OEM standards (**Figure 8.4**).

Lining Edge Codes

The side edge of a new brake lining is stamped with a code number established by the SAE (**Figure 8.5**). The intent of the SAE *edge code* is to provide a uniform means of identification and to rate the friction characteristics of different linings based on tests performed in a laboratory. The edge code does not indicate or guarantee quality. A lining with good stopping characteristics could wear out quickly or cause excessive wear to a rotor or drum.

There are three groups of numbers or letters in the edge code. The first is a letter that identifies the brake manufacturer. The second group is a manufacturer's combination of letters, numbers, or both that describes the materials used in the manufacture of the lining. It gives the manufacturer a way to identify the lining in case of a recall or warranty return.

The lining's coefficient of friction (CF) is described in the third group, which is two letters (**Figure 8.6**). The first letter is the cold coefficient of friction and the second letter is the hot coefficient of friction.

Figure 8.4 The Brake Manufacturers Council single-end brake dynamometer evaluation program, called BEEP for brake effectiveness evaluation procedure, tests brakes against OEM standards.

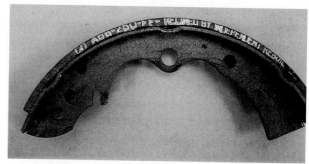

Figure 8.5 SAE friction codes are stamped on the edge of brake linings. *(Courtesy of Tim Gilles)*

Code C	0.00 to 0.15
Code D	0.15 to 0.25
Code E	0.25 to 0.35
Code F	0.35 to 0.45
Code G	0.45 to 0.55
Code H	0.55 and higher
Code Z	Ungraded

Figure 8.6 SAE coefficient of friction codes for brake linings. The first letter is the cold coefficient of friction (CF) and the second letter is the hot CF.

Figure 8.3 D3EA testing uses front and rear brake dynamometer tests. *(Courtesy of Raybestos)*

Most passenger cars will be EE or EF. Through the careful design of friction materials, brakes are designed to provide a good balance in performance between hot and cold stops. A narrow coefficient of friction change between the cold and hot temperature ratings is desirable.

A brake that does not maintain its original friction coefficient can grab or fade. A coefficient of friction lower than 0.15 (Code C) will result in a hard pedal. A high coefficient above 0.55 (Code G) will cause brakes to grab.

Breaking in New Linings

In the past, linings were not fully cured and had to be broken in easily as they finished curing. Today's linings are more fully cured but still require some break-in. To help brake linings finish curing and to seat them to the rotors or drums, accelerate to 30 mph and make twenty to thirty stops with medium to firm pressure. Do not overheat the brakes.

■ BRAKE DRUMS

Most brake drums are made of gray cast iron. Its high carbon content makes it very hard. (A diamond, which is pure carbon, is the hardest material on earth). The high carbon content of gray cast iron usually causes it to be highly resistant to wear from brake linings, but if a lining wears to the point where rivets or the steel rim of the shoe come into contact with the drum, rapid wear will occur.

Brake drums tend to trap the heat generated during braking. There are several design considerations that help rid them of heat, although heat conductivity from metal to air is relatively slow.

- Drums often have fins cast into their outer surfaces to help transfer heat to the surrounding air (**Figure 8.7**). The more surface area the drum has, the more heat it can conduct.

- Some drums are bimetal iron/aluminum to reduce weight and increase heat transfer.
- A heavier brake drum will not increase in temperature as much during a stop as a lighter one of the same material because it is able to absorb more heat. Wheel rims sometimes have finned center areas to increase the flow of air around the drum.

NOTE: *When 1 pound of cast iron absorbs 1 BTU of heat its temperature increases by approximately 10°F. During a stop, if 10 BTUs of heat are produced, a 1-pound brake drum will increase in temperature by about 100°F. If the drum only weighed 10 pounds, the temperature would only increase by about 10°F.*

■ BRAKE ROTORS

Rotors are made from a variety of materials including cast iron, composite, or aluminum composite. There are also stainless steel and plated iron rotors. Like brake drums, most rotors are made of gray cast iron because its high carbon content makes it very hard.

Disc brake rotors are either *solid* or *ventilated* (**Figure 8.8**). Disc brake components are part of a vehicle's unsprung weight. Because it affects vehicle handling, engineers try to keep unsprung weight as light as possible. Lightweight solid rotors are used in lighter cars. Ventilated rotors, used in larger cars and light trucks, are heavier, but they have abundant surface area so they can dissipate the extra heat generated in stopping heavier vehicles.

Sometimes the rotor and hub are one piece. This design is more costly, so a typical rotor and hub assembly has two pieces, which allows the rotor to be replaced without replacing the hub.

Directional Rotors

A ventilated rotor typically has straight fins extending from its center. Some ventilated rotors have fins that

Figure 8.7 Fins in the outer surface of the brake drum provide increased surface area to transfer heat to the surrounding air. *(Courtesy of Tim Gilles)*

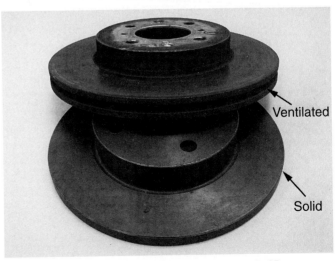

Figure 8.8 Rotors are solid or ventilated. *(Courtesy of Tim Gilles)*

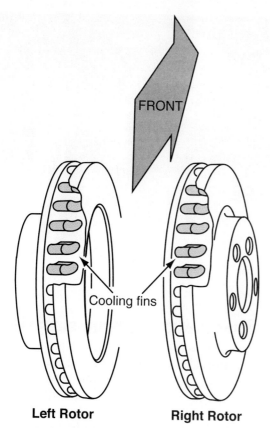

Figure 8.9 Directional rotors with curved fins.

are curved or angled to pump air more efficiently. These rotors are directional and are marked for installation on the left side or right side of a vehicle.

NOTE: *Left is viewed from the driver's seat, so a rotor marked R would be installed on the passenger side of the car.*

Some of these rotors are installed with angled fins pointing toward the rear of the car at the top of the rotor (**Figure 8.9**). Others are installed with the fins facing forward. It costs more money to produce directional rotors and the possibility of installation error is higher. Therefore, these rotors tend to be used in high-performance or luxury vehicles.

Ceramic Brake Rotors

Manufacturers have spent a good deal of money developing newer friction materials for brake pads. Rotor development is even more costly. In race cars and some high-end street sports cars, grey cast iron in rotors has been replaced with high-technology materials that combine ceramics in the casting. Cast iron rotors actually amplify noise because they vibrate at a frequency that is in the range of normal human hearing. The newer, less dense materials are not as likely to vibrate, and when they do, it is at a different frequency than cast iron.

Solid rotors can be made entirely of the same material, but vented rotors are of composite construction in which a steel hub is combined with a ceramic rotor. Ceramic rotors can weigh up to two-thirds less than a cast iron rotor of the same size. However, ceramic rotors are costly to make and manufacturers use different methods to produce them.

New design brake rotors made of carbon fiber, resins or sintered carbon powder, and ceramics are found on some high-end cars. They have all of the advantages of the other ceramic rotors and provide substantially reduced unsprung weight.

Composite Rotors

Composite rotors have been used on some vehicles since the 1980s. A composite rotor is lighter than a conventional cast iron rotor. Its friction surfaces are made of cast iron and the center of the rotor is thin stamped steel (**Figure 8.10**).

There are two noted advantages to composite rotors. They are about a pound lighter than a normal cast rotor, which reduces unsprung weight. This results in better steering feedback and reduced ride harshness. Also, a different type of iron that can reduce noise can be used in the rotor disc casting.

A disadvantage of composite rotors is that special adapters, which simulate a rotor being bolted to the hub and wheel, must be used when machining them. The alternative to this is to machine them on the vehicle. This is covered later in the chapter.

There were problems with some early composite rotors during the late 1980s which caused Ford to replace rotors on some of their cars with single-piece castings. Some brake companies have marketed one-piece cast rotors to replace composites in the aftermarket. This was popular with some technicians because they were easier to remachine. Replacing composites with solid rotors can be problematic, however.

When composite rotors need to be replaced, the best choice is to install new composite rotors. The stamped steel center section of the rotor can be up to

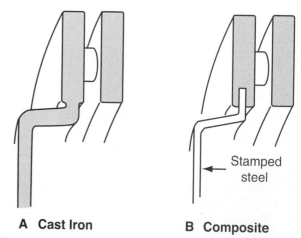

Figure 8.10 (A) A single-piece cast iron rotor. (B) A composite rotor is made up of a steel center section cast with an iron friction surface.

4 mm thinner than the same, thicker section of a single-piece cast rotor. The change in thickness results in the center of the front tires being farther apart. Two alignment considerations, scrub radius and camber (covered in Chapter 17), can be affected by this change, leading to increased steering harshness and tire wear, as well as more load on the outer wheel bearing. A cast rotor installed on only one side of a vehicle can produce a pull. If cast rotors are installed on both sides, the increase in scrub radius can result in more tire wear.

Metal matrix composite (MMC) rotors, used in the high-performance market, are made of an alloy of aluminum and magnesium and have a matrix reinforced with silicone carbide. MMC rotors combine aluminum's light weight and heat transfer advantages with the heat dissipation and wear properties of ceramics.

Alloys combine different metals into one. Alloys are more difficult to cast into an even structure, especially when they combine metals with different melting temperatures. Casting problems can be minimized by adding ceramic materials to the mix.

Early designs used expensive heat treatment methods to prepare the aluminum for its service as a braking material. A newer manufacturing method is alkali etching, which removes the top of the aluminum and exposes hard ceramic particles to the brake pads. When the rotor is new, the pad material embeds in its roughened surface. In theory, from that point the friction material of the pads never touches the metal surface of the rotor, so the rotor should not experience wear.

The friction surface of these rotors appears light gray in color, streaked with darker colors due to the transfer of friction material from the pads. The rotors are nonmagnetic so there is no rust.

Metal matrix rotors dissipate heat about four times as fast as cast iron rotors and are extremely resistant to distortion from heat. An increased amount of heat flows from the brakes to the wheels and engineers have discovered that aluminum wheels run about 200°F cooler than steel wheels. If stock wheels are replaced with aftermarket ones, be sure they include all the heat dissipation features of the originals.

Metal matrix rotors are very hard and diamond-cutting tools are recommended for machining them. They can be machined with carbide-cutting tools, but they will only work one time before they become dull.

Grooved and Drilled Rotors

A popular new rotor design in the high-performance sport aftermarket is disc brake rotors that are dimpled, drilled, or grooved (**Figure 8.11**). These design features help reduce brake fade caused when the pads give off gas under high heat operation, but they produce additional noise.

NOTE: *As the pads rub against the grooves and holes, they make a whirring noise that becomes less noticeable as the rotors and pads break in.*

Figure 8.11 This high-performance rotor has dimples and grooves. *(Courtesy of Tim Gilles)*

■ DRUM AND ROTOR SERVICE

The following section describes inspection and machining procedures for brake drums and rotors.

When rotors and drums are exposed to higher temperatures than they were designed to handle, many different changes result. Rotors and drums must be given a careful visual inspection for defects, which can include hard spots, cracks, and excessive wear that can call for replacement.

Hard Spots

Hard spots and heat checks (**Figure 8.12**) develop when a drum or rotor becomes overheated and carbon becomes unevenly dispersed in the cast iron. This can cause squeaks. Hard spots sometimes cannot be removed with a carbide tool bit because the high carbon area is as hard as the tool bit. In the days of four-wheel drum brakes, brake lathe fixtures were available for grinding hot spots. Grinding is not cost-effective. Machining will not "remove" the hard spots. Even when hard spots are machined, the metal that remains below the surface of the rotor has undergone a metallurgical change in its composition. The hard spot below the surface will not wear as fast as the rest of the rotor surface so it will reappear. Therefore, a drum rotor with hard spots must be replaced.

■ DRUM WEAR INSPECTION

Wear can be obvious (see **Figure 6.22A**) or it can be so minor in appearance that it must be measured to find a problem. Types of wear include bellmouth (**Figure 8.13**) and out-of-round wear (**Figure 8.14**).

Pedal pulsation results when hydraulic pistons move in and out during a stop. This happens when rear drum brakes have become out of round, which happens when overheated brakes have the parking brake set firmly against them.

Chapter 8 Friction Materials, Drums, and Rotors **171**

Drum Wear Limits

Limits to maximum brake drum diameter are set by law. Brake drums must be measured to be sure they are not worn beyond their maximum allowable diameter. An excessively worn drum will be too thin and can overheat.

NOTE: *A shop that reuses a drum that is worn or has been machined beyond the legal limit assumes liability in the event the vehicle is involved in an accident.*

In the past, drums could be machined to a diameter of 0.060 inch beyond the original size. This would still allow 0.030 inch wear before the drum reached its discard diameter (0.090 inch oversize). This older standard sometimes does not apply to today's vehicles.

Always read the outside of the drum or check the specification book for the discard diameter of a drum. This specification has been cast or stamped on brake drums since the 1972 model year (see **Figure 1.54**). According to General Motors service information, the discard diameter is the maximum dimension allowed *prior* to wear. It is *not* the maximum machining diameter. If the brake linings are to be able to wear out before the discard dimension is reached, the drum must have 0.030 inch left after machining to allow for future wear.

Drums are measured with a special brake drum micrometer as described in Chapter 1.

■ DISC ROTOR INSPECTION

The surfaces of rotors are inspected for heat and unusual wear in much the same way that drums are inspected.

Rotor Thickness and Runout Measurements

A minimum thickness, called a discard dimension, is stamped or cast into all rotors (**Figure 8.15**). A

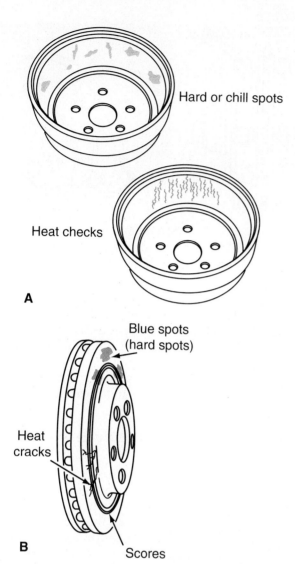

Figure 8.12 (A) Brake drum defects can include hard spots, heat checks, and cracks. (B) Possible rotor defects.

Figure 8.13 Bellmouth wear.

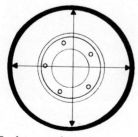

Figure 8.14 An out-of-round drum.

Figure 8.15 The minimum thickness specification is cast into the surface of disc brake rotors. *(Courtesy of Tim Gilles)*

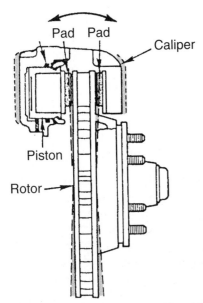

Figure 8.16 Rotor runout causes the piston to move in and out.

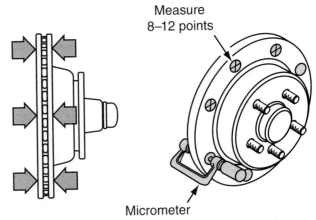

Figure 8.17 Check for thickness variations by measuring at six or more points around the rotor.

micrometer is used to make this measurement as described in Chapter 1.

Runout describes how much the rotor moves out of the vertical plane when rotated (**Figure 8.16**). In the past, maximum rotor runout was limited to 0.004 inch (0.1 mm). This is less than the thickness of two average human hairs. Standards developed in 2002 call for factory runout tolerance of less than 0.001 inch (0.025 mm). Runout is measured with a dial indicator as described in Chapter 7.

Cause and Effect of Rotor Runout and Variations in Thickness

In some vehicles, rotor runout can result in a pulsating brake pedal. Yet, runout is not the *original* cause of brake pedal pulsation. Runout causes the rotor to wear unevenly, resulting in variations in its thickness. As the warped rotor rotates, "high spots" on the rotor surface repeatedly touch the pads and the lower spots miss the pads. As miles of driving and braking accumulate, this wear to the rotors becomes measurable as *thickness variation*. Measure in six or more places around the rotor (**Figure 8.17**). When the rotor has variations in thickness and the driver applies the brake pedal, the caliper squeezes the pads against the thin areas. Then, the thicker area pushes the piston back into the caliper bore, returning fluid to the master cylinder and pushing back on the brake pedal. This cycle repeats itself and results in pulsation of the brake pedal. Maximum allowable thickness variation is 0.0005 inch (0.125 mm).

Pedal pulsation does not necessarily occur right after a brake job. Rather, it develops slowly. Rotor runout can be caused by incorrect rotor machining or improper wheel lug torque procedures. The latter is especially true with cast aluminum wheels, which are less forgiving of incorrect torque than stamped steel wheels. Be sure to tighten wheel lugs with a torque wrench. Torque is best done in at least two steps and in a star pattern. See Chapter 13 for more information on wheel torque.

When runout is not acceptable, sometimes changing the position of a nonintegral rotor on its hub will correct this.

■ MACHINING DRUMS AND ROTORS

When worn linings are replaced, drums and rotors in need of machining are serviced on a brake lathe. When done properly, this results in a fresh, true friction surface for the linings to apply against.

If a rotor or drum is excessively worn, it must be replaced rather than remachined (**Figure 8.18**). **Remember:** the discard diameter is cast into all drums produced since 1972, and brake rotors are stamped with a minimum thickness specification.

Figure 8.18 The rivets on this pad have worn so deeply into the rotor that the rotor must be replaced. *(Courtesy of Tim Gilles)*

Should Drums and Rotors Always Be Machined?

When inspecting a rotor, a technician must decide whether it needs to be refinished or replaced. When a rotor is in good condition, it will probably provide a better surface than a freshly machined rotor. A used rotor surface on properly operating brakes will be smooth and have some friction material embedded within the pores of the metal. A used rotor can provide better stopping performance than a new one.

When both drums are smooth, have normal wear, and are not out of round, they can be reused with new linings. If one drum requires machining, the other should be machined as well to provide surfaces with equal coefficient of friction. To avoid leverage problems, drums should be within 0.010 inch of each other in finished diameter.

NOTE: *When a used rotor is not going to be machined, it is important that the corresponding rotor on the other side of the vehicle not be machined. Otherwise, a brake pull can result. Rotors are always machined in pairs.*

Several years ago one manufacturer's bulletin specified that rotors did not need to be machined unless they were worn with a groove deeper than 0.060 inch. A later technical bulletin rectified this error, changing the specification to 0.006 inch. Another bulletin, from Wagner Brake, says machining the rotors is only necessary if there is pedal pulsation due to warpage. Score marks up to 0.050 inch deep are said not to affect brake operation. Some rotors are manufactured with a groove in the center of the friction surfaces. This groove is normal and does not require machining.

Some technicians resurface brand new rotors to ensure that they are true. If you purchase quality parts, a new rotor should have a superior surface finish and be straight and true. If your new rotors need to be machined, consider purchasing brake parts made by another manufacturer.

Federal brake safety standards apply only to new vehicles. Be sure to purchase rotors that are as good or better than original equipment. Poor-quality rotors can have inferior metallurgy and be prone to hot spots. Low-quality rotors can also have dimensional tolerance problems and might not fit the hub or lugs correctly. There are differences in the way rotors are cast. **Figure 8.19** shows cutaway rotors with differences in the castings of the cooling fins. There are also big differences in the metallurgy of iron castings. Poor-quality castings will distort more easily and are more prone to hot spots. They can also increase a vehicle's stopping distance.

Drums and Rotor Lathes

There are several common brake lathes manufactured. They all operate in a similar manner. Some lathes are dedicated machines, which means they do either

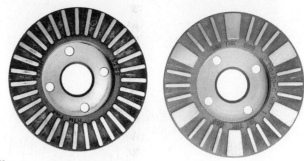

Figure 8.19 Cutaway rotors showing differences in the castings of the cooling fins. *(Courtesy of NAPA)*

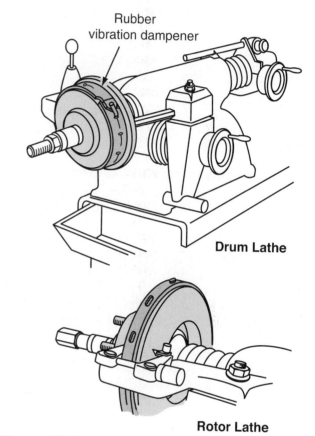

Figure 8.20 Typical multiuse brake lathe setups for machining drums and discs.

drums or discs but not both. Newer lathes are multiuse machines (**Figure 8.20**). They can be set up for either disc or drum machining. Making the changeover between the two is easy and can be done in a relatively short time.

Multiuse Brake Lathe. A multiuse brake lathe is not difficult to use, but you must pay attention to the differences between drum and rotor machining. Machining with a lathe is called "turning" a drum or rotor. Turning a drum requires the cutting tool to move from left to right, while cutting a rotor requires the tool to move in an opposite plane, from inside to outside.

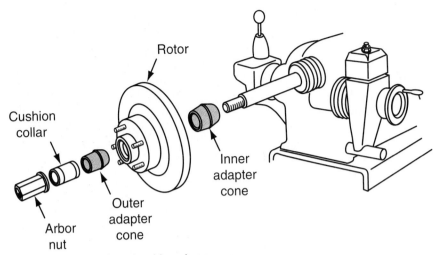

Figure 8.21 Cone adapters fit into the wheel bearing races.

NOTE: *To prevent vibration, lathe parts must be tight. There are two clamp screws on a typical multipurpose lathe. One must be tight while turning a rotor but not while turning a drum. The other must be tight while turning drums but not rotors. Common sense will help you determine which is which.*

Mounting Rotors and Drums

Whether machining a rotor or a drum, some parts of the setup are the same. For instance, mounting adapters for installing a drum or rotor on a brake lathe, called *mandrels* or *cones*, are the same for both. It is very important that the areas that contact the lathe's centering mandrels be free of rust, dirt, or nicks (refer to **Figure 7.21B**). Contaminants or nicks can produce a drum or rotor that, although its appearance is good, is untrue following machining. A pulsating brake pedal can result.

There are two standard mounting arrangements for drums and rotors. The hub is an integral part of some drums and rotors. With others, the drum or rotor can be removed after the wheel is removed.

Prior to mounting the hub assembly on the lathe, the bearings are removed from the hub and the hub is carefully cleaned. Select a mandrel that fits into the wheel bearing races (**Figure 8.21**). One or more spacers is installed on the outside so the arbor nut can be used to tighten the mandrels into their corresponding bearing races.

NOTE: *The arbor is sometimes called a* spindle.

The arbor nut is typically a left-hand thread and one end is relieved of threads so it can be tightened past the threads on the arbor of the lathe. Install the nut so it faces the correct direction. Most lathe manufacturers recommend that you help center the rotor or drum by slowly rotating it on the arbor adapters as you tighten the shaft nut.

Figure 8.22 shows how a drum is mounted when it has an integral hub with wheel bearing races. On nonintegral hubless drums or rotors, where the hub remains on the vehicle, floating adapters are used (**Figure 8.23**). A cone adapter fits part way into the hole in the center of the drum or rotor. A spring, installed before it, pushes the cone into the hole while the floating drum adapters are clamped around the flat area of the drum or rotor. It is important that this area

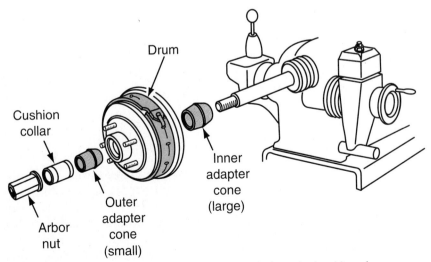

Figure 8.22 Typical lathe mounting for a drum with an integral hub and wheel bearing.

Chapter 8 Friction Materials, Drums, and Rotors

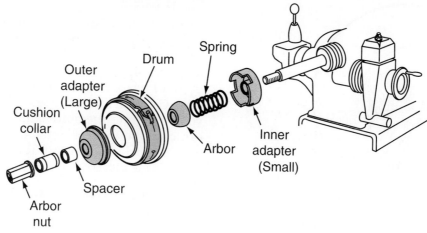

Figure 8.23 A floating adapter setup is used on drums or rotors that are hubless.

> **SHOP TIP** To ensure concentricity when machining, the drum or rotor should be secured to the hub whenever possible. Washers are installed under the lug nuts. Use all of the lug nuts and torque them equally.

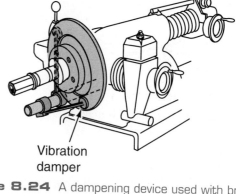

Figure 8.24 A dampening device used with brake rotors.

be clean and free of rust. If the hub is easily removable from the vehicle, a nonintegral rotor can be installed on it for machining. The lug nuts are installed upside down and tightened to hold the rotor to the hub.

Composite rotors require special mounting adapters or oversize bellcaps that simulate the torque of the wheel. They are more likely to vibrate, producing a harsh finish. Both sides of a composite rotor must be cut at the same time.

Silencing Band

To prevent vibration, use a sharp cutting tool, cut at the correct speed, and use a silencing device or band. Silencing devices are easy to install and prevent vibrations by dampening them out. One kind of dampener is a device that is a fixture on some brake lathes (**Figure 8.24**). It is adjusted to rub against the rotor during machining to dampen vibrations. One of its advantages is that it can be used on fixed or ventilated rotors. **Figure 8.25A** shows a popular silencing band used to dampen vibrations on ventilated rotors. It is easily installed using a clasp that fastens around metal buttons spaced at various intervals along the band to allow for installation on different diameter rotors (**Figure 8.25B**).

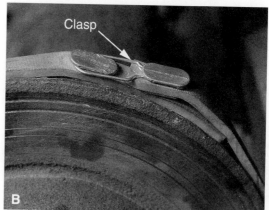

Figure 8.25 (A) A silencer band used to dampen vibrations on ventilated rotors. (B) A clasp fastens around metal buttons spaced at various intervals along the band to allow for installation on rotors of different diameters.

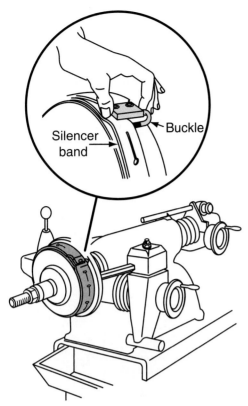

Figure 8.26 Wrap the silencer band around the drum with the buckle facing toward you.

The band used on a brake drum is wrapped snugly around the drum with the buckle finger coming toward you over the top of the drum (**Figure 8.26**). Do not make your wrap too neat. Part of the band should hang over the edge of the drum to prevent noise in that area, which is most likely to vibrate.

Noise when machining the rotor or drum is caused by vibration. If machining continues with the noise present, the surface of the machined drum or rotor will have a wavy pattern (**Figure 8.27**) that will result in a vibration during stopping.

Figure 8.27 When machined without a sound-deadening device, the surface of the machined drum or rotor will have a wavy pattern. *(Courtesy of Tim Gilles)*

Tool Bits

To achieve a surface finish smooth enough to prevent noise, tool bits must be sharp. If a silencing band is correctly installed on the drum or rotor and there is still noise, replace your tool bit.

Tool bits are made of carbide, a very high-carbon steel. Carbide is hard and brittle, so it can chip easily. If you bump into the brake lathe while it is cutting, this can chip the tool bit. Some popular premium-quality tool bits include ones with rounded superabrasive tips and carbide bits coated with titanium nitride.

Check and Adjust Arbor Speed

Choose the best arbor speed for the diameter of the drum or rotor. Larger diameters require slower speeds. (See **Figure 8.40** for recommended speeds.) The arbor speed of some brake lathes is adjusted by moving the belt to different sized V belt pulleys. Others are adjustable by selecting speeds electronically from a control panel. Some lathes have only one speed.

Scratch Cut

After mounting a drum or a rotor on a lathe, a **scratch cut** is made to verify that the setup does not have runout that will later show up on a newly machined part after it is reinstalled on the vehicle. To make a scratch cut, follow these procedures:

1. Take a small cut in the middle of the rotor or drum (**Figure 8.28**).
2. If the cutter does not touch all around the surface, stop the lathe, loosen the arbor nut, and

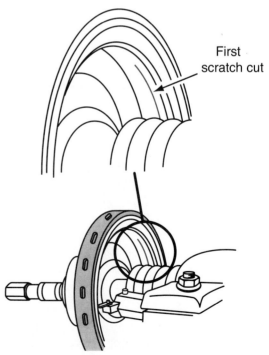

Figure 8.28 Making a scratch cut. Take a small cut in the middle of the drum or rotor.

rotate the drum or rotor 180 degrees on the lathe arbor (**Figure 8.29**). Be sure the inside adapter does not turn with the rotor when you reposition it.

3. Make another cut just to the side of the first cut.
4. If the two cuts are side by side (**Figure 8.30**), the rotor or drum has worn out of alignment and the problem is not with the setup. Go ahead and finish machining.
5. If the second cut is opposite the first cut on the other side of the rotor or drum, the problem is with the lathe setup (**Figure 8.31**). Change the setup by rotating each of the mounting mandrels again by 180 degrees.

Precut the Inner and Outer Ridges

When a rotor or drum wears, ridges or lips develop on both sides of the friction material contact points (**Figure 8.32**). When machining a rotor or drum, first

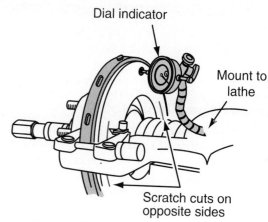

Figure 8.31 Scratch cuts that are on opposite sides of the rotor or drum indicate that the lathe setup is not correct.

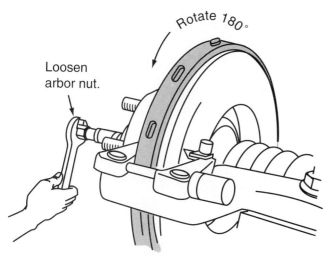

Figure 8.29 If the cutter does not touch all around the rotor surface, rotate the drum or rotor 180 degrees on the lathe arbor, while being sure the inside adapter does not turn with the rotor.

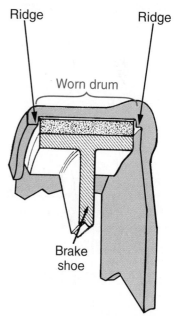

Figure 8.32 Ridges or lips develop on both sides of the friction material contact point.

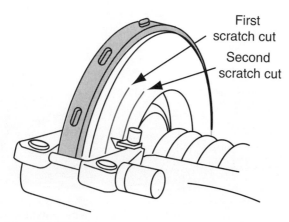

Figure 8.30 Make a second scratch cut off to the side of the first cut.

remove the inner and outer ridges by manually operating the lathe in both areas. A ridge is usually rusty and rust is very hard. When cutting the ridge manually, undercut the rust from the worn side. Because the ridges are unworn, they are probably the same diameter as a new drum. If the lathe happens to be doing a rough cut when cutting the friction surface, the cutter will be stressed when it encounters this unworn area and the tool bit could be chipped. Also, some lathes might stop turning when they encounter the increased turning effort presented by this area.

■ MACHINING BRAKE DRUMS

Drums that are within size specifications can be machined. Unless you are matching drum diameters, remove the smallest amount of metal possible. First, turn the drum with the most wear. Then, decide whether to turn the other to the same size or buy two new drums.

NOTE: *The industry standard is for rotors and drums to be within 0.010 inch of each other (from side to side on the vehicle) in thickness or diameter after machining.*

Position the Tool Holder

Adjust the tool holder so that it extends from the lathe as little as possible (**Figure 8.33**). A tool holder extending too far out of the lathe will tend to vibrate.

The position of the lathe bit is controlled by one of two handwheels. One handwheel positions the cutter on the surface of the drum, from the inside to the outside (left to right). The other adjusts the depth of the cut. **Figure 8.34** shows typical brake lathe controls.

Turning a Drum

Before cutting the drum, adjust the depth-of-cut handwheel to zero.

1. Turn on the lathe and advance the cutting tool until it makes a very light scratch in the surface of an evenly worn part of the drum. Reset the handwheel to zero.
2. Position the cutter over the deepest groove in the drum face and advance the cutter until it takes a light cut in the bottom of the groove. Check the dial to see how much metal will need to be removed in order to clean up the drum surface.

When you cut the surface of a drum by 0.015 inch, the increase in the diameter of the drum will actually be 0.030 inch because both sides of the drum rotate against the lathe cutting bit (**Figure 8.35**). The dial on a drum lathe shows "graduated" readings (**Figure 8.36**). What you read on the feedwheel dial is the actual change in diameter rather than the amount of the cut.

3. When cutting a drum, remember that a deeper cut makes a rougher finish. Most drums require

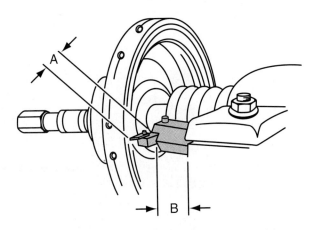

Keep distances A and B as short as possible.

Figure 8.33 The tool holder should extend from the lathe as little as possible. Keep distances A and B as short as possible.

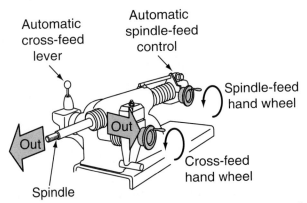

Figure 8.34 The position of the lathe bit is controlled by one of two handwheels. Turning the spindle-feed handwheel counterclockwise moves the spindle out from the lathe. Turning the cross-fed handwheel counterclockwise moves the tool bit away from the lathe spindle.

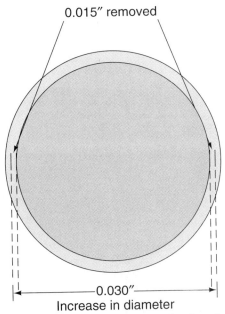

Figure 8.35 Removing 0.015 inch of surface metal results in an increase of 0.030 inch in drum diameter.

Figure 8.36 The dial on a drum lathe shows actual diameter "graduated" readings. This means when the tool bit is advanced 0.0145 inch, the drum size will increase by the 0.029 inch shown on the dial. *(Courtesy of Tim Gilles).*

a rough cut followed by a finish cut. A **rough cut** has a feed rate of 0.020 inch per revolution of the drum. This is at least three times as fast as the finish cut and leaves a rough thread in the surface of the drum. The maximum depth of cut on the rough cut is limited by the ability of the lathe to advance under load. If you try to cut too much on some lathes, the drum will stop turning when the belt slips. A **finish cut** should be about 0.004–0.006 inch deep with a feed rate of 0.002–0.006 inch per revolution of the drum.

4. Start your cut at the inside of the drum. Adjust the depth of cut, lock it down, and engage the automatic feed. Watch and listen carefully to see that the machine is operating correctly. If you have any doubt about the setup, shut the lathe off immediately and seek assistance.
5. **Remember:** *The finish cut feed rate is used for the final cut.*
6. When finished, remove the drum and remeasure it to verify the diameter is not over the machining limit.

■ ROTOR MACHINING

When machining a rotor it is best to cut both sides at once, maintaining equal force on both sides. This reduces the rotor's tendency to distort during machining. Some older lathes cut only one side at a time, but all newer machines cut both sides at once. Here is the procedure for turning a rotor:

1. After the rotor is mounted on the lathe, loosen the nut that holds the lathe cutting assembly and center it on the rotor with the two tool bits at an equal distance from the rotor surfaces (**Figure 8.37**).
2. Then, calibrate the handwheel setting to zero. To do this, turn on the lathe and advance the cutting tool until it makes a very light scratch in the surface of an evenly worn part of the rotor. Hold the handwheel while rotating the dial to zero (**Figure 8.38**).
3. Next, position the cutter over the deepest groove in the rotor face and advance the cutter until it takes a light cut in the bottom of the groove. Check the dial to see how much metal will need to be removed in order to clean up the rotor surface. If it is more than 0.008 inch on one side, make two passes with the cutter instead of attempting to finish the rotor in one pass. A deeper cut will result in a rougher surface finish. Most rotors can be machined with one finish cut. Remove the smallest amount of metal possible. A cut of 0.004 inch on each side is best because carbide tools can vibrate and/or break when making too small a cut.

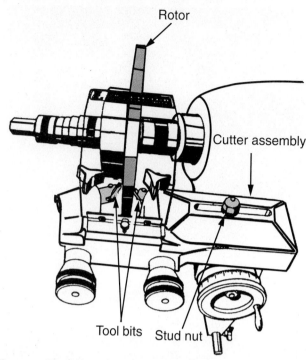

Figure 8.37 Loosen the stud nut and center the cutter assembly on the rotor with the two tool bits an equal distance from the rotor surfaces.

Figure 8.38 Hold the handwheel while rotating the dial to zero.

4. Make a scratch cut as described earlier to verify that your setup is correctly aligned.
5. Remove the inner and outer ledges from both rotor faces as described earlier.
6. The automatic feed progresses from the inside of the rotor to the outside. Position the cutting tools at the inside of the rotor faces.
7. Choose a fast or slow feed rate, depending on how much material needs to be removed. Then, turn on the feed to start cutting.

Sometimes one side of the rotor is more worn than the other. In this case, remove less metal from the unworn side. Rotors with fixed calipers, however,

Figure 8.39 The lathe cuts a thread in the surface of the rotor during refinishing. (Courtesy of Tim Gilles)

SHOP TIP To get the best possible finish, equal on the inside and outside surfaces of the left- and right-side rotors, use a fresh piece of sandpaper on each side every time.

should have the same amount of metal removed from each side.

Rotor Surface Finish

A lathe cuts a thread in the surface of the rotor as it is machined (**Figure 8.39**). The thread will be very fine if the lathe cross feed speed is slow, or it can be coarse if the cross feed speed is fast. The better the quality of the rotor surface finish, the less tendency there will be for the pads to make noise. This is true with all pads, not simply semimetallics that tend to be more prone to noise. Besides contributing to noise, a rough surface finish on a rotor can also cause the pads to be moved out from the center of the rotor.

The final surface finish should be quite smooth (10–50 microinches). Feel the finish of a resurfaced rotor with your fingernail. It should be at least as smooth as the finish on a new rotor. Most new rotors have a very high-quality surface finish of 20–25 microinches RA (roughness average). **Figure 8.40** lists recommended speeds and feeds.

A *nondirectional finish* (**Figure 8.41**) can be applied to the face of the finished rotor using a rotating sander (**Figure 8.42**) or a drill with a flex-hone (**Figure 8.43**). One special sanding tool is shown in **Figure 8.44**. Use #120–#150 grit sandpaper for about 60 seconds or until the surface is smooth.

Wash the Machined Surfaces

It is very important to wash machined surfaces after machining (**Figure 8.45**). If you fail to do this, small, loose metal dust grains left in the pores of the

ROTOR REFINISHING GUIDE

	Rough cut	Finish cut
Spindle speed		
10" and under	150–170 rpm	150–170 rpm
11" to 16"	100 rpm	100 rpm
17" and larger	60 rpm	60 rpm
Depth of cut (Per side)	0.005" to 0.010"	0.002"
Tool cross feed (Per rev.)	0.006" to 0.010"	0.002" Max.
Vibration dampener	Yes	Yes
Sand rotors	No	Yes
Sanding instructions	60 seconds per side with 150 grit sandpaper	
Cleaning instructions	Use brake parts cleaner and dry with paper towel	
Instructions for breaking in the new disc brake pads:		
Make ten stops from 30 mph to 5 mph under moderate braking pressure. Allow the brakes to cool between stops.		

Figure 8.40 Recommendations for rotor refinishing.

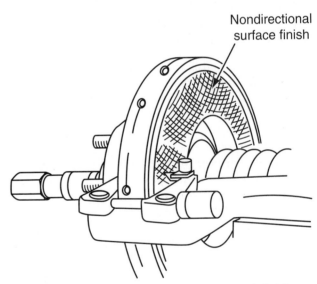

Figure 8.41 A nondirectional crosshatch finish smoothes the threads in the rotor surface.

machined surfaces will become embedded in the friction material. This will result in noise and wear and the only solution will be to replace the new friction materials once again. Use soap and water or a nonpetroleum solvent such as brake parts cleaner or denatured alcohol. Do not dry with compressed air because it contains small amounts of oil. Paper towels work well. Also, if

Chapter 8 Friction Materials, Drums, and Rotors **181**

Figure 8.42 Use an orbital sander to sand off the threaded finish while the lathe turns. *(Courtesy of Tim Gilles)*

Figure 8.43 A flex-hone can be used to finish the rotor surface. *(Courtesy of Tim Gilles)*

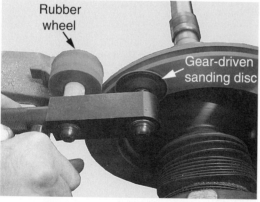

Figure 8.44 As the rubber wheel is turned by the rotor it rotates the sanding disc. *(Courtesy of Tim Gilles)*

Figure 8.45 Wash the rotor after machining it to remove metal dust so it cannot become embedded in the new pads.

SHOP TIP When machined surfaces have been freshly cleaned, do not touch the friction surfaces. Pick up a drum or rotor from the outer edge or from the center hole.

180 degrees on the arbor shaft as you remount the rotor. Then, use a dial indicator to check the runout on the freshly machined area of the rotor. Runout should not exceed 0.003 inch.

Some manufacturers use tapered shims to compensate for rotor runout. When runout has been measured on the vehicle, mark the rotor to show the place of greatest measurement. Maximum runout when mounted on the lathe should be in the same position. This can be checked with a dial indicator once the rotor has been mounted on the lathe. If runout is not in the same place, shims can be installed to adjust it so it is the same as it was on the vehicle. Also, a refinished rotor can be checked for runout after it is installed on the hub. If runout is present, mark the location of the high spot and install a shim with its thickest part on the other side, away from the high spot. With the correct size shim installed, runout can be eliminated. Many manufacturers recommend on-vehicle machining to prevent these problems.

ON-VEHICLE ROTOR MACHINING

Some earlier disc brakes had rotors that were difficult to remove from the vehicle, sometimes resulting in damaged front wheel bearings. In response to this, on-the-car lathes were developed (**Figure 8.46**). In recent years, runout tolerances have become tighter and on-the-car lathes are again increasing in popularity. On-vehicle machining is also a good choice when machining composite rotors.

you wash with hot water it will evaporate quickly on its own.

Checking Lathe Accuracy

After centering and tightening a rotor on the lathe, machine the outermost 1 inch of the rotor face. Remove the rotor and reposition all of the adapters

Figure 8.46 An on-the-car lathe. *(Courtesy of Tim Gilles)*

> **CAUTION** Rotors are made of iron, which is magnetic. When turning rotors on the vehicle, metal chips removed from the rotor as it is machined will stick to antilock brake wheel speed sensors. Be sure to clean any chips from the sensor before reassembly.

Because the rotor does not need to be removed, there is no chance for foreign material to become lodged between the rotor and hub. Also, machining the rotor while it is on the hub ensures a true rotor and hub assembly. Otherwise, *tolerance stack* can occur. An example of this is when a rotor has 0.002 inch runout and the hub has 0.002 inch runout. This might be within the tolerance allowed during manufacturing, but if the two amounts are combined for 0.004 inch runout, pedal pulsation could develop as the rotor wears unevenly.

There are three designs of on-vehicle lathes. One older type had the rotor turned by the vehicle's powertrain. Because the rotor is turned by the powertrain, this lathe cannot be used on rear-wheel drive front disc brakes. Of the two self-powered types, one has the lathe head mounted on the caliper mount. The other uses the wheel mounting surface. Caliper-mounted lathes were the first type to be in wide use. The newer, hub-mounted lathes have fewer adapters and are faster to use. Some of them compensate automatically for runout, which saves even more time.

■ HUB AND LUG REMOVAL

The hub is an integral part of some drums and rotors. Sometimes the two need to be separated, or damaged lug studs need to be replaced. Some rotors and drums have lug studs that are swaged to hold them to the wheel hub (**Figure 8.47**). *Swaged* means that a shoulder on the lug stud has been deformed with a press during installation to expand the stud, holding the two parts tightly together. Before the drum or rotor can be removed from the wheel hub, the lug studs must be removed. A cutter is used to remove the swaged area before the studs can be pressed out without damaging the hub (**Figure 8.48A**). If a repair shop does a good deal of this work, they will own the cutting tool. Otherwise, this work is sublet to a machine shop. **Figure 8.48B** shows how a stud is supported while its shoulder is pressed on during the swaging operation.

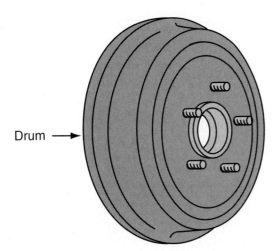

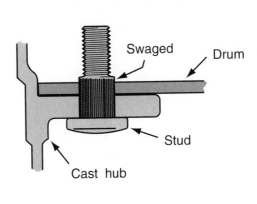

Figure 8.47 Some wheel studs are swaged.

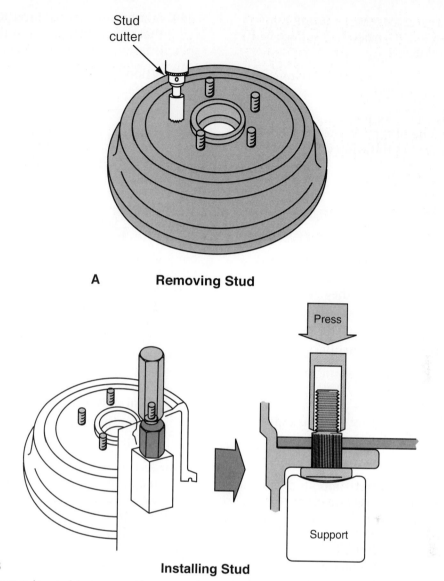

Figure 8.48 (A) A cutter is used to remove the swaged area before the studs can be pressed out without damaging the hub. (B) Each stud is supported while its shoulder is pressed against to expand it during the swaging operation.

REVIEW QUESTIONS

1. Three newer OE friction materials include semimetallic, _____ _____ (NAO), and ceramic.

2. Which linings have the best fade resistance of any lining but work very poorly when cold?

3. _____ fiber linings, premium linings used on police cars, race cars, and aircraft, are semimetallic composites, which use hollow fibers instead of powder.

4. List two aftermarket brake evaluation programs.

5. Which group of letters or numbers in the SAE edge code provides coefficient of friction?

6. What type of rotor has friction surfaces made of cast iron and a center made of thin stamped steel?

7. List two types of brake drum wear.

8. Which kind of wheels are most sensitive to improper wheel lug torque procedures?

9. To avoid leverage problems, drums should be within _____ of each other in finished diameter.

10. _____ rotors require special mounting adapters or oversize bellcaps that simulate the torque of the wheel.

11. When there is _____ during rotor machining, the rotor's machined surface will have a wavy pattern.

12. To prevent _____, use a sharp cutting tool, cut at the correct speed, and use a silencer band.

13. What is the name of the test cut that is done on a lathe to verify that the setup does not have runout?

14. A lathe cuts a thread in the surface of the rotor that is _____ if the cross feed speed is fast.

15. A nondirectional finish can be applied to the face of the finished rotor using a rotating _____.

PRACTICE TEST QUESTIONS

1. Technician A says semimetallic linings work better as they get hotter. Technician B says NAO linings work better as they get hotter. Who is right?
   A. Technician A   B. Technician B
   C. Both A and B   D. Neither A nor B

2. Technician A says that as ceramic disc pads wear, a dark brown dust develops on the wheels. Technician B says brake dust is a combination of friction compound and rust. Who is right?
   A. Technician A   B. Technician B
   C. Both A and B   D. Neither A nor B

3. All of the following lining compounds have less tendency to produce noise *except*:
   A. NAO   B. low-metallic
   C. semimetallic D. ceramic

4. Technician A says Federal Motor Vehicle Safety Standard (FMVSS) 135 applies only to new and aftermarket brakes. Technician B says SAE edge codes can be used to determine the quality of aftermarket linings. Who is right?
   A. Technician A   B. Technician B
   C. Both A and B   D. Neither A nor B

5. Technician A says it is best to replace a composite rotor with a single-piece casting. Technician B says to replacing a composite rotor with a cast rotor can cause a tire pull. Who is right?
   A. Technician A   B. Technician B
   C. Both A and B   D. Neither A nor B

6. A car has a pulsating brake pedal. Technician A says this could be due to out-of-round brake drums. Technician B says this can be caused by an incorrect front wheel lug torque procedure. Who is right?
   A. Technician A   B. Technician B
   C. Both A and B   D. Neither A nor B

7. Technician A says rotor runout can result in a pulsating brake pedal. Technician B says thickness variations in the rotor cause brake pedal pulsation. Who is right?
   A. Technician A   B. Technician B
   C. Both A and B   D. Neither A nor B

8. Technician A says rotors are always machined in pairs. Technician B says to resurface brand new rotors to ensure that they are true. Who is right?
   A. Technician A   B. Technician B
   C. Both A and B   D. Neither A nor B

9. Technician A says machining composite rotors requires special adapters that simulate a rotor being bolted to the hub and wheel. Technician B says it is best to machine these rotors on the vehicle. Who is right?
   A. Technician A   B. Technician B
   C. Both A and B   D. Neither A nor B

10. Technician A says following machining, a drum must be of sufficient thickness to allow for .0.030 inch of future wear. Technician B says a deeper cut will result in a smoother surface finish. Who is right?
    A. Technician A   B. Technician B
    C. Both A and B   D. Neither A nor B

CHAPTER 9

Other Brake Systems: Parking Brake and Power Brake

■ OBJECTIVES

Upon completion of this chapter, you should be able to:

✔ Explain disc and drum parking brake operation.
✔ Describe parking brake operation and service.
✔ Explain vacuum power brake booster operation.
✔ Explain the differences between the three types of vacuum power boosters.
✔ Describe the operation of hydraulic power boosters.
✔ Describe power brake diagnostic procedures.

■ KEY TERMS
bulkhead
drum-in-hat
firewall
Hydro-boost brake

■ INTRODUCTION

Disc and drum brakes—the service brakes—have auxiliary support systems. These include the brake power assist and the parking brake. This chapter familiarizes you with the support systems of the normal hydraulic brake system.

■ PARKING BRAKE FUNDAMENTALS

Vehicles are required by federal law to have a parking brake. The parking brake, sometimes called the *emergency brake,* must operate independently of the service brakes. The parking brake can share the same parts as the service brakes, but it must be able to be applied independently. For instance, service brakes are typically applied hydraulically by applying the foot brake pedal, and parking brakes are applied mechanically through a cable operated by a hand lever or foot pedal.

"Emergency brake" is not really an accurate name. Although a parking brake might help stop a moving car, most are not very effective at that. Federal requirements simply call for the parking brake to "hold a parked vehicle," not stop a moving vehicle. The federal safety standard that governs automotive brakes, FMVSS105, has different specifications that apply to separate areas of the brake system, including the parking brake.

The federal standard includes different specifications for vehicles with manual and automatic transmission. With a manual transmission, a parking brake must be able to lock the wheels of a parked, fully loaded vehicle and hold it on a 30 percent grade, while an automatic transmission brake must hold on a 20 percent grade.

Parking Brake Lever

A lever multiplies the force a driver applies to the parking brake. It can be either hand- or foot-operated (**Figure 9.1**). A ratchet or latch engages to hold the hand or foot lever in the applied position until it is released by the driver. To release a foot brake, the driver pulls on a lever by hand. On some types, pressing on an already engaged foot pedal releases the ratchet. On some luxury cars, a vacuum-operated automatic mechanism releases the parking brake in response to a vacuum signal (**Figure 9.2**). These systems also have a mechanical means of releasing the brake in the event of a vacuum system failure. Hand-operated parking brakes are released by rotating the handle or depressing a release button.

Federal law specifies that the force required to apply a hand brake cannot exceed 80 pounds. A foot brake must fully apply at less than 100 pounds.

Parking Brake Cables and Equalizer

Passenger car parking brakes are applied by pulling on a cable or a steel rod. The rod is connected from the

SECTION 2 BRAKES

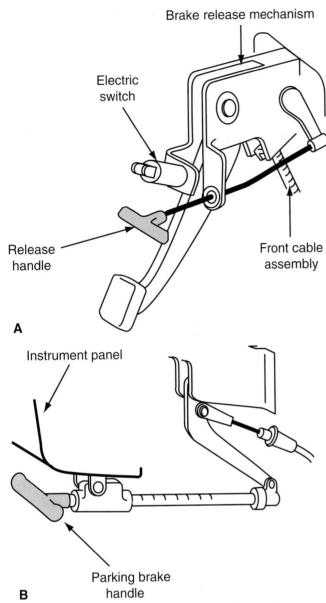

Figure 9.1 (A) A foot-operated parking brake with a release lever. (B) A hand-operated parking brake.

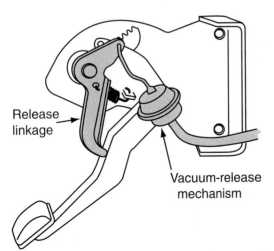

Figure 9.2 A vacuum-release mechanism on a foot-operated parking brake.

hand brake or foot pedal in the driver's compartment to an *equalizer* (**Figure 9.3**). The equalizer is usually located near the center of the car under the driver's compartment. A typical parking brake has a cable attached to a lever at each of the rear wheels (**Figure 9.4**); a few vehicles have had parking brakes on the front wheels. The front end of each cable is attached to one side of the equalizer. Because it pivots in the center, the equalizer applies the parking brake equally at each rear wheel.

Some cars have a single cable from the brake lever to one of the brakes on the rear axle and a different equalizing mechanism that applies pressure to a second cable connecting it to the brake on the other side of the axle (**Figure 9.5**). Another type of brake has the equalizer built into the brake lever under the dash.

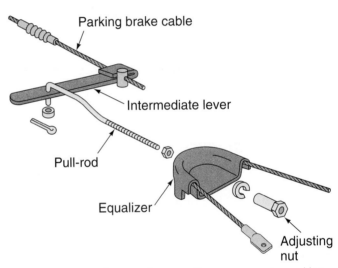

Figure 9.3 An equalizer applies pressure to parking brake cables from each side of the vehicle.

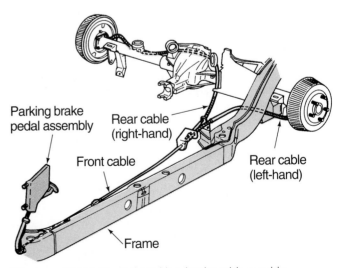

Figure 9.4 Typical parking brake with a cable attached to a lever at each of the rear wheels.

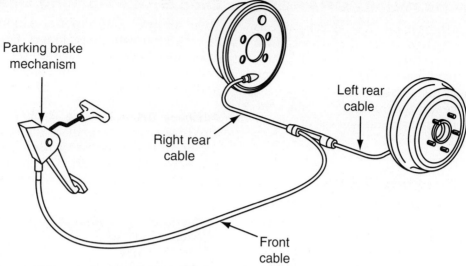

Figure 9.5 This equalizing mechanism uses the right rear cable to pull on the left cable.

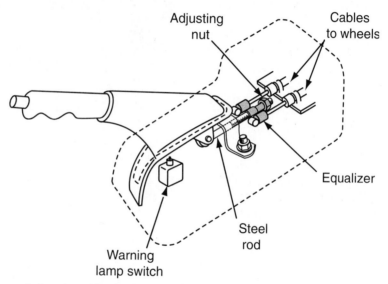

Figure 9.6 A center console hand parking brake control unit.

The lever is attached to the equalizer by the cable and housing. When a center console hand parking brake lever is used, it is often connected to the equalizer by a steel rod (**Figure 9.6**). Made of woven steel strands, a brake cable is usually exposed to air between the equalizer and a sheathed section of conduit that connects to the rear brakes (**Figure 9.7**). The conduit is flexible and the cable has enough extra length to allow the axle to move up and down without affecting the application of the parking brake. The cable is sometimes coated with plastic.

Parking Brake Warning Lamp

A red warning light illuminates on the instrument panel to alert the driver that the brake is applied. The warning light is designed to prevent damage to the parking and service brakes that would occur if the car were driven while the parking brake linings are in contact with the friction surfaces of the drum or rotor.

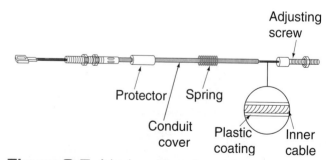

Figure 9.7 A brake cable and conduit.

The parking brake warning light can have its own separate circuit, but it is usually part of the brake warning light circuit that warns of a loss of hydraulic pressure. The warning light is activated by a switch on the parking brake lever. With the key on and the parking brake applied, the light should be illuminated. This is a good way to check the bulb for the hydraulic circuit as well.

TYPES OF PARKING BRAKES

Parking brakes are of three main types. Almost all cars use a parking brake located at the rear wheels. Cars with front disc brakes and rear drum brakes typically use a parking brake that is part of the rear drum brakes. Cars with four-wheel disc brakes use one of several types of parking brakes on the rear brakes. Some vehicles use a parking brake that operates on the drive line, not the rear brakes.

Drum Brake Parking Brake

Drum brakes usually use an *integral-type parking brake* in which a cable-actuated bar applies the drum-type emergency brake (**Figure 9.8**). The bar, called the *parking brake strut*, is a flat piece of steel attached to a lever that pivots on a pin connected to the secondary brake lining. A spring is installed at the end of the strut bar to prevent the strut from vibrating when the parking brake is not applied.

The strut has slots that fit into corresponding slots in the primary and secondary linings. The cable pulls on the pivot lever, which pushes on the strut. The strut bar separates the brake linings, wedging them into the brake drum with equal force.

NOTE: *When parking on a steep hill with the vehicle facing upward, a driver can maximize his or her parking brake effort by applying the service brakes while allowing the car to slowly sink back downhill until the turned front tire comes into contact with the curb (as prescribed by law). Using the service brakes in this manner while applying the parking brake does two things. First, it increases the effort that would normally be applied by the parking brake lever. Second, it takes advantage of servo action to wedge the linings tightly into the drum for greater holding ability. This is especially true with dual-servo brakes.*

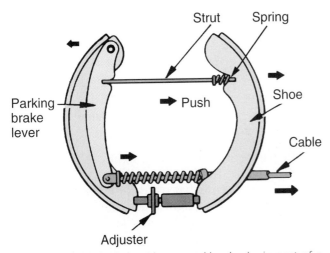

Figure 9.8 An integral-type parking brake is part of a rear drum brake.

Disc Brake Parking Brakes

There are two main types of parking brakes used on cars with four-wheel disc brakes. One uses a miniature drum and shoes housed within the inside of the rotor. The other uses the rear disc brakes. These are applied in different ways.

Auxiliary Drum Parking Brake. Some four-wheel disc brakes use an auxiliary internal drum-type parking brake that is housed within a section of the rear disc rotor (**Figure 9.9**). This parking brake is often referred to as a **drum-in-hat** brake (**Figure 9.10**).

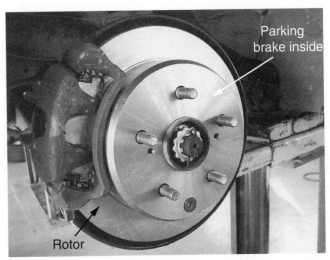

Figure 9.9 An auxiliary parking brake is housed within the rotor. *(Courtesy of Tim Gilles)*

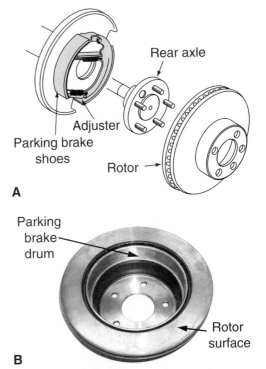

Figure 9.10 (A) Parts of an internal drum-type parking brake, often called a drum-in-hat brake. (B) The parking brake area is the drum inside the rotor.

The disc is the hat and the drum is the parking brake section found inside of it. This parking brake is found on Corvettes, Toyotas, Volvos, and other cars.

The drum-in-hat design costs more to produce and it weighs more than a conventional rear disc parking brake, but its operation is independent of the rear disc brakes and it provides a more positive parking brake than a conventional rear disc.

The drum-in-hat parking brake is applied in the same manner as a conventional drum parking brake. A pivot lever applies force to a strut between the front and rear shoe (see Figure 9.8). When the parking brake is released, return springs pull the shoes out of contact with the drum and the anti-rattle spring on the strut prevents it from vibrating. A return spring on the brake cable returns it to its unapplied position.

Parking brake shoes in the drum-in-hat brake wear very little, unless the car is driven with the parking brake partly applied.

Integral Disc Brake Parking Brake. Many cars have a parking brake that is an integral part of the rear disc service brakes. Most cars use a conventional parking brake cable arrangement, as described previously, to rotate a lever on the back of the caliper. There are two types of domestic car brakes that use screw mechanisms. Some imported cars use a simple cam and lever (**Figure 9.11**).

Screw and Nut Parking Brake Caliper

One type of brake caliper with a parking brake, developed by General Motors Delco Moraine, hydraulically

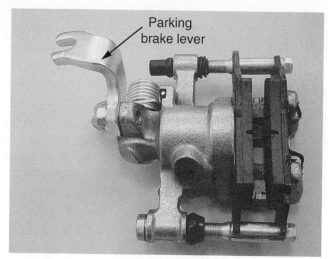

Figure 9.11 A simple cam and lever disc parking brake. *(Courtesy of Tim Gilles)*

self-adjusts like a normal disc brake. The parking brake self-adjusts using a lead screw with a very steep thread pitch attached to a lever (**Figure 9.12**). When the parking brake is applied, the lever threads rotate into an adjusting nut in the caliper piston. Rotation of the high-pitched screw causes the caliper piston to move against the brake pads to set the parking brake.

Self-adjustment of the parking brake occurs when the service brakes are applied and the piston travels farther than normal clearance would allow. This causes the piston to pull against the screw and adjusting nut. When a gap develops between the adjusting nut and cone as the brakes are applied, the nut rotates on the screw threads and makes the adjustment. During

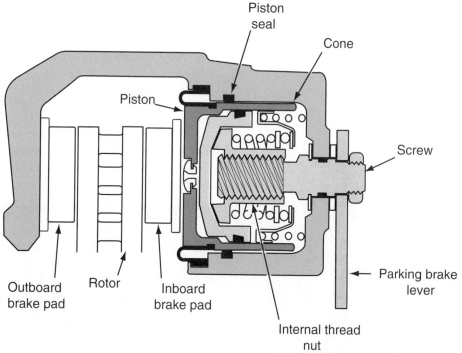

Figure 9.12 A lead screw rear disc self-adjusting parking brake has a steep thread pitch.

normal brake application, the nut and piston cannot rotate due to the high pressure between the nut and cone and the piston.

Ball and Ramp Parking Brake Caliper

Another manufacturer of brakes for domestic cars, Kelsey Hayes, uses a different design of disc parking brake. Like the Delco Moraine caliper, self-adjustment of this brake also occurs during normal operation of the service brakes, and self-adjustment of the parking brake occurs when the piston moves past a predetermined point in relation to the parking brake self-adjusting mechanism. The following is a description of that process.

When the lever on the back of the caliper is rotated, an operating shaft turns along with it. The operating shaft has three detent ramps on its inside face. There are three corresponding ramps on the outer face of the thrust screw as well. A ball fits into each of the three ramps. The thrust screw is prevented from rotating by a pin. Applying the parking brake moves the lever, which rotates the operating shaft. The three balls roll in their ramps in the faces of the thrust screw and operating shaft (**Figure 9.13**). This results in an inward force on the brake piston to apply the pads against the rotor.

During normal operation of the service brakes, self-adjustment occurs as the pads wear and the piston moves farther out of the caliper bore. When the piston moves past the point of normal parking brake operation, the force pulling between the thrust screw and the piston causes the adjuster assembly in the piston to rotate, taking up the extra clearance.

NOTE: *Some older integral disc brakes self-adjusted when using the parking brake rather than the service brake.*

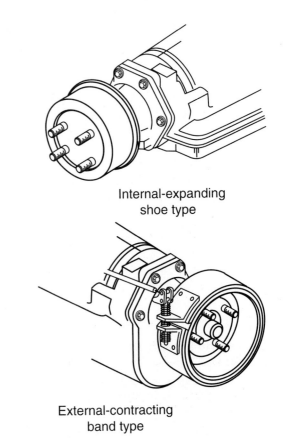

Figure 9.14 Two types of drive line or transmission parking brakes.

Auxiliary Transmission Parking Brake

On some older cars, an independent-type emergency brake was mounted behind the transmission on the front of the drive shaft. This design is still found today on heavier trucks and RVs. It can be either an internal-expanding or an external-contracting type brake. The internal-expanding brake resembles the drum-in-hat brake found on some four-wheel disc brake cars (**Figure 9.14**). A small, threaded starwheel adjuster is used to adjust the initial clearance between the lining and the drum. Because there is no lining-to-drum contact when the vehicle is rolling, wear is only caused by driving with the parking brake applied. There is no need for adjustment of the shoe-to-drum clearance as an ordinary maintenance item.

■ PARKING BRAKE SERVICE

The parking brake does not usually require service independent of the service brakes. Its operation can be checked by applying the parking brake and verifying that it engages before it is at half travel. Once the parking brake cable is correctly adjusted it should not require further adjustment unless the car is an older one whose service brakes do not have an automatic adjusting feature. A cable can stretch during its lifetime, so minor adjustment might be necessary.

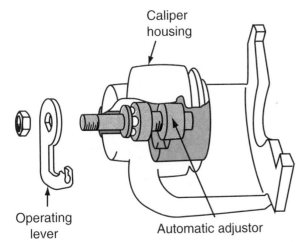

Figure 9.13 A ball and ramp self-adjusting rear disc parking brake.

Servicing the Parking Brake Cable

Inspect the parking brake cable by first checking to see that it applies easily. If not, check for damage or rust on the cable. During a rear brake job, the parking brake cables will be disconnected from the brake linings. This provides a good opportunity to inspect them. From the underside of the vehicle, pull the cables as far as possible and inspect them. Do not apply liquid lubricants to brake cables. Most cables are lubricated with dry lube, such as graphite. Liquid lubricant can run into the brakes, possibly contaminating the linings.

On drum parking brakes, the parking brake cable usually has a spring-loaded clip that fits into the brake backing plate. If a parking brake cable requires replacement, use a small hose clamp to help remove it from the backing plate (**Figure 9.15**). An offset wrench can also be used (**Figure 9.16**).

Figure 9.17 Installing the parking brake cable to the actuating lever. *(Courtesy of Tim Gilles)*

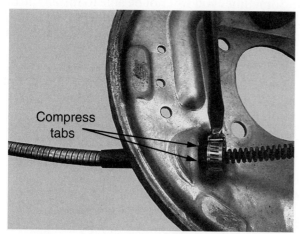

Figure 9.15 A small hose clamp can be used to depress the parking brake cable retaining tabs. *(Courtesy of Tim Gilles)*

SHOP TIP Unless there is a need to remove the brake cable, most technicians leave the lever attached to the brake cable during brake service.

Locking needle nose pliers or diagonal cutters can be used to pull the cable spring back during installation of the brake cable end to the actuating lever on the secondary brake shoe (**Figure 9.17**). There are also special tools available for this purpose.

Disc brake cables are often attached to the pivot lever at the caliper by a clevis pin. Remove the clip from the pin to remove the end of the cable. Then remove the clip that holds the cable (**Figure 9.18**).

Parking Brake Cable Adjustment

Service manuals often describe parking brake adjustment as the number of clicks that the brake handle or pedal makes before being fully applied. The specification might say "less than nine clicks." A good rule of thumb is that the brake should be fully applied at half travel.

On vehicles with drum brakes, changing the service brake adjustment will affect the parking brake adjustment. Do not adjust the parking brake unless the service brakes have been adjusted first.

Figure 9.19 shows a typical under-car adjustment for drum brake parking brakes. For the initial adjustment, apply the parking brake lever or pedal until its ratchet clicks once. Then tighten the nut at the equalizer until all slack is removed from the cable.

NOTE: *Some vehicles have the adjustment at the parking brake lever.*

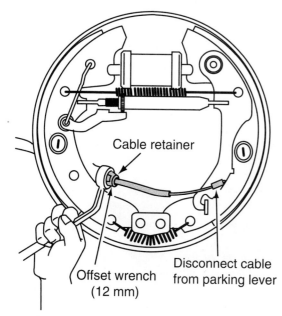

Figure 9.16 An offset wrench can be used to depress the retaining tabs.

The rear wheels should turn freely. As you turn a rear wheel, pull on the cable. You should feel the wheel start to drag as the linings contact the drum or

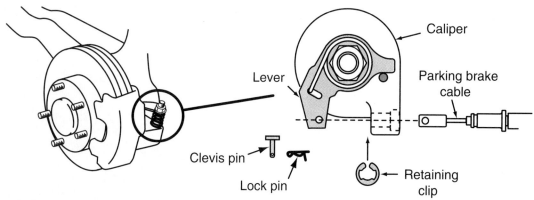

Figure 9.18 Removing a parking brake cable.

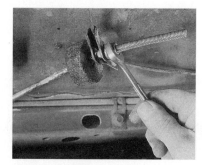

Figure 9.19 Typical parking brake adjustment.

rotor. Apply the parking brake tightly and check to see that it is at half travel. If not, adjust it in whichever direction is necessary to correct the pedal travel. Then release the pedal or lever all the way and verify that the wheels turn freely.

Some Chrysler cars have a self-adjusting parking brake pedal. A pin is positioned to prevent the cable tensioning spring from unwinding if the pedal or cable are removed. Check the service manual for information.

During a brake lining replacement, the correct adjustment of the parking brake cable should be verified. Sometimes you will not be able to install the brake drum over new linings because someone adjusted the parking brake cable to compensate for a lack of self-adjustment, or drum-in-hat parking brake linings might have become worn due to driving with the brake partly set. In either of these cases, the drum will not fit over new, thicker linings and you will have to back off the parking brake cable adjustment.

When parking brake shoes have been replaced, a drum brake adjusting gauge is used to set the lining-to-drum clearance (**Figure 9.20**). On some brakes, this is the only adjustment that is done. On others, there is an access hole in the backing plate to allow for a conventional adjustment (**Figure 9.21**).

When a vehicle has rear disc brakes and the parking brake height is low, apply the parking brake firmly, release it, and reapply it several times while watching the parking brake height. If the height does not improve, use a brake pedal depressor to firmly apply

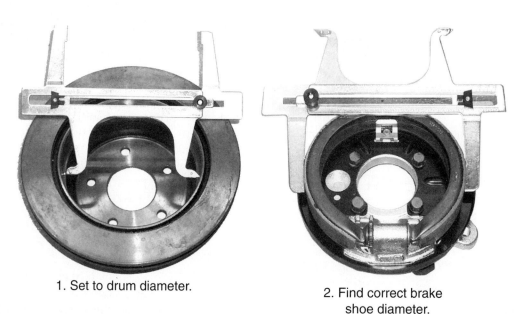

1. Set to drum diameter.

2. Find correct brake shoe diameter.

Figure 9.20 An adjusting gauge is used to adjust drum-in-hat parking brake clearance.

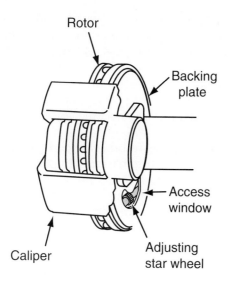

Figure 9.21 Some parking brakes have a traditional adjustment through the backing plate.

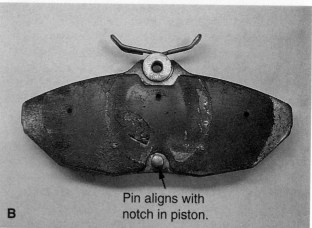

the service brakes. Then, tap on the caliper with a hammer to try to free it up. If the height is still low, disassemble the caliper.

Disc Brake Caliper Self-Adjustment Reset

Disc brake calipers automatically compensate for clearance during application of the service brakes. During installation of new pads on disc brakes with a self-adjusting parking brake, the caliper pistons must be rotated back to the point where they were before the old pads wore down and the self-adjuster moved out to compensate. This can be done using a special tool like the one shown in **Figure 9.22**.

When repairing disc caliper parking brakes, remember that as with other self-adjusting mechanisms, the left and right sides are not interchangeable. Due to the complexity of these calipers and the chance that some parts are not readily available at a reasonable cost, many shops install rebuilt units instead of servicing them themselves. If you use a rebuilt caliper, be sure to keep all of the old parts together with the old caliper so it will be acceptable as a core return.

Pedal travel on four-wheel disc brakes tends to be farther than with disc/drum combinations. This is because the rear pads have to travel farther during brake application. When there is a problem with the self-adjuster, pedal travel will become excessive more quickly.

After replacing a caliper, the initial clearance must be adjusted to within 1/16 inch or less. Adjustment can usually be done by repeatedly applying the parking brake. If the adjustment is too tight to install the caliper, the piston must be screwed back into its bore with the tool shown in **Figure 9.22C**.

Figure 9.22 (A) A threaded-type rear disc brake self-adjuster piston has notches that hold it aligned with the pad. (B) The back of a pad used with a threaded caliper piston has an aligning pin. (C) A tool used to retract the piston. *(Courtesy of Tim Gilles)*

■ PARKING BRAKE WARNING LIGHT SERVICE

Be sure that the warning light comes on when the parking brake is applied. The switch is located at the pedal or lever. When the parking brake is applied, the switch closes to illuminate the light. It opens when the brake is released. The mounting bracket controls the switch adjustment. Sometimes, bending the bracket is required to adjust the switch.

Accidents Caused by Incorrect Parking Brake Adjustment

A number of accidents are caused by driving vehicles with the parking brake applied or partly applied in a split diagonal brake system. If you adjust a parking brake too tightly the same thing can happen. Here is why this can be dangerous.

When the parking brake is partly applied and the car is driven on a highway for a period of time, a good deal of heat is generated. This heat is absorbed by the brake fluid, which can boil. When the fluid boils in the brake line, a large pocket of compressible air develops. As the driver leaves the highway and attempts to stop at the end of an off-ramp, the pedal moves to the floor as it compresses the pocket of air in the line. The car can then collide with a car in front of it or run a red light into oncoming traffic. The driver claims that the brakes failed, but the investigating traffic officer finds that the brakes are operational. This is because the brake fluid has cooled and the air pocket has vanished.

If the hydraulic system had not been a split diagonal, the front brakes would have had sufficient pedal travel remaining to stop the vehicle on the first pedal application, although braking would have been somewhat diminished. This problem is especially serious with drum-in-hat parking brakes because their drum cannot reject the heat, which migrates into the rotor and causes it to expand.

■ POWER BRAKES

Power brakes began to appear on some cars in America during the mid-1950s, when cars still had drum brakes on all four wheels. The reduced braking effort afforded by power brakes was popular with car buyers. However, in the 1950s power brakes were only available as an optional accessory on all but luxury automobiles. When disc brakes began to be installed on most new vehicles, power brakes became more common.

By the late 1970s, disc brakes were an OE installation on most vehicles. Disc brakes do not have the self-energizing action of dual-servo drum brakes. In fact, gripping a shiny piece of steel between two relatively smooth brake pads requires a substantial increase in braking effort. Disc brakes have the ability to apply more stopping energy at the wheels, but they also require more application force. The solution to this problem was the installation of power brakes on most vehicles. Power brakes reduce the pedal effort but still allow a reasonable feedback to the driver during braking.

Without power brake assist, the driver is able to sense braking effort by the "feel," or feedback, of his or her effort on the brake pedal. When you apply the brakes as hard as you can, the car stops. It is normal to feel resistance to the force you put into application of the brake pedal. Power brakes take that feedback away, so engineers have devised ways to put pedal feel back into the power brakes. Without this feedback, a driver would only feel the foot pressure applied to the pedal and not the result of the boosted force applied to the master cylinder.

One way to increase brake force is to change the size of the piston in the master cylinder and/or wheel cylinders. This method has its limits. A smaller piston in the master cylinder results in higher force at the wheel cylinders. Unfortunately, the pedal must move farther for the brakes to apply.

For modern brake systems, the most effective and economical means of boosting brake pressure is to use a vacuum-assist brake device called a *booster* installed behind the master cylinder.

There are also hydraulic-assist brake boosters, but these are not found in mainstream applications. They are typically used when the production of sufficient engine vacuum is a problem, such as in a diesel or turbo-charged engine.

A brake booster often allows the master cylinder to have a larger bore. Remember, a larger bore master cylinder exerts less pressure on the brake fluid. At the same time, it moves a larger volume of fluid with equal amounts of pedal movement. This means that the brakes can apply with less pedal travel.

Types of Power Brakes

There are two main types of power brakes. The type used on most vehicles is mounted behind the master cylinder (**Figure 9.23**). It uses the difference between atmospheric pressure and vacuum to apply force (**Figure 9.24**) from a brake booster powered by vacuum supplied from the intake manifold of the engine. The booster and master cylinder are mounted on the **firewall,** also called the **bulkhead,** which is the metal wall between the engine and passenger compartments. The pedal apply rod goes into the back of the

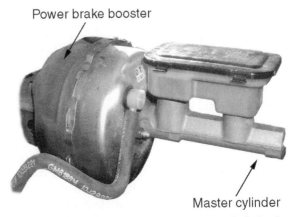

Figure 9.23 The power brake booster is bolted to the back of the master cylinder.

Chapter 9 Other Brake Systems: Parking Brake and Power Brake   195

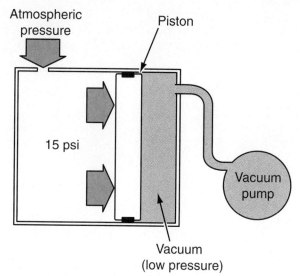

Figure 9.24 The difference between atmospheric pressure and vacuum can be used to apply force.

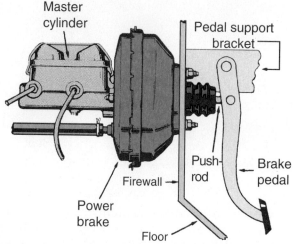

Figure 9.25 The power booster and master cylinder are bolted to the firewall.

Vintage Chassis

There are several methods of increasing the force applied by the brake pedal. The most obvious way is to step harder on the brake pedal, but this is not always practical due to variations in the size and weight of the driver. Other ways to increase pedal force have been used on cars for many years. The first method of increasing force to the brakes was to lengthen the brake lever. In the horse-and-buggy days, vehicles had a very long lever to apply a hand brake. Increasing the length of the pedal arm on today's brakes would result in a similar increase in application force, but the pedal would have to move farther before the brakes would apply, so this method is somewhat limited.

unit and the master cylinder is bolted to the front of it (**Figure 9.25**).

The second kind of power brake operates a brake booster using oil pressure supplied by the power steering pump. This system is covered later in the chapter.

■ VACUUM BRAKE BOOSTERS

The vacuum brake booster is mounted on the firewall (see **Figure 9.25**). The master cylinder is bolted to the front of it with the back of it sealed from the atmosphere. The brake booster provides an increase in the braking effort supplied by the driver. In the event of a power brake failure, the service brakes will still operate.

Power brakes depend on engine vacuum, but in the event of engine stall the power brake will still work. Federal standard FMVSS 105 calls for at least one power-assisted stop if engine power is lost. Pedal effort will increase substantially if power is lost, but the first stop will be normal. The second stop will have less assist than the first stop, the third stop will have less assist than the second stop, and so on until there is no further assist.

Brake Booster Check Valve

The booster is able to provide reserve braking because it has a check valve, which is typically mounted on the front of the brake booster at the end of the hose that supplies vacuum from the intake manifold (**Figure 9.26**). The check valve is a one-way valve that prevents manifold vacuum from leaving the power booster if engine vacuum drops below the tension of its spring or if the engine is shut off (**Figure 9.27**). When vacuum drops, the check valve moves against its seat to prevent air from entering the reservoir.

As the load on the engine increases or decreases, intake manifold vacuum changes. Without a check valve to allow the booster to act as a reservoir, a change in engine vacuum would produce a change in the amount of brake power assist. For instance, during

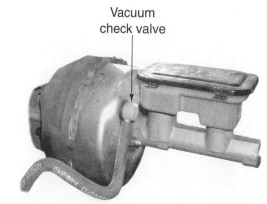

Figure 9.26 A check valve located on the vacuum booster.

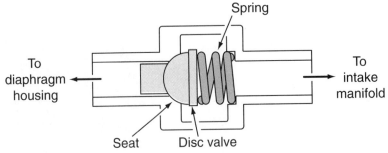

Figure 9.27 Cutaway of a typical vacuum brake booster check valve.

wide-open throttle, engine vacuum drops to zero. During heavy acceleration, vacuum stored in the booster would be depleted and the booster would not work if a quick stop were needed.

Charcoal Filter

Some vehicles use a charcoal filter (**Figure 9.28**) between the manifold and check valve to trap fuel vapors so they cannot enter the booster and damage the rubber diaphragm. Sometimes a fuel filter is installed in the line for the same reason. The vacuum check valve also serves this dual purpose of keeping fuel vapors from reaching the power booster.

Vacuum Booster Operation

The common kind of power brake booster in use today is the *vacuum-suspended* power brake. The booster is a metal chamber divided in half by a rubber diaphragm. Each side is isolated from the other by a valve at the end of the brake pedal apply rod in the back of the power booster.

In a vacuum-suspended power brake system, when the engine is running and the brakes are released, vacuum is present on both sides of a diaphragm inside the booster (**Figure 9.29**). Air movement in the power booster is controlled by valves that are part of the pushrod assembly. They are actuated by movement of the pushrod.

During a stop, the vacuum valve is closed and the air valve is opened in response to pedal movement. This uncovers a passageway for air to be drawn from the passenger compartment into the rear chamber of the booster (**Figure 9.30**). Another passageway is blocked at the same time, preventing the incoming air from entering the front chamber of the power booster. The result is a difference in pressure between the two chambers. The pressure is low in the front chamber compared to atmospheric pressure in the rear chamber.

With a pressure differential between the two chambers, there is a forward assist against the master cylinder apply piston. The amount of assist depends on the difference between the pressures in the front and rear chambers of the brake booster. When the pedal is held lightly, a spring-loaded valve prevents more pedal assist from occurring. During holding, the vacuum and atmospheric valves are both closed.

A power piston is attached to the middle of the rubber brake booster diaphragm. Force from the diaphragm goes from the power piston to a metal rod to apply force to the rear-most piston in the master cylinder. This mechanical connection ensures that a failure in vacuum assist will not result in a loss of braking.

The term *vacuum-suspended* comes from the fact that vacuum is present on both sides of the diaphragm when no braking is taking place. During a stop, atmospheric pressure is directed to the back side of the diaphragm (**Figure 9.31**), resulting in an increase in braking force. The power piston moves with the brake pedal rod, controlling the introduction of air to the back side of the booster diaphragm. The power piston valve assembly is called a floating valve because the valve bore moves with the piston. A large coil spring on the front housing returns the power piston/diaphragm assembly to the unapplied position when the brakes are released.

Figure 9.28 A charcoal filter is sometimes installed in the vacuum line to the booster.

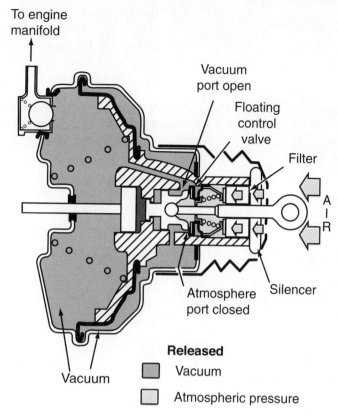

Figure 9.29 A power brake in the release position.

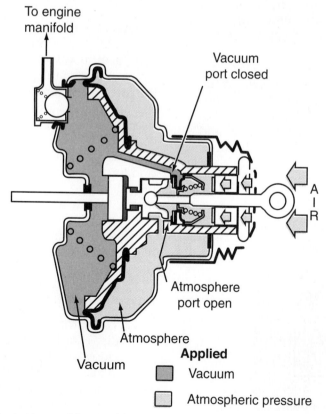

Figure 9.30 A power brake in the applying position.

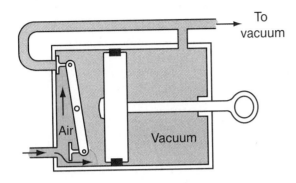

Figure 9.31 A vacuum-suspended power brake.

Power Booster Air Filter

During application of the brakes, air enters the power booster through a filter behind the pedal pushrod boot (refer to **Figure 9.30**). The filter cleans the air as it is drawn into the booster from the passenger compartment. In the same manner that the engine air filter silences the noise of the incoming air into the engine, the brake booster filter reduces the hissing sound of the air entering the power brake booster from the passenger compartment. In fact, if you listen carefully this noise can be heard while the engine is off and the brakes are applied, which can be useful in diagnosing problems with the power brake system.

Types of Vacuum-Suspended Brake Boosters

There are three main types of vacuum-suspended power brake boosters. Two of them use a single diaphragm, one with a reaction disc and the other with a reaction lever. The other design, called a tandem booster, uses two diaphragms to double the amount of force available from the same diameter reservoir.

Pedal Feel. To provide pedal feel, a reaction disc or a reaction plate and levers are used. The meaning of "reaction" is that as force is applied to the master cylinder, an equal force is developed in the opposite direction. The booster is capable of developing hundreds of pounds of force in addition to what is applied by the driver. In theory, the booster could force the pedal all the way back up to its unapplied position. The reaction disc or plate and levers are used to prevent this from happening and to provide a realistic brake feel. The amount of force returned to the driver's foot is only 20–40 percent of the force applied by the booster. Although the feedback force is less than the booster force, they are always proportional and the feedback is always the same percentage of the booster force.

Reaction Disc Booster. One single-diaphragm design has a rubber reaction disc applied by an input pushrod and vacuum valve plunger (**Figure 9.33**). The disc, which is part of the power piston, is compressed under the force of the brake pedal. The air valve moves into the application position (allowing air to enter the rear chamber) when there is sufficient pedal force to compress the reaction disc. During braking, the compressibility of the reaction disc allows it to absorb reaction force back from the master cylinder. In addition to providing pedal feel, as the disc is compressed it also moves the air and vacuum control valves to regulate the amount of pressure that is applied by the diaphragm. As more force is applied to the pedal, the reaction disc is compressed further and more feedback is returned to the pedal as increased pedal feel.

A booster has three stages of operation: applying, holding, and released.

1. When the brakes are being applied, the reaction disc is fully compressed. First, the vacuum valve closes and then the air valve opens.
2. When the brakes are in the holding position, the reaction disc is only partly compressed and the floating valve closes off both passageways to maintain the pressure that has already been applied.

Vintage Chassis

Very early power brakes used a system called atmospheric (air) suspended brakes (**Figure 9.32**). When the brakes were released, there was atmospheric pressure on both sides of the diaphragm. During a stop, vacuum was routed to the front side of the diaphragm. This system had some shortcomings not shared by vacuum-suspended brakes, including the necessity for a vacuum reservoir and inconsistent braking effort. These brake boosters were not used beyond the late 1950s.

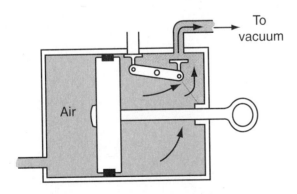

Figure 9.32 An atmospheric (air) suspended power brake.

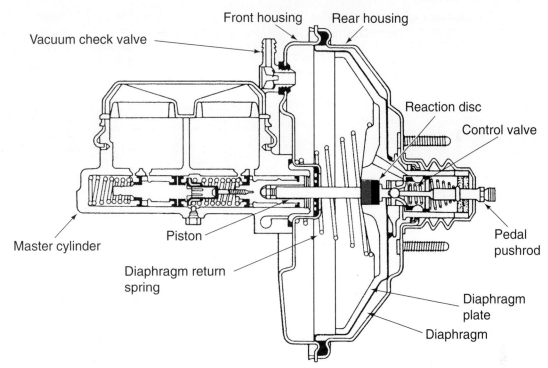

Figure 9.33 A reaction-disc vacuum booster.

3. When the brakes are released, the reaction disc is not compressed at all and the diaphragm and valves are returned to the back of the booster by the diaphragm return spring. The air valve is closed and the vacuum valve between the chambers is open.

NOTE: *With the vacuum valve open and the air valve closed, vacuum will equalize on each side of the diaphragm. Unless the booster is defective, this is the only stage of booster operation in which air will be able to enter the vacuum side. As the brakes are released, a minor increase in engine rpm might be noticed when the air from the back side of the diaphragm is drawn into the vacuum chamber.*

When there is enough pedal pressure to compress the reaction disc completely, maximum assist occurs. This point, called *runout,* occurs when there is full intake manifold vacuum on one side of the diaphragm and full atmospheric pressure is on the other side. When runout has occurred, there is no further power assist. At this point, a further increase in hydraulic pressure to the brakes can only happen due to extra pressure from the driver's foot.

Plate and Lever Booster. Another single-diaphragm booster design uses a reaction plate and levers to provide reaction force to the brake pedal (**Figure 9.34**) similar to the resistance and feel of nonpower brake systems. The fixed end of each lever contacts the power piston. The levers' other ends, which are spring-loaded, can move. During application of the brakes the springs are compressed, allowing the ends of the levers to contact the vacuum and air valves. Movement of the vacuum and air valves starts to bleed vacuum and add atmospheric pressure to the back side of the booster diaphragm, increasing reaction force to the brake pedal.

Tandem Brake Booster. The amount of assist that a booster can provide is directly proportional to the area of its diaphragm. A 9-inch-diameter diaphragm provides less assist than a 10-inch-diaphragm. The size of the diaphragm is limited by the position of the brake cylinder in the engine compartment. If a larger diaphragm is desired, the solution is to use a tandem booster. This booster has two diaphragms located in

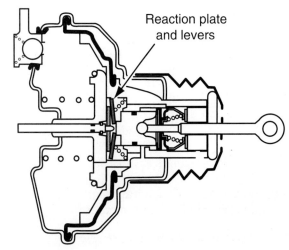

Figure 9.34 Reaction plate and lever vacuum booster.

the same housing. The amount of force available is proportional to the total area of the two diaphragms.

NOTE: *If the diaphragms are 9 inches in diameter and each diaphragm has 63.58 inches of area, the total amount of application force will be that of 127.16 square inches. A = pi × R² (Area of a circle using 9-inch-diameter diaphragm)*

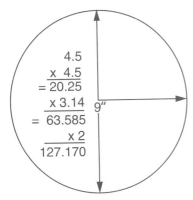

Figure 9.35 shows a tandem power booster. The two separate diaphragms have a support plate. Some tandem boosters use a housing divider between the diaphragms. Because it has an extra diaphragm and chamber, a tandem booster is slightly longer than a single diaphragm booster by about 20–25 percent.

The tandem booster works in the same manner as a single-diaphragm booster. However, the air and vacuum valves must control conditions on both sides of both diaphragms at the same time. With the brakes released, vacuum is present on both sides of

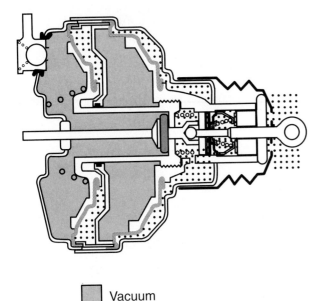

Figure 9.35 A tandem booster in the applied position. Vacuum is on the front side of each diaphragm.

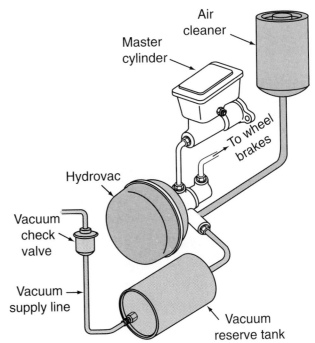

Figure 9.36 Pressure multiplier or Hydrovac booster.

each diaphragm. During a brake application, air is leaked into only the rear (brake pedal side) of each diaphragm.

Another style of booster is the *pressure multiplier*. This type is found on many motor homes and trucks. It is typically called a *Hydrovac* brake (**Figure 9.36**). The unit operates the same way as other vacuum-suspended boosters, but it has its own brake fluid piston. Pressure from a conventional master cylinder is routed to the Hydrovac unit, where it is multiplied and sent to the wheel cylinders. Since 1967 vehicles have tandem master cylinders, which requires that there be two Hydrovac units, one to serve each half of the master cylinder.

Some vehicles in the 1950s used a linkage-style booster. A linkage from the pedal applied the master cylinder and the boost was to the brake pedal.

Power Booster Mounting Location

Most boosters are mounted directly behind the master cylinder. The front of the booster has a seal that keeps vacuum from drawing brake fluid from the master cylinder into the booster and intake manifold. The master cylinder's rear-most piston is applied by a rod in the power booster that is carefully adjusted so the piston can return to the unapplied position behind the compensating port (see Chapter 4).

Function of Vacuum in the Brake Booster

A vacuum brake booster uses low pressure to provide an assist at the brake pedal.

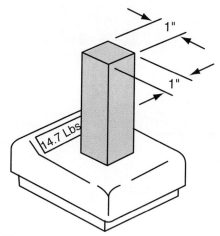

Figure 9.37 If temperature is 68°F, atmospheric pressure at sea level is 14.7 psi.

NOTE: *Low pressure is commonly referred to as vacuum. True vacuum is the complete absence of air, but this condition is only possible in space or in a laboratory. In the United States, pressure is measured in pounds per square inch (psi). Vacuum (low pressure) is measured in inches of mercury (in. Hg). When the temperature is 68°F, atmospheric pressure at sea level is 14.7 psi (**Figure 9.37**).*

An understanding of engine vacuum is important to your ability to diagnose power brake booster problems. The vehicle's engine is an air pump. Each intake stroke results in suction in the intake manifold that results in low pressure of about 16–20 in. Hg at idle. When the brakes are not in use, this *intake manifold vacuum* is applied to both sides of the diaphragm in the vacuum booster.

Atmospheric pressure is what is in the passenger compartment at the back side of the power brake booster. Remember, the power booster is mounted on the firewall and its back side extends into the passenger compartment. This air pressure is waiting to be used to assist in applying the brakes (**Figure 9.38**).

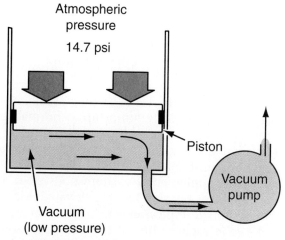

Figure 9.38 A difference between atmospheric pressure and vacuum is used to assist in applying the brakes.

NOTE: *A complete vacuum is either 29.9 in. Hg below atmospheric pressure or 14.7 psi below atmospheric pressure. Rounding off 29.9 to 30 and 14.7 to 15 makes it easier to compare the two.* **Figure 9.39** *compares vacuum and pressure equivalents. You can see that the ratio is approximately 2:1. There are two handy tips for conversions:*

1. *To convert psi to inches of mercury (in. Hg), multiply by 2*

2. *To convert inches of mercury (in. Hg) to psi, divide by 2*

For proper operation of the power brake, manufacturers typically specify a minimum vacuum level of 15 in. Hg at idle. If intake manifold vacuum is 15 in. Hg and atmospheric pressure is 15 psi, what is the pressure differential? The answer is 7.5 psi pressure differential:

Vacuum side pressure = 15 in. Hg ÷ 2 = 7.5 psi

So:

Atmospheric side pressure = 15 psi − 7.5 psi = 7.5 psi (pressure differential)

In this example, a power booster makes about 7.5 pounds of force for every square inch of diaphragm

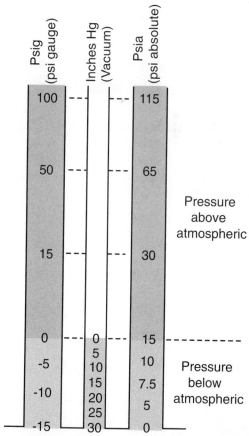

Figure 9.39 When pressure is measured from a theoretical perfect vacuum, which is 29.2 in. Hg, the pressure reading is expressed in pounds per square inch absolute (psia). Pressure measured on a gauge that measures vacuum (pressure less than atmospheric) is expressed in pounds per square inch gauge (psig).

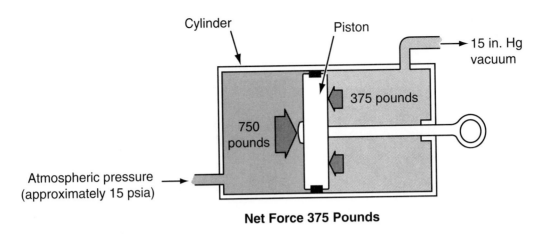

$$15" \text{ vacuum} = 7.5 \text{ psi absolute}$$
$$15 \text{ psi} \times 50 \text{ square inches} = 750 \text{ pounds force}$$
$$7.5 \text{ psi} \times 50 \text{ square inches} = 375 \text{ pounds force}$$
$$\overline{375 \text{ pounds pressure differential}}$$

Figure 9.40 Explanation of vacuum assist.

area. If the diaphragm has 50 square inches of area, it will develop 375 pounds of extra force to add to the effort of the driver's foot (7.5 psi × 50 square inches = 375 pounds). **Figure 9.40** explains how this works.

At higher altitudes, brake assist will be diminished. Engine vacuum is reduced by about 1 inch for every 1,000 feet in elevation above sea level (**Figure 9.41**).

Auxiliary Vacuum Reservoir

Some late-model cars and cars with high-performance engines use a separate vacuum reservoir to store vacuum so braking effort is always consistent (**Figure 9.42**). An auxiliary reservoir is necessary when an engine has been equipped with an aftermarket camshaft that has high valve overlap. This results in

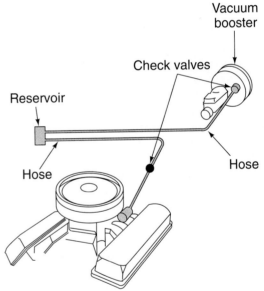

Figure 9.42 An auxiliary vacuum reservoir is used with diesel engines and those with turbochargers.

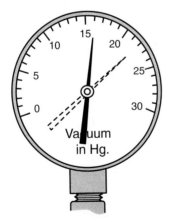

16 to 22 in. Hg of vacuum is considered normal for altitudes below 1,000 feet

Subtract 1 inch for every 1,000 feet higher.

Figure 9.41 Normal engine vacuum.

engine vacuum at idle that is far less than the 16–20 in. Hg typical of a stock engine. A high-performance engine might have 10 in. Hg of vacuum at idle. As rpm increases, vacuum quickly jumps up to normal levels as increasing engine speed allows the engine to run more efficiently. The auxiliary reservoir stores vacuum so it can help limit changes in braking. When a brake system has a supplemental vacuum booster in addition to the power booster, two check valves are used. In addition to the typical check valve installation at the end of the vacuum line at the power booster, an additional check valve is installed between the intake manifold and the reservoir (**Figure 9.43**).

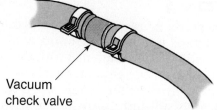

Figure 9.43 An in-line vacuum check valve.

Auxiliary Vacuum Pump

Diesels do not make engine vacuum so they need an auxiliary vacuum pump. This device uses a diaphragm to create a vacuum that can provide normal engine intake manifold vacuum. Some late-model cars with gasoline engines have high vacuum demands. Other than supplying the needs of the power brakes, vacuum devices are used to power many other auxiliary devices on vehicles. These include heating and air conditioning, cruise control, and emission devices. As more vacuum is used to supply these devices, especially at idle, less intake manifold is available for power brake assist.

An auxiliary vacuum pump can be driven by crankshaft-driven belt, the engine's camshaft (**Figure 9.44**), or an electric motor. Sometimes the pump is driven off the back of the engine's alternator (**Figure 9.45**).

An electrically controlled pump can be operated by the on-board computer. When it senses low vacuum in the brake booster, the computer controls a relay that supplies power to the pump's motor. Electrically driven pumps are either diaphragm or vane types. A diaphragm pump moves a diaphragm to create a vacuum. A vane pump has several vanes that rotate in an eccentric cavity to create a vacuum. As the vanes move through the eccentric cavity, air is pumped out and replaced by a vacuum.

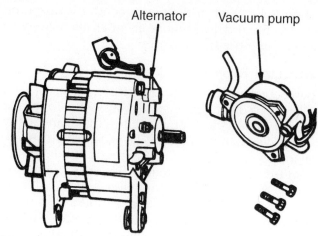

Figure 9.45 An auxiliary vacuum pump driven off the back of the alternator.

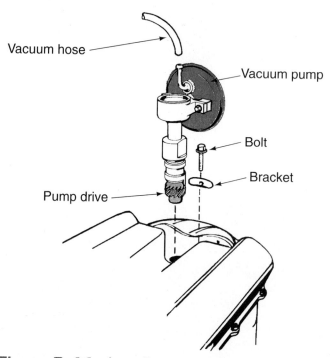

Figure 9.44 An auxiliary vacuum pump.

POWER BRAKE SERVICE

Repairs to power boosters include replacement of the hose, filter, or check valve. Power boosters can usually be disassembled, but they are not commonly repaired in the field any longer. New and rebuilt units are available. You can purchase a power booster by itself or with a rebuilt master cylinder attached.

Most complaints regarding power brakes have to do with increased pedal effort. This could be due to several things, not all of which are brake-related. According to brake manufacturers, a high percentage of power brake boosters returned during warranty are not defective but misdiagnosed. Manufacturers call these costly errors NCFR (no cause for removal). Many of the NCFR complaints are related to incorrect brake bleeding techniques.

Another NCFR complaint results from the installation of larger diameter tires and wheels on pickup trucks and SUVs. Large tires require more leverage to be exerted by the wheels against the brake system. Hard pedal effort is the result. A smaller bore master cylinder or larger bore wheel cylinders can help correct the problem.

A defective power booster results in a hard brake pedal. A low brake pedal or a soft brake pedal are *not* power brake–related complaints. A soft pedal is typically due to air in the hydraulic system and a low pedal can be related to brake adjustment or a failure in one side of the master cylinder's hydraulic system.

Vacuum Booster Operation Test

To test the operation of a power brake booster, exhaust all vacuum reserve from the power booster by applying the foot pedal several times with the engine off. You should hear the sound of the air rushing into the reservoir. Next, hold your foot on the pedal while starting the engine. If the power booster is operating correctly, the pedal will move about an inch closer to the floor after the engine starts.

Vacuum Booster Leak

A power brake booster can experience problems such as a hole in a diaphragm or a sticking valve. To test for a booster internal leak, shut off the engine and apply steady pressure to the brakes. The pedal height should remain constant for at least 30 seconds. If the booster has an internal leak, the pedal will slowly rise during this test.

To test for a leaking valve, connect a vacuum pump to the check valve inlet to the booster (**Figure 9.46**). Apply 20 inches of vacuum (20 in. Hg) while the brake pedal is released. Wait a few minutes to see that the vacuum remains steady. If vacuum drops, a control valve leak might be the cause. A front seal on the booster housing might also be the cause, but this could not be due to a leak in the diaphragm. The diaphragm should have equal pressure on both sides of it when the brakes are released, so a leaky diaphragm would not change the results of this test. A leaky air valve, on the other hand, would allow air to enter the vacuum chamber and lower the vacuum reading on the vacuum pump's gauge.

If vacuum holds steady, have a helper apply the brake pedal firmly. This should result in a rapid drop in the vacuum reading as air enters the rear chamber. While your helper maintains pressure on the pedal, apply 20 in. Hg once again. The vacuum reading should not drop more than 2 in. Hg in following 30 seconds. If it does, the booster could have a leaky diaphragm or vacuum control valve.

Brake Drag

A defective power booster can sometimes cause brakes to drag, resulting in rapid brake lining wear. To verify this, apply the brakes several times with the engine off. With the wheels off the ground, rotate the tires. They should turn freely. Start the engine and recheck to see that the wheels still turn freely. If not, the booster is attempting to apply the brakes. In the passenger compartment, verify that there is free travel between the brake pedal apply rod and the back of the power booster. If not, adjust the pedal apply rod. If there is free travel but the brakes still drag, replace the booster.

Vacuum Supply Checks

Check the hose that supplies vacuum to the power booster from the intake manifold. It should not be damaged, hard, or swollen. A replacement hose to a power booster must be reinforced and fuel-resistant, because fuel will damage the diaphragm in a power booster. Some hoses have an in-line filter in them to prevent fuel from entering the power booster during an intake manifold popback. The filter also prevents evaporating hydrocarbons from leaking into the passenger compartment.

To check for a restricted hose, disconnect the hose to the booster with the engine running. If flow is sufficient, an air leak (vacuum leak) this large should cause the engine to stall or stumble severely. Observe the inside of the hose to see that it is dry. Brake fluid in the hose can be due to a defective secondary seal on the rear piston in the master cylinder. In this instance, the rear master cylinder chamber will also experience unexplained fluid loss.

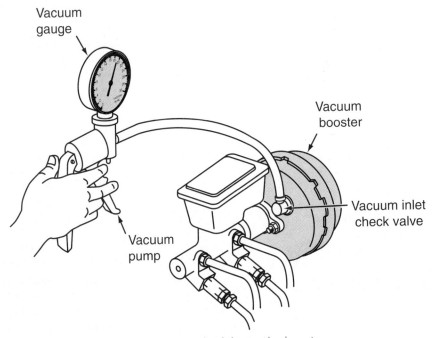

Figure 9.46 Connect a vacuum pump to the check valve inlet to the booster.

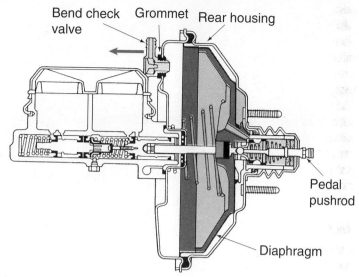

Figure 9.47 Bend the check valve against its rubber sealing grommet to make it leak air.

The brake booster has a one-way check valve that traps vacuum when the engine is off. If the check valve is bad, the brake effort will vary according to the load on the engine. Checking the power brake check valve is done with the engine off. Carefully bend the valve against its rubber grommet (**Figure 9.47**). If the check valve is holding, air will rush into the front part of the booster. With the valve removed, you should be able to blow through it in one direction and it should seal from the other side.

Oil in the hose can be an indication of a faulty vacuum check valve. A defective check valve can allow fumes from the crankcase to enter the hose and booster when the vacuum supply stops when the engine is shut off.

When you replace a check valve, use water or brake fluid to lubricate the grommet during installation. Check valves are plastic and can be damaged with careless handling (**Figure 9.48**).

Some manufacturers use an in-line check valve in the vacuum hose. To test check valve operation in these cars, remove the vacuum hose at the booster side and apply vacuum. If vacuum holds, the check valve is good.

Checking Vacuum

If there is a problem with braking effort, install a vacuum gauge into the line at the power booster (**Figure 9.49**). A typical *minimum* vacuum specification at idle is 15 inches of mercury. If vacuum is lower than that, several things could be the cause:

- The engine could have a high-performance camshaft
- The ignition timing could be retarded
- The car could be at high altitude
- There could be damage or a restriction in the vacuum supply line
- There could be an engine intake manifold vacuum leak

Before condemning a booster, drive the vehicle with a vacuum gauge connected to the engine. At cruising speed, apply the brakes several times and verify that the vacuum supply remains sufficient.

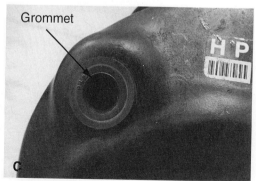

Figure 9.48 (A) A power brake check valve installed in its grommet. (B) A typical check valve is made of plastic and is fragile. (C) Lubricate the grommet with water or brake fluid before reinstalling the check valve. *(Courtesy of Tim Gilles)*

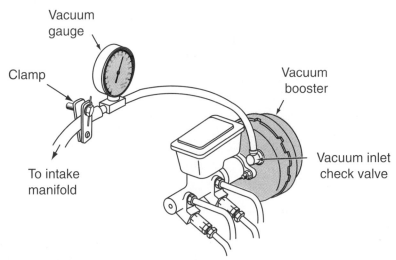

Figure 9.49 Use a vacuum gauge to check vacuum supply. You can check to see that it holds vacuum by clamping off the line.

With the engine running, pinch off the vacuum hose on the engine side of the T with pinch pliers. Vacuum should be maintained for at least 15 seconds without a loss of more than 1 in. Hg. If leakage is more than this, double-check to see that your pinch pliers are thoroughly clamping the hose. Otherwise, the check valve is defective or the booster has an internal leak.

Leaking Power Booster Front Seal

A leak in the seal in the front of a power booster can allow fluid to be drawn into the power booster and burned in the intake manifold. The rear chamber of the master cylinder will continually use fluid. The key to diagnosing this is recognizing that there is no evidence of external fluid leakage.

Replacing a Power Brake Booster

Most shops do not rebuild power brake boosters. Rebuilt units are available for most cars and trucks. Before removing the power booster, remove the nuts that fasten the master cylinder to it. The brake lines on most cars can be left on the master cylinder so the brakes will not require bleeding after booster replacement. Sometimes the lines must be removed to allow enough clearance to remove the booster.

A typical power booster has four studs that protrude through the bulkhead into the passenger compartment (**Figure 9.50**). When these studs are easily accessible, replacement of the booster is not a difficult job. On some cars, the pedal support will drop when the booster is removed. On these cars, remove the brake light fuse before removing the booster.

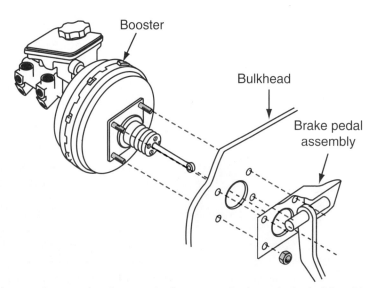

Figure 9.50 A typical power booster has four studs that protrude through the bulkhead into the passenger compartment.

Booster Pushrod Clearance Adjustment

When replacing a power booster, it is important that the pushrod depth be correct. If the pushrod is too high, the master cylinder compensating port will be covered, resulting in brake drag after the fluid warms. A pushrod that is too low will cause excessive brake pedal travel. There are special tools and different methods specified for making this adjustment (**Figure 9.51**). Complete rebuild units are available that include a rebuilt master cylinder assembled to them. These will already be correctly adjusted.

The following is a method of verifying that the booster pushrod adjustment is correct:

1. Remove the master cylinder reservoir cap and cover the opening with clear plastic wrap so you can observe the fluid. Have a helper apply the brakes. A slight movement of the fluid should be observed during both application and release of the brakes.
2. If there is no movement during pedal application, the pushrod is set too high. This causes the primary cup in the master cylinder to be positioned past the compensating port, which is why no fluid movement can be observed. Loosen the master cylinder mounting nuts and position it away from the booster with some cardboard shims. If the fluid movement is now normal as the pedal is depressed, shorten the adjustment on the pushrod and reinstall the master cylinder on the front of the power booster.
3. Excessive fluid movement indicates too low an adjustment, which allows the master cylinder pedal rod to move too far before the primary piston cup covers the compensating port. Lengthen the pushrod and check once again for the correct amount of fluid movement on pedal application and release.

■ HYDRAULIC POWER BRAKE BOOSTERS

Some power brake systems use fluid pressure instead of vacuum to produce the power assist. These are commonly found on cars with diesel engines because diesels do not have a throttle plate and do not produce sufficient vacuum to operate a vacuum booster. Turbocharged engines also have problems with vacuum supply because they have pressure in the intake manifold when the turbocharger is working. Hydraulic boost systems are found on these engines and also on some older gasoline engine vehicles that might have lower than normal engine vacuum available, usually due to emission control systems.

As mentioned earlier, some low vacuum engines use an auxiliary vacuum pump to accommodate a normal vacuum brake booster. When an auxiliary vacuum system is not used, there are two types of hydraulic power brakes that can replace it. One type uses the power steering system and the other uses an electric motor to create the pressure.

■ MECHANICAL HYDRAULIC POWER BRAKES

A power brake system powered off the power steering pump was developed by Bendix. This is called a **Hydro-boost brake** (**Figure 9.52**). A Hydro-boost power brake is located in the same place and fits into a similar amount of area as a conventional vacuum power booster.

In a Hydro-boost system, power steering fluid is also used to boost braking pressure. The hydraulic fluid used to assist steering and braking operates within the same system and returns to the same reservoir. Although the power steering pump also provides the hydraulic pressure for brake assist, the power steering and braking systems are totally isolated from one another and the power braking system uses its own fluid. Power steering fluid and brake fluid cannot be allowed to mix.

The power steering pump pressurizes the hydraulic brake booster fluid to the hydraulic booster assembly. The booster has a large spool valve that directs the flow of hydraulic fluid through the booster assembly. During application of the brakes, the spool valve directs the fluid to a power chamber where a boost piston reacts to the pressure and moves the master cylinder piston forward. The boost piston is the equivalent of the diaphragm and power piston in a vacuum booster. The master cylinder operates in the same manner as a conventional master cylinder.

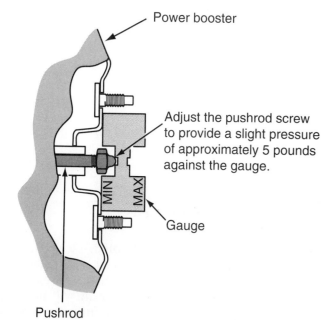

Figure 9.51 A tool for checking power booster pushrod depth.

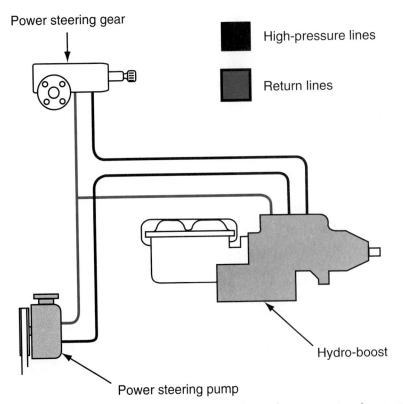

Figure 9.52 A Hydro-boost power brake operates off pressure from the power steering pump.

The Hydro-boost system includes an accumulator, a pressurized storage reservoir with a spring, or compressed gas behind a sealed diaphragm. The accumulator stores hydraulic pressure after the engine is off or in the event of a power steering hose or belt failure. Federal law requires reserve braking if there is a loss of vacuum or hydraulic pressure. During a loss of hydraulic pressure the accumulator allows sufficient reserve for one to three additional power-assisted stops. After that reserve is expended, there is still a mechanical connection between the pedal and the master cylinder, so the brakes will still operate but will require much more pedal pressure.

Prior to 1978, Hydro-boost systems used a spring-loaded accumulator. Systems made after that time used pressurized nitrogen gas. Since 1981, the accumulator has been built into the power piston and is housed within the booster assembly.

Hydro-Boost Inspection

The Hydro-boost system depends on a power steering pump that must be in good operating condition. Be sure the pump drive belt is in good condition and properly tensioned. Check to see that the fluid level is correct and that the fluid does not smell burnt or appear dirty.

NOTE: *Some units use a special Hydro-boost fluid that is not ATF or power steering fluid.*

If there is a leak in the hydraulic system, verify its location with the engine running while an assistant alternately turns the steering wheel against a steering lock and applies the brakes (**Figure 9.53**). Both of these conditions call for high pump pressure, which can cause a small leak to become apparent. Be sure the hydraulic section of the service brakes is free of air, which will prevent an accurate diagnosis of the Hydro-boost system.

If pump operation is in question, a pressure test can be performed. Follow the test equipment manufacturer's instructions and refer to power steering pressure specifications found in the service literature for the vehicle.

The following test can be done to check a Hydro-boost for proper function:

1. With the engine off, depressurize the accumulator by depressing the brake pedal several times.
2. With your foot backed slightly off the pedal from holding the brakes, start the engine.
3. The pedal should kick back and then return to its normal height.

Hydro-Boost Repair and Replacement

A Hydro-boost can be disassembled, cleaned, and reassembled with new seals. A special seal installer is needed to install seals on the input shaft. If you do not do these repairs regularly, a rebuilt unit will be your best option. If you disassemble the unit, be sure to relieve the accumulator pressure first and consult the appropriate shop library information.

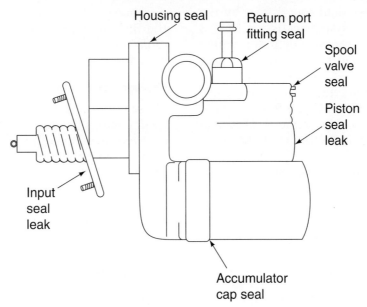

Figure 9.53 Possible sources of Hydro-boost leaks.

A Hydro-boost unit is installed behind the master cylinder, like a conventional vacuum power booster (**Figure 9.54**). To remove the unit, disconnect the lines from the power steering pump and remove the master cylinder without disconnecting the brake lines. Four studs and nuts fasten the unit to the bulkhead. Locate the nuts inside the passenger compartment below the brake pedal and remove them. Reverse the removal procedure to install the replacement unit.

Hydro-Boost Bleeding

After the hydraulic lines are reconnected to the power steering pump, air must be bled from the lines. Fill the reservoir and crank the engine for several seconds without starting it. Top off the reservoir with fluid. Then start the engine and turn the steering wheel twice from stop to stop. Shut the engine off and step on the brake pedal several times to bleed the accumulator of pressure before topping off the reservoir.

■ ELECTRIC HYDRAULIC POWER BRAKES

The hydraulic power brake system developed by General Motors is called the Powermaster (**Figure 9.55**). It uses an auxiliary electric motor–driven pump to develop hydraulic pressure for the brake system. Because the Hydro-boost system shares a hydraulic pump for both the brake and power steering systems, there is a potential for possible problems. The Powermaster system was developed in response to that potential.

There are other advantages to the Powermaster system. The electric pump only works when needed, so fuel efficiency is good. The need for hoses from the power steering pump is eliminated with this design.

A Powermaster has a self-contained power booster built into the master cylinder and uses a pump driven

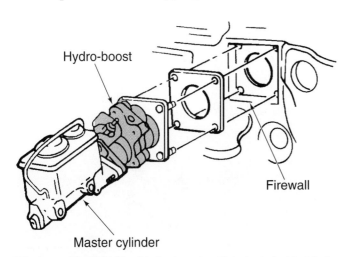

Figure 9.54 The Hydro-boost unit is installed behind the master cylinder.

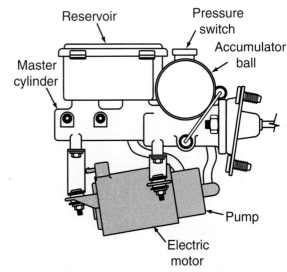

Figure 9.55 A Powermaster brake booster.

by an electric motor rather than one driven by the power steering pump. The system calls for DOT 3 brake fluid, supplied by the master cylinder, which lubricates the pump and provides for normal hydraulic brake application.

The Powermaster master cylinder has three partitions for fluid rather than two (**Figure 9.56**). Two of the reservoir chambers are for the service brakes, and a third, larger reservoir holds the fluid for the brake booster. The bottom of the reservoir has two ports. One of these supplies the fluid for the pump and the other is a return port.

Fluid flow from the reservoir to the booster and back is quite fast, faster than normal brake fluid movement. For this reason, the reservoir chamber for the booster is vented to the atmosphere rather than being sealed as in a normal master cylinder. The fluid for this part of the system is not intermingled with the normal brake fluid and its boiling point is not critical because high temperatures are not a factor in system operation. It will absorb moisture from the air, but this is not a concern because the fluid will not boil.

Connected to the power booster section of the master cylinder are an accumulator and a pressure switch. The accumulator is charged with nitrogen gas on the back side of a diaphragm. The pump fills the accumulator with fluid, against the nitrogen pressurized diaphragm.

Unlike the Hydro-boost system, which uses the accumulator for emergency stops only, "boost" pressure from the Powermaster accumulator is used during all braking operations. A pressure switch controls the pump, maintaining 510 to 675 psi in the accumulator. When the pump is off, a check valve maintains pressure in the accumulator. The pressure switch is also part of the instrument panel warning lamp circuit, which warns the driver if there is a problem with the system.

The fluid level normally appears to be low because the fluid is in the accumulator. Fluid should not be added to the chamber unless the accumulator has been discharged, which will fill the reservoir back up. Filling the fluid without discharging the accumulator can cause the reservoir to overflow when the accumulator does discharge.

Powermaster Service

For safety, before servicing Powermaster parts relieve accumulator pressure by depressing the brake pedal at least ten times with the key off. Hide the key in a safe place prior to this. Turning on the key repressurizes the system.

Powermaster service can include replacement of the return hose, pressure switch, accumulator, or pump assembly while the unit remains on the car (**Figure 9.57**). Removing and rebuilding the unit is also an option in a dealership. But in the aftermarket, rebuilt units are typically installed.

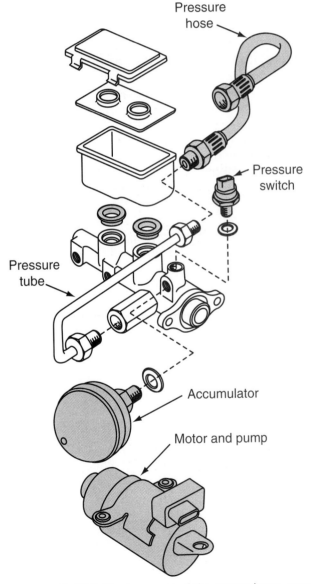

Figure 9.57 Replacement of the return hose, pressure switch, accumulator, or pump assembly can be done while the unit remains on the car.

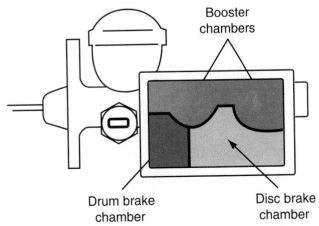

Figure 9.56 A Powermaster master cylinder has three partitions for fluid rather than two.

Diagnostic procedures include connecting a pressure gauge to the Powermaster unit. Excessively high pressure can point to problems with a pump, switch, accumulator, or booster.

Replacing a Powermaster Booster

Before replacing a Powermaster booster, disconnect the electrical connection for the pump motor. The brake lines will also need to be removed. Remember, the booster is integral with the master cylinder, so the pedal pushrod will have to be removed before unbolting the master cylinder from the bulkhead.

The master cylinder section of the Powermaster must be bench bled prior to installation (see Chapter 4). After installation, fill the third reservoir with brake fluid, connect the electrical connector to the pump, and turn on the ignition switch. The pump will run but should shut off within 20 seconds. If the pump operates beyond 20 seconds, consult the appropriate service information. Remember that the pump section will only appear to have the correct fluid level after the accumulator has been discharged.

REVIEW QUESTIONS

1. The parking brake must operate independently of the _____ brakes.
2. Which parking brake must hold a vehicle on a steeper hill, one on an automatic equipped vehicle or one on a manual transmission vehicle?
3. What is the name for the internal drum-type parking brake, housed within a section of the rear disc rotor?
4. Which has more pedal travel, four-wheel disc brakes or hybrid disc/drum brakes?
5. List two instances where a hydraulic assist brake booster might be needed.
6. With a brake booster, a master cylinder can have which type of bore—larger or smaller?
7. What is the name of the part that allows a vacuum brake booster to continue to be able to provide reserve braking after the engine is off?
8. Some vehicles have a filter between the engine's intake manifold and the power brake booster's check valve to keep _____ _____ from entering the booster and damaging the rubber diaphragm.
9. In a vacuum-suspended power brake system, when the brakes are released when the engine is running, what condition is found on both sides of the booster diaphragm?
10. What is the term for the point at which a power booster has reached maximum assist with full intake manifold vacuum on one side of the diaphragm and full atmospheric pressure on the other side?
11. What is another name for the pressure multiplier power brake system found on many motor homes and trucks?
12. To convert inches of mercury (in. Hg) to psi, do you divide or multiply by 2?
13. At higher altitudes, engine vacuum is reduced by about 1 in. Hg for every _____ in elevation above sea level.
14. What is the minimum engine vacuum level specified by most manufacturers for proper power brake operation?
15. What is the name of the power brake system that is powered off the power steering pump?

PRACTICE TEST QUESTIONS

1. Technician A says an integral rear disc parking brake hydraulically self-adjusts like a normal disc brake. Technician B says an integral rear disc parking brake requires a self-adjusting mechanism. Who is right?

    A. Technician A   B. Technician B
    C. Both A and B   D. Neither A nor B

2. Technician A says when new front disc pads are installed, a special tool is required to retract the pistons in their bores. Technician B says that for disc brakes with a self-adjusting parking brake, caliper pistons must be rotated during installation of new pads. Who is right?

    A. Technician A   B. Technician B
    C. Both A and B   D. Neither A nor B

3. Technician A says to apply liquid lubricant to brake cables. Technician B says on drum brake cars, changing the service brake adjustment will affect the parking brake adjustment. Who is right?
   A. Technician A   B. Technician B
   C. Both A and B   D. Neither A nor B

4. Two technicians are discussing vacuum booster operation during a stop. Technician A says engine intake manifold vacuum is directed to the back side of the diaphragm, resulting in an increase in braking force. Technician B says air enters the power booster through a filter behind the pedal pushrod boot. Who is right?
   A. Technician A   B. Technician B
   C. Both A and B   D. Neither A nor B

5. Which of the following is *not* a stage of vacuum brake booster operation?
   A. Released   B. Applying
   C. Holding   D. Boosting

6. Technician A says air enters the front side of the vacuum booster from the back side when the brakes are first released. Technician B says when the brakes are released, a minor increase in engine rpm might be noticed. Who is right?
   A. Technician A   B. Technician B
   C. Both A and B   D. Neither A nor B

7. Technician A says a low brake pedal can be caused by a defective power booster. Technician B says a soft brake pedal can be caused by a defective power booster. Who is right?
   A. Technician A   B. Technician B
   C. Both A and B   D. Neither A nor B

8. A power brake is being tested with a vacuum pump connected to the check valve inlet while the brake pedal is released. Technician A says if vacuum drops, a control valve leak could be the cause. Technician B says a leaky diaphragm could be the cause. Who is right?
   A. Technician A   B. Technician B
   C. Both A and B   D. Neither A nor B

9. Technician A says if a booster check valve is bad, braking effort will be low but constant. Technician B says braking effort will vary with the load on the engine. Who is right?
   A. Technician A   B. Technician B
   C. Both A and B   D. Neither A nor B

10. Technician A says a defective power booster can cause brakes to drag. Technician B says a defective power booster can cause rapid brake lining wear. Who is right?
    A. Technician A   B. Technician B
    C. Both A and B   D. Neither A nor B

CHAPTER 10

Antilock Brakes, Traction, and Stability Control

■ OBJECTIVES

Upon completion of this chapter, you should be able to:
- ✔ Describe the reason for ABS.
- ✔ Explain the theory of ABS operation.
- ✔ Describe the parts of two-, three-, and four-channel ABS.
- ✔ Explain the differences between integral and nonintegral ABS.
- ✔ Explain how ABS provides traction control and stability enhancement.
- ✔ Explain ABS and normal brake warning light operation.
- ✔ Describe how to bleed ABS brakes.
- ✔ Describe service procedures for ABS brakes.

■ KEY TERMS

acceleration slip regulation (ASR)
antilock brake controller
controller antilock brake (CAB)
dump
electromechanical hydraulic (EH) unit
electronic brake and traction control module (EBTCM)
electronic control module (ECM)
integral ABS
lateral acceleration sensor
nonintegral ABS
rear antilock brake system (RABS)
rear-wheel antilock (RWAL)
traction control system (TCS)
wheel speed sensor

■ INTRODUCTION

The ability of brakes to do their job is limited by the grip of a vehicle's tires to the road surface. If the tires do not slip, the vehicle will go in the direction it is steered. Once it loses traction, however, steering control is lost. Even with the best quality tires, when a car skids there is a loss of stopping ability and control. If a driver could release pressure on the brake pedal just before a wheel locked up, the skid could be avoided. When a wheel stops turning, friction between the tire and road generates heat. This softens the tire, causing it to lose traction.

Total traction loss is referred to as 100 percent slip. A slip rate of 50 percent means that the wheel is rolling at a 50 percent slower speed than a freely rolling tire at the same vehicle speed. Maximum traction occurs at about 10–20 percent slip (**Figure 10.1**).

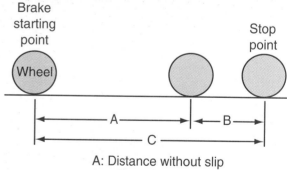

Figure 10.1 Slip rate.

A: Distance without slip
B: Slipped distance
C: Actual distance to stop

$$\text{Slip rate} = \frac{B}{C} = \frac{\text{Vehicle speed} - \text{wheel speed}}{\text{Vehicle speed}}$$

ANTILOCK BRAKES

Most newer cars are equipped with computerized *antilock brake systems (ABS)* to prevent the wheels from locking up (**Figure 10.2**). The ABS uses sensors and a computer to monitor wheel speed. During normal operation an antilock system works like a conventional brake system. It only functions differently when a wheel locks up. Wheel speed sensors measure the rotational speed of the wheel. If a wheel locks up, an **antilock brake controller** pulsates (modulates) the pressure to that wheel in much the same manner as when a race car driver pumps the brakes during a turn to avoid a skid. ABS systems can do this pulsation much faster than a human, however. A typical ABS system can pulsate the pressure to the brake system from ten to twenty times per second.

NOTE: *At some types of auto races, you can hear a pulsating chirp from ABS-equipped cars as the tires lock and unlock during turns. This noise results as the ABS dump solenoid ratchets.*

An antilock system helps avoid loss of control when a wheel loses traction. It does not necessarily result in a shorter stopping distance. One popular saying is, "ABS does not mean you will avoid an accident. It just means you can pick your target."

When a road is bumpy, when there is loose gravel, or when road is slick, less pedal force is needed to activate the ABS. Antilock brake systems are disabled below a certain speed. If the ABS starts working at low speed on an icy road, the vehicle will skid on the ice in a straight path, but still without traction. Stopping on snow is faster if the wheel locks, because the snow builds a mound in front of the tire as it skids.

When the ABS senses a failure, the brake system reverts to conventional-only braking. The ABS light on the dash comes on and a code is set in the vehicle computer, but braking is normal (except for the integral high-pressure systems, which can lose their rear brakes during a failure).

Pedal Feel

Cars with an ABS use the conventional braking system during normal stops. During an ABS stop the pedal feels different than with a conventional braking system. When the ABS becomes active, a small bump followed by rapid pulsation is felt in the brake pedal. The bump is because the pump is returning fluid to the master cylinder reservoir. The pulsation is noticed in some systems more than in others.

During normal stops, there is no pedal pulsation. A pulsating pedal during a non-ABS stop could be due to a warped brake drum or rotor. Some early ABS systems had an increase in pedal height during an ABS stop.

ANTILOCK BRAKE SYSTEM COMPONENTS

Various ABS system designs are covered later in this chapter. Some ABS system components are common to all of the ABS systems, whether they are integral or add-on, or one, three, or four channel. Some systems do not use all of the components listed here.

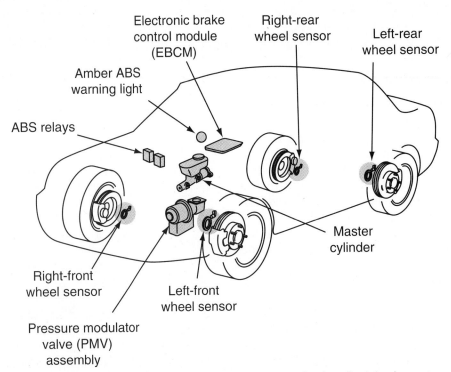

Figure 10.2 The basic electrical and hydraulic components of a four-wheel antilock brake system.

Vintage Chassis

The name *ABS* comes from the German word *antiblockiersystem*. ABS was patented in 1936 by Fritz Ostwald in Germany. It was first used on B-47 jet bombers in 1947 for better stopping on slick runways. Today, most large passenger and military planes have ABS.

In 1970, a prototype automobile with analog controls was developed, but it was the emergence of computerized circuits that finally made automobile ABS a reality. In 1978, Bosch manufactured the ABS 2S. Its cost would be the equivalent of about $5,000 today. During the early 1980s, many European luxury cars imported to the United States came equipped with ABS. With increased production, the cost of antilock brakes dropped. In 1985, Ford introduced the Teves system on some of its cars. Soon after that, other domestic vehicles were equipped with antilock brakes. Teves became known as ITT/Teves for a few years and is now known as Continental Teves.

Electronic Control Unit

The ABS computer is known by several different names and can be found in several places. Often referred to as "the controller," its official names include the **electronic control module (ECM)**, *electronic brake control module (EBCM)*, or **controller antilock brake (CAB)**. If the system includes traction control (covered later in this chapter), the name is **electronic brake and traction control module (EBTCM)**. The controller can be located inside the trunk, in the passenger compartment, or on the master cylinder, or it can be attached to the hydraulic control unit.

Inputs such as those from the wheel speed sensors, brake pedal sensor, and fluid level sensor provided data to the computer. The computer acts on those inputs to correct differences in wheel speed during a loss of traction. Brake fluid output from the master cylinder is interrupted by solenoid-operated pressure modulator valves (PMV). When the computer senses a wheel locking up, electrical current is directed to the solenoid. This energizes a magnetic field to operate the valve. Operation of the solenoid valves is covered later in this chapter. The computer also monitors electronic operation of the system with a self-test every time the ignition system is cycled, and the first time the vehicle is driven after a key cycle.

Wheel Speed Sensors

Each wheel with skid control has a sensor to detect the speed of wheel rotation. **Wheel speed sensors** operate in the same way that the magnetic trigger in the distributors of some electronic ignition systems

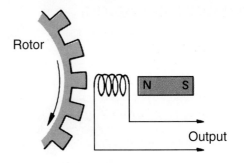

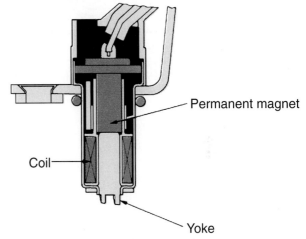

Figure 10.3 Most sensors are a coil of wire wrapped around a permanent magnet core.

operates. Most sensors are *permanent magnet (PM) generators*, with a coil of wire wound around a permanent magnet core (**Figure 10.3**). The sensor is positioned near a toothed ring, called a *tone ring* or *exciter ring*. The toothed ring spins with the wheel.

On front wheels, the tone ring is mounted to the inside hub of a rotor (**Figure 10.4A**) or on the outer CV joint housing (**Figure 10.4B**). As each tooth moves past the magnet, the magnetic field around the coil increases. As the tooth moves away, the strength of the magnetic field weakens. The space between the tooth and the magnet is called the *airgap*. Movement of the teeth causes a constantly changing airgap. These changes in the magnetic lines of flux around the sensor cause alternating voltage. **Figure 10.5A** shows the points where maximum positive and negative voltages are reached. **Figure 10.5B** shows conditions resulting in no voltage induction.

The sensor's frequency changes with the speed of wheel rotation (**Figure 10.6**). The faster the teeth pass the magnet, the higher the voltage output of the sensor and the higher the frequency of the voltage oscillations. When the brakes are applied, the electronic control unit "wakes up" and looks at this information to calculate the speed of the wheel and compare it to the speed of the other wheels.

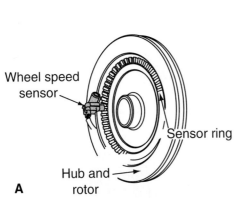

Figure 10.4 (A) An ABS sensor and sensor ring mounted inside the rotor. (B) A wheel speed sensor with the tone ring mounted on the outer CV joint. *(Courtesy of DaimlerChrysler Corporation)*

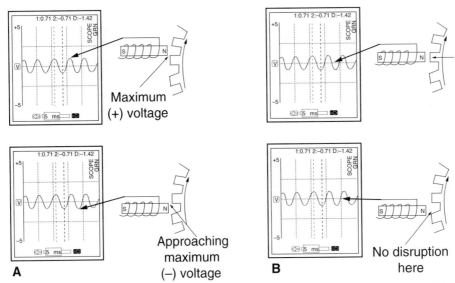

Figure 10.5 (A) Conditions that cause maximum voltage induction. (B) Both of these conditions result in zero volts, because the magnetic field is not being interrupted.

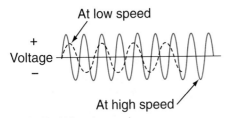

Figure 10.6 Wheel speed sensor.

A **lateral acceleration sensor** is found on some ABS systems. This sensor measures forces encountered while turning. The controller uses this information to change ABS control during hard cornering.

Wheel Speed Sensor Installations. There are many different ABS sensor installations. A rear drum brake ABS system for front-wheel drive cars has the tone ring located inside the brake drum, with the sensor installed through a hole in the backing plate (**Figure 10.7A**). Some have sensors in the hubs or on the back of the hubs (**Figure 10.7B**). Many single channel rear-wheel drive ABS vehicles have a sensor mounted in the differential (**Figure 10.8**). A tone ring on the differential ring gear provides the necessary speed information (**Figure 10.9**). Some systems have a sensor on the transmission tailshaft or on the transfer case in four-wheel drives. AC wave forms are prone to radio interference (RFI) with their signals. To avoid RFI, wheel speed sensor wiring is enclosed in a shielded (braided metal) tube or it is wound in twisted pairs. The latter are wires that are wound around each other about five to ten times per foot. Twisted pairs self-cancel RFI.

Chapter 10 Antilock Brakes, Traction, and Stability Control

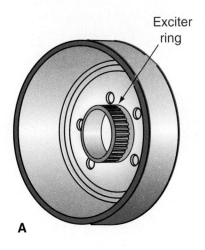
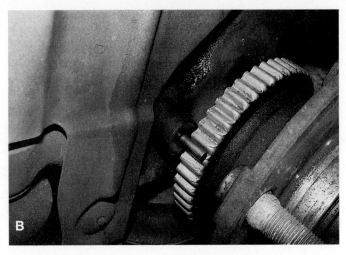

Figure 10.7 (A) Rear ABS with the tone ring inside the drum. (B) Rear ABS sensor and tone ring on the back of the hub. *(Courtesy of Tim Gilles)*

Figure 10.8 An ABS sensor mounted on the differential. *(Courtesy of Tim Gilles)*

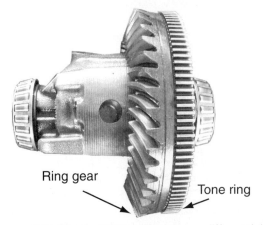

Figure 10.9 An ABS tone ring on a differential ring gear. *(Courtesy of Tim Gilles)*

Hydraulic Control Valve Assembly

A *hydraulic control valve assembly,* sometimes called an **electromechanical hydraulic (EH) unit** or *electrohydraulic control unit (EHCU),* has mechanical and electrical parts that cause hydraulic pressure to pulsate or modulate. It operates when one wheel's speed drops a certain amount below the speed of the other wheels on the vehicle.

NOTE: *When the wheels are rotating normally during braking, the ABS system is bypassed and the brake system operates in its normal base mode.*

■ TYPES OF ANTILOCK BRAKE SYSTEMS

The control valve assembly can be combined with the master cylinder and brake booster in an integral system (**Figure 10.10**) or it can be separate in a nonintegral system (**Figure 10.11**). Several manufacturers—Bendix, Bosch, Delco/Delphi (GM), Continental/Teves, Kelsey-Hayes, and Lucas Girling to name a few—produce antilock brake systems. You will need to know

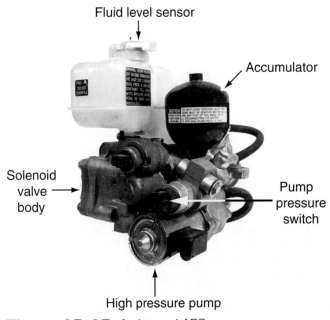

Figure 10.10 An integral ABS.

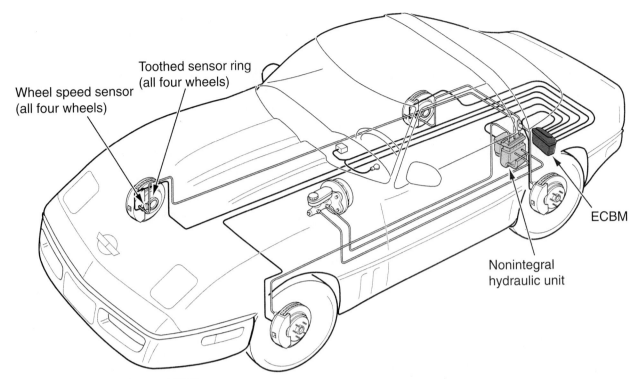

Figure 10.11 A nonintegral ABS.

which system you are working on before starting any repairs.

Integral ABS

Most of the early systems were **integral ABS** (see **Figure 10.10**). This means that the ABS combines the master cylinder, power brake booster, and hydraulic circuitry in one single hydraulic assembly. Hydraulic circuitry includes:

- The master cylinder and reservoir
- A brake booster that operates hydraulically rather than with vacuum
- A pressure pump and motor
- A pressure accumulator
- Switches that monitor pressure
- Pressure modulator valves
- A fluid level sensor

Notice that this type of ABS includes an integral hydraulic brake pressure booster. The booster works during both ordinary and ABS stops. It has a pump that provides the boost pressure. Hydraulic pressure to the rear brakes comes from the pump only, not from the master cylinder. If the pump fails, the pedal will be hard and the rear brakes will not work. These systems are often found on front-wheel drive cars so this is not as severe a problem as it would seem at first. Rear brakes on front-wheel drive cars only provide about 20 percent of the braking power during a stop.

Integral systems were used from 1985 until the early 1990s. Some of the integral antilock brake systems include Bosch III; Teves Mark II; Delco Moraine III and Bendix 4, 9, and 10. The Bendix names come from the number of solenoids contained in the hydraulic unit. Bendix 4, the last integral system used, stopped production with the 1995 Jeep.

Pumps and Motors. Earlier integral systems use the pump for a source of hydraulic pressure. They also have accumulators as part of the modulator (see **Figure 10.10**). An accumulator is filled with a charge of high-pressure nitrogen gas (**Figure 10.12**). It stores brake fluid under very high pressure to keep a ready

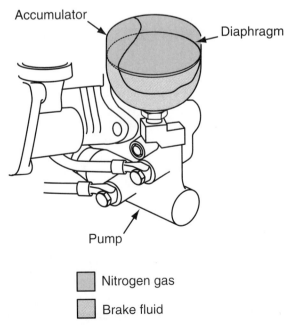

Figure 10.12 Pressure in an accumulator.

source of constant pressure to operate the ABS system when needed. Safety precautions regarding the service of these systems are covered later in this chapter.

Reservoir. This ABS reservoir is usually much larger than a normal brake system reservoir. Some systems use a second reservoir. The reservoir, which acts like a holding tank, collects pressurized fluid as it returns from the brakes at the wheels. Its location in the hydraulic circuit is before the pressure pump.

Pump Control Switches. Systems with a pump have a switch to turn the pump on and off. Some systems have a pressure-sensitive switch that monitors accumulator pressure. It turns on the pump when pressure drops to a certain level. If pressure drops below a certain level, an amber ABS warning light on the dash will come on. This light is in addition to the red brake hydraulic system warning light. Operation of these warning lights is covered later in this chapter.

Nonintegral ABS

Some of the earlier and all of the later-model antilock brake systems are **nonintegral ABS**, also called *nonintegrated, remote,* or *add-on ABS*. They have gained in popularity because of their lower cost and relative simplicity. Nonintegral ABS have a conventional power brake and master cylinder. The ABS unit is separate from the master cylinder, in series with its brake lines (see **Figure 10.11**). Since the early 1990s, nonintegral three- and four-channel ABS systems use a hydraulic pump to circulate fluid. They do not use the pump for power assist and do not have to be depressurized prior to repairing the brakes, like the integral system does.

Nonintegral ABS systems include:

- Bendix III (LC4 in Chryslers), 6, and Mecatronic (some Fords)
- Bosch 2, 2S, 2U, Micro, and ABSR
- Delphi (Delco Moraine) VI with and without traction control
- Kelsey-Hayes (purchased by Lucas-Varity) RWAL, RABS, 4WAL, and EBC
- Nippondenso
- Sumitomo 1 and 2 (based on Bosch)
- Teves Mark IV and Mark 20
- Toyota Rear Wheel

The method by which fluid pressure is controlled depends on the design of the system. ABS can be either *two-wheel* or *four-wheel*. They can be *one, two, three,* or *four channel*. Two-wheel systems are covered here first. Their components and operation are typical of more complicated systems, as well. A four-wheel system is basically a single channel times four.

Two-Wheel ABS

Two-wheel ABS systems only work on the rear wheels (**Figure 10.13**). Called *single channel,* they are found on sport utility vehicles (SUVs) and light trucks. ABS is especially important on the rear wheels of these vehicles. The brakes are designed to be able to stop a fully loaded truck. When the truck is empty, the rear brakes are prone to locking up. Single-channel systems offer a big improvement to an earlier addition to rear brakes,

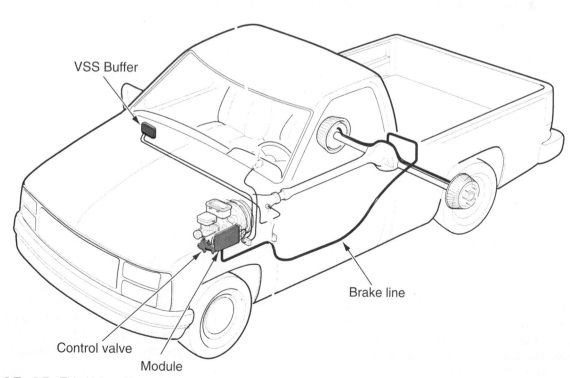

Figure 10.13 This Kelsey-Hayes two-wheel RWAL system acts only on the rear wheels.

the proportioning valve. Proportioning valves are still included in the rear brake system, along with ABS.

Two common two-wheel antilock systems are the Kelsey-Hayes (Lucas-Varity) **rear-wheel antilock (RWAL)** system and the **rear antilock brake system (RABS)**. RWAL is the GM name and RABS is the name Ford and Chrysler use. The difference in the two systems is that GM uses a single warning light, whereas Ford and Chrysler use two lights. ABS warning light operation is covered later in the chapter. Toyota has a two-wheel ABS system for their light trucks as well.

To control skids, the rear brakes are modulated at the same time. A centrally located speed sensor is normally found on the top of the differential, in the transmission, or in the four-wheel drive transfer case. The front brakes are normal brakes without any extra control.

The system needs only one sensor. With a standard differential, when one rear wheel locks the other wheel will try to speed up and the drive shaft changes speed rapidly. The signal goes to the ABS computer, which in turn shuts off flow to the rear brakes. There is no need for a separate vehicle speed sensor to compare the speed of the differential to vehicle speed. Speed simply changes and the sensor reacts to it. The system logic knows how fast it is possible for the wheels to slow down during braking. This is called the *deceleration factor*.

Some manufacturers offer an on-off switch for ABS brakes. At times it is desirable to shut off the ABS. When driving in snow, for instance, a locked-up wheel builds a wedge of snow in front of it that aids in stopping. The single-channel system is disabled on four-wheel drive vehicles when in four-wheel drive mode. Stopping power is more equalized in four-wheel drive. Also, the sensor has too much trouble making good decisions when it is getting input from all four wheels through the drive shaft and transfer case.

Four-Wheel ABS

Four-wheel ABS can be either three channel or four channel. Front wheels in both types are controlled separately. Most rear-wheel drive and some front-wheel drive vehicles use three-channel systems. A three-channel system has a sensor at each front wheel. The rear works like the single-channel two-wheel system. One sensor controls both rear wheels together. The rear sensor can be mounted in the differential housing, receiving signals from a notched ring mounted to the outside of the differential ring gear (see **Figure 10.8**). This sensor often doubles as the vehicle speed sensor (VSS).

The most effective ABS system is the four-channel system that has a sensor to monitor each wheel. It is the most expensive system because a wheel speed sensor is needed at all four wheels.

NOTE: *Having a sensor at each wheel does not necessarily mean that the ABS is four channel. Some three-channel systems have four-channel braking. They have four wheel sensors but control the rear brakes together hydraulically.*

Some of the newer systems use dynamic proportioning, which eliminates the traditional proportioning valve to the rear brakes (see Chapter 4). The brake pressure modulator valve's (BPMV) rear inlet valves are cycled during braking to maintain correct front-to-rear brake balance. If the rear brakes decelerate more than the front brakes, the system activates.

Unlike two-wheel ABS, four-channel systems continue operating when the vehicle is in four-wheel drive. On some systems, an electronic device that measures inertia, called a G-sensor, tells the controller the rate of stopping while in four-wheel drive.

ANTILOCK BRAKE SYSTEM OPERATION

Signals from each of the wheel sensors are pulsed to the ABS control module. When a wheel begins to lock up, the signal from its sensor drops off rapidly. To prevent lockup, the control module *blocks* further hydraulic fluid pressure to that wheel. If the frequency of the signal continues to drop off, fluid pressure will be *released* to that wheel. This is called pressure **dump** or *decay*. When the wheel begins to rotate freely once again, pressure is reapplied and braking can continue. If the wheel locks again, the cycle can repeat for up to ten to twenty times a second.

In most ABS systems, solenoid valves control the holding and releasing of hydraulic system pressure (**Figure 10.14**). Solenoid valves are small and weigh very little, allowing them to move fast. A typical ABS has two valves for each channel, an isolation valve, and a dump valve. An *isolation* valve, also called a *block* valve or a *hold* valve, is normally open, allowing brake fluid to flow. The isolation valve is always the first valve to operate during an ABS stop.

Single Channel Antilock Brake Operation

During a two-wheel ABS stop, the isolation valve closes to block further hydraulic pressure from reaching the rear brakes. When action by the isolation solenoid is not sufficient to prevent wheel lockup, the normally closed dump, or release, valve cycles open and closed rapidly to bleed system pressure. Excess brake fluid flows into a spring-loaded accumulator chamber (**Figure 10.15**). A drop in pedal height that is limited to the size of the accumulator reservoir can occur. Once the accumulator is full, pedal height cannot drop any further. The accumulator is very small because very little fluid needs to flow to achieve a drop in fluid pressure.

On some systems, the pedal can drop to the floor as the accumulator fills. The driver will have to pump

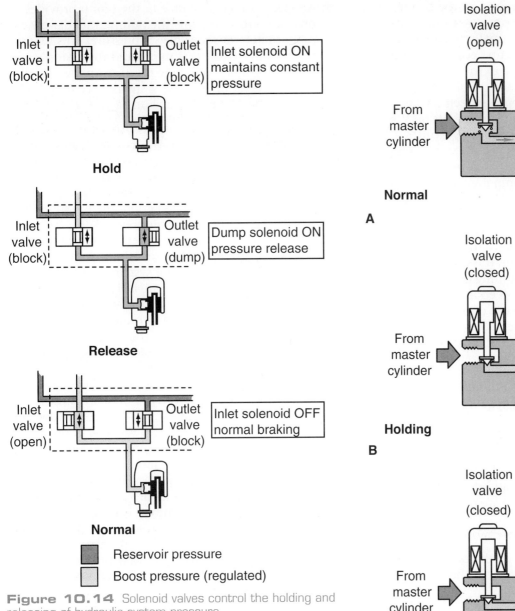

Figure 10.14 Solenoid valves control the holding and releasing of hydraulic system pressure.

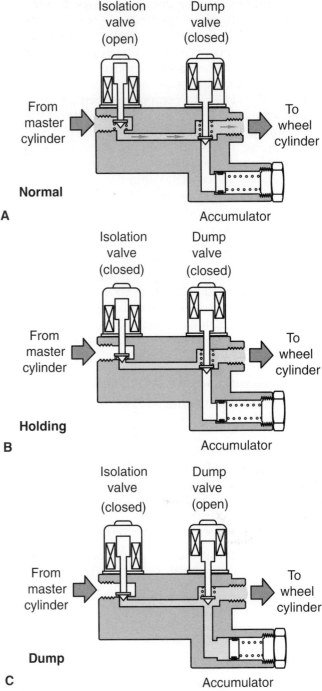

Figure 10.15 During normal braking of this RWAL system, the isolation valve is open and the dump valve is closed so there is a fluid flow between the master cylinder and the rear brakes (A). During the pressure holding stage, the isolation valve closes (B). During the pressure reduction stage, the dump valve opens to bleed pressure to the accumulator (C).

the pedal to get a fresh charge of fluid from the master cylinder to raise pedal height. Rear-wheel ABS does not have a pump to increase pressure or raise pedal height.

Once pressure to the rear brakes has been relieved and the computer senses the rear wheels turning again, the dump valve closes. The isolation valve remains closed and pressure to the rear is resumed as fluid pressure returns to the rear brakes from the accumulator. When the wheels are turning normally again, the isolation valve reopens to allow normal braking to resume. Remember, all of this happens very fast.

Three- and Four-Channel ABS Operation

Some three- and four-channel systems use a single combination valve. It has a three-position solenoid that controls normal fluid flow and the hold and release functions. These are the three stages of ABS solenoid operation:

1. *Pressure buildup,* also called *increase*—This is normal braking. Neither the inlet nor outlet

solenoid valves are energized by the computer. With both valves open, pressure from the master cylinder flows normally to the brakes. This mode is called buildup because the driver can still increase pressure to the brakes by increasing pedal pressure.

2. *Pressure hold*—The computer detects a wheel slowing rapidly during a stop. It shuts off flow from the master cylinder, blocking a further increase in pressure at the wheel. System pressure remains constant at that point.
3. *Pressure reduce*—The computer reduces pressure to a wheel by opening a solenoid valve. Pressure escapes to a low-pressure area in the system. This can result in a drop in pedal height. When pedal travel reaches about 40 percent, the electric pump increases pressure in the system. The computer energizes a relay, turning on the pump until the pedal height rises enough to close the switch. The pump is capable of more pressure than that required by the ABS. Relief valves return the excess fluid pressure to the master cylinder reservoir.

Electromagnetic Antilock Brakes

Some nonintegral systems, such as the Delco (Delphi) ABS-VI used on more cars than any other ABS system, do not use an accumulator or electric pressure pump. Instead, these systems use a motor pack (**Figure 10.16**). Three small, high-speed, bidirectional screw drive electric motors increase and decrease fluid pressure in each wheel circuit. They are positioned quickly and accurately in their bores.

The following describes system operation:

- One of the motors controls both rear brakes.
- A separate motor controls each of the front brakes.

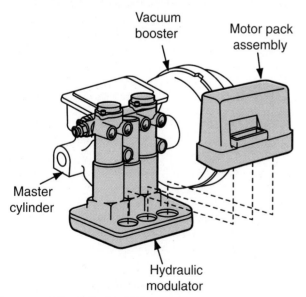

Figure 10.16 An ABS motor pack.

- An electromagnetic brake in the pump provides precise control. It stops the pump immediately as it alternately raises or bleeds pressure. This brake is important because it must hold the piston's position against hydraulic pressure applied by the foot brake. Some of these systems have nine brake pads in the ABS unit, three on each piston.

Conventional wheel speed sensors and a computer are used with this system. A conventional master cylinder and power booster are used as well. The system does not have a high-pressure pump, so system pressure is close to normal brake system pressure.

During normal brake operation, the modulator pistons are in their highest position, called "home." A pin extends from the end of each piston to hold its check valve open (**Figure 10.17**). This allows normal braking. The system is a split diagonal system. However, during an ABS stop, the rear pistons are isolated from the master cylinder and the system is no longer split diagonally.

During an ABS stop, one or more of the pistons are driven downward by the electric motor(s). If a front brake piston is moving, the solenoid must be closed. When the piston moves down, the pin moves away from the check valve closing it and isolating the brake cylinder that it controls from the master cylinder. Pressure remains the same as long as the piston does not move farther. There is no dump valve as in other antilock systems. If additional pressure reduction is needed, the computer commands the motor to move the piston farther downward. This drops, or dumps, the pressure. When the locked wheel and tire starts turning freely once again, the pistons are moved back up in their bores to unseat the check valve and resume normal braking.

Electrically operated pistons can cycle seven times per second, compared to ten to twenty times for a typical solenoid ABS system. The two isolation solenoids in an electromagnetic system are a safety backup provided only on the front brakes. They are normally open. During an ABS stop, they close to allow the system to be isolated from the master cylinder. If the ABS fails while one or both of the front pistons are down in their bores, the solenoids will provide an alternate circuit so the front brakes can still operate. There will be no rear brakes in this instance.

ABS System Safeguards

If a malfunction occurs in the operation of an ABS, the computer shuts the system off. The control module has a diagnostic procedure that starts when the vehicle starts and finishes at a speed of 4–10 mph. The ABS remains asleep until it receives a signal from the brake switch. When the ABS is shut off, the brake system operates as a normal system.

NOTE: *If an emergency spare is used, the differences in wheel rotation speed on that axle are sensed; the ABS system will not function and the ABS light will come on.*

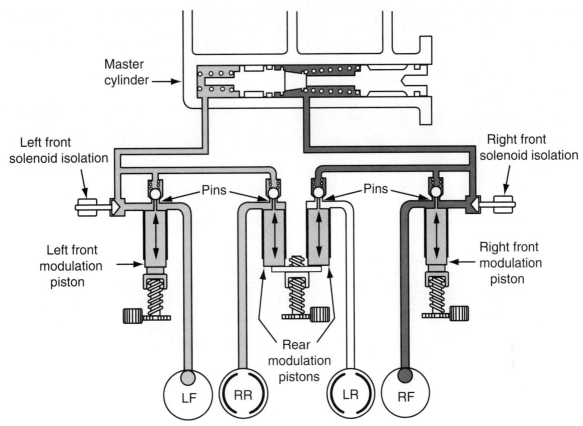

Figure 10.17 An electromagnetic ABS system. This is a split diagonal system. Pins at the tops of the pistons hold the check valves open during normal braking.

Brake Performance during an ABS Stop

When testing ABS in a hard stop, you should feel the solenoids or motors pulsating in the brake pedal. The amount varies by system.

During an ABS stop due to a loss of traction, the correct response by a driver should be to press even harder on the brake pedal. Even if only one wheel is locking, you will feel the pulses of the controller. ABS should allow the driver to steer away from a hazard, but harsh moves of the steering wheel during an ABS stop can result in spinout. ABS braking distances can be longer on some surfaces because a locked tire digs into a rough surface with greater adhesion.

Traction Control

The ABS system is sometimes used to limit the amount the wheels can spin during acceleration. This is called a **traction control system (TCS)** or **acceleration slip regulation (ASR)**. If a driving wheel starts to spin during acceleration, the controller applies hydraulic pressure to its brakes.

Electronic engine controls work with the ABS system to make traction control even more precise. Most systems limit slip by reducing the effect of accelerator pedal movement and engine power. Bosch, the company that developed the first ASR system in 1986, refers to the application of traction control as "throttle relaxation."

The computer's objective is to match traction with engine power. An engine's power output can be controlled by:

- Disabling fuel injectors (**Figure 10.18**)
- Retarding ignition timing
- Upshifting the transmission
- Closing the throttle (**Figure 10.19**)

Sensors provide the computer with information on vehicle speed and cornering forces. The computer makes a slip threshold calculation. Traction, or brake

CASE HISTORY

Emergency vehicles are equipped with a switch to turn the ABS system on and off, because sometimes a skid will get a car to spin. A highway patrol officer received notice of a burglary in progress. As he raced toward the address, the suspect's speeding vehicle passed him going in the opposite direction on the street. From habit, the officer slammed on the brakes while attempting to throw his vehicle into a skid in order to make a quick 180-degree turn. The ABS was on, however, and would not allow the inside wheel to stop turning, so he lost control of the vehicle.

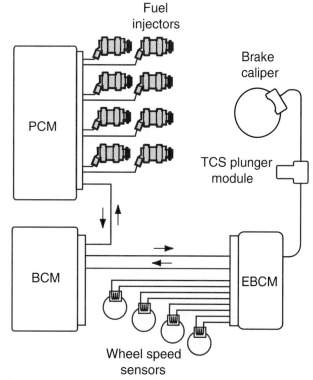

Figure 10.18 TCS controls for the Bosch III system. The PCM can cut off one to four fuel injectors on the V8 engine for torque management.

speeds, brake intervention is quick enough to prevent wheel spin, however.

The ABS controller disables traction control if it decides that the brakes are likely to overheat.

Another feature made possible by traction control is *electronic limited slip*. Limited slip differentials work mechanically. Electronic limited slip applies the brakes to a wheel that is spinning in mud or on ice, sending the torque to the wheel on the other side of the vehicle.

Electronic Stability Control

Stability programs use engine and ABS brake computers to stabilize the vehicle in the event a sudden evasive maneuver results in an unstable condition (**Figure 10.20**). Using a combination of the features from ABS, TCS, electronic brake-force distribution (EBD), and active yaw control (AYC), the computers determine if the car is actually traveling in the direction that the driver steered it. If understeer is sensed, hydraulic pressure is increased to the inside rear brakes to help correct it (**Figure 10.21**). Oversteer is corrected by applying the outside front brake. Previously available only on luxury vehicles, this system can now be found even on compact cars. The GM Delphi TRAXXAR vehicle stability enhancement and the Continental Teves electronic stability program (ESP) are examples.

The program can apply brakes at any one of the vehicle's wheels. The computers monitor the following sensors:

- Wheel speed sensors
- Steering angle sensor
- Yaw rate sensor
- Lateral acceleration sensor

intervention, has a higher priority at low speeds and directional control has priority at speeds above 50 mph. The system can either engage engine control or brake intervention separately or at the same time. It is possible to implement engine controls very quickly, while brake intervention is relatively slow. At lower

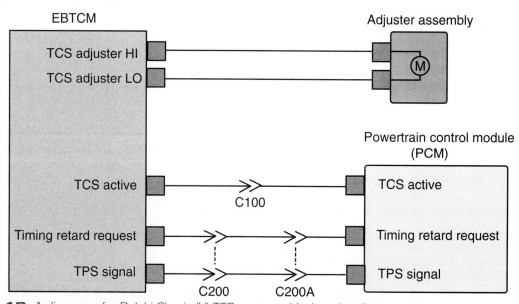

Figure 10.19 A diagram of a Delphi Chasis IVI TCS system with throttle adjuster.

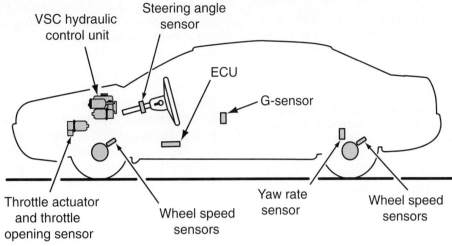

Figure 10.20 Parts of an electronic stability control system.

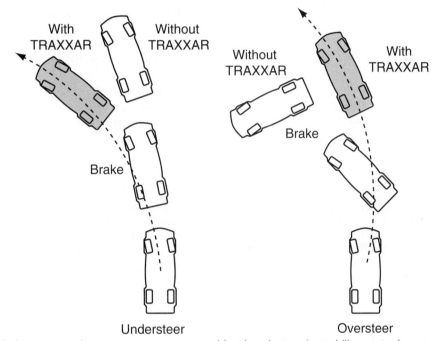

Figure 10.21 Understeer and oversteer are corrected by the electronic stability control program.

■ ANTILOCK BRAKE SYSTEM SERVICE

Antilock brake systems have proven to be very trouble-free. In fact, less than 1 percent of problems in the brake system arise from the ABS system. If there is a brake problem and the ABS light on the dash is not illuminated, the likelihood is that the conventional brake system is the cause of the problem. Be sure to check out the normal brakes first.

One way to test for ABS operation is to test-drive the car in an empty parking lot. Apply full pressure to the brakes during a hard stop at about 25 mph. If the ABS is functioning, the system should prevent a skid. Pulsation will be felt in the brake pedal and sometimes buzzing or clicking sounds will come from the modulator.

Electrical problems are diagnosed using the vehicle's on-board diagnostics. Detailed ABS service manuals are available from brake manufacturers.

Red Brake Warning Light

There are two kinds of brake warning lights. An amber light is only for ABS problems. A red light is the typical hydraulic system warning light. A low fluid level or low pressure in half of the hydraulic system will illuminate this lamp. The red light also serves the dual purpose of warning the driver that the parking brake is applied.

On integral antilock brake systems, the red light can come on when there is a failure. This results when the rear part of the system is not operating. On some ABS, there are three intensities of brightness to the

SHOP TIP: When there is erratic ABS operation, pull the ABS fuse and operate the vehicle. If the problem persists, the normal brakes are at fault.

light: The parking brake is the brightest; a hydraulic failure produces a light of medium brightness; and the red light is dim when there is an ABS failure. The warning signal passes through a resistance. To verify whether the light is at full illumination, apply the parking brake with the key on and look to see if it is brighter.

Amber Brake Warning Light

The amber (yellow) brake warning light is for ABS failures (**Figure 10.22**). It comes on after engine startup and remains on for a short time during the self-test. The amber light operates as follows:

- When the parking brake is applied and the key is on but the engine is off, both lights should come on. The amber light will go out in less than 4 seconds.
- When the engine starts, the amber light will come on for about 4 seconds before going out, if there are no problems. The self-test is taking place during this time.
- If both lights are on at the same time with the engine running, the master cylinder has run low on fluid.
- The pressure modulator valve has its own fluid reservoir. Low fluid in that system will light the amber light, but not the red light. A code will not set with this condition.

Diagnosing ABS Problems

When diagnosing a problem with the ABS, the first step is to inspect and repair the brake system as if it were non-ABS controlled. An example of a problem with the normal brake system that can mimic an ABS problem is false modulation. *False modulation* is when the ABS operates when it is not supposed to. Some early ABS cars had this problem during their warranty period because rear drum brake shoes had the wrong coefficient of friction, which caused the ABS to operate during moderate stops. Replacing the rear linings was the recommended repair that solved the problem. Glazed or oil soaked linings can result in a similar situation. When a pulsating pedal is the symptom, check for grabbing brakes.

Other things that can affect ABS operation include damaged or incorrect drum brake return springs or having two primary linings installed on one side of the car, with two secondaries on the other side.

A customer might complain that sometimes his car feels like it is accelerating during a stop. This can result when a problem with a wheel speed sensor causes the ABS to restrict hydraulic pressure to one wheel at low speeds.

Distortion of the disk brake caliper seal is what normally retracts its piston. If a rotor is warped or a wheel bearing is loose, the piston can be moved deeper into its bore, causing ABS to operate when stopping.

Check to see that all of the tires are the correct size. Next, check all fuses and connectors. ABS electrical problems share those common to all computer systems. Power and ground connections must all be tested prior to replacing components. The ground connection for the computer is especially important because the computer controls the valves by switching them to ground. Connectors are susceptible to resistance and there are often several connections. The computer will not allow much deviation from its expected measurement. Poor connections can cause multiple trouble codes. Follow these steps:

1. Use a scan tool to retrieve diagnostic trouble codes (DTCs), write them down, and then erase them. Some older systems use a flashing ABS light to signal trouble codes. Check the service literature for information on these.
2. Next, test-drive to see if a DTC resets.
3. Follow the recommended repair procedure. Always follow the detailed diagnostic troubleshooting charts in the specified order. Look over the entire chart first and note resistance specifications for electrical components. Experience with a particular repair will sometimes lead you to go right to the test of one part.
4. Erase stored DTCs and test-drive the vehicle again to verify the repair.

SHOP TIP: To check for a grabbing brake, drive the vehicle and operate the parking brake.

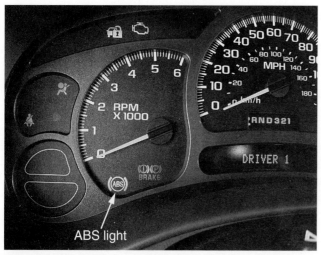

Figure 10.22 An amber warning light is used for ABS failures. *(Courtesy of Tim Gilles)*

Review the wiring diagram before testing an ABS electrical circuit. When measuring voltage and resistance use a high-impedance multimeter. An analog meter can damage an electronic circuit. Always turn off the ignition before disconnecting components.

ABS BRAKE FLUID SERVICE

There are many types of antilock brake systems, but they have a good deal of service similarities in common. Those items are covered here first, followed by some more specific service items. Always remember to locate the manufacturer's information for the vehicle you are working on before attempting to service an ABS.

Inspecting Brake Fluid Level

Follow the correct procedure for inspecting brake fluid level. Check the car under-hood label or the service manual. Integral ABS will appear to be low because fluid is in the accumulator. Some systems must be depressurized before checking. Others require the key to be on with the accumulator pressurized.

Depressurizing ABS Pressure

Some integral ABS systems operate under extremely high pressure. These systems have an accumulator that must be bled of pressure before attempting any service.

NOTE: *When bleeding the brakes, accumulator pressure will not be in the brake lines unless the brake pedal is depressed. Do not attempt to bleed brakes by depressing the brake pedal while opening and closing bleed screws.*

Before removing a caliper or master cylinder be sure the system is depressurized. Most systems are depressurized by applying the brake pedal at between twenty and forty times while the key is off. The pedal should feel like a normal power brake with no vacuum left in the booster. After restarting the vehicle, the ABS indicator light will go out when the system is repressurized.

Some Hondas and Acuras require special equipment for depressurizing.

Flushing and Bleeding ABS

Clean brake fluid is especially important in ABS systems. Contaminated brake fluid can result in costly repairs. Fluid can be checked for moisture contamination using an electronic moisture tester. The test strips change color when there is moisture contamination.

An annual fluid flush is a good recommendation for antilock brake systems. Fluid should be changed at least once every 2 years and certainly any time the brake pads are replaced. Some manufacturers have longer change interval recommendations because their factory fill is a premium, longer-life brake fluid.

NOTE: *Do not use DOT 5 (synthetic) brake fluid with ABS. Silicone brake fluids are not hygroscopic (do not accumulate water) like ordinary brake fluid. Water can accumulate in some areas of the system and cause problems. Use the recommended fluid, which is usually DOT 3 because it can flow easier. If the fluid does not flow easily air bubbles can be introduced, which can cause spongy brakes. The correct fluid type is usually specified on the master cylinder cap or reservoir body.*

Before bleeding an ABS, always check the service information. Brake part manufacturers provide brake bleeding sequence manuals. A typical bleed sequence for a front-to-rear hydraulic split system is shown in **Figure 10.23**.

There are many different procedures. Some integral systems require bleeding accumulator pressure and disconnecting the power to the ABS controller. Then the system is bled like conventional brakes.

Sometimes there is a special procedure for bleeding a brake pressure modulator valve. A scan tool is necessary to run the auto-bleed sequence (**Figure 10.24**).

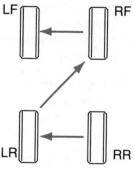

Figure 10.23 A typical bleeding sequence for a split hydraulic system.

CAUTION Extra care is necessary when servicing high-pressure antilock brake systems. Up to 2,700 psi can be available at the wheels even when the engine is off. Pressure this high is about twenty times as high as shop air pressure. Failing to bleed off pressure before opening a brake line or bleed fitting could result in serious injury to the technician. *It can force brake fluid to go right through your hand!*

Figure 10.24 A handheld antilock brake system scan tool. *(Courtesy of OTC/SPX Service Solutions)*

The ABS solenoids are cycled and the pump runs to purge air from the modulator. These circuits are normally closed during normal braking so the scanner is necessary to put the system into ABS mode.

If you do not need a scan tool to bleed the brakes, the system should be bled following procedures in the ABS repair manual. A good rule to follow is not to let the ABS system or master cylinder run dry. Then special bleeding procedures will not be required.

After bleeding the system, put the ABS into operation with a couple of hard stops during a test drive. This will help to get any remaining air out of the hydraulic control unit (HCU). It will not bleed the air from the device completely, but it can help after the correct procedures are followed.

Bleed antilock brakes with the key off or a false code can be set.

Dirty fluid can be forced into the ABS by careless repair practices. Always open a caliper bleed screw before pushing a piston into its bore during a brake pad replacement. Pushing the caliper pistons back in their bores without opening the bleed screws can result in one of several trouble codes caused by debris entering the hydraulic control unit.

Remove fluid from the master cylinder and look for sediment at the bottom. If there is sediment here, it is probably also in the control unit. It might be possible to clean the control unit as follows:

1. Flush the entire system with a pressure bleeder at 20–25 psi.
2. Be sure to clean the sediment from the bottom of the master cylinder first.
3. Use 2 to 3 quarts of DOT 3 fluid for the flush.
4. Follow the correct bleeding instructions.

Wheel Speed Sensor Service

The majority of ABS problems result from failure of a wheel speed sensor. Some of them fail because of exposure to harsh operating conditions and others fail due to abuse. Wheel speed sensors can become demagnetized or polarized by physical impact. This is caused most often by hammering during the removal or installation of front-end components, a CV joint service, or a MacPherson Strut repair.

SHOP TIP A safe and easy way to flush all brake systems, especially when you do not have a dedicated scan tool for the vehicle, is to use the gravity method. Simply place containers at all four wheels and open the bleed screws one turn each. Let the system drain slowly while you replenish the fluid supply. The fluid gets replaced but no air gets into the system.

The computer runs a self-test every time the ignition is cycled. It checks for low or high resistance. When the value is out of range, the computer turns on the dash light and disables the ABS. The computer senses the resistance at the sensor and at the computer and makes a judgement as to which code to set.

Verify the resistance of a suspected bad sensor:

1. Infinite resistance is usually due to an opening in the sensor's coil winding.
2. High resistance in the sensor is often due to corrosion in the harness connector due to exposure to water and salt.
3. Too little resistance could be due to a sensor coil that has shorted against another winding. A short like this lowers the total resistance of the coil.
4. If there is almost no resistance, look for sensor wires that are touching each other.

Testing a Sensor. A diagnostic flowchart will ask you to check a wheel speed sensor for AC voltage output. To perform the test, spin a wheel while reading its AC voltage output with a voltmeter or an oscilloscope. The flowchart will ask for a voltage reading at a specified speed. Typical voltage output ranges upward with speed from a minimum of at least 650 millivolts (0.65 volt). A faster turning wheel gives higher amplitude (voltage) and frequency. Notice in **Figure 10.25A** how far apart the oscillations are. The wheel is turning

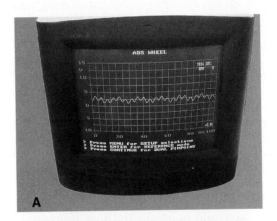

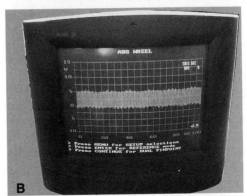

Figure 10.25 (A) A wheel speed sensor pattern from a slow turning wheel. (B) A wheel speed sensor pattern from a faster turning wheel. *(Courtesy of Tim Gilles)*

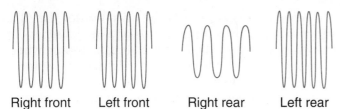

| Right front | Left front | Right rear | Left rear |

Figure 10.26 When one wheel begins to lock, the sensor sends a signal to the computer.

slowly and voltage is low. **Figure 10.25B** shows a faster turning wheel with higher voltage and frequency. If one wheel slows up (**Figure 10.26**) the computer will spot this and act to prevent it from locking. **Figure 10.27** shows a pattern from a sensor with a bad ground. In this case, the computer would set a code and turn on the ABS light.

NOTE: *Twisted pair wiring should not be pierced for testing. The wires are purposefully long, with few connections. Do not use butt connectors to repair them. Do not use a sensor with damage to a twisted wire pair.*

To connect a scope or multimeter, back probe the connector at the wheel. Do not attempt to pierce the insulation or shielding on the wiring loom. Sometimes a rear sensor can be hard to access. You can use the wiring diagram to go to the ABS computer. The sensor generates its own current so you need both of its wires to make the test. The signal is AC, so polarity does not matter as long as you have the correct two wires.

If there is no signal at a wheel, disconnect the connector and measure directly off the pins for the sensor. If there now is a signal, check for a short in the vehicle's wiring harness. If there is no signal with the wheel spinning, check the sensor for continuity with an ohmmeter. **Figure 10.28** shows a differential mounted sensor being tested for continuity.

Metal Shavings. Wheel speed sensors are magnetic so they attract metal shavings. An erratic sensor signal

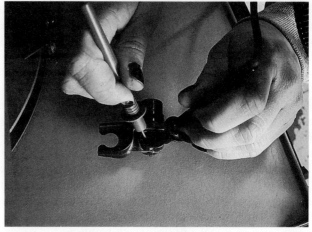

Figure 10.28 Testing a speed sensor for resistance. *(Courtesy of Tim Gilles)*

is the typical result. The following are conditions in which metal can be attracted to a sensor:

- When drum brake linings or disc brake pads wear down to their metal backings, shavings can be picked up by the sensor. Some semimetallic linings also shed magnetic dust as they wear.
- A drum brake sensor is most apt to collect metal because it is located in close proximity to the brake linings and drum.
- Sensors mounted on the transfer case or differential are also prone to picking up metallic debris from normal wear of the internal parts and metal dust shed from the clutch pack of a locking differential. **Figure 10.29** shows where metal collects on a differential mounted sensor.
- Turning the rotors with an on-the-car lathe can also result in metal shavings being attracted to sensors located near the rotor.

NOTE: *Some sensors have plastic housings. In addition to being fragile, they could be damaged by spray cleaners.*

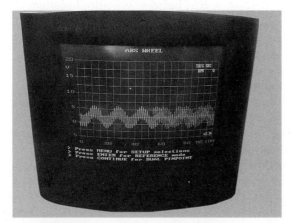

Figure 10.27 A wheel sensor pattern with a bad ground. *(Courtesy of Tim Gilles)*

Figure 10.29 A wheel speed sensor is magnetic. *(Courtesy of Tim Gilles)*

Replacing a Wheel Speed Sensor. Damage to the harness is the most common wheel sensor service problem. The wiring on a sensor is not serviceable and it is sensitive to abuse. Changing the resistance or configuration of the wiring could change the signal sent to the computer by the speed sensor.

Replacing a sensor is not difficult. Simply disconnect its wiring and unbolt it. Be sure to route wires where intended. Interference from ignition, the alternator, wiper motors or blower motors can result in electrical "noise" that the computer will try to interpret.

The sensor and harness are sold together as a unit. Sensors can be quite expensive, sometimes costing hundreds of dollars each. The wiring is routed like a brake hose so it has enough slack to allow for suspension deflection.

Sensor Air Gap. The gap affects the voltage produced by the sensor. The closer the gap, the higher the signal amplitude. Older systems had adjustable air gaps. Most newer ones do not. Air gaps range from 0.005–0.050 inch (**Figure 10.30**). Check the gap all around the sensor ring using a brass feeler gauge. The gauge will not stick to the sensor's magnet. If the gap is too large the steering knuckle could be damaged, requiring replacement. A gap that is too small can result in contact between the tone ring and sensor if the wheel bearing is loose, or when the suspension flexes. This can damage the tone ring or the sensor.

SHOP TIP When transmission or suspension work is being done, a safe practice is to disconnect the wiring from its connection at the chassis to prevent the chance of damaging it.

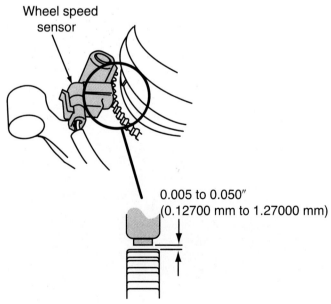

Figure 10.30 Wheel sensor gap measurement.

Wheel bearings can change operation of the ABS. When the sensor is on the rotor or axle, looseness in the wheel bearing will affect the air gap. Be sure wheel bearings are not loose or worn. Tighten the spindle nuts on front-wheel drive axles to the correct specification. A wheel bearing that is bad can also produce a trouble code.

Tone Ring Inspection. Some types of tone rings are susceptible to damage. Look for damaged teeth that can cause the ABS to stop functioning. Tone rings mounted externally on the CV joint live in the harshest environment. They are exposed to all sorts of road hazards in all kinds of weather. Look for tone ring damage if a new problem has surfaced after recent front-end service work, or after replacing a half shaft with a rebuilt one lacking a tone ring. Sometimes a loose wheel stud can move out of its mounting, damaging a tone ring.

The diameter and number of teeth on the tone ring vary with the application. This can result in a problem if the replacement axle has the wrong number of teeth on a tone ring mounted on its outer CV joint.

Use a press when replacing a tone ring. Do not hammer it into place. Besides the chance of bending or warping the ring, this can cause magnetism or change the polarity of the ring. A tone ring that has suffered an impact that causes it to be out of round can confuse the computer with variations in air gap. The computer sees this wheel going at different speeds and causes the system to go into antilock mode.

Some rear wheel sensors are integral with the wheel bearing. These cannot be replaced separately. An erratic sensor signal can also result from a bad wheel bearing. In either event, the entire wheel bearing assembly must be replaced.

Additional Precautions with ABS

There are several precautions common to all antilock brake systems:

- Do not fast charge the battery with the computer connected. A bad battery will allow voltage to rise too high.
- Do not jump-start an ABS vehicle with a battery charger on the fast charge/boost setting. Slow charge first or disconnect the negative battery cable before fast charging.
- Do not arc weld on the frame with the computer connected.
- Do not install an antenna (CB radio, cellular phone) near the ABS controller.
- Do not change the tire size, other than width. Taller or shorter tires have different rotational speeds. Using the minispare can disable the ABS.
- Do not disconnect or reconnect electrical ABS parts while the ignition switch is on. This can cause an

electrical current surge that can damage electronic parts.

Integral ABS Service

An integral ABS uses the same inputs as a nonintegral ABS. Sensor and fluid concerns are the same. Follow system depressurization instructions provided by the manufacturer.

Most manufacturers of integral ABS service their hydraulic assemblies as a unit. This can be very costly to the customer. This means that the motor, pump, and other parts are not readily available. Other manufacturers sell the pump, motor, and valve body assemblies as separate parts.

If the pump fails in an integral ABS, power-assisted braking does not work. Additional pedal effort will also be necessary because only the front brakes will operate.

Rear-Wheel Antilock Brake Service

Rear-wheel ABS is found on many light trucks and SUVs. Since 1990, RWAL has recorded soft codes but will only store one code at a time. If a code is not a hard fault, the controller shuts off the malfunction indicator light (MIL) and the codes remain in memory.

The following also apply to rear-wheel ABS systems:

- ABS is disabled in four-wheel drive.
- The red light will illuminate if there is no battery or ignition voltage to the ABS controller.
- Codes can be cleared by pulling a fuse with the key off.

ABS activation at low speeds can be caused when the sensor does not give enough of a signal for the processor. It can be caused by a defective sensor, metal shavings, and a sensor not fully seated after it was replaced.

One common problem in two-wheel ABS systems is that the dump valve can get some dirt in it. This results in a drop in pedal height, but with a firm pedal. At first the pedal feels okay, but with continued foot pressure it will drift lower toward the floor.

In a rear-wheel-only system, the brake pedal switch is a key input. The computer monitors vehicle speed all the time, but ABS only functions when the brakes are applied and voltage from the switch drops from system voltage to less than 1 volt. If the computer receives system voltage during a stop, it disables the ABS and sets a code.

CAUTION With some rear-wheel systems, a permanent code can be set if the fuse is pulled to clear codes in key-on position. The controller must be replaced to regain ABS.

A diagnostic trouble code can be activated by a two-footed driver. A code sets if the controller senses input from the brake light circuit while accelerating to about 40 mph. If the left foot rests lightly on the brake pedal while accelerating, the brake switch is activated. The ABS computer does not know the brakes are not being used and can set a code for a bad brake switch. Check the switch for correct adjustment.

Electromagnetic ABS Service

The Delco VI ABS sometimes suffers a failure that results in a low pedal. Follow the diagnostic chart in order to repair the problem. Part of the procedure calls for bleeding the modulator at various bleed screw locations.

One problem that sometimes occurs with this system is when a brake on a piston fails. When this happens the motor pack must be replaced. If the brake does not hold the piston from turning, it can move down in its bore, resulting in a drop in pedal height.

These antilock systems are driven by gears (**Figure 10.31**). Occasionally, one of these gear falls off. The gears must all be rotated counterclockwise as far as they can be turned. This is called "rehoming" the gears. They must remain in this position while installing the motor pack and cover.

The Delco VI ABS sometimes looks like an integral system because the master cylinder is mounted to the hydraulic modulator. They are connected by fluid transfer tubes and O-rings. The tubes are not reusable and replacement master cylinders often are supplied with new tubes and O-rings. Be sure to bench bleed the master cylinder before installing it.

Common ABS Electrical Problems

Vehicle speed sensor buffers are known to fail quite often. A blown fuse is a clue to this problem. A buffer is a solid-state (electronic) device. Check inputs and outputs first before condemning the buffer. See if its input wire has power or if it is open.

Figure 10.31 The ABS IV valves are driven by gears. *(Courtesy of Tim Gilles)*

Check the resistance of the VSS against specifications. A typical resistance would range from 900 to 2,000 ohms. Check for an AC output voltage from the VSS. The best way is to use an oscilloscope. When replacing a buffer, it will have to be calibrated for the tire size by an authorized parts supplier.

A failed ABS/TCS relay with bad electrical contacts is another common problem. This can be hard to pinpoint because it is often intermittent. This switch is inexpensive and is often replaced when no other test has pinpointed the problem.

On some cars, the use of an incorrect brake lamp or dash lamp can cause the ABS to shut off and the dash warning to illuminate. Some systems are so prone to voltage changes that a bad alternator can sometimes cause an ABS problem.

REVIEW QUESTIONS

1. At what percentage of slip does maximum traction occur?
2. How fast can a typical ABS pulsate the pressure to the brake system?
3. What is the name of the ABS with a conventional power brake and master cylinder that is also called remote or add-on ABS?
4. Which ABS can have up to 2,700 psi available at the wheels even when the engine is off?
5. Changes in the magnetic lines of flux around the wheel speed sensor cause a(n) _____ current voltage.
6. Why do manufacturers enclose wheel speed sensor wiring in a braided metal tube or wind it in twisted pairs?
7. What are two names for the wheel speed sensor's toothed ring that spins with the wheel?
8. What is the name of the high-pressure ABS that can lose its rear brakes and power assist during an ABS failure?
9. How many channels is an antilock brake system that has separate hydraulic controls for the front brakes but a common sensor and solenoids for the rear brakes?
10. What are the three stages of ABS solenoid operation in three- and four-channel systems?
11. When fluid pressure is released to a wheel, this is called pressure dump or _____.
12. In most antilock brake systems, _____ control the holding and releasing of hydraulic system pressure.
13. Which type of ABS uses a motor pack instead of solenoids?
14. Which can be implemented faster in a traction control system, engine controls or application of the brakes?
15. What is the name of the system that can control understeer and oversteer?

PRACTICE TEST QUESTIONS

1. Technician A says when the ABS senses a failure, the brake system reverts to conventional braking. Technician B says during normal operation the ABS works as a conventional brake system. Who is right?
   A. Technician A   B. Technician B
   C. Both A and B   D. Neither A nor B

2. Technician A says most wheel speed sensors are permanent magnet types. Technician B says some ABS systems use a Hall effect wheel speed sensor. Who is right?
   A. Technician A   B. Technician B
   C. Both A and B   D. Neither A nor B

3. Technician A says a wheel speed sensor's voltage output does not change with the speed of wheel rotation. Technician B says a wheel speed sensor's frequency changes with the speed of wheel rotation. Who is right?
   A. Technician A   B. Technician B
   C. Both A and B   D. Neither A nor B

4. Which of the following is/are locations for a wheel speed sensor tone ring?
   A. Inside the brake drum
   B. In the differential
   C. On the transmission tailshaft
   D. On a CV joint
   E. All of the above

5. Technician A says the purpose of ABS is to help avoid loss of steering control when a wheel loses

traction. Technician B says an ABS stop will result in a shorter stopping distance than a stop with normal braking. Who is right?
- **A.** Technician A
- **B.** Technician B
- **C.** Both A and B
- **D.** Neither A nor B

6. The amber (yellow) brake warning light is on. Technician A says there is an ABS failure and the ABS system is deactivated. Technician B says this can be due to low fluid level in one-half of the master cylinder. Who is right?
- **A.** Technician A
- **B.** Technician B
- **C.** Both A and B
- **D.** Neither A nor B

7. A left front wheel locks up during an ABS stop. Technician A says if the blocking solenoid for the left front brake is energized, the master cylinder is isolated from the left front wheel cylinder. Technician B says if the dump solenoid opens, normal braking will be reestablished. Who is right?
- **A.** Technician A
- **B.** Technician B
- **C.** Both A and B
- **D.** Neither A nor B

8. Which of the following could cause a code and illuminate the ABS warning light?
- **A.** An ABS solenoid problem
- **B.** A faulty wheel speed sensor signal
- **C.** A faulty ABS control module
- **D.** All of the above

9. Technician A says a scan tool is needed to bleed some hydraulic control units. Technician B says to use DOT 5 brake fluid with ABS systems. Who is right?
- **A.** Technician A
- **B.** Technician B
- **C.** Both A and B
- **D.** Neither A nor B

10. In a traction control system, the engine's power output can be controlled by:
- **A.** Disabling fuel injectors
- **B.** Retarding ignition timing
- **C.** Upshifting the transmission
- **D.** All of the above

SECTION Three

Tires and Wheels

■ **CHAPTER 11**
Bearings, Seals, and Greases

■ **CHAPTER 12**
Tire and Wheel Fundamentals

■ **CHAPTER 13**
Tire and Wheel Service

CHAPTER 11

Bearings, Seals, and Greases

■ OBJECTIVES

Upon completion of this chapter, you should be able to:
- ✔ Understand terms that relate to wheel bearings.
- ✔ Select the correct grease to use for a particular application.
- ✔ Describe the various wheel and axle bearing arrangements.
- ✔ Diagnose wheel bearing–related failures.
- ✔ Service wheel bearings on front and rear axles.

■ KEY TERMS

alloy
angular contact
auxiliary lip
axle bearing
bearing cage
bearing seal
brinelling
capillary action
double-row bearing
full-floating axles
grease
inner race
live axle
press-fit
pressure bearing packer
radial load
radial thrust
roller bearing
sealing lip
semi-floating axle
spalling
tapered roller bearing
wave seal
wheel bearing

■ INTRODUCTION

Bearings of different types are found throughout the automobile. The first part of this chapter deals with the fundamentals of bearings, seals, and lubricants; the last part covers bearing service.

■ PLAIN BEARINGS

Plain bearings are the kind used as engine crankshaft bearings. They do not use rolling parts and they provide sliding contact between two mating surfaces. Frictionless, or antifriction, bearings provide a rolling contact and allow easy rotation when any part moves.

■ FRICTIONLESS BEARINGS

Frictionless bearings are either ball, roller, or needle bearings (**Figure 11.1**). Bearings are made of hardened steel **alloys**. They are ground to a precise finish and size. Frictionless bearings must be lubricated.

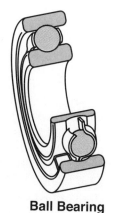
Ball Bearing

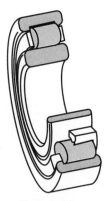
Roller Bearing

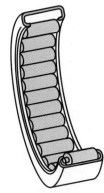
Needle Bearing

Figure 11.1 Types of frictionless bearings.

Some bearings are sealed on either side or both sides. The function of the seal is to keep grease in the bearing and contaminants out.

In a ball or roller bearing, the balls or rollers ride between an **inner race** (cone), or raceway, and an outer race, or cup, (**Figure 11.2**). The balls or rollers are usually held in position in a **bearing cage** or separator. The cage, usually made of stamped steel or plastic, keeps them from bunching up and rubbing against each other. It also serves to keep the bearings properly located so that the load on them is evenly distributed. When the bearing is the type that can be disassembled, the cage prevents the loss of the parts.

Direction of Bearing Load

Bearing manufacturers make bearings to handle different types of loads. A bearing load that is in an up-and-down direction is called a **radial load** (**Figure 11.3**). When the load is in a front-to-rear or side-to-side direction, this is called a *thrust load*, or axial thrust (**Figure 11.4**).

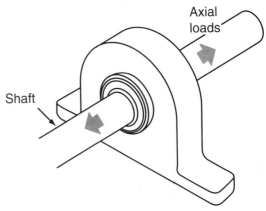

Figure 11.4 Direction of thrust loads on a bearing.

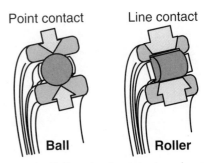

Figure 11.5 Ball and roller bearings have different contact areas.

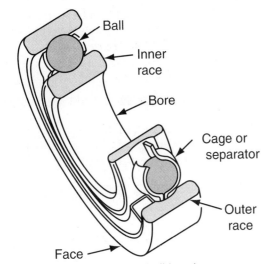

Figure 11.2 Parts of a ball bearing.

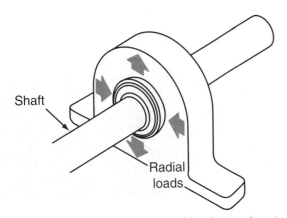

Figure 11.3 Direction of radial loads on a bearing.

▉ BALL BEARINGS

Ball bearings ride in grooves that are ground into the surfaces of the inner and outer races (see Figure 11.2). A big disadvantage to ball bearings is that most of the load of the vehicle is exerted on the bottom of the bearings on a very small surface area of the ball and race. The rest of the balls do not do much to support the load. On the other hand, an advantage of ball bearings is that they have less friction than roller bearings and can operate at higher speeds. **Figure 11.5** shows the difference between ball and roller contact areas.

When a ball bearing is at rest, its load is distributed equally wherever the balls and races are in contact with each other. When there is motion and the ball begins to roll, material in the race actually bulges out in front of the ball (**Figure 11.6**). Then it flattens out behind the ball. Metal-to-metal contact cannot be allowed, so lubrication is critical.

Ball bearings control both end thrust (side movement) and radial movement (up and down) (**Figure 11.7**). Single-row ball bearings are popular in trans-

Chapter 11 Bearings, Seals, and Greases 237

Figure 11.6 Metal bulges out in front of the ball when a load is applied to the bearing.

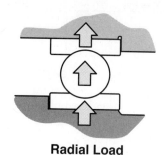

Radial Load

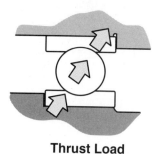

Thrust Load

Figure 11.7 Ball bearings control both end thrust and radial movement.

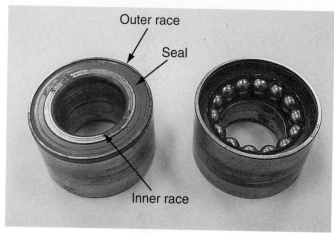

Figure 11.9 Double roller bearings. *(Courtesy of Tim Gilles)*

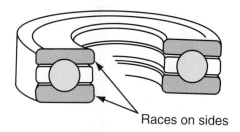

Ball Thrust Bearing

Figure 11.10 In a ball thrust bearing, the races are faced sideways rather than up and down.

missions, alternators, steering gears, and differentials. They are designed primarily for radial loads but can withstand thrust loads as well. When more load capacity is needed, extra balls are added to the bearing. This reduces the bearing's thrust capacity, however.

When a ball bearing must control thrust, the groove in the bearing race will be offset to one side. **Figure 11.8** shows a typical **radial thrust**, or **angu-**

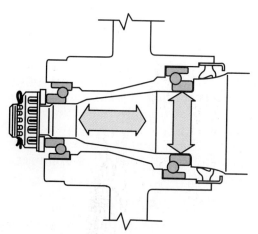

Figure 11.8 A radial thrust, or angular contact, ball wheel bearing set controls thrust in both directions.

lar contact, ball wheel bearing setup in which two radial thrust control bearings face each other. This controls thrust in both directions.

Single-row bearings are susceptible to damage when a shaft is misaligned. When a ball bearing is required to have more thrust capacity or when misalignment is likely to be a problem, a **double-row bearing** is used. This bearing is typically found in air-conditioning clutches and front-wheel drive front bearing hubs (**Figure 11.9**).

Ball bearings can also be designed primarily to control thrust. This is done by facing the races sideways rather than up and down (**Figure 11.10**).

ROLLER BEARINGS

When a greater load-carrying capacity is needed, **roller bearings** are used instead of ball bearings. They provide more surface area of contact with the race and can carry a greater load (see **Figure 11.5**). There are several types of roller bearings. Some roller bearings have only an outer or inner bearing race. The shaft or bore will be precision ground and hardened to serve as the other race. This is often the case with rear-wheel drive (RWD) rear axles in light vehicles (**Figure 11.11**). Rear-wheel drive cars have straight roller bearings on the rear axle. These bearings support radial loads well but cannot control thrust. Axles must be

238 SECTION 3 TIRES AND WHEELS

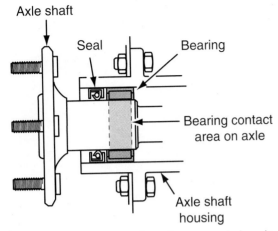

Figure 11.11 A straight roller rear axle bearing for a rear-wheel drive car.

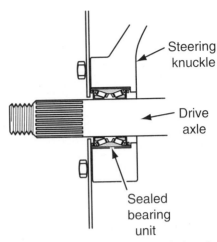

Figure 11.13 This FWD drive axle bearing has two tapered roller bearings.

retained or keyed so that end thrust cannot be applied to the bearing.

Roller bearings do not control end thrust. A thrust bearing (or a pair of tapered roller bearings) must be used to control end thrust. A thrust bearing or bushing is one that is mounted at 90 degrees to the wheel bearing's load.

Tapered Roller Bearings

The most popular type of roller bearing used in automobiles is the **tapered roller bearing**. Tapered roller bearings are used for front or rear wheel bearings on nondrive axles because they can control end thrust when two of them are installed with their tapers facing in opposite directions. The small diameters of the tapers face toward each other (**Figure 11.12**). In front wheel bearings, the larger of the two bearings is installed on the inside, where it supports the weight of the vehicle. The outer (smaller) bearing simply keeps the wheel aligned and provides end thrust control. Two outside bearing cups, or races, are pressed into the hub.

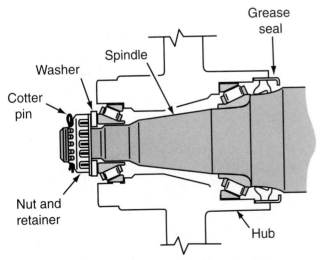

Figure 11.12 Tapered front wheel bearings are installed with their tapers facing in opposite directions.

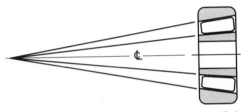

Figure 11.14 The angles of the tapers of the bearing rollers meet at a common point.

Sometimes, two tapered roller bearings are used in a single bearing assembly. These are found where thrust loads are greater, such as on front wheel hubs on front-wheel drive cars (**Figure 11.13**).

Tapered roller bearings tend to be self-aligning. The angles of the tapers of the bearing rollers are such that all of the angles meet at a common point (**Figure 11.14**). This makes each bearing roller align itself with the shaft.

Tapered roller bearings are assembled with the inner race, bearings, and cage as a single unit (**Figure 11.15A**). They have lips on the inside and outside edges of the inner bearing race (**Figure 11.15B**). This prevents the rollers from being able to slide out of the bearing.

Needle Bearings

When a roller bearing is very small, it is called a *needle bearing*. Needle bearings can be used to control thrust or radial loads (**Figure 11.16**), but they are not as good for high speeds as roller or ball bearings. They are most often used where space is limited. Needle bearings do not tolerate misalignment. One example of a common use of needle bearings is in universal joints (see Chapter 21).

■ WHEEL BEARINGS

Wheel bearings or axle bearings are found on all of the wheels of a vehicle. Some of them require service and others are simply replaced when they fail. The

Chapter 11 Bearings, Seals, and Greases 239

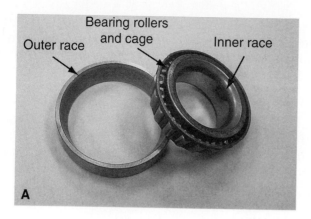

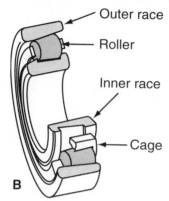

Figure 11.15 (A) Tapered roller bearing construction. (B) A tapered roller bearing has ledges on the inside and outside of the inner bearing race. (A, *Courtesy of Tim Gilles*)

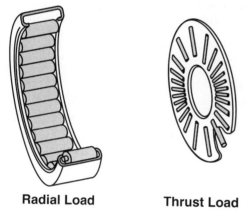

Figure 11.16 Needle bearings can be used to control thrust or radial loads.

term **wheel bearing** refers to nondrive front and rear wheel bearings. Bearings on **live axles**—those that drive wheels—are referred to as **axle bearings**.

Drive Axle Bearings

Drive axle bearings are located at the ends of the rear axle housing on a rear-wheel drive car or on the hub on a front-wheel drive car. Some are lubricated by differential lubricant and others are permanently lubricated and sealed.

Passenger car rear axles use a **semi-floating axle** design (**Figure 11.17**). This design has a bearing that rides on the axle. This design is unsafe for heavy loads. If the axle breaks, the wheel can fall off. This axle design is found on vehicles up to ½ ton pickup trucks. That is why they are not often used with large campers.

Full-Floating Axles

Full-floating axles are found on some light-duty and larger trucks, usually ¾ ton and larger. In this axle design, the bearings do not touch the axle. They are located on the outside of the axle housing (**Figure 11.18**). If the axle breaks, the rear brake drum and wheel will still be supported. This axle design is for heavy loads.

Front-Wheel Drive Bearings

Front-wheel drive bearings are compact because they must fit into a tight space. There is a pair of either ball or tapered roller bearings housed in the steering knuckle (see **Figure 11.13**). The small stub axle that the bearing supports is usually the end of the CV joint. It has splines that attach it to the front hub. A nut that holds the hub to the CV joint is tightened to keep tension (preload) against the bearings.

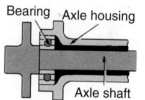

Semi-floating Type

Figure 11.17 In a semi-floating axle, the bearing is installed on the axle.

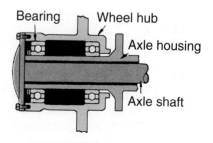

Full-floating Type

Figure 11.18 In a full-floating axle, the bearing is installed on the outside of the axle housing.

GREASES

Lubricants have several purposes. They reduce friction and wear. They must also dissipate heat and protect metal surfaces from rusting. The lip of the seals are lubricated by lubricants too. This helps them keep the lubricant in and dirt and water out.

Greases are used in wheel bearings, chassis joints, and universal joints. Some components are equipped with grease fittings for replenishing grease. Others must be disassembled to replace the grease.

Grease is a combination of oil and a thickening agent. In some situations, grease has certain advantages over oil. When grease is used, oil circulation systems are unnecessary, the tendency for leakage is reduced, and the lubricant remains in place after a long shutdown.

Grease does not leak or flow out of a bearing like oil would. It acts like a liquid lubricant when a shearing force is applied to it, but the direction of fluid flow is only in the direction that the bearing turns. It does not flow out of the bearing because it is not fluid at right angles to the direction that the bearing is turning.

NOTE: *The properties of a grease are limited by the quality of oil that it is made of. Greases are usually 70–90 percent liquid petroleum oils made semifluid or solid by adding a soap as a thickening agent at a concentration of from 5–25 percent.*

Because the rate of oil oxidation doubles with every 20°F increase in temperature, when a bearing runs hot its grease will require more frequent changing. A grease with a life expectancy of 1,000 hours at 100°F has a 500-hour life expectancy if used at 120°F. Grease made from petroleum should not be used at temperatures in excess of 325°F unless it is being changed every hour or two.

Grease Performance Classification

Automotive grease is classified by the National Lubricating Grease Institute (NLGI) for performance. This standard is based on an American Society for Testing and Materials (ASTM) standard grease specification approved in May 1990.

Automotive grease is classified in two general NLGI groups:

1. If the standard has an "L" prefix, the grease is for the lubrication of suspension and steering components such as ball joints and steering pivot points.
2. If the prefix is "G," the grease is intended for wheel bearings.

Figure 11.19 shows a further breakdown of the classifications.

Some greases might overlap and be suitable for both chassis and wheel bearings. The NLGI has a logo, which can be displayed on the product (**Figure 11.20**). Only the highest of each category of a combination of the two can be included in the logo.

Extreme Pressure Lubricants

Extreme pressure (EP) lubricants, which are added to some greases, are the same as those found in gear lubricants (see Chapter 21). Metallic soap thickening agents

LA	Service typical of chassis components and universal joints in passenger cars, trucks, and other vehicles under mild duty only. Mild duty will be encountered in vehicles operated with frequent relubrication in noncritical applications.
LB	Service typical of chassis components and universal joints in passenger cars, trucks, and other vehicles under mild to severe duty. Severe duty will be encountered in vehicles operated under conditions which may include prolonged relubrication intervals, or high loads, severe vibration, exposure to water or other contaminants, etc.
GA	Service typical of wheel bearings operating in passenger cars, trucks, and other vehicles under mild duty. Mild duty will be encountered in vehicles operated with frequent relubrication in noncritical applications.
GB	Service typical of wheel bearings operating in passenger cars, trucks, and other vehicles under mild to moderate duty. Moderate duty will be encountered in most vehicles operated under normal urban, highway, and off-highway service.
GC	Service typical of wheel bearings operating in passenger cars, trucks, and other vehicles under mild to severe duty. Severe duty will be encountered in certain vehicles operated under conditions resulting in high bearing temperatures (disc brakes). This includes vehicles operated under frequent stop-and-go service (buses, taxis, urban police cars, etc.) or under severe braking service (trailer towing, mountain driving, etc.).

Figure 11.19 Automotive grease classifications.

Figure 11.20 NLGI logos.

add a small amount of EP effect. Graphite or molybdenum do not add EP properties, but they increase antiwear and frictional modification characteristics.

Chassis Lubricants

A chassis lubricant is a grease of a consistency that allows it to be applied through a zerk fitting with a grease gun. It must adhere to the bearing surface and seal out dirt and water. Chassis lubricant is highly resistant to being washed away with water.

Chassis parts move up and down repeatedly, which can break down the structure of the grease. Grease that does not have adequate shear resistance will break down, become like oil, and flow away from the surfaces where it is needed.

Wheel Bearing Grease

Wheel bearing lubricant is grease with a high resistance to heat. It is usually thicker and specially formulated for use with various types of bearings. Because the grease is on a spinning part, it must resist the tendency to be thrown off. A poor-quality bearing grease can cause a safety hazard by leaking onto the brake linings when heat causes it to thin excessively. Excessive temperatures can also cause bearing failure when the lubricant breaks down.

Universal Joint Grease

Universal joint grease is made specifically for universal joints. Some universal joint designs require special lubricants. Be sure to follow the manufacturer's recommendations.

Multipurpose Grease

A multipurpose grease satisfies the requirements of chassis, wheel bearing, and universal joint lubricants. It is the most common type of grease used in service shops. However, multipurpose does not mean all-purpose. It meets only certain requirements. The lubricant most often used in chassis lubrication is a multipurpose lithium-based grease.

Solid Lubricant Greases

Some greases contain solid lubricant materials such as molybdenum (moly) or graphite. These greases are often used to lubricate speedometer cables, emergency brake cables, splines, and leaf springs.

■ WHEEL BEARING SEALS

Automobiles and equipment use seals for the following reasons:

- To seal in lubricants
- To keep different lubricants separated
- To keep out dirt
- To maintain vacuum or pressure

Seals are either *dynamic* for sealing moving parts or *static* for sealing fixed parts (**Figure 11.21**).

Wheel **bearing seals** are generally lip seals (**Figure 11.22**). There is usually a garter spring behind the

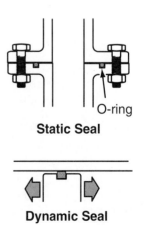

Figure 11.21 Static seals do not move. Dynamic seals work on moving parts.

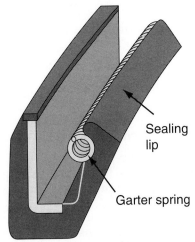

Figure 11.22 The garter spring increases the sealing lip's pressure on the shaft.

when there is a large amount of dirt because the auxiliary lip causes increased frictional heat.

Some bearings are sealed on one or both sides. When there is a seal on both sides, the bearing is not serviceable and must be replaced if there is a problem. If lubricant leaks from one of the seals, replace the bearing. Be sure not to immerse the bearing in solvent. If solvent gets into the bearing, it will dilute the lubricant. This can lead to bearing failure.

In some cases, there is a seal on only one side of a bearing. This type of bearing is commonly found on the rear axle of RWD vehicles. These bearings are lubricated by oil from the differential.

NOTE: *Seals are made of different types of materials depending on their intended use. The most popular seals are made of synthetic materials such as nitriles, Polyacrylates, silicones, and fluoroelastomers.*

Other Seals Designs

When a lubricant is to be directed back to its source, the lip of the seal is sometimes fluted (**Figure 11.25**). A newer seal design is the **wave seal**. This seal does not ride on the shaft in a straight line like a conventional seal. It oscillates, pumping oil back toward its source (**Figure 11.26**). It has less friction, lasts longer,

seal lip to further increase the pressure of the seal on the shaft. The spring can be separate or cast into the seal itself. The open side of the sealing lip always faces the lubricant (**Figure 11.23**). This is so that any pressure from the lubricant will increase the wiping pressure of the seal on its shaft.

Some seals have more than one lip; one is the **sealing lip** and the other is the **auxiliary lip** for sealing dust (**Figure 11.24**). Auxiliary seals are only used

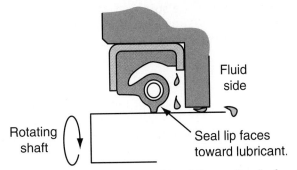

Figure 11.23 The open side of the sealing lip faces the lubricant.

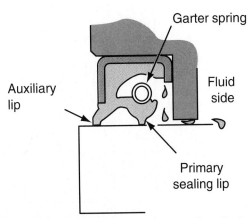

Figure 11.24 This seal has an auxiliary lip for dust and dirt.

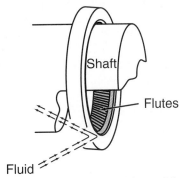

Figure 11.25 The flutes on this seal lip direct the oil back to its source.

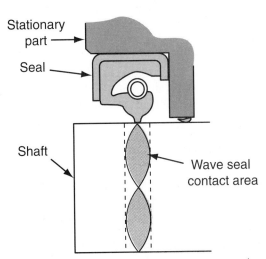

Figure 11.26 A wave seal oscillates, pumping oil back toward its source.

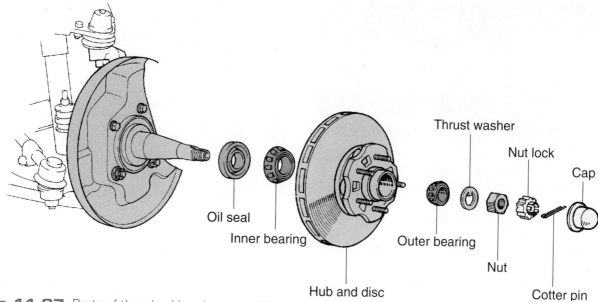

Figure 11.27 Parts of the wheel bearing assembly.

and works equally well no matter which direction the shaft turns.

SEAL TOLERANCE

Seals can usually accommodate a shaft that is undersized up to 1/64 inch (0.016 inch) but only if all parts are in perfect alignment. The Rubber Manufacturers Association (RMA) recommends that runout tolerance be held to + or −0.003 inch for shafts up to 4 inches in diameter. The surface finish should be smooth—10–20 microinches.

WHEEL BEARING SERVICE

If the lubricant can leak out, moisture can leak in. Also, once grease has leaked out, the bearing will soon suffer shock loads due to the lack of lubrication. Pieces will come off the bearing and races. The pieces will circulate around the bearing, resulting in heat buildup and failure of the bearing.

Periodic maintenance of serviceable bearings on nondrive axles consists of cleaning the bearings, repacking them with grease, and adjusting the clearance or preload after reinstallation. Parts of the wheel bearing assembly are shown in **Figure 11.27**. On drive axles, the only service done to a bearing is to replace it if it is bad. That service is covered in Chapter 21.

REPACKING WHEEL BEARINGS

Unsealed ball and roller wheel bearings are lubricated with grease, which also protects the metal from corrosion and helps carry away heat. These wheel bearings are sealed from the elements, but over a period of time moisture and brake and road dust can accumulate. It is customary to clean and repack the wheel bearings with grease whenever the brakes are relined or at 30,000 mile intervals.

On front-wheel drive (FWD) cars, the procedure for repacking the rear axle bearings (if they are not the sealed type) is similar to the procedure for repacking the front axle bearings on RWD cars. Do one side at a time so that parts are not accidentally interchanged from side to side.

A cotter pin is easily removed using diagonal cutters. Firmly grasp the cotter pin at its looped end and pry it against the spindle to pull it from its hole (**Figure 11.28**). After removing the dust cap and cotter pin, remove the spindle nut and the tabbed washer

> **SHOP TIP** Keep the bearing parts together so they can be reassembled in their original bearing races. Bearing parts become wear-mated to each other.

Figure 11.28 Removing a cotter pin.

244 SECTION 3 TIRES AND WHEELS

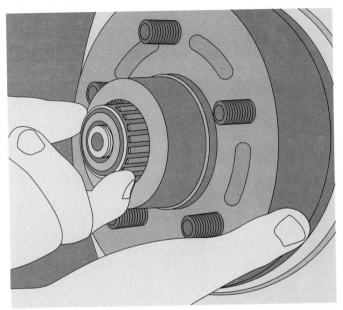

Figure 11.29 Tilting the rotor will help eject the outer wheel bearing.

that is under it. To easily remove the outer bearing, rock the tire back and forth at the top. The bearing will usually pop out on the spindle (**Figure 11.29**).

Remove the Seal

Pull the wheel and hub from the spindle. When working with the wheel bearings, be sure to keep grease and solvent off braking surfaces.

Seals are usually replaced during a bearing repack. Reusing the old seal means taking a chance that grease will leak out and contaminants will get into the bearing. To remove the seal, use a long dowel or drift to pound the bearing against the seal from the inside (**Figure 11.30**). The seal can also be removed with a screwdriver or special seal removal tool (**Figure 11.31**).

A favorite trick used by many chassis technicians is to *carefully* remove the bearing seal using the spindle nut. First, remove the outer bearing. Then, reinstall the spindle nut (**Figure 11.32**) and pull the rotor gently

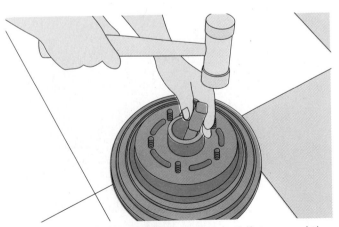

Figure 11.30 Use a long dowel or drift to pound the bearing from the inside.

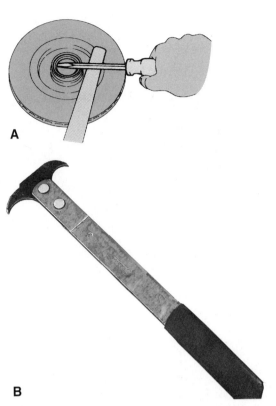

Figure 11.31 (A) Using a screwdriver to remove a wheel bearing seal. (B) A seal removal tool. *(Courtesy of Tim Gilles)*

Figure 11.32 After removing the outer bearing, install the spindle nut. *(Courtesy of Tim Gilles)*

and firmly against the back of the inner wheel bearing (**Figure 11.33**) to remove the seal (**Figure 11.34**). Be careful; rough handling can damage the bearing cage. Also, some seals are too tight in the hub to be removed by this method.

Clean out the old bearing; do not just add grease. Wipe all of the old bearing grease from the spindle, the bearing, the bearing cups, and the hub. Check the

Figure 11.33 Use the rotor to *gently* pull against the inner bearing and seal. *(Courtesy of Tim Gilles)*

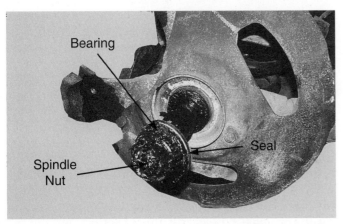

Figure 11.34 The nut, bearing, and seal after seal removal. *(Courtesy of Tim Gilles)*

 **CAUTION**
- Do not spin the bearing with air. Spinning a dry bearing can damage it and can also be dangerous if bearings come loose from the bearing cage.
- When blowing off parts, blow into the solvent tank to avoid making a mess.

grease on the shop towel for metal flakes, which would indicate bearing failure.

NOTE: *It is important to remove all grease from the bearings and hub because the ability of the new grease to lubricate can be compromised if it is mixed with a grease with an incompatible soap.*

Clean the bearing (**Figure 11.35**) and let it air dry on a paper towel or blow it dry from the ends, parallel to the rollers, using compressed air. Be careful not to

Figure 11.35 Clean the wheel bearing.

blow the old dirty grease into the bearing. Rinse the bearing off with alcohol or brake cleaner afterwards.

■ BEARING INSPECTION AND DIAGNOSIS

After the bearings are cleaned, inspect them for damage. If a bearing is damaged, the cause should be determined so the problem does not happen again. Check the bearing and bearing race for pitting and other signs of damage. If any damage is apparent, replace the bearing and its race. Save the old bearing to compare it with the new one. Bearing replacement is covered later in this chapter.

Bearings usually fail slowly. Noise is the clue that something is wrong. As metal is deformed or comes off the bearing races, the noise will become more pronounced. There are two main kinds of damage to a bearing, spalling and brinelling.

Spalling is when pieces break off the bearing metal (**Figure 11.36**). The noise from spalling is a random, high-pitched sound. **Brinelling** is when the bearing or race has indentations from shock loads

Figure 11.36 A spalled bearing. *(Courtesy of Tim Gilles)*

Figure 11.37 Brinelling is the name for dents in the bearing race that result from the rollers hammering against the race.

Figure 11.39 A bearing packer used with a pressurized grease gun. *(Courtesy of Tim Gilles)*

(**Figure 11.37**). The noise that results from brinelling damage is a regular, low-pitched sound.

The biggest shock load on a bearing is often during installation. Pressing should be done directly only on the race that is press-fit. Another cause of bearing damage can be due to an improperly grounded arc welder used when welding a trailer hitch or making body and frame repairs.

Adding Grease to Bearings

To pack bearings by hand, put a small amount of grease in the palm of your hand. Stroke the large open end of the bearing cage against the grease until ribbons of grease start to appear at the opposite ends of the bearing (**Figure 11.38**). Smear a fresh layer of grease all around the bearing and on the clean bearing races.

Most shops own a **pressure bearing packer**. A grease gun is applied to a grease fitting to pack the bearing (**Figure 11.39**).

After the bearings are packed with fresh grease, place them on a paper towel. Remember to reinstall them into the same bearing races. Leave a ring of grease below the bearing race to help keep the fresh grease inside the bearing area after it heats up and begins to flow.

Grease the Inside of the Hub

Put a small amount of grease in the cavity of the hub. Do not fill it up. According to bearing engineers, the amount of grease that is packed into a bearing determines how well it will work. If the cavity is full, there is no place for the excess grease or pressurized air to go when the bearing heats up. If the hub is full, there is no place for excess grease in the bearing to go. This results in excess heat and fluid friction, which can cause the bearing or seal to fail.

High-speed bearings, like those on race cars, are supposed to be filled to only 25 percent of their free space. When a bearing is totally full, it should be used only at low speeds. **Figure 11.40** shows the recommended amount of lubricant to be installed in the hubs of passenger cars.

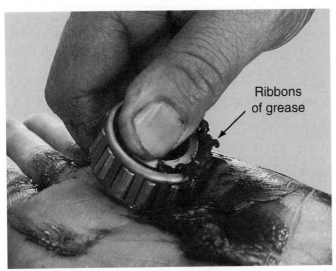

Figure 11.38 The bearing is thoroughly packed when ribbons of grease appear at the opposite end of the bearing. *(Courtesy of Tim Gilles)*

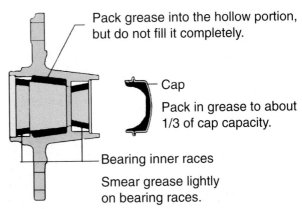

Figure 11.40 Partly fill the hub and dust cap with grease.

Inspect the Spindle

Just behind the area where the bearing rides on the spindle, there is a raised area that the seal rides on. Clean this area so that the new seal is not accidentally ruined (**Figure 11.41**). Inspect the condition of the spindle and check the fit of the large bearing on it.

NOTE: *A worn spindle is usually caused by a bearing seizure that forces the inner race of the bearing to spin on the spindle. The resulting heat softens the hardened spindle, which requires its replacement.*

A bearing is designed to creep on the spindle when loaded, so sometimes there are marks on the spindle. This is normal as long as the bearing feels snug and yet moves freely on the spindle. Inspect the bearing cups in the hub (**Figure 11.42**).

After the inner bearing is installed in the hub, the seal can be installed. Be careful to install it straight. A special tool is helpful for this (**Figure 11.43**). Install the seal with its open end, or lip, facing in toward the bearing (see Figure 11.23). Be sure to lubricate the seal lip so it does not burn up (**Figure 11.44**). Adjust the wheel bearing and install the cotter pin according to the instructions provided earlier.

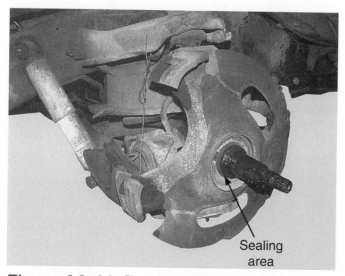

Figure 11.41 Clean the sealing area on the spindle so that the new seal is not accidentally ruined. *(Courtesy of Tim Gilles)*

Figure 11.42 Inspect the bearing races in the hub. *(Courtesy of Tim Gilles)*

Figure 11.43 (A) Select the correct adapter and position the seal over its bore. (B) Pound the seal flush with the top of the bore. *(Courtesy of Tim Gilles)*

Figure 11.44 To prevent damage to the new seal, be sure to lubricate the seal lip before installing the bearings on the spindle.

Pack some grease into the dust cap. (Do not fill it all the way, though.) Packing keeps out contaminants and provides a reservoir for fresh grease. As the grease in the bearing oxidizes, it can dry out. Oil in the fresh grease in the cap can replenish the old grease through capillary action.

■ REPACKING DISC BRAKE WHEEL BEARINGS

The procedure for repacking disc brake wheel bearings is the same as that followed for drum brakes, except that the disc caliper must be removed in order to gain access to the inside wheel bearing. The caliper must be supported or wired to the steering knuckle support. Do not let calipers hang on brake hoses. The inside of the brake hose can be damaged. When reinstalling the caliper, torque the caliper bolt to specifications. When the caliper bolts have an Allen head, a socket drive Allen head tool should be used with a torque wrench.

■ WHEEL BEARING ADJUSTMENT

Wheel bearings must be properly adjusted. If they are too loose, the wheel will be able to move. This can mimic problems that result from worn parts. Some of these problems are shake, noise, wander, steering wheel play, cupped tire wear, and an intermittent low brake pedal on disc brake cars.

The bearing is designed to operate with very little clearance. As a bearing rotates under a load, some heat is developed and the bearing expands. Manufacturers specify methods for adjusting their bearings (**Figure 11.45**). A common adjustment will result in the bearing having about 0.001–0.005 inch of running clearance, leaving enough room for a protective oil film between the bearing and race. This is somewhere around the thickness of a hair. A bearing that is adjusted too tightly will develop more friction, use more power, and may ultimately fail.

An easy method of adjusting a loose bearing can be done with the tire raised off the ground. There is a dust cap that is pressed into the hub at the end of the spindle. The dust cap is removed with a large slip-joint pliers or a special dust cover tool (**Figure 11.46**).

SHOP TIP As a test to see that the bearing is not too tight, insert a screwdriver under the tabbed washer that is under the spindle nut and pry the washer up and down. There is enough slack below the washer and the spindle that it should move freely. If the washer is difficult to move, the adjustment is too tight. Though this tip works on most cars, always check the service manual for the correct procedure when you are working on a make of automobile for the first time.

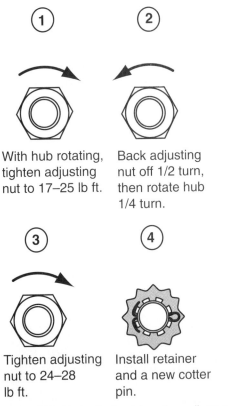

Figure 11.45 Typical wheel bearing adjustment directions.

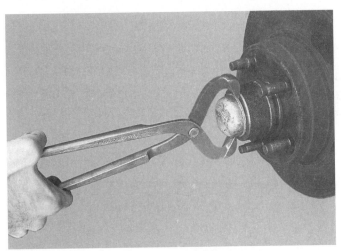

Figure 11.46 A special tool for removing a wheel bearing dust cap. (Courtesy of Tim Gilles)

Some spindle nuts have lock tabs or lock nuts, but most are kept in place with a cotter pin. Remove the cotter pin so the spindle nut can be tightened. Using both hands, grasp the top and bottom of the tire and try to rock it back and forth. As the spindle nut is tightened further, less and less movement will be felt.

The washer under the spindle nut has a tab that fits into a groove on the spindle (see Figure 11.27). Its function is to keep the bearing from trying to turn the spindle nut. It fits the spindle loosely so it should be easy to move.

Figure 11.47 A bearing retainer used to fine-tune the spindle nut adjustment. (Courtesy of Tim Gilles)

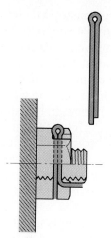

Figure 11.49 Cotter pin installation.

Wheel bearing nuts are typically hexagonal (six-sided). A popular feature used with wheel bearing adjusting nuts is a bearing retainer, a sheet metal cup that has twelve notches (**Figure 11.47**). This is used to fine-tune the adjustment so that the cotter pin that locates the nut can be installed closer to its intended position.

■ SELECTING AND INSTALLING A COTTER PIN

When selecting a new cotter pin, use the largest diameter one that will fit into the hole (**Figure 11.48**). To keep less inventory in stock, repair shops commonly purchase cotter pins of the same length (2 inches, for instance) and cut them with diagonal cutters during installation.

One of the ends of the cotter pin will be longer than the other. Pull on the longer end with the diagonal cutters to seat the cotter pin fully in its hole. Then, pull the long end out and over the end of the spindle. Cut it off. Then, cut off the remaining end of the cotter pin flush with the spindle (**Figure 11.49**). This provides a securely fastened cotter pin that will be easy to remove during the next service. **Figure 11.50** shows the right and wrong ways to install a cotter pin.

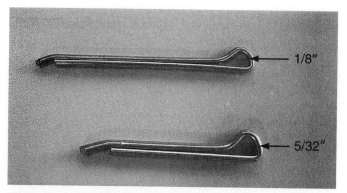

Figure 11.48 Select the largest diameter cotter pin that fits into the hole. (Courtesy of Tim Gilles)

Figure 11.50 The right (A) and wrong (B) ways to install a cotter pin. (Courtesy of Tim Gilles)

■ DIAGNOSING WHEEL BEARING NOISE

If a wheel bearing groan is heard, driving the car can sometimes help the technician pinpoint the problem. First, check the tires for damage and ensure that they are properly inflated. Find an empty parking lot or deserted road and make slow left and right turns. This shifts the weight of the vehicle from one side to the other. When the weight is increased on the bearing, the noise increases. The inside tire always turns at a lower rpm than the outside tire so this can also affect the noise level coming from the wheel.

When driving, changes in bearing noise can be evident with slight left or right steering pressure as weight transfers. The noise from a wheel bearing that is bad will change pitch as the wheel speeds up and slows down when turning from one side to the other. When the outside wheel has the bad bearing, the noise will become worse because that wheel is under more side-to-side load and is turning faster. Applying the brakes can also cause the noise level from a bad bearing to become less as they contact the drum or rotor with the bad bearing.

Check bearings in the service bay using the following procedure:

1. Check the clearance of a defective/noisy bearing. It will usually be loose. If looseness is found, clean serviceable bearings and check them visually. Defective sealed bearings must be replaced.
2. Raise the front of a front-wheel drive vehicle and power the wheels while listening for a suspected bad bearing in a solid area close to the bearing.
3. Spinning the wheels can be done using an on-the-car wheel balancer for nondrive wheels.

REPLACING BEARING RACES

Antifriction bearings usually have one race that is **press-fit**. The other race is a push-fit. A push-fit race slides into place by hand. The press-fit race is usually pressed onto or into the rotating part (outer race in brake hub). The push-fit race is usually pushed onto or into the stationary part (inner race on spindle).

When a damaged wheel bearing is replaced, the press-fit race must be removed from the bearing hub. The race might still look good to the eye but it has been subjected to loads for just as long as the failed bearing. Leaving the old race to be used with a new bearing is an invitation for failure. Bearings should last about 150,000 miles.

The old bearing race is removed by pounding it with a drift punch or a special tool. When using a drift punch there are recesses in at least two places in the side of the hub that allow for hammering on the back of the race (**Figure 11.51A**). Be sure to hammer a little bit from one side, then the other, so the race is not distorted during removal, damaging the hub. A special tool or a punch can be used to remove the race (**Figure 11.51B**).

The new race must fit the hub tightly. If not, replacement of the hub is usually required. A loose bearing race is often the result of a bearing seizure, which can be caused by a bent spindle. Drive the race into the

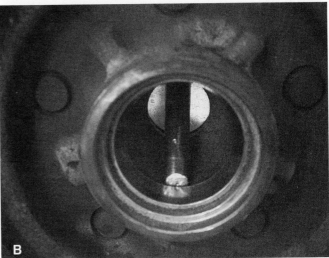

Figure 11.51 (A) This recess leaves room for a punch to remove the race. (B) Removing the bearing race with a brass punch. *(Courtesy of Tim Gilles)*

hub, being careful not to cock it off to one side. If using a soft punch (**Figure 11.52**), hit on one side of the cup and then on the other. Drive it all of the way into the bore until it seats solidly against the ridge at the bottom of the hub. When the race bottoms out, the sound from the pounding will change.

A special tool is handy for installing a bearing race (**Figure 11.53**). A new race can be chilled in a refrigerator to make it easier to install. If a bearing is allowed to run in a misaligned position, it will overload and fail (**Figure 11.54**).

SERVICING FRONT-WHEEL DRIVE BEARINGS

Unlike bearings found on typical rear-wheel drive vehicles, front axle bearings in most FWD vehicles are permanently sealed. Many FWD vehicles use a double-

Figure 11.52 Drive on one side and then the other to avoid cocking the bearing race. *(Courtesy of the Timken Company)*

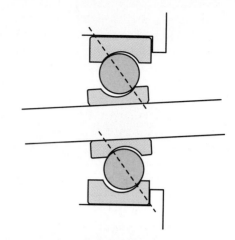

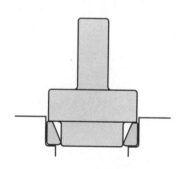

Figure 11.54 Misaligned bearing race wear. *(Courtesy of Tim Gilles)*

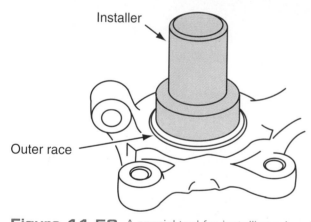

Figure 11.53 A special tool for installing a bearing race.

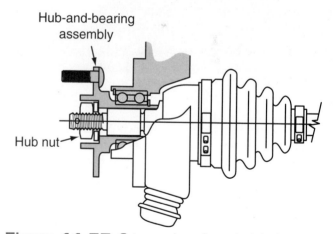

Figure 11.55 Cutaway view of a typical double-row ball bearing installation in the wheel hub.

row ball bearing (see **Figure 11.9**) with bearing races that prevent thrust movement when the spindle nut is correctly torqued. **Figure 11.55** shows a typical installation in the wheel hub. Some FWD vehicles use double tapered roller bearings (see **Figure 11.13**).

Front Bearing Inspection

When a bearing becomes worn or damaged, it will become loose and/or make noise. With the wheel raised off the ground, attempt to rock the tire in and out from top to bottom in the same manner as checking clearance in the front wheel bearing of a RWD car. If you can feel clearance in the bearing, it must be replaced.

Front Bearing Replacement

A sealed front wheel bearing is retained in the steering knuckle hub assembly in one of several ways. Some are

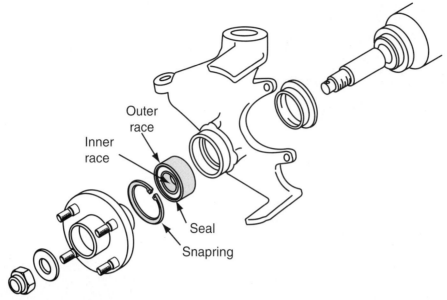

Figure 11.56 A snap ring installed to retain the front wheel bearing in the steering knuckle.

Figure 11.57 Remove the (A) hub retaining bolts and (B) spindle nut.

held in with a snapring as shown in **Figure 11.56**. Others are either pressed or bolted onto the hub assembly. There are also vehicles that have sealed front wheel bearings retained by the spindle nut, like conventional repackable bearings on rear-wheel drive cars.

Bearing Replacement—Bolt-on Type. Bearings that are bolted onto the steering knuckle are the easiest to replace. This type of installation is common on GM and some Chrysler vehicles. After removing the hub retaining bolts and spindle nut (**Figure 11.57**), the bearing hub assembly is removed from the steering knuckle of a GM vehicle using a slide hammer–type puller (**Figure 11.58**). After the old seal is removed from the knuckle, the parts are cleaned and inspected prior to reassembly with a new seal and bearing (**Figure 11.59**).

Bearing Replacement—Press-on Type. Press-fit bearings are found on Ford, some Chrysler, and most Asian

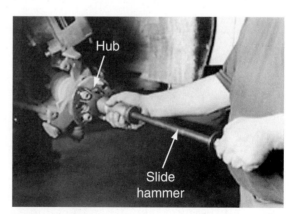

Figure 11.58 Removing the hub assembly with a slide hammer puller.

vehicles. The steering knuckle must be removed from the vehicle in order to press out the old bearing and install the new one.

Removal of most of these FWD front wheel bearings requires the axle to be removed from the hub. The

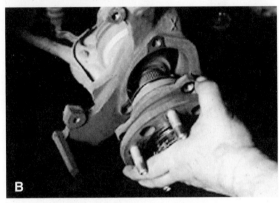

Figure 11.59 After the old seal is removed from the knuckle, (A) the parts are cleaned and inspected prior to (B) reassembly with a new seal and bearing.

end of front-wheel drive axle shaft, called a stub shaft, has drive splines that fit into splines in the rotor hub. The bearing supports the end of the axle just behind the splines. To get to the bearing requires removing the axle (half-shaft and CV joints). A puller is sometimes necessary (**Figure 11.60**). At other times, a tap with a soft-faced hammer is all that is needed to free the stub shaft from the bearing.

Figure 11.60 A puller is sometimes needed to remove the hub from a front-wheel drive car. *(Courtesy of Tim Gilles)*

Figure 11.61 Disassembling steering and suspension parts to gain access to a front wheel bearing. (A) Preliminary disassembly. (B) Stub shaft and steering knuckle removed. *(Courtesy of Tim Gilles)*

Some steering and suspension disassembly will be required prior to pressing the bearing from the hub. **Figure 11.61** shows a typical disassembly process. Steering linkage removal is covered in more detail in Chapter 17.

Disassembly of some suspensions can result in a change in wheel alignment settings. Be sure to mark camber eccentrics prior to disassembly (**Figure 11.62**). **Figure 11.63** shows a typical removal sequence for this steering knuckle design.

Removing a Press-Fit Bearing

When the steering knuckle has been removed from the vehicle, a hydraulic press or a special puller set is used to replace the bearing. The following section describes bearing removal and replacement using a press.

Remove the Hub from the Bearing

Using a press mandrel that is slightly smaller than the inner bearing race, press the hub out of the steering knuckle and bearing (**Figure 11.64**). On some cars, the inner bearing race will come apart from the bear-

Figure 11.62 Mark the location of the camber eccentric on the strut prior to disassembly.

Figure 11.63 (A) Remove the bolts holding the strut. (B) Slide the knuckle assembly off of the axle stub shaft.

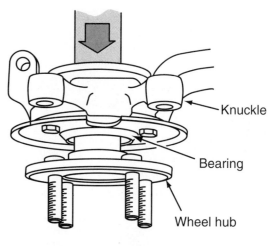

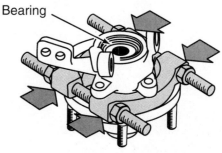

Support bearing separator at arrow locations.

Figure 11.64 Press the hub out of the bearing.

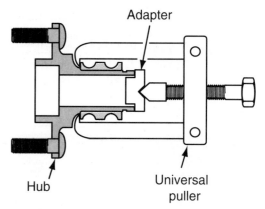

Figure 11.65 Using a puller to remove the inner bearing race from the hub.

ing, remaining on the hub. Use a puller to remove it (**Figure 11.65**).

Remove the Bearing from the Knuckle

Remove the bearing retainer (**Figure 11.66**). It can be a snapring, plate, or collar. Press the outer part of the bearing from the knuckle (**Figure 11.67**). Use a mandrel that is slightly smaller than the outer bearing race.

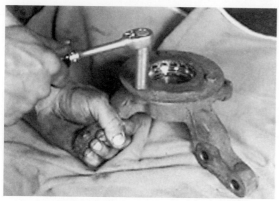

Figure 11.66 Removing the bearing retaining plate. Some use a snap ring.

Figure 11.67 Press against the outer bearing race to remove it from the steering knuckle.

> **SHOP TIP** An old bearing made slightly smaller by grinding its outside diameter will work nicely for removing and installing bearings.

Install the New Bearing in the Knuckle

Lubricate the housing bore and the outside of the bearing. Then use a large mandrel to press the bearing into the knuckle (**Figure 11.68**).

NOTE: *Be sure to press on the outside bearing race. Pressing on the inner race will damage the bearing.*

When the bearing is flush with the outside edge of the steering knuckle, the old bearing with its outside diameter ground smaller can be used to finish pressing the bearing into its bore. A special puller set like the one shown in **Figure 11.69** can also be used.

Install the Hub

Position the bearing retaining plate (**Figure 11.70**) and install the hub into the bearing (**Figure 11.71**).

NOTE: *The bottom of the bearing must be supported on its inner race during this process or the bearing will be damaged.*

Figure 11.68 Press against the bearing outer race using a large mandrel to install it into the knuckle.

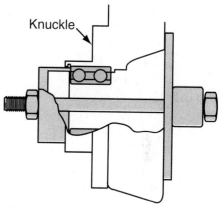

Figure 11.69 Pressing the bearing into its bore using a special puller.

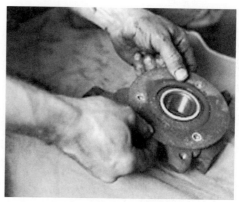

Figure 11.70 Install the bearing retaining plate before pressing the hub into the bearing.

Figure 11.71 When pressing the hub into the bearing, the bottom of the bearing must be supported on its inner race or the bearing will be damaged.

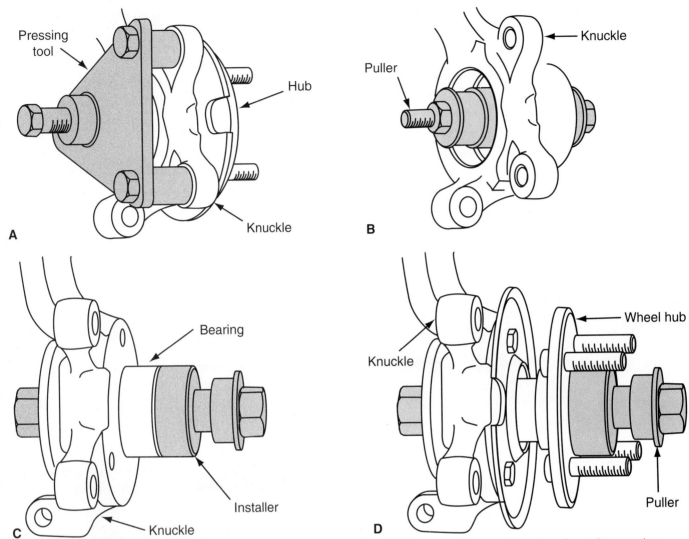

Figure 11.72 (A) Removing the wheel hub from the bearing. (B) Removing the wheel bearing from the steering knuckle. (C) Installing the wheel bearing in the steering knuckle. (D) Installing the wheel hub in the wheel bearing.

Removing and Replacing a Bearing Using Pullers

When removing and replacing a bearing using a puller, follow similar precautions to avoid bearing damage as described earlier for using a hydraulic press. Be sure to apply pressure to a point that will not result in damage to the bearing. The sequence when using a puller set is shown in **Figure 11.72**.

■ REASSEMBLY OF AXLE STUB SHAFT TO BEARING

Some manufacturers recommend that the bearing be replaced any time a front-wheel drive assembly is disassembled, that is, when the front hub is removed from the spindle. End play is controlled by the size of the parts. When they are tightened together, the end play should be correct.

 **CAUTION** The vehicle should not be rolled on its wheels with the axle removed or the bearings can be damaged.

Special care is required during reassembly. When a puller was needed to disassemble an axle from the hub, a certain degree of force will be required to reinstall it in the hub. Be certain that the parts are aligned before applying any force. There are usually two rows of ball bearings with their races tapered toward each other to avoid end thrust. Be sure that the bearing assembly is held together as a unit before forcing it onto the axle shaft. The bearing can be ruined during this process.

On ball bearings, which is what most front-wheel drive cars are now equipped with, the bearings are preloaded. Torque for the axle nut ranges up to 200 pounds. Always check the manufacturer's specifications.

Although it is common practice, it is not recommended to torque the axle nut with the vehicle on the ground. Instead, have an assistant hold pressure on the brakes while you torque the nut to specifications, or use a brake pedal lock. Then, stake the nut into the groove in the spindle, if required (**Figure 11.73**). Remember:

- Many vehicles have self-locking nuts that should not be staked. Reusing axle nuts is not recommended. Use a new axle nut.
- Do not overtighten and then back off.
- Do not use an impact wrench on the axle nut.

Figure 11.73 After some front-wheel drive bearing retaining nuts are torqued, staking is needed to keep them in place. *(Courtesy of Tim Gilles)*

REVIEW QUESTIONS

1. What kind of bearing is a ball or roller, antifriction or plain?
2. The separator that holds roller or ball bearings properly spaced in the bearing assembly is called a _____.
3. An up-and-down load on a bearing is called a _____ load.
4. A front-to-rear or side-to-side load on a bearing is called a _____ load.
5. Which bearing design has greater load-carrying capacity, roller or ball?
6. A _____ bearing is a very small roller bearing.
7. Which grease would be intended for use in wheel bearings, L or G?
8. Which kind of thickener is most often used for multipurpose grease?
9. What kind of oil can damage a nitrile seal?
10. How much running clearance would a correctly adjusted tapered roller front wheel bearing have?
11. What is it called when the bearing or race has indentations from shock loads?
12. On a race car, how far should the wheel bearing hub cavity be filled with grease?

PRACTICE TEST QUESTIONS

1. Two technicians are discussing nondrive tapered front wheel bearings. Technician A says that the outside wheel bearing supports the load of the vehicle. Technician B says that the inside bearing holds the wheel in alignment. Who is right?

   A. Technician A B. Technician B
   C. Both A and B D. Neither A nor B

2. Technician A says that tapered roller bearings tend to be self-aligning. Technician B says that ball bearings tend to be self-aligning. Who is right?

   A. Technician A B. Technician B
   C. Both A and B D. Neither A nor B

3. Technician A says that tapered roller bearings have lips on the edges of the inside and outside races. Technician B says that the angles on tapered rollers are what make them self-aligning. Who is right?

   A. Technician A B. Technician B
   C. Both A and B D. Neither A nor B

4. Technician A says that *axle bearing* is the term for those bearings found on nondrive front and rear wheels. Technician B says that *wheel bearing* is the term for those bearings found on live axles. Who is right?

   A. Technician A B. Technician B
   C. Both A and B D. Neither A nor B

5. Technician A says that heavy-duty trucks use semi-floating axles. Technician B says that if an axle breaks on a full-floating axle, the rear brake drum and wheel will still be supported. Who is right?
   A. Technician A
   B. Technician B
   C. Both A and B
   D. Neither A nor B

6. Technician A says that grease is made up mostly of lubricating oil. Technician B says that soaps are what make grease thick. Who is right?
   A. Technician A
   B. Technician B
   C. Both A and B
   D. Neither A nor B

7. Technician A says that the open side of the sealing lip always faces the lubricant. Technician B says to clean a sealed bearing before reassembly by soaking it in solvent. Who is right?
   A. Technician A
   B. Technician B
   C. Both A and B
   D. Neither A nor B

8. Technician A says during a bearing repack to fill up the hub cavity with grease. Technician B says that a wheel bearing should fit the spindle tightly. Who is right?
   A. Technician A
   B. Technician B
   C. Both A and B
   D. Neither A nor B

9. When selecting a new cotter pin, use the _____ diameter one that will fit into the hole.
   A. largest
   B. smallest

10. Technician A says that the rate of oil oxidation doubles with every 20°F increase in temperature. Technician B says that even if a used bearing race looks good to the eye, it should not be reused with a new bearing. Who is right?
    A. Technician A
    B. Technician B
    C. Both A and B
    D. Neither A nor B

CHAPTER 12

Tire and Wheel Fundamentals

■ OBJECTIVES

Upon completion of this chapter, you should be able to:
- ✔ Describe how a tire is constructed.
- ✔ Understand the various size designations of tires.
- ✔ Explain the design differences among tires.
- ✔ Be able to select the best replacement tire for a vehicle.

■ KEY TERMS

all-season tires
aspect ratio
bead
bead seats
bead wire
belt
bias-ply
carcass
chafing strips
compact spare tire
corporate average
   fuel economy
   (CAFE)
drop center
footprint
hydroplaning
hysteresis
inner liner
isoprene
kPa
load index
placard
plus sizing
ply
P-metric radial tire
profile
psi
racing slick
radial-ply
retreads
ribs
rim well
rim width
run-flat tire
safety beads
sidewall deflection
sipes
size equivalents
speed rating
stud-centric
temperature grade
tire chains
traction
traction grade
tread
tread wear ratings
tubeless tires
tube-type tire
uniform tire quality
   grade (UTQG)
wear bars
wheel offset
wheel rim

■ INTRODUCTION

A service technician should be able to advise customers about tires, discuss aspects of tire design, and help customers make the safest (and best) choice when purchasing new tires and/or wheels. Tires and wheels are an important automotive safety and service specialty area. In-depth information about them is presented in this chapter and in Chapter 13.

■ TIRE CONSTRUCTION

Tires are constructed of several layers of rubber materials, cords, and two rings of wire, called beads (**Figure 12.1**). The *casing,* or **carcass**, is the internal structure of the tire. The **ply** is metal or fabric cord that is rubberized (covered with a layer of rubber). The plies provide strength to the tire to support the load of the vehicle.

The ends of the plies wrap around the steel **bead** before being bonded to the side of the tire. The beads

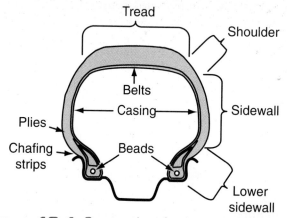

Figure 12.1 Construction of a tire.

are coils of wire at the side edges of the tire. These give the tire the strength to stay firmly attached to the wheel. Some tires use **bead wire** bundles that are cut

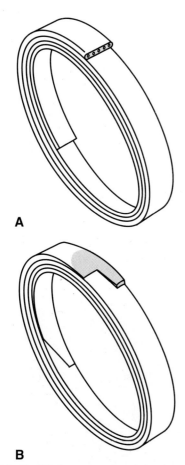

Figure 12.2 (A) A conventional tire bead bundle with an overlapped joint. (B) Bead cable that eliminates the overlapped joint.

to length and butted together with an overlap joint. This can produce a tire that is not as round or as well balanced as premium tires, which have a single strand of bead wire wrapped in loops (**Figure 12.2**). This eliminates the balance and roundness problems of bundled beads. **Chafing strips** are hard strips of rubber that protect the beads from damage that could result from chafing against the rim.

A **belt** is a cord structure made up of plies. It is located only in the area of a tire under the tread and does not extend under the sidewalls.

The **tread** is the section of the tire that rides on the road. A sidewall covering of rubber protects the casing plies between the tire tread and the tire bead.

■ TUBELESS TIRES

Because of safety considerations and ease of servicing, car manufacturers in the 1950s began to put **tubeless tires** on all of their cars. Almost all passenger car tires sold since the early 1960s are of the tubeless design. Some imported cars still had tube-type tires until the middle 1970s, and wire wheels have tubes to prevent leakage from the spoke holes.

The inside of a tubeless tire found on a passenger car has an **inner liner** bonded to it that seals air into the tire. The liner is thicker than the liner on a **tube-type tire**. Tubeless tires are actually safer than tube tires. When a tubeless tire is punctured, it will usually not go flat immediately. A nail tends to be held in the tire by the inner liner, allowing air to escape more slowly. A tube-type tire tends to go flat instantly when punctured because the walls of the inner tube tend to tear.

■ TRACTION

A tire's **traction** is defined as how well it grips the road. Traction is affected by the road surface and contaminants such as water, ice, or debris. It is also affected by the tire's tread, the tread material, inflation pressure, width of the tread, cord ply design, and wheel alignment, among other things.

Figure 12.3 shows the coefficient of friction of different stopping surfaces. Concrete has a better coefficient of friction than asphalt. Stopping distance is affected by the speed of the vehicle. Driver reaction time also contributes to the extension of stopping distances that come with higher speeds (**Figure 12.4**).

■ TIRE TREAD

The tread is a band made of a rubber compound designed to have various traction and wear characteristics. A federal grading standard (discussed later in this chapter) that is cast into the sidewall of the tire describes a tire's traction and wear characteristics. Grooves in the tread allow traction on wet surfaces, giving the water a place to go. They also allow the tire to flex without squirming, which would cause wear.

Treads are designed for specific types of weather and conditions. The design selected is always a compromise. The best traction on a dry paved road would be with a **racing slick**, or a bald tire. That same tire would be dangerous in the rain. Water forms a wedge under a tire that can actually float a vehicle. This is called **hydroplaning**, or *aquaplaning*. A deep tread pattern will break through a water film and grip the road at low speeds, but at high speed the tire can hydroplane (**Figure 12.5**). Beyond a certain speed, the tread pattern loses its ability to remove water from the tread. The only solution to hydroplaning is to slow down. Weight is

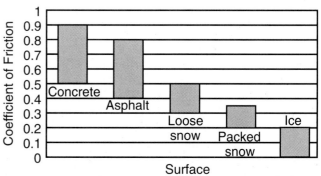

Figure 12.3 Coefficient of friction of different stopping surfaces. *(Courtesy of Bridgestone/Firestone)*

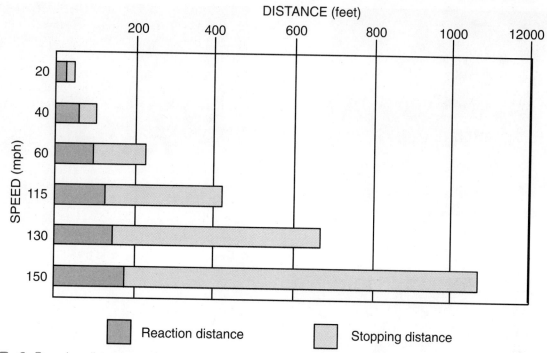

Figure 12.4 Stopping distance and reaction time of the driver increase dramatically with higher speed.

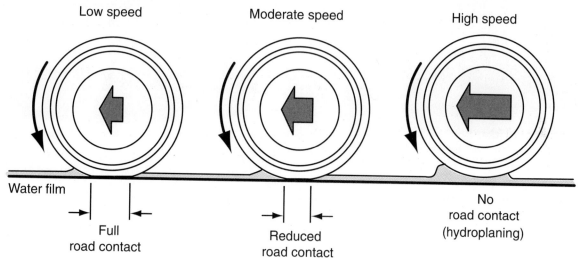

Figure 12.5 Hydroplaning.

also a factor in hydroplaning. The heavier the vehicle, the more resistant it is to hydroplaning.

Although worn tires are more likely to hydroplane, they are better on dry pavement than new tires. Imagine a pencil eraser. As the tire wears, the tread rubber gets stiffer.

Tires with large grooves are designed for use in mud and snow. But the large tread pattern can result in noise on the highway. Treads are often spaced at random intervals to minimize noise.

Sipes are small grooves in the tire tread that look like knife cuts (**Figure 12.6**). They allow extra gripping as the tire flexes. Sipes also clear water off the

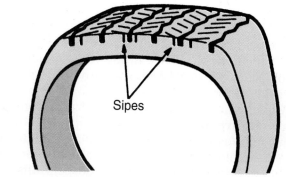

Figure 12.6 Tread sipes.

Figure 12.7 Tread wear indicators. (Courtesy of Tim Gilles)

Figure 12.9 Symmetric tread pattern. (Courtesy of Butler Engineering)

Figure 12.8 Marks on the edge of the tread indicate the location of the wear bars. (Courtesy of Tim Gilles)

Figure 12.10 An asymmetric tread pattern is different from side to side. (Courtesy of Tim Gilles)

road, wiping the contact area to provide a better grip. **Ribs** in the tire tread are designed to pump water from the road through the grooves to the back of the tire, where it is thrown out onto the road.

At the bottom of the tread are **wear bars**. These are raised areas cast into the bottom of the tire tread area to indicate when the tread has become worn to its safe limit (**Figure 12.7**). Marks on the edge of the tread indicate the location of the wear bars (**Figure 12.8**).

Tread Pattern Designs

Tires have tread patterns for different driving conditions. Symmetric tires have treads of the same design on both sides (**Figure 12.9**). They can be installed on either side of the vehicle.

Asymmetric tread patterns are different from side to side (**Figure 12.10**). When cornering, the force is on the outside of the tire so the outside of the tire has larger tread blocks to provide extra stability. Grooves and blocks in the inside of the tread help dissipate water. Asymmetric tires must be mounted in one direction only, which makes them more expensive, and their position on the vehicle cannot be rotated except from front to rear.

Tread grooves for water diversion are located on the inside and a harder sidewall is located on the outside. Directional tread patterns move water out to the sides of the tire, but asymmetric tread designs only improve wet performance. Performance is unchanged in dry conditions.

Some tire treads are unidirectional (**Figure 12.11**). They allow a vehicle to accelerate faster because they have less rolling resistance. They also allow faster stopping. When they are rotated, they must remain on the same side of the vehicle.

Tire Tread Material

Tread materials also call for compromise. Hard materials may wear longer but not provide sufficient traction. Materials for mud and snow tires must remain soft in cold weather. Soft materials must provide sufficient

Figure 12.11 A tire with a unidirectional tread pattern. *(Courtesy of Tim Gilles)*

> ### Vintage Chassis
>
> Charles Goodyear patented the vulcanization process in 1842. **Isoprene**, the core substance of natural rubber, was synthesized in 1910 in Germany. Natural and/or synthetic rubbers have chemicals such as carbon black and antioxidants added to them to improve grip, abrasion resistance, flexibility, and oxidation resistance.

wear. Natural rubber is compounded in different proportions with synthetic rubber to achieve the desired characteristics. Synthetic rubbers are more resistant to heat and solvents than natural rubbers.

Rubber. Rubber trees are grown in all of the subtropical areas of the world. Pure rubber freezes at only 40°F and becomes sticky at 86°F. It swells when contacted by many liquids and is damaged by sunlight. For rubber to be useful, it must be vulcanized, or heated, to make it stable.

Hysteresis is the term used by chemical engineers to describe a rubber's energy absorption characteristics. A high hysteresis compound results in quiet running, a comfortable ride, and better wet and dry grip. A low hysteresis compound has good lateral stability, low rolling resistance, and minimized tread wear.

Tire Cord

Because rubber is elastic and not very strong, it must be reinforced with material such as fabric, fiber, or steel cords. Without these materials, a tire would blow up like a balloon. The most common cord material until World War II was cotton. Today, cord material in the casing is made of either rayon, nylon, or polyester. Cord material for belts can be steel, rayon, nylon, fiberglass, or aramid (Kevlar), which was developed specifically for the tire industry.

■ TIRE PLY DESIGN

Older vehicles used **bias-ply** tires. Modern vehicles use **radial-ply** tires, although some trucks and RVs still use bias-ply tires that have reinforcing belts under the tread area. **Figure 12.12** shows the difference between radial and bias tire construction.

Radial tires have casing plies that run across the tire from bead seat to bead seat in the "radial" direction of the wheel. The outside circumference of the tire is held together by reinforcing belt rings of slightly angled cord material (**Figure 12.13**).

Bias-ply, diagonal, or *cross-ply* tires have casing plies that cross each other at angles of 35–45 degrees. They ride softer than radials, but their tread tends to squirm when rolling. This results in tire wear. Belts beneath the tread give the tire stability. Bias tires with belts under the tread last longer than unbelted bias tires because the belts keep the tread from squirming (**Figure 12.14**).

Radial tires provide longer tread life, better grip to the road surface, and improved fuel economy. They ride rougher at low speeds than bias tires but can actually be smoother at faster speeds while travelling over

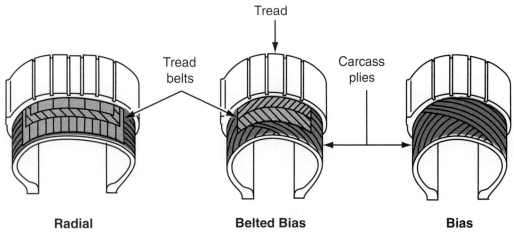

Figure 12.12 Comparison of radial and bias tire construction.

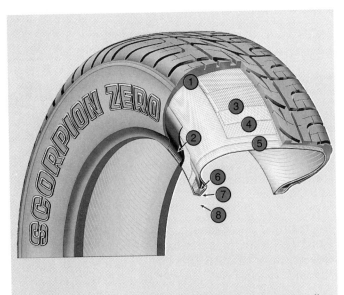

1. Tread
2. Tread base
3. Two-ply nylon wound breaker
4. Two steel-cord plies
5. Two rayon-carcass plies
6. Double nylon bead reinforcement
7. Bead filler
8. Bead core

Figure 12.13 Typical tubeless tire. *(Courtesy of Pirelli Tire of North America, LLC)*

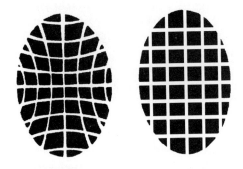

No belts | With belts

Figure 12.14 Belts stabilize the tread in the tread contact area.

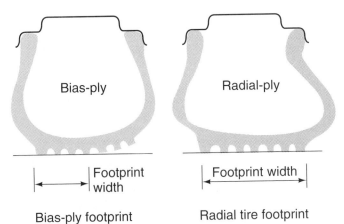

Bias-ply footprint | Radial tire footprint

Figure 12.15 The flexible radial sidewall allows more of the tread to remain in contact with the road, even when cornering.

highway expansion joints. A radial is more expensive to construct than a bias tire because more labor is required during manufacturing.

A bias tire is made in a full circle mold that has two halves, like a bagel cut in half. The parting line for a tire made in this type of mold will run down the center of the tire's tread. A segmented mold is often used for constructing radial tires. A radial tire that has been made in a segmented mold will have several radial parting lines running from bead to bead across its tire tread.

A tire acts like a part of the suspension system as it supports the load of the vehicle, isolating the passengers from road shock as its sidewall deflects. **Sidewall deflection** allows more of the tread to actually be in contact with the road surface. The larger area of contact, called the tire's **footprint**, allows the load on the tire to be spread across a wider area of the tire. A larger footprint also enables the tire to grip better so it can transmit forces of the engine and brakes to the road surface.

A radial tire flexes on its sidewall and is more resistant to wear because its tread surface stays flat on the ground, even when cornering (**Figure 12.15**). Because of the bulging sidewall, a properly inflated radial will appear to be low on air when compared to a bias-ply tire.

Vintage Chassis

Classic car owners with bias tires often wish to upgrade to radial tires. Cars built prior to 1972 usually do not have suspension systems designed for radial tires. Installing radials on these cars can result in a somewhat harsher ride at slow speeds. Installing radials on wheels that were designed only for bias tires can result in a dangerous wheel failure because radials exert more pressure against the sides of the rim. Cars produced after 1975 have numbers on the rims that designate their use with radial tires. If the number includes an R, the rim is designed to be used with radials.

The difference in handling characteristics between radials and diagonal bias tires makes it best not to mix them on the same vehicle. Sometimes, a customer's car will have two belted-bias tires that are in good condition. The customer might want to begin to make the switch to radials without buying all four new tires. The Rubber Manufacturers Association (RMA) recommends that the two new radial tires be installed on the rear.

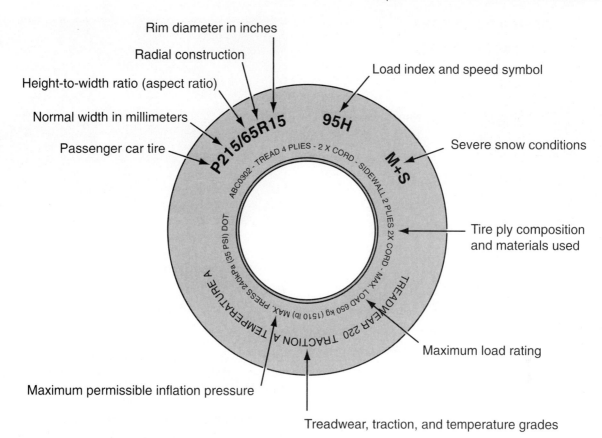

Figure 12.16 Information found on a typical passenger tire sidewall.

Radial tires have less resistance to rolling, which improves a vehicle's gas mileage. Fuel economy standards have been mandated by the U.S. Congress for several years. Each vehicle manufacturer must meet a **corporate average fuel economy (CAFE)** standard or pay a "gas-guzzler" penalty to the government. Because radials help vehicles achieve better fuel economy, these tires are included on all new vehicles as original equipment.

See Chapter 13 for more information on radial tire service.

TIRE SIDEWALL MARKINGS

The U.S. Department of Transportation (DOT) requires the listing of certain information on the tire (**Figure 12.16**). Included on the tire sidewall for a typical passenger car are:

- Tire size
- Maximum permissible cold air pressure
- Load rating—an indication of the load limit for each of the vehicle's tires under cold inflation
- The name of the material that the cords of the tire are made of
- The number of plies in the tread and sidewall areas
- If the tire is a radial tire
- Whether the tire is tube type or tubeless
- DOT manufacturing code
- M+S—this indicates that the tire meets the RMA definition for a mud and snow tire
- Uniform tire quality grade (UTQG) standard grading for traction, treadwear, and temperature

Tire Size

A tire information sticker, sometimes called a **placard** (**Figure 12.17**), has been required on cars sold in the United States since 1968. It is located on the door post, the edge of the door, the gas filler door, or the glove box door. The placard indicates the correct original equipment (OE) tire size, the cold inflation pressure, and the gross axle weight (for commercial vehicles). If

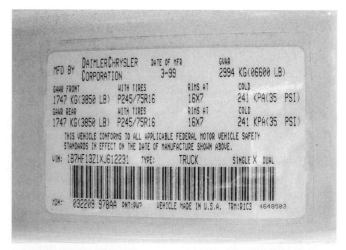

Figure 12.17 Door jamb decal showing gross vehicle weight rating, and tire size. *(Courtesy of Tim Gilles)*

266 SECTION 3 TIRES AND WHEELS

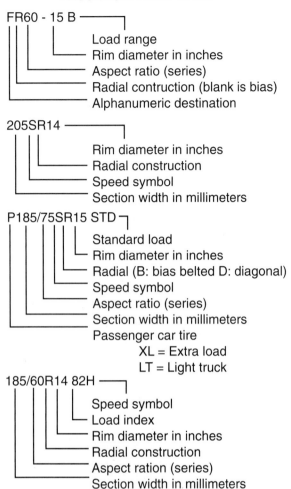

Figure 12.18 Different ways of measuring tire size.

Figure 12.19 A tire size label for a light truck tire. *(Courtesy of Tim Gilles)*

Figure 12.20 A European tire size label. Notice there is no P in the size designation. *(Courtesy of Tim Gilles)*

there is no placard, check the owner's manual for the information.

The tire's size is listed on the sidewall, using one of several ratings (**Figure 12.18**):

- P-metric (P205-75 R15)
- Numeric (6.70-15)
- Alphanumeric (FR78-15)
- Light trucks (LT 235/75R-15; **Figure 12.19**)
- European metric (205/40 R17)—Note that European tires do not list a P or an LT at the front of the rating (**Figure 12.20**).

The alphanumeric rating was commonly used until the early 1970s, but metric cross-sectional measurement is now universal. **Figure 12.21** shows how the tire size designation is interpreted for the **P-metric radial tire**, the most common tire in use today. The first letter tells the type of tire it is:

- P means passenger car
- LT means light trucks, or C means commercial
- T means temporary spare (covered in more detail later)

The tire's cross-sectional width (215 mm) is listed next. With each size increase (from 205 to 215, for instance) the width of the tire increases by 10 millimeters.

There is usually a letter in the size designation. Some of the possible letters are:

- R—radial
- B—belted bias construction (this is sometimes left blank)
- D—diagonal bias construction

A tire's height is called its **profile**. A low profile tire is shorter than a normal tire. The number that comes after the cross-sectional width of the tire is the **aspect ratio**, which is a measurement of the height-to-width ratio (**Figure 12.22**). In Figure 12.21, aspect ratio is expressed as the number 65. An example of a higher profile tire would be a 70. A 60 would have a

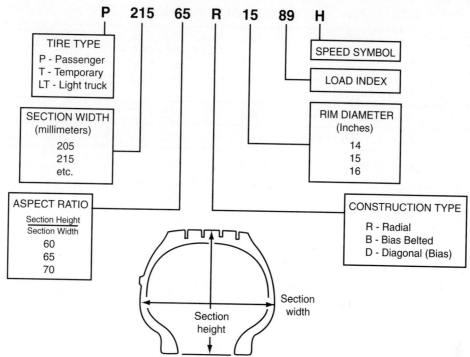

Figure 12.21 Tire size designation on a P-metric radial.

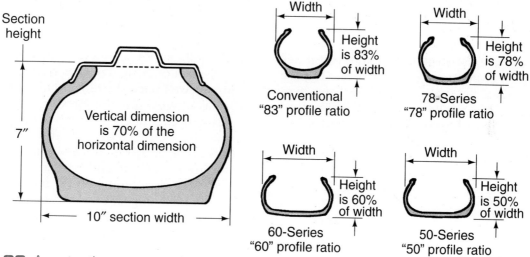

Figure 12.22 Aspect ratio compares a tire's height and width.

lower profile. When a designation does not include an aspect ratio, the tire is an 80 series (P215 R15).

The last part of the sequence, R15, shows that the tire is a radial tire to be mounted on a 15-inch diameter rim. This is sometimes followed by 90H—an optional load/speed index.

Speed Rating

Sometimes a **speed rating** (**Figure 12.23**) is listed as part of the size designation. This rating was originally developed in Europe, where higher freeway speeds are legal. Tires made for use in the United States have speed ratings based on tests that meet SAE J1561 standards.

Vintage Chassis

Some older tires include a speed rating on their labels: P 215-65 HR15. The H is the speed designation for 130 mph. This means that a properly inflated tire with an H designation has been designed to operate at up to 130 mph for short periods of time, such as when passing cars. Sustained high speeds can damage the tire.

SPEED SYMBOLS / RATINGS

SPEED SYMBOL	MAXIMUM SPEED	APPLIES TO PASSENGER CAR TIRES	APPLIES TO LIGHT TRUCK TIRES
ZR	Above 149 mph (240km/h)	Yes	No
Y	186 mph (300km/h)	Yes	No
W	168 mph (270km/h)	Yes	No
V	(with service description) 149 mph / (240km/h)	Yes	Yes
H	130 mph (210km/h)	Yes	Yes
U	124 mph (200km/h)	Yes	Yes
T	118 mph (190km/h)	Yes	Yes
S	112 mph (180km/h)	Yes	Yes
R	106mph (170km/h)	No	Yes
Q	99mph (160km/h)	(winter tires only)	Yes
P	93mph (150km/h)	No	Yes
N	87 mph (140km/h)	No	Yes
M	81 mph (130km/h)	Temporary spare tires	No

Figure 12.23 Speed ratings.

Newer tires have the speed rating and load index listed separately, after the size designation (**Figure 12.24**). The new designation is a two- or three-digit **load index** followed by the speed symbol. The load index, developed by the International Standards Organization (ISO), provides an industry standard for a tire's maximum load at the designated speed rating. Individual load capacities are listed in **Figure 12.25**. A tire's load-carrying ability is related to the strength of its sidewall plies. A tire with a higher load capacity also has a higher inflation pressure.

The speed rating also serves as an indicator of better handling characteristics that result from improvements to the tire. At high speeds, non-speed-rated tires distort in the sidewall and tread areas (**Figure 12.26**). Speed-rated tires require extra reinforcement in the sidewall, including sidewall bead stiffeners and nylon cap plies or belt edge strips (**Figure 12.27**). Bead stiffeners are made of extra hard rubber that prevents the sidewall from bulging. Tires with these additions provide a quicker steering response time. You can feel the difference at as little as 35 mph.

Low profile tires often have sections of sidewall that extend beyond the flange of the rim to protect the rim from damage when the tires rub against a curb.

Nylon cap plies are extra plies on the sides of the normal tire plies. Additional centrifugal force results from higher speed, but nylon shrinks when heated, pulling back on the sidewall and flattening the tire footprint. This helps to keep the rear end from breaking loose. Tires used for extreme high speeds have a sidewall reinforced with a band of steel.

Speed symbols for passenger cars range from the L rating (74.5 mph/120 km/h) to ZR (over 149 mph/240 km/h). The letters that denote changes in speed ratings change in 20 kilometer per hour (km/h) increments.

Figure 12.24 This tire has a Y speed rating and a 95 load rating. It is rated for 186 mph and can carry a load of 1,521 pounds. The ZR is required in the size designation on Y rated tires. *(Courtesy of Tim Gilles)*

LOAD CAPACITY INDEX CHART

LOAD INDEX	MAX. LOAD PER TIRE Kg.	Lbs.	LOAD INDEX	MAX. LOAD PER TIRE Kg.	Lbs.
70	335	738	93	650	1,433
71	345	760	94	670	1,477
72	355	783	95	690	1,521
73	365	804	96	710	1,565
74	375	826	97	730	1,609
75	387	853	98	750	1,653
76	400	882	99	775	1,708
77	412	908	100	800	1,764
78	425	937	101	825	1,819
79	437	963	102	850	1,874
80	450	992	103	875	1,919
81	462	1,018	104	900	1,984
82	475	1,047	105	925	2,039
83	487	1,074	106	950	2,094
84	500	1,102	107	975	2,149
85	515	1,135	108	1,000	2,205
86	530	1,168	109	1,030	2,271
87	545	1,201	110	1,060	2,337
88	560	1,234	111	1,090	2,403
89	580	1,278	112	1,120	2,469
90	600	1,323	113	1,160	2,535
91	615	1,356	114	1,180	2,601
92	630	1,389			

Figure 12.25 Individual tire load capacities.

Figure 12.26 This tire does not have a speed rating. Under high speed, it distorts in the tread and sidewall areas. (Courtesy of Bridgestone/Firestone)

H-rated tires were the first speed-rated tires, hence, the rating does not follow alphabetical order. These are some of the most common ratings:

Q—99 mph (160 km/h)—winter tires

R—106 mph (170 km/h)—heavy-duty light truck tires

S—112 mph (180 km/h)—family cars and vans

T—118 mph (190 km/h)—family cars and vans

H—130 mph (210 km/h)—high-performance passenger cars

Figure 12.27 Speed-rated tires have stiffer sidewalls and more belts. (Courtesy of Bridgestone/Firestone)

V—149 mph (240 km/h)—high-performance sports cars

NOTE: *Good winter tires tend to have only a Q rating because they generate a great deal more heat than conventional tires.*

Light truck tires run hotter and do not dissipate heat as well as passenger car tires. Some of the newer truck tires carry an R or S speed rating.

Vintage Chassis

Under the original speed rating system, V-rated tires (130 mph/210km/h) had the highest achievable speed rating. This category was originally defined as "unlimited." As tire manufacturing abilities improved and safety considerations increased, the V rating became a "limited" rating. The Z rating was added as the top speed rating, for speeds in excess of 149 mph, with the exact speed rating determined by the manufacturer and varying with size.

New High-Speed Ratings. Because some new vehicles are capable of very high speeds, new tire speed ratings have been developed. A W-rated tire is rated at 168 mph (270 km/h) and a Y-rated tire is rated at 186 mph (300 km/H). These tires still carry a Z in their tire number designation, with the W or the Y listed after with the load rating, such as *275/35/ZR19 (99Y)*. When the load and Y speed rating are enclosed in parentheses, the speed rating is in excess of 186 mph (300 km/h).

Load Rating

A tire's load rating tells how much weight it can safely support at a specified air pressure. It is very important not to use a tire that has too low a load rating for the weight of the vehicle.

P-metric radial tires found on today's passenger cars are all of a uniform standard load rating or an extra load rating. The amount of load one of these tires can support is determined by the area of the tire and the amount of air pressure in it. Standard load tires reach their maximum load carrying capacity when inflated to 35 psi. Extra load tires achieve maximum load at 41 psi. An extra load tire is labeled with "XL," as in P205 70R15**XL**.

Although a P-metric standard load tire has a normal maximum inflation of 35 psi, its tire sidewall might be branded with a maximum pressure of 44 psi. This means that even though the tire can be safely inflated to this pressure, its maximum load-carrying capacity is not increased.

Tire companies recommend that heavy RVs be weighed separately at each wheel. This is to determine the load rating and the air pressure that should be used to safely support the load at each corner of the vehicle.

GROSS VEHICLE WEIGHT

The vehicle's gross vehicle weight rating (GVW or GVWR) includes the weight of the vehicle, the weight of the passengers it has seats for (estimated at 150 pounds each), and the maximum amount of luggage load. The GVWR can be found on a plate or sticker on the door jamb (see **Figure 12.17**). It is sometimes listed on the vehicle's registration as well.

Curb weight is the weight of the vehicle without passengers or luggage, but it includes a full tank of fuel and all fluids filled in the vehicle.

Vehicles, especially pickup trucks, are often overloaded. It is important that tires, brakes, and axles be of sufficient size or capacity to support the load and that a vehicle not be loaded beyond its weight rating.

When towing, be sure that the weight of the trailer is within the maximum capacity of the vehicle. The best way to prevent overloading is to weigh each axle of the vehicle on platform scales. The RMA (PO Box 3147, Medina, Ohio 44258) provides free information on vehicle weighing procedures.

DOT Codes

The DOT symbol (**Figure 12.28**) signifies that the tire meets DOT safety standards. Before the year 2000, there were ten characters (a combination of numbers and letters) in the DOT code. Today there are up to twelve, but usually eleven. All of the characters except for the fifth, sixth, and seventh are regulated by the DOT.

Vintage Chassis

Before P-metric tire sizes, tire sizes were alphanumeric. Passenger car tires were "Load Range B," which had a maximum inflation pressure of 32 psi. Light truck tires carried higher inflation pressures with load ranges of C, D, or E. Used on passenger car tires in the past, load ratings are now used only on light truck tires. Light truck tires are any tires that have the letters LT in their size callout.

Figure 12.28 DOT symbol on a tire. *(Courtesy of Tim Gilles)*

DOT 8X72 WL1 1603

8X	The location of the plant where the tire was manufactured
72	The tire size (regulated by the U.S. government)
WL1	The manufacturer's option code, used to tell the difference between tires
16	The week of the year the tire was manufactured
03	The year the tire was manufactured

Figure 12.29 A typical DOT code and the meaning of its numbers and letters.

The first two characters list the plant and manufacturer where the tire was made. There is a separate code for the same manufacturer in different countries, so the country of origin can be determined. DOT tire manufacturers and countries of origin are listed in the appendix at the back of this book.

The second set of characters in the code tells the size and type of the tire. The third set consists of three characters that are not regulated and can be used by manufacturers as they choose. The final four digits tell the week and year the tire was made. The first two digits are the week of the year and the last two are the year. Prior to the year 2000, the date of manufacture was listed as three digits, but since then four digits have been used so it is possible to tell the decade when the tire was manufactured. **Figure 12.29** shows a typical DOT code and the meaning of its numbers and letters.

■ ALL-SEASON TIRES

All-season tires have a tread designed to improve traction on snow or ice, while producing acceptable noise levels on dry, smooth roads. When radial tires became common on vehicles, they were found to have more traction on snow than the bias-ply tires they replaced. When a tire has specially designed pockets and slots in at least one tread edge, it can be labeled with a mud and snow designation. Any combination of the letters M and S (M+S, MS, M&S, or M/S) on the tire sidewall means that the snow tire meets definitions set by the RMA (**Figure 12.30**).

Snow Tires

M&S tires do not guarantee winter driving performance or safety, however. A snow, or winter, tire is specially designed for winter performance. In 1999, the RMA and the Rubber Association of Canada decided on a performance standard for snow tires that would allow consumers to identify tires designed to enhance traction in harsh winter conditions. Winter tires must meet traction tests on packed snow specified by the American Society for Testing and Materials (ASTM). Tires that meet this standard are labeled with a "snowflake on the mountain" symbol on the tire sidewall next to the M&S symbol (**Figure 12.31**).

Summer tires become harder in the winter. Winter tires are made with a tire rubber compound that remains soft in the winter. If these tires are used during the summer, they will be excessively soft and will experience rapid tire wear and generate a good deal of heat. As mentioned earlier in the chapter, they have low Q-speed rating.

Snow tires have deeper tread grooves designed to provide a better grip when driving on snow-covered roads (**Figure 12.32**). Tread depth for snow tires is

Figure 12.31 This winter-rated tire has a snowflake symbol showing it is designed for driving in snow. *(Courtesy of Tim Gilles)*

Figure 12.30 This tire is rated for snow.

Figure 12.32 A snow tire has an aggressive tread pattern. *(Courtesy of Tim Gilles)*

13/32–15/32 inch deep, compared to new passenger car tire depth of 10/32 inch.

Most manufacturers recommend that snow tires be installed on all four wheels to prevent handling problems. Wide tires do not cut through snow as easily as narrow tires. Snow tires can be fitted in the original equipment size. But when the vehicle has wide tires, narrower tires of the same load capacity are often recommended by the tire dealer. Changing tire sizes is covered later in this chapter.

Some states allow tires with the severe snow symbol to be used in place of studded tires or snow chains. Winter tires can provide better handling in snow than is afforded by four-wheel drive without winter tires.

Snow Chains

Snow chains are used in some mountain areas during severe winter weather when roads have become covered with ice and snow. Most manufacturers recommend against the use of chains. In fact, some new vehicle sales procedures include having the buyer sign a paper saying that only cable chains can be used. When a vehicle has sufficient clearance between the tires, fenders, and suspension components, **tire chains** provide a viable means of achieving traction on snow-covered roads.

Tire chains can be of either the cable or chain type. Cable chains (**Figure 12.33A**) are not as effective as conventional chains, but they work well in low-clearance applications and are not as apt to cause damage to the tire or vehicle from incorrect installation. **Figure 12.33B** shows the different conventional types of tire chain. Heavy-duty conventional chains have reinforced lugs to better bite into ice and snow.

Run-Flat Tires

Some new cars do not carry a spare but use one of several methods that allow them to run with little or no tire pressure. Because an underinflated tire will develop heat that will damage the tire, a run-flat tire must be used with a low-pressure detection system.

A typical **run-flat tire** has a stiffer sidewall and a tighter tire bead (**Figure 12.34**). The stiff reinforced sidewalls are four to six times as thick as a normal tire's sidewalls (**Figure 12.35**). When a conventional tire is driven without air its beads tend to fall out of the rim's safety bead areas into the drop center. The vehicle can lose control and cause a serious accident. Run-flat tires have a special bead design to prevent this from happening. A run-flat tire can partly support the vehicle even when the tire is completely empty of air and can be driven up to 70 miles without air before suffering damage. When driving over expansion joints on the freeway, a driver should expect the reinforced sidewalls to contribute to a rougher ride.

Many modern vehicles have high torque engines and high-performance brakes that can cause the tire to slip on the rim. When this happens, the wheel weights will move to a different place in relation to the tire, resulting in vibration due to imbalance. High-

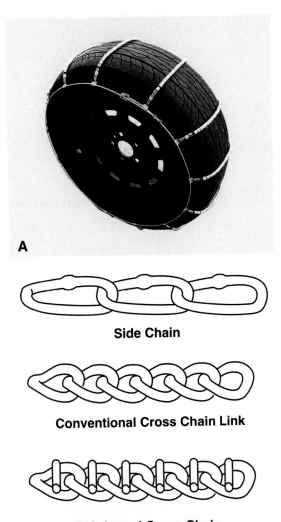

Figure 12.33 (A) Cable snow "chains" are installed when the clearance between the tire and fenderwell is reduced. (B) Conventional and heavy duty-chains. (A, *Courtesy of Tim Gilles*)

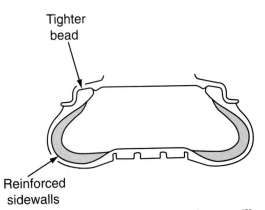

Figure 12.34 A typical run-flat tire has a stiffer sidewall and a tighter tire bead.

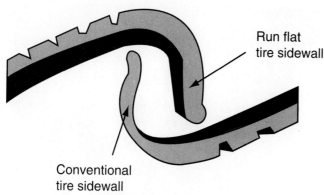

Figure 12.35 The stiff reinforced sidewalls of a run-flat tire are four to six times as thick as a normal tire's sidewalls. (Courtesy of Hunter Engineering Company)

performance and run-flat tires with tighter fitting beads prevent the tire from slipping on the rim. Some luxury performance cars and SUVs also have "bead lock" rims designed to hold the tire tightly to the rim.

Other run-flat designs use bead retention systems with special tires, special rims, and sometimes a third part (**Figure 12.36**). One method, which uses an ordinary tire and rim, has one or two rings or a foam filler. The rings fill the drop center of the wheel to prevent the beads from moving into that area if there is a loss of air. The foam filler design maintains the original shape of the tire in case of an air loss.

Compact Spare Tire

Most new cars come equipped with a **compact spare tire**. The compact tire is considerably smaller than a regular tire and is to be used temporarily only. Many have a limited speed of 31 mph (50 kilometers per hour) and a distance of 31 miles (50 kilometers). The speed and distance warning will be printed on the sidewall of the tire.

Retreads

Retreads were very popular in the past but are not very common any longer except for the rear tires of large trucks. A retread is a worn tire that has had its entire tread removed. A new tread area is vulcanized (bonded) to the tire carcass.

Low Pressure Warning

The Tread Act, legislated by the U.S. Congress in 2000, requires new vehicle manufacturers to install a tire low pressure warning system on all cars. Tire pressure monitoring will be phased in until all vehicles have it by the year 2006. Sensors transmit a radio frequency (RF) to the receiver (**Figure 12.37**). In some pressure monitor systems, when the pressure in a tire drops below a predetermined point, 25 psi for instance, a warning

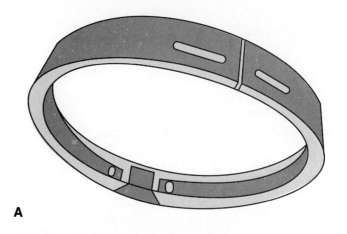

A

B

Figure 12.36 Run flat devices. (A) This device fills the inside of the rim and prevents a tire from coming off of the bead seats. (B) A cutaway of a run-flat tire with an insert to support the tire in case of air loss (B, Courtesy of Goodyear Tire Company)

light on the instrument panel illuminates. Other systems monitor tire pressure continuously.

Direct and Indirect Pressure Monitors

Low tire pressure can be monitored using a direct or indirect method. Direct tire pressure monitoring,

274 SECTION 3 TIRES AND WHEELS

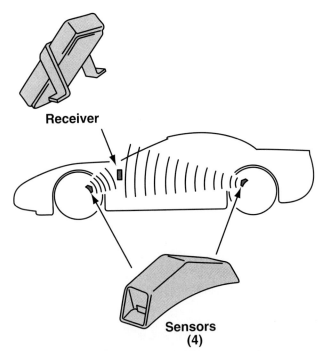

Figure 12.37 Direct tire pressure sensing uses pressure sensors and a receiver.

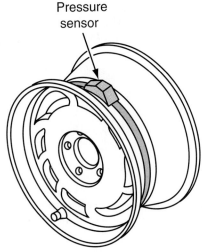

Figure 12.38 Early sensors were strapped in the dropwell of the rim, opposite the valve stem.

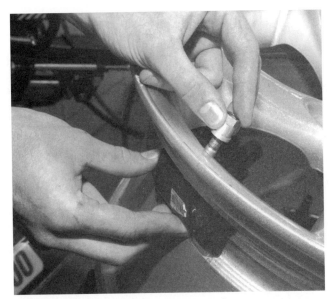

Figure 12.39 Late-model direct tire sensors are part of the valve stem. *(Courtesy of Hunter Engineering Company)*

which uses individual wheel sensors and a computer, is more costly. Early sensors (prior to 1997) were strapped in the dropwell of the rim, opposite the valve stem (**Figure 12.38**). Later sensors, called *integral sensors,* are part of the valve stem (**Figure 12.39**). Each wheel's sensor sends a different signal to the monitor so it can determine which wheel has a pressure problem. A typical sensor is powered by a lithium battery with a ten-year service life. To save the battery, the sensor operates only at speeds over 20 mph (32 kph) and sends a signal once an hour when parked. Pressure below 24 **psi** (168 **kPa**) or above 39 psi (272 kPa) illuminates a warning light on the instrument panel.

An indirect tire pressure monitor system uses the antilock brake system to compare the speed of all of the vehicle's wheels, allowing a 10-psi difference in pressure, but if all four tires are low it does not detect a problem. This presents a new concern for tire makers: If they believe their tires are correctly inflated, consumers might have a false sense of security because no instrument panel light is illuminated.

■ TIRE QUALITY GRADING

American manufacturers use the **uniform tire quality grade (UTQG) system**, which rates tread wear, traction, and temperature resistance with numbers and letters. The UTQG is printed on the label of a new tire (**Figure 12.40**) and also on the tire sidewall. The rating of the tire shown in **Figure 12.41** is 420AA.

Figure 12.40 A new tire label lists the DOT quality grades. *(Courtesy of Tim Gilles)*

Figure 12.41 Uniform tire quality grade (UTGC) markings on a tire sidewall. *(Courtesy of Tim Gilles)*

Tread Wear

The amount a tire's tread wears will vary with wheel alignment, road surface texture, tire rotation maintenance, vehicle speed and braking practices, the weight of the vehicle, and the size of the tire. Tire manufacturers test their tires according to government-specified procedures. The government prescribes a test procedure and course. A convoy of less than four test vehicles drives the same 400-mile test course on public roads in Texas. The test sequence lasts 7,200 miles, with tread depth being measured every 800 miles. Tire wear results are compared with those done on a control group of tires. **Tread wear ratings** are simply an overall indicator of performance, and ratings are not identical between manufacturers.

Tread wear ratings range from under 100 to over 500, increasing in increments of 20. The number 100 represents a standard tire. A 200 would be expected to last twice as long on the government test course, and a 150 would last about one and one-half times as long. The actual life of a tire will vary based on road conditions, climate, air pressure, alignment, driving habits, vehicle loading, and other factors.

Traction Grade

With regard to the actual ratings, AA is the highest, whereas C is the lowest. Prior to 1997, A was the highest rating for wet braking traction. The first letter is the **traction grade**, which indicates stopping ability on wet asphalt pavement and concrete. This rating, done on specified government test surfaces, covers only braking in a straight ahead direction, not cornering.

Temperature Grade

The second letter is the **temperature grade**. It indicates a properly inflated tire's resistance to generating heat and its ability to dissipate heat at highway speeds. Temperature ratings are determined using specified government tests in a laboratory on a test wheel. Grade C is the minimum standard required by law. Standards B and A exceed this standard. Continuous high-speed driving can result in deterioration of the tire's material. Coupled with excessive temperatures, this can lead to sudden tire failure.

■ CHANGING TIRE SIZE

Customers often want to make a change in the size of an original equipment tire. It is very important that the intended use of the tire be known before making a change from original equipment tires. Selecting a replacement tire that is the exact same size as shown on the placard is not always possible. If tire size is changed, be sure to substitute a tire that has an equal or greater load-carrying capacity. Tire companies provide charts that give the maximum load various sizes of their tires can accomodate at listed cold pressures (see **Figure 12.25**).

Tire Size Equivalents

A change in tire size can usually be accomplished without sacrificing safety or design considerations when the correct **size equivalent** is used. At first, this can be confusing. Tires with different bead-to-bead diameters (13", 14", 15", etc.) can all have the same outside diameter. For instance, the following five tire sizes all have the same diameter *and* load capacity:

1. 175/70SR13
2. 205/60R13 H
3. 185/60R14 H
4. 205/55VR14
5. 195/50VR15

When changing tire sizes, the following five things need to be considered in the replacement tire:

1. The intended use of the tire
2. The width of the wheel rim
3. The overall diameter of the replacement tire assembly
4. A speed rating equal to or greater than that of the original tire
5. An overall load-carrying capacity equal to or greater than the load index number listed on the original tire

Usually, as the diameter of the tire increases, the load capacity of the tire increases.

Tire manufacturers publish application handbooks that give information such as tire dimensions,

Changing from a P185/65R14 (85 load index/1124-pound maximum carrying capacity) to a P185/60R14 (82 load index/1047-pound capacity) will result in almost 7 percent less load capacity.

revolutions per mile, diameter, acceptable rim sizes for each tire size, and the cross section of the tire. When unsure about a possible change in tire size, consult the technical assistance available from all of the major tire manufacturers.

Since the early 1970s, tires with lower profiles have become increasingly popular. When a tire with a lower profile is being installed, a wider tire and a wheel with a larger diameter will be used to make up the difference in overall tire assembly height. This method, called **plus sizing**, produces the same overall diameter as the original equipment tire (**Figure 12.42**). The new combination has less sidewall flexibility but has the same load capacity. It also has a larger footprint.

Lower profile tires grip better and they are more responsive. A high-profile tire provides better loose snow or mud traction because it does a better job of cutting through these materials. Narrow wheels and tires with the proper load capacity can be a better choice for customers who find their snow traction to be inadequate.

Wheel rim width must also be considered when changing tire size. A wider wheel provides more support to the tire sidewall. A narrower wheel allows the sidewall to flex more easily, providing a softer ride.

When changing tire size, tire load capacity and diameter can be maintained by using the following formula: When changing to a 5 percent lower profile (reducing the aspect ratio by 5), choose a tire with a 10-mm wider cross section. For instance, change from 215/75 R15 to 225/70 R15.

Tire manufacturers recommend against changing from a lower profile to a higher profile tire, which results in reduced vehicle performance. When making a change in the size of tires on a vehicle, the recommendation is to replace tires in sets of four.

Overall Tire Diameter

DOT standards require the overall diameter of a replacement tire to be within +2 percent to −3 percent of what the tire was as original equipment. Changing tire diameter can affect antilock brakes, the speedometer, gear ratios, and four-wheel drive.

Many newer vehicles are equipped with antilock brake systems (see Chapter 10). A small pickup coil at each wheel measures wheel rotation speed and generates a signal that is sent to the ABS computer. Tires of a different height than original can cause excessive tire chirping or erratic system operation during a panic stop. When the size difference is large enough, the ABS warning light may come on. A small amount of tire height difference between tires on different sides of the vehicle is enough to cause the computer to turn on the ABS warning light and set a trouble code. In this case, the brakes will continue to operate normally but ABS function will be disabled. On some computer-controlled vehicles, changing the tire size will require a change in the computer's program.

When the diameter of a tire is changed, front end geometry is altered and the speedometer will need to be recalibrated. Installing higher diameter tires to provide increased load is called *oversizing*. Higher diameter tires raise the vehicle's center of gravity. This reduces a vehicle's ability to hold the road and maneuver quickly in an emergency.

SAFETY NOTE: A tire should never be installed on a wheel that is narrower or wider than the manufacturer's recommended wheel width. Installing wider tires on a vehicle will probably mean that wider wheels will have to be installed.

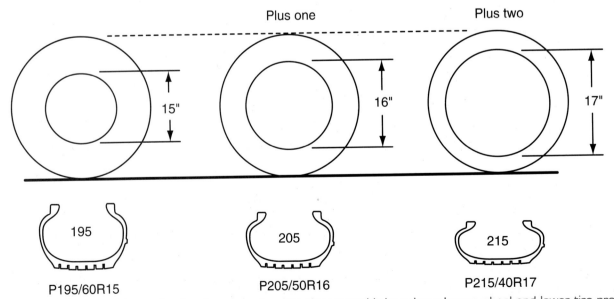

Figure 12.42 Plus sizing maintains the same diameter tire assembly by using a larger wheel and lower tire profile.

Undersizing is when a smaller sized tire is installed, often because it is less expensive. Never install undersized tires on a vehicle. They will wear faster and have less load-carrying ability. The vehicle will be lower, the speedometer will no longer be accurate, and the increase in engine rpm for a given speed will result in a decrease in fuel economy.

WHEELS

Wheels are made either of steel, aluminum, or an alloy of aluminum. Original equipment wheels on less expensive passenger cars are made of steel. Wheels have two parts, the center flange, and the **wheel rim** (**Figure 12.43**). The center flange of steel wheels is stamped because this is the least expensive production method. A strip of steel is rolled and butt welded at the ends to form the rim, which is then spot welded to the center flange.

The raised sections on either side of the drop center area are called **bead seats**. This is where the tire actually seals. There are raised sections on the inside edges of the bead seats called **safety beads** (**Figure 12.44**). These help to keep the tire bead on the bead seat in case of an "air out" until the vehicle can be safely stopped. Rims for tubeless tires must have safety beads in order to be DOT approved.

A **drop center** or **rim well** provides a means of removing and installing a tire from the wheel. The tire bead is reinforced with wire and it will not stretch.

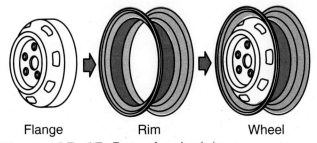

Figure 12.43 Parts of a wheel rim.

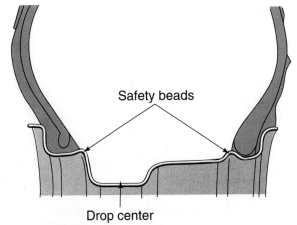

Figure 12.44 Safety beads keep the tire on the wheel in the event of a flat tire.

When a tire is installed on the wheel, one side of the bead is pushed into the drop center so that the other side of the bead can be pulled over the edge of the rim (see Chapter 13).

Wheels are centered by one of two methods:

1. **Hub-centric**—This means that the center of the wheel has a machined counterbore that pilots on a machined area of the hub. This is the most precise method of centering the wheel to the axle. Most original equipment wheels are hub-centric.
2. **Stud-centric**, or lug-centric, wheels located on the wheel studs. Many aftermarket or custom wheels are stud-centric. Custom wheels that are made for a specific model of vehicle can be hub-centric, but they are more expensive because they are not universal to many makes of vehicle. When stud-centric wheels, which have a bigger center hole, are installed on hub-centric vehicles the result can be an improperly centered wheel, which causes vibration.

Service information related to balancing hub- and stud-centered wheels is found in Chapter 13.

Custom Wheels

Customers sometimes purchase custom wheels when they want to have a cosmetic change in the appearance of the vehicle or when different sized tires are installed. Aftermarket wheel quality is rated by the Specialty Equipment Manufacturers Association (SEMA) and by their affiliate the SEMA Foundation (SFI). Wheels carrying their certification are manufactured to SFI standards.

Aluminum wheels can be cast, forged, or rolled. They can be either a single-piece casting or they can have lighter rolled rim halves bolted to a cast center section. Race cars use alloy wheels.

An alloy occurs when two or more metals are combined to make one.

Custom wheels for street use are single-piece castings of light alloy aluminum with a weather-resistant coating. The more costly custom wheels fit a single application only. Less expensive wheels are made of

> ### Vintage Chassis
>
> In the fifties, sixties, and seventies, aluminum wheels were commonly called "mags," due to the magnesium part of the alloy. The alloy used for aluminum mags combines aluminum with magnesium and silicon. These wheels are strong and light but are not practical for passenger cars because they are expensive and do not resist corrosion.

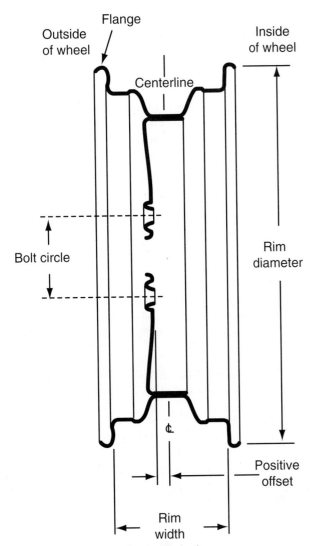

Figure 12.45 Wheel terminology.

weaker materials. They are stud-centric and are made to fit a variety of applications.

Rim width is the measurement from bead seat to bead seat (**Figure 12.45**). It is usually about 80 percent of the cross-sectional width of the tire. The centerline is at one-half of the rim width.

Wheel Offset

Wheel offset is the difference between the rim centerline and the mounting surface of the wheel. Sometimes a certain amount of offset is designed into a wheel to allow it to clear the fender well. The offset of the wheel is also important to proper brake cooling because it affects the distance between the brake caliper and the wheel.

Wheel clearance is not included in application handbooks. This is something that must be carefully checked on the vehicle. The use of wider tires and wheels, or wheels that are offset a different amount than stock, can result in interference between the tire and fender well or suspension components.

Wheel offset is described in various ways. The most common offset classification is as follows (**Figure 12.46**):

- Negative wheel offset increases the track width of the tires.
- Positive offset (the opposite of negative offset) is found often on FWD cars.

When a wheel is replaced, the new wheel should be of the same offset to maintain the proper scrub radius.

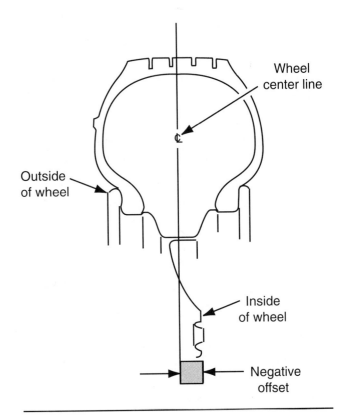

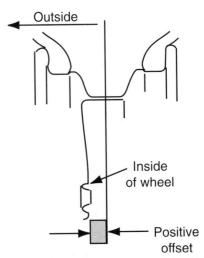

Figure 12.46 Positive and negative wheel offset.

CAUTION: Scrub radius (see Chapter 18) has an effect on the handling and steering effort of the vehicle. Changing the height or centerline of a wheel from that designed by the manufacturer can produce a change in scrub radius from positive to negative or vice versa. The result can seriously affect vehicle handling. This is just one reason why replacing stock tires and wheels with ones of a different size or offset is something that many shops avoid.

■ LUG STUDS

Wheels have different numbers of lug studs, between three and eight, depending on the load on the vehicle. Most passenger cars use four or five lug studs, while light trucks usually use six or eight. Heavier trucks and some RVs sometimes use fewer lug bolts or studs, but they are larger in diameter and are tightened to a much higher torque.

There are also various bolt patterns used. A bolt pattern that is listed in a catalog as being 6–5½ is a six-bolt pattern spaced around a 5½-inch circle. Bolt patterns with an even number of bolt holes are easy to measure. Simply measure the distance from the center of one bolt to the center of the one across from it. Five-bolt patterns are more difficult to measure. Templates are available to help determine the size of a bolt pattern.

Lug studs have a serrated shank so that they will remain tight in the hole in the hub during tightening.

Lug Nuts

Lug nuts can use either metric or standard screw threads. Always check stud threads when installing a replacement nut.

■ TIRE VALVE STEMS

Passenger car tire valve stems are usually rubber and are designed to be used at pressures of less than 62 psi (4.2 bar). A spring-loaded valve core is screwed into the valve stem (**Figure 12.47**). Valve stems have a screw-on dust cap. Some of these dust caps have a gasket that prevents air loss past the valve core.

> *Vintage Chassis*
>
> Lug nuts for cast wheels (mags) are long and thick and must fit into a large, deep hole. These lug nuts must be used with a washer to avoid damaging the wheel. Most lug nuts are made with the washer permanently installed on them (see Chapter 13).

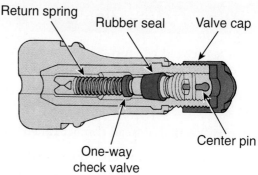

Figure 12.47 Parts of a rubber valve stem.

NOTE: *According to the Pirelli Armstrong Tire Corporation, an imperceptible leak of one bubble per minute can result in the loss of 1.5 psi (1 bar) per month.*

There are two common valve stem lengths available. A short stem is used when there is a hubcap, and a long stem accommodates the use of full wheel covers (**Figure 12.48**). Rubber valve stems come in common diameters of 11.3 mm or 15.7 mm to match the two common wheel rim hole sizes. When a longer stem is required, tire valve extensions can be screwed onto an existing valve stem. Extensions have a spring-loaded check valve so that air pressure can be checked and adjusted as needed.

Light trucks and custom wheels are usually equipped with threaded metal stems that have a rubber bushing and are fastened to the wheel with a nut (**Figure 12.49**). The common wheel hole size for metal stems is 11.5 mm.

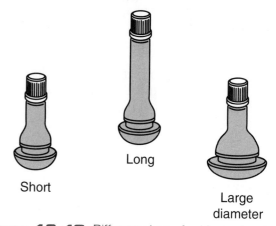

Figure 12.48 Different sizes of rubber valve stems.

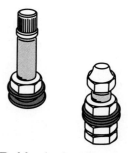

Figure 12.49 Metal valve stems.

REVIEW QUESTIONS

1. What is it called when water forms a wedge under the tire, causing it to lose traction?
2. Which tire allows better fuel economy, radial or bias?
3. What is the area of tread contact called?
4. What is another name for a tire's height?
5. What is the mph limit for a speed rating of H?
6. What are the two letters that are found on snow or all-season tires?
7. When changing a tire sized 195/75 R15 to a 5 percent lower profile, the correct tire size would be _____.
8. What is the lower area between the bead seats of a wheel rim called?
9. When a wheel is offset to increase the track width of the front tires, this is called _____ offset.
10. A small leak of one bubble per minute can result in the loss of _____ psi per month.

PRACTICE TEST QUESTIONS

1. Technician A says that a properly inflated bias tire will have a bulging tire sidewall. Technician B says when installing two radials with two bias tires, install the radials on the front. Who is right?
   - A. Technician A
   - B. Technician B
   - C. Both A and B
   - D. Neither A nor B

2. Technician A says that a tire with a tread wear rating of 200 would be expected to last twice as long under the same conditions as one with a 100 rating. Technician B says when the diameter of a tire is changed, front end geometry is altered. Who is right?
   - A. Technician A
   - B. Technician B
   - C. Both A and B
   - D. Neither A nor B

3. Technician A says that an alloy is produced when two or more metals are combined. Technician B says that custom wheels are usually hub-centric. Who is right?
   - A. Technician A
   - B. Technician B
   - C. Both A and B
   - D. Neither A nor B

4. Technician A says that the tire ply design that has the cords intersecting at an angle is called a radial tire. Technician B says that the P in a P-metric radial's size means performance. Who is right?
   - A. Technician A
   - B. Technician B
   - C. Both A and B
   - D. Neither A nor B

5. Technician A says that the UTQG standard rates a tire's traction. Technician B says that the UTQG standard rates a tire's tread wear rate. Who is right?
   - A. Technician A
   - B. Technician B
   - C. Both A and B
   - D. Neither A nor B

6. Technician A says that a drop center is when the center of the wheel has a machined counterbore that pilots on a machined area of the hub. Technician B says that a tubeless tire will deflate faster than a tube-type tire. Who is right?
   - A. Technician A
   - B. Technician B
   - C. Both A and B
   - D. Neither A nor B

7. Technician A says that radial tires ride smoother at low speeds than bias tires. Technician B says that radial tires ride smoother at faster speeds while travelling over highway expansion joints. Who is right?
   - A. Technician A
   - B. Technician B
   - C. Both A and B
   - D. Neither A nor B

8. Technician A says the amount of load a tire can support is determined by the area of the tire. Technician B says the amount of load a tire can support is determined by the amount of air pressure in it. Who is right?
   - A. Technician A
   - B. Technician B
   - C. Both A and B
   - D. Neither A nor B

9. Technician A says that a 70 is a lower profile tire than a 75. Technician B says that an H rating is for higher speeds than a Z rating. Who is right?
   - A. Technician A
   - B. Technician B
   - C. Both A and B
   - D. Neither A nor B

10. Technician A says that a tire's height is called its footprint. Technician B says that a leak of one bubble per minute from a tire can cause the tire to go flat overnight. Who is right?
    - A. Technician A
    - B. Technician B
    - C. Both A and B
    - D. Neither A nor B

CHAPTER 13

Tire and Wheel Service

■ OBJECTIVES

Upon completion of this chapter, you should be able to:
- ✓ Understand the importance of correct tire pressure.
- ✓ Rotate tires on all passenger and light-duty vehicles.
- ✓ Repair tire punctures in the correct manner.
- ✓ Determine the causes of tire-related vibration.
- ✓ Understand and perform tire balancing.

■ KEY TERMS

dynamic balance
gross vehicle weight (GVW)
hub-centric
inflation pressure
lug-centric
shimmy
static balance
tire rotation

■ INTRODUCTION

Tire service is a large area of automobile repair. The average owner can expect to replace at least one set of tires on his or her car. Tire life depends on tire quality, air pressure, vehicle weight, driving conditions, suspension condition, and wheel alignment.

■ TIRE INFLATION

Tire wear can be caused by incorrect **inflation pressure** (**Figure 13.1**). Tires typically lose about 1–2 psi of pressure a month through permeation of the sidewall, and maybe more during hot weather. This is a normal occurrence. Think about what a balloon looks like a few days after a birthday party. Like tires, balloons hold air—they leak slowly.

Maintaining correct air pressure is the most important factor in the safety, performance, and life expectancy of a tire. Underinflation is the most common cause of radial tire failures. Low tire pressure in a radial tire changes the normal deflection of its sidewall. This raises the amount of heat generated within the tire's carcass. The following are results of low tire pressure:

- Temperature of the tire rises
- Load-carrying capacity of the tire is lowered
- Tire tread life is reduced
- Fuel consumption increases
- Outside edges of the tire wear excessively

A car will usually pull to the side that has a low tire, especially when it is a front tire. However, a radial tire can often be run at a tire pressure that is below specifications without exhibiting noticeable handling symptoms.

High tire pressure can cause the center of the tread to wear excessively. A rough ride can result because the tire is actually the first part of the suspension and spring system.

■ CHECKING AIR PRESSURE

Vehicle owners should be encouraged to check tire air pressure at least once a month and prior to a long trip.

NOTE: *According to the Rubber Manufacturers Association (RMA), customers should be advised to check their tire pressures prior to driving while the tires are still cold. In the event they find a low tire, after a short drive to the service facility that has an air compressor air can be added in the amount that the tire was underinflated.*

The following list details the effects of temperature on a tire's pressure:

- As the air in the tire expands due to heat, pressure normally increases. Air pressure should be checked when the tires are cold.

Vintage Chassis

When an older car with bias tires had low air pressure the driver would usually notice it in the handling of the vehicle because bias-type tires experience a strong pull to the side of a tire with lower inflation.

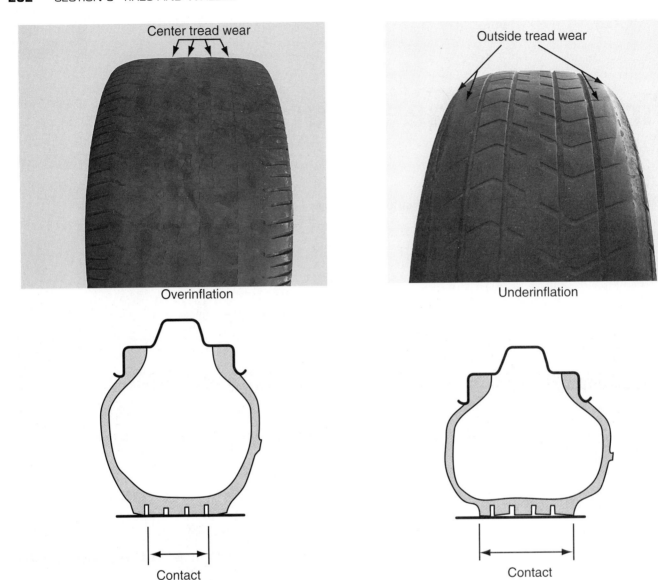

Figure 13.1 (A) Overinflation wears the center of the tread. (B) Underinflation results in excessive outer tread wear. (Photographs courtesy of Tim Gilles)

- It takes less than 3 minutes, or one mile of driving at moderate speeds, to make tires too hot to check accurately. According to Michelin Tires, this can increase air pressure by 4 psi or more. So when adjusting pressure in a hot tire, add 4 psi to the maximum gauge reading desired. For instance, if the recommended cold pressure is 24 psi and the gauge reads 26 psi, fill the tire to 28 psi. Be sure to recheck it when cold the next day.
- Air should not be let out of a tire when it is hot.
- Each change of 10°F in outside temperature will result in about 1 psi change in tire pressure.
- Tire pressures taken in direct sunlight can read higher than those taken in the shade, even when both check the same when cold.
- A hot tire that has lower pressure than the recommended cold pressure is seriously underinflated.
- Cold tire pressure should never be higher than the maximum pressure molded into the tire sidewall.

Tire inflation pressures should always be the same for both tires on one axle so ride and handling are not affected.

An accurate air pressure gauge must be used to check tire pressures (**Figure 13.2**). A normally inflated radial tire has a bulging sidewall and tire pressure must be dangerously low before there is a visible difference. **Figure 13.3** compares two tires with 10 pounds' difference in pressure.

CHECKING AND ADJUSTING TIRE PRESSURE

All valve stems should have screw caps on them. These keep out dirt and moisture and provide a backup in case the valve core leaks. Before adding air to a tire,

Chapter 13 Tire and Wheel Service 283

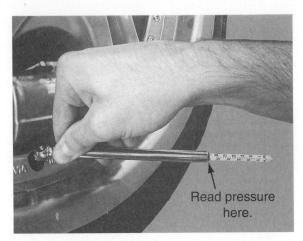

Figure 13.2 A tire pressure gauge showing 34 psi. *(Courtesy of Tim Gilles)*

Figure 13.3 The appearance of a radial tire sidewall changes very little with inflation pressure. *(Courtesy of Tim Gilles)*

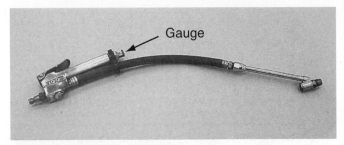

Figure 13.4 This type of tire pressure gauge is rarely accurate. *(Courtesy of Tim Gilles)*

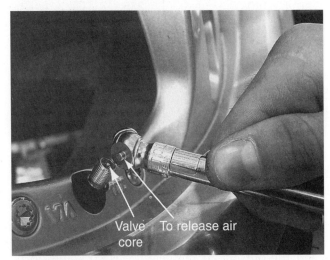

Figure 13.5 A pin on a typical tire pressure gauge used to deflate the tire after too much pressure is added.

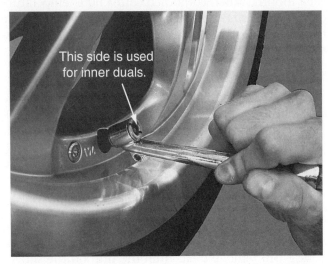

Figure 13.6 Either side of this tool can be used for easier access to the valve stem. *(Courtesy of Tim Gilles)*

Vintage Chassis

Older, bias-type tires had relatively stiff tire sidewalls, compared to modern radial tires. When they were low on pressure, there was a visible difference in the sidewall appearance.

blow air through the air chuck to clear it to prevent dirt from being forced into the valve core.

Always use a high-quality tire gauge. Inexpensive gauges are often inaccurate. One type of gauge is a part of the air chuck (**Figure 13.4**). These air gauges are often abused and become inaccurate when left installed on an air hose.

Tire pressure gauges are usually equipped with a pin that can be used to release air (**Figure 13.5**). Some tire air chucks and gauges have two sides that can be used to inflate and check tire pressures (**Figure 13.6**). The top side is handy to use when checking and inflating inner tires on trucks and motorhomes with dual wheels.

Tire pressure specifications are listed in metric and inch equivalents. **Figure 13.7** is a conversion chart from metric kilopascals to pounds per square inch of pressure.

SECTION 3 TIRES AND WHEELS

TIRE INFLATION PRESSURE CONVERSION CHART			
Inflation Pressure Conversion Chart (Kilopascals to PSI)			
kPa	psi	kPa	psi
140	20	215	31
145	21	220	32
155	22	230	33
160	23	235	34
165	24	240	35
170	25	250	36
180	36	275	40
185	37	310	45
190	38	345	50
200	39	380	55
205	30	415	60
Conversion: 6.9 kPa=1 psi			

Figure 13.7 A conversion chart from the metric system (kilopascals) to the English system (PSI).

Let us examine the effects of not checking tire pressures regularly. If you fill your tires during the summer when the outside air temperature is 90°F and do not check them again for six months, the pressure will be considerably lower. A typical tire loses 1 psi in pressure each month. When combined with a 60°F temperature change in seasonal climates, the pressure change in the tire can be substantial. **Figure 13.8** shows typical pressure changes under three conditions.

■ TIRE WEAR

According to Goodyear, a 4-psi decrease in pressure below the recommended amount can result in a 10-percent loss of tread life. Additional decreases of air pressure can result in even more wear and the possibility of serious damage to the tire. Underinflation can also cause the edges of the tire to wear (see **Figure 13.1**).

The fastest tire wear occurs during hard cornering, braking, and acceleration. Rough pavement also contributes to accelerated tire wear. Slow-speed sharp cornering wears the front tires, while high-speed cornering will remove tread from the tires on the side of the vehicle to which the weight is transferred.

When a tire wears to within 1/16 inch (2/32 inch) of the bottom of its tread, wear bars begin to become more obvious at regularly spaced areas around the tread circumference (see **Figure 12.7**). The wear bars are raised areas cast into the bottom of the tire tread area to indicate when the tread has become worn beyond its safe limit. The RMA recommends that tires with 1/16 inch of remaining tread depth be replaced. These tires are unsafe in wet weather and more apt to be damaged by road hazards. Federal regulations require vehicles in excess of 10,000 pounds **gross vehicle weight (GVW)** to have 4/32 inch minimum tread depth on front tires. **Figure 13.9** shows a gauge used to measure tread depth.

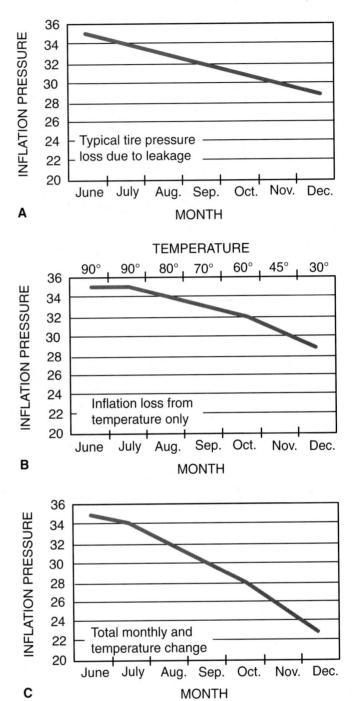

Figure 13.8 Changes in tire inflation due to time and variations in temperature.

When a tire exhibits scalloped or cupped wear, the cause is usually that the tire has been hopping up and down on the road. This movement, known as wheel tramp (**Figure 13.10**), results from bad shock absorbers, worn parts (such as ball joints or control arm bushings), out-of-balance tires, or too much tire runout.

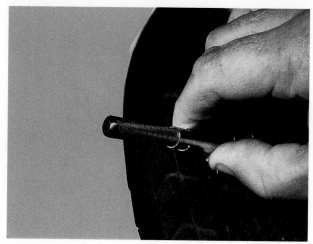

Figure 13.9 A tread depth gauge. *(Courtesy of Tim Gilles)*

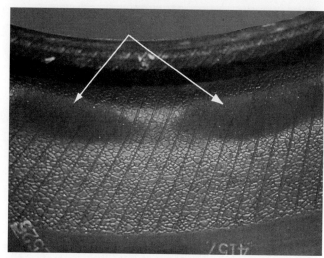

Figure 13.11 The inside of this tire is bruised from driving with low air pressure. *(Courtesy of Tim Gilles)*

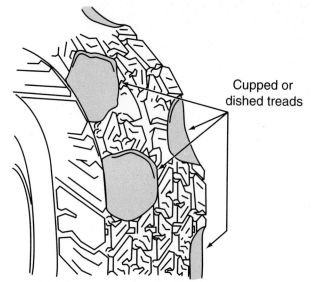

Figure 13.10 Tire wear resulting from wheel tramp.

Front tires on some front-wheel drive cars develop abnormal tire wear. Some manufacturers specify tire rotation at very low mileage intervals to compensate for this. Tire rotation is covered in more detail later in this chapter.

Vintage Chassis

Older, belted bias tires often experience step wear, which is wear on the second row of tread from the outside. This is a normal condition with these tires. It occurs because the inside and outside rows of tire treads on these cars carry most of the weight, while the second and sixth ribs carry the least. This allows them to scrub slightly as the tire rotates. Changing the inflation pressure of the tire will not affect this wear.

Inspect the tire for physical damage. When a car is driven with a tire that is flat or underinflated, the tire can become damaged (**Figure 13.11**). Look for evidence of tread or sidewall separation. This might show up as an out-of-round tread or visible deformities on the outside of the tire. But sometimes damage from underinflation might not be visible from the outside of the tire and can only be determined when the tire is removed from the wheel.

■ SIDEWALL CHECKS

Sidewall cracks are usually caused by age and years of exposure to the sun and ozone (**Figure 13.12**). Cracks on the outside of sidewalls often develop on recreational vehicles because RV driving is seasonal. Motor homes are used mostly on the highway (very low tire wear). Tires can be many years old before the tread wears out. When an RV is parked in the same spot for long periods of time and the tires have not been regularly rotated, only the tires on the side of the vehicle

Figure 13.12 Sidewall cracks are caused by age and exposure to sunlight. *(Courtesy of Tim Gilles)*

Figure 13.13 A slight indentation in the sidewall of a radial tire is normal. *(Courtesy of Tim Gilles)*

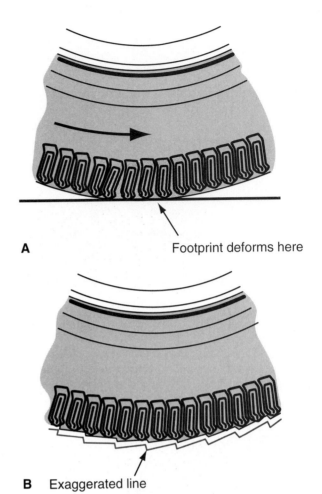

Figure 13.14 (A) Drive cornering scrub wear that occurs as the footprint of a radial tire squirms against the pavement. (B) This results in a sawtooth heel-and-toe wear pattern on the tread.

that have been exposed to more sunlight may be cracked.

According to the RMA, a slight sidewall indentation, also known as sidewall undulation, is a common construction characteristic of radial tires (**Figure 13.13**). These indentations are due to tire construction and are purely a visual characteristic that will not affect tire performance. The indentations are due to overlaps in the cord material of about ⅜ inch. Indentations are normal, but if the tire bulges, there is no supporting cord and the tire must be replaced. If there is any question concerning sidewall appearance, the tire should be removed from service and inspected by a knowledgeable tire dealer, or the tire manufacturer's representative should be contacted. Cuts or cracks in the sidewall that allow cords to be exposed are cause for replacement of the tire.

Recreational vehicles and light trucks with dual rear tires often experience uneven wear on the rear tires. The diameters of dual wheel tires must be within ¼ inch of each other.

■ TIRE ROTATION

Front wheels on all cars experience the most wear because they turn at an angle during steering and also because weight transfers forward during a stop. Moving tires to different locations on the vehicle is called **tire rotation**. Regular tire rotation allows the tires to wear more evenly, allowing them to be replaced in complete sets. Most manufacturers specify regular rotation intervals. When there is no recommendation, rotation is recommended at 6,000 to 8,000 miles or before that if tire wear is evident.

In the past, moving radial tires to the other side of the car was not recommended, but this is no longer true as long as there is not a specified rotation pattern to the tire. If one front tire shows uneven wear, it can be switched with the other front tire.

Front-wheel drive cars experience far more wear on the front tires. Rear tires on these cars last much longer. A typical rotation pattern for front-wheel drive cars with radial tires involves moving the front tires to the rear on the same side of the car. The rear tires are then moved to the front, but to opposite sides of the vehicle. This is said to even out and minimize *drive cornering scrub* wear that occurs as the footprint of a radial tire squirms against the pavement (**Figure 13.14A**). This results in a sawtooth heel-and-toe wear pattern on the tread (**Figure 13.14B**). Rotating the tires to another side of the vehicle smooths this wear. If it is allowed to develop, noise and vibration can result, along with accelerated tread wear.

Figure 13.15 shows typical rotation patterns for front- and rear-wheel drive vehicles. Four-wheel drive vehicles use the same pattern as rear-wheel drive. Notice how the front-wheel drive rotation pattern has rear tires crisscrossed to the front. The front tires are then moved directly to the rear on the same side. The

> **SHOP TIP** "X to the drive wheels"; with front-wheel drive, "X-to-the-front"; with rear-wheel drive, "X-to-the-rear."

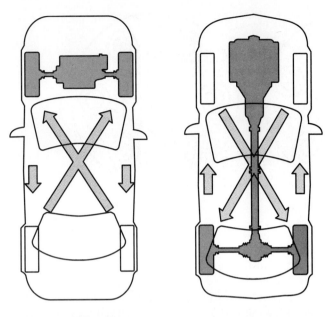

Front-wheel Drive   **Rear-wheel Drive**

Figure 13.15 A typical tire rotation pattern. With front-wheel drive, "X-to-the-front"; with rear wheel drive, "X-to-the-rear."

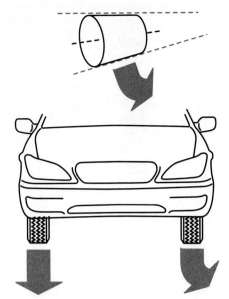

Figure 13.16 Tire conicity results in a pull.

opposite is true with rear-wheel drive and four-wheel drive rotation patterns.

Follow the manufacturer's recommended tire rotation pattern.

- Some rotation patterns include rotation of the spare tire. In this case, insert the spare into the rotation pattern at the right rear and reposition the tire that would normally have been installed on the right rear as the new spare. Other cars are equipped with only an emergency compact spare, which is not rotated onto the vehicle.
- Some vehicles have tires of a smaller size on the front. These should not be rotated to the rear.
- Some tires are designed to be mounted and run in a specified direction of rotation only. These tires are rotated in the front-to-rear pattern and are cross-rotated only if they are removed from the rim and remounted so that they will still rotate in the designated direction.
- Paired tires should be of the same size designation, construction, and tread design. According to the RMA, if radial and nonradial tires are used on the same vehicle, put radials on the rear.

Sometimes after a tire rotation, the car can exhibit a pull to one side or the other. This can be due to an inherent pull within a radial tire caused by tire conicity (when the tread is tapered like a cone) or off-center belts (**Figure 13.16**). If the offending tire was previously installed on the rear of the car (nonsteering wheels), it can cause a problem when installed on the front of the car. **Figure 13.17** shows some recommendations for isolating the problem. Off-center belts can also produce outside shoulder wear on the tire.

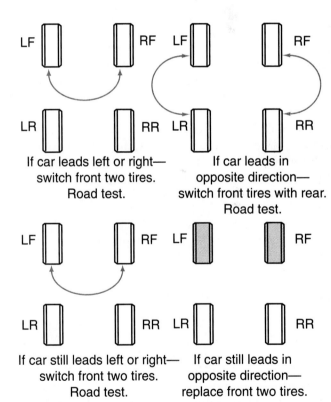

Figure 13.17 One manufacturer's recommendation for diagnosing tire pull caused by conicity.

■ REMOVING AND TIGHTENING LUG NUTS

Most lug nuts have right-hand threads and are loosened by turning counterclockwise. Loosening and removing lug nuts is easiest when done with an impact wrench. When the wheel is in the air, it will not have to be held from turning.

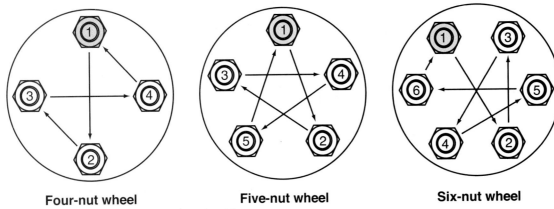

Figure 13.18 The torque sequence for wheel lugs.

Tighten lug nuts evenly in a crisscross pattern (**Figure 13.18**). Use a torque wrench to avoid warping a disc brake rotor. Specifications are available for the different makes of cars. Just as with other fasteners, lug bolts are stretched when they are properly installed so that they maintain clamping force. Lug nuts that are loose allow the wheel to exert all of the weight of the vehicle on the lug bolts rather than on the wheel. Be sure to install a socket extension between the socket and the torque wrench as needed so your hand and the handle of the torque wrench will be able to clear the tire sidewall (**Figure 13.19**). Use a ½-inch drive extension only. Inaccurate torque readings can result from the use of excessively long or ⅜-inch drive extensions.

Lug nuts for steel wheels (and some aluminum wheels) are tapered on the side that faces the wheel

SHOP TIP Remove lugs using the tightening pattern, to avoid warping a hot rotor.

CAUTION Do not loosen the lug nuts excessively when the wheel is on the ground. The car can fall down. If using a hand wrench, have someone hold the brakes applied while you loosen them with the wheel off the ground or loosen them one-half turn only before raising the vehicle.

Vintage Chassis

A few vehicles have left-hand threads on the lug nuts on the left side of the car. These are labeled with an *L* on the end of the lug stud (**Figure 13.20**). Most of these left-hand threads will be found on Chrysler vehicles from the 1950s to the 1970s.

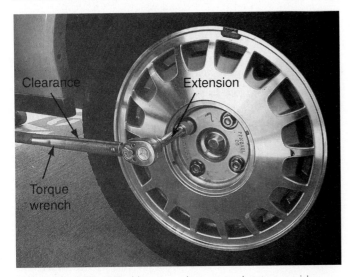

Figure 13.19 Use a socket extension to provide clearance to the tire sidewall. *(Courtesy of Tim Gilles)*

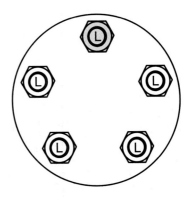

Five-nut wheel

Figure 13.20 Lug studs with left-hand threads are labeled. *(Courtesy of Tim Gilles)*

Chapter 13 Tire and Wheel Service 289

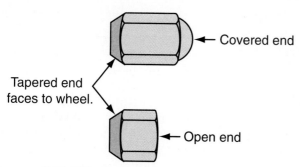

Figure 13.21 Lug nuts for steel wheels are tapered on the side that faces the wheel.

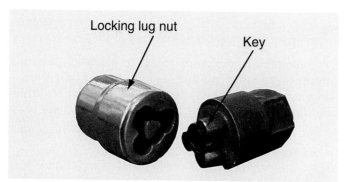

Figure 13.22 A locking lug nut and key. *(Courtesy of Tim Gilles)*

> **CASE HISTORY**
>
> *A student complained of a clunking sound and poor vehicle handling. He asked the teacher to inspect the vehicle. The teacher was amazed to find all of the lug nuts on inside out. All of the holes in the rims were damaged so badly that new wheels were required.*

(**Figure 13.21**). Some wheels use special lug nuts that cover the ends of the wheel lugs, so these nuts can only be installed in one direction.

Antitheft lug nuts are popular on custom wheels. They require a special key (**Figure 13.22**). Sometimes, the customer loses the key or forgets to bring it to the shop when the car is due for a tire or brake repair. Most shops that do tire repair have a special tool kit with several adapters for removing antitheft lug nuts without the key (**Figure 13.23**). The kit has several adapters that are wedged against the outside of the lug nut. The inside of each adapter is tapered (**Figure 13.24A**). A hammer is used to force the adapter onto the outside of the wheel lock (**Figure 13.24B**). To loosen the wheel lock, the removal tool is driven rapidly by an impact wrench lock while holding it firmly against the wheel lock (**Figure 13.24C**).

A special type of lug nut is used with aluminum wheels. It is usually equipped with a washer that cannot be removed. The shank on the nut must not be too

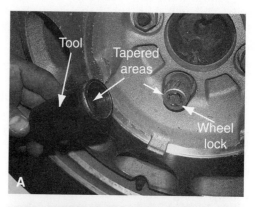

Figure 13.24 (A) The inside of the adapter is tapered. (B) A hammer is used to force the adapter over the lug nut. (C) The lug nut is removed using an impact wrench. *(Courtesy of Tim Gilles)*

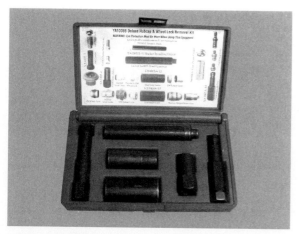

Figure 13.23 Special tool kit with adapters for removing antitheft lug nuts. *(Courtesy of Tim Gilles)*

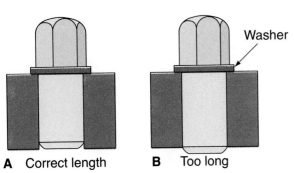

Figure 13.25 Lug nuts in case aluminum wheels: (A) is correct and (B) would allow the wheel to be loose.

> **SHOP TIP** Put an anti-seize lubricant on the outside of aluminum wheel lug nuts before installation. This will help avoid the electrolysis that can occur between the steel lug nuts and the aluminum wheel.

long or it can bottom out, leaving the wheel loose (**Figure 13.25**). Also, the lug nut must have a small amount of clearance so that it is free to turn inside of the hole in the wheel.

■ REPAIRING WHEEL STUDS

Occasionally, a lug bolt will be stripped or broken. If only a couple of threads are damaged, they can be cleaned up with a thread chaser. Broken lug bolts must be replaced.

If a rotor or drum separates easily from the hub, the lugs can usually be driven from the hub with a brass hammer or punch. If excessive effort is needed, a tie-rod press can be used to force the lug bolt out of its hole. The new lug stud can be installed with an inverted lug nut and washers as shown in **Figure 13.26**.

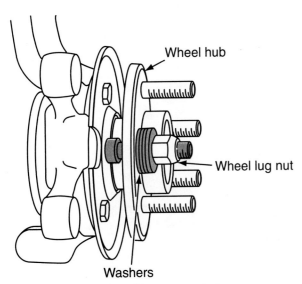

Figure 13.26 Installing a new stud.

Vintage Chassis

On drum brake cars, when the drum is removed with the hub, the studs are swaged, which means that they are deformed to keep them tight. The swaged area can be cut off with a special tool to make them easier to remove (see Chapter 8). Drum brakes are no longer found on the fronts of newer vehicles.

■ REMOVING AND MOUNTING TIRES ON RIMS

There are some important points to be aware of before attempting to remove and install tires on wheel rims. Tires can explode and fingers can be cut off if proper caution is not observed:

- Be sure to use the proper size and construction of tire to match the wheel rating.
- Be sure that the rim diameter matches the diameter molded on the tire sidewall. Be especially careful that a tire or wheel is not a metric size that was manufactured for the overseas market. Although you might be able to mount a metric tire on a standard wheel, it will not fit properly and can come off.

Tire problems such as leaks and vibrations can sometimes be traced to improperly mounted tires.

The following information applies to either of the two most common tire changing machines.

Deflate the Tire

Before removing the tire from the wheel, remove the valve core to deflate the tire completely. **Figure 13.27** shows a tool used to remove the valve core. If the tire is to be patched and reinstalled without rebalancing, be sure to mark the locations of the valve stem and any wheel weights with a marking crayon. If another

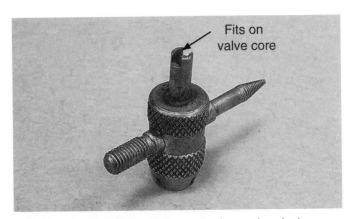

Figure 13.27 A tool for removing and replacing valve cores. *(Courtesy of Tim Gilles)*

Figure 13.28 Remove wheel weights from the rim. *(Courtesy of Tim Gilles)*

tire is to be installed on the rim, first remove any balance weights from the rim (**Figure 13.28**).

Tire Changers

There are three types of tire changers in common use. Two of them, the rim clamp and the center post, have been used for many years (**Figure 13.29**). The rim clamp changer was originally developed in Europe and has become the predominant tire changer in North

Figure 13.30 Tire changer. *(Courtesy of Hunter Engineering Company)*

America in recent years. Tire changing requirements for high-performance and specialty aftermarket wheels and tires have presented new challenges. A new tire changer design has emerged in response to these needs. **Figure 13.30** shows this new tire changer.

In the following section, the procedures for mounting and dismounting conventional tires are similar and are covered first. High-performance tire service, which is very similar, is covered later.

Unseating the Beads

Before a tire can be removed from a wheel, its beads must be unseated from their bead seats (**Figure 13.31**). This is called "breaking the beads." Lubricate the tire bead with rubber lube. Rubber lube reduces friction between the tire and wheel, preventing damage to the tire and making removal easier.

Figure 13.29 (A) A center post tire changer. (B) A rim clamp tire changer. *(Courtesy of Tim Gilles)*

- Do not attempt to unseat the beads of an inflated tire.
- Do not hit the tire or rim with a hammer. A hammer should not be necessary when removing a passenger car tire.
- Never let go of the tire iron when in use. It can flip up and hit you.

Figure 13.31 Breaking the bead on each type of tire changer. (A, *Courtesy of Tim Gilles*)

Removing the Tire from the Wheel

The bead on the bottom side of the tire is forced into the drop center of the wheel so that the upper bead can be pulled over the edge of the rim (**Figure 13.32**). On either type of tire machine, removal of the tire from the wheel requires the use of a bar called a tire iron (**Figure 13.33**). First, the upper bead is pulled over the outside of the rim; then, the lower one. Do not try to remove both beads at the same time. Be careful not to tear the bead. This will ruin the tire. If the bead does not come off the rim without binding:

- Be sure there is enough rubber lube on the bead.
- Double-check to see that the lower bead is totally forced into the drop center of the wheel.

■ INSPECTING THE TIRE AND WHEEL

After the tire is removed from the rim, inspect its inside for cuts, carcass damage, penetrating objects, loose cords, dirt, and liquid. Inspect the condition of the bead by pulling out on it in several places around its circumference. If there is a sharp bend in the bead, do not mount the tire. The bead seat could be broken. It is not worth taking a chance with someone's safety.

Inspect the condition of the wheel rim for sharp edges, dents, cracks, and other damage. Small dents in the rim flange can be straightened. When there is a larger dent in the wheel, it must be checked for straightness (runout) before it can be used. This procedure is covered later in this chapter.

Check the Bead Seat

It is quite common for rust to damage the bead seat on a wheel. If the rust is only on the surface, it can be removed with a wire brush (**Figure 13.34**). If the bead

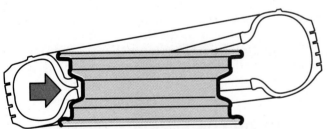

Figure 13.32 Both beads are in the drop center of the rim while the bead is pulled over the edge of the tire.

Figure 13.33 A tire iron is used to pull the top bead over the edge of the rim. Be sure that the other side of the tire bead is in the drop center of the rim. (*Courtesy of Tim Gilles*)

> **SAFETY NOTE** Never mount a tire or wheel that is damaged. This could result in injury or death to the occupant(s) of the vehicle. In addition, the shop can be held liable in court for negligence and the technician can face criminal charges.

Figure 13.34 Prepare the wheel for the mounting of a tire by using a wire brush to remove all dirt and rust from the sealing surface. Apply rubber compound to the bead area of the tire.

Figure 13.35 (A) Cut the bottom of the old valve stem. (B) The valve stem is installed with a special tool. (Courtesy of Tim Gilles)

seat is not smooth, the tire will have a slow leak and the wheel will need to be replaced.

Valve Stem Service

Rubber valve stems are customarily replaced when new tires are installed and the tire is off the rim. The valve stem can also be replaced without removing the tire from the rim. This is commonly done when an old valve stem becomes cracked or starts to leak. Leaks can be found by prying the stem to the side and observing any air leakage.

To remove a valve stem, it is cut off with a knife (**Figure 13.35A**) or it is forced through the hole in the rim using the valve stem installing tool. To install the new stem, thread it into the installing tool, lubricate it with rubber lube, and pull it into the hole (**Figure 13.35B**). Be sure that it is pulled all of the way into place and is properly seated in the hole in the rim.

Rubber Lube

Lubricate both tire beads with an appropriate rubber lube (**Figure 13.36**). Good quality rubber lube is slippery and fast drying. Lubricate the bead seats on the rim with rubber lube also. Using rubber lube provides the following advantages:

- Rubber lube reduces friction between the tire beads and the edge of the rim during mounting.
- Rubber lube helps to seal around the bead during initial inflation of the tire.
- Friction between the bead seats and the tire bead will be reduced when inflating the tire. This is

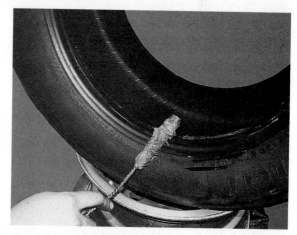

Figure 13.36 Lubricate both tire beads with rubber lube. (Courtesy of Tim Gilles)

important so that the beads will be fully seated and the tire tread will not be distorted. (Always wear safety glasses when inflating the tire.)

The following are some cautions to observe regarding rubber lubricants:

- Do not use petroleum products, which will damage the rubber in the tire.

> **SHOP TIP** A new valve stem comes with a new valve core. Remove the valve core before attempting to inflate the tire.

- Rubber lube should not be diluted with water, which can rust the rim.
- Do not use silicone lubricants or liquid soaps, which will allow the tire to spin on the rim.

Directional Tires

Some tires are designed to be run in only one direction. Check the sidewall of the tire to see if there are direction arrows. Be certain the correct side of the tire is facing out. Some tires have whitewalls or special lettering. All tires have a serial number, which is on the back side of the tire (**Figure 13.37**). Other tires are designed to rotate in one direction only. These will have an arrow indicating the direction of rotation (**Figure 13.38**).

■ INSTALL THE TIRE

Clamp the wheel to the tire machine with the narrow bead ledge up. Install the inside bead of the tire over

Figure 13.39 The narrow bead ledge is up and the lower bead is in the drop center of the rim. *(Courtesy of Tim Gilles)*

Figure 13.37 A tire's serial number is usually found on the back of the tire. *(Courtesy of Tim Gilles)*

the flange of the rim (**Figure 13.39**). As more and more of the bead is passed over the edge of the rim flange, force the bead down into the drop center well of the wheel. This is important! If one side of the bead is not in the drop center, the diagonal mounting distance will be excessive and the other side will not be able to be stretched over the flange (**Figure 13.40**). When the first bead is totally installed and positioned in the drop center of the wheel, install the other bead over the flange of the rim (**Figure 13.41**).

Radial tire sidewalls are more flexible and their tire beads are not forced into position as readily as those of older bias-type tires. If the beads are not seated properly, the tire can have an out-of-round condition that will make it difficult to balance.

Inflating the Tire

Many tire machines have a provision for tire inflation.

Seating the Beads

Tubeless tires require a substantial volume of airflow for the beads to start to seat on the rim. Occasionally, getting the tire to take on air can be quite difficult.

Figure 13.38 Some tires are directional. *(Courtesy of Tim Gilles)*

SAFETY NOTE
- Inflation can be the most dangerous part of tire installation as tires can explode during inflation.
- If a machine is not equipped with a lock-down clamp to hold the wheel, inflate the tire in a safety cage. Use an extension hose with a clip-on air valve. This is so the technician can stand away from the tire during inflation.
- Do not leave tools on the tire sidewall when inflating the tire.

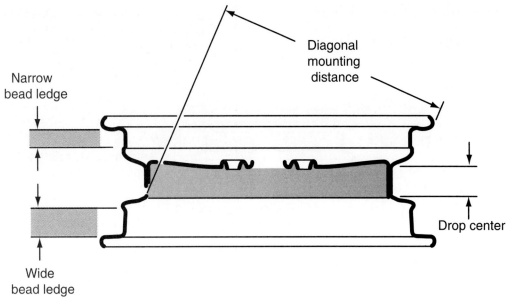

Figure 13.40 The bead must be all the way past the edge of the narrow bead ledge or the diagonal mounting distance will not be sufficient to allow the bead to pass over the edge of the wheel.

Figure 13.41 As the bead is drawn over the edge of the wheel rim, it is important that the other side of the bead stay in the drop center. *(Courtesy of Tim Gilles)*

SHOP TIP Removing the valve core from the valve stem during filling allows a greater volume of air to enter the tire.

Most tire machines have an inflation chamber with a sealing ring. The sealing ring forces a good deal of air from the air chamber into the area surrounding the lower bead of the tire (**Figure 13.42**). The air chamber ensures that there is a tremendous amount of airflow at once. If air was simply pumped through the normal shop air lines, the volume of air could not be maintained.

NOTE: *Be sure there is an adequate supply of air to the tire machine. A regulator or fitting that is too small will restrict the air supply line.*

If the tire is pulled against the upper bead seat while inflating it, both beads will usually seat. It is more difficult to seat the beads on wide rims. If the beads will not seat, sometimes the outer bead can be pushed against the inner bead using the bead breaking attachment to force it down over the safety hump. Another trick is to bounce the tire vertically on the floor so that the air in it can force the beads outward.

When inflating a tire on a center post tire machine, the holddown clamp should be in place. When the tire begins to inflate, loosen the holddown clamp one turn. Otherwise as the tire expands, the lower sidewall can wedge against the tire machine, making it difficult to loosen the holddown clamp.

The last bead to seat is usually the top one. As the tire beads slip over the safety humps on the bead seats of the rim, a loud pop is usually heard. This is normal.

Many tire machines include an in-line dial indicator-type tire gauge. Do not put more pressure than 40 psi, or the maximum pressure listed on the tire sidewall, when attempting to seat the beads. If the beads will not seat with 40 psi, break the beads again and thoroughly lubricate the tire and rim again. Reposition the tire on the rim and reinflate it. When a tire is inflated in a safety cage, it can be inflated to a maximum of about 50 psi (3.5 bar) to seat the beads. Run-flat tires (covered later) often require higher pressures to seat the beads.

For especially difficult tire inflations, such as with wider wheels, some tire specialty shops have special bead seaters (**Figure 13.43**). A different size bead seater is needed for each rim diameter.

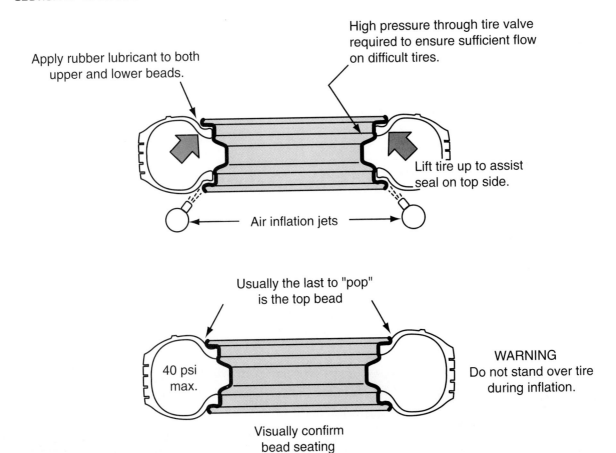

Figure 13.42 Hold the tire against the top bead seat and use the air jets to seal the bottom bead.

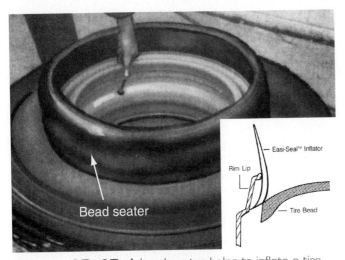

Figure 13.43 A bead seater helps to inflate a tire. *(Courtesy of Hardline International, Inc.)*

SAFETY NOTE
- When inflating a conventional tire to a pressure higher than 40 psi, be sure it is in a safety cage.
- Be careful that the direction the valve stem is pointing is away from you when inflating a tire. It can shoot out of the wheel.

Several potential problems can occur with conventional tire changers. Center post tire changers were originally designed to be used with steel wheels. The pilot hole in the wheel is clamped against the tire changer and the wheel can be overly stressed at the center hole during bead breaking, resulting in a bent rim.

Rim clamp tire changers are sometimes referred to as *tabletop "steel jaw"* tire changers. Custom rims can be damaged on the inside by the teeth of the steel jaw clamps. Some of today's rim designs allow much of the inside of the rim to be visible. Also, the anticorrosion coating that was applied during manufacture can be damaged. Custom rims should be clamped from the outside. If the rim does not remain tight on the clamping table it can spin, resulting in scuffing to the outside of the inner rim edge. This can be caused by difficult installations or because of a problem with the tire changer.

There are some potential cautions to be aware of when using these machines:

- If shop air pressure drops due to high air demand, the clamps can slip.
- With repeated use, jaws wear out and no longer grip tightly.
- The jaws on the clamping table are not all applied in the same manner, and as parts of the apply assembly wear, clamping force can become uneven.

CAUTION Inflating a tire while it is clamped from the outside is dangerous.

Low profile tires often have sections of sidewall that extend beyond the flange of the rim to protect the rim from damage when it rubs against a curb. This can cause problems when using a shovel-type bead breaker (**Figure 13.44**). This area is where the shovel would normally be applied. If the shovel slips toward the rim, it can scratch or chip the rim. If the shovel is applied to the sidewall closer to the tread area, it can damage the tire sidewall (**Figure 13.45**). There are other types of tire changers designed to prevent this.

Mounting High-Performance Tires

A vehicle can have many thousands of dollars invested in the tires and wheels. Aftermarket wheels can cost between $100 and $1,500 each and speed-rated tires can cost up to $600 or more each. An incorrectly serviced tire might cost the shop more than $1000.

Servicing high-performance tires with tighter tire beads can produce a tear in the bead or damage the tire sidewall if done incorrectly. Removing these tires is done with a tire changer that breaks the bead loose with rollers instead of the typical shovel-type bead breaker used on conventional tire changers.

NOTE: *Tire service should not be performed without proper equipment. Tires with aspect ratios less than 60 will typically call for more sophisticated tire equipment.*

Low-profile tires are more apt to suffer bead damage during removal from the rim. Always apply rubber lubricant to both beads during removal and installation of tires.

Bead Roller and Tulip Clamp Tire Changer

The newest tire changer design (see **Figure 13.30**) was developed for use with high-performance wheels and tires. It uses hydraulic bead rollers (**Figure 13.46**) on the top and bottom beads to loosen the beads with a

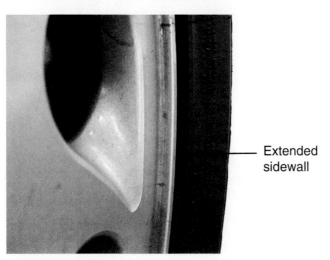

Figure 13.44 Low-profile tires often have an extended tire sidewall that protects the rim from curb scrapes. *(Courtesy of Tim Gilles)*

Figure 13.45 This tire sidewall can be damaged if the bead breaker is positioned too far away from the edge of the wheel rim. *(Courtesy of Tim Gilles)*

Figure 13.46 Hydraulic bead rollers break the bead seat on the top and bottom. *(Courtesy of Tim Gilles)*

CAUTION: Tires that require more than 40 psi to inflate must always be installed in an inflation cage. When reinflating these tires, 60–80 psi could be required to seat the beads so it is critical to install them in a cage for safety reasons.

force of up to 4,000 lbs. The tulip clamping system uses pads protected by rubber, which allows the wheel to shift without damage to its surface. A spring-loaded device centers the wheel by its hub mount. Practically any size wheel can be mounted in this changer, from 5 inches to 23 inches in diameter and up to 19 inches wide.

In the event a tire is incorrectly mounted and binding occurs, the mounting head will break. This saves expensive tires and wheels from damage. Mounting heads for this tire changer are easy to replace and inexpensive.

Removing a tire is similar to the rim clamp, except that the bead is automatically pushed into the drop center while the bead lever tool pulls the bead over the edge of the rim (**Figure 13.47**).

When installing a tire on a rim with an integral tire pressure sensor, the sensor must be positioned correctly in relation to the tire bead (**Figure 13.48**) or it can be damaged. The tire is held in position on the rim using blocks (**Figure 13.49**).

Install the Valve Core

The valve core should not extend above the top of the valve stem. If it does, the valve cap could contact the top of the valve core and let air out. A shorter valve core should be installed. After the valve core is installed, inflate the tire to the amount recommended on the vehicle placard. Check to see that the valve core is sealing by applying a drop of rubber lube or water to

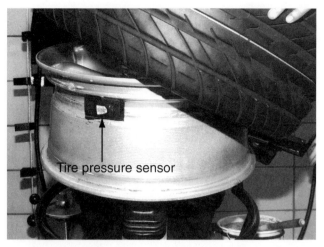

Figure 13.48 When mounting the bottom bead, be sure the tire pressure sensor is in this position. *(Courtesy of Hunter Engineering Company)*

Figure 13.49 The tire is held in position using blocks. *(Courtesy of Tim Gilles)*

the top of the valve stem and watching for bubbles. Then install the valve stem cap.

TIRE AND WHEEL RUNOUT

When a tire is not correctly mounted on a rim or is out of round it will have runout. Runout is when the tire is not round as it spins. If the tire is higher on one side than the other, handling problems will occur. Excessive runout can cause a car to shake at highway speeds.

NOTE: *A favorite technicians' saying is, "You can balance a square tire, but the ride will be unacceptable."*

Spinning the wheel while holding an object like a screwdriver or a pen near it for reference will give an indication of whether a tire or wheel is badly bent.

Pirelli Tires recommends that tires be inflated initially to 40 psi to thoroughly seat the beads. Sometimes, a tire is not properly seated on the rim. The sidewall has centering ribs or locating rings (**Figure 13.50**)

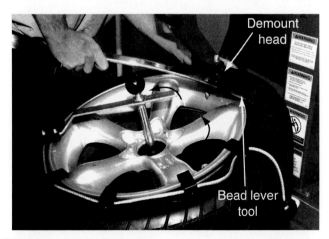

Figure 13.47 As the wheel rotates counterclockwise, the demount head rolls the bead into the drop center while the bead lever pulls the bead over the top of the rim. *(Courtesy of Hunter Engineering Company)*

Figure 13.50 Check the locating ring to see that the tire is evenly mounted.

that can be inspected to see that they are even with the edge of the rim flange all the way around the tire. If not, the beads need to be broken down again to center the tire. Dismounting and remounting the tire on the rim 180 degrees from its original position can sometimes result in improvement.

Another possible cause of runout is that a second-quality or blemished tire might be out of round. Runout can be either lateral (sidewall wobble) or radial (tread up and down) (**Figure 13.51**). To check runout, raise the vehicle until the tires clear the ground by about 1 inch. Place a tire runout gauge against the sidewall or tire tread while rotating the tire (**Figure 13.52**). Combined tire and wheel radial runout should be less than 0.060 inch and lateral runout less than 0.045 inch (1.1 mm).

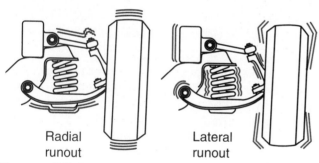

Figure 13.51 Tire runout can be either radial or lateral.

Figure 13.52 Checking radial runout with a tire runout gauge. *(Courtesy of Tim Gilles)*

When radial runout is excessive, verify that the wheel is not the source of the runout. Then try moving the tire to a new position on the rim to correct the problem. If this does not correct the problem, the tire should be replaced.

NOTE: *New road force tire balancers, covered later in this chapter, can match the high spot on a rim to the low spot on a tire to minimize runout.*

Checking Wheel Runout

Checking runout is best done with a dial indicator. Radial runout is measured in an up-and-down direction (**Figure 13.53A**) and lateral runout is measured from side to side (**Figure 13.53B**). Raise the vehicle and mount a dial indicator to read off the correct place on the wheel rim for each of those measurements. Once again, radial runout should be less than 0.060

Vintage Chassis

Many shops used tire-truing machines on the bias-type tires found on older cars and trucks. Rubber was shaved off the tire to correct excessive radial runout. This corrected the problem but significantly reduced tire life.

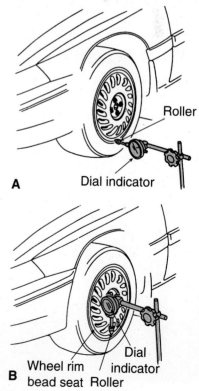

Figure 13.53 (A) Checking lateral runout. (B) Checking radial runout.

inch and lateral runout should be less than 0.045 inch (1.1 mm). The lateral runout limit on aluminum wheels is 0.035 inch.

TIRE REPAIR

When a tire is losing air, it is checked for leaks using a water bucket. Roll the inflated tire slowly around in the tank while looking at the area where the tread meets the water (**Figure 13.54**). Small bubbles from a slow leak can usually be found in this manner. If a leak is not evident, push the valve stem from side to side. Also, check carefully where the tire bead meets the rim. If a soak tank is not available, apply soapy water to the outside of the tire.

Mark the location of the leak and remove the nail or screw if it is still there (**Figure 13.55**). Notice whether it comes out at an angle or not. During the repair, the hole will be reamed at the same angle as the injury to the tire.

Figure 13.54 Check for bubbles at the water line as you spin the tire slowly in the tank. *(Courtesy of Tim Gilles)*

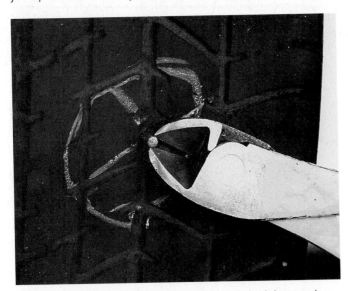

Figure 13.55 Mark the location of the injury and remove the nail or screw. *(Courtesy of Tim Gilles)*

Inspecting the Tire

When inspecting a tire before repairing it, remove it from the wheel. A special tool called a bead spreader can be used to hold the beads apart during the tire inspection and repair. Check the entire outside and inside of the tire carefully to see that it is not damaged from having been run flat. Use a bright light. The tire must be dry prior to inspection. Inspect the outside of the tire for anything that can allow moisture in, like weather checking, cracks, or tread separation. Tires that should not be repaired include those that:

- Smell of burned rubber or have bluish discolored rubber on the sidewall flex area
- Have evidence of loose cords
- Have an inner liner that has blisters, bubbles, cracks, or a visible casing cord pattern
- Are worn beyond the wear bars

REPAIRING A TIRE

Tire puncture repairs can be done with a rubber tire plug, a patch, or both. Pinhole repairs can be made using a patch only. A pinhole is a very small hole that closes up visually when the nail is removed.

Tire repairs can be made when they are within the tread area, or crown, (**Figure 13.56**) as long as the puncture is less than ¼ inch in diameter. Most shops will *not* perform repairs to tire sidewalls and repairs to the bead area can *never* be safely done.

NOTE: *With expensive off-road or farm tires, repairs to holes in the sidewall area or holes larger than ¼ inch in the tread area can sometimes be made, but only by a special full-service tire facility (not traditional retail tire outlets).*

The RMA publishes guidelines for the proper repair of all types of tires. Tires are sometimes repaired improperly. Rubber plugs are often installed without removing the tire from the wheel. This is a substandard repair because the tire should be removed from the wheel and carefully inspected to see that it is in

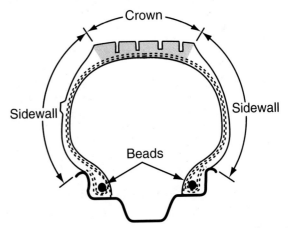

Figure 13.56 Repairs are usually made only to the crown area.

good shape. Damage can happen when a repair is done to a tire that has been run flat, has a hole that is too large to be repaired, or has suffered damage from an impact. **Figure 13.57** shows types of tire damage that are not visible from the outside.

There are differing opinions about the repairability of tires. When the cost of a quality repair is considered against the cost of a new tire, the correct decision is often to replace the tire with a new one. Sometimes the puncture appears to be in an acceptable area of the

Figure 13.58 This puncture is outside of the crown area. It should not be repaired.

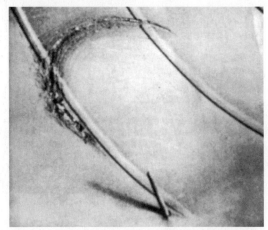

A   Interior damage caused by nail

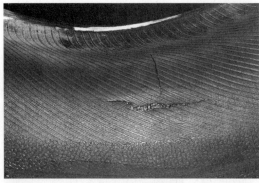

B   Impact damage

C   Rim bruise break

Figure 13.57 Tire damage unseen from the outside: (A) The damage to the interior of this tire was caused by a nail. (B) A rim bruise break is caused by impact against something like a curb or pothole. (C) This damage was caused by driving for a short distance on a severely under-inflated tire. (A, Used with permission of Michelin North America Inc. All rights reserved. (B and C, *Courtesy of Rubber Manufacturers Association*)

tread. Here is a good way to tell if a tire can be repaired. During the tire repair process, a puncture is reamed. If the resistance of the steel belts is not felt while reaming, then a passenger car tire should be replaced. The puncture shown in **Figure 13.58** turned out to be outside of the area supported by the tire's belts. Although the tire is in very good condition, it was replaced with a new one.

Each tire manufacturer publishes repairability standards that they distribute to their dealers. Some manufacturers say that high-speed tires are not repairable because these tires are used on powerful cars at high rates of speed. Others say that a high-speed tire will lose its speed rating if repaired. Still others give repair procedures for these tires. In any event, the tire should be removed from the rim if a repair is to be made.

A proper repair consists of the following steps:

1. Remove the tire from the wheel.
2. Inspect the tire carefully.
3. Run a reamer or drill through the hole.
4. Install a tire plug.
5. Smooth off the inside of the plug.
6. Install a patch on the inside of the tire.

Figure 13.59 shows a sequence of photos of the tire repair process.

Preparing a Tire for Repair

First locate the puncture on the tire tread and mark it with a crayon (**Figure 13.59-1**). Before attempting to repair a tire, the inside of the tire liner must be cleaned. The rough inner surface of a tire is covered with a material that will prevent the adhesion of a patch unless the surface is properly prepared. Using rubber cleaning solvent, clean the area where the new patch will be installed (**Figure 13.59-2**). A scraper is used to clean the liner because it is more effective in removing oils, grease, and silicone (**Figure 13.59-3**).

Figure 13.59-1 Mark the puncture with a crayon. *(Courtesy of Tim Gilles)*

Figure 13.59-2 Clean the area with rubber cleaning solvent. *(Courtesy of Tim Gilles)*

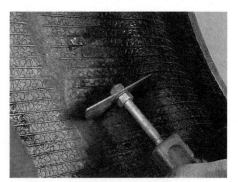

Figure 13.59-3 Clean the area to be repaired with a scraper. *(Courtesy of Tim Gilles)*

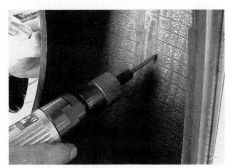

Figure 13.59-4 Ream the hole to remove metal burrs from the belts. *(Courtesy of Tim Gilles)*

Figure 13.59-5 Remove the plastic coating. *(Courtesy of Tim Gilles)*

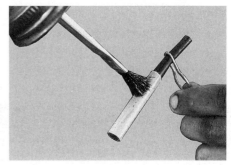

Figure 13.59-6 Apply vulcanizing cement to the plug. *(Courtesy of Tim Gilles)*

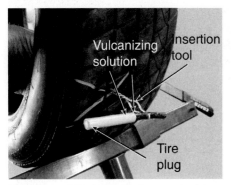

Figure 13.59-7 Insert the plug into the hole. *(Courtesy of Tim Gilles)*

Figure 13.59-8 Pull the plug until ½ inch extends above the tread surface of the tire. *(Courtesy of Tim Gilles)*

Figure 13.59-9 Cut the end of the plug ⅛ inch to ¹⁄₁₆ inch above the inner liner. *(Courtesy of Tim Gilles)*

Plugging the Hole

A steel-belted radial tire that has a puncture through a steel belt probably has exposed metal. Moisture can damage the inside of the belt area of the tire so this puncture is best repaired by filling it with a rubber plug and then applying a patch.

Ream the Hole. To prepare a puncture for the installation of a plug, the hole is first drilled, or reamed, (**Figure 13.59-4**). Holes that go through steel belts beneath the tread can have sharp pieces of steel sticking into them. When the tire deflects during normal driving, a newly installed plug can be cut into small pieces if the hole is not prepared properly by reaming. When a drill is used instead of an abrasive reamer specifically designed for this purpose, it should have a speed of less than 5,000 rpm to avoid excess heat. Before drilling or reaming, probe the hole with an awl to determine the direction of the puncture.

Install the Plug. A rubber plug is inserted into the hole in the tire. The plug comes with a plastic coating that protects the soft vulcanizing surface. This must be carefully removed, being sure not to touch its sealing

Chapter 13 Tire and Wheel Service   303

Figure 13.59-10 Cut the plug flush with the tread surface. *(Courtesy of Tim Gilles)*

Figure 13.59-11 Buff the area where the patch will be placed. *(Courtesy of Tim Gilles)*

Figure 13.59-12 Vacuum the dust. *(Courtesy of Tim Gilles)*

Figure 13.59-13 Apply vulcanizing cement to the buffed area. *(Courtesy of Tim Gilles)*

Figure 13.59-14 Remove the backing from the adhesive side of the patch. *(Courtesy of Tim Gilles)*

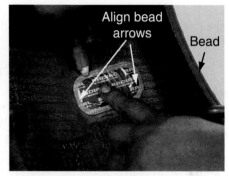

Figure 13.59-15 Center the patch over the hole. *(Courtesy of Tim Gilles)*

surface (**Figure 13.59-5**). Apply the vulcanizing cement to the plug (**Figure 13.59-6**). The cement is the same one used to vulcanize a tire patch to the tire liner.

Insert the pulling wire into the hole (**Figure 13.59-7**). Pull the plug until ½ inch extends above the tread surface of the tire (**Figure 13.59-8**). Use a flexible knife to cut the end of the plug ⅛ inch to 1/16 inch above the inner liner (**Figure 13.59-9**). Cut the plug flush with the tread surface on the outside of the tire (**Figure 13.59-10**). Be careful not to stretch the plug while cutting it.

Buff the Inner Liner. Lightly buff the liner with a fine buffing stone and a low-speed buffer motor (less than 5,000 rpm) to complete the preparation of the surface (**Figure 13.59-11**). First, outline an area ½ inch around the outside of the patch with a marking crayon. Be careful not to buff through the liner into the casing plies. Just clean the surface of the rubber. If a portion of steel belt extends into the inside of the tire, carefully grind it away. Because compressed air contains oils and moisture, it should not be used to remove the dust. Vacuum the dust from the buffing operation (**Figure 13.59-12**).

Patching a Tire

After the plug is cut flush with the inside of the tire liner, clean and scrape the area once again with rubber cleaning solvent and a scraper. When the cleaned area has thoroughly dried, apply an even amount of the vulcanizing cement to the buffed area where the patch will be installed (**Figure 13.59-13**). Vulcanizing cement is liquid when it is in good condition. If the cement has solidified or is like jelly, it should be discarded. It is a good idea that the cement be the same brand as the patch so that the two are compatible.

Allow the vulcanizing cement to dry completely before applying the patch. Drying time varies with temperature and humidity.

Install the Patch. Remove the backing from the adhesive side of the patch. Be careful not to touch the sticky adhesive (**Figure 13.59-14**).

NOTE: *The patch should be installed while the tire is in its natural shape. A bead spreading tool is commonly used by many shops when affixing a patch to the inside of the tire. A bead spreading tool is not recommended for radial tires when applying the patch.*

Center the patch over the hole, making sure that the bead arrows on the patch point to the beads of the tire (**Figure 13.59-15**). Roll the patch into place with a corrugated tire stitcher (**Figure 13.59-16**). Roll firmly from the center outward, using as much hand pressure as possible. After stitching is complete, remove the thin plastic covering from the top of the patch (**Figure 13.59-17**). To complete the repair,

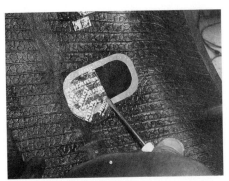

Figure 13.59-16 Use the stitcher to seat the patch to the tire casing. *(Courtesy of Tim Gilles)*

Figure 13.59-17 Remove the thin plastic covering from the top of the patch. *(Courtesy of Tim Gilles)*

Figure 13.59-18 Apply tire sealer around the outside of the patch. *(Courtesy of Tim Gilles)*

apply tire repair sealer around the outside of the patch (**Figure 13.59-18**).

Combination Plug-Patches

Some plugs are a combination of plug and patch (**Figure 13.60A**). These can only be used when the hole goes straight into the tire. If the hole is slanted, use a plug and a patch. Some combination plug-patches are mushroom-shaped and must be installed using a special plug installation tool (**Figure 13.60B**). Be sure to dip a blunt probe in vulcanizing cement and push it through the hole to thoroughly

SHOP TIP Moving the puncture to the up position will allow the cement to dry more quickly and evenly. Solvents are heavier than air.

Figure 13.60 (A) A combination plug-patch. (B) One type of combination patch-plug installed with an insertion tool. *(Courtesy of Tim Gilles)*

Figure 13.61 Tire sealant was the cause of this tire's vibration problem. The tire could not be balanced. *(Courtesy of Tim Gilles)*

SAFETY NOTE Some puncture sealants are flammable and can explode if drilling through the steel belts of the tire during an external plug repair produces a spark. Removing the tire from the wheel reduces the danger of an explosion.

coat the walls of the hole. After reinstalling the tire on the wheel and inflating it, use water to double-check that the leak has been fixed.

Liquid Puncture Sealants

Liquid puncture sealants are commonly available in automotive stores. While these products may be handy when a flat tire occurs, they are not recommended by tire manufacturers. Liquid tire sealants can also make it impossible to balance a tire, due to the shifting weight of the sealant within the tire (**Figure 13.61**).

■ TIRE AND WHEEL BALANCE

Tire imbalance is one of several possible causes of vehicle vibration. Some other possibilities are a bent axle or wheel rim, an out-of-phase drive shaft, bad universal joints, or misaligned drivetrain components. Tire imbalance can result in a cupped tire, a loss of traction, and premature wear to steering and suspension parts.

Although tire imbalance is not always the cause of vehicle vibration, it is a major cause. Front tires are the most prone to exhibit symptoms from imbalance. Vibration from rear tire imbalance may not be as obvious, but tire wear can still occur.

Imbalance results when the weight of a tire's materials is not equally distributed around the tire. The result is that one side of the tire is heavier than the other. This can be due to errors in manufacturing or it can result from tire wear. Sometimes new tires have a minor imbalance that can be corrected by adding specified amounts of weight to the wheel rim at certain points.

When an excessive amount of weight is required to achieve balance, a defective tire could be the cause or the tire might be incorrectly seated on the wheel rim. Remember, out-of-round tires can be balanced but will still cause vibration from wheel tramp.

Wheel Weights

Wheel weights are attached to the rim to correct imbalance (**Figure 13.62**). Clip-on wheel weights are attached to the rim using a special wheel weight hammer (see **Figure 13.28**).

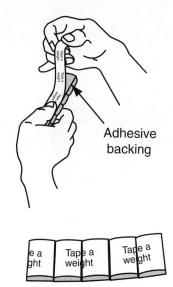

Figure 13.63 Lead tape weights.

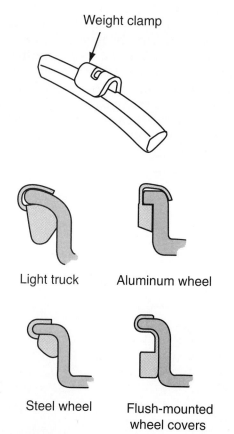

Figure 13.62 Lead wheel weights are installed on the wheel to counterbalance a tire. There are several styles for different applications.

Several styles of lead clip-on weights are available in increments of ¼ ounce. Some have an extra long clip for fitting under wheel covers. Another style has a wider, larger clip to fit on pickup truck wheels.

Aluminum wheels require alloy or coated clips if they have a flange that will accept a clip-on weight. The coated weight will not corrode the wheel. Lead strips that are cut to length are also available for aluminum wheels (**Figure 13.63**). These are attached to the wheel with double-sided tape after cleaning a place on the inside of the wheel with sandpaper. Sometimes they are installed on the inside and outside of the wheel, but this is unsightly. Chrome weight strips are available also.

NOTE: *Be careful when working around aluminum wheels. They are easily scratched. Plastic shields that cover tire tools can be used to prevent damage.*

■ TYPES OF WHEEL BALANCE

The two types of wheel balancing methods used on cars are static and dynamic.

Static Balance

If a wheel with static imbalance were mounted on a spindle with the heavy spot at the top, the heavy spot would rotate to the lowest possible position on its own. **Static balance** means that when the wheel is balanced, it does not have a tendency to rotate by itself.

Static imbalance subjects the wheel to vertical impacts that become worse at higher speeds. The impacts occur because the tire has a heavy spot on one end of its tread (**Figure 13.64**). A small amount of imbalance when the wheel is at rest can amount to a great deal of pounding force when the wheel is spinning at highway speeds (**Figure 13.65**). For instance, if a 15-inch wheel is 1 ounce out of balance, when the

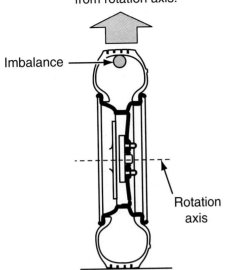

Figure 13.64 When the heavy spot is rotated, force is directed out away from the center of the wheel.

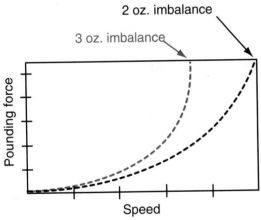

Figure 13.65 Pounding force increases as speed and the amount of imbalance increase.

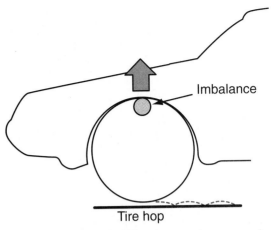

Figure 13.66 Wheel tramp happens when the speed and force of the imbalance become large enough to cause a tire to leave the road surface.

Vintage Chassis

Older, narrower bias tires were static-balanced. Static means an object is stationary. Although today's modern wheel balancers are capable of measuring the static imbalance of a spinning wheel, the name remains from when imbalance was measured with the wheel at rest on a bubble balancer. Static balancing, also called single plane balancing, is done in a single plane where compensating weight is added on the opposite side of the wheel (**Figure 13.67**). If it were possible to put the compensating weight at exactly the same place on the opposite side of the tire, the amount of the weight would be exactly the same as the imbalance. Unfortunately, tires are sometimes heavier on one side of the tread than the other. Modern wheel balancers correct this imbalance.

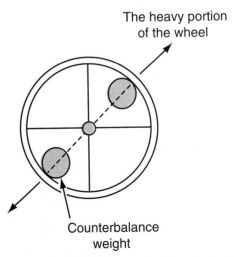

Figure 13.67 To balance a tire in a single plane, compensating weight is added on the opposite side of the wheel.

wheel rotates at 60 mph the pounding force will be 4.6 pounds.

Static imbalance causes wear to mechanical parts, vibration, and gouged tire tread wear, called cupping (see Figure 13.10) In severe cases, especially when accompanied by a bad shock absorber, a tire can actually hop so badly that it leaves the road surface. This is called wheel tramp (**Figure 13.66**).

The part of the tire that is contacting the ground is actually traveling at 0 mph. This is called static friction or static contact. During wheel tramp when the tire recontacts the road, a small amount of the tire's rubber is scrubbed off. This is because the tire was traveling at the same speed as the vehicle when it left the road surface. This scrubbing happens all around the tire, accounting for the cupped wear all around the tire's tread surface.

Figure 13.68 Location of weights for static balance. *(Courtesy of Tim Gilles)*

Static balancing is an option when using a computerized spin balancer. Although there are better ways of balancing aluminum wheels, this option may be used when unsightly weights are undesirable on the outside of the wheel. When static balancing an aluminum wheel, weights with tape on their back sides are fixed to the inside of the wheel at the center plane (**Figure 13.68**).

On-the-Car Balancers

Wheels can be balanced on the car using a spin balancer (**Figure 13.69**). When a wheel is spun with a balancer, *static balance* is not an accurate term. The correct term is *single-plane balance* or *kinetic balance*. This method is popular for exotic and high-speed cars because the tire, wheel, and spinning brake parts are all balanced as a unit.

NOTE: *When a tire has been balanced on the vehicle and the wheel is removed, the balance can be changed if it is reinstalled in a different position. It should be marked for both the wheel location on the car and the lug position.*

Figure 13.69 An on-the-car spin balancer. *(Courtesy of Hunter Engineering Company)*

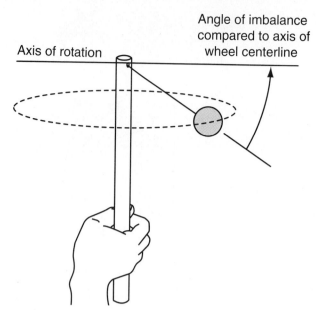

Figure 13.70 As the speed of rotation increases, the imbalance moves toward the axis of rotation.

Couple Imbalance

When a tire tread is lopsided, with more of its weight on the outside or the inside of its tread, this is known as *couple imbalance*. Couple imbalance only shows up when the wheel is spinning. The wheel will have a tendency to **shimmy**—wobble from side to side—which also results in faster tire wear. **Figure 13.70** shows how a spinning heavy spot will seek the centerline of the tire. This is why the tire shimmies. When the heavy spot is in the front, it causes the wheel to move one way. After the tire rotates one half revolution, the heavy spot causes the wheel to try to turn in the opposite direction (**Figure 13.71**).

The previous example is of an imbalance on only one side of the tire. Usually there are imbalances on both sides of the tire, requiring counterbalancing weights to be installed on both sides of the wheel.

A wheel may be in static balance but not couple balance. If a tire has been statically balanced by adding all of the weight to one side of the offset wheel rim, that weight will be thrown side to side in a different direction as the tire spins, resulting in shimmy. The proper way to balance a wheel statically is to split the amount of weight to be used in half, placing an equal amount of weight on both sides of the rim (**Figure 13.72**). Pirelli Tires recommends this procedure any time the amount of weight to be added is in excess of 0.71 oz. (20 grams). Tape weights are often added in two places on the inside of the wheel (**Figure 13.73**).

Dynamic Balance

It is unlikely that a tire and wheel will have only static imbalance. Dynamic imbalance is the combination of both static and couple imbalance. **Dynamic balance** means balance in motion. It is also called two-plane

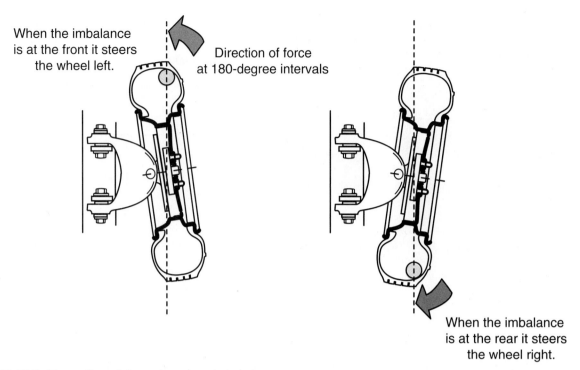

Figure 13.71 The action of dynamic or couple imbalance.

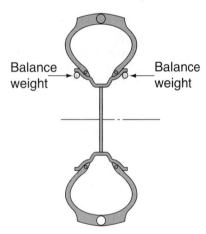

Figure 13.72 The amount of weight is split in half and put on both sides of the rim to avoid creating a coupling imbalance.

balance because it measures side-to-side (lateral) force as well as up-and-down (axial) force. Lateral forces are felt when a steering wheel moves back and forth.

Dynamic wheel balancers spin the wheel and locate vibration in it. The computer splits the tire into two halves and measures lateral and radial (axial) forces on each side of the tire's center. Weights are added to the proper side of the rim to correct the imbalance. A tire that is dynamically balanced will be statically balanced as well. Dynamic balancing can be done with a computer balancer or with an on-the-car spin balancer.

Computer Balancers. Off-the-car computerized wheel balancers (**Figure 13.74**) are very popular and easy to use. Computer balancers balance in both the

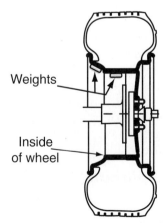

Figure 13.73 Tape weights are added to the inside of the wheel for cosmetic purposes.

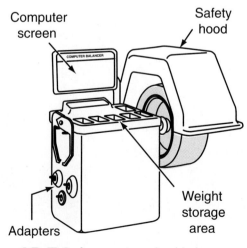

Figure 13.74 A computer wheel balancer.

static and dynamic planes. When the wheel is off the car for tire replacement, computer balancing is usually performed on the new tire. The wheel is mounted on either a horizontal or a vertical threaded shaft using adapters that are supplied with the machine.

Centering a Wheel on the Balancer

Imbalance can also be caused when a wheel is not properly installed on a wheel balancer. According to the FMC Corporation, a 36-pound tire and wheel assembly that is off-center by only 0.006 inch will result in a ½-ounce imbalance error.

When mounting a wheel on the wheel balancer, the best method is to center the wheel by the same method used by the manufacturer. There are two ways that wheels are centered:

1. **Hub-centric,** or *pilot-hole-centric,* is when the hole at the center of the wheel locates the hub (**Figure 13.75**).
2. **Lug-centric** is when the lug nuts center the wheel.

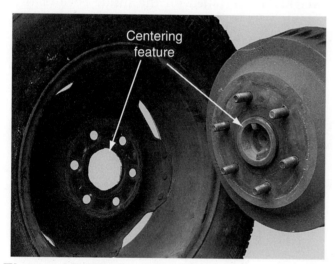

Figure 13.75 The center hole of a hub-centric wheel fits on the hub. This is a rear wheel and hub. *(Courtesy of Tim Gilles)*

To determine whether a wheel is hub-centric, see if the wheel fits the hub snugly with the lug nuts removed.

Preparation for Mounting the Wheel. The backing plate and wheel lug flange must be clean and undamaged. Be sure to remove all foreign objects from the wheel and tire before attempting tire balance. When the wheel and tire assembly is first spun, look for signs of obvious runout and double-check the mounting.

Mounting Hub-Centric Wheels

Use a centering cone when mounting a hub-centric stamped steel wheel. The most accurate centering system is to install it from the back side of the wheel. This is because the center hole was originally stamped from the back. **Figure 13.76** shows the recommended mounting arrangement on the wheel balancer.

It is important that the centering cone fit the shaft snugly. When new, a centering cone has about 0.001 inch of clearance. Expandable collets are available to provide the most accuracy in centering. This is because they expand when tightened, eliminating all clearance between the collet and the shaft.

Mounting Lug-Centric Wheels

To mount lug-centric wheels, use a special lug centering adapter (**Figure 13.77**). The arms on the adapter are mounted in different places, depending on the number of lug holes. There are only five adapter arms. With six-

CASE HISTORY

A man purchased a used Suburban that had aftermarket aluminum wheels and oversized tires. The tires appeared to be in good condition, with less than half of the tread worn. Unfortunately, they had been balanced incorrectly, using adapters in the center hole instead of a lug-centric adapter. The tires wore unevenly after the purchase, developing cupped tread wear. They were now too far out of balance and had to be replaced.

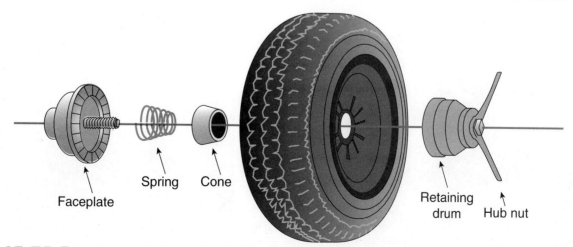

Figure 13.76 The recommended computer balancer mounting arrangement for hub-centric wheels.

Figure 13.77 The lug-centric adapter installed on a wheel balancer. It can be configured with three, four, or five adjustable arms. (Courtesy of Tim Gilles)

lug wheels, use three arms. With eight-lug wheels, use four arms. After the arms are installed and tightened on the adapter, it is held against the mounting flange on the wheel balancer while screws are installed and tightened (**Figure 13.78**). Special lug nuts hold the wheel against the adapter (**Figure 13.79**).

Figure 13.78 A lug-centric adapter is held against the balancer mounting flange while retaining screws are installed and tightened. (Courtesy of Tim Gilles)

Figure 13.79 Special lug nuts hold the wheel against the lug-centric adapter. (Courtesy of Tim Gilles)

Program the Wheel Balancer

After the wheel is mounted, the technician needs to program three references into the computer:

1. The width of the rim, measured by a rim caliper (**Figure 13.80**)
2. The location of the flange on the wheel, measured by a gauge (**Figure 13.81**)
3. The diameter of the wheel (for instance, 13", 14", or 15")

Balance the Wheel and Tire

To balance a tire, the safety hood is lowered over the wheel. The wheel is spun for a short time and then

> **SHOP TIP** Weights for aluminum wheels have double-sided tape attached to their backs. When balancing aluminum wheels with tape weights, use duct tape to temporarily attach the weights to the wheel. Install one weight on the wheel as near to the outside as possible. The other weights are installed on the inside edge of the wheel.

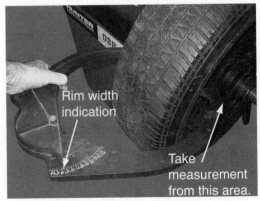

Figure 13.80 Use a special caliper to measure the width of the rim. Apply its ends to the places on the inside and outside of the rim where wheel weights would be applied. (Courtesy of Tim Gilles)

Figure 13.81 This measurement tells the computer where the wheel is located on the balance shaft. (Courtesy of Tim Gilles)

stopped. The machine gives a readout telling how much weight to use and where to put it. After installing the weights, the wheel is spun once again to check the accuracy of the balance job. A reading of zero on both sides of the wheel means that the wheel is ready for installation on the car.

NOTE: *Make certain there is enough clearance between the wheel weights and the disc brake caliper.*

Match Mounting

Some computer balancers have an extra feature that matches a tire's imbalance to a wheel's imbalance. This procedure is used when a weight of more than 2 ounces is required on one side of a wheel. The tire is given an initial spin on the balancer. Then it is deflated and installed on a tire machine so that the beads can be broken down. After rotating the tire on the rim 180 degrees, the tire is reinflated and rebalanced. The computer can now tell how much the wheel is out of balance and how much the tire is out of balance. If neither is unacceptable, the tire is once again rotated on the rim to complete the match mount.

■ FORCE VARIATION

Today's lighter vehicle designs are more sensitive to road feel. When a load is placed on a tire, there is a change in the stiffness of the sidewall and the tire's footprint. There can be stiffer or weaker areas of the tire when under load (**Figure 13.82**). Manufacturers have specified acceptable limits for this change, which is called *force variation*. A wheel and tire might be free of runout when measured with a dial indicator. Yet, the tire might vibrate under load due to excessive force variation.

Some tires come with a mark or a paint dot on their sidewall to indicate the high or low side of force variation. Unfortunately, this marking is not uniform among manufacturers, so simply lining the valve stem up with the mark could result in an exaggeration of the problem. Also, some manufacturers use the mark to label the high point, whereas others use it to label the low point. There are tire balancers that use a special device to measure and correct force variation (**Figure 13.83**). These tire balancers locate tire and wheel problems caused by force variation, runout in the wheel, and the random position between the force variation and runout when the tire is mounted on the rim. The balancer detects variations in wheel runout (**Figure 13.84**) and the

Figure 13.83 (A) A tire balancer with a force roller. (B) When the hood is lowered over the tire, the force roller presses against the treads as the wheel spins. *(Courtesy of Tim Gilles)*

Figure 13.82 There can be stiffer or weaker areas of the tire when under load. *(Courtesy of Hunter Engineering Company)*

Figure 13.84 This wheel balancer detects wheel runout. *(Courtesy of Tim Gilles)*

312 SECTION 3 TIRES AND WHEELS

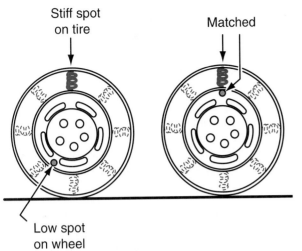

Figure 13.85 Matching the high or stiff spot on the tire with the low spot on the wheel reduces vibration.

computer recommends a change in the tire and wheel positions to match the stiff or high spot on the tire with the low spot on the rim (**Figure 13.85**). This minimizes problems caused by force variation.

■ MOUNTING THE WHEEL ON THE CAR

Before installing a wheel on a car, double-check to see that the bolt holes in the wheel center are in good condition. It is not uncommon to find steel wheels where the bolt holes are worn from a wheel that was loose, lug nuts that were installed upside down, or lug nuts that were overtightened. Finger-tighten the lug nuts. Shake the wheel to center it over the centering area on the hub. Cross-tighten the lug nuts with a torque wrench.

■ TIRE CHAIN SERVICE

Chains must be installed correctly to avoid damaging the tire or underbody of the car. Loose chain ends must be secured so they do not swing out and cut a brake hose, damage an ABS sensor wire, or pop a tire. Driving speed must be reduced markedly when driving with chains. If you have ever ventured to the mountains during or after heavy snow, you have probably witnessed accidents caused by ignorant drivers who attempted to navigate slick, dangerous roads as if they were driving on dry pavement.

Chains must be relatively tight so they cannot twist during driving and puncture the tire sidewall. It is easier to install chains if the tires are off the ground. There are specialty installation ramps available for this purpose.

Chains can be safely installed on drive wheels using the following procedure:

1. Lay the chain out carefully on the ground in front of one of the two drive tires (**Figure 13.86**). Be sure the loops of the cross chains are facing toward the ground and the chains are not twisted. To prevent puncture of the sidewall, the outer loops must face out, with the flat side of the chain toward the tire sidewall (**Figure 13.87**).
2. Tire chains have a provision for adjustment only on one side (**Figure 13.88**). Be sure the side that is not adjustable is positioned toward the inside of the vehicle.
3. Roll the vehicle forward onto the chain, being sure that the connecting chain links are near ground level at the front or back of the tire so you can easily connect the inner chain hook (**Figure 13.89**).

Figure 13.86 Lay out the tire chains prior to installing them on the drive wheels. *(Courtesy of Tim Gilles)*

Figure 13.87 The flat side of the chain faces the tire sidewall. The loops must face outward. *(Courtesy of Tim Gilles)*

4. Connect the inner loop to a chain link that provides the tightest possible connection.
5. Connect the front adjustable link and pull it tight.
6. Drive the vehicle forward a short distance, allowing the chain to even out on the surface of the tire tread.
7. If the chain is still loose, reposition the adjusting hook to the tightest link location possible.
8. Use a spring-type or rubber slack adjuster. Connect its hooks to evenly spaced areas around the circumference of the chain to pull it tight (**Figure 13.90**).

Figure 13.89 Roll the vehicle onto the chain, with the connecting links near ground level at the front or back of the tire. *(Courtesy of Tim Gilles)*

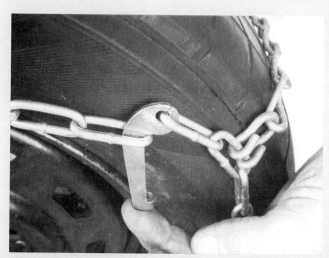

Figure 13.88 One side of the chain has a provision for slack adjustment. *(Courtesy of Tim Gilles)*

Figure 13.90 Use a slack adjuster to remove slack in the chain. *(Courtesy of Tim Gilles)*

REVIEW QUESTIONS

1. What kind of tire has a bulging sidewall when properly inflated?
2. About how much does pressure in a tire increase as the tire warms up?
3. Wear bars show up around the tire tread at what tread depth?
4. What is the name for the up-and-down action of the tire that results in scalloped tire wear?
5. What is applied to the tire bead before installing a tire on a rim?
6. Should the valve core be in or out of the valve stem when filling the tire with air to seat the bead?
7. What kind of hole can be repaired using only a patch?
8. A hole in the tire tread area of a steel-belted tire must be _____ or _____ before installing a plug in it.
9. What kind of tool is used to clean the inside of a tire liner before installing a patch?
10. What type of wheel balance is measured with the wheel stationary?
11. If a 15-inch wheel is 1 ounce out of balance, when the wheel rotates at 60 mph, the pounding force will be _____ pounds.

12. How fast is the part of the tire that is contacting the ground traveling?
13. When a tire is lopsided, with more weight on one side than the other, what kind of imbalance results?
14. What is the combination of both static and couple imbalance called?
15. To mount _____ centric wheels, use a special lug centering adapter.

PRACTICE TEST QUESTIONS

1. Technician A says when static balancing a wheel, it is better to place the weight on the inside of the wheel. Technician B says tire balance problems become less noticeable as speed increases. Who is right?
   A. Technician A   B. Technician B
   C. Both A and B   D. Neither A nor B

2. Technician A says that a computer tire balancer balances the tire statically. Technician B says that the computer tire balancer balances a tire dynamically. Who is right?
   A. Technician A   B. Technician B
   C. Both A and B   D. Neither A nor B

3. Wheel lug nuts on front wheels are being installed using an impact wrench. Technician A says that the lug nuts can be loosened while the car is off the ground without holding the wheel from turning. Technician B says that tightness must be double-checked with a torque wrench. Who is right?
   A. Technician A   B. Technician B
   C. Both A and B   D. Neither A nor B

4. Technician A says that wheel lugs on most cars are loosened by turning them clockwise. Technician B says that lug nuts are installed with the tapered side away from the rim. Who is right?
   A. Technician A   B. Technician B
   C. Both A and B   D. Neither A nor B

5. Technician A says a tire's inflation pressure drops as the car is driven. Technician B says the tire bead must not be lubricated for assembly or the rubber will rot. Who is right?
   A. Technician A   B. Technician B
   C. Both A and B   D. Neither A nor B

6. Technician A says the valve core should be installed before inflating a tire to seat its beads. Technician B says to use a shop towel and solvent to clean a tire's inner liner before applying cement. Who is right?
   A. Technician A   B. Technician B
   C. Both A and B   D. Neither A nor B

7. Technician A says holes larger than a pinhole must be plugged. Technician B says holes larger than a pinhole must be patched. Who is right?
   A. Technician A   B. Technician B
   C. Both A and B   D. Neither A nor B

8. Technician A says the part of the tire that is in contact with the ground is traveling at the same speed as the car. Technician B says the combination of static and couple balance is called dynamic balance. Who is right?
   A. Technician A   B. Technician B
   C. Both A and B   D. Neither A nor B

9. Technician A says to unseat tire beads while the tire is inflated. Technician B says to pound on the beads of a passenger car tire with a special tire hammer. Who is right?
   A. Technician A   B. Technician B
   C. Both A and B   D. Neither A nor B

10. Technician A says not to inflate a tire over 25 pounds when seating a bead. Technician B says if a hot tire has a lower pressure than the recommended cold pressure, it is seriously underinflated. Who is right?
    A. Technician A   B. Technician B
    C. Both A and B   D. Neither A nor B

SECTION Four

SUSPENSION AND STEERING

■ **CHAPTER 14**
Suspension Fundamentals

■ **CHAPTER 15**
Suspension System Service

■ **CHAPTER 16**
Steering Fundamentals

■ **CHAPTER 17**
Steering Service

■ **CHAPTER 18**
Wheel Alignment Fundamentals

■ **CHAPTER 19**
Wheel Alignment Service

CHAPTER 14

Suspension Fundamentals

■ OBJECTIVES

Upon completion of this chapter, you should be able to:
- ✓ Identify parts of typical suspension systems.
- ✓ Describe the function of each suspension system component.
- ✓ Compare the various types of suspension systems.

■ KEY TERMS

active suspension system
adaptive suspension systems
aluminum space frame (ASF)
backbone chassis
ball joints
body-over-frame
bounce
carbon fiber monocoque
center bolt
coil spring
composite springs
compression
control arms
dive
double wishbone suspension
independent suspension
jounce
leaf spring
load-carrying ball joints
MacPherson strut
monoleaf spring
multilink suspension
overload leaves
passive suspension systems
rebound
rigid axle
semi-elliptical spring
shackle
shock absorbers
short/long arm (SLA) sususpension
space frame
spring rate
sprung weight
squat
torsion bar
unibody
unsprung weight
wheel rate
wishbone

■ INTRODUCTION

The vehicle chassis includes the frame, suspension system, steering parts, tires, brakes, and wheels (**Figure 14.1**). The suspension supports the vehicle, cushioning the ride while holding the tires and wheels correctly positioned in relation to the road. Several frame and suspension designs are in use today. Over the years, many nonstandardized part names have been associated with these designs. The names used in this chapter are the ones that have become standard. As you read the chapter, familiarize yourself with the names of these parts. Suspension system parts include the springs, shock absorbers, control arms, ball joints, steering knuckle and spindle, or axle.

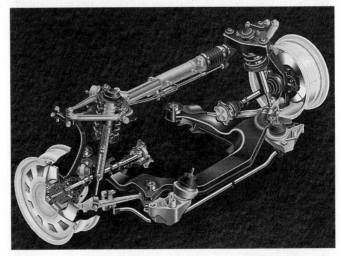

Figure 14.1 Suspension and steering systems.

■ SPRUNG AND UNSPRUNG WEIGHT

Sprung weight is the weight that is supported by the car springs. Sprung weight includes the powertrain,

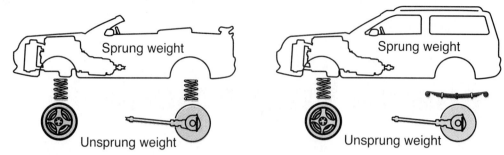

Figure 14.2 A comparison of sprung and unsprung weight.

body and frame, and anything else carried by the weight of the springs. Anything *not* supported by the springs is **unsprung weight**, which includes such things as the tires and wheels, brakes, bearings, axles, and differential.

Vehicle control increases with the reduction of unsprung weight. The tire and wheel assemblies react to irregularities in the surface of the road, whereas the sprung components are relatively insulated from those effects (**Figure 14.2**). Having low unsprung weight means that inertia has less of an effect on those parts as they react to potholes and bumps in the road. This means that the springs have less work to do.

FRAME AND SUSPENSION DESIGNS

Cars are designed to be as lightweight as possible to improve handling and fuel economy. Many newer cars have front-wheel drive. Almost all older, heavier cars had rear-wheel drive.

Various types of frame designs are used in cars and trucks. Almost all passenger cars made today have unitized bodies. Sport utility vehicles (SUVs) and light trucks usually have a body-over-frame design. Sports and exotic vehicles have several different frame designs. Limited production vehicles often use different frame designs than those manufactured in very large production numbers. This is due to economics. Modern passenger vehicle frames are made in huge stamping plants and welded together by robots. Limited production manufacturers cannot afford these manufacturing practices.

Body Over Frame

The **body-over-frame** design has a separate frame with all of the powertrain parts mounted in it. A separate body assembly is bolted to the frame after the suspension and drivetrain are constructed with the chassis. Rubber insulating bushings separate the frame from the car body (**Figure 14.3**).

There are two types of body-over-frame designs. The very first frame design, the ladder frame (**Figure 14.4**), was used by nearly all of the world's auto manufacturers until the early 1960s and is still used in trucks and large vans. As the name implies, the frame resembles a ladder, with longitudinal side rails and cross members. This frame design is simple and inexpensive to manufacture, but it is heavy and has a tendency to twist more than other frame designs. Because of the under-vehicle space requirement of the frame, a high roof line is a characteristic of these vehicles.

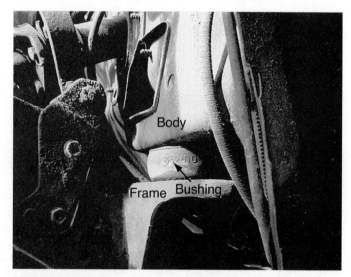

Figure 14.3 Insulating bushings separate the body from the frame. *(Courtesy of Tim Gilles)*

The other body-over-frame design is the *perimeter frame* used by most manufacturers from the mid-1960s to the mid-1970s. Perimeter frames are still used in today's SUVs and in large passenger cars like the Ford Crown Victoria or Lincoln Town Car (**Figure 14.5**). The perimeter frame is a redesign of the ladder frame. The perimeter frame still has narrow integral subframes at each end, but its midsection of steel frame is moved outward to outline the perimeter of the passenger compartment. This change allows the body to be positioned lower on the frame, but these vehicles are too heavy to meet today's fuel efficiency requirements.

Unibody

The vast majority of today's passenger cars have a unitized body design called **unibody**, or *monocoque*. When mass-produced in large quantities, this design is inexpensive. Newer unibodies also provide very good crash protection.

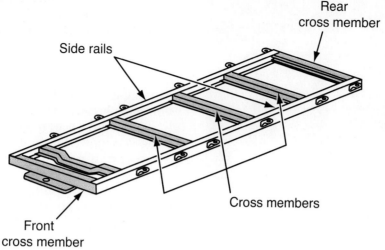

Figure 14.4 A ladder frame.

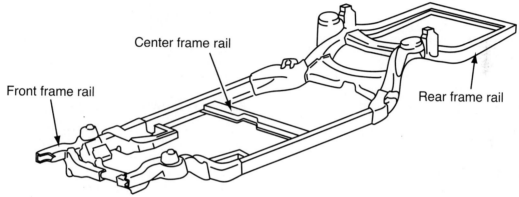

Figure 14.5 A perimeter frame.

A sheet metal floor pan has small sections of frame at the front and rear. A subframe bolted to the front includes the engine, transaxle, and steering/suspension system (**Figure 14.6**). The unitized design incorporates a good deal of the body into the frame. The single-piece unibody is made up of several parts welded together. The floor pan is the largest part. Other stamped sections are held in jigs while robots weld them to the floor pan. Body parts, such as the doors, hood, trunk (deck lid), quarter panels, and roof are added to the unibody to form the completed car body.

Figure 14.6 A subframe supports the powertrain and suspension and steering systems.

Several variations of the unibody design have been used:

- A *single-box* design is used in some small cars and vans. It has extra space in front, which can make it more difficult to park because the driver's visibility of the front of the car is reduced.
- A *two-box* body is used for hatchbacks.
- A *cab-forward* design creates more room for front passengers by using a shorter engine compartment. The bottom of the windshield is considerably closer to the front wheels, and the front wheels are located farther forward. The cab-forward design is used in many European cars. Chrysler was the first manufacturer to use this design in North American automobiles (**Figure 14.7**).
- An example of a small car design that is used to increase passenger safety in the event of a crash is the *sandwich body* used in the A-class Mercedes (**Figure 14.8**). The passenger compartment is noticeably higher (200 mm) in this vehicle. The location of its transverse engine is higher than the floor pan but beneath the passenger compartment.
- A *three-box* body is used for increased passenger safety in a front-end collision. With this kind of

Figure 14.7 A cab-forward body design.

Figure 14.8 Some cars use a sandwich body that increases passenger safety. *(Courtesy of Daimler Chrysler)*

design, the engine is pushed under the driver in a front-end collision. This type of design provides crash protection without increasing weight by adding more steel.

Crash Protection

Passenger vehicles sold in the United States must pass a crash test into a wall at 30 mph. Since 1978, the federal government has been comparing the safety performance of vehicles in flat-barrier crashes at 35 mph, because the majority of serious real-world collisions are the kind simulated in these tests. Crash tests are recorded in slow motion, giving designers a true picture of a vehicle's *crumple zone*. Designers adjust the crumple zone according to the size of the vehicle. Large vehicles have a softer crumple zone than small cars because they can absorb more crash energy, which gives better protection to the occupants of both vehicles involved in a collision.

As crash standards were further refined during the 1990s, more metal was required to meet the standards. Vehicle manufacturers are experimenting with materials other than steel to better meet these standards, while maintaining or improving strength and cutting weight. Designers are always looking for ways to improve crash protection and fuel economy. Aluminum is lighter than steel and is often used in body panels like the trunk or hood. It is also used in suspension parts because it lessens unsprung weight.

Auto body sheet metal parts are typically stamped using very powerful presses. A new technology, called hydroform, is replacing the press in a newer unibody manufacturing process called ultralight steel auto body (ULSAB). Hydroforming improves body metal because pressed or stamped parts are not of uniform thickness. The corners and edges of a stamped part are thinner than the rest of the part. This means that parts must sometimes be constructed of a thicker gauge of steel to ensure that the edges and corners of the metal part are of sufficient thickness following press forming.

Sheet metal is not used in the hydroforming process. Instead, thin steel tubes are shaped in a die. High-pressure fluid pumped into the steel tube forces it to conform to the inside of the die, resulting in a steel part with equal thickness throughout.

Other Frame Types

Many different frame designs have been used in sport and exotic vehicles that are not mass-produced in large quantities. A tubular **space frame** looks like a birdcage made up of many round and square tubes that are welded together. This frame was first used in the 1950s on the Mercedes 300SL Gullwing. Round tubes are used when possible because they are stronger. Square tubes are used where body panels can be better mounted to the frame. To achieve maximum strength, there is usually a high frame section under the doors that makes access to the vehicle more difficult (**Figure 14.9**). This is why the 300SL had "gull wings" to make access to the interior easier.

Figure 14.9 A tubular space frame usually has a higher frame section under the doors as shown in this gull-wing Mercedes-Benz SLR McLaren. *(Courtesy of Daimler Chrysler)*

Vintage Chassis

An important early change in aerodynamics came with the fastback design of the 1960s. Engineers discovered that lowering the slope of the rear roof line to under 20 degrees resulted in a steep drop in resistance to airflow.

The shape of a vehicle influences fuel efficiency and handling. A drop of water has the most efficient aerodynamic shape—0.05 CD. CD is the co-efficient of drag. Newer aerodynamic cars have a CD of about 0.30. The amount of aerodynamic drag on a car goes up by four times as vehicle speed doubles. For racing purposes four times the engine power would also be required to increase top speed by twice as much. If a racer wanted to raise a car's top speed from 200 mph to 220 mph without changing the race car's shape, horsepower would have to increase by 10 percent. Engineers do research in wind tunnels to produce vehicles that have higher top speeds and improved fuel economy.

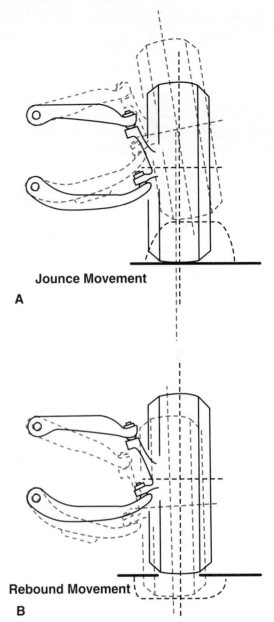

Figure 14.10 (A) Upward motion is called jounce; (B) downward motion is called rebound.

Variations of the space frame design have been used in mass production vehicles like the Corvette, the Camaro, and some minivans. While they do not look like birdcage tubed construction, their monocoque bodies provide very good crash protection, even without the fastened-in-place body panels of earlier vehicles.

An **aluminum space frame (ASF)** chassis developed by Alcoa Aluminum is being used in mass production of the Audi A8. It is said to be 40 percent stiffer and 40 percent lighter than steel monocoque. Extruded aluminum frame sections are connected by die-cast aluminum nodes. This chassis is expensive.

The **backbone chassis** is found in some light sports cars with low production numbers. Originally developed by Lotus, this chassis uses a long tube backbone to connect the front and rear axles.

The **carbon fiber monocoque** racing chassis used on some exotic cars, like Porsche and Ferrari, comes from the aircraft industry. This design uses different carbon fibers, with DuPont's Kevlar being the strongest and best known. The carbon fiber monocoque is a very expensive chassis to manufacture.

■ SPRINGS

Springs support the load of the car, absorbing the up-and-down motion of the wheels that would normally be transmitted directly to the frame and the car body. Up or down motion is referred to as **bounce**. When the wheel moves up as the spring compresses, this is called **compression** or **jounce** (**Figure 14.10A**). When the wheel moves back down, it is called **rebound** (**Figure 14.10B**).

A tire can absorb some of the shock caused when it hits a bump in the road, but the tire passes over the bump so fast that a spring is needed to absorb the rest of the shock. There are four types of springs used in vehicles: the coil spring, the torsion bar, the leaf spring, and the air spring.

Coil Springs

The **coil spring** is the most common type of spring used in the front and rear of passenger cars (**Figure 14.11**). It is constructed of spring steel rod wound into a coil. The coil spring is dependable and relatively inexpensive. It can carry a heavy load but is relatively light in weight, especially when compared to leaf springs.

Coil springs are made from an alloy of different types of steel, usually mixed with silicon or chromium.

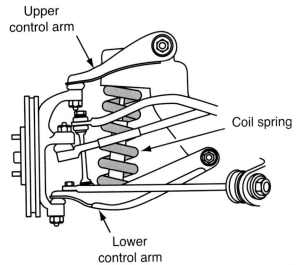

Figure 14.11 Typical front suspension with a coil spring.

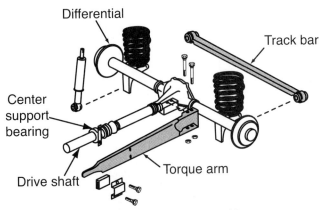

Figure 14.13 A rear suspension with a torque arm, also called a traction bar.

An alloy is a metal mixed with other metals. Springs are *tempered,* which hardens the steel alloy. Tempering is a very precise process in which the spring is heated to a specific temperature and then cooled at a precise rate. If the spring is cooled too slowly, it will anneal, or become soft. Cooling the spring too quickly will make it brittle.

Coil springs are painted or coated with vinyl or epoxy to reduce noise and prevent rust or nicks that could raise stress in the spring and result in breakage.

When a coil spring is used with a rigid rear axle, the axle wants to move out of its correct position so upper and lower control arms are used to control fore and aft movement (**Figure 14.12**). A reinforcement bar, or track bar, is sometimes needed to keep the coils in a stable side-to-side position. The track bar is also called a transverse link, a Watts rod, or a Panhard rod. This is shown in **Figure 14.13**, along with a *torque arm,* sometimes called a *traction bar.* The one shown in this figure is a coil spring rear suspension with a single arm attached between the differential and the drive shaft center support bearing.

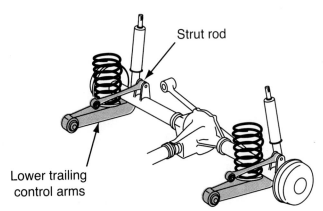

Figure 14.12 A coil spring used in a rear suspension with the axle held in alignment by a strut rod.

Vintage Chassis

Traction bars were common aftermarket additions to the muscle cars of the 1960s. They were installed on both sides of high-powered vehicles to keep the differential from wrapping itself in a circle when heavy torque was applied.

Spring Rate

Spring rate is the load required to deflect the spring 1 inch. If 200 pounds can deflect a spring 1 inch, 400 pounds will deflect it 2 inches. The resulting spring rate would be 200 pounds per inch. This constant rate of change is called *linear rate*. The linear rate would continue until the spring ran out of travel, bottoming out.

Wheel Rate

Wheel rate is a term used to describe the spring's effect at the wheel rather than at the base of the spring. Sometimes a spring's load is different than the spring rate because the leverage of a control arm compresses the spring. If you study **Figure 14.12**, which is a type of suspension covered later in this chapter, you will see that the bottom of the coil spring is mounted on the lower control arm about halfway toward the wheel from its inner pivot point. This results in approximately a 2 to 1 leverage ratio, which causes two changes in relation to the spring:

1. If the tire moves 2 inches, the spring will only deflect 1 inch.
2. If the load on the tire was 1,000 pounds, the spring would have to be able to support 2,000 pounds.

MacPherson strut suspensions (covered later in this chapter) have a more direct relationship between spring rate and wheel rate. If you compare MacPherson strut springs and SLA springs, you find that MacPherson strut springs are longer but narrower (**Figure 14.14**).

Figure 14.14 A coil spring for a MacPherson strut is longer than a conventional coil.

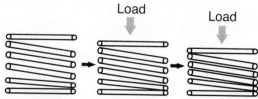

Figure 14.15 One type of coil spring is made from a tapered rod that is thinner at the ends.

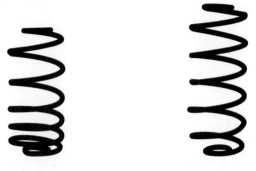

Figure 14.16 The most common variable-rate coil springs have a consistent diameter and unequally spaced coils.

NOTE: *Tire track width also affects the wheel ratio. When aftermarket offset wheels that increase the distance between the tires are installed, the spring rate is reduced and the vehicle height drops. This presents a problem for engineers because it changes wheel alignment geometry.*

Springs are also rated by the frequency at which they oscillate. Softer springs with low spring rates bounce at a slower frequency than stiffer springs.

The *natural frequency* describes how fast a car body would bounce if shock absorbers were not installed to prevent this. This is typically about one cycle per second for a passenger car and twice as fast for a race car. Tires have a natural frequency of about ten cycles per second. This is how fast the car would bounce if it did not have a suspension system. Manufacturers tune their suspension systems so that the front suspension system has a frequency that is about 80 percent of the rear. This is to prevent the vehicle from pitching forward and backward as the front springs react to a bump, followed by the rear springs hitting the same bump. If the front and rear had the same natural frequency, the vehicle might react to the equally spaced expansion joints on the freeway with the front end in jounce and the rear in rebound, and vice versa. Differing the natural frequencies of the front and rear suspensions prevents this pitching from becoming an annoyance.

Linear-Rate Coil Spring. Linear-rate coil springs are constructed from a single thickness of wire, with equal spacing between the coils. They have a constant spring rate, regardless of the load. Heavier duty springs have a thicker wire diameter and can carry higher loads. Linear-rate coil springs are used for regular and heavy-duty applications and in sport suspension packages. Heavy-duty springs are installed in vehicles that are designed to tow trailers and perform other similar tasks.

Variable-Rate Coil Spring. More load-carrying capacity in a vehicle results in a harsher ride, especially when the vehicle is not carrying an extra load. Variable-rate springs automatically maintain ride quality as the load varies, with an increased load-carrying capacity comparable to heavy-duty springs with the same maximum load rating.

There are different varieties of variable-rate springs. One type of spring is made from a tapered rod that is thinner in diameter at its ends. The ends of the spring do not carry as much load as the center coils. When the spring is being compressed, its load-carrying capacity increases and it becomes stiffer (**Figure 14.15**). The result is a smoother ride when going over smaller bumps, while still allowing for a heavier carrying capacity. This type of variable rate spring is not as widely used because it is more costly to produce.

The most commonly used variable-rate spring has a consistent diameter and unequally spaced coils wound in a cylindrical shape. The coils are more closely spaced on one end than on the other (**Figure 14.16**). The more closely spaced coils at one end of the spring do not function until the spring is compressed sufficiently at the other end and in the middle to cause them to produce force. The spring's *active* coils work throughout the complete range of spring compression. The spring's *transitional* coils (more widely spaced) progressively bottom out, becoming inactive once they are compressed to their maximum capacity.

Torsion Bar

A **torsion bar** is a straight rod that twists when working as a spring (**Figure 14.17**). When the wheel moves up during jounce, the torsion bar twists in one

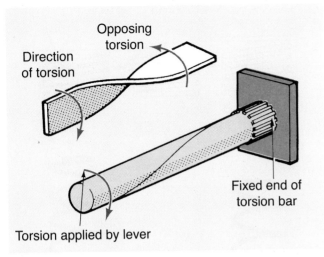

Figure 14.17 A torsion bar is a straight rod that twists when working as a spring.

direction. When the wheel rebounds, the torsion bar unwinds.

Torsion bars are made of heat-treated alloy steel with a hex head or splines at each end. One end of the torsion bar fits into a mating surface at the frame (**Figure 14.18**). The other end attaches to the moveable lower control arm of the vehicle's suspension system.

A torsion bar can be mounted in the chassis to run either front to rear or side to side. Light-duty trucks and SUVs use longitudinal torsion bars. Transversely mounted torsion bars were used on some older cars. Almost all torsion bar installations are on the front.

Some vehicles use torsion bars because they do not require much vertical space and the car can be designed to be lower on the front. Compared to coil and leaf springs, torsion bars can store a higher amount of maximum energy. A shorter, thicker torsion bar can carry more load than a longer, thinner one.

Spring tension can be adjusted by turning a screw against a bracket mounted on one of the torsion bar ends. This allows restoration of the vehicle's correct ride height prior to an alignment.

Leaf Spring

Most **leaf springs** are mounted at a right angle to the axle (**Figure 14.19**). They are very resistant to lateral movement, so control arms or struts are not needed. A leaf spring is made of a long, flat strip of spring steel rolled at both ends to accept a press-fit rubber insulating bushing. The front end of the spring is attached directly to the frame and the rear of the spring is connected to the frame with a spring **shackle** (**Figure 14.20**). The spring shortens and lengthens as it compresses and rebounds and the shackle compensates for these changes.

As a leaf spring is deflected, it becomes progressively stiffer. To provide a variable spring rate, extra springs, called *leaves*, of varying lengths are added to the *master* leaf (**Figure 14.21**). Each leaf is curved more than the one above it. Only the master leaf is rolled at the ends. A **center bolt** extends through a hole in the center of all of the leaves to maintain their position in the spring. In addition to the center bolt, metal clips keep the leaves centered. The main leaf is the strongest leaf in the spring pack. The rest of the leaves are progressively shorter as they are positioned further from the main leaf.

Leaf springs are curved, or arched, during manufacture. **Semi-elliptical springs** are so-named because if the curve of the spring were doubled it

Figure 14.19 A leaf spring suspension.

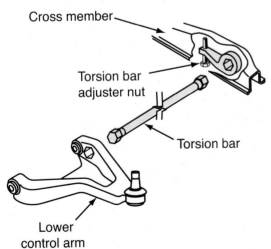

Figure 14.18 A typical torsion bar suspension.

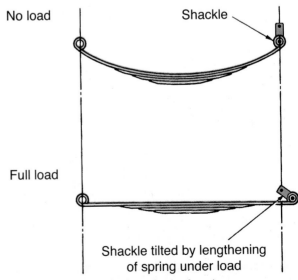

Figure 14.20 Action of a leaf spring.

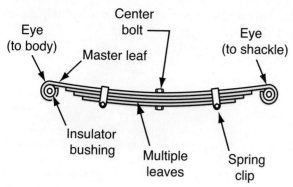

Figure 14.21 Leaves of shorter lengths are added to the master leaf to make a variable spring rate.

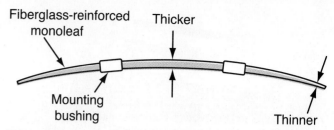

Figure 14.23 A transversely mounted fiberglass composite spring.

would form an ellipse. The name describes the amount of the ellipse the spring is curved to. For instance, there are also springs that are *quarter elliptical.*

When a multiple leaf spring deflects, the lengths of the leaves change. Because each leaf slides on the one next to it, friction and noise can result. The friction between the leaves dampens spring oscillations, but it also causes a rougher ride than a coil spring. Passenger cars have insulating strips made of plastic and zinc or synthetic rubber between the leaves. These strips provide a softer ride and minimize squeaks.

Some trucks have one or more **overload leaves** that do not work until the other leaves have deflected enough under load to allow them come into contact (**Figure 14.22**).

A **monoleaf spring** is a single strip of steel that is thicker in the center and gradually tapers toward the thinner outside ends. Tapering a leaf gives it a variable spring rate, providing for a better ride. A monoleaf spring also has no problem with friction and noise. Monoleaf springs can be mounted longitudinally or transversely and can be used in front or rear suspensions.

Some vehicles made since the early 1980s have **composite springs** made of reinforced fiberglass or graphite-reinforced plastic. Corvettes, for instance, use composite leaf springs mounted transversely in the rear (**Figure 14.23**). Weight is about 30 pounds less per spring, and the springs do not rust. They are also easily manufactured to be tapered across their length. A drawback to these springs is that they are expensive.

Air Springs

An air spring has a rubber air chamber attached by tubing to an air compressor. Air springs are found on suspension systems that control ride height. Some systems can control spring rate as well.

Some vehicles use only air springs. Other vehicles use coil springs to support the weight of the vehicle. An auxiliary air spring enclosed within the coil spring adjusts ride height. Some light trucks and SUVs have air suspension systems used in conjunction with the regular leaf springs that act as adjustable overload springs. The different air spring designs are covered in more detail later in this chapter.

SUSPENSION CONSTRUCTION

There are different suspension designs. Some parts are common to different suspension types, while others are unique to one suspension design. Different suspension designs and parts are covered in the next section.

Independent and Solid Axle Suspensions

A **rigid axle**, called a *straight axle* or *solid axle,* is found on the rear of most RWD vehicles, some FWD vehicles, and some heavy truck front ends. On truck front ends, a rigid axle is called an *I beam* (**Figure 14.24**). I-beam axles are very strong and can support a great deal of

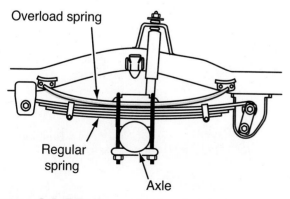

Figure 14.22 An overload spring does not work until the other leaves have deflected enough under load to contact it.

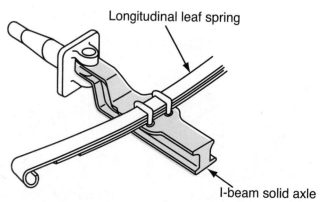

Figure 14.24 An I-beam axle.

SECTION 4 SUSPENSION AND STEERING

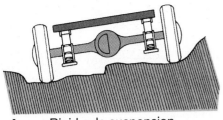

A Rigid axle suspension

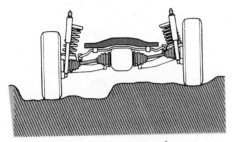

B Independent suspension

Figure 14.25 (A) When a rigid axle goes over a bump, the other wheel is affected as well. (B) When an independent suspension goes over a bump, only one wheel deflects.

weight. When a wheel attached to a rigid axle goes over a bump, the wheel on the other side is affected by that movement as well (**Figure 14.25A**). Because the rigid axles on I-beam front ends are heavy, they increase the vehicle's unsprung weight, which results in a rougher ride.

Independent suspensions are found on most passenger car front ends and some rear ends. When a wheel on a car with independent suspension goes over a bump, only that wheel will move up and down (**Figure 14.25B**). Independent suspension systems have less unsprung weight than rigid axle suspensions, which provides improved ride quality as well.

Control Arms

Control arms are used on independent suspensions to allow the springs to deflect (move up or down). When they are A-shaped like the ones shown in **Figure 14.26**, they are called *A-arms* or **wishbones**.

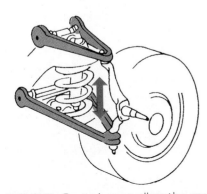

Figure 14.26 Control arms allow the springs to deflect. *(Courtesy of Federal-Mogul Corporation)*

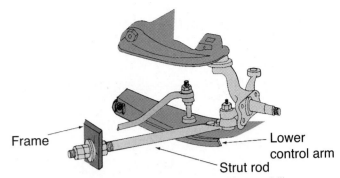

Figure 14.27 The strut rod provides stability to a narrow control arm. *(Courtesy of Federal-Mogul Corporation)*

Another type of control arm commonly found on lighter cars has a single bushing, with a *strut rod* (also called a *radius rod*) for stability (**Figure 14.27**).

Bushings

Rubber bushings are used to separate many suspension parts. The bushing has an outer and an inner metal shell (**Figure 14.28**). The outer part of the bushing is pressed into the control arm and the inner part fits against a pivot shaft (**Figure 14.29**).

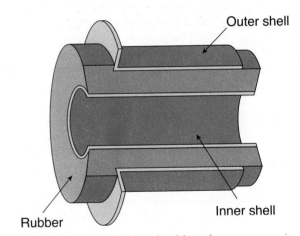

Figure 14.28 Rubber bushings keep suspension parts separated.

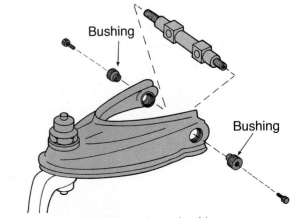

Figure 14.29 Control arm bushings.

Chapter 14 Suspension Fundamentals 327

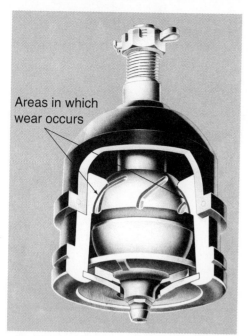

Figure 14.30 A ball joint cutaway. *(Courtesy of Federal-Mogul Corporation)*

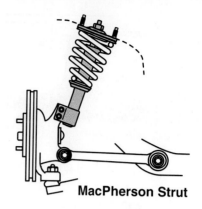

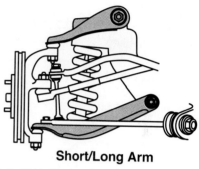

Figure 14.31 Two popular suspension designs.

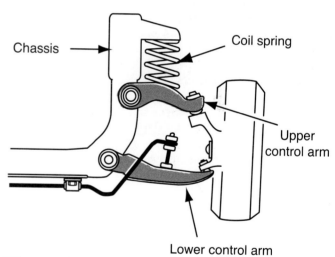

Figure 14.32 An SLA suspension with the coil spring above the upper control arm.

During compression (jounce) or rebound, the control arm moves up or down, twisting the rubber bushing and allowing parts to pivot without any metal to metal contact, somewhat like a vulcanized motor mount. Some of the suspension's resistance to body roll comes from the resistance of the bushings to twisting.

Ball Joints

Ball joints (**Figure 14.30**) are used to attach the control arm to the spindle. A ball joint allows motion in two directions, moving with the same up-and-down motion as the bushings on the other end of the control arm. A ball joint also allows the spindle to pivot for steering. Types of ball joints are covered later in this chapter.

■ SUSPENSION TYPES

Several front suspension designs have been used on vehicles. Two of the most popular ones are the **short/long arm (SLA) suspension** and the MacPherson strut (**Figure 14.31**). Rear suspensions on cars are sometimes independent, but most SUVs have rigid axles. Some are used with leaf springs, but most have coil springs.

Some SLA suspensions on unibody cars have the coil spring above the upper control arm (**Figure 14.32**). On vehicles with frames, the spring is situated below the upper control arm (see **Figure 14.26**).

SLA Double Wishbone Suspension

Double wishbone suspensions are used on both front and rear wheels. Because both upper and lower control arms resemble the shape of a wishbone (**Figure 14.33**), this suspension system is called a double wishbone. The SLA double wishbone suspension uses two control arms of unequal length that are not parallel to one another. The shorter one, located on top, slants downward at the outside. This causes the wheel to tilt *inward* at the top when the spring is compressed, which allows the tread width of the front tires to remain constant (**Figure 14.34**). If the control arms were both the same length, the tire would slide from side to side as it goes over bumps (**Figure 14.35**).

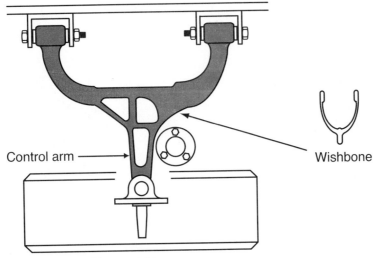

Figure 14.33 The control arm resembles a wishbone.

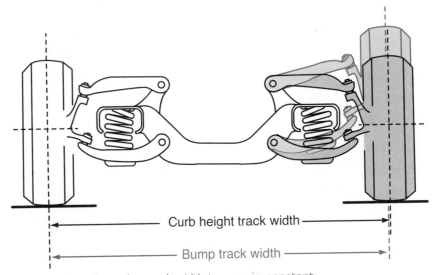

Figure 14.34 SLA suspension allows the track width to remain constant.

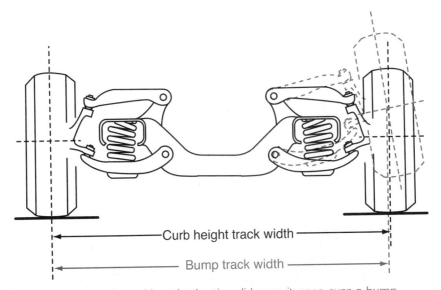

Figure 14.35 With control arms of equal length, the tire slides as it goes over a bump.

> ### Vintage Chassis
>
> Earl Steele MacPherson designed the MacPherson strut suspension during the late 1940s. MacPherson, who was born in Britain, became an engineer, first working for Chevrolet and later with the European division of Ford. The MacPherson strut was first used on the 1949 French Ford Vedette.

Double wishbone suspensions are popular for several reasons:

- They have improved directional stability and steering control.
- They react to body roll less than any other type of independent suspension.
- They maintain precise wheel position under all driving conditions and can have a large amount of vertical travel without a camber change. For instance, the C5 Corvette, which uses a double wishbone suspension, has zero camber change through 95 percent of its vertical travel.
- They are good for absorbing bumps and the tires maintain more surface area on the road.
- The suspension is more rigid because the ball joints and steering knuckle are positioned within the profile of the wheel.

High-performance double wishbone suspensions use upper and lower control arms made of lightweight, high-strength aluminum alloys. This results in less unsprung weight, which improves traction and ride quality. Sometimes a double wishbone suspension is not an option because it requires more space than MacPherson struts or some multilink designs.

MacPherson Strut Suspensions

Many smaller cars use the **MacPherson strut** design. The SLA suspension is a better suspension, but weight, space-saving, and cost considerations account for the MacPherson strut's popularity.

A MacPherson strut incorporates the coil spring and shock absorber into its front suspension (**Figure 14.36**), using only a single control arm on the bottom. It does not have an upper control arm and upper ball joint. The spindle is part of the strut casting and a pivot bearing at the top allows the entire unit to rotate when steering. Shock absorbers are covered later in this chapter.

Modified Strut. Some older vehicles used a modified MacPherson strut without coil springs mounted on them (**Figure 14.37**). The lower control arm is the same as on an SLA suspension. Conventional coil

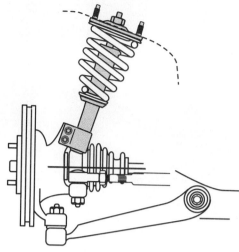

Figure 14.36 A MacPherson strut assembly.

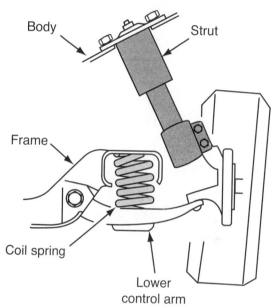

Figure 14.37 A modified MacPherson strut suspension.

springs are positioned between the lower control arms and the frame.

High-Performance Suspensions

Several suspension types have been used in sports and racing vehicles. The multilink and the double wishbone are the most popular ones.

Multilink Suspensions. When an independent suspension has more than two control arms, it is considered a multilink suspension. Extra links are used to keep the wheel in a more precise position during cornering and on bumps. Steering control is improved and tire wear is minimized. A **multilink suspension** is basically a double wishbone suspension that has each arm of the wishbone as a separate part. **Figure 14.38** shows a simple multilink rear suspension. Other

Figure 14.38 A simple multilink suspension. *(Courtesy of International Iron and Steel Institute)*

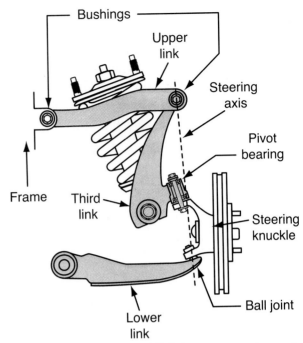

Figure 14.40 One multilink front suspension system, sometimes called a long knuckle SLA.

the front cross member of the frame. The steering pivot axis runs through the lower ball joint and the bottom of the pivot bearing and is independent of the upper and third links.

The shock absorber fits between the bottom of the third link and the unibody fender reinforcement (**Figure 14.41**). The shock, which has a coil spring attached to it, resembles a MacPherson strut. But the

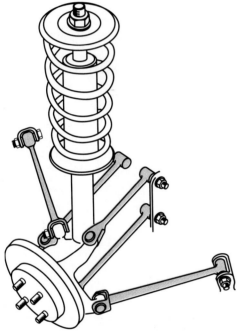

Figure 14.39 A more complicated five-link suspension.

designs, like Honda's five-link suspension (**Figure 14.39**), are more complicated. This suspension is a double wishbone with a fifth control arm.

Multilink suspensions give the designer more options in tuning the suspension. Several variations of these front suspensions have been used on different luxury and sport vehicles since the late 1980s. One variation of the SLA suspension, the multilink shown in **Figure 14.40**, is sometimes called a long knuckle SLA. An upper link attaches to a bracket on the frame on one end, and another link attaches to the steering knuckle and spindle by a pivot bearing rather than a ball joint. Rubber insulating bushings are installed in each end of the upper link.

The lower link is like a conventional lower control arm with rubber bushings connecting its inner end to

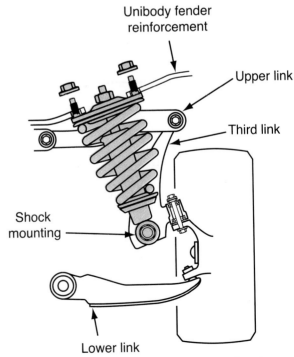

Figure 14.41 The shock absorber and spring resemble a MacPherson strut.

coil spring and shock do not rotate with the wheel as a MacPherson strut does.

■ SHOCK ABSORBERS

Cars have four **shock absorbers**, commonly called shocks or dampers, one at each corner of the vehicle. Shock absorbers do not actually absorb shock; their function is to dampen spring oscillations by converting the energy from spring movement into heat energy. Normal motion of the springs is uncomfortable to passengers, causes cupped tire wear, and is unsafe. This motion can be controlled using shock absorbers.

Shock absorbers are designed to resist, or damp out, excess and unwanted motion in the suspension. When a spring is compressed, it absorbs energy. During spring rebound, this energy is released. The first cycle is followed by more compression-rebound cycles called *oscillations* (**Figure 14.42A**). Shock absorbers convert the kinetic energy of this motion into heat energy as they damp out the excess motion. Without this damping, the vehicle would bound about in an unsafe and uncomfortable manner. In fact, poor shock absorbers are known to aggravate situations that lead to potential rollovers, especially with top-heavy vehicles like SUVs and vans.

Shock absorbers are also critical for proper tire-to-road contact (**Figure 14.42B**). This affects braking, steering, cornering, and overall stability. The effectiveness of antilock brakes is also determined in part by the good tire-to-road contact provided by shock absorbers.

NOTE: *Although shock absorbers absorb much of the road shock, the vehicle's tires are actually the primary shock absorbers.*

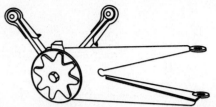

Figure 14.43 An early friction shock absorber.

Vintage Chassis

Early shock absorbers were mechanical friction devices. Friction material similar to that found on a clutch or brake was fitted between two levers. One of the levers was attached to the frame and the other to the spring mount or spring (**Figure 14.43**). They required frequent adjustment because they were prone to wear.

Hydraulic Shock Absorber Operation

In a modern *direct-acting* hydraulic shock absorber, one end is attached to the suspension and the other is attached to the car body or frame (**Figure 14.44**). The shock is mounted on rubber bushings to allow for slight changes in its angle of installation as it is compressed and extended.

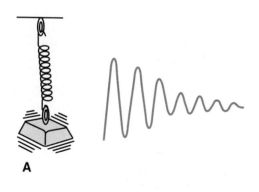
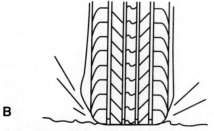

Figure 14.42 (A) When a compressed spring rebounds, it begins to oscillate. (B) Under load, the tire forms a mechanical interlock with the road surface.

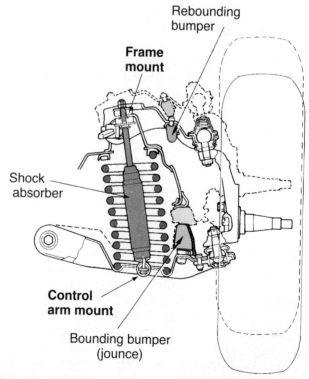

Figure 14.44 One end of the shock absorber is attached to the control arm, and the other end is attached to the frame.

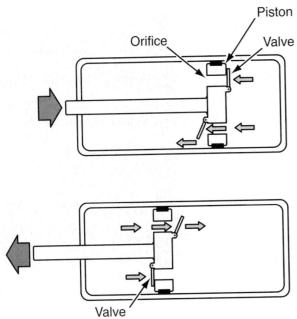

Figure 14.45 Fluid is forced through an orifice to dampen spring action.

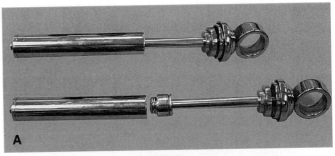

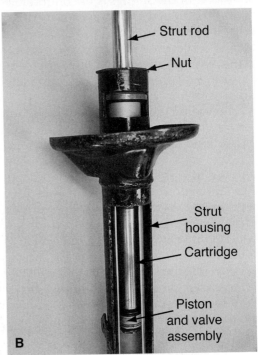

Figure 14.46 (A) Parts of a conventional shock absorber. (B) Cutaway of a MacPherson strut shock absorber. *(Courtesy of Tim Gilles)*

To damp the action of a spring, a hydraulic shock absorber forces oil through small openings, called valves, like water through a squirt gun (**Figure 14.45**). This action generates hydraulic friction, which converts kinetic energy to heat energy as it reduces unwanted motion. Using energy in this fashion lessens the oscillations of the springs.

NOTE: *The shock only converts unwanted motion to heat. If it were to remove all of a spring's energy, all motion would stop.*

A shock absorber has two chambers with a piston that forces fluid through the valves from one chamber to the other. The faster the piston moves, the more resistance it encounters. **Figure 14.46A** shows the parts of a conventional shock absorber. **Figure 14.46B** shows a cutaway MacPherson strut shock absorber.

Today's shock absorbers are either twin-tube or monotube (single-tube) designs (**Figure 14.47A**). Monotube shocks are not as common as the twin-tube design but are still found in many applications. As a monotube shock is compressed, the piston rod displaces oil. There must be a reserve space to accept the extra oil (**Figure 14.47B**). In twin-tube designs, the outer tube provides the reserve chamber. A monotube shock usually has the reserve chamber in the same column as the working piston. When the shock extends, oil returns once again to the pressure chamber.

Front and rear shocks are not the same. Usually they have chambers and valves of different sizes to control the differences in weight between the front and rear of the car. Also, the center of gravity shifts when the vehicle stops or is thrown into a turn, causing the front shocks to do more of the work.

The up-and-down movements of a shock are called compression (or jounce) and extension (or rebound). Shock absorbers are considered *double-acting* because they control motion when moving both up and down. Early friction shocks had a 50:50 ratio, controlling the motion of the springs equally on compression and rebound. A typical modern shock provides different resistance on compression and rebound (70:30, for instance).

Compression and Rebound Resistance. Unlike springs, which are sensitive to loads, hydraulic shocks are sensitive to velocity. The faster the piston moves through the oil in the shock, the higher the resistance.

Shocks must deal with a wide variety of suspension motions and velocities. Resistance can be very high, so a series of orifices and valves are necessary to properly manage fluid flow. Either too little or too much resistance can result in poor adhesion to the road. Too much resistance can cause a harsh ride. Too little resistance allows excess body motion, poor control of unsprung weight, and wheel bounce. A defective

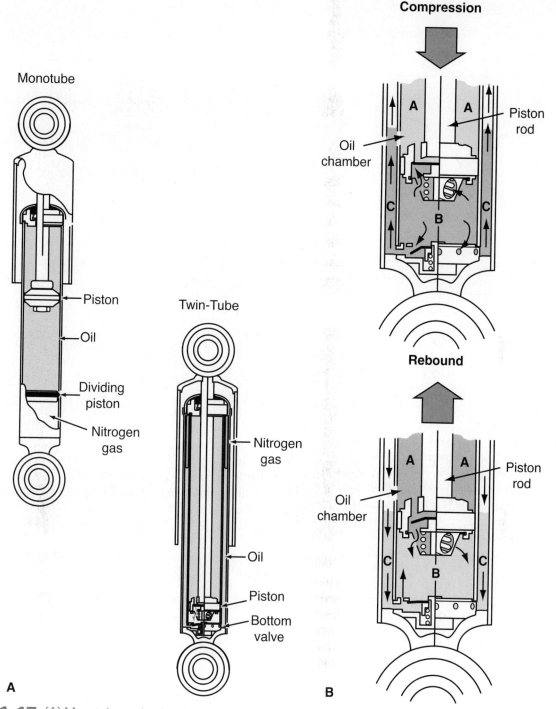

Figure 14.47 (A) Monotube and twin-tube shock absorbers. (B) Shock action during the compression and rebound cycles.

shock will have too little resistance. A graph of resistances provided at various piston velocities is called a damping force curve (**Figure 14.48**).

NOTE: *Most modern shocks compress more easily than they extend.*

Compression damping is used to control the relatively light unsprung weight of the tires, wheels, and brakes. It works with the spring to keep the tire in contact with the road surface. A car with a wheel that hops off the pavement has a compression control problem with its shock absorber.

Rebound damping controls excess chassis motion as the shock extends when the heavier weight of the car body is in motion. A car that floats as it travels down the road has a rebound control problem with its shock absorber.

Because piston velocities on the rebound side (body motion) tend to be much lower than on the compression side (for example, while hitting a chuckhole),

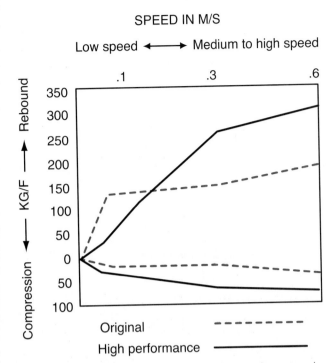

Figure 14.48 A dampening force graph comparing two different shocks for the rear of a pickup truck. Notice how rebound dampening increases dramatically at medium and high fluid speed.

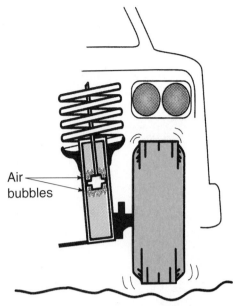

Figure 14.49 Air bubbles form around the valves during bumpy rides. *(Courtesy of Tokico)*

compression damping force is usually much lower than rebound resistance. Fluid movement changes depending on whether the car hits a hole (fast fluid movement) or whether the car leans over (called body roll) as it goes through a turn (slow fluid movement).

Virtually all shock absorbers use three stages of valving. The initial valve opening (first damping stage) is a small hole because little force is generated by low piston velocities. As the piston moves faster, an additional (second stage) valve opens at a level of pressure determined by the manufacturer. The final stage of valving (third stage) is called the high-speed restriction.

Bump Stops and Limiters. Shock absorber movement follows the travel of the rest of the vehicle's suspension. If the shocks or struts reach their extreme travel limit by topping or bottoming, damage can result. Some vehicles use rubber bump stops (sometimes called snubbers) on control arms to limit excessive travel. Internal bump stops are generally found only in monotube struts or strut cartridges.

Normal full shock absorber travel is considered to be a good thing, but altered vehicle height is a typical cause of reduced total travel that allows a shock to top out or bottom out. Raising or lowering a vehicle excessively in either direction can force shocks and struts to their extreme travel points. This can result in damage to the shock, as well as to other suspension parts.

Aeration of Fluid. Aeration, or cavitation, occurs when hydraulic fluid becomes mixed with air. A shock must be installed in a nearly vertical position so that when it extends air will not be drawn in place of fluid from the outer reservoir. This would result in a "skip," as the shock moves through its range of motion.

As the shock piston moves, pressure builds up in the fluid in front of it. A drop in pressure happens behind the piston, causing air bubbles and making the fluid foamy. When driving on rough roads fluid is rapidly forced through the check valves. The entire area of fluid around the piston becomes aerated and shock operation suffers (**Figure 14.49**). During normal operation, the air will usually work its way back to the air chamber.

There are two ways that designers avoid the tendency toward aeration. One is to use spiral grooves or a flat spiral ring around the outside of the reservoir tube. Another way to avoid aeration of the shock fluid is to use gas shocks.

Gas Shocks

Gas shocks were invented to control cavitation or foaming of the oil. All oil has some air or gas bound up in solution. Pressurizing the oil keeps the air in solution so the piston works in clear oil and can provide consistent damping.

Pressurizing the oil column in the shock absorber keeps the bubbles in the solution. Compare this to what happens when a carbonated beverage is first opened. When the top is removed, the pressure drops and the bubbles come out of the soft drink. Pressurizing the shock keeps the shock oil clear and eliminates the skips that occur in the action of a regular shock absorber.

Some gas shocks have a pressurized gas-filled cell. This is a plastic bag that takes the place of the free air in a normal shock. These bags usually have low

Vintage Chassis

In 1953, French physicist Christian Bourcier de Carbon designed the monotube high-pressure gas shock absorber and founded the DeCarbon company the same year. A short time later, a license was sold to the German company Bilstein. The shocks became original equipment on the 1957 Mercedes Benz. The patent has expired, and now many top companies use monotube technology.

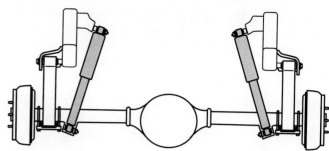

Figure 14.51 Shock absorbers mounted to the rear of the axle.

pressure (10–20 psi of refrigerant gas) in them. True gas shocks have a reservoir pressurized with nitrogen gas at 100–200 psi (**Figure 14.50**). Nitrogen is used because it is dry (no water vapor), chemically inert at normal shock temperatures, and fairly inexpensive.

Twin-tube gas shocks are pressurized with nitrogen gas at 90 to 140 psi, depending on the manufacturer and the damping force desired. They are usually less expensive to manufacture than monotube shocks.

Monotube gas shocks have pressures between 250 psi and 400 psi. Since the monotube has no base or bottom valve, it needs the high pressure to support the compression valving, as well as to control cavitation.

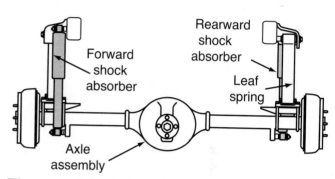

Figure 14.52 Shock absorbers staggered on each side of the rear axle.

NOTE: *Gas shocks are packaged with a strap to hold them in the compressed position. They extend on their own when the strap is removed for installation.*

Monotube gas shock absorbers, also called DeCarbon shocks after their inventor, offer better cooling of the shock fluid and can be mounted upside down. Monotube shocks are more susceptible to physical damage than twin-tube shocks.

Rear Shock Mounts

To minimize vibration and improve ride quality, rear shock absorbers on rear-wheel drive vehicles are mounted in one of two ways. Some of them have the shocks slanted inward and toward the rear at the top (**Figure 14.51**). Others have one shock mounted in front of the axle and the other one mounted behind it (**Figure 14.52**).

■ AIR SHOCKS/LEVELING DEVICES

Shock absorbers are not normally designed to carry the weight of the vehicle. If they were, the height of the vehicle would be affected when they were removed or when they wore out.

There are aftermarket devices that use the shock absorber as a way to correct or adjust the height of the vehicle. The two common ones are air shocks and coil springs that are mounted on the outside of the shock body. There is a disadvantage to leveling the vehicle in either of these ways. When shocks support the weight

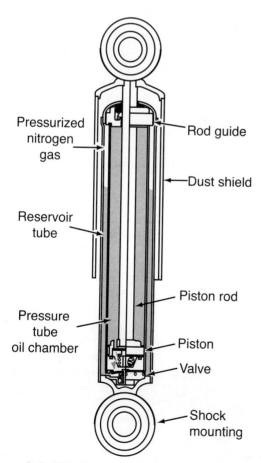

Figure 14.50 Parts of a pressurized gas shock.

of the vehicle, the shocks and shock mounts are prone to breakage. Air shocks (**Figure 14.53**) have a rubber bladder that, when filled with varying amounts of air, raises the rear of the vehicle to compensate for a heavier than normal load.

NOTE: *A small amount of air must be kept in the bladder at all times. If the vehicle is driven while the bladder is empty, the bladder can be folded into the shock and tear.*

A *coil spring shock* (**Figure 14.54**) has a constant rate spring that works the same way during both extension and compression. Its action in preventing body roll is similar to that of a stabilizer bar (covered later in this chapter).

■ OTHER FRONT END PARTS

Other parts are attached to the suspension to help control the ride. Parts like stabilizers and strut rods are insulated from front suspension parts and the frame with rubber bushings (**Figure 14.55**).

Stabilizer Bar

A stabilizer bar, also called a *sway bar* or an *anti-roll bar*, is used on the front or rear of many suspensions (**Figure 14.56**). It connects the lower control arms on both sides of the vehicle together, reducing sway and functioning as a spring when the car leans to one side.

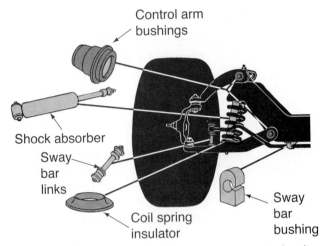

Figure 14.55 Parts are insulated from each other with rubber bushings. *(Courtesy of Federal-Mogul Corporation)*

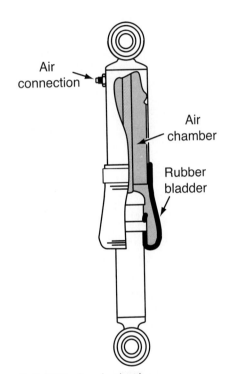

Figure 14.53 An air shock.

Figure 14.54 A coil spring shock.

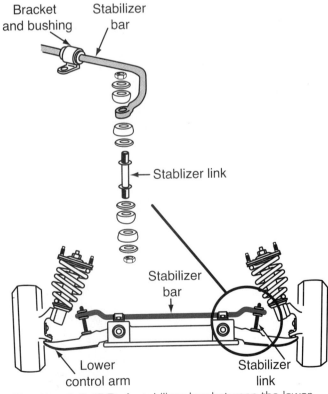

Figure 14.56 A stabilizer bar between the lower control arms reduces sway.

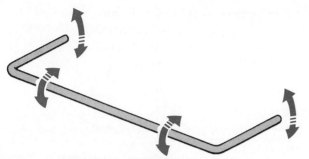

Figure 14.57 The action of a stabilizer.

When both tires move up or down an equal amount, the stabilizer simply rotates in its bushings (**Figure 14.57**). If one of the wheels moves up, the bar twists as it tries to move the other wheel along with it.

Stabilizer links and bushings provide some flexibility and softness to the sway control so the suspension can still operate somewhat independently during minor bumps.

Spindle and Ball Joints

The *spindle support arm*, also called the *steering knuckle*, includes the axle on which the wheel bearing is mounted. Ball joints (see **Figure 14.30**) attach the control arm to the spindle. Depending on the design, ball joints are located on top of or under the control arm.

On suspensions with two control arms, ball joints function as either **load carrying ball joints** or followers (**Figure 14.58**). The ball joint on the control arm that has the spring mounted on it is the load carrier. The function of the follower ball joint is to maintain parts in the proper position.

There are two styles of ball joints. They are either compressed all the time—*compression type*—or always pulling apart—*tension type* (**Figure 14.59**).

The MacPherson strut suspension system uses only one ball joint because it has only one control arm. It has a pivot bearing at the top of the strut that allows the strut to rotate for steering (**Figure 14.60**). On a strut suspension, the pivot bearing carries the load and the ball joint is a follower.

■ SUSPENSION LEVELING SYSTEMS

Normal suspension systems are called **passive suspension systems**. Passive systems have either a firm or a soft ride. Their height varies according to mechanical forces on the suspension and they do not adjust to these changes. Early leveling systems were manual, using air shocks and a compressor. A manual switch was used to change the height of the car body. Today's systems are automatic and they are called **adaptive suspension systems**.

Electronically controlled suspension systems are used by manufacturers on some of their luxury vehicles. They keep the vehicle at the same height when weight is added to different parts of the car. Some of the advanced systems can vary the damping capability

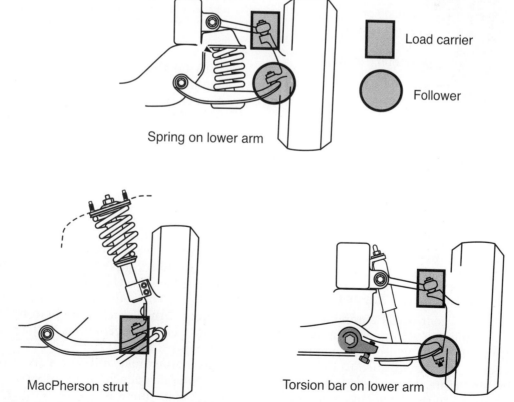

Figure 14.58 The location of load carrier and follower ball joints.

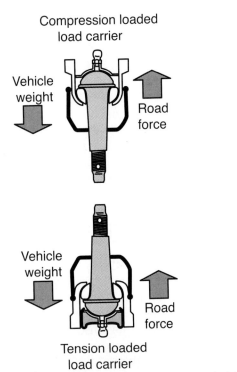

Figure 14.59 Tension and compression loaded ball joints.

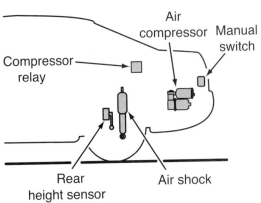

Figure 14.61 An automatic leveling system that uses air shocks.

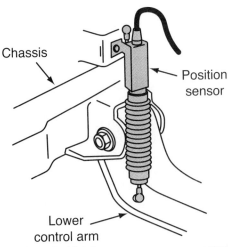

Figure 14.62 Height sensors are connected between the body, and a suspension member sends a signal to the control module if a height change occurs.

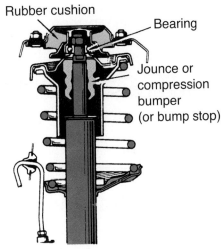

Figure 14.60 A pivot bearing allows the strut to rotate.

of the shock absorbers as well. Most of the systems have air or conventional springs.

Automatic Suspension Leveling

There are two-wheel and four-wheel automatic leveling systems. The simplest leveling systems use air shocks or air springs filled by air from a compressor (**Figure 14.61**). An air dryer is attached to the pump to condition the air before it enters the shocks. A height sensor connected to the frame and axle housing is used for vehicle height input (**Figure 14.62**). It can turn on the compressor or bleed air to correct changes in height.

With electronically controlled systems, a computer reacts to signals from sensors at all four wheels to change the amount of air in air springs at each wheel (**Figure 14.63**). The aim is to keep the vehicle level to the road from side to side and front to rear.

Several sensors are found with different systems. Some of these sensors and their functions include:

- Three or four height sensors located at the wheels. They are often rotary Hall effect sensors. When there are three sensors, one is for the solid rear axle. Height sensors electronically measure the distance between the control arm or axle and the frame. The computer can also use their signal to prevent the vehicle from bottoming out when going over major variations in the road surface, like railroad tracks.
- Signals from brake and door sensors that help the computer decide whether to disable automatic height adjustment when the car is stopping or when passengers are getting in or out.

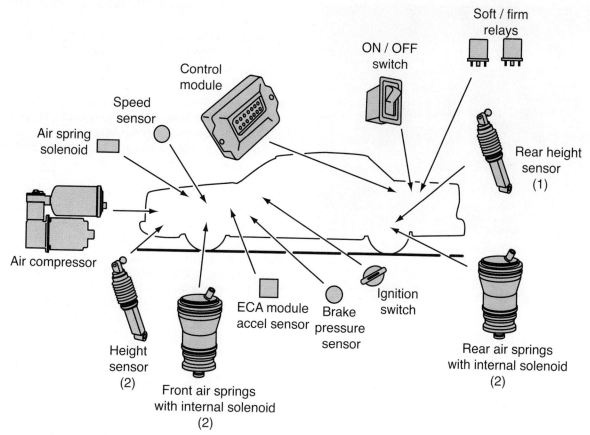

Figure 14.63 A computer controls the air springs at each wheel.

- Speed sensors that are used in some systems to lower the vehicle in the front or in both the front and rear for high-speed aerodynamics. In some systems, spring rate is increased in response to higher speed as well.
- A photo diode and shutter in the steering columns of some advanced systems that cause the spring rate to change when the vehicle is turning.
- A G sensor on some cars that helps the system accommodate severe maneuvers with a change in spring rate.
- An acceleration sensor that senses the rate of acceleration using the throttle position sensor or mercury switches.
- A mode switch installed in the dash of some cars that lets the driver select the degree of ride harshness desired.

Some light trucks and SUVs use air suspension systems in conjunction with the regular leaf springs. The air spring, positioned between the truck frame and the leaf spring, acts as an adjustable and overload spring (**Figure 14.64**).

Electronically Controlled Shock Absorbers

Electronically controlled shock absorbers have variable control. The amount of shock damping changes as the size of a metered orifice within the shock is manipulated by the computer. An actuating motor on the top of the shock (**Figure 14.65**) turns a control rod that changes the size of the orifice (**Figure 14.66**).

The newest adaptive systems use solenoid-actuated shock valves that allow almost instantaneous changes in the size of a shock's damping orifice. The suspension can react to changes in body height in 0.010 second

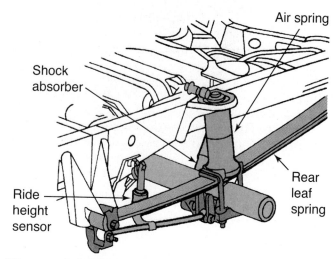

Figure 14.64 An air spring used with a conventional spring.

340 SECTION 4 SUSPENSION AND STEERING

Figure 14.65 An electronically controlled shock absorber.

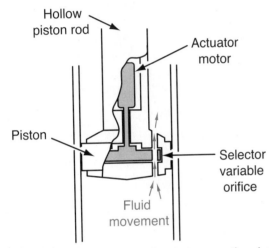

Figure 14.66 The control rod changes the size of the orifice.

(10 milliseconds). These devices can be found on shocks and struts.

Some electronically controlled shocks use a variable orifice controlled by gas or air. Normal shock absorber valves are used. Shock absorbers deflect in one direction to control fluid flow:

- At lower speeds, they operate normally (the valve is fully open). This is the comfort mode and fluid flows unhampered through the orifice. The valves work as normal.
- The orifice is partly restricted between 40 mph and 60 mph. Fluid flow is balanced between the small orifice and the normal shock valve. This provides a medium degree of control.
- At speeds above 60 mph, or during acceleration or braking, the orifice shifts to a firm ride position. Flow through the orifice is restricted to increase the effort required to move oil through the shock absorber.

Magneto-Rheological Fluid Shock Absorbers. Another type of electronically controlled shock absorber that does not use electromechanical solenoids or valves is the magneto-rheological (MR) shock absorber. It uses a fluid that rapidly changes its viscosity in response to computer-controlled signals. The fluid is a synthetic oil with iron particles in suspension. The iron is dispersed throughout the fluid, allowing the shock to operate as a normal shock under ordinary driving conditions. Each shock has an electrical winding that can be energized by the control module in response to a signal from a wheel position sensor when a large bump in the road is encountered. This causes the iron particles in the shock fluid to align themselves (**Figure 14.67**), turning the fluid into a viscous, gooey mass. The current is supplied up to 1,000 times per second, so it can very quickly vary the damping characteristics of the shock from firm to normal.

During hard braking, a vehicle wants to **dive**. The front of the car is pushed down and the rear of the car slides up. During hard acceleration, the front of the vehicle lifts and the rear lowers. This is called **squat**. Advanced active suspensions help control some of the forces normally encountered in driving. They can reduce pitch and body roll as well as help to control squat during acceleration and dive during braking.

Active Suspensions

An **active suspension system** works with sensors, a computer, and activators to solve these problems. Active suspensions were first developed by Lotus in England. A few luxury cars have active suspensions.

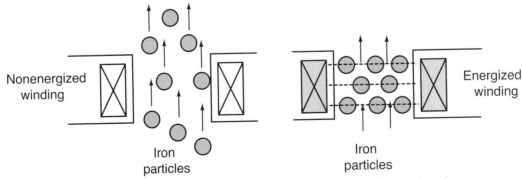

Figure 14.67 Magneto-rheological fluid changes in viscosity to cause a firm shock absorber.

An active suspension does not require conventional shock absorbers or springs. Each wheel has a double-acting hydraulic cylinder (high-speed actuator) to keep the car body level during all driving conditions. Shock absorbers and springs are not necessary on these vehicles because the actuators replace them. They provide a smooth ride and yet have the handling characteristics of a vehicle with a stiff suspension. The suspension can be programmed by the driver as to the type of ride desired.

The active suspension is powered by a hydraulic pump driven by the engine. It requires about the same amount of power as a typical power steering pump during a turn (3–5 horsepower). The high-speed actuators can raise or lower the vehicle in .003 second. As soon as a bump has been absorbed by an actuator, pressure is reestablished to keep the wheel in contact with the road and maintain ride height. During hard stops, an active suspension reduces the tendency for the vehicle to dive or lean. Power consumed by the system is least when the vehicle is riding flat.

The computer uses signals received from sensors to track the position of each actuator. It can sense whether a wheel is in jounce or rebound. It also senses how heavily each wheel is loaded and whether the wheel is turning or pointing straight ahead. The computer sees a bump and immediately releases pressure from a control valve. The valve can release pressure instantly or relatively slowly, depending on what the computer program specifies. After the suspension absorbs the shock, pressure is forced back into the actuator to keep the tire in contact with the road and maintain ride height. Sensors provide the computer with necessary information regarding extension and compression of each actuator and how heavily the vehicle is loaded. Different sensors were covered earlier in this chapter.

NOTE: *In case of a flat tire, the system can be told to raise the tire so a jack is not needed.*

REVIEW QUESTIONS

1. List three items that are part of a vehicle's unsprung weight.
2. What type of body-over-frame design do the vast majority of today's passenger cars use?
3. What is the name of the frame section that includes the engine, transaxle, and steering and suspension systems and that is bolted to the front of a unibody vehicle?
4. Which kind of spring is adjustable?
5. What is the name of the manufacturing process that hardens the steel alloy of a spring by heating it to a specific temperature and then cooling it at a precise rate?
6. If a coil spring is compressed 1 inch with 200 pounds and 2 inches with 400 pounds, this is a _____ rate spring.
7. What is the name of the suspension type that uses two different length control arms?
8. What is the name of the part that is used to control the oscillations of a spring?
9. What is the difference between the amount of shock control on compression and extension called?
10. An original equipment 70:30 shock gives _____ percent of its resistance during extension and _____ percent during compression.
11. What is the term that describes the softness or harshness of a vehicle's ride?
12. What kind of shock can be mounted upside down?
13. If a small amount of air is not kept in an air shock, what can happen?
14. What is the name of the kind of front suspension that has a built-in shock absorber and only one ball joint?
15. What is the name of the ball joint design that is always trying to pull apart?

PRACTICE TEST QUESTIONS

1. When one wheel of a rigid axle vehicle goes into a pothole, the wheel on the other side of the vehicle moves as well.
   A. True
   B. False

2. Which of the following is true about MacPherson strut suspensions?
   A. The ball joint is a follower.
   B. It has only one control arm.
   C. Both A and B
   D. Neither A nor B

3. Technician A says that a typical shock absorber will compress more easily than it extends. Technician B says that some shocks have gas under pressure to prevent cavitation of the fluid. Who is right?
   A. Technician A
   B. Technician B
   C. Both A and B
   D. Neither A nor B

4. Technician A says that shock absorbers are designed to carry the weight of a vehicle. Technician B says shock absorbers absorb shock. Who is right?
   A. Technician A
   B. Technician B
   C. Both A and B
   D. Neither A nor B

5. Technician A says that the MacPherson strut is a better front suspension design than the SLA. Technician B says a MacPherson strut suspension is often called a double wishbone suspension. Who is right?
   A. Technician A
   B. Technician B
   C. Both A and B
   D. Neither A nor B

6. Technician A says that some types of ball joints carry a load and others do not. Technician B says that a load-carrying ball joint is one that has the spring attached to it. Who is right?
   A. Technician A
   B. Technician B
   C. Both A and B
   D. Neither A nor B

7. All of the following are true about unsprung weight except:
   A. Lower unsprung weight is desirable for better handling.
   B. Tires are unsprung weight.
   C. The powertrain is unsprung weight.
   D. Independent suspension systems have less unsprung weight than rigid axle suspensions.

8. The kind of frame design that has a high frame section under both doors is called a
   A. Space frame
   B. Body-over-frame
   C. Unibody
   D. None of the above

9. Technician A says independent suspensions with more than two control arms are called multilink suspensions. Technician B says dive is when the front of the vehicle lifts and the rear lowers during hard acceleration. Who is right?
   A. Technician A
   B. Technician B
   C. Both A and B
   D. Neither A nor B

10. Technician A says shock absorber rebound damping controls the lighter weight of unsprung weight. Technician B says a car that floats as it travels down the road has a shock absorber rebound control problem. Who is right?
    A. Technician A
    B. Technician B
    C. Both A and B
    D. Neither A nor B

CHAPTER 15

Suspension System Service

■ OBJECTIVES

Upon completion of this chapter, you should be able to:
- ✔ Diagnose suspension system problems.
- ✔ Service suspension system components.
- ✔ Describe suspension system repairs.
- ✔ Replace MacPherson strut cartridges.
- ✔ Replace suspension bushings.

■ KEY TERMS

diagnostic trouble code (DTC)
gas-filled shock absorber
pinch bolt
steering knuckle

■ INTRODUCTION

Emphasis in this chapter is on the diagnosis and service of suspension system parts. The reader will become familiar with commonly performed chassis diagnosis and repair procedures. When suspension parts are in good condition and properly aligned, they are subjected to two forces. These forces are the weight of the car on the springs, and the force of the road on the tires (**Figure 15.1**).

Suspension parts take a large amount of abuse as the car is regularly driven over potholes, speed bumps, and concrete expansion joints. Forces are transmitted through the suspension parts, causing wear over time (**Figure 15.2**).

■ DIAGNOSING SUSPENSION SYSTEM PROBLEMS

When diagnosing suspension problems, carefully question a customer about the symptoms. Problems with the suspension system usually come to light when the customer complains of noises such as a clunk or squeak, vibration, tire wear, or a handling problem. All of the

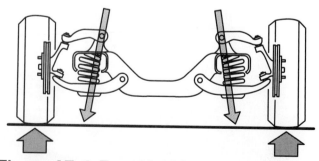

Figure 15.1 The weight of the car and the force of the road act on the suspension system.

Figure 15.2 Locations of wear.

suspension system parts depend on the others for their performance. The rest of the system is affected when one part becomes worn.

Wheel alignment settings can change when a suspension component wears. Check all ball joints and bushings for looseness. Look for wear or cracking on rubber bushings and the sway bar links. Bushing wear can cause suspension parts to change position. This can cause tire wear or steering pull, either all the time or just when braking.

Spring sag can result in tire wear. On most cars, the camber angle will change when the ride height drops (**Figure 15.3**). Ride height measurement is covered in Chapter 19.

■ SHOCK ABSORBER SERVICE

Shock absorbers can be functioning poorly before the appearance of obvious signs of failure, such as tire cupping or excessive front end float. Signs of inadequate

343

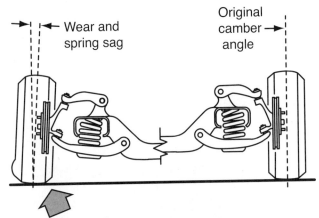

Figure 15.3 The camber angle will go toward negative on most cars when the ride height drops.

shock absorbers include excessive body motion, road wander (especially in wind), poor adhesion of the tire to the surface of rough roads, a harsh ride, suspension bottoming, and poor braking. A bad shock absorber on one corner of the vehicle can cause the entire vehicle to feel strange.

A defective shock absorber can cause a tire to "hop." Feel the tire tread around the total circumference of the tire. When a scalloped or gouged wear pattern develops on a tire (see **Figure 14.5**), look for a bad shock absorber. Tires that are unbalanced or out-of-round can also be a contributing factor.

A shock can be defective because it is leaking or physically damaged. When one damaged shock is found, both shocks on that end of the vehicle (front or rear) are replaced. A leaking shock can result from a defective seal, an improper dust cover, or lowering a vehicle too far. A damaged piston rod on a MacPherson strut can result from holding the rod with a pliers during installation of the top nut. This can cause a leak as well.

Some shocks still work well after 50,000 miles. Others may be weak after 4,000 miles. Original equipment shock absorbers are designed to give a satisfactory (soft) ride when used with tight, new suspension bushings. By the time a vehicle has accumulated over 20,000 miles and the suspension has loosened up, some owners choose to replace the shocks simply to stiffen the ride on the highway. Internal parts can also experience wear. Shocks should be checked to see that they still provide good control.

Testing a Shock

A preliminary shock test with the car on the ground is the bounce test. Push down hard two or three times on the fender at each corner of the car. After the fender is pushed down, it should oscillate only about 1.5 cycles and then settle. Hand-testing shocks tests only the very lowest piston speed so it does not fully indicate the condition of the shock.

When the operation of the shock is in question, it can be unbolted from its lower mount. Then the shock can be moved to see that it moves slowly and with equal resistance through its normal range of motion. A skip or lag in a shock as its direction of motion is changed indicates a defective shock.

NOTE: *The bounce test only checks the first stage of shock absorber operation. A failure in the second or third stage valves will not be evident from this test. So even if a shock passes the bounce test it still might be defective.*

The bounce test or looking for cupped tires only highlights worst case shock absorber failures. Evaluating worn shocks is best done by driving the vehicle over a variety of roads.

Shocks do not often fail at the same rate. Usually one shock has a problem, requiring the replacement of both.

Sometimes a customer complains of noise, the location of which can be determined during the bounce test. The sound of the fluid being forced through the valves in the shock is normal.

Perform a visual inspection of the shock:

1. Inspect the condition of the shock mounts and rubber cushions.
2. Look to see if any fluid has leaked out of the shock, indicating a bad seal. Fluid cannot be replenished and the shock will have to be replaced. However, it is normal for a slight amount of moisture to be on the seal (**Figure 15.4**).
3. If the outside of the shock body is damaged (**Figure 15.5**), replace both shocks.
4. If rubber bump stops show signs of contact, a shock could be damaged or the vehicle might have been lowered or raised excessively.

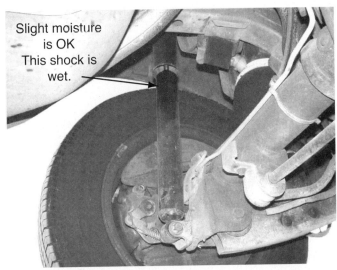

Figure 15.4 A slight amount of moisture on the seal is normal. However, the body of the shock should not be wet. *(Courtesy of Tim Gilles)*

Chapter 15 Suspension System Service **345**

Figure 15.5 The outside of this shock absorber is damaged. *(Courtesy of Tim Gilles)*

Shocks sometimes limit spring travel. In fact, if the shocks are removed on some coil spring cars, the spring can actually fall out. When the shock is removed, the control arm can drop quickly downward in response to spring pressure. If the shocks are to be replaced, the car should be raised on a hoist that supports the wheels. If this is not possible, support the axle with a high lift stand while removing the shock (**Figure 15.6**).

Figure 15.6 (A) Support the axle; (B) remove the bottom of the shock from its mount. *(Courtesy of Tim Gilles)*

The top of a shock absorber is often hidden within the trunk (**Figure 15.7**) or under a rear seat. A special tool is available to help with shock absorber removal and installation. One end is used to hold the shock rod while loosening or tightening the retaining nut (**Figure 15.8**). The other end of the tool is installed through the hole in the trunk and threaded onto the end of the shock rod to assist in pulling it into its socket in the trunk (**Figure 15.9**).

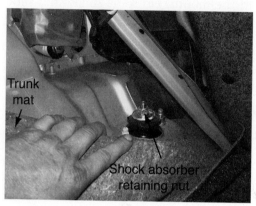

Figure 15.7 The top of this shock absorber is located under the floor mat in the trunk. *(Courtesy of Tim Gilles)*

Figure 15.8 This end of the tool is used to hold the shock rod while loosening or tightening the retaining nut. *(Courtesy of Tim Gilles)*

Figure 15.9 This end of the tool is installed through the hole in the trunk and threaded onto the end of the shock rod to assist in pulling it into its socket in the trunk. *(Courtesy of Tim Gilles)*

Shock Mounts

Check shock mounts to see that the shock is secured to the vehicle. A common problem with shock absorbers is that they become loose. Shocks are mounted to the chassis with rubber cushions. Shock mounts for passenger cars are usually the single stud (bayonet) or ring mount. Some of the ring mounts have a bar (crosspin), stud, or bolt pressed into them (**Figure 15.10**).

Rubber cushions allow the shock to have some flexibility in the mount. On some single-stud shocks, the tightness of the nut can determine how well the shock cushion functions. Some stud mount nuts are properly tightened when torqued against a shoulder on the stud. Others are tightened only until the rubber bulges out even with the end of its metal retainer (**Figure 15.11**).

Bleeding/Purging Shocks

When a conventional shock absorber has been stored on a shelf in a horizontal position, an air space can develop within it. This will cause a skip in the middle of the shock's travel. If the air is not purged or bled, the

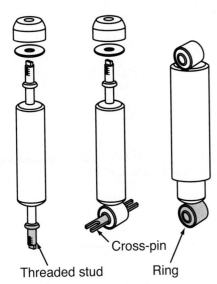

Figure 15.10 Different shock mounts.

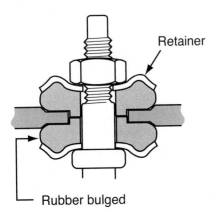

Figure 15.11 Tighten the nut until the cushions bulge almost to the outer edge of the retainers.

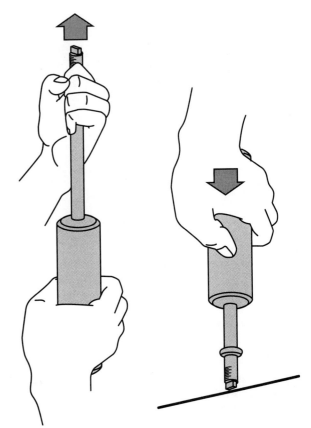

Figure 15.12 Purging air from a shock.

shock can be misdiagnosed as defective. Bleeding/purging shocks of air is done as follows (**Figure 15.12**):

1. Extend the shock while it is in its normal vertical position.
2. Turn the shock over so that the top is down and fully collapse the shock.
3. Repeat the process four or five times while working out any skips caused by air.

■ AIR SHOCKS

An air shock can sometimes fail because of a hole in its rubber bladder. Sometimes the rubber will rot if oil comes into contact with it. If the shock is installed in a tight location, the rubber can rub on suspension parts when the air shock is inflated. Air shocks can be checked for leaks using a soapy water solution. Rub the solution over the lines, fittings, and bladder while the shocks are inflated. Look for bubbles, which would indicate a leak.

Gas Shocks

A **gas-filled shock absorber** will expand to its fully extended position if not restrained. When gas shocks are removed from the box, there is a band around them to hold them compressed (**Figure 15.13**). If a gas shock has lost its gas charge, it will no longer expand on its own. It may still have oil in it, but the gas charge will no longer be there to prevent foaming when the oil becomes heated.

Figure 15.13 A band around a gas shock holds it compressed. *(Courtesy of Tim Gilles)*

MacPHERSON STRUT SERVICE

A large majority of vehicles have MacPherson struts. When a MacPherson strut shock fails, there are two common repair procedures:

1. Replace the entire strut assembly
2. Install a strut cartridge into the original shock housing (**Figure 15.14**)

Removing the Strut

The entire strut assembly is removed from the car (**Figure 15.15**). First, mark one of the bolts and its location at the top of the strut tower (**Figure 15.16**). The

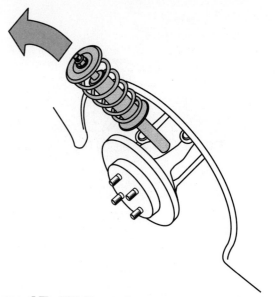

Figure 15.15 The entire strut assembly is removed from the car.

- Hammering on a pinch bolt can damage the bolt and the ball joint. If the pinch bolt does not come out easily, pry on the lower control arm to release any tension on it.
- To minimize the danger of the compressed spring, unload it prior to removing the strut. Raise the vehicle by the frame and allow the spring to extend as far as possible.

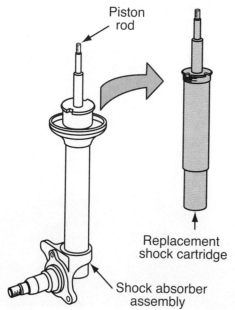

Figure 15.14 A replacement cartridge is installed in the old strut.

Figure 15.16 Prior to loosening the strut to chassis bolts, make alignment marks on the strut bolts and the chassis.

strut is easily removed from the top. Removing the bottom part of the strut sometimes requires that the lower ball joint be disconnected from the spindle support, or **steering knuckle**. This will allow the lower control arm to be pried down so the strut can be removed. Some lower ball joints have a **pinch bolt** (**Figure 15.17**) that holds the ball joint. Others will have a cotter pin and castle nut that will need to be removed.

348 SECTION 4 SUSPENSION AND STEERING

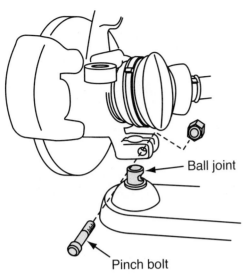

Figure 15.17 A pinch bolt holds the steering knuckle to the ball joint.

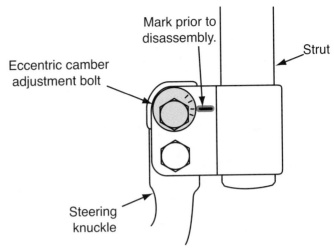

Figure 15.19 Mark an eccentric strut bolt before removal.

Some struts are easily removed at the bottom by removing two bolts that fasten the strut to the spindle support (**Figure 15.18**). On some imports, removing a brake line from a caliper is necessary. Brake bleeding will be required following the strut repair or replacement. Sometimes one of the two bolts has a slot that allows for camber adjustment during a wheel alignment. If the vehicle has this adjustment feature, be sure to mark the location of the bolt so it can be replaced in the same position (**Figure 15.19**).

A spring compressor is used to compress the coil spring (**Figure 15.20**). With the spring compressed, the nut at the top of the strut can be removed from the strut rod (**Figure 15.21**). Once the nut has been removed, the parts can be disassembled from the strut (**Figure 15.22**).

■ The center nut must not be removed from the top of the strut before the strut assembly is removed from the vehicle. This would allow the spring to be decompressed during strut removal.

■ Be careful with the compressed spring. Dropping it can cause it to dislodge from the spring compressor with dangerous results.

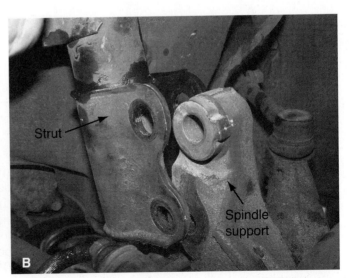

Figure 15.18 (A) Remove these bolts to remove the spindle support and caliper from the strut. (B) Pull the strut away from the spindle support. *(Courtesy of Tim Gilles)*

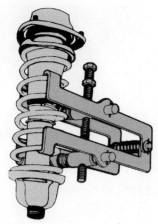

Figure 15.20 A MacPherson strut compressor. *(Courtesy of Federal-Mogul Corporation)*

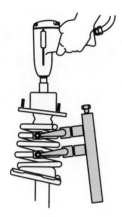

Figure 15.21 Removal of a nut from the strut piston rod.

> **SHOP TIP** A tight nut can be removed using an impact wrench, squeezing the trigger rapidly to produce short bursts of power and avoid spinning the shaft.

Replacing a Strut Cartridge

Struts that are serviceable with a shock cartridge have a nut at the top. Sometimes a special spanner wrench is required to remove it (**Figure 15.23**). At other times, a conventional hex nut is used. One of these is shown in **Figure 15.24**, which provides an exploded view of the parts being removed from a strut housing.

Adding Oil to the Strut Body

Unlike a replacement cartridge, an original equipment strut is not enclosed in a cartridge. The old shock fluid will need to be poured out of the strut. Some lightweight oil is poured into the strut body to help conduct heat between the new strut cartridge and the outside of the strut housing. The old shock oil can be used for this. Two or three ounces of oil is usually enough. The amount is not important, but it should almost reach the top of the strut tube with the cartridge replaced.

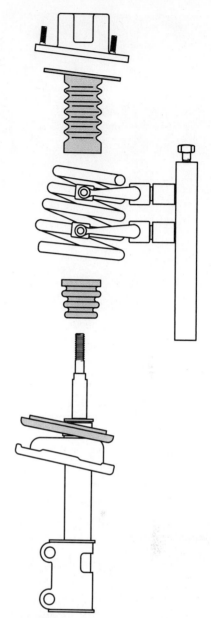

Figure 15.22 Removal of upper strut mount, upper insulator, spring, spring bumper, and lower insulator.

Figure 15.23 Using a spanner wrench to remove a nut from the strut. *(Courtesy of Tim Gilles)*

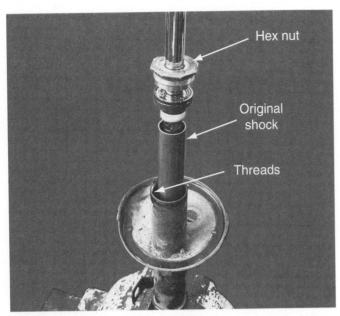

Figure 15.24 Removing parts from a strut housing. *(Courtesy of Tim Gilles)*

SHOP TIP ATF is a good choice of oil to add to the strut tube because it is red in color. If the shock leaks later on, you will be able to determine if the oil is from the sealed strut cartridge within the strut tube, or is ATF leaking past the threads of the retaining nut at the top of the tube.

Center the strut cartridge in the strut body before installing the locknut.

Inspecting the Upper Strut Bearing

Inspect the condition of the upper strut bearing (see **Figure 14.60**) while the strut assembly is disassembled. A questionable bearing can easily be replaced at this time.

Installing the Coil Spring

Install the coil spring and tighten the locknut. Be sure both ends of the spring are correctly seated before removing the compressor (**Figure 15.25**).

Reinstalling the Strut Assembly

Reinstall the strut assembly on the car in the same position it was in before. A wheel alignment should always be performed after a strut replacement. If the

Do not use an impact wrench to tighten the locknut. The piston rod can spin rapidly, damaging the new seal.

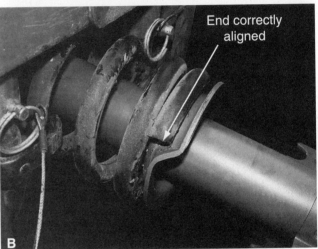

Figure 15.25 (A) The end of the spring is not aligned with its seat. (B) Correct installation. *(Courtesy of Tim Gilles)*

brake caliper hose was disconnected, brakes will need to be bled as well.

■ BUSHING SERVICE

Bushings are made of synthetic rubber. They insulate suspension parts from noise and road shock. Rubber bushings deteriorate with age. They are also susceptible to heat damage. Upper control arm bushings are especially prone to heat damage because of their close proximity to engine exhaust manifolds. Driving on bumpy roads can also result in heat and fatigue of the bushing's rubber.

Bushings should not be lubricated because petroleum attacks rubber and can ruin it. Rubber lubricant (the kind used to aid in mounting tires) or brake fluid can be used to help quiet a squeaky, hardened bushing, but the fix is only temporary.

Control Arm Bushings

Damage or distortion to a control arm bushing can produce changes in wheel alignment settings (**Figure 15.26**). Inspect the control arm bushings for deterio-

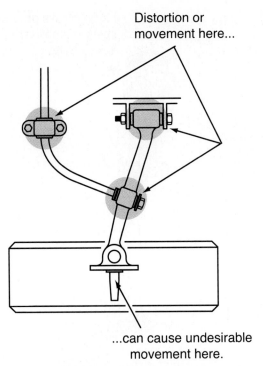

Figure 15.26 Defective bushings can change alignment settings.

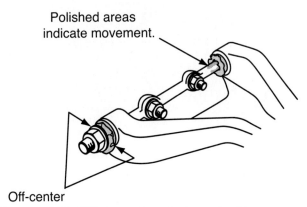

Figure 15.27 Inspect control arm bushings.

Figure 15.28 Using an air chisel to remove a control arm bushing. *(Courtesy of Federal-Mogul Corporation)*

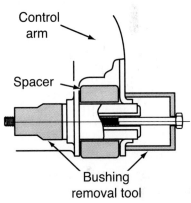

Figure 15.29 Control arm bushing removal with a puller.

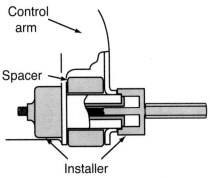

Figure 15.30 Control arm bushing installation with a puller.

ration and splits in the rubber (**Figure 15.27**). See if the bushing is off-center. Sometimes bushings are in a position where visual inspection is difficult. Using a flashlight and mirror will sometimes help.

Push on the fenders of the vehicle while listening for noise. To inspect bushings for looseness, use a prybar to see if the control arm can be moved.

Special offset replacement control and bushing sets are available that compensate for sagged frame components (such as strut towers leaning together) that can cause front end alignment problems.

Control arm bushings are pressed or driven out. It is important that the control arm holes not be damaged during the process. If an air chisel is used (**Figure 15.28**), be sure that the chisel bit is wide and dull. If it is sharp, it will cut the bushing rather than push it out. The new bushing is pressed or pounded in with a special driving tool. **Figure 15.29** shows removal of a bushing using a bushing puller. The puller kit also includes adapters for bushing installation (**Figure 15.30**).

NOTE: *Tighten bolts or nuts that hold bushings in place only when the suspension is in its normal ride position.*

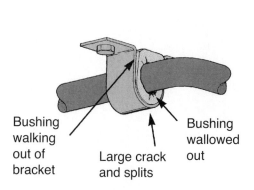
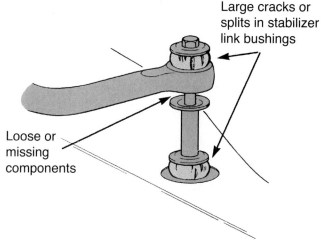

Figure 15.31 Inspect stabilizer bushings. *(Courtesy of Federal-Mogul Corporation)*

Strut Rod Bushing Service

Refer to **Figure 14.27** for the relationship between the strut rod, lower control arm, and bushing. When replacing a strut rod bushing, the nut is removed. This allows the lower control arm to move fore or aft. Next, remove the fasteners that hold the strut rod to the control arm and slide the strut rod out of the bushing. Most bushings have a center spacer that will only allow the bushing to be compressed a certain amount during reinstallation when the strut rod nut is tightened against it.

Stabilizer Bar Service

Inspect the bushings at both ends of the stabilizer bar (**Figure 15.31**). Stabilizer links do not have to be removed to replace the inner bushings. Replacement stabilizer bushings have a split in them. Some original bushings do not have a split; these can be cut off with a razor blade. Then install the new bushing.

■ SPINDLE SERVICE

The steering knuckle and spindle are not serviced. If the spindle is damaged, it is replaced. Damage from a collision will be noticed during a check of the steering axis inclination (SAI) alignment angle. If a wheel bearing fails it can wear or heat the spindle. This changes the metallurgy of the spindle and it must be replaced.

Figure 15.32 This spindle broke after a burned wheel bearing was replaced. *(Courtesy of Tim Gilles)*

CASE HISTORY
A front wheel bearing failed on a 1965 Mustang. There was wear on the spindle that the technician failed to notice. He replaced the wheel bearing and returned the car to the customer. Several months later, the customer's daughter was driving the car when she drove over a small pothole in the road. The spindle broke off just behind the area where the old bearing had burned (**Figure 15.32**). She was not traveling fast and, luckily, was not hurt when the front wheel fell off the car.

CAUTION When replacing stabilizer bar bushings, be sure that both front wheels are either on the ground or in the air. Otherwise, the sway bar will be spring-loaded. With one wheel jacked up, removal of the nut on the top of the stabilizer link could lead to a serious injury.

■ BALL JOINT SERVICE

Ball joints are relatively trouble-free, but occasionally they wear out. The ball joint is sealed within a rubber boot filled with grease. Many ball joints are equipped with a threaded plug (**Figure 15.33A**). Many shops install a zerk fitting in place of the plug (**Figure 15.33B**). Grease is injected into the ball joint using a grease gun whenever grease fittings are found.

Chapter 15 Suspension System Service

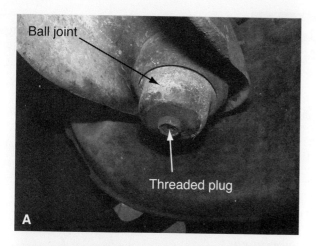

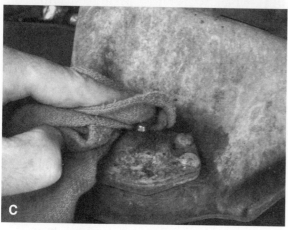

Figure 15.33 (A) The threaded plug is removed. (B) A zerk fitting is installed. (C) Clean a grease fitting before injecting grease into it. *(Courtesy of Tim Gilles)*

NOTE: *Do not inject too much grease into the joint. Simply refill it until you see grease movement.*

A small bleed hole in the ball joint boot allows for grease movement during lubrication. Other cuts or tears in the boot will allow water and dirt into the joint, causing it to wear. Be sure to clean the outside of a dirty fitting before injecting it with grease (**Figure 15.33C**).

Feel around the outside of the boot to see if it is torn. If there is a hole in the boot, the joint will proba-

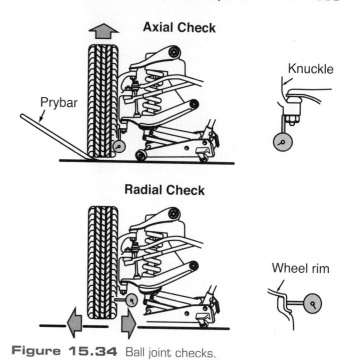

Figure 15.34 Ball joint checks.

bly fail soon and should be replaced. Also inspect for signs of rust or cracks on the control arm near the joint.

Manufacturers list specifications for movement of ball joints. Always check the specifications and be sure that the proper procedure is followed. Vertical, or axial, movement is usually specified, although some manufacturers specify horizontal, or radial, movement (**Figure 15.34**).

A *load-carrying ball joint* usually has some movement when unloaded. In the past, the unnecessary replacement of good ball joints was a common occurrence. As a result, some states now require measurement and documentation of the amount of clearance found on a worn ball joint before it can be replaced. Most load-carrying ball joints have a wear limit of 0.060 inch of vertical movement, but some joints can have as much as 0.200 inch of movement and still be within specified limits.

Before checking ball joint wear, determine whether the ball joint is a load carrier or follower (see Figure 14.58). It is important to know this when testing ball joints for wear because the load-carrying joint must be unloaded to test it. Load-carrying ball joints are either tension or compression types (see **Figure 14.59**). Tension-loaded ball joints are more common. The ball joint on the control arm that has the spring mounted on it is the load carrier. The ball joint on a strut suspension is the follower joint and therefore carries no load.

Measuring Ball Joint Wear

Check the manufacturer's recommended procedures before checking ball joints. For an accurate wear check, the ball joint must not support the weight of the vehicle. Some SLA suspension systems have the spring above the upper control arm. A jack is placed under the frame when checking ball joint clearance on these vehicles (**Figure 15.35**). A piece of wood or a special tool made of aluminum is used to hold tension off the ball joint. Prying between the tire and ground will show any clearance.

When the spring bears against the lower control arm, a different procedure is used. The jack is positioned directly under the lower control arm to unload the ball joint while prying between the tire and the ground (**Figure 15.36**).

The follower joint should be checked at the same time that the loaded ball joint is unloaded for testing. Follower ball joints hold the steering knuckle in the correct position and allow it to pivot during bumps and turns. Recommendations for most follower ball joints call for replacement if there is "any perceptible movement" in the joint. Always check specifications.

To check a follower ball joint for movement, unload the joint and try to move the tire back and forth while looking for movement. For this test to be accurate, wheel bearing clearance cannot allow the wheel to move.

> **SHOP TIP**
> - A quick way to eliminate wheel bearing clearance from the measurement is to apply the brakes during the test. When a helper is not available, brakes can be applied using a pedal depressor.
> - To measure ball joint clearance without a dial indicator, measure the length of the ball joint with a caliper and then measure it again when the joint has been unloaded.

Wear Indicator Ball Joints. Some ball joints have a wear indicator built into them. The most common type of wear indicator has a shoulder that sticks out of the bottom of the joint about 0.050 inch when it is new (**Figure 15.37**). If the ball joint has worn, this shoulder will recede into the ball joint housing. When it is flush, the ball joint should be replaced. A wear indicator ball joint must be loaded and at normal ride height to read the indicator.

Separating Ball Joint Tapers

The ball joint is connected to the steering knuckle with a tapered connection called a *steering taper* (**Figure 15.38A**), which is also used in other steering connections. Removing a ball joint can be done with a special tool (**Figure 15.38B**). Another method of ball joint removal is to "break" the taper using a large hammer:

1. First, remove the cotter pin from the ball joint nut and loosen the nut several turns.
2. Position the vehicle so that the coil spring is pushing on the ball joint. This could require lifting the vehicle or allowing its weight to rest on the wheels.

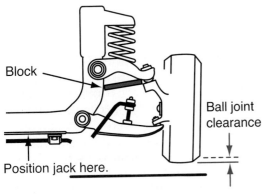

Figure 15.35 Floor jack position to check ball joint wear when the spring is above the upper control arm.

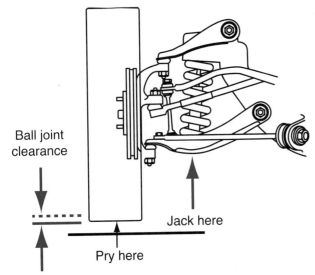

Figure 15.36 Floor jack position to check ball joint wear when the spring is on the lower control arm.

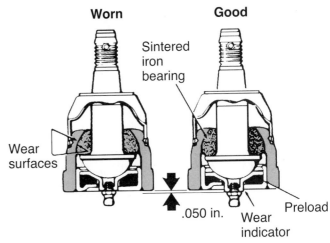

Figure 15.37 Checking a wear indicator ball joint. (Courtesy of Federal-Mogul Corporation)

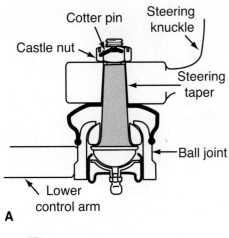

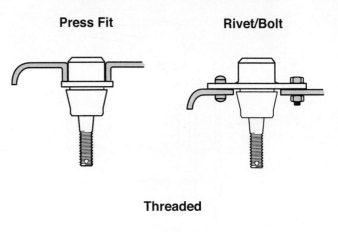

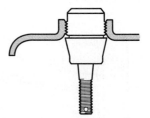

Figure 15.39 Three common ball joint retaining methods.

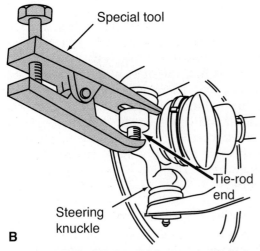

Figure 15.38 (A) A ball joint taper. (B) A tie-rod end and ball joint puller.

3. Use a hammer to pound sharply on the steering knuckle on the outside of the taper. This will deform the taper and spring pressure will separate the ball joint from the steering knuckle.

NOTE: *A pickle fork (see **Figure 17.9**) will probably ruin the rubber seal on the ball joint. If the ball joints are not going to be replaced, this will be a costly subtraction from the profit of the job.*

Replacing the Ball Joint

A ball joint can be retained in one of several ways. **Figure 15.39** shows three of the most common methods.

Some original equipment ball joints are fastened to the control arm with rivets. These rivets must be removed in order to extract the ball joint (**Figure 15.40**). First, a small drill bit is used to bore a hole about halfway into the rivet. Then, the rivet head is drilled using a large bit until the rivet head falls off. Finally, bolts and nuts are used to hold the replacement joint in place.

NOTE: *Many technicians use an air chisel to cut the heads off the rivets, but this calls for ear protection for the technician and others in the shop.*

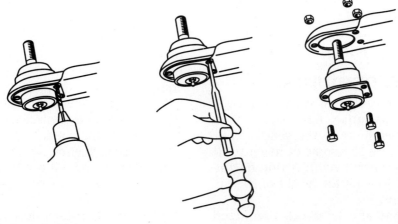

Figure 15.40 Replacing a ball joint.

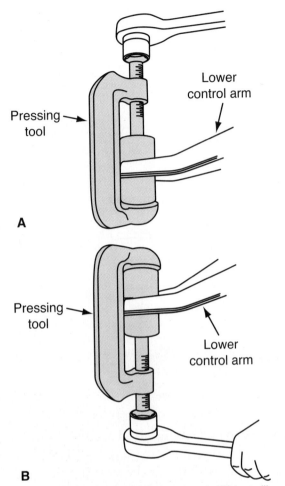

Figure 15.41 (A) Ball joint removal. (B) Installing a pressed-in ball joint in the lower control arm.

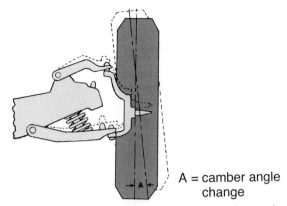

Figure 15.42 The position of the control arm changes in its arc due to spring sag. (Courtesy of McQuay Norris)

Ball joints can also be pressed or threaded into the hole in the control arm. If the control arm has been removed from the vehicle, a press-fit ball joint can be removed using a standard hydraulic press. While the control arm is still on the car, a special press set allows the removal and replacement of these ball joints. **Figure 15.41** shows this process.

Some press-fit ball joints have a spot weld that holds them in place. This must be carefully removed. When a replacement ball joint is installed, a snapring often takes the place of the weld.

NOTE: *Be sure the grease release hole in the new ball joint's rubber boot is aimed away from the brakes.*

■ COIL SPRING SERVICE

A coil spring will rarely break unless it has been constantly overloaded or if its surface has a stress raiser from a nick or defect. Over the years, the weight of the car causes the coil spring to lose some of its tension. This results in a lowered ride height. A ride height check can be performed to see if the height of the car conforms to specifications.

Incorrect ride height affects wheel alignment angles, changing camber and toe. With an SLA suspension system, the car is designed to ride at a height that minimizes changes in tread width. When the suspension system drops due to sagged springs, the upper control arm is in a different place in its arc of travel. This means that there will be even more camber change during bumps (**Figure 15.42**). Tires will wear faster and handling will be affected.

A vehicle that is too low cannot be aligned properly, so this test should be done before attempting a wheel alignment. The ride height check is covered in Chapter 19.

Adjusting Spring Height

Coil springs must be replaced when they have sagged beyond specifications. An advantage of the torsion bar spring used on a few vehicles is that spring height can be adjusted. A screw is turned against a bracket mounted on one of the torsion bar ends. The vehicle's correct ride height can be restored prior to an alignment.

Air shocks or shock absorbers with coil springs around them are designed to be used only for temporary overload conditions. The weight of the vehicle will rest on the shock mounts instead of the spring seat. Shock mounts are not designed to continually support the vehicle.

Coil Spring Replacement

Replacement springs are purchased from a dealer or aftermarket parts supplier. They are replaced in front or rear pairs. Some aftermarket springs have a part number stamped on the end of the coil. Original equipment springs are often tagged with a part number wrapped to one of the coils. The tag might be missing or unreadable. Replacement springs must be of the same kind as the one in the vehicle. Coil springs have different shaped ends. Square-end coils can be tapered or untapered. The full wire ends are cut off cleanly on *tangential springs*. **Figure 15.43** shows different spring ends.

Aftermarket regular duty coil springs are similar to original equipment springs but are sometimes designed

Figure 15.43 Different spring ends.

to replace several different springs. See Chapter 14 for more information on springs.

SLA Coil Spring Replacement

Most passenger cars require the use of a coil spring compressor for coil spring removal and replacement. When a spring is installed in a vehicle, it is held in a compressed position. When it is fully extended, it is much longer. Before removing a spring, the wheel, shock absorber, and stabilizer links are removed. The outer tie-rod is disconnected from the steering arm.

During a coil spring replacement, only the lower ball joint needs to be removed. The upper ball joint can remain in place. The upper control arm and steering knuckle will be moved out of the way during the coil spring replacement. Sometimes, both upper and lower ball joint tapers must be broken to be able to get the disc brake splash shield and spindle to clear the lower control arm. Remove the cotter pin from the ball joint nut and back the nut off several turns.

When using the special ball joint press to loosen a taper connection, unload the ball joint with a floor jack. For maximum leverage, position the jack so it is as close to the outer ball joint as possible (**Figure 15.44**). A piece of wood will help get more leverage and keep the jack level.

Figure 15.44 Position the jack so it is as close to the outer ball joint as possible. (Courtesy of Federal-Mogul Corporation)

 Be sure the spring is secured before the ball joint is released.

Reach across to jack up the control arm from the opposite side of the vehicle (**Figure 15.45**). This will allow the jack to roll toward the inside of the vehicle as the control arm is lowered, releasing spring pressure. Raise the jack to compress the spring until the vehicle begins to lift.

Remove the lower ball joint nut and lift the steering knuckle assembly away from the ball joint stud. Support it out of the way. A coil spring compressor can also be used to keep the spring compressed while repairs are performed to the spindle or upper control arm (**Figure 15.46**).

Use two clips to hold the spring compressed. Clips are either short or long; use the one that fits best. Short clips are used to hold four rungs of the spring. Long clips are used to clip five rungs. Position the two clips

Figure 15.45 The correct way to use a jack when doing SLA coil spring work. (Courtesy of Tim Gilles)

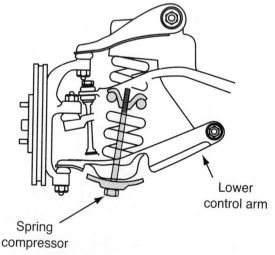

Figure 15.46 A spring compressing tool installed on spring in a short/long arm suspension system.

SAFETY NOTE: Never compress the spring far enough for the coils to stack up.

side by side on the same spring rungs and as low as possible on the inboard side of the spring.

Pry down on the lower control arm and pry the coil spring from its place. Mark the locations of the clips. Use a coil spring compressor to compress the spring and remove the clips.

Lay the old spring next to the new spring. Align the dominant ends (**Figure 15.47**). The dominant end is the end that aligns in the coil spring seat in the frame or lower control arm. Mark the new spring in the same location of the old spring. Compress the spring and install the spring clips (**Figure 15.48**).

Because of the spring clips, the spring will be bowed. This will make aligning the upper and lower spring seats easier during installation. If there are insulators that fit into the spring seats (**Figure 15.49**), tape them onto the spring to make installation easier.

The spring seats must be accurately aligned. During installation, align the upper end first (**Figure 15.50**). Then, push the bottom of the spring into place. Be sure the lower spring end fits into the pocket in the spring seat (**Figure 15.51**).

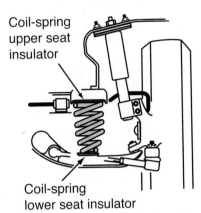

Figure 15.49 Insulators that fit into the spring seats.

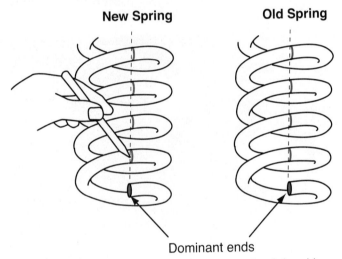

Figure 15.47 Align the dominant ends of the old and new springs.

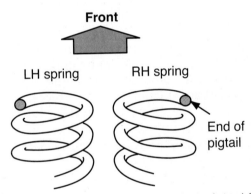

Figure 15.50 Upper ends of springs viewed from the top.

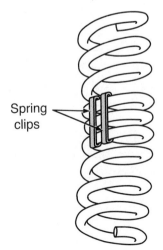

Figure 15.48 Compress the spring and install the spring clips.

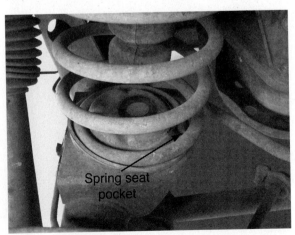

Figure 15.51 The end of the spring fits into a pocket in the spring seat.

Figure 15.52 Align the cotter pin hole and install the ball joint in the steering knuckle. *(Courtesy of Federal Mogul Corporation)*

Jack the control arm to compress the spring and install the ball joint stud in the tapered hole (**Figure 15.52**). Be sure to align the cotter pin hole so it is parallel to the brake backing plate. Otherwise, it will be difficult to install the new cotter pin in the castle nut. Torque the castle nut to specifications.

NOTE: *If the cotter pin hole does not line up, tighten until the next hole does. Do not back off on the nut to align the hole.*

After connecting the ball joint(s), remove the spring clips from the spring. A small prybar may be needed for this.

Complete the assembly of all of the components. Before lowering the wheels onto the ground, loosen the bolts that compress the control arm bushings. Drive the vehicle a short distance and jounce the suspension several times. Before doing a wheel alignment, retorque to specifications all fasteners that hold bushings.

Torsion Bar Removal and Replacement

The torsion bar adjusting bolt must be loosened before removing a torsion bar (**Figure 15.53**). First, measure how far its adjusting bolt extends above the surface of the bracket (**Figure 15.54**). This will save time when readjusting ride height after reinstallation. There are tools available that can remove tension from the torsion bar while the adjusting bolt is loosened (**Figure 15.55**). If you do not have this tool, be sure the wheels are lifted off the ground before attempting to loosen the adjusting bolt.

If the torsion bar is to be reinstalled, mark its end and its adjusting arm prior to removal so it can be reinstalled in the same position (**Figure 15.56**).

 **CAUTION** Some SUVs with torsion bars have air suspension systems. To avoid injury, disable the air suspension system before working on the system.

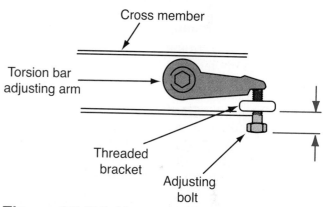

Figure 15.54 Measure how far the adjusting bolt extends above the surface of the bracket.

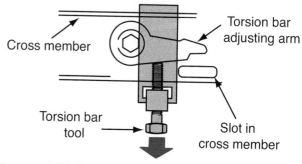

Figure 15.55 A tool used to remove tension from the torsion bar.

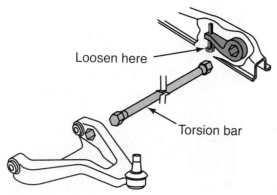

Figure 15.53 Remove tension from the torsion bar adjusting bolt.

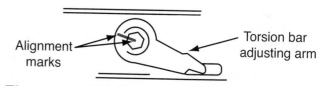

Figure 15.56 Mark the end of the torsion bar before removing it.

Some torsion bars are tight in their sockets. A tool is available that clamps on the torsion bar. Hammering on it removes the torsion bar from its socket.

Before reinstalling a torsion bar, be sure it is going into the correct side of the vehicle. Torsion bars are directional. They are marked left or right, as viewed from the driver's seat, and must be installed on the proper side of vehicle.

After the torsion bar is installed, check and adjust ride height as needed (see **Figure 19.7**).

Leaf Spring Service

Leaf spring problems include broken leaves, spring sag and differences in ride height from side to side, squeaks, worn bushings, and broken center bolts. Bushings and shackles can be replaced without removing the spring from the vehicle.

When leaf springs have sagged, they are removed and sent to a facility where they can be "re-arced." This reestablishes the original ride height.

Multiple leaf spring leaves have holes through their centers. A center bolt runs through the holes of all of the springs, with the head of the bolt fitting into an opening in the spring mount on the axle. The center bolt holds the position of the axle in relationship to the springs so the rear wheels can track behind the front wheels. Sometimes the center bolt breaks. This can alter the position of the rear axle, changing vehicle tracking and alignment.

When insulators need replacing or there is a broken spring leaf, secure the spring with a C-clamp or in a large vise. Then, remove the center bolt and disassemble the spring pack. Spring leaves are different lengths. Before disassembly, use a crayon to number the springs so they can be easily replaced in their original positions.

Removing leaf springs is easiest when done on a wheel contact lift with the vehicle frame supported on jack stands or the lift's air jack. If a wheel contact lift is not available, the job is performed with the vehicle on the ground. Raise the body of the vehicle up until the wheels just start to leave the wheel contact surface. Then, disconnect the spring from one side at a time. When possible, leave the vehicle in the same position so the spring can be easily reinstalled. When positioning the spring, be sure that the head of the center bolt fits into the hole or recess in the spring seat. This correctly positions the spring fore and aft.

■ WHEEL ALIGNMENT

Following completion of suspension repair jobs, a wheel alignment will be required. This will reposition suspension components so that the car will be safe to drive and will go straight without unusual tire wear. Remember to torque bolts that go through bushings only after the vehicle is resting on its tires.

■ ELECTRONIC SUSPENSION SERVICE

Electronic suspension system problems are often related to mechanical failures, including leaks in the system. A leak can occur in a rubber spring bladder or nylon tubing can be damaged.

Electrical failures can also occur in the air system. The length of time a compressor can run is limited to a few minutes in the event of a solenoid malfunction. Compressor failures do occur, however.

If the battery is being charged, the ignition switch must be in the off position when the air suspension switch is on. If not, damage to the compressor relay or motor may occur.

Wiring connections are also a common source of problems. Sometimes a problem can be corrected simply by disconnecting and reconnecting wires.

Electronic failures will set a computer code and cause an instrument panel light to flash or illuminate continuously. The source of electronic problems can usually be traced using a scan tool to read the **diagnostic trouble codes (DTCs)** stored in the vehicle computer's memory. Manufacturer scan tools are best for diagnosing the most troublesome problems on specific makes of vehicles. But the majority of aftermarket scan tools are able to diagnose the most probable electronic failures.

Scan tools were originally designed to diagnose emission-related problems. Today there are many other computer-controlled systems on automobiles. Scan tool cartridges are available to help in the diagnosis of electronic system problems in these systems (**Figure 15.57**).

Figure 15.57 An interchangeable program cartridge for the vehicle is inserted into the scan tool.

PRC Diagnostic Trouble Codes (DTCs)

Code	Defect
6	No system problem
1	Left rear actuator circuit
2	Right rear actuator circuit
3	Right front actuator circuit
4	Left front actuator circuit
5	Soft relay control circuit shorted
7	PRC control module
13	Firm relay control circuit shorted
14	Relay control circuit

Figure 15.58 PRC system diagnostic trouble codes (DTCs).

CAUTION Be sure the leveling system is disabled before raising a vehicle on a lift, jacking it up, or towing it. Some computer systems are not activated by the ignition switch and stay on all the time. The ignition switch must be off, and the leveling system switch—often located in the trunk—must be turned off before lifting or towing the vehicle.

There are several sensors for electronic suspension systems (see Chapter 14). Service literature contains detailed charts and diagrams for each manufacturer's system. **Figure 15.58** shows a typical list of diagnostic trouble codes for a programmed ride control (PRC) system.

When raising a vehicle with an electronically controlled air suspension, a frame contact hoist (sometimes called a body hoist) is recommended by most manufacturers. Lifting by the frame allows air springs to be in their extended position, where the rubber bladder (**Figure 15.59**) will not be damaged. With the suspension hanging free, height sensors lengthen to their maximum position. If the system is on, the computer sees this as a "high car" condition and responds by bleeding pressure from the air springs. If the vehicle is lowered to the ground while the springs are in this position, the rubber can be damaged if the air spring bags happen to fold inward.

When lifting a vehicle with a hydraulic jack, lift by the front cross member and at the manufacturer's specified rear lift points in front of the rear tires. Before towing a vehicle with an automatic leveling system, be sure to check the service manual for precautions.

Prior to doing a wheel alignment, the vehicle must be set at regular curb height. Failing to do this will result in a faulty wheel alignment.

Electronic-Controlled Shock Absorbers

Some shock absorbers are electronically controlled to provide a ride that is more firm or more soft than normal. When a firm ride is selected, an actuator motor rotates in response to an electronic signal to reposition a valve in the shock, restricting oil movement and causing a rougher ride.

To remove the actuator, squeeze the plastic retainers and lift the motor from the top of the strut (**Figure 15.60**). To test the actuator, turn on the ignition while

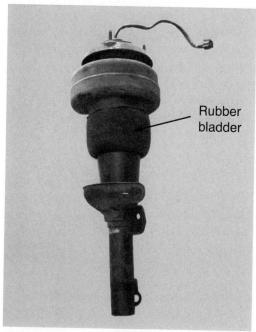

Figure 15.59 An air spring strut assembly from an electronic suspension system. *(Courtesy of Tim Gilles)*

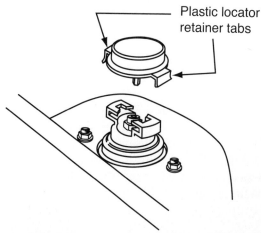

Figure 15.60 Squeeze the plastic retainers and lift the motor from the top of the strut.

the actuator wiring is still connected. When the ride control switch is activated to the firm or soft position, the motor should rotate the control tube on the bottom of the actuator within a few seconds. If it rotates, the actuator is operating. If not, refer to manufacturer's instructions before testing the unit.

To diagnose the actuator system, you will need wiring color codes and specific instructions. A typical diagnostic procedure will have you rotate the actuator motor to where it is normally positioned under different commands and make identifying marks on the actuator and the control tube. With the electrical wiring disconnected, the actuator control tube can be turned with a small screwdriver to put it into the desired position. Electrical resistance is tested in different positions using an ohmmeter, and the results are compared with specifications.

REVIEW QUESTIONS

1. A _____ shock expands to its full travel when not installed.
2. What is added when installing a MacPherson strut cartridge to help it conduct heat to the outside of the strut?
3. If a wheel bearing burns and damages the spindle, what part should be replaced?
4. Are there any holes in a good ball joint boot?
5. In which two directions are ball joints checked for wear?
6. Which kind of ball joint is more common, tension or compression?
7. When a wear indicator ball joint wears, does the shoulder move into or out of the ball joint?
8. When a replacement ball joint is held in place with nuts and bolts, what held the previous ball joint in position?
9. What is used with a spring compressor to aid in the removal and replacement of coil springs?
10. What is the name of the end of a coil spring that aligns in the coil spring seat in the frame or lower control arm?

PRACTICE TEST QUESTIONS

1. Technician A says that a leaking shock should have the fluid refilled. Technician B says it is normal for a small amount of moisture to be on the outside of the shock body. Who is right?
   A. Technician A   B. Technician B
   C. Both A and B   D. Neither A nor B

2. Technician A says that a shock mount should be tightened until the rubber bumper is totally compressed. Technician B says to lubricate shock absorber bushings with light oil. Who is right?
   A. Technician A   B. Technician B
   C. Both A and B   D. Neither A nor B

3. Technician A says when replacing control arm bushings with an air chisel, the tool must be sharp. Technician B says when replacing stabilizer bar bushings, be sure that both front wheels are raised or on the ground together. Who is right?
   A. Technician A   B. Technician B
   C. Both A and B   D. Neither A nor B

4. Technician A says a load-carrying ball joint must be unloaded to test it. Technician B says load-carrying ball joints must be replaced if there is any perceptible movement. Who is right?
   A. Technician A   B. Technician B
   C. Both A and B   D. Neither A nor B

5. Technician A says applying the brakes will temporarily eliminate wheel bearing clearance. Technician B says to install a new ball joint with its rubber boot grease hole facing against the brake backing plate. Who is right?
   A. Technician A   B. Technician B
   C. Both A and B   D. Neither A nor B

6. Technician A says air shocks are designed to permanently raise a vehicle's height. Technician B says when jacking on a control arm to compress a coil spring, the jack should be 90 degrees (perpendicular) to the tires. Who is right?
   A. Technician A   B. Technician B
   C. Both A and B   D. Neither A nor B

7. Technician A says to align the cotter pin hole in a ball joint so that it is parallel to the brake backing plate. Technician B says after reaching torque

specifications, back off on the ball joint locknut to align the cotter pin hole. Who is right?

- A. Technician A
- B. Technician B
- C. Both A and B
- D. Neither A nor B

8. Technician A says that wear on one side of a tire could result from a bad shock absorber. Technician B says that when one shock absorber is bad, all four shocks should be replaced. Who is right?

- A. Technician A
- B. Technician B
- C. Both A and B
- D. Neither A nor B

9. Technician A says that worn bushings can change alignment settings. Technician B says that the sound of the fluid being forced through the valves in the shock is normal. Who is right?

- A. Technician A
- B. Technician B
- C. Both A and B
- D. Neither A nor B

10. A MacPherson strut shock absorber is leaking badly. Technician A says that sometimes an entire strut assembly must be replaced. Technician B says some vehicles use a strut cartridge installed into the original shock housing. Who is right?

- A. Technician A
- B. Technician B
- C. Both A and B
- D. Neither A nor B

CHAPTER 16

Steering Fundamentals

■ OBJECTIVES

Upon completion of this chapter, you should be able to:
- ✔ List the parts of steering systems.
- ✔ Describe the principles of operation of steering systems.
- ✔ Compare linkage systems to rack-and-pinion systems.
- ✔ Describe how power steering systems operate.
- ✔ Understand the operation of four-wheel steering systems.

■ KEY TERMS

ball socket
belt
center link
collapsible steering column
fast steering ratio
flow control valve
high cordline belt
idler arm
integral power steering
linkage-type power steering
lock to lock
Magnasteer
neoprene
parallelogram steering
pitman arm
pressure relief valve
rack
rack-and-pinion steering
recirculating ball and nut steering gear
side slip
steering ratio
tensile cords
tie-rod
toe-out on turns
turnbuckle
wormshaft

■ INTRODUCTION

The steering system works with the suspension system. It allows the driver to steer the car with a comfortable amount of effort. Steering system parts include the *steering gear*, the *steering linkage*, the *steering wheel*, and the *steering column*. There are two styles of steering. One has a gearbox and parallelogram linkage (**Figure 16.1**). The other is a simple long rack with linkage extending from its ends (**Figure 16.2**).

■ STEERING GEARS

The two common types of automotive steering gears are the **recirculating ball and nut steering gear** and the **rack-and-pinion steering**. There are other

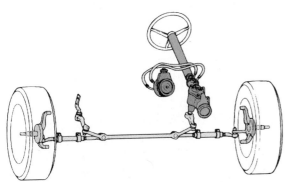

Figure 16.1 A steering gearbox and parallelogram linkage. *(Courtesy of Federal-Mogul Corporation)*

Figure 16.2 Rack-and-pinion steering with linkage. *(Courtesy of Federal-Mogul Corporation)*

types of steering gearbox designs, but these are the primary ones used in automobiles and light trucks.

The number of teeth on the driving gear compared to the number of teeth on the driven gear help determine the **steering ratio**. The length of the steering arms, pitman arms, and idler arms (parts of the parallelogram steering linkage) also play a part in determining steering ratio.

When the steering wheel is turned all the way in one direction, it stops against a lock. Turning the wheel all the way from one lock to the other is called **lock to lock**. The steering ratio refers to how much the steering wheel must be rotated for the wheel to turn a given amount. A **fast steering ratio** is about three turns lock to lock. Slower ratios require the wheel to be turned about four times. The ratio of a steering gear varies, depending on whether the car has a power assist. Cars with power steering usually have faster ratios. A 15:1 ratio means that when the steering wheel is turned 15 degrees the front wheels will turn 1 degree.

Recirculating Ball Steering Gear

In a recirculating ball steering gear (**Figure 16.3**), a *sector gear* (part of the *pitman* or *sector shaft*) meshes with a *ball nut* that rides on bearings on the **wormshaft** or steering shaft, to provide a smooth steering feel. The ball nut has curved channels for ball bearings to ride in. The steering shaft also has bearing channels. The balls rotate and recirculate through tubes (ball returns).

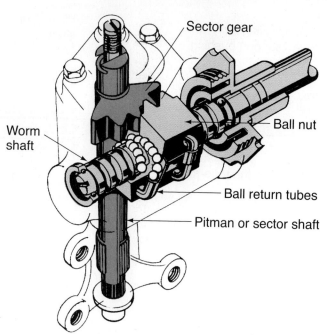

Figure 16.3 A recirculating ball steering gear. *(Courtesy of Federal-Mogul Corporation)*

Rack-and-Pinion Steering

On a rack-and-pinion steering system (**Figure 16.4**), the end of the steering shaft has a *pinion gear* that meshes with the *rack gear* (**Figure 16.5**). They are used on many cars because they are lighter than standard

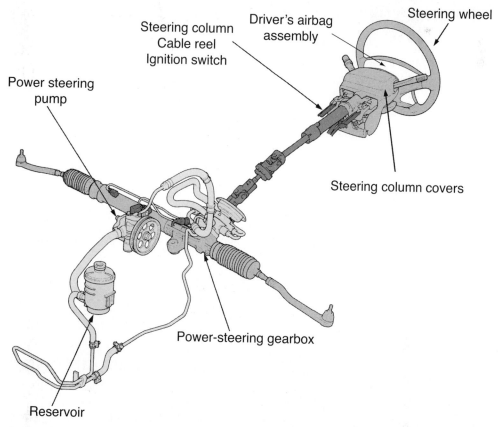

Figure 16.4 A complete power-assisted rack-and-pinion steering system. *(Courtesy of American Honda Motor Co., Inc.)*

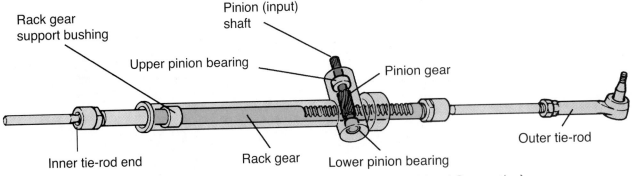

Figure 16.5 Parts of a rack-and-pinion steering gear. *(Courtesy of Federal-Mogul Corporation)*

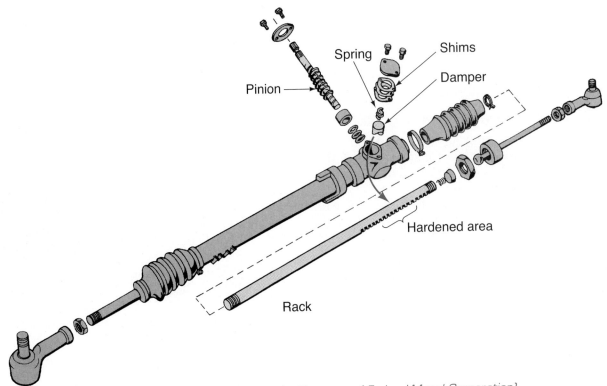

Figure 16.6 A damper holds tension against the rack. *(Courtesy of Federal-Mogul Corporation)*

steering gears and are easier to assemble to the vehicle at the factory. A typical rack-and-pinion often has a faster ratio.

A spring-loaded damper preloads the rack gear to prevent the **rack** from flexing, which would cause gear backlash (**Figure 16.6**). An adjustment screw or shims are used to adjust the amount of tension against the rack. The pinion shaft is usually supported by needle bearings. The rack is most often supported by plain bushings.

Rack-and-pinion steering systems are more easily damaged when the front wheels hit a curb or rock. They transmit more road shock that can be felt back through the steering wheel. The rack is mounted on rubber bushings to help cushion shocks (**Figure 16.7**). The rack-and-pinion unit can be mounted in several locations. Some units are on the subframe and others are on the fire wall.

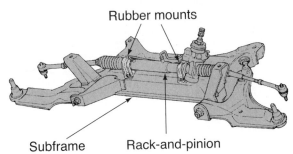

Figure 16.7 The rack is mounted on rubber bushings. *(Courtesy of Federal-Mogul Corporation)*

■ STEERING LINKAGE

The steering gear is connected to the wheels by the steering linkage. Steering linkage parts vary depending on the design used, but all designs include *tie-rods*, *steering arms*, and a *steering knuckle*. A comparison of

Chapter 16 Steering Fundamentals **367**

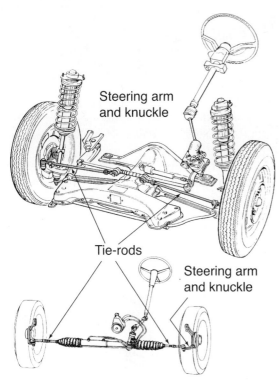

Figure 16.8 Both types of steering have tie-rods, steering arms, and a steering knuckle. *(Courtesy of Federal-Mogul Corporation)*

the steering linkage parts of the two major steering systems is shown in **Figure 16.8**.

When a recirculating ball and nut steering gearbox is used, there can be a number of different linkage designs used, depending on the suspension design. The most popular steering design in use with the short/long arm suspension is the parallelogram.

Parallelogram Steering Linkage

On passenger cars, a recirculating ball gear usually uses a **parallelogram steering** system (**Figure 16.9**). The name comes from the parallelogram shape made by the steering linkage during a turn. Parallelogram steering parts are shown in **Figure 16.10**. Tie-rods on each side are connected by the **center link**. The **pitman arm** is the part that connects the steering box to the center link. An **idler arm** supports the center link on the passenger side (**Figure 16.11**).

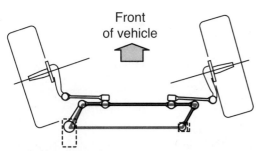

Figure 16.9 The shape of a parallelogram is made by the steering linkage during a turn.

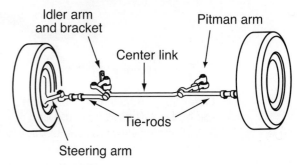

Figure 16.10 Parts of a parallelogram steering system during a turn.

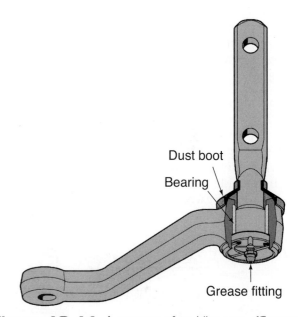

Figure 16.11 A cutaway of an idler arm. *(Courtesy of Federal-Mogul Corporation)*

Ball Sockets. **Ball sockets** connect the steering linkage parts. They allow parts to rotate during a turn and pivot as the steering deflects during a bump.

Tie-Rods. Tie-rod (**Figure 16.12**) ends are attached to pivot points at the front wheels. Not only do they transmit motion from the steering wheel to the front

Figure 16.12 A cutaway of a tie-rod end.

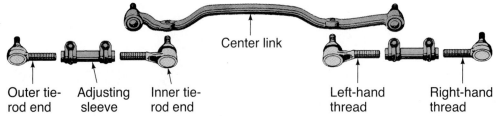

Figure 16.13 A threaded sleeve connects the inner and outer tie-rods. *(Courtesy of Federal-Mogul Corporation)*

wheels, they maintain the correct front wheel toe (the amount that the front tires are aimed inward or outward at the front). There is a threaded adjusting sleeve that connects the inner and outer tie rods (**Figure 16.13**). It has a right-hand thread on one end and a left-hand thread on the other end. During an adjustment, when it is turned it acts like a **turnbuckle**. Turning it one way shortens the tie-rod assembly. Turning it the other way lengthens it.

Steering Arm. The tie-rods attach to the front wheels at the steering arms (**Figure 16.14**). The steering arm is attached to the steering knuckle, which includes the *spindle* (**Figure 16.15**).

During a turn, the inside wheel must turn sharper than the outside wheel because they follow different circular paths (**Figure 16.16A**). The steering arms are angled inward, rather than being parallel to the frame. This angle, called the Ackerman angle, provides an important steering angle, **toe-out on turns** (**Figure 16.16B**).

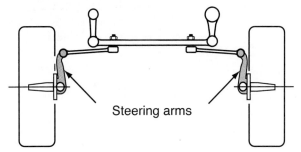

Figure 16.14 The tie-rods attach to the front wheels at the steering arms.

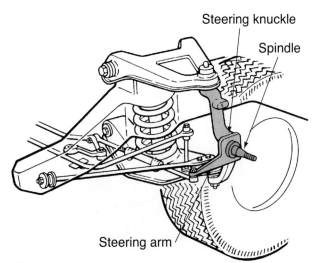

Figure 16.15 A steering arm and knuckle. *(Courtesy of McQuay Norris)*

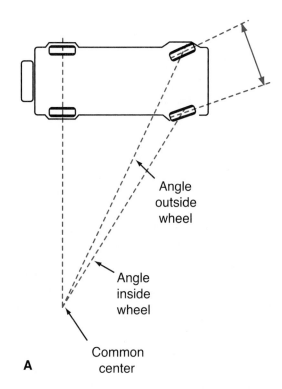

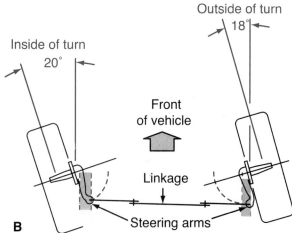

Figure 16.16 (A) The inside wheel turns sharper than the outside wheel because they follow different circular paths. (B) The steering arms are bent at an angle to allow the fronts of the tires to toe-out during a turn.

Rack-and-Pinion Steering Linkage

Rack and pinion does not have the complicated steering linkage used with a recirculating ball and nut steering box. In most systems, two tie-rods come out of the steering rack (see **Figure 16.6**). They have conventional tie-rod end ball sockets only on the outer ends (**Figure 16.17**). These attach to the steering knuckles. The inner tie rod ends are ball sockets, enclosed within *rubber bellows* or *boots* on the rack (**Figure 16.18**). In another type of rack-and-pinion system, the inner tie-rods attach to the center of the rack gear (**Figure 16.19**).

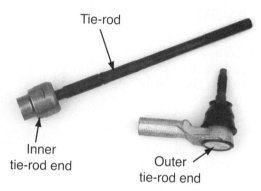

Figure 16.17 A rack-and-pinion tie-rod with inner and outer ends.

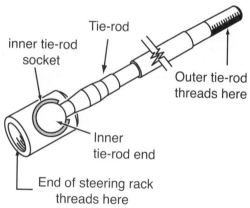

Figure 16.18 The inner pivots on a rack tie-rod are ball sockets.

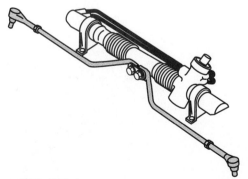

Figure 16.19 Inner tie-rods attach at the center of the rack.

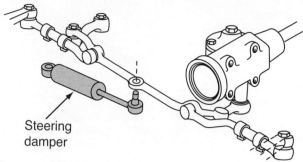

Figure 16.20 A steering damper absorbs road shocks to the steering linkage.

Steering Damper

Some steering linkages use a *steering damper*, a horizontally mounted shock absorber to minimize the effect of road shocks to the steering wheel (**Figure 16.20**).

■ STEERING COLUMN

The steering wheel is splined to a steering shaft located in the center of the steering column. A locknut retains the steering wheel to the shaft. The shaft is supported by bearings at the top and bottom of the steering column.

Included in the steering column is the turn signal switch and horn control. Sometimes a headlight and dimmer switch, windshield wiper and washer controls, transmission shift selector, cruise control, and ignition switch and lock (between the shift selector and the steering wheel) are built into the steering column. This is called the *European style* of controls.

Tilt and Collapsible Steering Columns

Tilt steering wheels and **collapsible steering columns** are also available. Tilt columns provide the driver with a means of adjusting the angle of the steering wheel. A tilt column includes a short section of shaft beneath the steering wheel connected to the steering column using gears or a universal joint. A typical tilt column uses a spring-loaded ratcheting mechanism to hold the steering wheel in position. A tilt release lever compresses the spring to release tension on the ratchet when steering wheel position adjustment is desired. Some columns are telescopic, allowing the height of the steering wheel in relation to the instrument panel to be adjusted as well.

Air bags are installed on the steering wheel in most vehicles and, by law, steering columns and shafts are required to be collapsible in case of an accident. There are different designs of collapsible steering columns and shafts. One type uses a two-piece outer section, retained with plastic shear pins (**Figure 16.21A**). It is combined with a two-piece inner steering shaft, also retained with plastic shear pins.

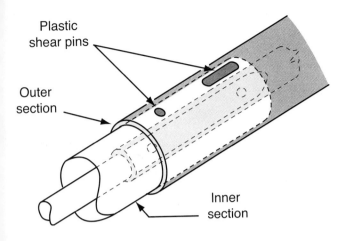

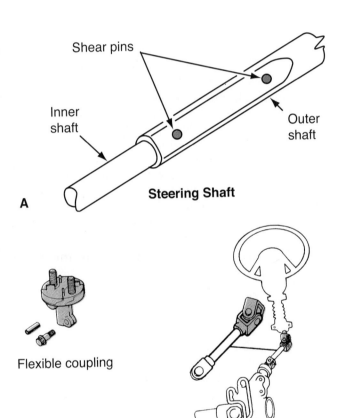

Figure 16.21 (A) A collapsible steering column and shaft. (B) A flexible coupling or a universal joint provides the attachment between the steering shaft and the splined input shaft of the steering gear. *(B Courtesy of Federal-Mogul Corporation)*

Service to the steering column and air bags, called supplemental inflatable restraints (SIR), is covered in Chapter 17.

The steering shaft connects the steering wheel to the steering gear. Because the steering column is mounted to the body of the car and the steering gearbox is mounted to the frame, there is movement or flex between the two. There is a *flexible coupling* or a *universal joint* between the steering shaft and the splined input shaft of the steering gear (**Figure 16.21B**). It allows a small amount of misalignment between the steering column and the steering gear. The flex coupling also keeps shock from transferring from the road to the steering wheel. When there is a greater angle, a universal joint is used.

When the transmission shift selector is located in the steering column, a slotted lock plate attached to the upper steering shaft is engaged with a lever to lock the steering wheel when the shift lever is in park and the ignition switch is off.

■ POWER STEERING

Steering systems found on most cars today are power assisted, although there are still some manual units made. Most power steering is hydraulic, with pressure supplied from the crankshaft by a belt-driven pump (**Figure 16.22**).

Power Steering Pump

A steering pump driven by a crankshaft belt supplies hydraulic power to assist steering effort. Three main types of power steering pumps have been used on cars (**Figure 16.23**). They are the *roller*, *vane*, and *slipper* types. Of these, the vane design is the most common (**Figure 16.24**). Though there are several types of power steering pumps, they all work in the same way.

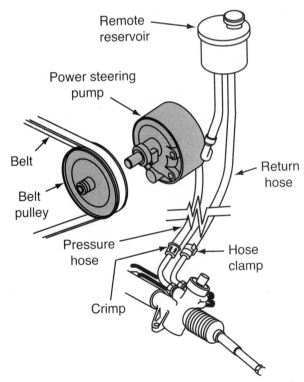

Figure 16.22 Parts of a hydraulic power steering system.

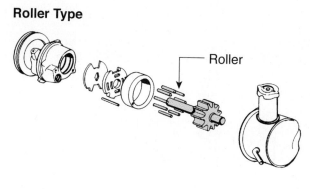

Roller Type

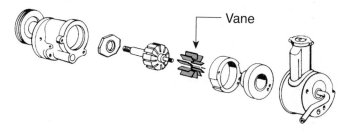

Vane Type

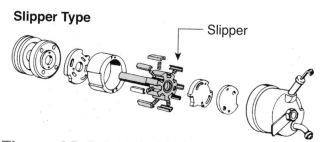

Slipper Type

Figure 16.23 Types of power steering pumps. *(Courtesy of Federal-Mogul Corporation)*

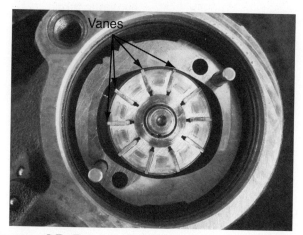

Figure 16.24 Internal parts of a vane-type power steering pump. *(Courtesy of Tim Gilles)*

As the pump shaft is turned, oil is drawn into the pump. The oil is squished into a smaller area, which traps it and pressurizes it for delivery to the steering gear.

Power Steering Pump Operation. The power steering pump works under difficult conditions. Steering assist is needed most when the car is stopped or nearly stopped. So a steering pump must deliver sufficient

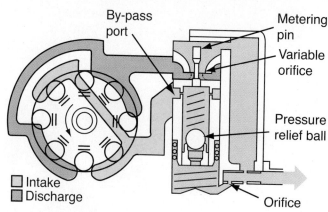

Figure 16.25 A control valve regulates fluid flow in the power steering pump.

fluid flow to be able to provide steering assist at low engine rpm and idle. A pump develops more flow as it is driven at higher speeds. When the pump causes excess fluid to flow at cruising rpm, it must divert this fluid back to the inlet side of the pump so it can be returned to the reservoir without creating pressure.

There is a two-stage relief valve in the pump. A control valve monitors the turning effort on the wheel to provide the correct amount of assist (**Figure 16.25**). Its two functions are to control flow and limit maximum pressure. The **flow control valve** is almost always working. The pressure relief part of the valve hardly ever opens. It opens when the steering wheel is held all the way to one side against a steering lock. This causes noise as pressure is released.

When the pump turns at low speed, fluid action is as shown in the drawing. At higher speeds, the pump by-pass port opens, allowing the excess fluid to return to the pump intake. When pressure in the steering system becomes too high, the **pressure relief valve** opens to allow the pressure to bleed off to the fluid intake side of the pump.

Steering Power Consumption. Power steering pumps require a considerable amount of horsepower to operate. When steering effort is high on small vehicles, sometimes the air-conditioning compressor will be shut off to compensate for the draw of the power steering system. Many late-model cars have computer-controlled charging systems that also shut off the alternator at idle when the power steering system is under load.

Power assist on very few later-model rack-and-pinion steering units is supplied by an electric motor. These use the alternator to supply power to the steering unit. Some late-model General Motors vehicles have a specifically electronically controlled electric unit located inside the vehicle under the dash.

Power Steering Hose

There are usually two power steering hoses. One is a pressure line and the other is a return line (see **Figure 16.22**). The return hose is not under high pressure like

372 SECTION 4 SUSPENSION AND STEERING

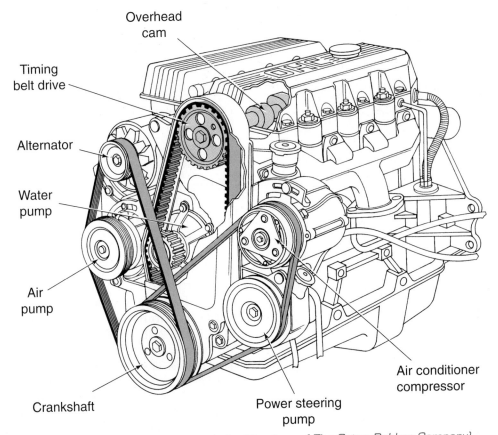

Figure 16.26 Belt-driven accessories and camshaft. *(Courtesy of The Gates Rubber Company)*

the pressure side is. The hose on the pressure side is made of reinforced oil-resistant synthetic rubber, crimped to metal tubing. It must be able to withstand temperatures of 300°F and handle pressures up to 1,500 psi. Some pressure hoses also have a flow reducer, an orifice to help control flow. The return line is usually connected to the power steering pump with a hose clamp. It must be able to withstand high temperatures, but operates under low pressure. It must also be resistant to damage by power steering fluid.

Drive Belts

Power steering pumps are usually driven with a belt from the crankshaft (**Figure 16.26**). The drive belt can be either a V belt or a V-ribbed belt.

Belt Material. Belts are very strong and flexible with **tensile cords** to provide strength (**Figure 16.27**). The overcord material on the top of the belt is made of **neoprene** or another kind of oil-resistant artificial rubber. The undercord of the belt is the area beneath the tensile cords. It supports the cord and transfers loads to the pulleys. Sometimes the undercords have a cord support platform with textile cords running perpendicular to the tensile cords. Tensile cords prevent the belt from sagging in the middle, which results in uneven load distribution and early belt failure.

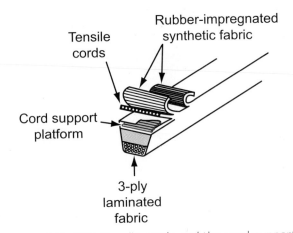

Figure 16.27 Tensile cords and the cord support platform.

V Belts. A V belt has more surface area in contact with the pulley groove than a flat belt of the same width. Sometimes notched belts are used on small diameter pulleys that would cause severe bending stress on the belt.

Belt Cords. Most V belts have polyester tensile cords. The strength of a belt is determined by the placement of the belt's tensile cords. **High cordline**

Chapter 16 Steering Fundamentals 373

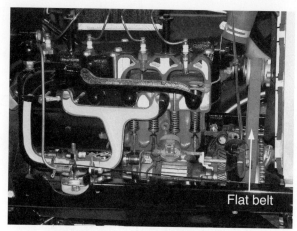

Figure 16.28 This early Ford engine used a flat belt. *(Courtesy of Tim Gilles)*

> **Vintage Chassis**
>
> Early cars used flat drive belts (**Figure 16.28**) until the V belt was invented in 1917.

belts (**Figure 16.29**) are stronger but require more material to manufacture. Center cord belts are cheaper but do not last as long.

High-cord belts have about 40 percent more cords because the cord is at a wider part of the belt. In Society of Automotive Engineers (SAE) tests, high-cord belts lasted about four times longer than center cord belts.

In past years, the edges of premium belts were covered with a fabric cover to protect them from the elements. Today's belts have no cover and do not show wear as easily. Sometimes they can appear to be in good shape, even though they are ready to fail.

Some engines use dual belts to drive accessories; these must be replaced in pairs when they are worn.

V-Ribbed Belts. V-ribbed belts are ribbed on one side (**Figure 16.30**) and flat on the other. **Figure 16.31** compares the pulleys for typical V belts and V-ribbed belts. Some vehicles use a short V-ribbed belt to drive a single accessory.

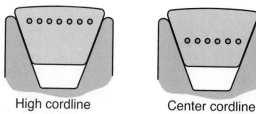

Figure 16.29 High cordline belts are stronger but require more material to manufacture. The higher cord position is evident when viewing the belt from the side.

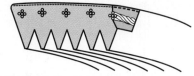

Figure 16.30 Construction of a V-ribbed belt.

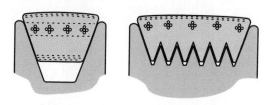

Figure 16.31 Comparison of a V-ribbed belt and a V-belt in their pulleys.

Serpentine Belt Drive. Serpentine belt drives first appeared in the late 1970s. Today, most newer vehicles use a single V-ribbed serpentine belt to drive all accessories (**Figure 16.32**). The thinness of a V-ribbed belt makes it more flexible so it can bend around smaller pulleys. It can also be bent backward so both sides of the belt can be used to transmit power. Usually the ribbed side matches the pulley grooves of accessories and the flat side goes against a spring-loaded tensioning roller, but the flat side is also capable of transmitting power.

The belt arrangement is called *serpentine* because the belt follows a snakelike path, weaving around the various pulleys. Compared to V belts, serpentine belts are easier to install, take up less space, transmit power more efficiently, and last longer. **Figure 16.33** compares the pulley space required for typical V belts and V-ribbed belts.

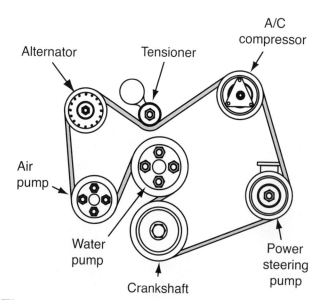

Figure 16.32 A serpentine V-ribbed belt drive.

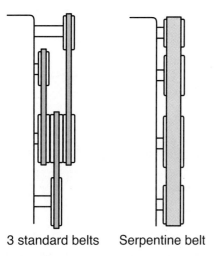

Figure 16.33 Serpentine belts save space.

■ TYPES OF POWER STEERING

Power steering systems are either rack-and-pinion or recirculating ball and nut units with a hydraulic control system added. Most are **integral power steering** systems (see **Figure 16.22**). *Integral* means "part of." This means the power steering components are within the steering gear. Integral units can be either recirculating ball or rack-and-pinion (**Figure 16.34**). Integral recirculating ball gearboxes have a ball nut and sector gear, just like a manual gearbox. The ball nut is housed within a *power piston*. Pressurized oil enters a chamber on either side of the power piston to provide steering assist (**Figure 16.35**).

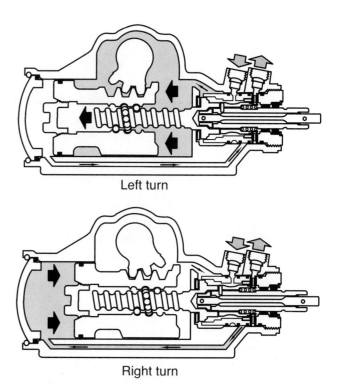

Figure 16.35 Pressurized oil enters a chamber on either side of the power piston to provide steering assist.

To sense and control power steering assist, gearboxes use either a pivot lever or a torsion bar acting on a spool valve. On the *pivot lever type*, turning effort causes a spool valve to be moved by a pivot lever (**Figure 16.36**). When the spool valve moves, fluid is

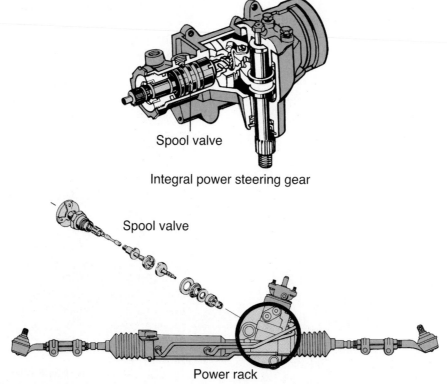

Figure 16.34 Types of integral power steering. *(Courtesy of Federal-Mogul Corporation)*

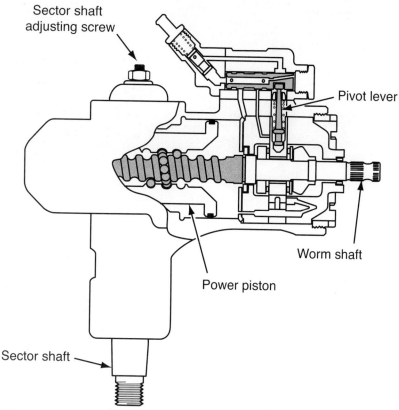

Figure 16.36 A pivot lever–type power steering gear.

directed to one side or the other of the power piston to provide the assist.

On the *torsion bar type,* also called a *rotary valve type,* a small sensitive, torsion bar twists in response to steering effort. This turns a rotary spool valve to direct pressure to the correct side of the power piston (**Figure 16.37**).

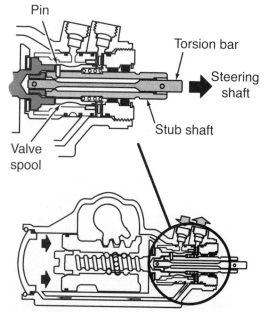

Figure 16.37 A torsion bar–type power steering gear.

A *linkage-type* gearbox, found on older cars and light trucks, uses a standard gearbox with a piston attached to the steering linkage. **Figure 16.38** shows fluid assist during left and right turns.

In a rack-and-pinion system, fluid is directed to a chamber on either side of the rack in a manner similar to the **linkage-type power steering**. **Figure 16.39** shows how power rack-and-pinion steering works during left and right turns. To sense and control assist, some racks use a torsion bar attached to the input shaft like in recirculating ball and nut steering. Other units use a spool valve that moves in response to the up-and-down movement of the pinion gear as it tries to drive the rack.

Electronically Controlled Variable Effort Power Steering

Power steering is installed on vehicles so that a car can be easily steered at low speeds. Once the vehicle achieves a reasonable speed, a fixed level of power assist is no longer necessary and can interfere with a driver's "feel" for the road.

On some late-model vehicles, the speed of the vehicle determines how much power assist is given. One system controls fluid output from the pump. Another type controls the amount of fluid pressure available in the power steering gear.

In pump-controlled units, an electronically managed actuator solenoid changes fluid flow in the power

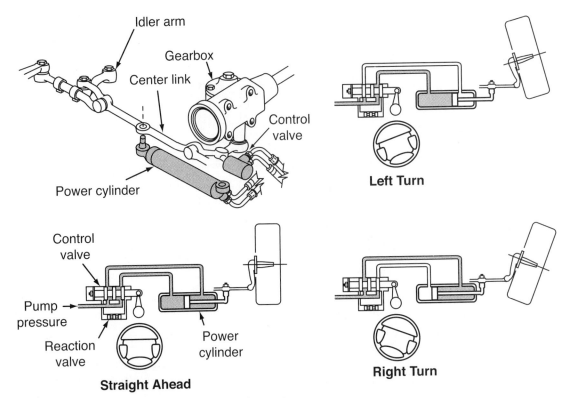

Figure 16.38 Fluid flow in a linkage power steering system.

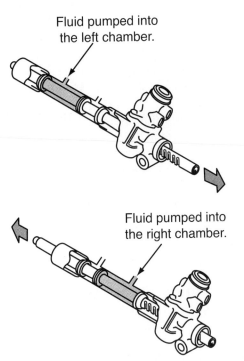

Figure 16.39 Fluid power assist in a rack-and-pinion unit.

pump flow is decreased. This increases the steering effort and gives the driver a better feel of the road.

In the steering gear–controlled units (**Figure 16.40**), the amount of boost available at the power steering gear is sensed and a module responds by changing fluid flow in the pump control valve to provide the correct amount of assist.

General Motors has an electronic variable assist steering system called **Magnasteer**. This system has a rotary actuator attached to the input shaft of the hydraulically powered rack-and-pinion steering gear (**Figure 16.41**). A control module varies a supply of electrical current to the actuator, which uses electromagnetic force on the steering gear input shaft to increase or decrease steering effort.

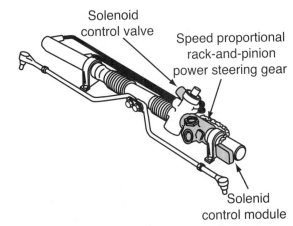

Figure 16.40 A speed proportional power steering unit.

steering pump control valve. This is done with a pulse width modulated signal, like in fuel injection systems. Maximum power assist occurs at 1,500 rpm (fast idle) while the vehicle is at rest. The computer varies the solenoid on-time, which allows pump pressure to be higher. As the speed of the vehicle increases, the amount of

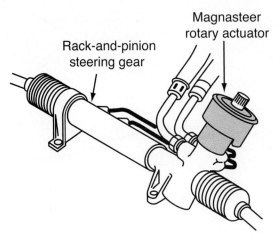

Figure 16.41 The Magnasteer system uses a magnetic actuator.

Turning the steering wheel rotates the input shaft/spool valve with its sixteen permanent magnet segments. The rotary actuator solenoid that surrounds the input shaft has sixteen matching segments, electromagnetically powered by a coil. At high speeds the control module causes the electromagnetic segments to attract the permanent magnet segments, increasing steering effort to improve road feel. At low speeds, the module reverses the polarity on the actuator's magnetic segments, reducing steering effort.

More advanced variable effort steering systems can adjust steering effort in response to lateral forces. A computer senses g-forces from a lateral accelerometer. It compares this signal with vehicle speed and a signal from the digital steering angle sensor that tells how fast the steering wheel is being turned (**Figure 16.42**). The steering effort is adjusted to fit the condition.

FOUR-WHEEL STEERING

Some manufacturers produce four-wheel steering systems. In four-wheel steering all four wheels turn, improving handling and helping the vehicle make tighter turns.

Generally, at low to medium speeds the rear wheels steer in the opposite direction as the front wheels to reduce turning radius, helping with parking. At high speeds they turn in the same direction as the front wheels to improve maneuverability during lane changes.

The front wheels do most of the steering. Rear-wheel turning is generally limited to 5 to 6 degrees during an opposite direction turn. During a same direction turn, rear-wheel steering is limited to about 1 to 1.5 degrees. If the rear-wheel turning radius was not limited, the wheels would bump into the curb as the vehicle attempted to maneuver out of a parallel parking spot (**Figure 16.43**).

Why Four-Wheel Steering?

During a turn, the rear wheels are always trying to catch up with the change in direction of the front wheels. This is called lag. At higher speeds, the result is sway. As the front wheels return to straight ahead, the rear wheels must compensate again (more sway).

Centrifugal force moves the rear of the vehicle sideways, causing the tires to slip. This can cause the car to spin out. During a high-speed turn, when the rear wheels are turned the same direction as the front wheels, the entire vehicle moves one way. **Side slip** is reduced and stability is improved. Steering response is faster because the lag between the front and rear wheels is eliminated.

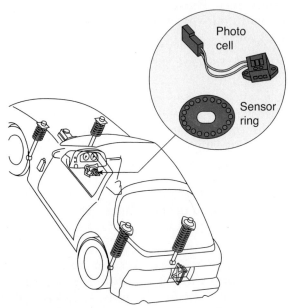

Figure 16.42 A digital steering angle sensor tells the computer how fast and how far the steering wheel is turned.

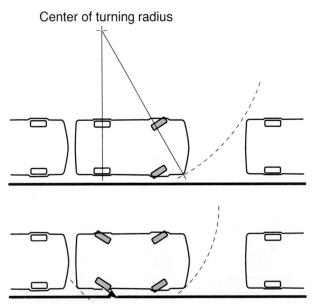

Figure 16.43 The action of conventional front wheel steering and four-wheel steering when parallel parking.

TYPES OF FOUR-WHEEL STEERING

There are several types of four-wheel steering: mechanical, hydraulic, and electric/hydraulic. The mechanical four-wheel steering system used by Honda has two steering gears, one in the front and the other in the rear. The conventional steering found on their two-wheel steering vehicles is used in the front. There is a shaft connecting the front steering box to the one in the rear.

The steering shaft from the front turns a planetary gear inside of a stationary internal gear (**Figure 16.44**). An eccentric pin on the outside of the planetary gear moves the steering rod. The wheels are turned with or against the front wheels (**Figure 16.45**). As the steering wheel is first turned, the front and rear wheels turn the same direction. After the steering wheel has turned past 120 degrees, the wheels start to straighten out again. At 240 degrees they are straight. Then, the wheels turn in the opposite direction with further steering wheel turning (for slow speed parking assistance). The maximum amount of opposite direction turning is limited to 5.3 degrees.

Hydraulic Four-Wheel Steering

In a simple hydraulic four-wheel steering system, the rear wheels are turned in only the same direction as the front wheels. Turning is limited to 1.5 degrees. The system only operates at speeds in excess of 30 mph in forward gears. A two-way hydraulic cylinder mounted in the rear turns the wheels. Fluid pressure comes from a pump driven by the differential. Fluid is supplied by a fluid storage reservoir in the engine compartment. As

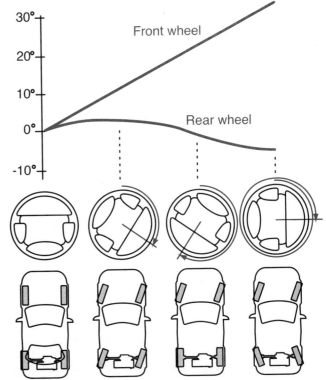

Figure 16.45 As the steering wheel is turned, the rear wheels are turned with or against the front wheels.

the front wheels are turned, pressure is increased to the rear-wheel hydraulic cylinder to turn the rear wheels.

Electric Four-Wheel Steering

Since the early 1990s, Honda has also used electronically controlled rear-wheel steering. A sketch of that system is shown in **Figure 16.46**. It is similar to the four-wheel steering system used in General Motors trucks, which use an electric rack-and-pinion steering unit in the rear (**Figure 16.47**). The turning radius of a truck with four-wheel steering is similar to that of a compact car.

NOTE: *Trucks equipped with these steering systems cannot use tire chains.*

Rear-wheel steering increases stability during freeway speed lane changes. It also offers much better control when towing a trailer. A mode select switch on the instrument panel gives the driver a choice of applications, including a selection for towing. The amount and direction of steering is based on an algorithm that considers the mode selection, the steering wheel position, and the speed of the vehicle. An electronic controller reacts to inputs and operates the system in one of three phases.

1. Negative phase, used at speeds from 0 to 45 mph, turns the rear wheels in the opposite direction of the front wheels. They steer at 25 percent of the steering angle of the front wheels. Twelve degrees is the maximum turning amount.

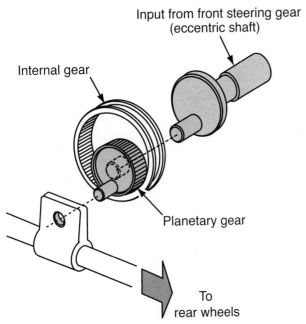

Figure 16.44 A steering shaft from the front turns a planetary gear inside a stationary internal gear.

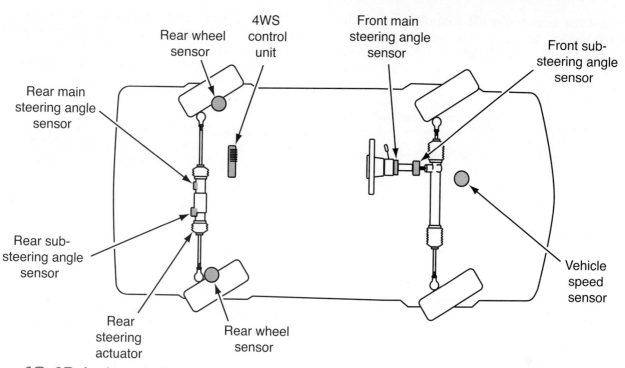

Figure 16.46 An electronically controlled four-wheel steering system.

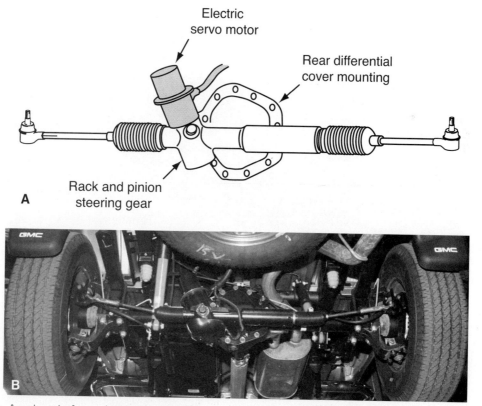

Figure 16.47 An electric four-wheel steering gear.

2. Neutral phase occurs when only the front wheels steer. This is the fail-safe mode, in which a powerful "return to center spring" keeps the rear wheels steering straight ahead. This occurs at approximately 45 mph.

3. Positive phase occurs during turns at more than 45 mph. The rear wheels steer in the same direction as the front wheels. When the selector is in the "four-wheel tow" position, positive steering starts at 25 mph.

Parts of the system include a handwheel position sensor located at the base of the steering column, a vehicle speed sensor, the electrically driven rack-and-pinion steering box, and the electronic control module.

The handwheel sensor determines how fast the steering wheel is turning, the number of degrees it is turned, and when it is at zero. A sensor also tells when the rear steering is centered.

NOTE: *The steering wheel is called a handwheel because in an electronically controlled steering system it is similar to a joystick.*

ELECTRONICALLY CONTROLLED ELECTRIC STEERING

Electronically controlled electric rack-and-pinion steering systems use computer controls to control an electric motor–driven steering gear (**Figure 16.48**). The computer considers three factors: road speed, amount of steering wheel turn, and how fast the wheel is turned. Antilock brake speed sensors and a steering angle sensor provide these inputs.

The motor-driven steering unit must be able to deliver a sufficient amount of current to the motor for steering. It must also be able to rapidly change direction. The motor can receive up to 75 amps. More current means more steering force. A sensor monitors steering wheel movement and load. Based on that signal, the computer regulates current flow to the steering gear motor.

There are several advantages to this system:

- It provides steering assistance even when the engine is not running.
- Steering feel is adjustable for given driving conditions.
- Because there is no power steering fluid, there are no leaks.
- It can be used in electronically controlled four-wheel steering systems.

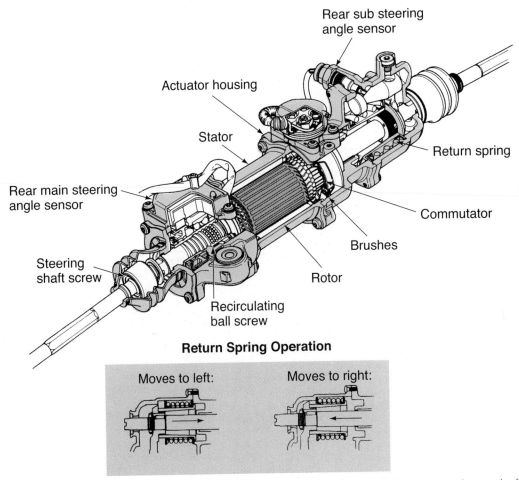

Figure 16.48 An electronically controlled electric motor–driven recirculating ball rear steering rack. *(Courtesy of American Honda Motor Co., Inc.)*

REVIEW QUESTIONS

1. One of the two most common types of automotive steering gears is the recirculating ball and nut. What is the name of the other one?
2. What is the term that compares the number of teeth on the driving gear to the number of teeth on the driven gear?
3. When the steering wheel is turned all the way in one direction, it stops against a _____.
4. What is the name of the steering part that is angled so that the front wheels toe-out during a turn?
5. What is another type of steering shaft coupling besides a flex coupling?
6. What are the names of the three main types of steering pumps that have been used on automobiles?
7. Which is the most common type of power steering pump design?
8. What is the name of the power steering design that has the power steering components within the steering gear?
9. What is the name of the power steering design that uses a standard gearbox with a piston attached to the steering linkage?
10. In most four-wheel steering systems, when the steering wheel is turned all the way in one direction, do the rear wheels turn the same way?

PRACTICE TEST QUESTIONS

1. Which of the following are true about turning radius?
   A. The outer wheel turns sharper
   B. The inner wheel turns sharper
   C. Both wheels turn the same amount

2. Front tires are supposed to toe outward during a turn.
   A. True   B. False

3. Technician A says that steering assist is needed most when the car is stopped or nearly stopped. Technician B says that the power steering must be able to provide more steering assist at low engine rpm and idle. Who is right?
   A. Technician A   B. Technician B
   C. Both A and B   D. Neither A nor B

4. On some power steering systems, the faster the vehicle is driven, the _____ power assist is given.
   A. More   B. Less

5. Which is/are true about four-wheel steering systems?
   A. The front and rear wheels turn the same amount
   B. The front wheels turn more than the rear
   C. The rear wheels turn more than the front

6. On most four-wheel steering systems, the rear wheels steer opposite the front wheels to reduce turning radius:
   A. At low to medium speeds
   B. At high speeds
   C. Never

7. The most popular steering design in use with the short/long arm suspension is the:
   A. Rack-and-pinion
   B. Recirculating ball and nut
   C. Worm and roller

8. A "fast" steering ratio is about _____ turns lock to lock.
   A. Three   B. Four
   C. Five   D. Ten

9. In the figure shown below, the part indicated by the arrow is a
   A. Idler arm   B. Tie rod
   C. Center link   D. Pitman arm

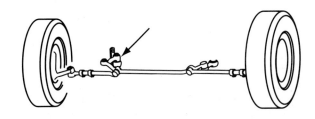

10. An electronically variable power steering system
    A. Increases assist at lower vehicle speeds.
    B. Decreases assist as vehicle speed increases.
    C. Both of the above.
    D. Neither of the above.

CHAPTER 17

Steering Service

■ OBJECTIVES

Upon completion of this chapter, you should be able to:
- Describe problems with steering system components.
- Inspect the condition of the steering system.
- Explain different types of repairs that can be made to the steering system.
- Describe repair procedures for rack-and-pinion and parallelogram steering systems.
- Perform steering system repairs in a safe and professional manner.

■ KEY TERMS

click-type tension gauge
clock spring
deploy
high point
integral reservoir
jam nut
morning sickness
preload adjustment
pyrotechnics
serpentine belt
steering free play
torque wrench

■ INTRODUCTION

In normal use, with the correct, clean lubricant and good seals, steering parts do not wear appreciably. When a seal is damaged and a lubricant leaks out, parts can rub against each other and dirt can leak in. Also, road hazards such as potholes can cause hard impacts to steering parts. This result in immediate damage or problems down the line from further wear. Belts and hoses can fail, causing operating problems with the steering system.

■ FLUID LEVEL CHECKS

Power steering fluid should be hot when checking its level (**Figure 17.1**). With the engine idling, turn the wheel several times in each direction to raise the temperature before shutting the engine off to check the

Figure 17.1 Check power steering fluid level when hot.

SAFETY NOTE The engine should not be running when checking power steering fluid.

fluid level. Sometimes a system has a reservoir located in a remote location from the pump.

Type of Fluid

Power steering fluid is recommended in power steering systems. Some manufacturers specify a special type of power steering fluid. Installing the wrong type of fluid in these vehicles can damage seals within the system.

■ DIAGNOSING STEERING PROBLEMS

Steering and suspension problems can be caused by looseness, incorrect height, and wheels that are not aligned to specifications. Wheel alignment theory is discussed in Chapter 18, but some of the terms will

Vintage Chassis

In the past, many manufacturers allowed the use of automatic transmission fluid instead of power steering fluid.

382

be discussed here when they apply to a particular problem.

Noise

Noise from the steering system can be due to several things. A loose power steering belt can result in a jerky feeling during a turn. A loose belt can also squeal when the wheel is turned, especially when the wheel is turned all the way to the end of its travel against a steering lock. If allowed to continue slipping without a belt tension adjustment, the belt will become glazed. The pulley can also wear out.

When the steering wheel is turned against a lock, the pump must bypass the extra pressure that develops. A "whirring" sound is normal as bypass occurs.

Noise can also occur because of a lack of fluid in the power steering reservoir. This will cause a growl that is louder when the steering wheel is turned, especially under load.

Loose parts will clunk when going over bumps or when turning the steering wheel from side to side.

Hard Steering

Hard steering can be caused by binding in the steering linkage, steering box problems, low tires, or incorrect wheel alignment settings. Always check power steering belt tension and fluid level. These are common causes of hard steering.

Tire Wear

Steering and suspension problems often result in unusual tire wear. Worn parts can cause scalloped or gouged tire wear (see **Figure 13.10**). Tire wear can also be due to improper alignment adjustments (see Chapter 19).

■ STEERING PART INSPECTION

The condition of the steering system must be inspected before doing a wheel alignment. Excessive vibration of front end parts can allow the wheels to move on their own. This can cause vibration, noise, tire wear, and unsafe driving conditions.

Remember to inspect the wheel bearings to ensure that they are not too loose or too tight. Also, check to see if they feel rough before performing other tests on the front end. If a wheel bearing is loose, the wheel alignment equipment will not be able to make accurate readings.

■ STEERING LINKAGE INSPECTION

The fastest way to discover obvious looseness is by performing a "dry park check." Have an assistant turn the wheel from side to side with the tires on the ground or supported by a wheel contact lift. Look for loose parts while observing the steering and suspension as the wheels turn.

NOTE: *A vehicle with power steering must have its engine running for this test to prevent looseness that might be present in the steering gear when there is no fluid pressure.*

If you can feel looseness at one steering position more than others, perform the check with the steering wheel in that position. When looseness is difficult to diagnose, some technicians prefer to jack up only one wheel at a time and check for looseness on the opposite side. This puts more effort against the loaded side, making loose parts more apparent. Switch sides to test the other wheel.

While the vehicle is in the air during a lube inspection, the steering linkage can be manually inspected without assistance. With the tires raised and hanging free, turn the steering wheel through its full range of travel. Check to see that there is no binding in the steering linkage or gearbox.

Steering Gear Looseness

Excessive **steering free play** can be caused by a worn steering column flexible coupling, worn steering linkage, worn rack mount bushings, loose steering box bolts, or a worn or misadjusted steering gear. Steering gearbox looseness is usually adjustable.

Check steering wheel free play with the engine off. Power steering vehicles should have the engine on. Move the wheel back while checking to see how far it moves before the wheels start to move.

NOTE: *Check the steering gearbox to see if it feels smooth throughout its entire travel. Check it in the straight-ahead position for binding at the center position and from lock to lock to see that there is no looseness or rough feel.*

Parallelogram Inspection

Tie-rods are held in position by the pitman arm, idler arm, and center link. Parts should allow pivoting and turning but prevent movement. They are tested by attempting to move them.

Pivot sockets are filled with lubricant and sealed with rubber seals. Check sockets for looseness or damage to the seal.

NOTE: *When checking for steering looseness, be sure that the steering wheel is in the straight-ahead position. This is where the most wear would tend to be.*

Idler Arm Inspection. To check the idler arm, grasp the center link at a point that is as near to the idler arm as possible. Push it firmly up and down while looking for movement (**Figure 17.2**). Movement here can result in a change in toe-in of the front wheels, causing tire wear.

Pitman Arm and Center Link Inspection. A small amount of movement where the pitman arm attaches to the steering linkage is sometimes considered normal. Check the manufacturer's specifications.

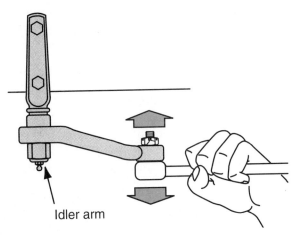

Figure 17.2 Checking an idler arm.

Tie-Rod Inspection. To inspect tie-rods for wear, push them by hand to see if there is any perceptible movement (**Figure 17.3**). There should be no movement. There are two types of tie-rod sockets. One is preloaded with no clearance. The other is spring loaded and can be compressed vertically. One way to test a tie-rod is to compress it (**Figure 17.4**). Spring-loaded sockets should have firm spring pressure when compressed all the way. Preloaded sockets should not give at all.

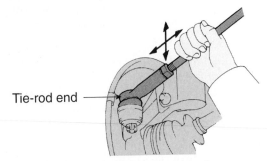

Figure 17.3 Checking for tie-rod end wear.

Figure 17.4 Testing a compressible tie-rod end.
(Courtesy of Tim Gilles)

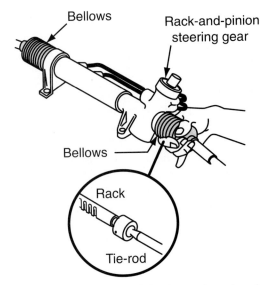

Figure 17.5 Location of the inner tie-rod end.

Rack-and-Pinion Steering Linkage Inspection

The inside tie-rod is housed within the bellows boot (**Figure 17.5**). Feel for looseness by grasping the tie-rod through the bellows boot. There should not be any looseness in the tie-rod and its socket. The steering linkage must be in its usual position for this test. If the tie-rod is hanging down in its socket, it could be bound up, preventing an accurate check.

■ STEERING LINKAGE REPAIRS

Steering linkage parts are locked by wedging them against each other with beveled connections called "tapers" (**Figure 17.6**). Locknuts, usually castle nuts with cotter pins, provide additional protection. When disassembling tapers to service or replace parts, there are several ways to get them apart, which are listed here in order of preference:

1. Tapers can usually be separated using a puller (**Figure 17.7**).
2. Linkage parts can also be separated by deforming the outside of the taper using two hammers (**Figure 17.8**).

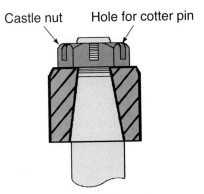

Figure 17.6 A steering linkage taper connection.

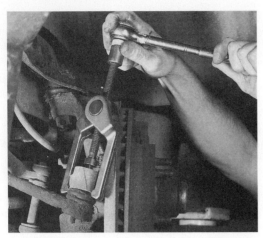

Figure 17.7 Using a tie-rod end puller. *(Courtesy of Federal-Mogul Corporation)*

Figure 17.10 These slotted holes are for mounting an adjustable idler arm. *(Courtesy of Federal-Mogul Corporation)*

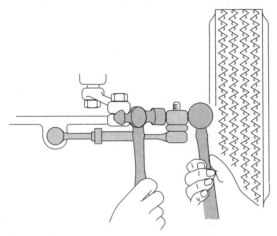

Figure 17.8 Using two hammers to loosen a steering taper connection.

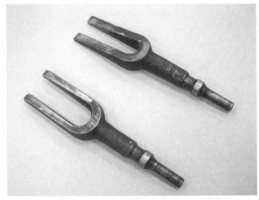

Figure 17.9 Two sizes of pickle forks for use with an air chisel. *(Courtesy of Tim Gilles)*

3. A *pickle fork* can be used if the part will not be reused (**Figure 17.9**). The pickle fork often cuts the seal and ruins the part so be very careful when performing this procedure.

Idler Arm Replacement

During replacement of an idler arm, check to see if it is one of the adjustable types (**Figure 17.10**). If the idler arm and pitman arm are not level to each other, the vehicle will experience bump steer. Bump steer causes toe to change when going over bumps (see **Figure 19.8**), which will also result in tire wear.

Pitman Arm Replacement

Removal of the pitman arm requires a special puller (**Figure 17.11**). It must be replaced in the same position on the splines of the steering gear. Be sure to check its position before removing it. Pitman arms often have a "blind spline." One wide spline ensures that the pitman arm can only be installed in one position.

Tie-Rod End Replacement

Tie-rods ends are replaceable, whether the steering system is a rack-and-pinion or parallelogram steering system. Tie-rod ends are threaded to provide a means of adjusting toe. When a tie-rod is replaced, measure

Figure 17.11 A pitman arm puller. *(Courtesy of Federal-Mogul Corporation)*

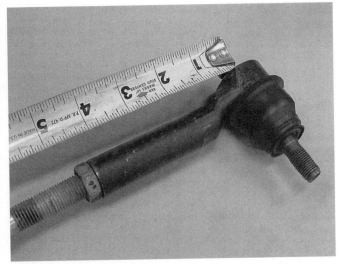

Figure 17.12 Measure the tie-rod before disassembling it. *(Courtesy of Tim Gilles)*

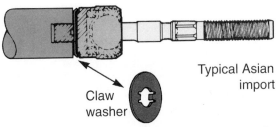

Figure 17.13 An inner tie-rod end held by a claw washer. *(Courtesy of Federal-Mogul Corporation)*

the old tie-rod assembly before disassembling it. An approximate toe-in adjustment can be made to the new one prior to installation (**Figure 17.12**). When removing the old tie-rod end, count the number of turns it takes to remove it. Then, turn the new tie-rod end onto its threads the same number of turns.

Some vehicles have tie-rods that appear to be the same on both ends. The difference is in the threads. One end has left-hand threads and the other has right-hand threads. It is possible to install the entire shaft backwards (which will work). Mark the tie-rod to identify the inner or outer end before removing it.

Before tightening tie-rod clamps, check to see that they are in good condition and are positioned properly so they can be clamped tightly (see **Figure 19.39** and **Figure 19.40**). Before doing a front end adjustment, spray penetrating oil on the threads of the tie-rods. Do this during the steering linkage inspection so the lubricant has time to soak in.

Rack-and-Pinion Tie-Rods

Outer rack-and-pinion tie-rod ends are serviced the same way as other tie-rod ends. There are several ways that the inner tie-rod socket is held to the ends of the rack. Be sure to follow the instructions that come with the replacement part.

A typical Asian import uses a *claw washer retainer*. The washer has inner tabs that fit into a recessed area of the rack (**Figure 17.13**). Deform the outside of the washer with a hammer and punch so that it conforms to the flats on the outside of the tie-rod socket.

Three other types of retaining methods are shown in **Figure 17.14**.

1. One method uses a **jam nut**. Some of these types use a thread adhesive. The rack must be held from turning while tightening the jam nut (**Figure 17.15**).
2. Another method uses a special washer that is staked into a flat area in the end of the rack.

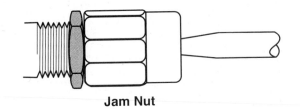

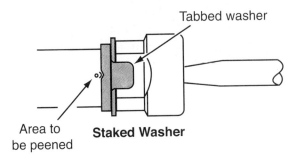

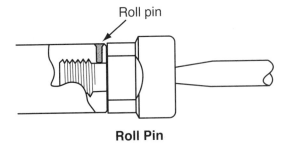

Figure 17.14 Three types of tie-rod retaining methods.

> **CASE HISTORY**
>
> A technician ordered a tie-rod end for a small pickup truck. She specified that the tie-rod end was for the outside left. When the new tie-rod was delivered from the parts store, she attempted to install it. The threads did not match up. The parts store double-checked and determined that they had sent the correct tie-rod. Further investigation showed it was an inside tie-rod end. Previously, the entire tie-rod assembly had been installed backwards. This had no effect on the operation of the steering but resulted in some confusion during service.

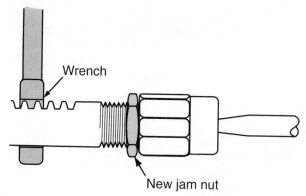

Figure 17.15 Hold the rack with a wrench while tightening the jam nut.

3. When a roll pin is used to hold the tie-rod to the rack, it must be drilled to remove it. After installing the new tie-rod end, a new hole is drilled for a new pin 90 degrees to the original one.

New bellows boot kits are available. Urethane boots are used by many domestic manufacturers. These are superior to neoprene boots.

CAUTION Accidentally deploying (activating) an air bag can result in an injury or death.

■ STEERING WHEEL, COLUMN, AND AIR BAG SERVICE

Removal of the steering wheel is sometimes necessary. This could be because a turn signal switch requires replacement or a horn does not work.

NOTE: *On most steering systems, when the steering wheel is off-center, adjustment is usually made by turning the tie-rods. It is not done by removing the steering wheel.*

Air Bags

Many newer vehicles come equipped with air bags, also called supplemental inflatable restraints (SIR) or a supplemental restraint system (SRS), to protect occupants of a vehicle during a crash (**Figure 17.16**). These parts are expensive and can be damaged during careless or uninformed service. Before attempting to remove a steering wheel on one of these vehicles, consult a service manual.

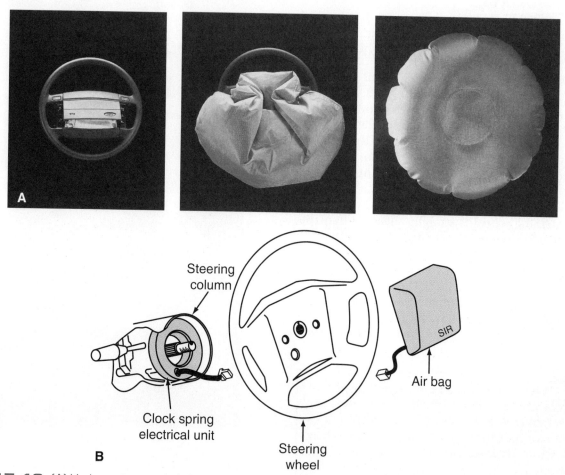

Figure 17.16 (A) Various stages of air bag inflation. (B) Steering column parts of an air bag. (A, *Courtesy of TRW*)

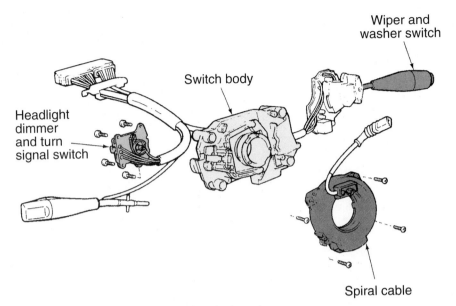

Figure 17.17 The spiral cable is commonly called a clock spring.

Air bags contain rocket propellant. They are shipped in cartons that say **pyrotechnics** on the label. When an air bag **deploys**, it opens so fast that only a very good video camera can freeze the action enough to slow it down for frame-by-frame investigation. It is not unusual for a windshield to be broken by the passenger side air bag.

The driver's side air bag is located in the steering wheel. Usually, it can be easily removed by loosening two or more screws beneath the steering wheel.

Air Bag Clock Spring. The driver's side air bag has a **clock spring** or spiral cable (**Figure 17.17**). It is called a clock spring because of its resemblance to the spring in a mechanical clock. The clock spring delivers the electrical signal from the air bag module to the steering wheel to deploy the air bag in the event of an accident.

The clock spring has a certain amount of travel from fully wound to fully unwound (usually four to five turns of the wheel). It is in its midway position when the steering wheel is straight ahead. When the wheel is turned one way against a steering lock, it unwinds. Turning the wheel the other way winds it up.

If the clock spring is not installed in the correct position, it will break as soon as the wheel has been turned in both directions. In that case, the air bag will not operate, the SRS dash light will come on, and the clock spring will have to be replaced. A new clock spring comes with an aligning mark so that it can be installed properly (**Figure 17.18**). Put the steering wheel in the straight-ahead position before removing a steering column. The arrows should be lined up on the clock spring.

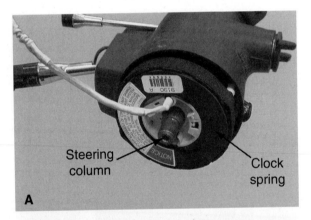

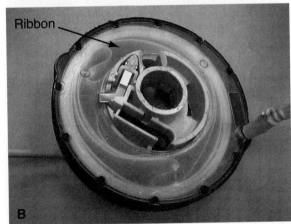

Figure 17.18 (A) The clock spring is located under the steering wheel. (B) A clock spring, viewed from the back side. The ribbon is made of nylon with a gold filament. *(Courtesy of Tim Gilles)*

SHOP TIP When removing an entire steering gear or column assembly, a quick tip is to turn the wheel all of the way against the steering lock and remove the ignition key, locking the steering column. During reinstallation, be sure the wheels are positioned in the same full travel position.

SAFETY NOTE: When removing a steering wheel on a vehicle with an air bag always check the manufacturer's service information for any specific procedures.

SHOP TIP: Stepping on the brake pedal is an OBDII procedure driveability technicians use to clear computer memory.

Simply disconnecting the battery will not ensure that the bag will not deploy, because there is a backup capacitor that will deploy the air bag in the event the battery becomes disconnected during a crash. A typical precautionary procedure before disconnecting an air bag would be to short the battery terminal ends together or depress the brake pedal after the battery has been disconnected.

With the battery disconnected and the key off, step on the brakes. Since the taillights are normally energized without the key on, they try to light. This works to discharge the air bag capacitor. Be certain the electrical connection is the correct one before disconnecting it. Air bag connectors are colored yellow (**Figure 17.19**).

Always store an air bag facing up on the workbench. If it accidentally deploys, an entire steering column could be shot into the ceiling. If the air bag is facing upward, there is less chance of injury to bystanders.

Steering Wheel Service

Some precautions are necessary when removing and reinstalling a steering wheel. Do not hammer on the end of the steering shaft when removing the steering wheel. This can damage the column. Install a puller in the

Figure 17.20 Removing a steering wheel with a steering wheel puller.

threaded holes in the steering wheel (**Figure 17.20**). If there are no alignment marks, make some with a centerpunch and hammer. When reinstalling a steering wheel, double-check to see that it is in the neutral position (halfway between the steering locks).

Steering Column Service

Steering column service can be complicated and the proper service literature should be consulted before repairs are attempted. In addition to disassembly information, a troubleshooting chart is typically included in this service material.

Noises in the steering column can be due to a loose or damaged steering coupling, misalignment, or a bearing or horn ring that lacks lubricant. Shaft bearings or bushings are typically held in place with snaprings. When bearings are not permanently lubricated, an O-ring retains lubricant.

Fixing mechanical problems with the ignition lock is also part of steering column service. These problems include failure to lock or unlock and excessive effort needed to turn the switch, which can be due to a defective lock cylinder or bent or misaligned parts.

Figure 17.19 Air bag electrical connections are yellow. *(Courtesy of Tim Gilles)*

390 SECTION 4 SUSPENSION AND STEERING

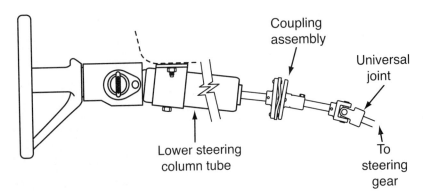

Figure 17.21 Parts of a steering column.

Flexible Joint Replacement. A worn flex coupling (see Chapter 16) can cause looseness in the steering wheel. To replace a flex coupling, loosen its bolt at the steering gear. **Figure 17.21** shows the locations of the parts. Remove the steering column mounting bolts where the column mounts to the instrument panel inside the passenger compartment. Move the column far enough toward the rear of the vehicle for the coupling to be removed from the steering gear.

■ STEERING GEAR SERVICE

Steering gears are generally reliable. The most common service is the repair of fluid leaks. Steering gear damage can be also be caused by collisions or deep potholes. Steering gears are occasionally rebuilt by a technician in the repair shop. When rebuilt steering gears are available at a reasonable price, they are often the repair of choice.

Recirculating Ball and Nut Steering Gear Service

Recirculating ball and nut manual steering gears need two adjustments (**Figure 17.22**). One of them is a **preload adjustment** to the bearings at the ends of the steering shaft (wormshaft). These are ball bearings, which call for a small amount of load to be on them. The load adjustment is measured with an inch-pound **torque wrench** (**Figure 17.23**).

The second adjustment is to the **high point**. The sector gear teeth in a steering gear have a high point where the gear teeth come closer together when the vehicle is traveling straight ahead (**Figure 17.24**). The high point must be centered when the vehicle travels straight ahead so the steering response is not faster in one direction than in the other.

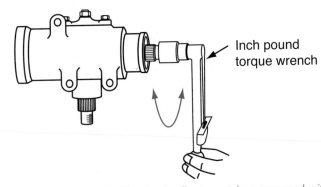

Figure 17.23 The load adjustment is measured with an inch-pound torque wrench.

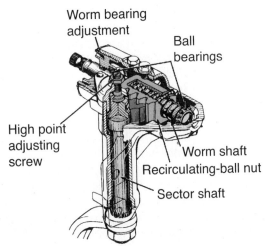

Figure 17.22 High point and worm bearing preload adjustments on a recirculating ball and nut steering gear. (Courtesy of DaimlerChrysler Corporation)

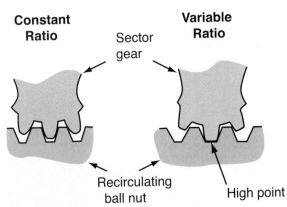

Figure 17.24 The high point is where the closest mesh between the teeth happens when the steering wheel is centered.

Manual Rack-and-Pinion Steering Service

It is important that the steering wheel on a rack-and-pinion be properly centered. This is for two reasons:

1. The center of the rack is hardened.
2. If the tie-rods are of different lengths, the vehicle can steer to one side farther than it can to the other.

Rack-and-Pinion Steering Looseness. A steering damper holds tension against the steering rack (see **Figure 16.6**). A rack gear preload adjustment can be made, but be sure to check the service manual for the correct procedure and specification. If the rack is worn in the center, tightening this adjustment can result in binding steering at the outsides during turns.

■ POWER STEERING SYSTEM SERVICE

Power steering system service includes repairs and replacement of the pump, flushing of the hydraulic system, seal replacement, belt and hose service, and service and repairs to the steering gear.

Reservoir Service

The fluid reservoir is made of sheet metal or plastic and is either attached to the pump or remotely located (**Figure 17.25**). An attached unit is called an **integral reservoir**. Some vehicles have a stand-alone pump (**Figure 17.26**) that is attached by hoses to a reservoir mounted in a remote location (**Figure 17.27**).

Power Steering System Flushing

Be sure to check the condition of the hoses. Power steering pumps or gears often fail because the inside of the power steering hose has deteriorated.

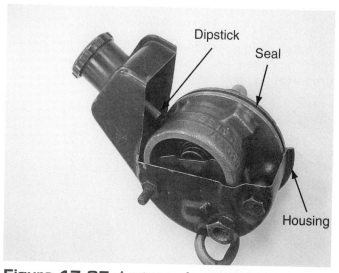

Figure 17.25 A cutaway of a typical sheet metal power steering pump and integral reservoir. *(Courtesy of Tim Gilles)*

Figure 17.26 This power steering pump has a remote reservoir. *(Courtesy of Tim Gilles)*

Figure 17.27 A power steering reservoir located in a remote location from the pump. *(Courtesy of Tim Gilles)*

In the past, power steering fluid was considered to be a long-life item. Today, rack and pinion steering is common on modern vehicles and the housing and valve that control steering are susceptible to wear. When the fluid becomes contaminated with metal particles, more wear results. The service life of today's rack-and-pinion steering systems can be lengthened considerably by changing the steering fluid at regular intervals. The most popular vehicle models, that is, those vehicles produced in the largest quantities, will have rebuilt steering units available at competitive prices. Low-volume vehicles do not enjoy the same availability of rebuilt parts and the replacement cost for a steering unit can be substantially higher.

When replacing a rack-and-pinion steering unit, the system must be thoroughly flushed. A common power steering problem is **morning sickness**—a loss of power assist on cold startups, often in one direction only. Failures from morning sickness result in contaminants being circulated throughout the fluid system. These contaminants can ruin a new rack if they are not flushed from the system. Some manufacturers recommend the installation of a filter when a rack is replaced. Service to power rack-and-pinion steering gears is covered later in this chapter.

Power Steering Flush Methods. Commercial fluid exchange units are available (**Figure 17.28**). These machines provide a convenience to the fast lube service market. Fluid can be exchanged easily and cleanly, without disconnecting hoses.

When a shop does not have a steering flusher, a technician can flush a steering system using the following procedure (**Figure 17.29**):

1. Raise the vehicle and remove the return hose from the reservoir. When the pump has an integral reservoir the return hose often has an ordinary hose clamp, rather than a crimped connection.
2. Plug the reservoir outlet.
3. Put the end of the hose in a drain pan.
4. Run the engine at idle and turn the steering wheel from lock to lock. Then, shut the engine off.
5. Refill the reservoir with clean fluid.
6. Start the engine and wait until fluid begins to flow from the hose. Then, shut the engine off.
7. Repeat the cycle of refilling the reservoir and running the engine until the fluid coming from the return hose is clean and free of air bubbles.

Bleeding the System of Air

Air in the power steering system will cause erratic operation and/or a growling sound, especially when the steering wheel is turned.

Check that the reservoir is full and remove air from the system as follows:

1. Be sure the tires are lifted off the ground so you do not cause a flat spot on the tire tread as you cycle the steering wheel back and forth.
2. When the fluid is warm, run the engine at idle and cycle the steering wheel from lock to lock several times, holding it against the steering lock at each side for 2 or 3 seconds. Then, inspect the fluid in the reservoir for bubbles. If the fluid appears to be foamy after bleeding is attempted, allow it to sit for several minutes until the foam disappears.

NOTE: *Ford recommends turning the wheel from lock to lock 20 to 25 times.*

Test-drive the vehicle. If there is still a groan from the power steering system, there is probably air remaining in the fluid. Do the following to check for air:

1. With the engine running at 1,000 rpm, verify that the fluid level is at the hot/full mark.
2. Shut off the engine and check the fluid level again. If the fluid is bled of air, the fluid level should not increase by more than 0.020 inch (5 mm) (**Figure 17.30**).

Some technicians like to bleed power steering air with the engine off. This prevents larger air bubbles from turning into many more smaller ones, which are

Figure 17.28 A commercial power steering fluid exchange machine.

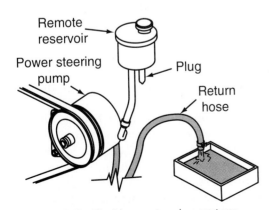

Figure 17.29 Flushing a steering system.

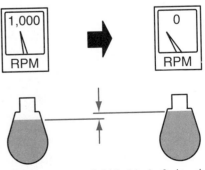

Figure 17.30 If the fluid is bled of air, when the engine is shut off the level of the fluid should rise less than 0.020 inch (5 mm). This is less than 1/32 inch.

SHOP TIP The best practice when replacing a power steering component is to avoid allowing air to enter the system. Do not let the pump reservoir become empty. Letting the fluid level drop too low allows air to be pumped into the system. When refilling the system after a steering gear or hose replacement, fill the reservoir, then start the engine and shut it off immediately. When the engine is started, the reservoir will be sucked empty in as little as a second as the pump fills empty voids in the hose or steering gear with fluid. Refill the reservoir and repeat the procedure until the fluid level no longer drops.

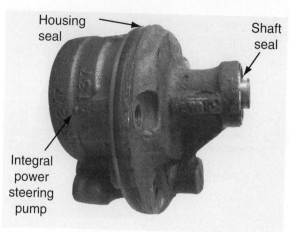

Figure 17.31 Power steering pump seals. *(Courtesy of Tim Gilles)*

more difficult to expel. Sometimes an external vacuum source is needed to remove air from the fluid. A rubber stopper is installed in the reservoir filler neck. Vacuum is applied through the stopper for about 15 minutes.

NOTE: A few vehicles have hydraulically assisted power brakes, known in the industry as hydro-boost, which are operated off the power steering pump. On these systems, depress the brakes several times while turning the steering wheel to help rid the system of air.

Power Steering Pump Replacement

Prior to removing a power steering pump, disconnect the return hose to drain the fluid. Loosen and remove the belt. On some vehicles it is easier to access power steering pump bracket fasteners with the vehicle raised on a lift. Sometimes a pump has more than one bracket. If bracket removal from the pump is necessary, make a careful observation of their locations.

Repairing Power Steering Pump Oil Leaks

An integral reservoir is sealed to the pump with a large O-ring (**Figure 17.31**). Seals are also located at the pump shaft and the fittings for the hoses (**Figure 17.32**). Occasionally there are seals at the mounting bolts.

Pulley and Seal Removal and Replacement

Replacement of the pump shaft seal requires removal of the pulley. The pulley or pump can be easily damaged with a conventional puller, so special pullers and installers are used for this purpose. **Figure 17.33** shows a popular puller used to remove the pulley from the steering pump. Another tool used to reinstall the pulley is usually included with a replacement pump. It

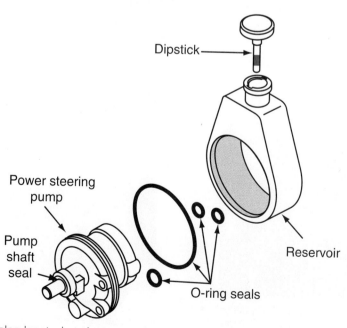

Figure 17.32 Seals are also located at the pump shaft and the fittings for the hoses.

Figure 17.33 Removing a power steering pump pulley. *(Courtesy of Tim Gilles)*

is turned into internal threads on the pump shaft (**Figure 17.34**). When there is a leaking front seal and space permits, the pulley can be removed without removing the pump (**Figure 17.35**).

Power Steering Pressure Diagnosis

When power steering system hydraulic pressure is low, there will probably be noise and it will be harder to steer the vehicle. When pressure varies, steering effort will be erratic, causing the steering wheel to jerk during a turn. Belt tension should be checked first because this is the most common cause of erratic steering effort.

Steering system pressure can be tested using a pressure gauge. Pressure testing varies among manufacturers. Check the service information for the vehicle. The following is a typical test procedure:

1. First, remove the pressure hose from the pump and install the pressure gauge between the pressure hose and pump (**Figure 17.36**).
2. Start the engine and cycle the steering wheel from lock to lock several times to bleed the system of air and heat the fluid.
3. Close the valve on the pressure gauge and note changes in the gauge reading (**Figure 17.37**).
4. Open the valve and observe the pressure reading on the gauge. It should equal the manufacturer's specifications. If not, repair or replace the pump.
5. While the valve is open, check pump pressures between 1,000 and 3,000 rpm and compare them with the manufacturer's specifications. If they do not match, look for a problem in the power steering pump's flow control valve assembly.
6. Hold the steering wheel against a lock while observing system pressure (**Figure 17.38**). If pressure is lower than specified, the steering gear is leaking internally.

Steering effort can be measured using a spring scale (**Figure 17.39**). Comparing the reading with the

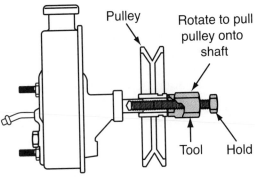

Figure 17.34 A tool used for reinstalling a power steering pump pulley is turned into threads in the pump shaft.

CAUTION Closing the valve for more than 10 seconds can result in excessive system pressure, which can cause a hose to burst.

Figure 17.35 When space permits, the pulley can be removed without removing the pump from its mounting. *(Courtesy of Tim Gilles)*

Figure 17.36 Connect the pressure gauge in series between the pump and the pressure hose.

Chapter 17 Steering Service **395**

manufacturer's specifications can highlight a steering system problem.

Power Steering Pump Service

When a power steering pump has a problem, it is typically replaced with a rebuilt pump. Some shops still perform power steering pump repairs. Typical repairs are covered in this section.

Flow Control Valve Service. The high-pressure hose connects to a fitting that also serves as a part of the flow control and pressure relief valve assembly (**Figure 17.40**). Removal of this fitting is often required before the pump can be removed from the housing or reservoir. The flow control valve piston and spring will fall out after the fitting is removed. The seat for the pressure relief valve ball is threaded into the control valve assembly (**Figure 17.41**).

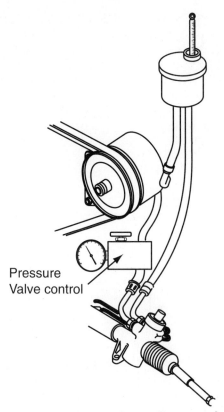

Figure 17.37 The pressure gauge has a valve that is positioned either open or closed during diagnosis.

Figure 17.38 Hold the steering wheel against a lock while observing system pressure.

Figure 17.39 Steering effort can be measured using a spring scale.

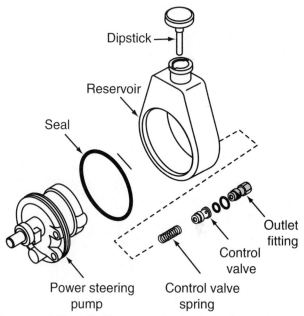

Figure 17.40 The control valve is located behind the power steering pump outlet fitting that connects to the pressure hose.

Disassemble and inspect all of the parts. Crocus cloth can be used to remove minor scratches from the control valve. Clean the housing and valve, and lubricate them with power steering fluid before reassembly. When the relief valve is mounted on the outside of the pump housing, be sure to use a new O-ring during reinstallation to prevent leaks.

Pump Disassembly. Typical pump disassembly requires removal of a retaining ring (**Figure 17.42A**). The pump housing can then be removed from the reservoir housing (**Figure 17.42B**).

Pulley Shaft Seal Service. Replacement of the power steering shaft seal requires removal of the pulley

396 SECTION 4 SUSPENSION AND STEERING

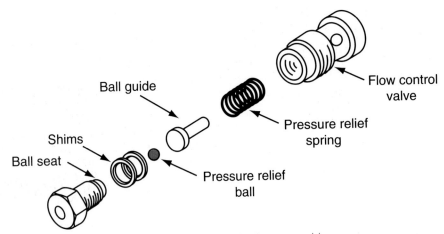

Figure 17.41 A typical flow control and pressure relief valve assembly.

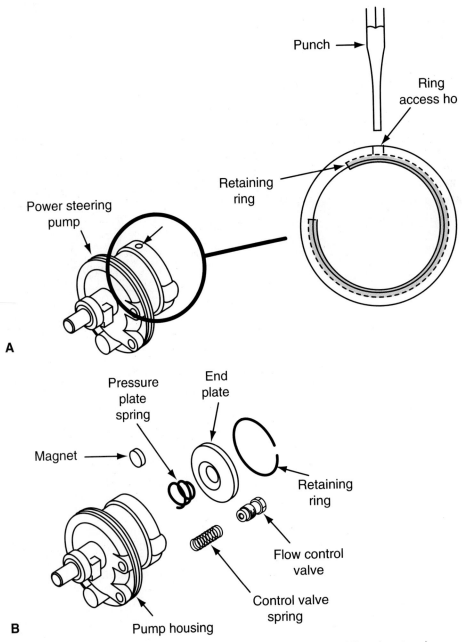

Figure 17.42 (A) Use a punch to remove the retaining ring when disassembling the steering pump. (B) The flow control valve and end plate are removed from the pump for inspection.

Chapter 17 Steering Service **397**

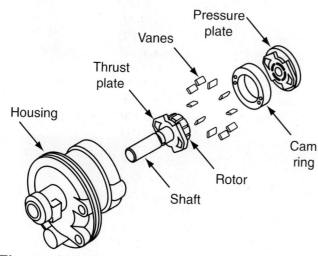

Figure 17.43 Parts of a typical power steering pump.

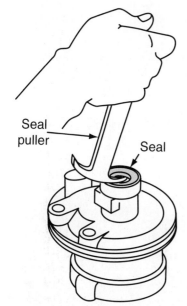

Figure 17.44 Removing a power steering pump shaft seal.

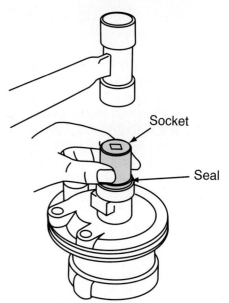

Figure 17.45 Replacing a power steering pump shaft seal.

(see **Figure 17.34**). **Figure 17.43** shows an exploded view of typical pump parts. When the pump is disassembled, the seal is easily removed (**Figure 17.44**). There are special tools available for installing the new seal, but when these are not available a socket and soft faced hammer can be used to install the new seal (**Figure 17.45**).

■ DRIVE BELT SERVICE

Drive belts are very strong and dependable when they are replaced at reasonable intervals. It is advisable to change them periodically *before* they fail, regardless of appearance. Belt failures rise significantly after 4 years of use, so that is the recommended interval for replacement.

Belt Inspection and Adjustment. Belts often appear good to the eye, even though they are about to fail.

Belt failure can result in a loss of power steering. Check the drive belt for wear, cracks, or damage. A belt that has any of these conditions should be replaced. Be certain that belt tension is sufficient.

When a belt slips on the pulley, it becomes glazed. A glazed belt must be replaced because it will not grip the pulley tightly, even if tension is at specification.

NOTE: *A slipping power steering pump drive belt will squeal loudly when the steering wheel is held against a steering lock.*

V-Belt Inspection. V belts must be the correct size. Normally, a V belt contacts only the sides of the pulley groove. When a V belt is the wrong size or has become worn from slipping, the bottom of the belt can contact the pulley groove bottom. A V belt should extend slightly out of the pulley groove (**Figure 17.46**). The

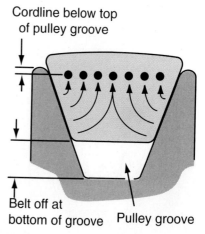

Figure 17.46 This V belt fits well in the pulley groove.

belt has a cord member to provide strength. If the belt rides too high in the groove, the belt will wear below the cord member.

When a V belt rides below the edge of the groove this indicates either a worn belt, a worn pulley groove, or a belt that is too small. A V belt that is too small for the pulley will bottom out in the pulley groove, and the sides of the belt will not grab. This results in a belt that slips, even when tensioned correctly.

Inspect the pulley grooves for oil, rust, or wear. Cracks on the face of an old V-ribbed belt are normal, but replacement is suggested. Smaller pulleys usually show wear first.

NOTE: *Surface finish of the pulley is more important with a V-ribbed pulley than with a V-type pulley.*

Some shops use belt dressings to help a belt to grip. The problem with these substances is that they are sticky and attract dirt, which wears the pulley groove. Some belt dressings actually attack and damage the belt material.

Belt Alignment. Inspect belt alignment before disassembly. V-belt pulleys must be in alignment within 1/16 inch for each foot of the distance between the pulleys. Misalignment can be due to the pulley shafts being unparallel or due to an accessory being located improperly. Misalignment can cause rapid belt and pulley wear and thrown belts. When pulleys are out of alignment, the belt will chirp. Noise is a result of vibration. In this case, the vibration results because the belt must continually slide into the groove as it moves against the pulley (**Figure 17.47**).

A misaligned V-ribbed belt can walk off the pulley (**Figure 17.48**) or can tear off a rib. The accessory can

Figure 17.48 A V-ribbed belt can walk off the pulley if it is misaligned.

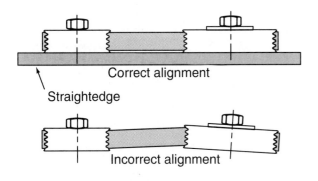

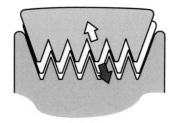

Figure 17.47 Noise happens when a misaligned belt vibrates in the pulley groove.

Figure 17.49 (A) A laser belt alignment tool installed in the grooves of a V-ribbed pulley. (B) The laser beam is focused on a groove in another pulley. *(Courtesy of Tim Gilles)*

sometimes be realigned using 0.030-inch front end alignment shims. **Figure 17.49** shows a laser tool used to check belt alignment between pulleys.

Replacing Belts. Loosen bolts on any accessories that act as drive belt adjusters and remove the belt. Compare the new belt to the old belt. If no belt is available, use a piece of string in the pulley groove to estimate the replacement belt size. Belts usually change in size in half-inch increments, which is reflected in the part number of the belt. Be sure the

SAFETY NOTE: Before replacing a belt, disconnect the battery to prevent an electric cooling fan from accidentally coming on. Also, the output terminal on the alternator will be energized (hot) when the battery is connected. Prying against the outside of the alternator can result in an accidental short circuit.

new belt is the right size. A belt that is too long can rub on a radiator hose or fuel line after tension is adjusted.

V-Ribbed Belt Replacement. Most V-ribbed belts have constantly spring-loaded tensioning idler pulleys that contact the smooth backside of the belt (**Figure 17.50**). A few others have "locked center" drives. In these, tension is controlled in one of the following ways:

- A tensioner pulley with an off center bolt (**Figure 17.51**)
- An adjustable jackscrew (**Figure 17.52**)
- An accessory such as the alternator (**Figure 17.53**)

Before removing a **serpentine belt** make a sketch of how it is installed. It can be complicated figuring out the path of one of these belts. Most vehicles have an underhood diagram label. Belt manufacturers produce catalogs with belt routing diagrams for the various makes of vehicles. These are available from parts

SHOP TIP If a belt seems too small to install, try installing it in a different order. For instance, if the belt will not go over the water pump pulley last, try installing it over the alternator pulley last.

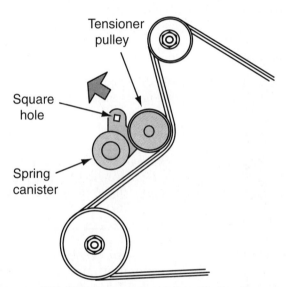

Figure 17.50 To install a V-ribbed belt with a spring-loaded belt tensioner, pull the pulley in the direction of the arrow to release belt tension.

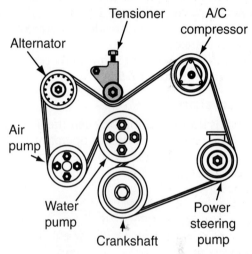

Figure 17.52 This belt is tensioned by an adjustable screw.

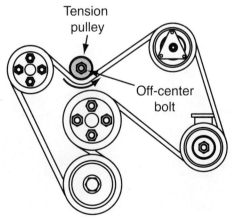

Figure 17.51 To loosen tension on this belt, turn the off-center bolt in the direction of the arrow.

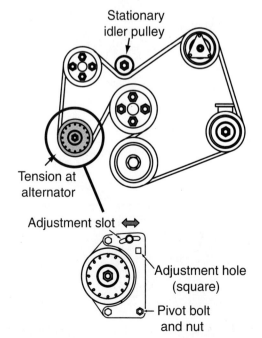

Figure 17.53 This belt is tensioned by the alternator.

SHOP TIP When a belt is replaced, the old belt can be stored in the trunk for emergencies. If a serpentine belt fails, power to all accessories and the coolant pump will be lost.

stores and are usually free of charge to their customers. Belt routing information is sometimes available in the owner's manual as well.

When the belt information is not available, the following tips may be useful:

- Remember that both sides of the belt can be used to drive accessories. The V-grooved side of the belt will mesh with pulley grooves. The flat side of the belt will go against a flat pulley.
- The belt is routed around the outside of the pulleys and is drawn in and looped around the smooth pulleys (**Figure 17.54**).
- Only one pulley can be threaded incorrectly. It is always near the center and is a smooth pulley. If the belt does not seem to fit properly one way, try another way (**Figure 17.55**).

Belt Tension

Belt tension is important for long belt life. Belts stretch slightly in the first few minutes of operation, and then remain constant in length. If they are overtightened, parts can be overloaded.

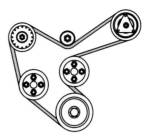

If this routing does not work...

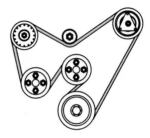

...try this.

Figure 17.55 Different ways to position a serpentine belt.

CASE HISTORY A technician replaced a damaged power steering pump pulley. After the new pulley was installed, the pump was mounted on the engine. When the technician tried to install the new belt, he discovered that the belt would not fit in the pulley groove. The new pulley had five grooves and the old one had six. Always count the grooves in a pulley before replacing it.

Too much belt tension can cause failure of the coolant pump bearing, the alternator, or the upper front main bearing (**Figure 17.56**). Improperly tightened belts can also cause water pump and alternator bearing noise. Loose belts can cause overheating and abnormal combustion, noise, or a dead battery.

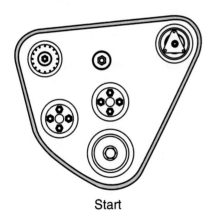

Start

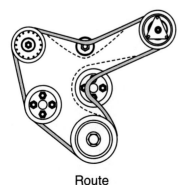

Route

Figure 17.54 The belt is drawn in from the outside and looped around smooth pulleys.

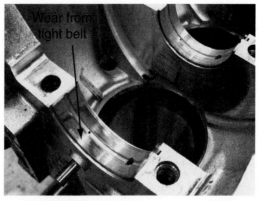

Figure 17.56 This front upper main bearing shows wear caused by too tight a belt. *(Courtesy of Tim Gilles)*

Tightening a Belt. There are several ways that belts are tensioned:

- Some accessories are provided with a place to pry against (**Figure 17.57**).
- Brackets sometimes have a hole that a prybar can be inserted into.
- Manufacturers often provide some provision for tightening, such as a ½-inch-square hole in the mounting bracket for a ½-inch-breaker bar to be inserted into (**Figure 17.58**).
- Sometimes a jackscrew adjustment is provided (see **Figure 17.52**).

When there is no provision for tightening the belt, pry against a strong area of the accessory. Be sure that the outside of an alternator or power steering pump is not accidentally damaged. A power steering reservoir is made of sheet metal or plastic. Prying against it can damage the pump, often resulting in a leak. Do not pry on the power steering reservoir or any other delicate part (see **Figure 17.25**).

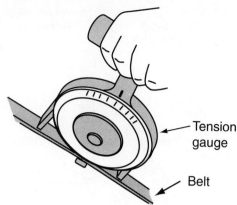

Figure 17.59 Use a belt tension gauge when adjusting the belt.

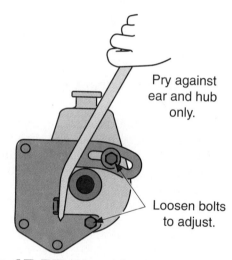

Figure 17.57 When tightening the power steering belt, pry only the designated areas.

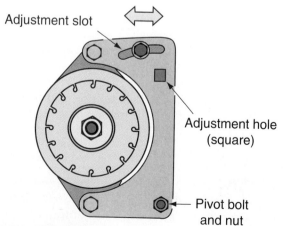

Figure 17.58 Use a breaker bar or ratchet in the square hole to tighten the belt.

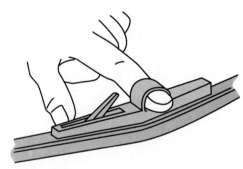

Figure 17.60 A click-type belt tension gauge is used for measuring tension on V-ribbed and timing belts.

V Belt Tension. New belts stretch slightly during the first few minutes of operation. After that, their length will remain constant. When adjusting a new V belt, set the tension about 15 pounds higher than the recommended specification. After running the engine for about 15 or 20 minutes, recheck and adjust the belt tension using a belt tension gauge (**Figure 17.59**).

V-Ribbed Belt Tension. V-ribbed belts usually require more tension than V belts. Maximum tension should be limited to 30 pounds per rib, checked at a splice free area.

Use a **click-type tension gauge** for V-ribbed and timing belts (**Figure 17.60**). After a new belt is installed, run the engine. Then, loosen the tensioner bolt and retighten it.

After some initial tension is lost, V-ribbed belts will maintain 20 pounds of tension per rib for a long time. Used belts should have 15 to 20 pounds of tension per rib.

■ POWER STEERING HOSES

The system has two hoses: a pressure hose and a return hose (**Figure 17.61**). The pressure hose is made of reinforced oil-resistant synthetic rubber, crimped to the metal tubing. It must be able to withstand temperatures of 300°F and handle pressures up to 1,500 psi.

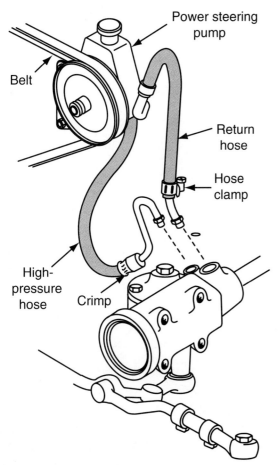

Figure 17.61 Power steering hoses. Pressure hoses are crimped. Return hoses, under less pressure, usually have hose clamps.

Sometimes the pressure hose has two different ends. The larger end fits on the pump side and acts as an "accumulator," which absorbs pulsations in the hose from variations in pressure. The ends of some pressure hoses have flared connections. Others have captured seals that allow the tubing to be able to turn even when tight. The return hose is usually attached only with hose clamps because it is not under high pressure.

Power Steering Hose Service

When checking a power steering hose, look for signs of leakage or dampness at the connections. Also look for signs of deterioration, such as cracks, signs of rubbing, or swelling. A failed pressure hose can leak, blowing oil onto an exhaust manifold where it can cause a fire.

 **CAUTION** Power steering fluid, which is not flammable at engine operating temperature, is flammable at temperatures a little above 300°F—lower than the temperature of an exhaust manifold.

SHOP TIP Sometimes a replacement hose is not available. Some businesses have special equipment for making hydraulic hoses. Check with a business that has heavy truck, industrial, or farm implement accounts.

Get the correct replacement hose and replace the old one. Parts stores have catalogs that have extensive illustrated listings for power steering hoses. It is also a good idea to compare the new hose to the old hose to be sure that the length and fittings are identical.

Refilling the Power Steering System

Be sure to use the correct fluid when refilling the system. Earlier systems traditionally used automatic transmission fluid, which is red in color. Many late-model vehicles use power steering fluid, which has a lower viscosity. It is either a yellow or yellow/green color.

After refilling the system, run the engine and check for leaks. On hoses that use O-rings, new O-rings should be installed. Bleed the system of air as discussed earlier in this chapter.

■ POWER STEERING GEAR SERVICE

Power steering failures include a lack of power assist in one direction and leakage from the steering gear. Power steering seal kits are available. Some shops rebuild power steering gears, but most buy rebuilt ones that carry a warranty.

A majority of today's new cars are equipped with power rack-and-pinion steering. Rack-and-pinion steering is more prone to problems because its weight has been cut over the years. Earlier power rack assemblies weighed as much as 54 pounds. Some newer racks weigh as little as 8–10 pounds. Weight reductions of this magnitude have had an effect on the durability of the steering gears. Some of the lighter units have not been very durable, requiring replacement while the vehicle still has relatively low mileage.

A rack-and-pinion steering gear can have an external leak at the pinion shaft (attached to the steering column) or from the rubber bellows at either end of the rack (**Figure 17.62**). Rack failures can be due to a torsion bar bent from a hard impact or serious internal or external leaks.

SHOP TIP If a flared fitting leaks, try loosening the fitting and twisting the tubing to seat it against the flare seat. Then retighten the fitting.

Chapter 17 Steering Service **403**

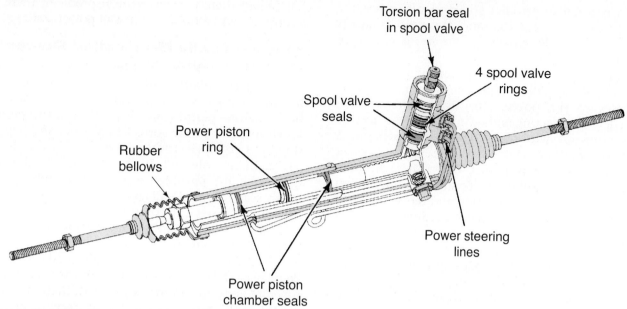

Figure 17.62 Rack-and-pinion sealing points. (Courtesy of Federal-Mogul Corporation)

Replacing Rack-and-Pinion Units

Rack and pinion gears are often replaced with rebuilt units. Some units have a good record of success when rebuilt by technicians; others do not. Sometimes grooves wear in the control valve housing (**Figure 17.63**). The symptom of this problem is usually morning sickness (loss of power assist on cold start-ups). Rebuilt units usually have a nickel-plated sleeve installed to correct this and prevent it from happening again. When deciding whether to rebuild a unit or buy a replacement, an auto parts supplier is a good reference source.

NOTE: *To prevent damage to an air bag clock spring, lock the steering column and remove the key, or wrap the seat belt through the steering wheel to prevent it from turning while the steering unit is removed.*

Before removing a rack-and-pinion unit from the car, the steering shaft coupler must first be removed. There are two main types of coupler designs (**Figure 17.64**). Both types use a roll pin or threaded shoulder bolt that fits into a groove on the pinion stub shaft. One type uses a roll pin or bolt at the top as well. The other uses a D-shaped spring-loaded slip fit coupler. The spring clip can fall out on the interior carpet of the car as the steering gear is lowered for removal. It must be reused.

Roll pins must be removed with a punch and chisel. On the type that has two roll pins, the lower

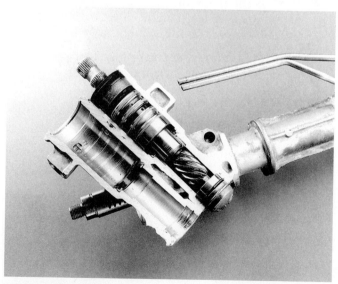

Figure 17.63 Grooves worn in the control valve housing. (Courtesy of Federal-Mogul Corporation)

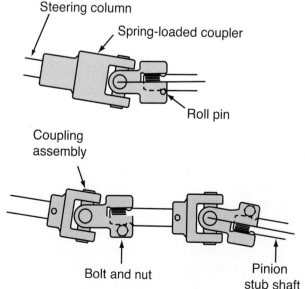

Figure 17.64 Two designs of steering shaft couplers.

roll pin is usually corroded and difficult to remove while the rack is on the car. Once the steering rack is on the workbench, this pin can more easily be removed.

There are usually four tubing connections into the top of a rack-and-pinion unit. The two that are from the hoses to the power steering pump must be removed. Disconnecting and reconnecting the hoses is often not possible with ordinary flare nut wrenches. A special crowfoot flare socket can be used. Some power steering lines have swivel ends. When these are fully tightened, they can still rotate. Be sure to tighten them to the correct torque specification and replace the two seals that seal them to the tube fitting. Sometimes, it is necessary to lower the cross member under the engine to gain access to the rack-and-pinion.

When replacing an entire rack-and-pinion unit, be sure that the rubber mounts are in good condition and that the brackets are installed properly. If the rack is not level, bump steer can result.

Variable Power Steering Service

Problems with variable power steering found in some newer vehicles can be either mechanical or electrical. When there is an electrical system failure, the power steering system goes back to operating on full-time power assist. In other words, steering becomes easier at higher speeds. If this is the case, refer to the electrical troubleshooting procedure in the vehicle service manual. When there is a mechanical problem, there will almost always be noise or a loss of power assist.

Electronically Controlled Power Steering System Service

A typical complaint with electronically controlled power steering is steering effort that is too easy or too hard. Before performing any other tests, perform visual inspections for correct fluid level and belt tension, and check electrical connections at the steering gear actuator solenoid. If the visual inspection does not locate any possible causes for the problem, connect a scan tool to the diagnostic connector under the dash near the steering column.

Turn the ignition to the on position and select the correct item from the scan tool menu. If there is an electrical defect, a diagnostic trouble code (DTC) will be displayed. Follow instructions in the service literature and perform required tests with a high-impedance digital multimeter. Service literature will include a detailed diagnostic chart.

CAUTION Be certain that the ignition key is off when connecting or disconnecting a scan tool.

REVIEW QUESTIONS

1. What happens to power steering temperature when the wheel is turned several times?
2. What is the name of the test for looseness in which an assistant turns the wheel from side to side with the tires on the ground?
3. How much looseness should there be between a rack-and-pinion inner tie-rod and its socket?
4. What is used to remove a pitman arm from a steering gear?
5. What measuring tool is used to make an approximate toe setting when replacing a tie-rod end?
6. What is another name for the spiral cable that sends an electrical signal from the steering column to the steering wheel?
7. What can happen if the clock spring is not centered before steering linkage is reconnected to a steering gear?
8. What is the name of the point in a steering gear where the gear teeth come closer together when the vehicle is traveling straight ahead?
9. Which kind of power steering system is more prone to problems, rack-and-pinion or conventional?
10. What is it called when there is a loss of steering power assist on cold startups?

PRACTICE TEST QUESTIONS

1. Technician A says that automatic transmission fluid can be used in all power steering systems. Technician B says that the engine must be running when checking steering looseness with a dry park check on a power steering vehicle. Who is right?
   - **A.** Technician A
   - **B.** Technician B
   - **C.** Both A and B
   - **D.** Neither A nor B

2. Technician A says that a loose belt can cause the steering wheel to jerk during a turn. Technician B says that a loose steering pump belt is more likely to squeal when driving straight. Who is right?
   - **A.** Technician A
   - **B.** Technician B
   - **C.** Both A and B
   - **D.** Neither A nor B

3. Which of the following is/are true about a vehicle that has been driven for a long time with a loose belt?
   - **A.** The belt has probably become glazed
   - **B.** The pulley groove could be worn out
   - **C.** All of the above
   - **D.** None of the above

4. Technician A says when testing looseness on a rack-and-pinion tie-rod, the wheels should be hanging free. Technician B says that inner and outer parallelogram tie-rod ends are identical. Who is right?
   - **A.** Technician A
   - **B.** Technician B
   - **C.** Both A and B
   - **D.** Neither A nor B

5. Technician A says if the steering wheel is off-center when driving on the highway, remove the steering wheel using a special puller and replace it in the straight-ahead position. Technician B says when removing a steering wheel, loosen the nut and then pound on the end of the shaft until the wheel pops loose. Who is right?
   - **A.** Technician A
   - **B.** Technician B
   - **C.** Both A and B
   - **D.** Neither A nor B

6. Technician A says to store an air bag face down on a workbench. Technician B says the air bag wiring connector must be disconnected before removing an air bag. Who is right?
   - **A.** Technician A
   - **B.** Technician B
   - **C.** Both A and B
   - **D.** Neither A nor B

7. Which of the following is/are true about rack-and-pinion steering?
   - **A.** The center of a rack-and-pinion steering rack is hardened
   - **B.** If a rack is not centered, the vehicle can steer farther to one side than to the other
   - **C.** All of the above
   - **D.** None of the above

8. Which type of steering gear must be centered so the steering response is not faster in one direction than the other?
   - **A.** Recirculating ball
   - **B.** Rack and pinion
   - **C.** Both
   - **D.** Neither

9. If the ends of the idler arm and pitman arm are not at equal height, bump steer will result.
   - **A.** True
   - **B.** False

10. Which power steering hose has a hose clamp on its connection to the pump?
    - **A.** The pressure hose
    - **B.** The return hose
    - **C.** They both have hose clamps

CHAPTER 18

Wheel Alignment Fundamentals

■ OBJECTIVES

Upon completion of this chapter, you should be able to:
- ✔ Describe each wheel alignment angle.
- ✔ Tell which alignment angles can result in tire wear when adjusted incorrectly.
- ✔ Explain which alignment angles can cause a vehicle to pull to one side.
- ✔ Understand how steering axis inclination works.
- ✔ Describe the effects of incorrect scrub radius.
- ✔ Explain why the Ackerman angle causes tires to toe-out when turning.
- ✔ Explain the results of understeer and oversteer.

■ KEY TERMS

Ackerman angle
camber
camber roll
caster
directional control
directional stability
dog tracking
included angle
king pin inclination (KPI)
lead point

negative camber
negative caster
negative scrub radius
oversteer
positive camber
positive caster
positive scrub radius
scrub radius
scuff
setback

slip angle
steering axis inclination (SAI)
toe-in
toe-out
tracking
turning radius
understeer
wheelbase

■ INTRODUCTION

Correct alignment of the wheels provides a vehicle with the ability to run straight on the highway with very little steering effort. Correct alignment also results in minimized tire wear. This chapter deals with the principles of, and relationships between, the different alignment angles.

There are five wheel alignment angles:

1. Toe
2. Camber
3. Caster
4. Steering axis inclination (SAI)
5. Turning radius

■ TOE

The alignment angle most responsible for tire wear is toe. The toe measurement compares the distances between a pair of tires from the front and from the rear (**Figure 18.1**). When the tires are closer together at the front, this is called **toe-in**. When the tires are farther apart at the front, this is **toe-out**. If the measure-

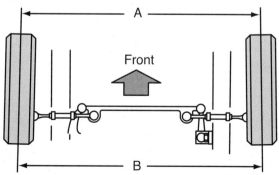

Figure 18.1 Toe is a comparison of the distances between the fronts and the rears of a pair of tires.

ment at the front of a pair of tires is 1/16 inch shorter than at the rear, this is 1/16 inch toe-in.

Every 1/16 inch of toe-in or toe-out results in 11 feet per mile of **scuff**. This means that the tires actually move sideways for 11 feet out of every mile traveled. If a car has 1 inch of toe, the tire is dragged 182 feet sideways every mile. The result is severe tire wear and decreased fuel economy.

Several things can cause incorrect toe:

- An improper adjustment
- Looseness in the steering linkage due to wear
- A collision with a substantial road hazard, like a curb, pothole, rock, or debris
- A change in either the caster or camber adjustment

Tie-rods and other steering linkage parts are built to be flexible if there is an impact. If steering linkage parts were brittle they could break during an impact, causing a dangerous loss of vehicle control. Because they are somewhat flexible, steering linkage parts will bend, rather than break.

NOTE: *Bent steering linkage will result in a change in the toe setting and the steering wheel will usually be off-center when traveling straight down the road.*

Front toe is adjustable on all vehicles and rear toe is adjustable on some. When a car is driven, the toe adjustment changes. Whether the tires deflect inward or outward when rolling depends on if the car has front- or rear-wheel drive. On rear-wheel drive (RWD) cars, rolling tires tend to toe-out when driving. This happens as the steering linkage deflects due to the rolling resistance of the tires, taking up any clearance that exists.

During an alignment, compensation is made so that front tire toe will be as close as possible to zero while rolling. The adjustment specification for RWD cars usually calls for the tires to be slightly toed-in, closer together at the front, while on the alignment rack.

Front-wheel drive (FWD) cars react in an opposite fashion. Wheels on these cars tend to toe inward as they are pushed by engine torque. Specifications for FWD cars usually call for a slight toe-out setting or zero.

NOTE: *A change in toe affects the position of the steering wheel.*

Information on front and rear toe adjustment is found in Chapter 19.

■ CAMBER

Camber, an adjustable angle on most vehicles, is the inward or outward tilt of a tire at the top. A tire that tilts out at the top has **positive camber** (**Figure 18.2**). A tire tilted in at the top has **negative camber**.

Camber is a tire wearing angle. The inside and outside edges of the tread on a cambered tire actually have two different radii (**Figure 18.3**), so they rotate at different speeds. When the tire leans to one side, the

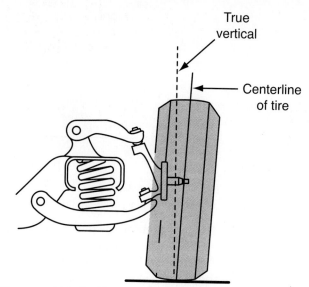

Figure 18.2 This wheel has positive camber.

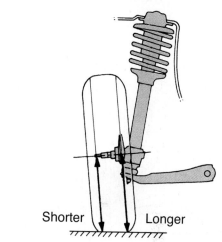

Figure 18.3 The tread will wear on the shorter side as the smaller diameter of the tire squirms against the road surface.

tread will wear on that side as the smaller diameter of the tire squirms against the road surface.

The camber angle is controlled by the position of the control arms or struts. The control arms are quite strong and are not usually affected by an impact, such as running into a curb. With SLA suspensions, negative camber typically develops as springs sag with age. Sagged springs are often accompanied by wear on the inside edges of the front tires from the resulting negative camber.

Camber is a **directional control** angle. A cambered tire tends to roll in a circle, as if it were at the large end of a cone (**Figure 18.4**). Think about what happens if you put an ice cream cone on a table and attempt to roll it. This is **camber roll**. Unequal camber between tires on opposite sides of the vehicle can cause a steering pull to the side with the most camber.

The inner wheel bearing is larger because it is designed to support the vehicle's load. Camber is usually positive, which loads the inner wheel bearing

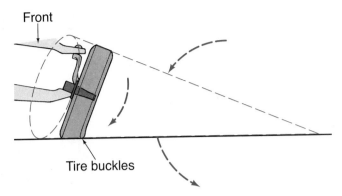

Figure 18.4 A cambered tire tends to roll in a circle.

(**Figure 18.5**). Negative camber causes the weight of the vehicle to be placed on the outer wheel bearing (**Figure 18.6**).

The angle of the spindle exerts leverage on the suspension. Negative camber increases the leverage on the spindle (**Figure 18.7**). When camber is negative, the angle of the spindle can cause instability when the wheel encounters a bump. The increased leverage caused by negative camber can cause a wheel to move out of the straight-ahead position if it goes into a pothole or over a bump (**Figure 18.8**). Positive camber diminishes the spindle's leverage, increasing steering

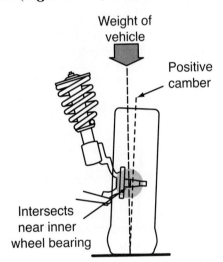

Figure 18.5 Positive camber loads the inner wheel bearing.

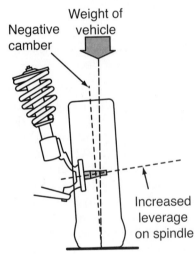

Figure 18.7 Negative camber increases the leverage on the spindle.

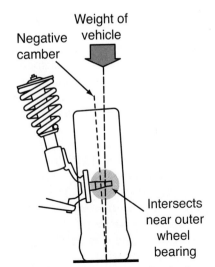

Figure 18.6 Negative camber loads the outer wheel bearing.

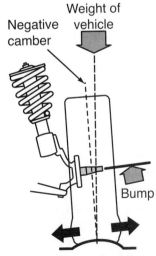

Figure 18.8 Negative camber's increased leverage can cause the wheel to move out of the straight-ahead position if it goes into a hole or over a bump.

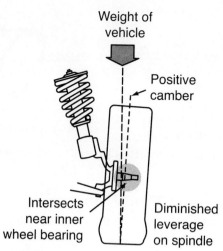

Figure 18.9 Positive camber decreases the spindle leverage, increasing steering effort.

effort (**Figure 18.9**). Sometimes negative camber is specified in order to improve directional stability or cornering ability.

■ CASTER

Caster is the angle that describes the forward or rearward tilt of the spindle support arm (**Figure 18.10**). When the top is tilted to the rear as shown in **Figure 18.11**, the wheel has **positive caster**. Its **lead point**, or *point of load*, is in front of true vertical. **Negative**

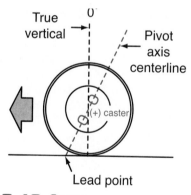

Figure 18.10 Caster.

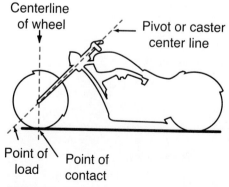

Figure 18.11 Bicycles and motorcycles have positive caster.

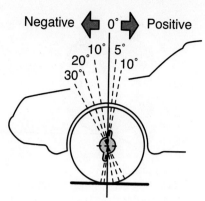

Figure 18.12 Comparison between positive and negative caster.

caster is the forward tilt of the steering axis. **Figure 18.12** compares positive and negative caster.

Caster is a directional control angle but does not cause tire wear when the car is going straight. A bicycle has positive caster, which helps it to continue in a straight-ahead direction without holding the handle bars. If the handle bars are aimed in the wrong direction (negative caster), the bicycle loses its stability.

When the front wheels on a car have different caster settings, the car will pull toward the side with the most negative caster. That wheel will have its point of load behind the wheel on the other side of the car. Caster causes the spindle to move either toward the ground or away from the ground during a turn, depending on whether it is negative or positive. Positive caster causes the spindle to move toward the ground when the wheel is steered to the left. Because the tire cannot move closer to the ground, the vehicle must lift (**Figure 18.13**). When turning effort is removed from the steering wheel after the turn, the tire attempts to move back toward the center. Steering a positive castered wheel to the right causes the spindle to move up, with the opposite results. Wheels with equal caster will cancel each other out as a return to straight-ahead force.

NOTE: *Steering axis inclination (SAI) exerts a similar force in terms of returnability to straight ahead, but in a different and consistent direction. SAI is covered later in the chapter.*

Figure 18.13 Because the tire cannot move closer to the ground, the vehicle must lift. This vehicle has negative caster.

410 SECTION 4 SUSPENSION AND STEERING

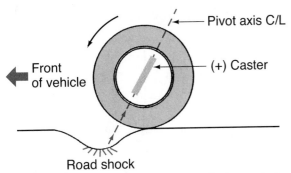

Figure 18.14 Positive caster results in road shock being transmitted through the steering column.

Positive caster aligns the steering axis with bumps encountered in the road so road shock is more likely to be felt in the steering wheel (**Figure 18.14**). Too much positive caster can result in a front wheel shimmy. Some vehicles that have a large amount of positive caster use a steering damper to prevent this (**Figure 18.15**). When the damper is worn out, the front wheels will shimmy uncontrollably when a front tire hits a small bump.

NOTE: *Excessive positive caster can cause cupped tire wear if the resulting shimmy is allowed to continue.*

Negative caster makes a vehicle easier to steer. It can also cause a car to wander and weave on the highway. Old cars had narrow tires. Their alignment specification called for positive caster to keep the car going straight without holding the steering wheel. Newer cars often have wider tires, which tend to keep rolling straight (**Figure 18.16**). Caster specifications on high-performance cars can range as high as 11 degrees, in order to enhance directional stability.

NOTE: *Light trucks have different caster settings depending on their ride height. Changing the caster setting can have a drastic effect on the driveability of the vehicle.*

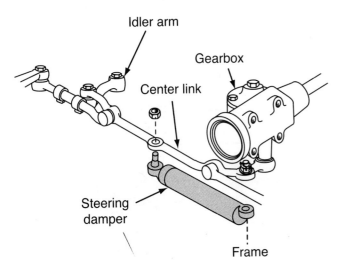

Figure 18.15 When there is a high amount of caster, a steering damper is often used to prevent wheel shimmy.

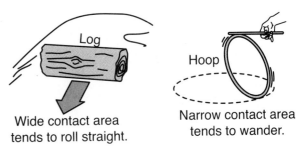

Figure 18.16 Wider tires tend to roll straight.

Caster and Tire Wear

Caster is actually a measurement of the camber angle as it changes during a turn. A high amount of caster can result in tire wear if the car is driven excessively on winding roads or in the city. This is because a high amount of caster influences camber roll. Although caster is known as primarily a directional control angle, it can cause tire wear, too. Alignment equipment manufacturers teach this differently. Some say that tire wear is not a result of caster.

■ STEERING AXIS INCLINATION

Steering axis inclination (SAI) is the amount that the spindle support arm leans in at the top (**Figure 18.17**). SAI is also known as *ball joint inclination* (BJI) or **king pin inclination (KPI)** for older vehicles or trucks with king pins. It is not a tire wearing angle.

NOTE: *SAI is the angle most responsible for helping the vehicle steer in the straight-ahead direction.*

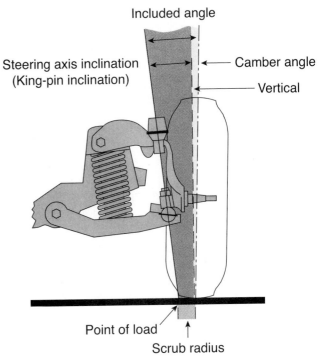

Figure 18.17 Steering axis inclination puts the pivot point under the tire.

How SAI Works

Although the spindle is horizontal to the ground (except for the camber angle) when the wheels are pointed straight ahead, it attempts to move closer to the ground during a turn (see **Figure 18.13**). The spindle cannot move any closer to the ground because of the tire, so the car must lift. During a turn, the weight of the car puts pressure on the wheel, causing it to want to return to the straight-ahead position.

SAI has three functions:

1. After a turn, SAI helps the vehicle return to a straight-ahead position as the weight of the vehicle causes the spindle's return to neutral position (**Figure 18.18**).
2. SAI keeps the vehicle going straight down the road. Positive caster does this, too, but the more positive the caster, the harder it is to steer the car.
3. SAI allows the car to have less positive caster (for easier steering), while still having good **directional stability**.

The combination of SAI and camber is called the **included angle**.

Some cars with a large amount of SAI will wear the outsides of the tires (especially on nonradial tires) because of the excessive amount that the wheel cambers during turns (see **Figure 18.13**). These cars do not require as much positive caster to keep the car from wandering.

In city driving, right-hand turns are made about 80 percent of the time. Delivery trucks are especially hard on right-hand springs, bearings, and tires, which must turn more sharply.

NOTE: *SAI can be used as a diagnostic angle in determining whether parts are bent.*

Scrub Radius

Scrub radius is a factor of steering axis inclination. It is the pivot point for the front tire's footprint. Scrub radius is the distance at the road surface between the centerline of true vertical at the center of the tire tread and the steering axis pivot centerline.

The junction of the steering axis and centerline pivot point is normally below the surface of the road (see **Figure 18.17**). How far below the surface of the road determines how much scrub radius there is.

More scrub radius makes it harder to steer the car. Positive camber reduces scrub radius, but tire life is best when the running camber angle is 0 degrees.

Positive and Negative Scrub Radius. Scrub radius on rear-wheel drive cars is called **positive scrub radius**. The front wheels toe-out when rolling (**Figure 18.19**). Front-wheel drive cars usually have **negative scrub radius**. The tires toe-in when rolling (**Figure 18.20**).

NOTE: *If a left front tire on a front-wheel drive vehicle with positive scrub radius blows out, the tire will pull hard to the left. The still-inflated right tire would pull inward.*

Vintage Chassis

Older vehicles had narrow, tall tires. Their alignment specifications often included more positive camber, which would result in less scrub radius.

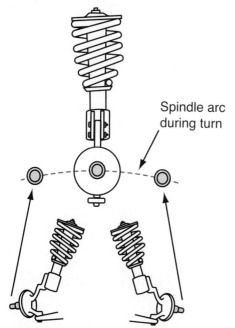

Figure 18.18 SAI aids in returning to a straight-ahead position after a turn when vehicle weight causes the spindles to aim straight ahead.

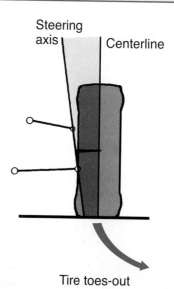

Figure 18.19 With positive scrub radius, the tire toes-out when rolling.

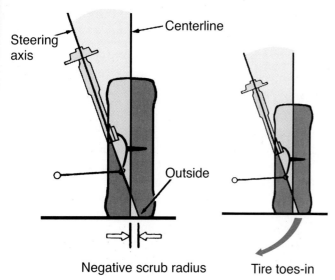

Figure 18.20 With negative scrub radius, the tire toes-in when rolling.

To prevent this dangerous situation, SAI has been increased on FWD cars. This results in negative scrub radius, which causes the blown tire to pull inward. The car will continue to go straight because this direction of motion counteracts the motion of the right front tire.

Incorrect Scrub Radius. Installing lower profile tires and offset wheel rims on a vehicle will result in a big increase in scrub radius (**Figure 18.21**). The pivot point can actually move outside of the tire's footprint area. The result is much harder steering, wheel shimmy, and a tendency to wander.

Negative scrub radius can result when tires and wheels that are too tall are installed on a rear-wheel drive vehicle. The result can be instability. When a tire on an SLA suspension system goes over a bump, the scrub radius can change from positive to negative as the camber changes. Vehicle handling can become dangerous.

NOTE: *When changing to tires and wheels of a different size, modifications to the suspension must be made. Wheel alignment cannot compensate for this.*

Other causes of incorrect scrub radius are a bent front suspension part or damage to the frame at the cross member or *strut tower*. The cross member, or cradle, is the large steel part of the frame beneath the engine and between the front wheels. The strut tower is the area inside of the front fenders that supports a MacPherson strut.

■ TURNING RADIUS

When a car makes a turn, the outside wheel must travel in a wider arc than the inside wheel (**Figure 18.22**). The alignment angle that controls this is called **turning radius**, toe-out-on-turns, or the **Ackerman**

> **Vintage Chassis**
>
> Horse-drawn wagons used a steering system that pivoted at the center. In 1884, Rudolf Ackerman patented a unique steering system with both of the wheels pivoting at the outside of the axle. The angle he designed into the steering arms caused the front wheels to toe-out when turning.

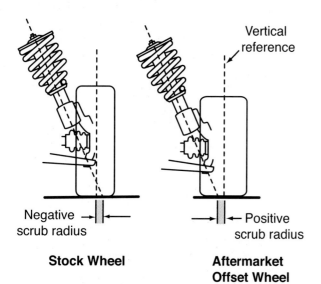

Figure 18.21 Installing lower profile tires and offset wheel rims on a vehicle will change scrub radius.

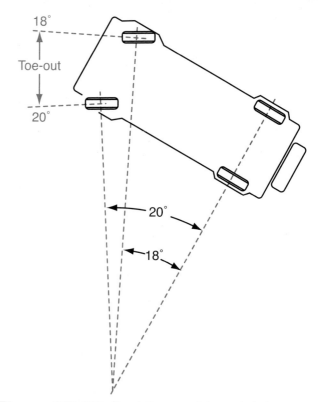

Figure 18.22 Front tires must toe-out during a turn.

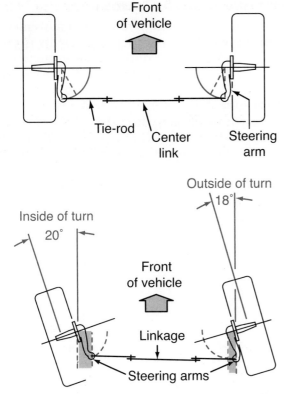

Figure 18.23 The tires turn different amounts as the steering arms move through their arcs of travel.

angle. The tires actually toe out during a turn because the steering arms are angled inward or outward (see **Figure 16.14**). When the wheels are turned, they move at different amounts as they move through their arc of travel (**Figure 18.23**).

■ TRACKING

The distance between the front and rear tires is called the **wheelbase**. The side-to-side distance between an axle's tires is called the *track*. For minimum tire wear and good fuel economy, the wheels must run on track. The rear tires are supposed to follow in the tracks of the front tires. All four wheels should form an exact rectangle.

Tracking is a term that refers to the relationship between the average direction that the rear tires point when compared to the front tires. When tracking is off, the front wheels do not follow the rear wheels (**Figure 18.24**). This causes the car to try to steer to the side, increasing tire wear and upsetting handling.

If the rear axle is out of line to the right, it will cause the steering wheel to be aimed to the right. This is often referred to as **dog tracking** due to its resemblance to the way that some dogs walk, with their rear feet not in alignment with their front feet. This situation will cause right front toe-out and left front toe-in. The result will be inside or outside toe wear. Toe wear is described in the prealignment section of Chapter 19.

■ SETBACK

Setback is the amount that one front wheel is behind the front wheel on the other side of the car (**Figure 18.25**). It is measured in degrees. A negative angle indicates that the wheel on the left side is set back. This will cause the vehicle to steer to the left. It can also cause a brake pull. A collision can cause incorrect setback.

■ SPECIAL HANDLING CHARACTERISTICS

The handling characteristics of a vehicle are described by terms such as *slip angle, understeer,* and *oversteer*.

Slip Angle

Tire manufacturers sometimes recommend different inflation pressures or alignment angles to change a vehicle's handling characteristics. A tire with lower inflation has a greater **slip angle**. Slip angle is the tendency during a turn for a tire to continue to go in the direction it was going, even though the rim has turned in response to steering wheel movement (**Figure 18.26**). Besides tire pressure, the amount of slip angle a tire has depends on the weight exerted vertically on it, the structure and type of tire, and the wheel alignment setting on the wheel. Positive camber causes a tire to have a greater slip angle.

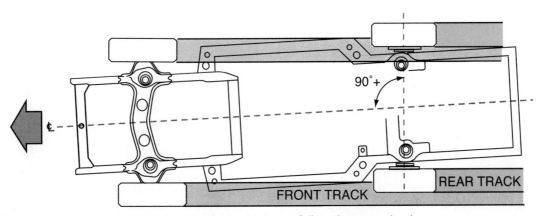

Figure 18.24 When tracking is off, the front wheels do not follow the rear wheels.

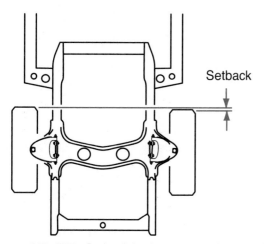

Figure 18.25 Setback is the amount that one front wheel is behind the one on the other side of the car.

Understeer and Oversteer

When a car does not seem to respond to movement of the steering wheel during a hard turn, this is called **understeer**. When a car turns too far in response to steering wheel movement, this is called **oversteer** (**Figure 18.27**). Desirable steering is said to be *neutral*.

Lower inflation pressures are sometimes specified for the rear tires of rear-wheel drive cars to prevent them from understeering. That way, when a car begins to slide and the driver lets up on the accelerator, the driver regains control. If the car had oversteer, easing up on the gas would result in a spinout.

When a car has four-wheel alignment capability, setting rear-wheel camber more negative than the front tires results in less tendency to oversteer. When tires of different profiles are installed on the same vehicle, the ones of the lower profile are installed on the rear to prevent oversteer.

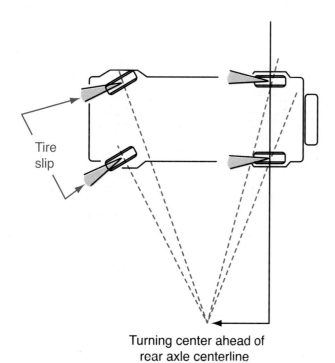

Figure 18.26 Slip angle.

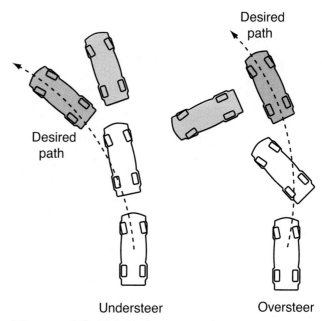

Figure 18.27 Understeer and oversteer.

REVIEW QUESTIONS

1. List the five wheel alignment angles.
2. Which alignment angle is most responsible for excessive tire wear?
3. If a tire has 1 inch of toe-in, how far is it dragged sideways in 1 mile of driving?
4. Which kind of car has its toe set inward when not rolling, front-wheel drive or rear-wheel drive?
5. When the top of a tire tilts inward toward the center of the vehicle, what positive or negative angle is this?
6. The angle that keeps a bicycle going straight when your hands are off the handle bars is called _____.
7. What is the name of the alignment angle that keeps the car going straight while allowing for easy steering?

8. When tires of a different height or wheel offset are installed on a vehicle, what alignment factor changes?

9. During a turn, which front wheel turns more sharply, the outside or the inside?

10. When a car does not seem to respond to the movement of the steering wheel during a hard turn, what is this called?

PRACTICE TEST QUESTIONS

1. Technician A says that camber can cause a car to pull to one side. Technician B says that camber can cause a tire to wear on one side of its tread. Who is right?
   A. Technician A
   B. Technician B
   C. Both A and B
   D. Neither A nor B

2. Which of the following is/are true about caster?
   A. When springs sag with age, caster changes.
   B. Caster is a measurement of camber change as the wheel is turned.
   C. Caster is a directional control angle.
   D. All of the above
   E. None of the above

3. Technician A says that scrub radius is what causes the tires to toe-in or toe-out when rolling. Technician B says that front-wheel drive cars usually have positive scrub radius. Who is right?
   A. Technician A
   B. Technician B
   C. Both A and B
   D. Neither A nor B

4. Technician A says that front tires toe-out during a turn. Technician B says that turning radius is caused by the toe setting. Who is right?
   A. Technician A
   B. Technician B
   C. Both A and B
   D. Neither A nor B

5. Which of the following will NOT cause a vehicle to pull to one side?
   A. Setback
   B. Tracking
   C. A low tire
   D. Toe-in

6. Two technicians are discussing setback. Technician A says setback can cause a brake pull. Technician B says incorrect setback can be due to a collision. Who is right?
   A. Technician A
   B. Technician B
   C. Both A and B
   D. Neither A nor B

7. Which of the following is/are true about front wheel toe?
   A. RWD cars are usually toed-in
   B. FWD cars are usually toed-out
   C. Wheels on FWD cars tend to toe inward when driving.
   D. Wheels on RWD cars tend to toe outward when driving.
   E. All of the above

8. Technician A says uneven caster can cause a car to pull to the side. Technician B says excessive caster will cause a tire to wear on one side of its tread. Who is right?
   A. Technician A
   B. Technician B
   C. Both A and B
   D. Neither A nor B

9. Which of the following is/are true about camber?
   A. Positive camber loads the inner wheel bearing
   B. Negative camber loads the outer wheel bearing
   C. Camber is also called the spindle angle
   D. Negative camber increases the leverage on the spindle
   E. All of the above

10. When a car makes a turn, the outside wheel must travel in a wider arc than the inside wheel. The alignment angle that controls this is called:
    A. Turning radius
    B. Toe-out-on-turns
    C. The Ackerman angle
    D. All of the above
    E. None of the above

CHAPTER 19

Wheel Alignment Service

OBJECTIVES

Upon completion of this chapter, you should be able to:
- ✔ Perform a prealignment inspection of the steering and suspension.
- ✔ Describe how to adjust caster, camber, and toe.
- ✔ Understand the different ways of adjusting alignment.

■ KEY TERMS

bump steer
crowned
eccentric cam adjustment
four-wheel alignment
geometric centerline
individual toe

orbital steer
radius plates
ride height
shims
slip plates
spread

steering pull
thrust angle
thrustline
toe angle
toe change
torque steer

■ INTRODUCTION

Before performing a wheel alignment, a thorough inspection of the steering and suspension systems must be performed. Loose parts will prevent an accurate and lasting adjustment. Looseness of suspension or steering parts can result in slack in the steering wheel, tire wear, shimmy, or an intermittent pull to one side or the other. For an alignment to be successful, suspension and steering components must be in good working order. A wheel alignment *should not be* attempted on a vehicle with worn or loose parts.

NOTE: *Incorrect alignment settings can result in pull, instability, and tire wear.*

Unusual tire wear or a vehicle handling problem are usually what cause a driver to bring a vehicle in for a wheel alignment. Front axles experience far more stress than rear axles because they support the weight of the engine and provide for steering. Although the rear axle is usually not the cause of problems, the technician must consider the entire frame and the steering and suspension systems to be able to properly align a vehicle.

This chapter first deals with suspension and steering system wear and inspection procedures. The latter part of the chapter deals with wheel alignment measuring and adjusting procedures.

■ PREALIGNMENT INSPECTION

Prior to a wheel alignment, the suspension and steering systems must be inspected. If parts are loose or worn, an alignment will not be successful:

- Tire pressures must be adjusted to specifications.
- The frame must be at the correct height.
- Worn bushings and pivoting parts must not allow excessive movement of suspension and steering parts.
- Steering gear and linkage coupling points must not have excessive clearance.
- Tires must be worn evenly from side to side for the vehicle to be level during alignment measurement.

■ TIRE WEAR INSPECTION

Unusual tire wear can be caused by worn parts, incorrect inflation, hard cornering, or incorrect wheel alignment. Camber and toe are the two adjustable wheel alignment angles that often cause wear. It is better if the alignment technician can look at worn tires before they are replaced. If tires are evenly worn and there is no pull or wandering, caster and camber should not require adjustment.

Tire Wear from Camber

Wear from incorrect camber shows up on either the outside or inside of the tire tread (**Figure 19.1**). Camber causes the inside and outside of the tire to have different diameters (see **Figure 18.3**). Outside wear is due to positive camber. It usually results from incorrect settings or when a vehicle with a high amount of steering axis inclination (SAI), does a lot of cornering.

NOTE: *Tire wear from camber results as springs sag over time, changing the height of suspension components.*

Tire Wear from Toe

Driving a vehicle with excessive toe is dangerous because the front tires are sliding. When toe is incorrect, a tire will develop a feathered edge, also called a *sawtooth pattern* (**Figure 19.2**). The feathered edge can be especially severe on bias tires, where it is often apparent to the eye. Feathered wear also occurs on radial tires and can often be felt by hand, although it is more difficult to detect visually on radials unless the toe is severely out of spec. Tire wear resulting from incorrect toe on radial tires often appears as wear to one side of the tire, similar to camber wear.

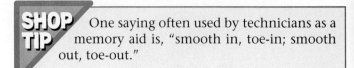

SHOP TIP One saying often used by technicians as a memory aid is, "smooth in, toe-in; smooth out, toe-out."

Before moving your hand across the tread surface, be sure the tire is not worn to the point where steel wire is exposed (**Figure 19.3**). To diagnose excessive toe on a worn tire:

1. Move your hand across the tire's footprint area from the outside to the inside. If you feel a feathered edge, the vehicle has excessive toe-out (see **Figure 19.2**).
2. If you feel a feathered edge when moving from the inside to the outside, the vehicle has excessive toe-in (**Figure 19.4**).

Other Toe Wear Factors

On a rear-wheel drive vehicle with toe-in, the right front tire tends to toe in more than the left and its outside edge wears because the tire is rolling under at that edge. With toe-out, the inside edge of the left tire will roll under, resulting in wear on the inside edge of the left front tire.

Radial tires with toe-in will both roll under, resulting in wear that looks like positive camber. Toe on the

Figure 19.1 Excess camber results in wear to one side of the tire tread. *(Courtesy of Tim Gilles)*

Figure 19.3 Be careful not to cut your hand on exposed steel belts. *(Courtesy of Tim Gilles)*

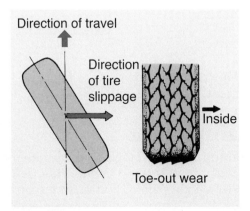

Figure 19.2 Toe-out wear.

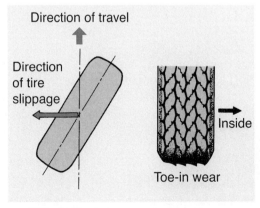

Figure 19.4 Toe-in wear.

418 SECTION 4 SUSPENSION AND STEERING

Figure 19.5 Wear from incorrect toe on a rear wheel. *(Courtesy of Tim Gilles)*

rear of a vehicle does not equalize like toe on the front does. If the tires are not rotated, they will develop diagonal wear (**Figure 19.5**).

■ RIDE HEIGHT CHECK

Because the weight of the vehicle is always resting on them, springs tend to sag as a vehicle ages. Alignment specifications are based on the assumption that the **ride height**, or *curb riding height,* of a vehicle is correct. Although it is possible to adjust alignment when the springs have sagged beyond specifications, tire wear and unusual handling can result. Prior to making any wheel alignment adjustments, ride height must be measured and compared to specifications (**Figure 19.6**). **Figure 19.7** shows the computer screen of a modern alignment machine that describes how to perform this measurement.

NOTE: *A vehicle that is either too high or too low will probably have a camber setting that is outside of specified limits.*

A vehicle with a short/long arm (SLA) suspension with springs that have sagged will not function within its desired alignment range as the springs deflect. This can cause changes in the way the vehicle reacts to

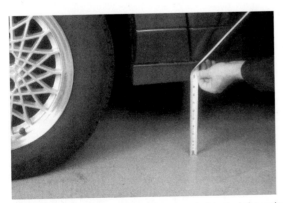

Figure 19.6 Measure and record ride height prior to aligning wheels.

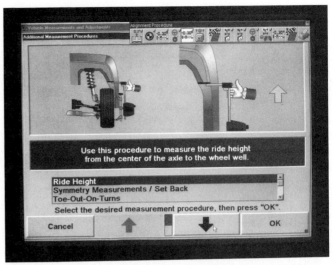

Figure 19.7 A screen from an alignment machine with a ride height feature.

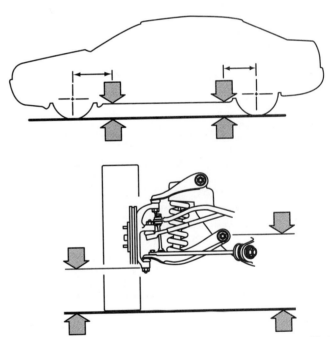

Figure 19.8 Typical ride height measurements. The lower sketch shows the typical measurement locations on an SLA system.

bumps. **Figure 19.8** shows where typical ride height measurements would be checked on an SLA suspension.

■ TOE CHANGE

As suspension height changes, toe measurement can change with it (**Figure 19.9**). The tie-rods are designed to remain parallel to the lower control arms when they pivot (**Figure 19.10**). When springs have sagged, **toe change** can result. Toe change causes the

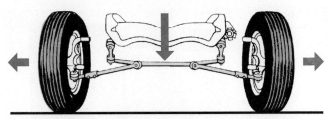

Figure 19.9 Toe can change with changes in ride height. *(Courtesy of Hunter Engineering Company)*

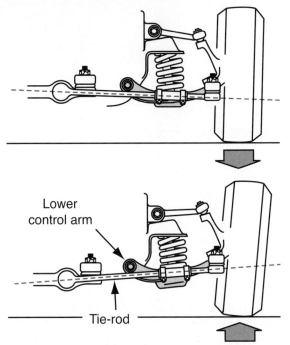

Figure 19.10 The tie-rods are designed to remain parallel to the lower control arms when they pivot.

tire to move on the road surface, scrubbing away tread. A small amount of toe change can be absorbed by the flexing of a properly inflated tire. Loose parts or incorrect wheel alignment can move the amount of allowable toe change out of normal limits.

When toe change is confined to one side of the vehicle, **bump steer**, or **orbital steer**, can result. Bump steer happens when a wheel with tie-rods at unequal heights goes over a bump. The vehicle momentarily steers in the direction that the wheel turns as the toe changes (**Figure 19.11**).

NOTE: *The height of the steering arm changes as caster changes. Toe change can be caused when there are different caster angles from side to side.*

Toe change measurement and correction is covered later in this chapter.

■ TORQUE STEER

Sometimes a vehicle will turn abruptly to the side during initial acceleration. This action, called **torque steer**, usually occurs on front-wheel drive vehicles with axles of unequal lengths. It can also result from anything that causes the axles to be at different heights, which produces unequal CV joint angles. The height difference could be due to a loose subframe or a problem with unequal spring height.

Vintage Chassis

Torque steer was not a problem on rear-wheel drive cars or front-wheel drive cars with small engines. When front-wheel drive became popular and horsepower increased, torque steer became noticeable. Older front-wheel drive cars sometimes had front driving axles of different lengths. During acceleration the long axle would twist slightly, delaying power transfer to its front wheel. During quick deceleration it would recoil. Some of the solutions attempted by automakers were to use a thicker axle, to change scrub radius, or to use an intermediate axle shaft so that drive axles would both be the same length.

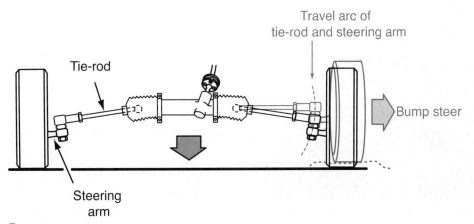

Figure 19.11 Bump steer.

SUSPENSION AND STEERING LOOSENESS

The motto of one suspension and steering part manufacturer is, "You can't align looseness." Steering and suspension components are designed to pivot without allowing any change in the positions of parts.

Perform a dry park check for steering and suspension looseness as described in Chapter 17. Always check the adjustment of the wheel bearings to see that they are not loose before attempting a wheel alignment.

TEST DRIVE

Unless a vehicle is unsafe to drive, a test drive should be done before performing repairs. Before driving the vehicle, a visual inspection must be performed to check the following items:

- Suspension bushings, visually and with a prybar
- Steering linkage pivot connections—firmly grasp the part and rock it to check for looseness
- Rubber grease boots on tie-rod ends and ball joints
- Shock absorbers

Also check to see if the vehicle has any signs of collision damage.

Tire Checks

To perform a tire check, follow these steps:

1. Adjust tire pressures to specifications. Check the condition of the valve stems and look for signs of impact damage that might have resulted in a bent rim (**Figure 19.12**).
2. Look for unusual wear on the tire tread.
3. Check for signs of damage to the sidewalls or tread area (**Figure 19.13**).
4. Be sure that tires of the correct size are used.
5. Radials and bias tires should not be mixed. Front tires should be of the same brand and tread pattern.

Figure 19.12 Inspect wheels and tires for impact damage. *(Courtesy of Tim Gilles)*

Figure 19.13 Distorted tread like this will affect a wheel alignment. *(Courtesy of Tim Gilles)*

Power Steering Checks

To perform a power steering check, follow these steps:

1. Check power steering fluid level. Look for evidence of leaking fluid.
2. Check drive belt tension.
3. Check the power assist to see that it works with equal ease in both directions.

During the test drive, the vehicle is checked for several conditions:

- Hard steering can be caused by binding parts, incorrect alignment, low tires, or a failure in the power steering system. Be sure to note whether the vehicle is easier to steer at higher speeds.
- Do tires squeal on turns? This could be due to a bent steering arm or low tire pressures.
- Are there squeaks and clunks? These noises can be the result of bad bushings, which can also cause changes in camber that produce brake pull.
- Is there a shimmy or tramp (the steering wheel shakes from side to side)? This could indicate a bent or out-of-balance wheel or excessive caster.
- A vehicle that wanders might have an incorrect caster angle setting.
- Does the vehicle pull to one side? When the steering wheel wants to go to the side by itself, especially when you let go of the wheel, this is called **steering pull**.

NOTE: *Sometimes pull can be the result of the crown of the road. Be sure to perform the test on a flat surface.*

- Does the vehicle pull to one side or the other? Does it always pull, or only pull during braking? This could be due either to alignment or brakes.
- If the vehicle pulls, does the direction or amount of pull change when the tires are rotated? A defective or damaged tire can cause a pull (usually to the side with the bad tire).

- Is it possible that a power steering problem could be causing the pull? Some vehicles have adjustable spool valves. With others, the steering box might require service. With the wheels off the ground, start the engine and verify that the wheels do not self-steer in either direction.
- Does the owner complain of a rough ride? This could be due to tire pressures that are too high, a bent or frozen shock absorber, or to the installation of radial tires on an older vehicle.
- Does the vehicle have a *body roll,* where it leans excessively to one side or the other during fast turns? The shock absorbers could be worn out.
- Is there a noise during turns that changes in pitch as the vehicle weaves to the left and then to the right? The wheel bearings on the outside wheels spin faster during a turn. Because of this and weight shift (which loads bearings), a bad bearing will make more noise when turning to one side than to the other.
- A wheel bearing noise is often accompanied by looseness.

Before attempting a wheel alignment, check for looseness in any related parts. With the vehicle raised, grasp the tires with your hands at the six o'clock and twelve o'clock positions and check for movement. If there is movement, readjust the wheel bearing. If there is still movement, locate its source in the suspension system.

NOTE: *A check for similar movement at the three o'clock and six o'clock positions verifies steering linkage condition.*

Test ball joints for looseness as described in Chapter 15. It is best to consult the appropriate service manual because procedures vary. Some ball joints are designed to have a specified clearance in them. Several part manufacturers provide specification tables.

Inspection Checklist

A checklist that can be used by technicians to ensure that no steps are accidentally forgotten during a suspension inspection and test drive is shown in **Figure 19.14.**

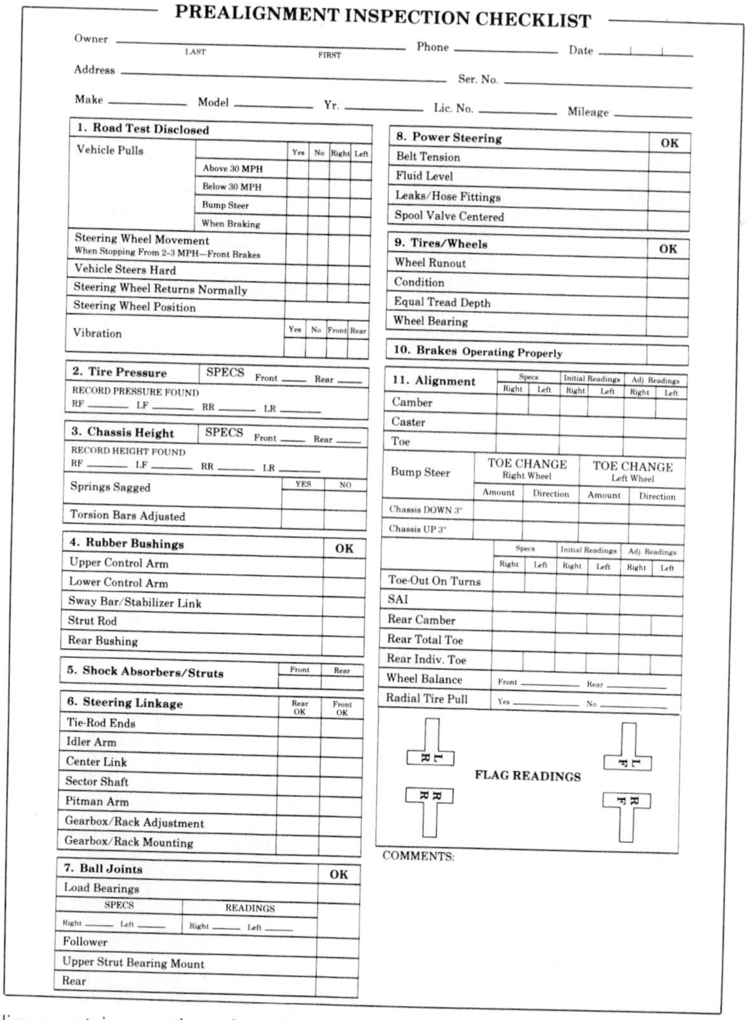
PREALIGNMENT INSPECTION CHECKLIST

Owner ______ LAST ______ FIRST ______ Phone ______ Date ______

Address ______ Ser. No. ______

Make ______ Model ______ Yr. ______ Lic. No. ______ Mileage ______

1. Road Test Disclosed		Yes	No	Right	Left
Vehicle Pulls	Above 30 MPH				
	Below 30 MPH				
	Bump Steer				
	When Braking				
Steering Wheel Movement When Stopping From 2-3 MPH—Front Brakes					
Vehicle Steers Hard					
Steering Wheel Returns Normally					
Steering Wheel Position					
Vibration		Yes	No	Front	Rear

2. Tire Pressure — SPECS Front ____ Rear ____
RECORD PRESSURE FOUND
RF ______ LF ______ RR ______ LR ______

3. Chassis Height — SPECS Front ____ Rear ____
RECORD HEIGHT FOUND
RF ______ LF ______ RR ______ LR ______

	YES	NO
Springs Sagged		
Torsion Bars Adjusted		

4. Rubber Bushings	OK
Upper Control Arm	
Lower Control Arm	
Sway Bar/Stabilizer Link	
Strut Rod	
Rear Bushing	

5. Shock Absorbers/Struts	Front	Rear

6. Steering Linkage	Rear OK	Front OK
Tie-Rod Ends		
Idler Arm		
Center Link		
Sector Shaft		
Pitman Arm		
Gearbox/Rack Adjustment		
Gearbox/Rack Mounting		

7. Ball Joints	OK
Load Bearings	
SPECS Right ____ Left ____ / READINGS Right ____ Left ____	
Follower	
Upper Strut Bearing Mount	
Rear	

8. Power Steering	OK
Belt Tension	
Fluid Level	
Leaks/Hose Fittings	
Spool Valve Centered	

9. Tires/Wheels	OK
Wheel Runout	
Condition	
Equal Tread Depth	
Wheel Bearing	

10. Brakes Operating Properly	

11. Alignment	Specs Right	Specs Left	Initial Readings Right	Initial Readings Left	Adj. Readings Right	Adj. Readings Left
Camber						
Caster						
Toe						

Bump Steer	TOE CHANGE Right Wheel Amount	TOE CHANGE Right Wheel Direction	TOE CHANGE Left Wheel Amount	TOE CHANGE Left Wheel Direction
Chassis DOWN 3"				
Chassis UP 3"				

	Specs Right	Specs Left	Initial Readings Right	Initial Readings Left	Adj. Readings Right	Adj. Readings Left
Toe-Out On Turns						
SAI						
Rear Camber						
Rear Total Toe						
Rear Indiv. Toe						

Wheel Balance — Front ______ Rear ______

Radial Tire Pull — Yes ______ No ______

FLAG READINGS
LR LF RR RF

COMMENTS:

Figure 19.14 A prealignment inspection checklist.

WHEEL ALIGNMENT PROCEDURES

The front suspension is designed to keep the wheels in the best possible position while rolling. Wheels must roll freely with as little tire scuff (side-to-side wear) as possible. Alignment settings can change according to such factors as vehicle speed, the roughness of the road surface, acceleration, braking, weight distribution, and cornering. Specifications are developed by manufacturers so that alignment can be adjusted with the vehicle at rest on a level alignment rack (**Figure 19.15**). If tire wear is experienced when the settings are within specifications, the technician should make adjustments to compensate.

Adjustments to original alignment settings might be needed because of previous misadjustment, minor wear of pivoting parts, collision damage, or spring sag. A new vehicle might need an alignment because it came from the factory with adjustments outside of the normal range of specifications. Design engineers list a range of adjustment limits for wheel alignment angles. If the adjustments fall within those limits and the vehicle "tracks" straight (goes straight on a level road), little tire wear should occur.

Only three of the five alignment angles are normally adjustable:

1. Caster
2. Camber
3. Toe

The other two angles are measured to check for damage to suspension and steering parts. Of the three adjustable angles, tire wear when traveling straight only results from incorrect camber and toe. Directional pull can be caused by camber and caster.

NOTE: *On some MacPherson strut systems, only toe is adjustable.*

MEASURING ALIGNMENT

Most alignment measurements are read in degrees and parts of degrees. Degrees represent part of a 360-degree circle in both the metric and inch systems. In the United States, remaining portions of degrees are either fractional or decimal. In the metric system, as in the world of science, portions of degrees are read in minutes (') and seconds ("). Like a clock, there are 60 minutes in a degree and 60 seconds in a minute.

When taking alignment measurements, ball bearing–supported plates are placed beneath the tires. These allow the tires to assume a relaxed position (see **Figure 19.15**).

On most alignment racks, the front wheels are positioned on **radius plates** (**Figure 19.16**). A radius plate has a gauge that measures in degrees how far a wheel is turned to the right or left. On four-wheel alignment racks, **slip plates** are under the rear tires.

NOTE: *Some of the newer alignment racks no longer require the use of radius plates (see **Figure 19.58**).*

Pins hold the upper plate from moving on the bearings. After the vehicle is driven onto the plates, the pins are pulled out of the plates. Pushing down on the bumpers lets the wheels creep into a relaxed position, as they would be in when rolling on the road. Air or hydraulic jacks are used to raise the vehicle off the lift during alignment adjustments or to reposition the radius plates.

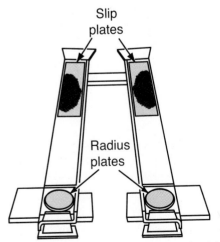

Figure 19.15 A wheel alignment rack with ball bearing–supported plates for the front and rear wheels.

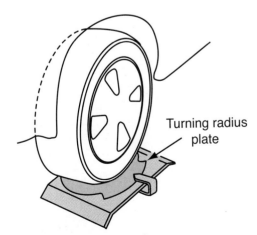

Figure 19.16 A radius plate.

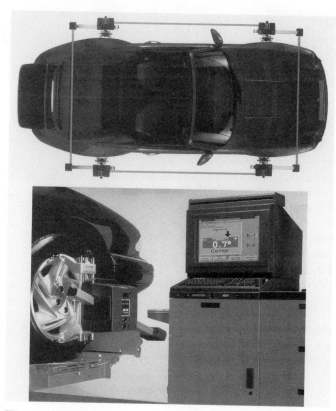

Figure 19.17 A modern four-wheel alignment machine looks at all four wheels. *(Courtesy of Hunter Engineering Co.)*

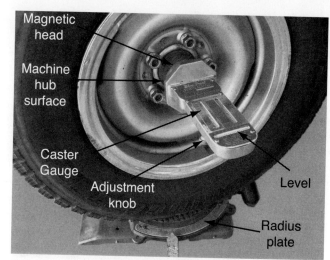

Figure 19.18 A portable alignment gauge attached to the hub with a magnet. *(Courtesy of Tim Gilles)*

Computerized alignment machines are used to do **four-wheel alignment** inspection (**Figure 19.17**). Many newer vehicles are equipped with rear-wheel adjustment, which requires four-wheel alignment capability. Mechanical measuring systems are covered in this section because it is important to understand what you are actually measuring when taking alignment readings.

To get accurate measurements, the vehicle must be level. Toe is measured in inches, millimeters, or in angle of toe. The remaining angles are measured in degrees of a circle.

When adjustment is possible, caster and camber are usually adjusted together since adjusting caster affects the camber reading. Toe is adjusted last, after caster and camber. When other alignment angles are changed or when parts are replaced, toe changes.

NOTE: *If tires are to be replaced, the new tires should be installed before attempting an alignment. An alignment rack provides a level location for checking suspension angles. The heights of worn tires will affect alignment measurements.*

Measuring Camber

Camber is simply a comparison measurement to true vertical, using a level. To measure camber, position the wheels straight ahead while reading the gauge or computer screen. If the wheels are not straight ahead, the caster angle can cause an incorrect camber reading.

Measuring Caster

Caster causes the wheel's camber angle to change during a turn. In fact, the caster measurement is actually a reading of camber change while turning.

To measure the caster setting, the wheel is first turned a specified amount either inward or outward. Whether you turn the wheel in or out depends where the vehicle manufacturer positioned + and − on the face of the gauge. The positions are not all the same.

Alignment settings for caster and camber are usually within a range, but adjusting alignment to within that range will not always ensure that the vehicle will go straight.

Road Crown and Pull. Roads are **crowned** (higher at the center than the outside) so that rain will run off (**Figure 19.19**). A vehicle aligned with equal settings

> ### Vintage Chassis
>
> The tools used for measuring caster and camber in older and portable mechanical alignment measuring systems have bubble levels. These are used for making comparisons to an exact level position. To attach the gauges to the wheels, a wheel rim adapter clamp is used or a large magnet holds the gauge to the wheel hub (**Figure 19.18**). The tool is mounted against a machined area of the hub. The wheel bearing dust cap is removed so that a pilot can center the gauge on the center hole in the end of the spindle.

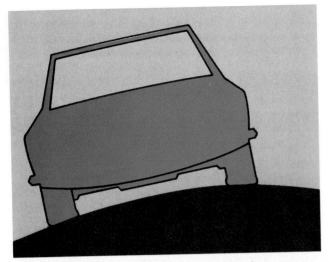

Figure 19.19 Roads are crowned so that water can run off them.

> **SHOP TIP**
> - Before making an alignment adjustment, decide if it would be helpful to unload the suspension components so the weight of the vehicle is not resting on the wheels. Understanding the suspension design can make removing and replacing shims or turning an adjustment eccentric easier.
> - Before making an alignment adjustment, note the current camber angle while the vehicle is raised. Then make the required amount of change to the "new" camber reading.

adjustment methods. On SLA suspensions, the camber adjustment is done with shims or eccentrics. **Figure 19.20** shows a common SLA adjustment in which **shims** are removed or installed to reposition the upper control arm.

Plan Ahead. When there are shims, caster and camber are changed together. Removing and replacing shims is sometimes a fair amount of work. Plan ahead. Think about the effect the shim will have. Look at the control arm inner shaft to see what the effect of adding

Vintage Chassis

Caster is measured using the same gauge that measures camber. On older, portable alignment heads, there are two levels on the gauge. One of the levels has a thumbscrew that allows it to be adjusted. At the 20-degree point on the radius plate, set the level to 0 degrees. Turn the wheel through a 40-degree sweep until the wheel is 20 degrees in the opposite direction. At this point, the reading on the gauge is the caster reading. It will read in degrees, either positive or negative.

from side to side will drift to the outer edge of a crowned road.

To compensate for road crown, two methods can be used:

1. Camber can be set slightly more positive on the driver's side.
2. Caster can be set so that it is slightly more negative on the driver's side of the vehicle.

Many alignment technicians like to correct for road crown using caster because camber that is too far positive will result in wear to the outer edge of the tire.

NOTE:

- *When correcting for road crown, caster creates less of a pull than camber.*
- *Caster must be within ½ degree of the caster setting on the other side of the vehicle.*

Adjusting Caster and Camber

There are many methods of adjustment for caster and camber. Alignment equipment manufacturers provide detailed charts and catalogs that describe specific

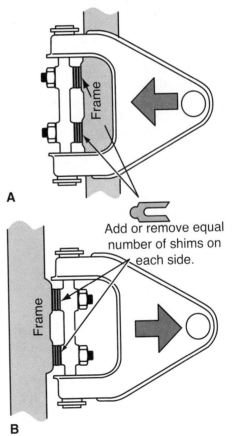

Figure 19.20 Changing camber with shims on inboard and outboard pivot shafts. (A) With this suspension design, it is easier to remove shims with the weight of the vehicle on the tire. (B) On this suspension, removing shims is easiest with the weight off the tire.

or removing shims will be. Some vehicles (usually light trucks) have an outboard pivot shaft on the frame. Shims have the opposite effect as the normal control arm with the pivot shaft inboard of the frame. **Figure 19.20** shows how to change camber with inboard and outboard pivot shaft locations. Changing only the camber angle is accomplished by removing or adding an equal number of shims to each side of the pivot shaft.

To change caster with shims requires removing or adding shims at either end of the control arm pivot shaft (**Figure 19.21**). This will change camber because it moves the control arm out, or in, on one side. To keep camber from changing much during a caster adjustment, remove a shim from one side of the pivot shaft and install it at the other end.

Service literature sometimes states the effect a particular shim thickness will have on both the caster and camber readings. More often, this information is not available. When a technician works repeatedly on one make of vehicle, experimenting with one size of shim will show the effect that shim has on alignment settings.

An example of the amount a shim will change a vehicle's alignment is as follows:

$1/16$ inch shim on one side only = $1/2$ degree caster change

or

Remove a $1/32$-inch shim from one side and add it to the other side for $1/2$ degree of caster change

Service literature sometimes includes a chart showing the amount of shim change to make to get a desired setting. Some computerized four-wheel alignment machines do the calculations for you (**Figure 19.22**).

Some vehicles use an **eccentric cam adjustment** on the upper or lower control arm or strut (**Figure 19.23**). Turning the adjustment repositions the camber and caster angle. Less common alignment adjustment methods include slotted holes in the upper control arm

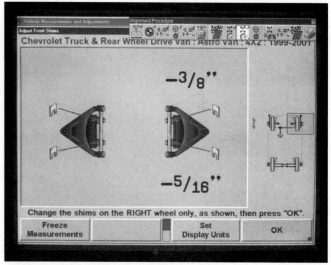

Figure 19.22 A computer screen from a wheel alignment machine showing recommended shim thicknesses.

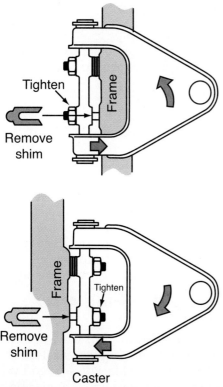

Figure 19.21 Removing shims from one side or adding them to the other moves the position of the upper ball joint toward the front or rear, depending on the position of the frame.

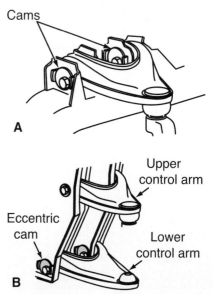

Figure 19.23 (A) An eccentric adjustment on an upper control arm. (B) An eccentric adjustment on a lower control arm.

shaft (**Figure 19.24**) and an eccentric bushing installed under a ball joint (**Figure 19.25**). Some computer alignment machines have a feature that calculates the position of the bushing (**Figure 19.26**).

When MacPherson struts are adjustable (not all of them are) there are slots in the upper bearing bracket

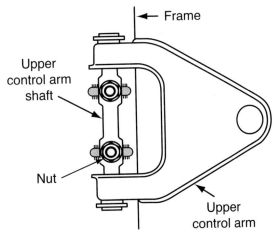

Figure 19.24 Slotted holes in this upper control arm shaft allow for caster and camber adjustment.

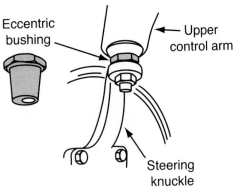

Figure 19.25 An eccentric bushing installed under the ball joint for alignment adjustment.

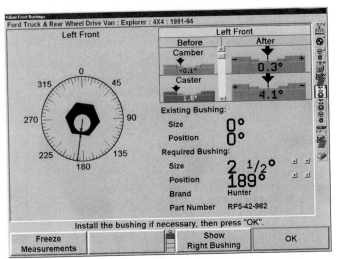

Figure 19.26 A computer alignment screen that calculates offset bushing position. *(Courtesy of Hunter Engineering Company)*

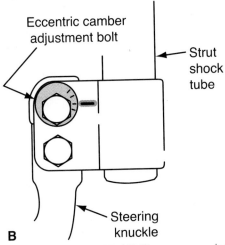

Figure 19.27 (A) The arrow points to a slot that allows movement of the top of the strut to adjust alignment. (B) An eccentric adjuster on a strut suspension. *(A, Courtesy of Tim Gilles)*

for adjusting caster (**Figure 19.27A**) or an eccentric between the shock tube and steering knuckle (**Figure 19.27B**).

When a vehicle has a narrow lower control arm, there is a strut rod running from it to the frame for strength. There are often threads on one end of the strut. Repositioning the nuts on the threaded area moves the control arm in an arc (**Figure 19.28**). The result is that caster changes. Moving the lower control arm toward the rear makes caster more negative.

Measuring Steering Axis Inclination

Steering axis inclination (SAI) does not change. It is not adjustable and some equipment does not measure it. On front-wheel drive MacPherson strut vehicles, SAI is the primary directional control angle and adjustable angles can be used to correct vehicle aim toward straight ahead. FWD MacPherson strut vehicles have lighter suspensions that are more easily damaged than

Chapter 19 Wheel Alignment Service 427

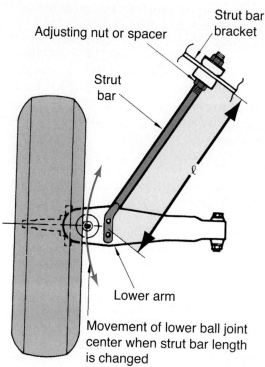

Figure 19.28 Adjust the length of the strut bar to change caster.

heavier RWD vehicles. A change in SAI only occurs if the spindle has been bent or there has been body damage, resulting in a bent strut tower. When this happens, the camber angle will also change. If the included angle is correct but the camber angle is not, the spindle/steering knuckle is not bent, so replacing it will not correct the problem. If the included angle is not correct, parts must be replaced.

Look for a bent spindle when a front wheel bearing wears out. Front wheel bearings on rear-wheel drive vehicles do not wear out very often, but a bent spindle can cause the bearings to be misaligned, increasing the load on it. If the cradle (the subframe) has shifted to one side, camber will change on both front wheels (**Figure 19.29**) but the included angle will remain the same. Some vehicles have an alignment hole that is used to verify correct cradle position in relation to the body (**Figure 19.30**).

To measure SAI on computerized alignment equipment, special procedures are followed. For example, during the caster measurement the steering wheel is locked and the front alignment sensors are leveled and locked in place before the wheels are turned, first to one side and then the other. The machine measures caster at the same time that it reads SAI.

NOTE: *Unless there is a symptom pointing to the possibility of incorrect SAI as a contributing cause, most technicians do not go through the extra procedures required to check SAI.*

Because it is quicker for the technician to measure only caster without SAI, the alignment machine offers that option as a default choice. When measuring caster

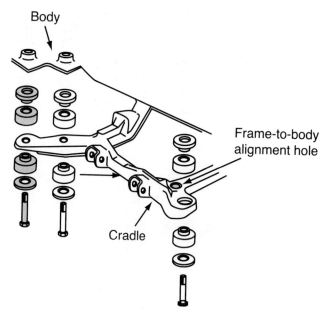

Figure 19.30 Some vehicles have an alignment hole that is used to verify correct cradle position in relation to the body.

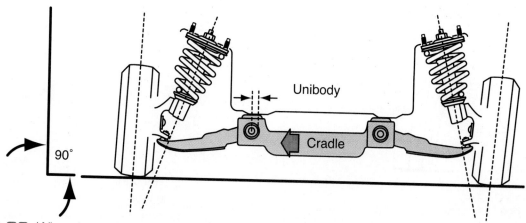

Figure 19.29 When the cradle shifts to one side, camber will change but the included angle will remain the same.

only, the alignment head is loose on its shaft and free to pivot.

SAI specifications are not usually provided by manufacturers. When they are available, some manufacturers provide separate camber and SAI specifications. Others list camber and the included angle, leaving the technician to calculate the SAI. Camber must be set correctly first. Comparing SAI from side-to-side is a good indicator of correct SAI angles.

The typical range of SAI specifications is:

- SLA/RWD   6–8 degrees
- MacPherson/FWD   12–16 degrees
- Maximum side-to-side variation 2 degrees

A typical vehicle with an SLA suspension might have 7.5 degrees SAI and 2.8 degrees positive caster. A MacPherson strut front-wheel drive vehicle might have 14 degrees SAI and 1.6 degrees positive caster.

Measuring Included Angle. The included angle is the amount of SAI minus camber. If camber is negative, subtract it from the SAI reading. If camber is positive, add it to the SAI reading (**Figure 19.31**). Included angle is usually within ½ degree from side to side. Double-check the specification.

If camber cannot be adjusted, check the included angle. When the included angle is off, the spindle, strut, or steering knuckle is bent, which requires replacement of the part. If the included angle is as specified and the camber angle is off, a damaged strut or steering knuckle is not the cause and replacing the part will not solve the problem.

Measuring Toe

Checking and adjusting toe after replacing a tie-rod end or other steering linkage component is an important part of the job. When measuring toe, the distances between the fronts and the rears of the front tires are compared (see **Figure 18.1**). An optical measuring gauge used to measure toe projects an image to a gauge on the opposite side of the vehicle (**Figure 19.32**).

Vintage Chassis

Measuring toe without an optical device required an accurate line to be scribed on the tread's footprint. With the tire raised off the ground, the tire was spun while the technician applied chalk to the center of its tread. Next, a narrow line was scribed in the middle of the chalked area. This was done using a special scribing tool or a board with a nail pounded through it. After the vehicle was lowered to the ground, the distances between the scribed lines on the tread surfaces at the front and rear of the tire were compared. Prior to the development of optical toe measurement, a tape measure or a trammel bar, also called a tram gauge, was used (**Figure 19.33**). The trammel bar was adjusted to measure at the same height as the spindles at the centers of the wheel hubs, which resulted in toe being measured at the largest circumference of the tire tread.

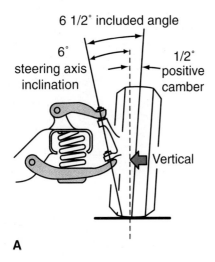
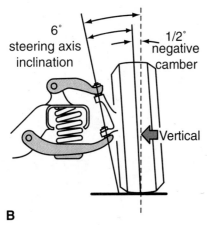

Figure 19.31 Calculating the included angle. (A) When camber is positive, add it to SAI. (B) When camber is negative, subtract it from SAI.

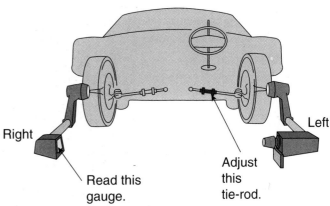

Figure 19.32 Adjust the tie-rod on the opposite side of the gauge.

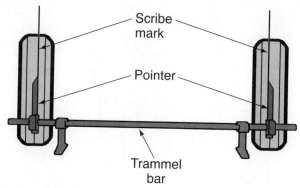

Figure 19.33 A trammel bar for measuring toe.

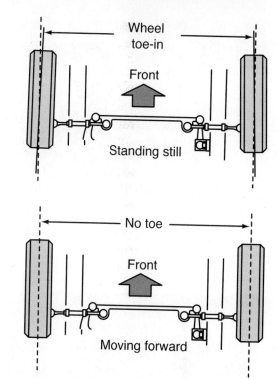

Figure 19.35 On a rear-wheel drive car, tires tend to toe-out when rolling.

Methods of Calculating Toe.
Toe has traditionally been measured as a distance in inches or millimeters, called total toe or toe distance. A more recent trend is to measure the **toe angle**, which is the angle between the two front tires. Toe angle is equal between the two front tires when the tires are rolling forward. The diameter of the tire does not affect the toe angle reading as it does with traditional toe distance measurement (**Figure 19.34**). This is an advantage, especially when oversized tires are used. When toe angle is measured in degrees, toe-in is a positive (+) angle and toe-out is a negative (−) angle. Zero toe is when the wheels are parallel. A chart in Appendix E at the back of the book lists conversions of lengths and angles used in wheel alignment.

Toe in FWD and RWD Vehicles.
Front- and rear-wheel drive vehicles have different toe specifications because they react differently when their tires are in motion. The objective is to have zero toe when the vehicle is in motion.

- Rear-wheel drive vehicles are usually set slightly toed-in (about 1/16 inch on new vehicles). This is because the front tires tend to push outward at the front when rolling due to the rolling resistance of the tires and slack (also called compliance) in steering linkages and bushings (**Figure 19.35**).
- Front-wheel drive vehicles are usually set with zero toe or a slight amount of toe-out. This is because the driving front tires tend to push inward at the front as they pull the vehicle forward on the road.

The running toe setting can undergo more of a change in response to scrub radius and to increased rolling resistance caused by low tire pressures or wider tires. Scrub radius changes when wheels with more negative offset than original equipment are installed.

Adjusting Toe
Steering linkages on most vehicles have either two or four tie-rod ends. On conventional parallelogram steering, the toe adjustment is made by turning the threaded "turnbuckle" sleeves between the tie-rod ends (**Figure 19.36**). Because one of the tie-rod ends has a left-hand thread and the other has a right-hand thread, as the sleeve is turned the tie-rod assembly will become either longer or shorter. This changes the toe setting.

First center the steering wheel and hold it in place with a steering wheel clamp (**Figure 19.37**). Then make the adjustment.

After turning the adjusting sleeves, the clamp must be properly positioned before tightening it. It must not come into contact with anything when the wheels are turned from side to side. Be sure that the opening of the clamp is not positioned over the split in the adjusting

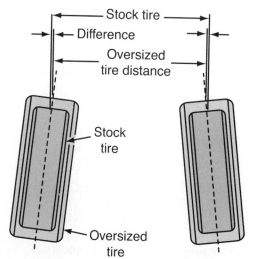

Figure 19.34 Oversized tires will affect toe distance measurement but not toe angle.

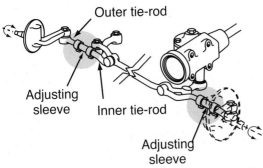

Figure 19.36 Turn the adjusting sleeves to adjust toe.

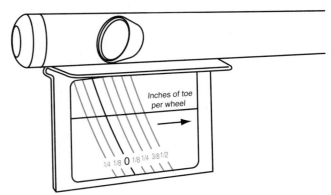

Figure 19.38 An optical toe gauge screen.

sleeve (**Figure 19.39**). On parallelogram systems, if one tie-rod happens to be tilted one way and the other is tilted the other way, the entire tie-rod will not be able to pivot (**Figure 19.40**). The tie-rods are positioned so that they are not binding up. This is most easily done by turning both tie-rods in the same direction before tightening the clamp.

Rack-and-pinion steering systems have a conventional outer tie-rod and an inner tie-rod end with a

Figure 19.37 A steering wheel holder and a brake pedal depressor in use. *(Courtesy of Tim Gilles)*

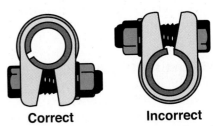

Figure 19.39 Be sure that the opening of the clamp is not positioned over the split in the adjusting sleeve.

Vintage Chassis

Before computer alignment machines, a trammel bar or optical toe gauge was used. When making a mechanical measurement with a trammel bar, a technician would sight down the side of the front tire to the rear tire and make an adjustment so that each front wheel looked to be in line with the rear one on the same side of the vehicle. When using an optical toe gauge, the gauge on one side was read while adjusting the tie-rod on the other side (**Figure 19.38**). Nowadays, computer alignment machines calculate toe and project it on the computer screen.

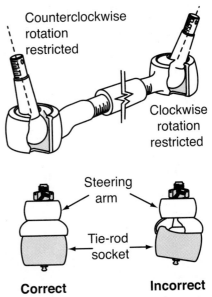

Figure 19.40 Position the tie-rods so that they will not bind.

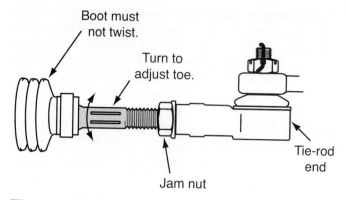

Figure 19.41 Adjusting toe by turning a rack-and-pinion tie-rod.

jam nut on each side (**Figure 19.41**). Hold the tie-rod end with a wrench or pliers while loosening and tightening the jam nut (there is usually a provision on the tie-rod to accommodate a wrench). The inner end of the tie-rod is a ball socket that can be rotated to adjust toe. It will not bind if a tie-rod is turned off-center like the parallelogram type shown in **Figure 19.40**.

NOTE: *Loosen the clamp on the boot as needed to be sure the boot does not twist while you adjust toe.*

Centering the Steering Wheel

Adjusting one of the tie-rods more than the other will cause the steering wheel position to move.

NOTE: *It is important that the steering wheel be centered using the tie-rods, not by removing the steering wheel and putting it back on straight.*

If the steering wheel is not straight ahead when driving, follow this procedure to center it:

1. Count the number of turns of the steering wheel while turning it from lock to lock.
2. Position the steering wheel so that it is halfway between the locks. It should be straight. If not, remove it and put it back on straight.
3. Use a steering wheel lock to hold the steering wheel in the centered position (see **Figure 19.37**).

This procedure will allow turn signals to cancel correctly during turns and minimize looseness in recirculating ball/nut steering gears.

When making a rough adjustment, turn each tie-rod an equal amount in opposite directions. This will approximately maintain the current toe setting. A small amount of change in the toe setting can occur when tightening the tie-rod clamps or jam nuts. This is because of slack between the threads.

NOTE: *Correct adjustment will often require that the initial setting be slightly off in anticipation of the change that will occur when the jam nut is completely tight. This can be compared to the fine-tuning that is done when adjusting valve clearance in an engine.*

Test-drive the vehicle on a straight, level road. If the steering wheel is off-center and points to the left, adjust the tie-rods so that the tires also point to the left (**Figure 19.42**). Adjust the wheels to face the opposite direction if the wheel points to the right.

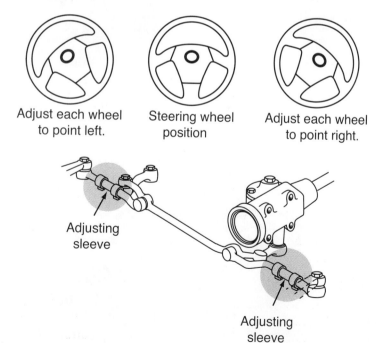

Figure 19.42 Centering a steering wheel by adjusting the tie-rods. Rotate the tie-rod sleeves equally in opposite directions.

SHOP TIP From the front of the vehicle, sight down the tires on each side of the vehicle to see if the front one aligns with the rear one. This will give a rough estimate of what direction the tie-rods need to be turned.

REMEMBER: *People often perceive things differently from each other. A customer might describe a steering pull because the steering wheel is off-center. The customer sees the wheel off-center and straightens it, causing the vehicle to steer to the side. Manufacturers' specifications for maximum allowable steering wheel angle variation are typically plus or minus 3 degrees.*

Toe Change Check. Toe change was discussed earlier in this chapter. Because toe can change with bumps, toe is accurate only when the vehicle is at the correct ride height. Use the jack on the alignment rack to raise the vehicle an equal amount on each side (**Figure 19.43**). Check to see that toe changes equally on each wheel. If not, look to see if one end of the steering linkage is out of level.

When a rack-and-pinion steering gear is mounted in an out-of-level position, its tie-rods will be at unequal angles. First check the mounting of the rack-and-pinion steering gear. Sometimes one or both of its holding brackets is loose or a bushing has become damaged. Measure and compare the height of the steering gear and linkage at several locations (**Figure 19.44**). Sometimes a long straightedge is helpful in spotting irregularities (**Figure 19.45**). Some vehicles use shims to adjust rack-and-pinion height to correct for toe change (**Figure 19.46**).

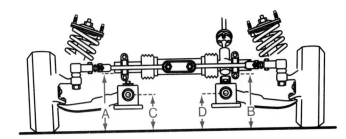

Figure 19.44 Measure and compare the height of the rack-and-pinion steering gear and linkage at several locations.

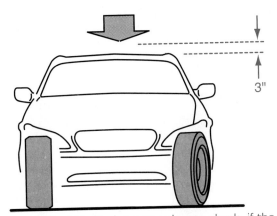

Figure 19.43 During a toe change check, if the wheels do not change an equal amount look to see if one end of the steering linkage is out of level.

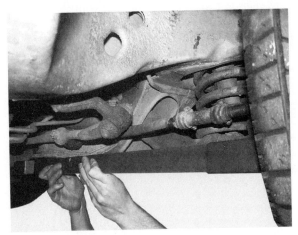

Figure 19.45 Sometimes a long straightedge is helpful in spotting irregularities in steering linkage angles. *(Courtesy of Tim Gilles)*

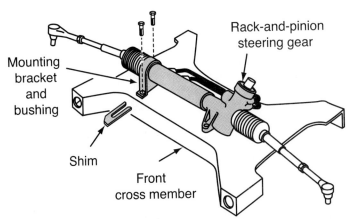

Figure 19.46 Some vehicles use shims to adjust rack-and-pinion height to correct for toe change.

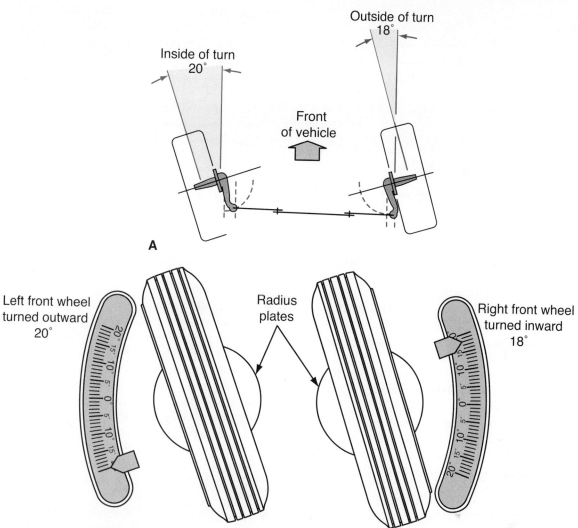

Figure 19.47 (A) The inner wheel turns at a sharper angle than the outer wheel. (B) Observe the difference in outer and inner wheel angles on the radius plates.

Sometimes, an idler arm will have slotted mounting holes so that its height can be adjusted. The vehicle might have had an accident and require the services of a frame straightening shop.

NOTE: *Aftermarket suspension parts sometimes relocate steering linkage connections, which can cause extreme changes in toe as the suspension system moves. After installation of new parts, it is always a good idea to check for toe change during suspension system travel.*

Measuring Turning Radius

Measure turning radius after other alignment adjustments have been made. To measure turning radius, observe the pointer on the radius plate while making a caster measurement. As the wheels are turned from side to side, the outer wheel should make a turn that is 2 or 3 degrees less than the inside wheel. For example, when the outside wheel is turned 18 degrees, the inside wheel's radius plate gauge should read about 20 degrees (**Figure 19.47**). The reading should be within 1.5 degrees of specifications found in the service manual.

The steering arms are angled to point to the center of the rear axle (**Figure 19.48**). This is called the Ackerman angle.

NOTE: *Turning radius is not an adjustable angle, but if a steering arm becomes bent the turning radius is affected and the tires can squeal on turns.*

■ GENERAL WHEEL ALIGNMENT RULES

A suspension/steering specialist will experiment with different vehicles to see what changes result from various adjustments. The following general rules are provided as a starting point. Due to variations in vehicle design, the rules will not always be true.

Caster and Camber

■ The vehicle will pull to the side with the least (more negative) caster.

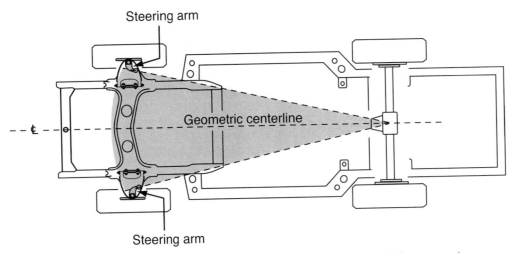

Figure 19.48 The Ackerman angle has the steering arms point to the center of the rear axle.

- The vehicle will pull to the side with the most (more positive) camber (**Figure 19.49**).
- Adjusting caster more negatively results in easier steering.
- The driver's weight will usually cause camber to increase on the left front wheel and decrease on the right front wheel.
- On shim-type vehicles, changing camber will not affect caster but changing caster can affect camber.
- Caster for both wheels should be set either positive or negative; not one wheel positive and the other one negative.
- Caster **spread** between the front wheel settings should not be more than ½ degree. Camber is sometimes specified with more than ½ degree of spread. Always follow the manufacturer's specifications.

NOTE: *Spread is the difference between alignment settings from side to side. It is also called cross caster or cross camber.*

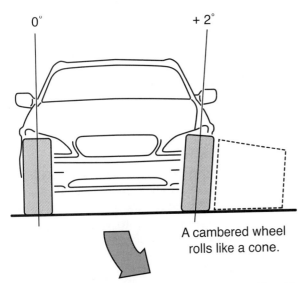

Figure 19.49 The vehicle will pull to the side with the most positive camber.

- Make caster equal from side to side. Use camber to compensate for road crown. Set the left side a quarter degree more positive than the right (if caster is equal).
- Though uncommon today, vehicles with manual steering have caster that is generally 0 to 1 degree negative. Power steering vehicles can have caster as high as 10 degrees (Mercedes).
- On a MacPherson strut vehicle, jounce the vehicle while measuring camber. If camber changes drastically on either front wheel, the strut is bent.

Toe

- Before driving on the alignment rack, be sure the pins are in the radius plates.
- Every ¼-inch turn of the adjusting sleeve results in about ¹⁄₁₆ inch of change in toe.
- Changes in caster and camber affect toe, so toe is adjusted last.
- Toe is measured in either inches, millimeters, or degrees.

■ FOUR-WHEEL ALIGNMENT

Modern wheel alignment equipment measures wheel alignment of all four wheels. Rear wheel toe and camber are often adjustable. When they are not, the front wheels are adjusted, with reference to the settings in the rear. That concept will be covered later in this chapter.

The **geometric centerline** of the vehicle is a line drawn between the center of the front axle and the center of the rear axle. When individual toe is measured at all four-wheels, the geometric centerline is used. Several measurement factors are taken in relation to the geometric centerline. **Figure 19.50**, a computer screen from a four-wheel alignment machine, shows several of these factors. The following section describes some of these factors. Other geometric centerline factors were covered in Chapter 18.

Chapter 19 Wheel Alignment Service   **435**

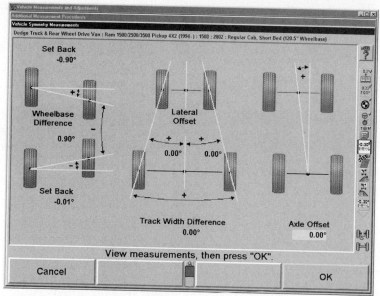

Figure 19.50 A computer alignment screen showing several factors related to the geometric centerline of the vehicle. (Courtesy of Hunter Engineering Company)

Thrustline is the direction that the rear wheels are pointing (**Figure 19.51**). When the rear wheels are held in a fixed position, the thrustline defines the wheels' true straight-ahead position. If the rear wheels are aimed to the right, the thrustline is to the right.

Thrust angle is the angle formed by the thrustline and the geometric centerline (**Figure 19.52**). A thrust angle is created when the rear wheels are not parallel to the vehicle centerline. This is most often a result of an accident or an impact that moves the position of the axle away from perpendicular to the centerline. The thrust angle is positive when off to the right and negative when off to the left. If the left rear wheel had excessive toe-out, the thrust angle would be left of the geometric centerline. Therefore, thrust angle would be negative. Full four-wheel alignment involves adjusting rear-wheel toe to factory specifications with the thrust angle at or near zero.

Independent rear suspensions typically have accommodations for toe adjustment. When rear toe is adjustable, adjusting individual rear toe will compensate for incorrect thrust angle. Aftermarket shims are available for some applications (see **Figure 19.67**).

Individual toe is measured at each wheel in reference to the geometric centerline of the vehicle. In

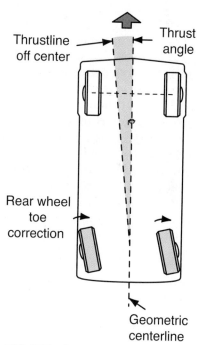

Figure 19.52 Correct rear-wheel alignment adjusts the rear wheel toe to specifications with the thrust angle near zero.

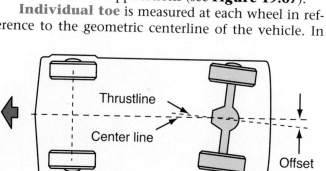

Figure 19.51 Thrustline is the direction that the rear wheels are pointing.

order to center the steering wheel, it is necessary to adjust the front individual toe for each wheel to compensate for the following:

- If rear toe is not adjustable and the thrust angle is not zero degrees
- If rear toe is adjustable, but the thrust angle cannot be brought to zero degrees using individual rear toe adjustment (which could be indicative of frame misalignment)

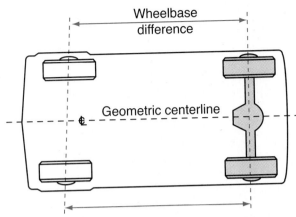

Figure 19.53 The wheels on the right side are closer together than the wheels on the left side.

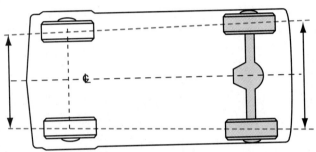

Figure 19.54 The front wheels often have a narrower track width than the rear wheels.

When the wheels are closer together on one side of the vehicle than the other, this is called wheelbase difference (**Figure 19.53**).

Track width difference is when the distance between the two front wheels is different than the distance between the two rear wheels (**Figure 19.54**). Front wheels on some vehicles have a narrower track width than rear wheels. In winter weather, this can pose a problem when the rear tires try to follow tracks made in snow or mud by the front tires.

Setback is the amount that one front wheel is behind the one on the other side of the car (see Figure 18.25). Setback is not adjustable, but it is measured during a four-wheel alignment. Maximum setback is not typically specified but .5 inch (13 mm) is an allowable amount. Setback is measured in degrees so this amount would be 0.50 degree. Incorrect setback is often caused by a collision.

> **SHOP TIP** Sight across the sidewall bulges of the front tires as you look at the rear tires. If the rear wheels have a wider track width, you will most likely see a small amount of tread on each side. If the rear-wheel thrustline is straight, you can determine if the steering wheel will be centered by observing an equal amount of tread on each side.

PERFORMING A FOUR-WHEEL ALIGNMENT

During a computer wheel alignment:

- Alignment sensors are installed on all four wheels. Some of the newest alignment machines use mirrors, rather than gauges on the wheels.
- With all four wheels being checked at the same time, the thrust angle is calculated.
- When toe on all four wheels has been adjusted correctly and thrust angle is the same as the geometric centerline, the steering wheel will be correctly centered.

Compensating the Alignment Heads

Many older wheel alignment machines required some time and a higher degree of skill to adjust the alignment heads that mounted on the wheels. Modern alignment machines have much quicker and easier compensation procedures. Here is a typical compensation procedure: First the wheel head is mounted and centered on the wheel (**Figure 19.55**). The wheel head is leveled with a bubble and locked in position (**Figure 19.56A**). The wheel is raised off the lift and rotated according to instructions until a light glows or an alarm sounds. At this point, a button on the aligner is pressed (**Figure 19.56B**). When the process is complete, the computer has determined the position of the wheel and alignment can proceed. **Figure 19.57** shows a wheel head compensation screen from a computer aligner.

Some of today's machines automatically compensate by simply rolling the vehicle on the alignment rack (**Figure 19.58**).

Measuring Caster and Camber Check

Caster and camber are measured in the same manner as with manual alignment equipment. The amount

Figure 19.55 Mount the wheel alignment head on the wheel. *(Courtesy of Tim Gilles)*

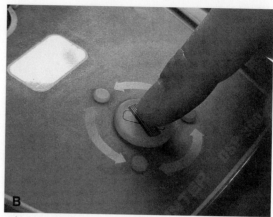

Figure 19.56 (A) The wheel head is leveled with a bubble and locked in position. (B) Pressing this button is part of the compensating process. *(Courtesy of Tim Gilles)*

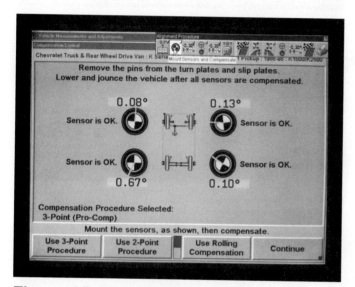

Figure 19.57 A wheel head compensation screen from a computer wheel alignment machine.

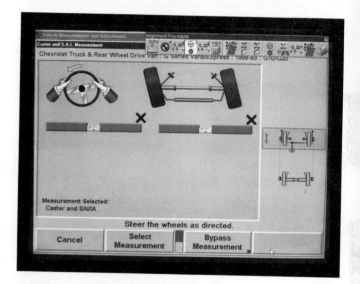

Figure 19.59 Using a computer alignment machine to make a caster reading.

Figure 19.58 A modern alignment machine that automatically compensates the wheels by simply rolling the vehicle on the alignment rack. *(Courtesy of Tim Gilles)*

of wheel sweep during a caster check is determined by the alignment program. The technician watches the screen (**Figure 19.59**) while slowly turning the wheels a few degrees in one direction on the radius plates. Then a turn in the opposite direction is required. When the wheels have been positioned correctly and the computer has the information it needs to make the calculations, alignment readings are displayed on the screen. **Figure 19.60** shows typical alignment readings.

One popular feature on some computer aligners is the presentation of still photo and video instructions for various procedures (**Figure 19.61**).

Adjusting Rear-Wheel Alignment

Camber and toe adjustments are possible on some vehicles.

Adjusting Rear Camber. **Figure 19.62** shows a typical camber eccentric adjuster on a vehicle with a double wishbone rear suspension. Another camber adjustment used on rear strut suspensions calls for the installation of a tapered wedge between the top of the rear knuckle and the strut (**Figure 19.63**).

Adjusting Rear Toe. Rear-wheel toe is adjusted in several ways, depending on the manufacturer. One

438 SECTION 4 SUSPENSION AND STEERING

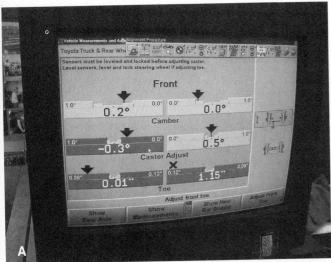

Figure 19.60 (A) Front-wheel alignment readings. (B) Front and rear alignment readings. (A, *courtesy of Tim Gilles*)

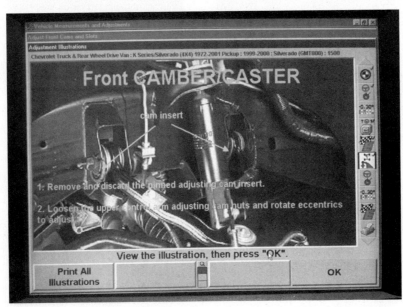

Figure 19.61 Some computer alignment machines have photo and video instructional presentations.

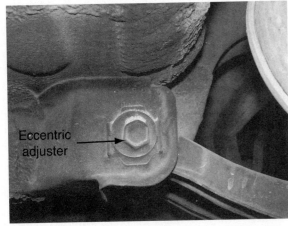

Figure 19.62 A rear camber eccentric adjuster. (*Courtesy of Tim Gilles*)

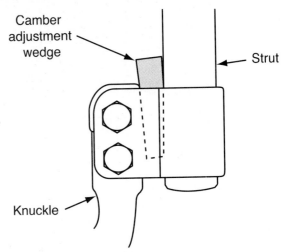

Figure 19.63 A rear strut suspension camber adjustment using a tapered wedge between the top of the rear knuckle and the strut.

common method involves moving the lower control arm. An adjustable linkage is attached to the knuckle. **Figure 19.64** shows a top view of a rear adjustment linkage on a wishbone suspension. On the suspension shown in **Figure 19.65,** one link of the lower wishbone is adjustable. Some manufacturers have a toe link with a slotted hole that allows the wheel to be moved to adjust toe when its holddown screw is loosened (**Figure 19.66**).

Another rear-wheel alignment adjustment method requires the installation of a tapered shim, typical of that used to adjust front-wheel alignment on four-wheel drive vehicles (**Figure 19.67**). Camber and/or toe can be changed depending on the installation position of the shim. Instructions are included with the shim or are given in detail when using the alignment machine (**Figure 19.68**). Some aftermarket

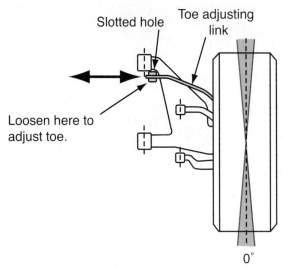

Figure 19.66 A slotted hole in the toe adjusting link allows the wheel to be moved to adjust toe.

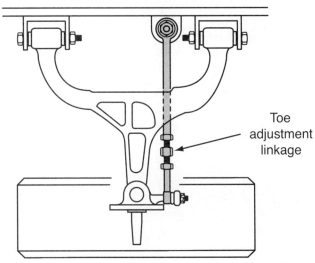

Figure 19.64 Top view of rear toe adjustment linkage.

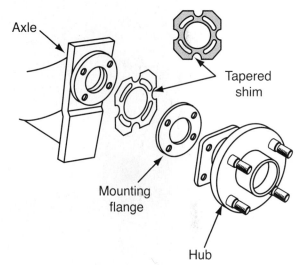

Figure 19.67 A tapered shim installed to correct rear-wheel alignment.

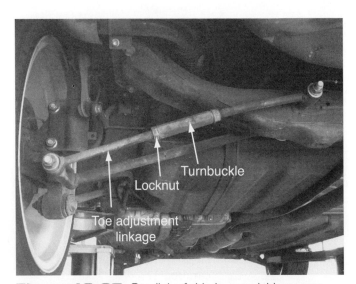

Figure 19.65 One link of this lower wishbone provides a toe adjustment. *(Courtesy of Tim Gilles)*

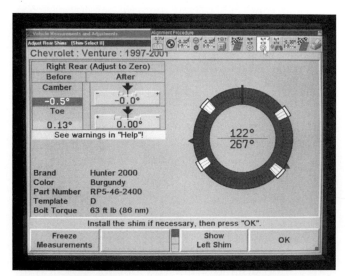

Figure 19.68 This computer aligner's screen gives instructions for selecting a shim.

SECTION 4 SUSPENSION AND STEERING

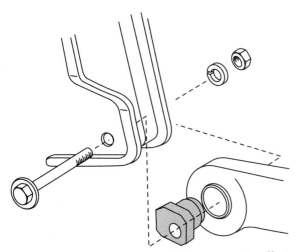

Figure 19.69 Some manufacturers supply offset bushings to adjust rear-wheel toe and camber.

manufacturers supply offset bushings to adjust rear wheel toe and camber (**Figure 19.69**).

More Wheel Alignment Rules

- Be sure there are no heavy loads in the vehicle when reading alignment angles. The rear axles of smaller front-wheel drive cars are so light that they can flex under load, resulting in camber change.
- Ideally, the fuel tank should be full.
- The vehicle, for example, a light truck normally operated with a camper or a car with a trunk full of sales literature, should be aligned in the condition in which it is normally driven.

One extra feature included in modern alignment equipment is a remote display that can be carried to the rear of the vehicle where it can be observed while making adjustments (**Figure 19.70**).

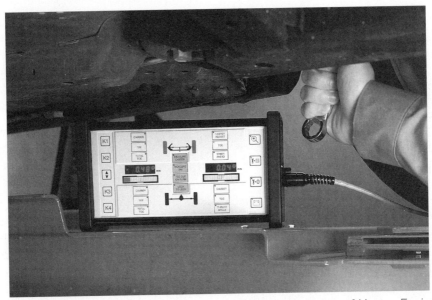

Figure 19.70 A remote display is useful when aligning rear wheels. *(Courtesy of Hunter Engineering Company)*

REVIEW QUESTIONS

1. When a front-wheel drive vehicle has two drive axles of different lengths, what can happen during acceleration?

2. If you move your hand across the tire's footprint area from the outside to the inside and feel a _____ edge, there is excessive toe-in.

3. On a rear-wheel drive vehicle with excessive toe-in, the _____ edge of the right front tire tends to wear.

4. What is the name of the test in which an assistant turns the steering wheel back and forth a short distance while you look for looseness in the steering linkage?

5. Which of the alignment angles is normally adjustable?

6. Which alignment angle is usually measured in inches or millimeters rather than degrees of a circle?

7. Caster results in _____ steering.

8. When the road is higher in the center than on the outside edges, what is this called?

9. Is SAI adjustable?

10. The included angle is the amount of SAI _____ camber.

PRACTICE TEST QUESTIONS

1. A vehicle's steering wheel is off-center when driving on a straight highway. The required repair is:
   A. Remove the steering wheel and put it back on in the correct position
   B. Adjust the tie rod ends during a toe adjustment
   C. Adjust the camber and caster
   D. Adjust the steering gear box

2. Which of the following will cause a change in turning radius?
   A. A bent steering arm
   B. A change in camber
   C. A change in caster
   D. Adjusting the steering axis inclination

3. Technician A says that a change in camber can cause a change in toe. Technician B says that a change in toe will cause a change in camber. Who is right?
   A. Technician A   B. Technician B
   C. Both A and B   D. Neither A nor B

4. Technician A says that toe can change when springs sag. Technician B says that centering the steering wheel is done by adjusting tie rod length. Who is right?
   A. Technician A   B. Technician B
   C. Both A and B   D. Neither A nor B

5. Technician A says that caster is read with the wheels facing straight ahead. Technician B says that caster spread between the front wheel settings can be up to 2 degrees. Who is right?
   A. Technician A   B. Technician B
   C. Both A and B   D. Neither A nor B

6. Which of the following is not true regarding camber?
   A. The car will pull to the side with the most positive camber
   B. The car will pull to the side with the most negative camber
   C. Excessive camber can cause a tire to wear on one side
   D. Sagged springs can cause negative camber

7. Which of the following is true regarding caster?
   A. The car will pull to the side with the most positive caster
   B. The car will pull to the side with the most negative caster
   C. Excessive caster can cause a tire to wear on one side
   D. Caster does not cause a tire to pull

8. Which of the following is/are true about toe change?
   A. It can cause bump steer
   B. It can happen when tie rods are at unequal heights and a wheel goes over a bump
   C. It can be caused by having different caster angles from side to side
   D. All of the above
   E. None of the above

9. A vehicle turns abruptly to the side during initial acceleration. Technician A says this problem can be caused by unequal spring height. Technician B says this could be due to a loose subframe. Who is right?
   A. Technician A   B. Technician B
   C. Both A and B   D. Neither A nor B

10. Two technicians are discussing how to compensate for the crown designed into a typical road. Technician A says camber can be set slightly more positive on the driver's side. Technician B says caster can be set so that it is slightly more negative on the driver's side of the vehicle. Who is right?
    A. Technician A   B. Technician B
    C. Both A and B   D. Neither A nor B

SECTION Five

Axles and Joints

■ **CHAPTER 20**
Front Wheel Drive CV Joint Fundamentals and Service

■ **CHAPTER 21**
Rear-Wheel Drive Shafts, Universal Joints, and Axles

■ **CHAPTER 22**
Drive Line Vibration Service

CHAPTER 20

Front-Wheel Drive CV Joint Fundamentals and Service

■ OBJECTIVES

Upon completion of this chapter, you should be able to:
- ✔ Describe differences between front- and rear-wheel drivetrains.
- ✔ Diagnose CV joint problems.
- ✔ Service CV joints.
- ✔ Replace CV joint boots.

■ KEY TERMS

bridge-type clamp
constant velocity (CV) joint
cross groove plunge joint
double offset plunge joint
earless clamp

fixed joint
half shaft
inboard side
knockoff-type CV joint
outboard side

plunge joint
Rzeppa CV joint
stub shaft
transaxle
tripod tulip plunge joint

■ INTRODUCTION

This chapter covers the theory of front-wheel drive axles, as well as the removal, service, and replacement of front drive axles and CV joints. Until the early 1980s, most vehicles came equipped with rear-wheel drive.

A front-wheel drive vehicle has a **transaxle**, which combines a transmission and differential in one unit. Drive axles extend to the front wheels out of each side of the transaxle. At each end of the drive axle is a constant velocity universal joint, called a CV joint.

Advantages of front-wheel drive include reduced weight and a more efficient drivetrain, resulting in better fuel economy. When combined with MacPherson struts there is less unsprung weight for better handling as well. The transmission hump necessary in rear-wheel drives is also eliminated.

A few front-wheel drive engines have been mounted in the engine compartment longitudinally, from front to rear with the transaxle (**Figure 20.1**).

Vintage Chassis

Front-wheel drive (FWD) first appeared on an American car in 1930, on the Auburn. The Cord, built later in the 1930s, also used front-wheel drive. The next American-made car to be built with front-wheel drive was the 1966 Oldsmobile Toronado. Front-wheel drive showed up on a few more vehicles but remained limited until the mid 1980s. Today, a majority of cars come with front-wheel drive.

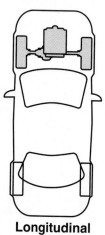

Longitudinal

Figure 20.1 A front-to-rear engine and transaxle.

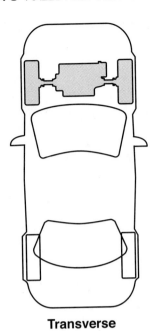

Transverse

Figure 20.2 A side-to-side engine and transaxle.

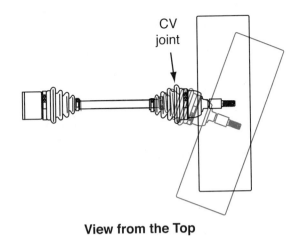

View from the Top

Figure 20.4 CV joints pivot during steering.

Most transaxles are mounted in a sideways (transverse) fashion (**Figure 20.2**). The transaxle is bolted to the engine.

Front-wheel drive axles represent a lucrative service area. The related area of vibration analysis is covered in Chapter 22.

■ FRONT DRIVE AXLES

The inner ends of front- and rear-wheel drive axles have splines that engage side gears in the differential. However, there is a major difference between rear-wheel drive axles and front-wheel drive axles. At each end of a front-wheel drive axle is a **constant velocity (CV) joint** (**Figure 20.3**). CV joints are necessary because the axles are driven at sharper angles than can be tolerated by ordinary universal joints (see Chapter 21). In addition, they allow steering of the front wheels to take place during power transmission (**Figure 20.4**). They also allow the axles to change length due to suspension deflection.

An ordinary universal joint changes its output speed twice in every revolution when run at an angle. A true CV joint provides a constant output speed no matter what the shaft angle.

In a rear-wheel drive vehicle, the drive shaft turns very fast because it is positioned before the gear reduction of the differential. Front-wheel drive axles turn slower than a rear-wheel drive shaft by the difference of the axle ratio (about one-third slower).

■ AXLE SHAFT PARTS

Each drive axle is called a **half shaft** or an *axle shaft*. There is a short shaft at the outside end called a **stub shaft** or *stub axle*. It is splined to the front hub so it can drive the front wheels. Some axle shafts also have a stub shaft on the inside. Others have a female-type splined yoke that slides over male splines similar to a rear-wheel drive transmission output shaft.

CV Joints

At each end of a front-wheel drive axle is a CV joint (see **Figure 20.3**). There are three ways of classifying CV joints:

1. Inboard and outboard
2. Fixed and plunge
3. Ball and tripod

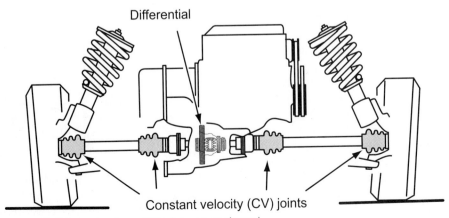

Figure 20.3 Front-wheel drive axles have CV joints at each end.

The inside end of the axle shaft is commonly called the **inboard side** and the outside end nearest the wheel is called the **outboard side**.

Plunge CV Joints. The inboard CV joint is called a **plunge joint**. When the suspension system deflects, the axle's length must change (**Figure 20.5**). The joint that takes care of this change in length is the plunge joint. It acts like a slip yoke does in a rear-wheel drive vehicle. As its name implies, it plunges in and out to compensate for changes in axle length. Because of its design, it can also transmit motion through an angle.

Fixed CV Joints. The outboard joint is called a **fixed joint**. This means that it does not allow for a change in length like the inboard joint. It allows for a change in the angle of the axle as the suspension moves over bumps in the road. A fixed joint also allows for severe pivoting during steering. It can accommodate angles up to 50 degrees, although the maximum amount of travel is usually limited to 46 degrees. The large angle that the outer joint is subjected to is why they wear out more often than inner ones. Plunge joints cannot change angles as effectively as fixed joints because they are limited to about 20 degrees (**Figure 20.6**).

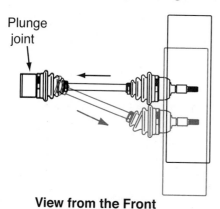

Figure 20.5 A plunge joint allows the axle to change length.

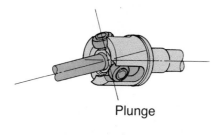

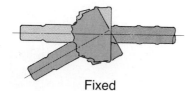

Figure 20.6 A plunge joint does not allow as much angle change as a fixed joint. *(Courtesy of Federal-Mogul Corporation)*

■ CV JOINT CONSTRUCTION

There are different types of plunge and fixed joints. Different combinations of joints are found on axles. The most common combination uses a **Rzeppa CV joint** for the outside fixed joint and a tripod for the inside plunge joint. The tripod has become more popular because it costs less to manufacture.

Fixed Joints

Almost all newer vehicles use a Rzeppa joint for the fixed outer joint. Named after Alfred Rzeppa, the Ford engineer who invented it in 1929, the Rzeppa joint is a ball-and-socket-type joint (**Figure 20.7**) with the inner race splined to the axle shaft. The outer race is either part of the stub shaft or it is splined to the stub shaft. Splined parts are held to their shafts with snaprings. Six ball bearings fit in between the inner and outer races, housed in a cage. The joint flexes in tracks in the inner race and outer housing. The balls act as both bearings and a medium for transferring power to the wheels.

A Rzeppa joint is like a bevel gear with balls instead of teeth. It provides even power transmission at all angles. As the inner and outer races move back and forth compensating for the angle change, the balls remain in a constant perpendicular plane (**Figure 20.8**). The cage holds the balls in this position while the inner and outer races move.

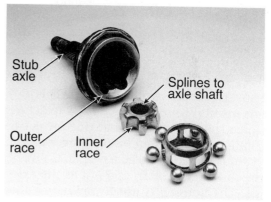

Figure 20.7 Parts of a Rzeppa CV joint. *(Courtesy of Federal-Mogul Corporation)*

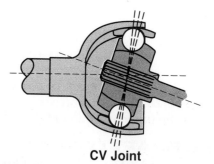

Figure 20.8 As the inner and outer races move back and forth to compensate for the angle change, the balls remain in a constant perpendicular plane.

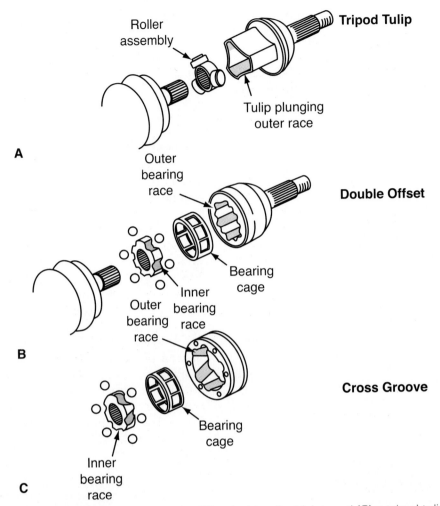

Figure 20.9 Plunge joints: (A) a cross groove joint; (B) a double offset joint; and (C) a tripod tulip joint.

Types of Plunge Joints

There are three types of plunge joints. A **tripod tulip plunge joint** (**Figure 20.9A**) has a spider assembly with three spherical rollers that ride on needle bearings, similar to a cross and yoke universal joint. The yoke or housing, often called a tulip, has three grooves that the rollers can roll in and out as the axle length changes. The housing is usually spline mounted, but can also be bolted to a flange.

A **double offset plunge joint** (**Figure 20.9B**) uses six balls evenly spaced by a cage. Most double offset joints are spline mounted to the transaxle. The outer housing is considerably deeper than a cross groove so it can plunge further. The races have straight grooves.

A **cross groove plunge joint** (**Figure 20.9C**), also called a pancake joint, has grooves in the bearing races that would cross each other if they extended far enough (**Figure 20.10**). There are six balls and a cage in addition to the inner and outer races.

■ AXLE SHAFT DESIGN

Axle shafts can be either solid or hollow. Some of them have damper weights to absorb vibration (**Figure 20.11**). Because the axle shafts turn so much

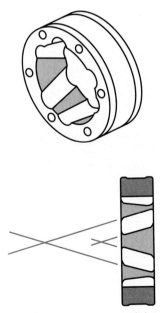

Figure 20.10 In a cross groove joint, grooves in the bearing races would cross each other if they extended far enough.

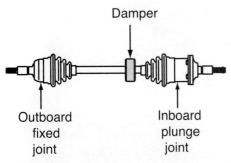

Figure 20.11 A damper on some axles absorbs vibrations.

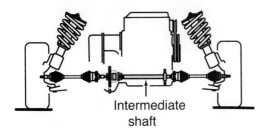

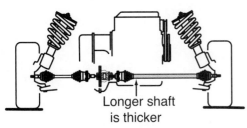

Figure 20.12 When axle shafts are the same length, there is an intermediate shaft.

slower than a rear-wheel drive shaft, balance is not as important.

Axle shafts can be the same length or they can be different lengths (**Figure 20.12**). Earlier vehicles had unequal length axle shafts because the transaxle was mounted off to one side of the engine. When there are different shaft lengths, the long one wraps up and there is a slight lag before it puts its torque to the wheel. The more horsepower the engine has, the greater the effect of this torque steer. One way of preventing torque steer is to make the longer axle shaft of larger diameter tubing.

The trend is to have half shafts that are of equal length on both sides of the vehicle. An intermediate shaft is used on the equal length system so that the axle shafts can be the same length. The Ford Taurus actually has the intermediate shaft as part of the transaxle.

CV Joint Boots

Boots are used to contain grease and protect the joint from the elements. The CV joint boot is attached to the axle and stub shafts with plastic or steel bands or straps. Boots are made of natural rubber, neoprene, silicone, or urethane.

■ FRONT-WHEEL DRIVE SERVICE AND REPAIR

Front-wheel drive axles often require service as a vehicle ages. Some checks and repairs of axle assemblies can be made with the axle on the vehicle. Others require removal and disassembly of the components.

■ CV JOINT BOOT SERVICE

Over a period of time, boots age and suffer the effects of the elements and repeated flexing. When a boot fails, grease is thrown from it. If the boot is not replaced before contaminants can enter it, the entire joint will fail. Failure of the boot is the most common problem with CV joints. They are designed to last about 75,000 miles. Boot replacement is a common service. The outboard boot is the one that fails most of the time.

Check the boots for damage. Grease will begin to spray from a tear in the boot. Check to see if a part has been rubbing on the boot. If a boot is torn and you can see evidence of contamination, the joint will probably have to be replaced.

NOTE: *CV joints are lubricated for life. If a boot tears or the clamp falls off, the joint will fail if not serviced within a short time. The estimated life of a joint is 8 hours without lubricant.*

The action of an operating CV joint is like that of a vacuum cleaner, causing suction of water into the joint. Storm runoff water carries a good deal of mud and grit that will destroy the bearing areas inside of a CV joint. When a boot has failed, the correct repair is to disassemble and inspect the CV joint before replacing the boot and grease. Be sure to check the seal at the transaxle and plunge joint for leakage as well.

NOTE: *A noisy joint will not get quieter with the installation of a new boot and grease.*

Boot Kits

Replacement boot kits come with new clamps and the correct amount of high temperature grease required to refill the boot and CV joint (**Figure 20.13**).

The half shaft is removed to replace the boot. A new nut is required for the stub shaft, but it does not come in the boot kit because too many different nuts are used. Just removing the stub shaft from the hub is all that is required to replace the outer boot.

There are a few axle shafts where the fixed joint cannot be disassembled. These can have the boot installed from the plunge joint end.

Axle Inspection and Diagnosis

Check the half shaft for obvious looseness. Noises coming from the front axle can be listened for during a test drive. To test for noise, weight must be on the suspension. Otherwise the joint is not in the same position where all of the wear occurs. Simply feeling the

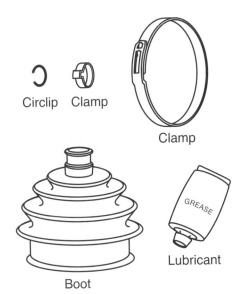

Figure 20.13 A CV joint boot kit.

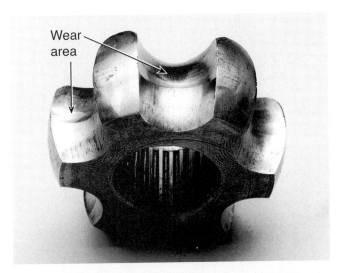

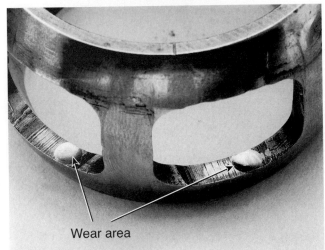

Figure 20.14 A worn cage or race can cause a clicking sound during a turn. *(Courtesy of Federal-Mogul Corporation)*

joint does not work for diagnosing noise unless the joint is extremely damaged.

Questions to ask the customer include:

- What does the noise sound like?
- Does it occur during acceleration, deceleration, or when turning?

■ CV JOINT DIAGNOSIS

This section explains diagnostic procedures for fixed and plunge joints.

Fixed Joint Problems

Test the vehicle during turns as you would when diagnosing axle bearings. A clicking sound during a turn could be due to a bad outboard joint. The clicking will increase and decrease with wheel speed. If the cage becomes worn, the result is a clicking sound when turning as the races move back and forth in their planes (**Figure 20.14**). Opposing balls in a Rzeppa joint always work together in a pair. When one ball wears its track, the opposite side will have identical wear.

NOTE: *Always check the splines on a shaft. They could be the cause of a clicking sound as well.*

Plunge Joint Problems

A bad plunge joint will make a clunking sound when starting from a stop, or during deceleration and when braking. If an accurate diagnosis cannot be made with the axle on the vehicle, remove it and disassemble it to find the cause of the problem. Excess wear is hard to detect visually and slight, almost imperceptible wear can result in a problem.

■ AXLE SHAFT REMOVAL

There are some differences in the methods by which axle shafts are removed and serviced. Check the service library for the procedure before attempting a job for the first time.

Loosen the Axle Nut

To remove the axle, first loosen the stub shaft nut in the center of the wheel while the wheel is still on the ground.

NOTE: *Most service information advises against using an impact wrench on the nut with the vehicle raised on a lift because of possible damage to parts, including the transaxle.*

SHOP TIP When the vehicle has custom wheels or if the wheel is already off the vehicle, you can block a ventilated rotor from turning. Wedge a large screwdriver in one of the vent holes. If you have a helper, you can have that person apply the foot brake tightly while you loosen the nut.

Figure 20.15 Some axle nuts are staked to lock them in place. *(Courtesy of Federal-Mogul Corporation)*

Some axle nuts are staked to lock them in place (**Figure 20.15**). A new nut will be required when these are disassembled.

Remove the Stub Shaft from the Hub

On some vehicles it is easier to move the hub away from the stub shaft if the tie-rod is removed first (see Figures 17.8 and 17.9).

Front-wheel drive vehicles typically have MacPherson strut suspensions. A stub shaft is located at the end of the fixed-type outer CV joint. The inner plunge-type joint cannot move far enough inward to allow removal of the stub shaft from the hub (see Figure 20.6 for the difference between fixed and plunge joints). First, the ball joint or the strut must be disconnected so the hub can be moved away from the stub shaft.

Some vehicles have a conventional ball joint with a tapered connection (**Figure 20.16**). Others use a pinch bolt–type ball joint connection. A pinch bolt fits through a groove in the ball joint (**Figure 20.17**). When it is tightened, it "pinches" the slotted connection to secure it to the ball joint.

After the ball joint is disconnected, the control arm can be pried down to allow removal of the stub shaft from the hub (**Figure 20.18**). On some suspensions, removal of a link bolt might also be required.

Removing the stub shaft from the hub can sometimes be done by tapping it with a brass hammer. Other times a puller is required (**Figure 20.19**). When

Figure 20.17 This ball joint design uses a pinch bolt.

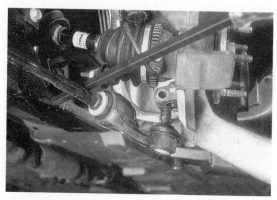

Figure 20.18 Prying against the strut to force the control arm down allows the ball joint to be separated from its connection.

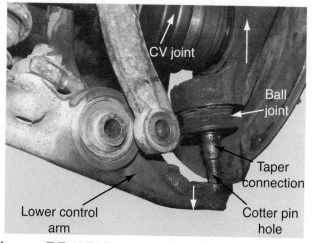

Figure 20.16 Pry the control arm down to release the ball joint. *(Courtesy of Tim Gilles)*

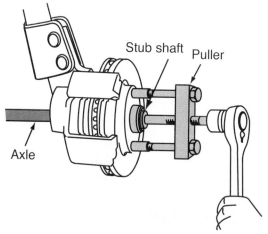

Figure 20.19 A puller is needed to remove the stub shaft from the hub on some vehicles.

450 SECTION 5 AXLES AND JOINTS

> **SHOP TIP**
> - Disconnecting the sway bar at both ends will make it easier to pry down on the control arm to separate it from the ball joint.
> - If you remove a strut, mark it either at the top or the bottom so that it can be reinstalled in the same position. Otherwise the wheel alignment will be off after the job is completed.

using the puller, the ball joint needs to be removed after the shaft is loosened part way or the plunge joint will bottom out.

When removing the stub shaft, be careful not to tear the front-wheel bearing seal on the back of the hub. After the outer joint is removed from the hub, be careful not to tear it on the suspension parts (**Figure 20.20**). Watch that the plunge joint does not come out of its housing. The boot can stretch enough to allow this.

NOTE: *Be careful not to nick the splines on the outboard axle stub shaft. Doing so will make it difficult to reinstall the splines in the hub.*

Remove the Axle

Be sure to note which kind of axle retention method is used at the transaxle. On a few vehicles, the inner end of the axle is simply unscrewed from a flange (**Figure 20.21**). A slide hammer is used to remove some types of splined inner joints (**Figure 20.22**). Other types might simply be pried away with a screwdriver (**Figure 20.23**). If you are not certain how the inner end of the axle is restrained, check the service manual. You might damage an aluminum transaxle if you use excessive force.

If the plunge joint has a spline (**Figure 20.24**), oil may leak from the differential when the axle shafts are pulled from the transaxle housing. If you did not drain the transaxle first, place a drain pan under the axle

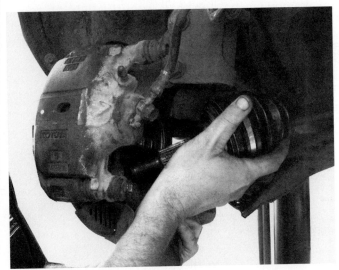

Figure 20.20 Move the stub shaft out of the way. Be careful not to damage the boot. *(Courtesy of Tim Gilles)*

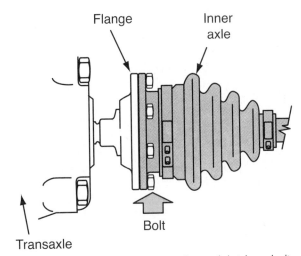

Figure 20.21 Sometimes a plunge joint is unbolted from a flange.

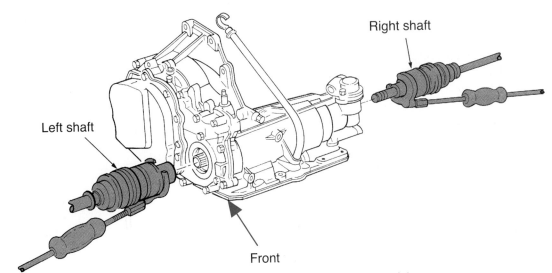

Figure 20.22 A slide hammer is used to remove some types of splined inner joints.

Figure 20.23 Prying a plunge joint out of a transaxle. *(Courtesy of Tim Gilles)*

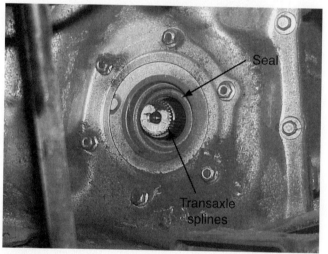

Figure 20.24 Appearance of transaxle splines following FWD axle removal. Oil will leak when the plunge joint is removed. *(Courtesy of Tim Gilles)*

Figure 20.25 Cut off the boot clamps.

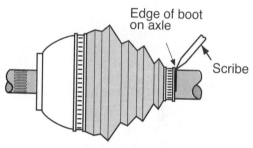

Figure 20.26 Mark the location of the edge of the boot on the axle shaft prior to disassembly.

tape or you can scribe a line (**Figure 20.26**). When the new boot is installed, you will know where to position it.

Clips and Snaprings. Wipe off the joint and inspect it to see what kind of retaining method is used on it. Several types of clips are used to retain CV joints on the axle shafts or to the transaxle (**Figure 20.27**). Clips are not reused, so they must be replaced.

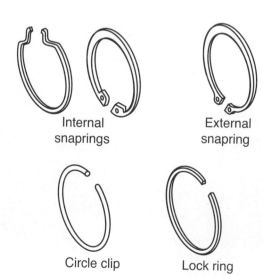

Figure 20.27 Several types of retaining rings.

before removing it. Replacing the axle seal in the transaxle should be done if it is hard or was leaking.

Boot Removal

Position the axle in a vise. Cut off the boot clamps and remove the boot (**Figure 20.25**).

Before removing the clamps, always mark the location of the end of the old boot on the axle. You can use

SHOP TIP Use a clean pan to collect all of the fluid that spills from the transmission. The fluid can be reinstalled. This way you do not have to locate the correct type of transmission fluid or measure the quantity you install. Be absolutely certain not to allow the fluid to drain over parts contaminated with road grime.

Circlips are commonly used. They come in various thicknesses and sizes. Measure the wire diameter before selecting a replacement. *Internal snaprings* and *external snaprings* are also common. They require snaping pliers for removal. *Lock rings* are sometimes installed in a groove to locate or limit movement of a part on a shaft.

CV JOINT REPLACEMENT

An internal circlip is the more common way of retaining a fixed joint to the shaft. The clip rides in a groove in the shaft. Its outer diameter goes against the inner race (**Figure 20.28**). At the other side of the inner race there is often a lock ring or a thrust washer. It will be the only thing visible when this type of joint is wiped off during an inspection. This joint design is called the **knockoff-type CV joint**. It is removed using a brass punch and hammer. There is a chamfer on the inner race that helps compress the clip during removal.

Some Japanese vehicles use another version of the knockoff-type joint. A circlip rides in a groove in the axle and fits into a corresponding groove inside the inner race (**Figure 20.29**). To remove this joint, pressure must be applied to the inner race. There is a special tool for doing this (**Figure 20.30**). If you use this tool to knock off another type of fixed joint that has a lock ring, be sure you are not driving against the lock ring. When the CV joint has been removed from the axle, the boot is removed (**Figure 20.31**).

Fixed Rzeppa joints are held to their shafts by one of two methods. For one type, when the joint is wiped off during an inspection an external snapring will be visible. There is a recess in the inner race where the tabs on the snapring fit (**Figure 20.32**). After removing the snapring with snaping pliers, the CV joint can be removed from the axle shaft (**Figure 20.33**). The other type of retaining method uses an internal clip that fits into a groove in the shaft. It expands into the joint's inner race. Rzeppa joints retained in this

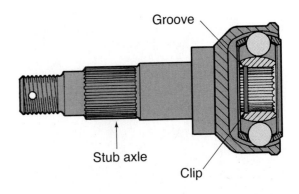

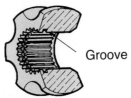

Figure 20.29 Some CV joints have a circlip that fits into a groove inside the inner race. *(Courtesy of Federal-Mogul Corporation)*

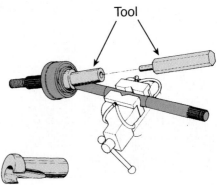

Figure 20.30 A special tool for removing knockoff-type CV joints. *(Courtesy of Federal-Mogul Corporation)*

Figure 20.31 After the CV joint is removed, the boot is removed.

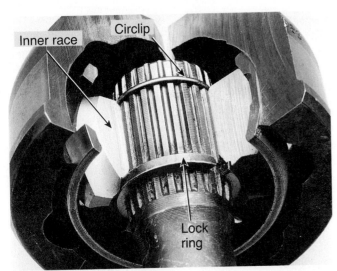

Figure 20.28 This inner race has a lock ring on one side of it and a snapring on the other side. *(Courtesy of Federal-Mogul Corporation)*

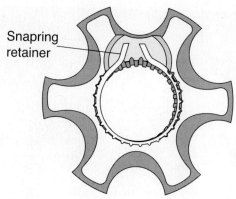

Figure 20.32 The tabs of the snapring fit into a recess in the inner race. *(Courtesy of DaimlerChrysler Corporation)*

Figure 20.33 Removing the CV joint from the axle.

manner are sometimes called *knock-off joints* because they are driven off using a soft hammer.

■ FIXED JOINT DISASSEMBLY AND INSPECTION

After the boot is removed, use a drift punch or a special tool to move the inner race to the side for a more complete inspection (**Figure 20.34**). Wipe off the parts and look for evidence of wear (**Figure 20.35**). Sometimes the metal is colored blue, which is acceptable as long as the bearing surfaces are all smooth and unworn.

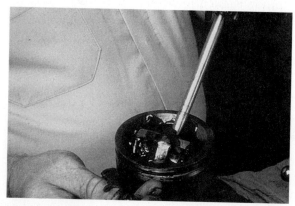

Figure 20.34 Move the CV joint to the side for cleaning and inspection. *(Courtesy of Federal-Mogul Corporation)*

Figure 20.35 These CV joint balls are pitted. The joint must be replaced. *(Courtesy of Federal-Mogul Corporation)*

To disassemble a Rzeppa joint, remove the balls (**Figure 20.36**). First mark all the major components with a felt marker so they can be reassembled in their original positions. To remove the balls, the inner race must be turned sideways. This can be done using a brass punch or a special tool. After removing the balls, the inner race can be turned so that it can be removed from the stub shaft (**Figure 20.37**). All of the parts are cleaned and inspected once again for wear.

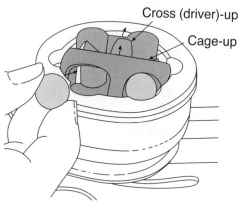

Figure 20.36 With the cage turned to the side, the balls can be removed. *(Courtesy of DaimlerChrysler Corporation)*

Figure 20.37 After removing the balls, the inner race can be turned so that it can be removed from the stub shaft.

Grease in a CV joint has a very difficult job to do. Parts of the CV joint should *not* be cleaned in solvent. The solvent film left behind can ruin the grease. Some technicians like to remove the balls one at a time. Then they clean and inspect them and put them back in. This makes reassembly easier. After wiping the joint clean, lightly grease all of the parts. Reassemble the joint, fill it with grease, and install it on the axle using a new clip.

Install the Boot

The boot kit includes one or more tubes with the correct amount of CV joint grease. Be sure to follow any instructions that come with the kit (**Figure 20.38**). Part of the grease is applied to the CV joint itself and the rest goes into the boot (**Figure 20.39A**).

The lock ring can be temporarily removed from the shaft to make installation of the boot easier. After the boot is slipped over the end of the axle, install a new circlip. Be careful not to overstretch the circlip. After the joint is pounded onto the shaft, pull on it to see that its snapring or circlip is seated in its groove.

Install the large end of the boot and its clamp. A large screwdriver with a dull tip is handy for installing

Figure 20.40 A latex glove is handy for keeping grease contained and clean after assembling a CV joint. *(Courtesy of Tim Gilles)*

the large end of the boot over the CV joint (**Figure 20.39B**).

SHOP TIP A good trick for keeping the grease inside of an inner CV boot and free from contamination is to temporarily put a latex shop glove over the end of the axle (**Figure 20.40**).

■ CV JOINT BOOT CLAMPS

There are several types of clamps (**Figure 20.41**):

- A *universal clamp* is a continuous strap made of stainless steel. It comes in two sizes. A special tool pulls it tight around the boot. A universal clamp will work on any boot except a urethane boot. Spray the clamp with penetrating oil to reduce friction during installation.
- The **bridge-type clamp** is the most common. It requires a special crimping tool (**Figure 20.42**).
- Urethane boots are stiffer. When the boot is urethane a tighter crimp is necessary so a heavy-duty crimping tool is needed. The tool is used with a torque wrench to ensure the correct clamping

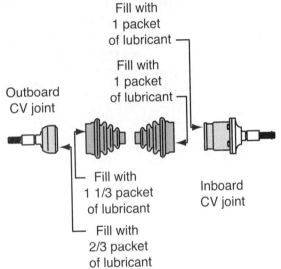

Figure 20.38 Follow instructions for lubricating the joint. Inboard and outboard instructions can be different.

Figure 20.39 (A) After installing the correct amount of grease in the joint, fill the boot with the remaining grease. (B) A large screwdriver with a dull tip is handy for installing the large end of the boot.

Chapter 20 Front-Wheel Drive CV Joint Fundamentals and Service 455

Bridge clamp

Earless clamp

Universal clamp

Press-on ring clamp

Figure 20.41 Types of boot clamps.

Figure 20.43 Crimping a bridge clamp on a urethane boot using a torque wrench. *(Courtesy of Federal-Mogul Corporation)*

Figure 20.42 A special crimping tool is used to tighten a bridge clamp.

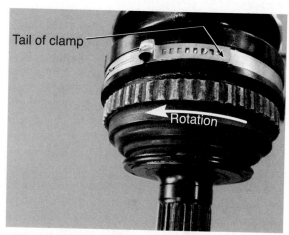

Figure 20.44 Install the clamp with its tail facing away from the axle's direction of rotation. *(Courtesy of Tim Gilles)*

load on the boot (**Figure 20.43**). Bridge-type universal clamps are available that can be cut to length.

- An **earless clamp**, also called a *low profile* or *locking clamp*, is used when there is a problem with the axle clearing other parts. Another type of clamp might accidentally get knocked off.
- Another type of clamp is a press-fit ring. It must be installed in perfect alignment or it will become overexpanded. These clamps are not reusable.

NOTE: *Do not try to use a regular hose clamp to replace a boot clamp. It will probably ruin the boot. In a pinch, a nylon tie strap can be used for a day or two until a clamp can be purchased.*

Installing the Small-End Clamp

After pulling the big end of the boot over the CV joint and installing its retaining clamp, vent the small end of the boot on the axle (**Figure 20.45**). Slide the small end of the boot until it is even with the mark you made on the shaft before disassembly and install the clamp in that position.

SHOP TIP Universal strap- or bridge-type retaining clamps (**Figure 20.44**) can be installed in either of two directions. Position the clamp so that its tail (the portion that sticks out) will not be able to wrap itself up on road debris. If it faces so the tail follows the band on the clamp as it rotates toward the front, debris will not catch on it as easily.

Figure 20.45 After installing the large clamp on the boot, let the air out of the small end before installing the clamp. *(Courtesy of Federal-Mogul Corporation)*

■ SERVICING AN INNER TRIPOD JOINT

Before removing the inner boot, tape the shaft and joint to show where the clamps should be reinstalled (**Figure 20.46**). This is especially important on plunge joints because they must be able to move in and out. Mark the location of the housing to the shaft so that parts can be reassembled in their original locations. Cut away the clamps and boot. Some tripods will simply come out of the tulip. Others will have retaining tabs, which must be bent out of the way, or a wire ring retainer.

Check to see if the tripod rollers can come off the spider. Put tape around a tripod so the balls and rollers do not come out (**Figure 20.47**). It can be difficult to get all of the rollers back in. Mark the tape to locate the tripod to the shaft on reassembly. It is not necessary to disassemble these. Simply roll them to feel that they are free and smooth.

Inspect the condition of the tulip. Feel its grooves for depressions in the metal. A polished appearance in

Figure 20.46 Put tape next to the ends of the boot and make marks to make reassembly of a plunge joint easier. *(Courtesy of Federal-Mogul Corporation)*

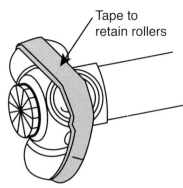

Figure 20.47 Put tape around the rollers so they cannot come off and spill out their needle bearings.

SHOP TIP On some axles, the boot can be installed from the other side while the fixed joint is off.

the center of travel is normal. The tripod will have to be removed in order to install the boot.

When reinstalling the joint onto the axle shaft, be sure to align any marks made during disassembly. Use a soft-faced hammer to tap the joint back onto the shaft. If it has a circlip, you will hear it click when it reaches the correct position.

After the boot and tripod are reinstalled, grease the rollers and fill the boot with the rest of the grease. Put the tulip back over the tripod and bend any retaining tabs as needed. Position the joint at mid travel and crimp the clamps.

Remove the retaining ring that holds the tripod to the shaft and remove the tripod. Some have different types of circlips and others have to be driven off.

■ DOUBLE OFFSET PLUNGE JOINTS

Double offset plunge joints vary in how they are connected to the axle shaft. Most of them are held by an internal circlip. To remove the joint requires that it first be disassembled. This kind of joint is identified by a large round retaining ring in a groove in the joint's outer housing (**Figure 20.48**). The other kind of double offset joint has a circlip that can be removed by pounding on the outer housing. A metal retaining ring holds the parts in the housing.

■ CROSS GROOVE JOINT SERVICE

A cross groove joint is serviced in a similar manner to Rzeppa joints. The joint is disassembled by turning the inner race and cage perpendicular to the outer cage and removing them.

Figure 20.48 This double offset plunge joint is disassembled by removing a retaining ring. *(Courtesy of Federal-Mogul Corporation)*

■ REBUILT HALF SHAFTS

Installing a complete rebuilt half shaft has become a popular repair. Although both the outer and inner joints are replaced, a rebuilt half shaft is often less expensive than just one joint kit would be. The determining factor in whether a technician will rebuild a shaft or buy a rebuilt half shaft is often whether there is other work to be done. With the reduced cost of parts and labor, installing rebuilt half shafts is often in the best interest of the customer.

■ INSTALLING THE AXLE

To install a spline-type joint into the transaxle, push in on the axle while turning it to align the splines. Push the splines all of the way in. If there is a circlip, it will click when it is in its groove. Pull on the housing of the joint to see if the snapring is seated. Do not pull on the axle shaft. The plunge joint could come apart.

After the inner end is installed, position the stub shaft into the hub as far as it can go. Install the washer and a new axle nut and tighten the nut until it is snug. Reinstall the front end components. Do not forget to torque the ball joint stud bolt if it was reassembled. A new pinch bolt is recommended. Leaving pinch bolts loose can cause serious accidents. Some pinch bolts have a knurled area under the head to keep them from turning. Be sure to only turn the nut.

When the vehicle is back on the ground, torque the new stub shaft nut. The torque on the spindle nut can be substantial. A planetary torque multiplier works well for this (**Figure 20.49**). Use a dull chisel to stake the nut into the slot in the spindle (see **Figure 20.15**). Always check the clearance of the boot clamps before test-driving the vehicle.

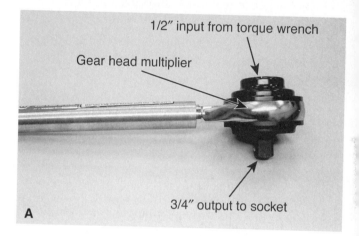

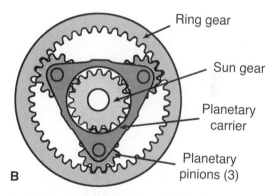

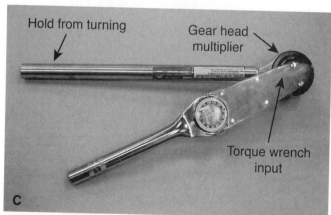

Figure 20.49 (A) A gear head multiplier is helpful when a high tightening torque is specified. (B) Parts of a planetary gearset. (C) The multiplier is kept from turning as torque is applied. *(A and C, courtesy of Tim Gilles)*

CASE HISTORY

A staff member brought her car into a school shop to have the students replace her CV joint boots. After the job was completed, she drove the car home with no problem. But when she backed down her driveway the next morning the wheel fell off to one side and turned 90 degrees to the other. She left a big skid mark on her driveway and ruined her tire before she realized there was a problem. The car was towed to the shop where it was discovered that the pinch bolt had come loose and the ball joint had fallen out of its hole. Had the wheel fallen off while she was driving on the highway, a deadly accident could have occurred.

REVIEW QUESTIONS

1. A FWD axle is also called a _____ shaft.
2. What is the name of the short splined shaft at the outside end of a FWD axle?
3. The two kinds of CV joints are fixed and _____.
4. A _____ joint is like a bevel gear with balls instead of teeth.
5. What kind of sound does a Rzeppa joint make if its cage becomes worn, allowing the races to move back and forth in their planes?
6. What is it called when a vehicle with unequal length half shafts steers one way during acceleration?
7. If one half shaft is thicker than the other, will it probably be the longer one or the shorter one?
8. What is the estimated life of a CV joint that has lost its lubricant?
9. What is the most common repair to CV joints?
10. Of what material is a universal clamp made?
11. Before pulling a splined axle from a transaxle, what do you position under it?
12. Before removing a boot from a CV joint, what do you do to the axle?
13. Which end of a CV joint boot is installed first, the big end or the little end?
14. What do you do to a plunge joint's tripod rollers to keep them from coming apart?
15. What do you do to the stub shaft nut after it is torqued to specification?

PRACTICE TEST QUESTIONS

1. Technician A says that ordinary universal joints cannot handle as steep an angle as a CV joint. Technician B says that a rear-wheel drive shaft turns at a faster rpm than a FWD axle. Who is right?
   - **A.** Technician A
   - **B.** Technician B
   - **C.** Both A and B
   - **D.** Neither A nor Bk

2. Technician A says that a plunge joint cannot change angles as effectively as a fixed joint. Technician B says that the outboard joint is a fixed joint. Who is right?
   - **A.** Technician A
   - **B.** Technician B
   - **C.** Both A and B
   - **D.** Neither A nor B

3. Technician A says that outboard CV joints wear out more often than inboard ones. Technician B says that almost all front-wheel drive vehicles use a Rzeppa inboard joint. Who is right?
   - **A.** Technician A
   - **B.** Technician B
   - **C.** Both A and B
   - **D.** Neither A nor B

4. Which CV joint boot fails most often?
   - **A.** The outboard boot
   - **B.** The inboard boot
   - **C.** They both fail at about the same rate

5. A new stub shaft nut is sometimes required when replacing CV joint boots.
   - **A.** True
   - **B.** False

6. Technician A says that to test an axle shaft for noise, the weight must be off it. Technician B says that a clicking sound during a turn could be due to a bad plunge joint. Who is right?
   - **A.** Technician A
   - **B.** Technician B
   - **C.** Both A and B
   - **D.** Neither A nor B

7. Technician A says that a bad plunge joint will make a clunking sound when starting from a stop. Technician B says that a bad plunge joint will make a clunking sound when braking. Who is right?
   - **A.** Technician A
   - **B.** Technician B
   - **C.** Both A and B
   - **D.** Neither A nor B

8. An impact wrench is the best tool for installing the stub shaft nut with the vehicle in the air.
   - **A.** True
   - **B.** False

9. Technician A says that removing a stub shaft from a hub always requires a puller. Technician B says that sometimes a stub shaft can be removed from the hub with a brass hammer. Who is right?
   - **A.** Technician A
   - **B.** Technician B
   - **C.** Both A and B
   - **D.** Neither A nor B

10. Technician A says to clean all CV joint metal parts in solvent. Technician B says to bottom out the tripod in its bore before clamping the boot to the axle shaft. Who is right?
    - **A.** Technician A
    - **B.** Technician B
    - **C.** Both A and B
    - **D.** Neither A nor B

CHAPTER 21

Rear-Wheel Drive Shafts, Universal Joints, and Axles

■ OBJECTIVES

Upon completion of this chapter, you should be able to:
- Describe the operation of universal joints.
- Describe the different types of rear axles and bearings.
- Diagnose and repair universal joints.
- Remove and replace axle bearings and seals.

■ KEY TERMS

American Petroleum Institute (API)
bearing-retained axle
Cardan
center support bearing
C-lock/C-clip axle
cross and yoke
drive line

drive shaft
EP additives
flange
galling
Hotchkiss design
in phase
propeller shaft
slip yoke

Society of Automotive Engineers (SAE)
spider
swing axles
torque tube
trunnions
U-joints

■ INTRODUCTION

Until the early 1980s, most vehicles were equipped with rear-wheel drive (RWD). Today, pickup trucks, sport utility vehicles (SUVs), and some higher-end vehicles still use rear-wheel drive. On RWD vehicles, **drive line** is a term that describes the parts that transfer power from the transmission to the rear wheels. The drive line includes the drive shaft and universal joints (U-joints), axles and axle bearings, and differential. RWD vehicles have a long drive shaft between the transmission and differential (**Figure 21.1**). This chapter covers parts of the drive shaft and also live axles. Live axles are ones that turn with the wheels.

Differentials are sometimes also referred to as *axles*. They are not covered in this text. Front-wheel drive vehicles have a transaxle with two half shafts that deliver power to the front wheels. Information related to FWD theory and service is found in Chapter 20.

As you study the drive line, you will learn more than just drive line service and repair procedures. Pay

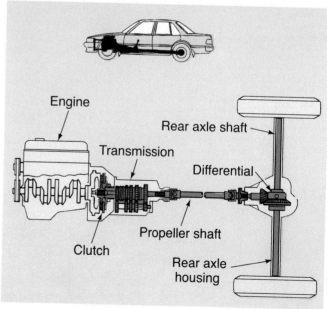

Figure 21.1 Rear-wheel drive.

460 SECTION 5 AXLES AND JOINTS

attention to the methods used in pressing bearings and other parts. Press skills are used in many other areas of automotive repair.

■ DRIVE SHAFT (RWD)

The **drive shaft**, or **propeller shaft**, not only transfers power, it allows for changes in drive line length and angle when springs deflect as a vehicle goes over bumps. It is usually made of steel tubing, although some late-model drive shafts are aluminum. Drive shafts are strong and light. They must be balanced and straight. Universal joints at both ends attach the drive shaft to other components. Yokes to accept the U-joints are welded to the shaft at both ends. A typical drive shaft (**Figure 21.2**) includes two U-joints and a slip yoke. Sometimes a rear yoke bolts to a flange on the differential.

> ### Vintage Chassis
>
> The open drive shaft, or **Hotchkiss design** (named after its inventor), has been in use since the 1950s and is the only one found on rear-wheel drive vehicles since that time. The older design that it replaced was an enclosed type called the **torque tube**.

Slip Yoke

As the vehicle goes over bumps, the rear springs allow the rear axle assembly to go up and down (**Figure 21.3**). The distance between the differential and the transmission changes, so the drive shaft requires a

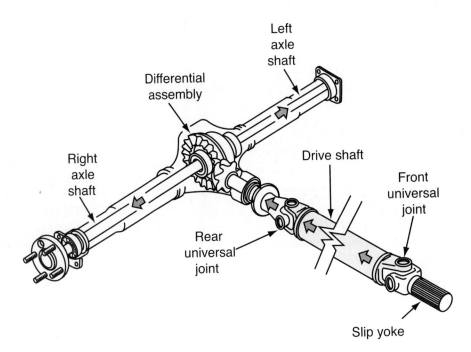

Figure 21.2 A typical drive shaft.

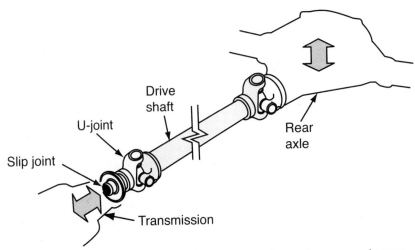

Figure 21.3 A slip yoke allows the drive shaft length to change as the car goes over bumps.

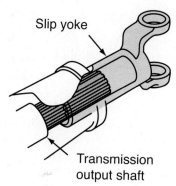

Figure 21.4 The slip yoke fits over splines on the output shaft.

means to change length by moving in and out of the transmission. A **slip yoke** is attached to a U-joint on the front end of the drive shaft. The other end of the slip yoke fits over splines on the output shaft or main shaft of the transmission (**Figure 21.4**). It slides in and out of the transmission as the distance between the transmission and differential changes.

The slip yoke is machined smooth on its outside diameter. This provides a sealing surface for the extension housing seal. It also provides a bearing surface for the extension housing bushing to act upon.

When a slip yoke is used with an automatic transmission, there is often a seal that fits over the output shaft. It rides against the *inside* of the slip yoke (**Figure 21.5**). Its purpose is to keep ATF (automatic transmission fluid), which is as thin as SAE 10 engine oil, from leaking out of the slip yoke through its vent hole (**Figure 21.6**). The output shaft splines are lubricated by the ATF. There must be a vent hole to allow the slip yoke to move in and out. Some yokes are greased and have a grease fitting. This design must be sealed at the end of the yoke.

If the rear yoke is not attached to the drive shaft, there will be a **flange** that is bolted to the front of the differential pinion shaft—the splined shaft that comes out of the front of the differential.

Universal Joints

Universal joints, often called **U-joints**, are located at both ends of the drive shaft. They transmit power at an angle (**Figure 21.7**). When the axle moves up or down, the U-joint allows for the necessary changes in angle at the ends of the drive shaft.

The most popular universal joint design is called a **cross and yoke**, or **Cardan**. It is two Y-shaped yokes connected by a cross, called a **spider** (**Figure 21.8**).

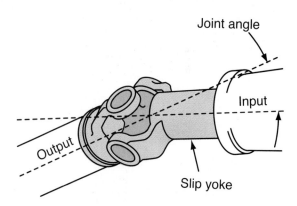

Figure 21.7 Universal joints allow power to be transmitted at an angle.

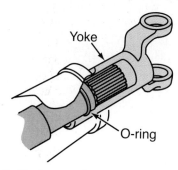

Figure 21.5 Automatic transmission slip yokes sometimes have seals.

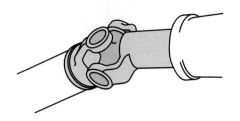

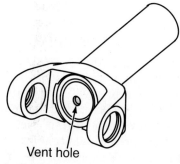

Figure 21.6 A slip yoke with a vent hole.

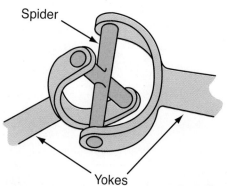

Figure 21.8 A universal joint is two Y-shaped yokes connected by a cross called a spider.

Most U-joints are made of forged, carburized steel. At the ends of the U-joints are four bearing caps with needle bearings (**Figure 21.9**). The needle bearings ride on **trunnions**, which are bearing areas ground on the ends of the cross or spider. The caps and bearings allow the joint to swivel as its angle changes. The caps and the ends of the trunnions must have a groove for grease.

Grease seals fit onto the ends of the bearing caps to keep dirt out and lubricant in. Replacement joints usually have a zerk fitting for lubrication during service (**Figure 21.10**). Original equipment joints on passenger cars do not have this feature.

There are several methods of holding the U-joints to the yokes (**Figure 21.11**). Universal joints are usually press fit into the yokes on the drive shaft. Snaprings fit into grooves in the yoke to keep the joint centered and retained. The most popular mounting found on vehicles is one in which U-bolts hold the U-joint in place between tabs in the rear flange. On the rear flange, there must be a feature that aligns the joint on center. Alignment is done either by tabs on the outsides or snaprings on the ends. Instead of a snapring, some original equipment joints are held in place with injected molded plastic. If the drive shaft is not installed exactly on center, serious vibration will result.

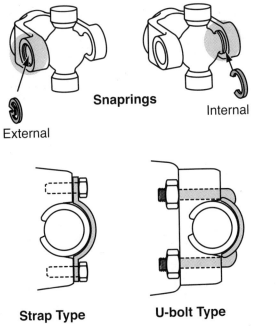

Figure 21.11 Several methods of holding universal joints to yokes.

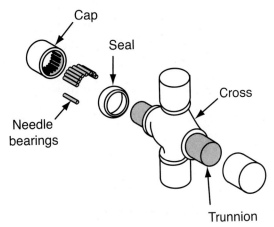

Figure 21.9 Parts of a universal joint.

Figure 21.10 Replacement joints usually have a zerk fitting for lubrication during service. (*Courtesy of Federal-Mogul Corporation*)

Two-Piece Drive Shaft

Balance is very important on a drive shaft. During balancing done at the factory, weights are spot-welded to the drive shaft. When a drive shaft is too long, it can flex, upsetting balance. Most vehicles have a single drive shaft, but when a vehicle has a longer wheel base, it will have a two-piece drive shaft. These are often found on large RWD cars, SUVs, and light trucks.

A properly assembled two-piece drive shaft will have its U-joint cups in alignment with one another. An improperly assembled two-piece drive shaft will be severely out-of-balance, causing major vibration. This condition, called *out-of-phase,* is discussed later in the chapter.

On a two-piece drive shaft, a **center support bearing** is used to hold the center of the shaft where the two shafts attach to each other. It bolts to the frame, cross member, or underside of the vehicle. The sealed bearing is supported in a rubber mount that dampens out noise and vibration (**Figure 21.12**).

■ DRIVE SHAFT ANGLE

Cardan U-joints have a problem that limits their use. When they are operated at an angle, the speed of the driven shaft varies as it revolves. The front of the drive shaft is attached to the yoke splined to the transmission output shaft, so they both turn at the same average speed. The yoke and drive shaft are coupled together by a U-joint, however, so the speed of the driven yoke at the opposite end of the drive shaft increases twice and

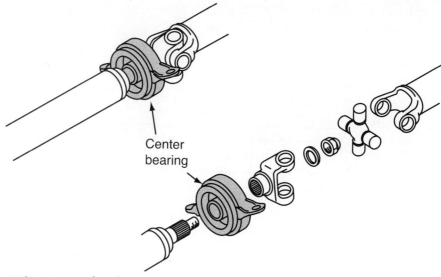

Figure 21.12 A center support bearing.

decreases twice during each drive shaft revolution. The elliptical path that the input and output yokes take can be compared to the face on a clock (**Figure 21.13**). At twelve and six o'clock the speeds of the input and output yokes are the same. In between, the output yoke turns at a speed that is not constant. The arrow in the

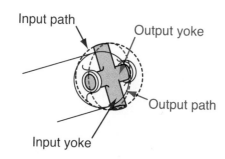

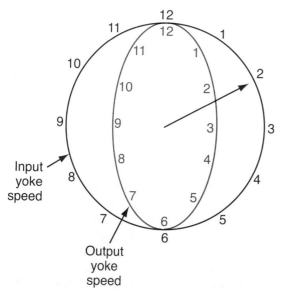

Figure 21.13 The elliptical path that the input and output yokes take can be compared to the face of a clock.

sketch shows that the output yoke has traveled to approximately the 2.2 o'clock position compared to the two o'clock position of the input yoke.

If the angle is increased, the change in velocity during each revolution increases. This can cause the drive shaft to whip like a jump rope. Most vehicle manufacturers do not use angles greater than 3 to 4 degrees with single U-joints because they cause excessive vibration. **Figure 21.14** compares the speed change of the U-joints with the shaft at both 10-degree and 30-degree angles. Study the illustration. If the drive shaft angle were set at 30 degrees at an input speed of 1,000 rpm, the speed of the driven yoke would vary from 700 rpm to 1,300 rpm during 90 degrees of drive shaft movement. Then it would change back. Imagine the vibration that would result from this situation.

With a one-piece drive shaft, arranging the U-joints so that they can cancel out each other's angle solves the problem (**Figure 21.15**). The angle is canceled out by the U-joint at the opposite end of the drive shaft. Both ends of the drive shaft must be **in phase** (**Figure 21.16**) for the canceling action to take place. This means that the trunnions of the front and rear U-joints are in the same plane (parallel). On a two-piece drive shaft, the front and rear halves can be assembled out-of-phase, resulting in vibration.

It is important that the angles at each end of the drive shaft be almost equal or vibration will result. When an angle is steeper, a constant velocity (CV) joint is used. The type used in front-wheel drive axles is covered in Chapter 19.

■ CONSTANT VELOCITY JOINTS

The vibration caused by the speeding up and slowing down of the drive shaft can be canceled by putting two Cardan U-joints next to each other. These CV universal

464 SECTION 5 AXLES AND JOINTS

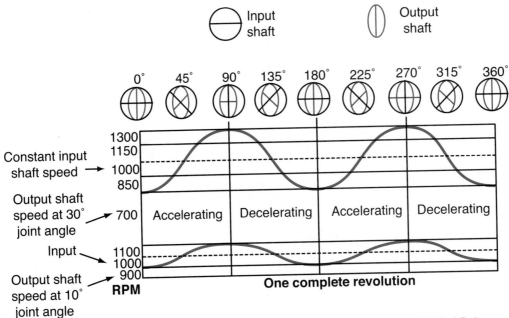

Figure 21.14 A comparison of the speed change of universal joints with the shaft at both 10-degree and 30-degree angles.

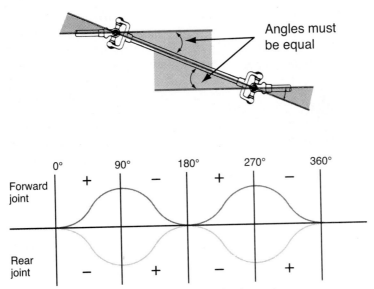

Figure 21.15 U-joints can be positioned to cancel out each other's angle.

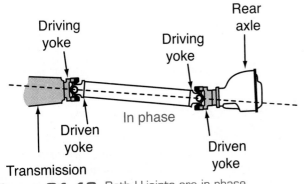

Figure 21.16 Both U-joints are in phase.

joints are called *double cardan* U-joints. **Figure 21.17** compares single and double cardan U-joints. The two joints are connected by a centering socket and yoke (**Figure 21.18**). The two joints are phased to cancel out each other's angle before the change in speed can go to the drive shaft. This is not a true constant velocity joint, but the speed change never leaves the joint. Its output is at a constant speed so the drive shaft does not get any of the vibrations it would get from a single joint. Because of their lack of vibration, constant velocity joints are found on larger luxury cars and pickup trucks.

Chapter 21 Rear-Wheel Drive Shafts, Universal Joints, and Axles **465**

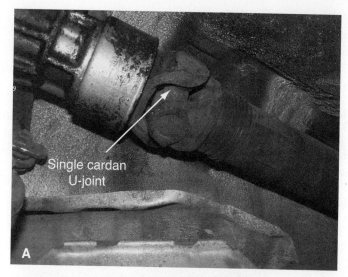

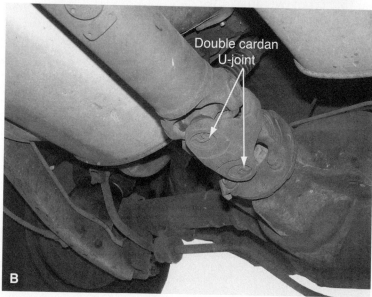

Figure 21.17 Comparison of single and double cardan U-joints. *(Courtesy of Tim Gilles)*

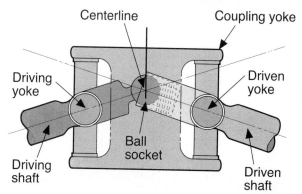

Figure 21.18 A double Cardan U-joint has two single Cardan U-joints connected by a centering socket and yoke.

Another type of constant velocity universal joint is the ball and trunnion, also called tripod tulip (see **Figure 20.9C**). Though not commonly used on rear-wheel drive vehicles, it is used on front-wheel drives because it is a true CV joint.

■ DRIVE SHAFT DIAGNOSIS

Drive shaft problems can result in noise or vibration from worn or rusted U-joints, a worn slip yoke, or a bad center support bearing. Worn U-joints can cause squeaking, clunking, knocking, or grinding sounds.

Sometimes a vehicle will have a clunking sound when changing from acceleration to deceleration. This can be due to worn slip yoke splines or a bad extension housing bushing. It can also be caused by problems in the differential or a very worn U-joint. Sometimes leaf springs can be loose at the differential, allowing the housing to "wind up." This means the input shaft to

SHOP TIP Before disassembling a universal joint, check to see that a wheel cover or hub cap is not making the noise. Remove all of them and drive the vehicle to see if the noise goes away.

the differential can move higher or lower in relation to its normal position, changing the drive shaft angle.

A worn center support bearing can cause a whining sound that varies with vehicle speed. The noise is constant in pitch, rather than changing or intermittent like U-joint noise. A U-joint noise changes pitch because of the changing angle of the U-joint.

NOTE: *The drive shaft is not always the source of noises that seem to be coming from it. A ringing sound, heard in the drive shaft, sometimes results from a bad clutch disc damper. Replacing the clutch disc usually solves the problem. If the vehicle has an automatic transmission, the problem can be due to a bad lockup converter.*

Drive Shaft Balance

Drive shafts are a possible source of vibration. In the highest transmission gear that is not overdrive, the drive shaft spins at engine rpm. If the shaft is bent or a U-joint is worn, a vibration can occur. Drive shaft balance is covered in Chapter 22.

Some drive shafts are built in two pieces with rubber dampening rings inside them. Another drive shaft style has a damper like a crankshaft vibration damper mounted on its outside. This absorbs torsional vibration.

Drive Shaft Inspection

Several checks are made when inspecting a drive shaft.

- Look for undercoating, missing balance weights, or obvious physical damage.
- Move the drive shaft up and down while watching the U-joints for looseness.
- Look for a rusty appearance around U-joint cups.
- Move the slip yoke up and down against the extension housing bushing while looking for excessive movement.
- Check to see that the motor mount on the transmission cross member is in good condition.

A dent in the drive shaft tubing will weaken it, which can cause it to kink easily under load. The strength of a drive shaft is longitudinal. Try standing on an empty soda can that is stood on end. Then quickly touch both sides of the can. Although it might support your weight, it will immediately collapse when you touch it. Cracks in a drive shaft result from physical damage. They always start on the surface, never at the inside.

■ UNIVERSAL JOINT DIAGNOSIS

When a universal joint begins to fail, a squeaking sound is often noticed just when the vehicle begins to go forward. The most common cause of U-joint failure is when its grease dries out. This often happens because the seal on the U-joint has failed, allowing moisture in. A vibration can also occur when a U-joint starts to fail.

With a worn U-joint, a sharp, one-time click will be heard when the vehicle direction is changed from forward to reverse or when the vehicle first takes off.

■ DRIVE SHAFT SERVICE

To remove the drive shaft, follow this procedure:

1. Mark the drive shaft so that it can be replaced in the same position. Use a crayon to mark the rear differential yoke and the companion flange (**Figure 21.19**).
2. Unbolt the rear U-joint from the differential companion flange.
3. Pry the rear U-joint forward away from the differential.
4. Wrap tape around the U-joint cups so that they cannot fall off the U-joint cross (**Figure 21.20**).

NOTE: *If one of the cups falls off the U-joint, one or more of the small needle bearings might fall out. If one gets lost, the entire U-joint must be replaced.*

5. On a two-piece drive shaft, unbolt the center support bearing.

NOTE: *Be sure to mark both halves of the shaft in the center where the splines are. The shaft will be out-of-phase if it comes apart and is not reassembled correctly (see Figure 21.16). Serious vibration will result.*

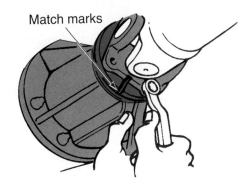

Figure 21.19 Mark the rear differential yoke and the companion flange before removing the drive shaft.

Figure 21.20 Tape wrapped around the U-joint cups to keep them from falling off. *(Courtesy of Tim Gilles)*

Figure 21.21 When the drive shaft is removed, a slip yoke installed in its place prevents oil from leaking from the transmission. *(Courtesy of Tim Gilles)*

> **SHOP TIP** To prevent an excessive amount of oil from leaking out of the transmission, install an old slip yoke onto the main shaft splines (**Figure 21.21**). If an old yoke is not available, a plastic plug or a bushing installation tool can be used.

6. The drive shaft will now slip out of the transmission. When it is removed, oil will probably come out of the transmission.

■ UNIVERSAL JOINT DISASSEMBLY

The procedure described here is for single cross and yoke (Cardan) U-joints. If the U-joint has any snaprings, remove them (**Figure 21.22**). Some snaprings are on the inside of the yoke; others are on the outside. When the snapring is on the outside a sturdy pair of pliers can be used.

NOTE: *Smaller needle nose pliers can become damaged because they are not sturdy enough.*

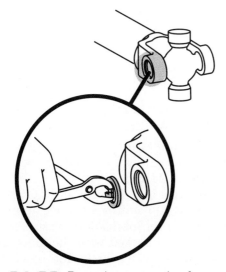

Figure 21.22 Removing a snapring from a U-joint.

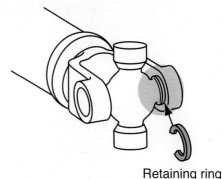

Figure 21.23 Sometimes snaprings go on the inside of the yoke.

Figure 21.24 Removing a U-joint with sockets and a vise.

When snaprings are on the inside, a punch or a special tool can be used to remove them (**Figure 21.23**). If the U-joint is retained by plastic resin, follow the manufacturer's service manual instructions. A small tube of resin is usually used.

There are several ways U-joints are commonly disassembled. The most common method is to use a bench vise. When pressing a U-joint out of its yoke, the bearing cup on one side must be pushed into a socket or pipe that is slightly larger in diameter (**Figure 21.24**). After the cup is removed from one side, the cross is forced against the other cup to push the cup out of the other side. The cross can be removed at this time. Then the cup is pounded out with a punch (**Figure 21.25**). The process is repeated on the other side of the cross to complete the removal of the U-joint.

NOTE: *One problem with using a vise is that it is weak when opened as far as it takes to accommodate the U-joint. A vise can be broken if excessive force is used when its jaws are far apart.*

Many shops have a special U-joint tool. It is used in the same manner (**Figure 21.26**). The hole in the opposite side of the tool is larger than the U-joint bearing cup that will be pressed into it during removal. A third method of U-joint removal is to use a press with a special tool.

468 SECTION 5 AXLES AND JOINTS

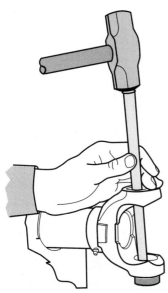

Figure 21.25 Pound out the bearing cup with a punch.

Figure 21.27 Brinelling on a trunnion. (Courtesy of Tim Gilles)

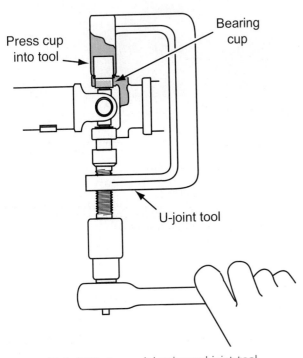

Figure 21.26 A special universal joint tool.

Figure 21.28 Galling on a trunnion. (Courtesy of Tim Gilles)

Some shops pound the U-joint out with a hammer. This process works, but you must be careful not to bend a yoke. Even a very small amount of misalignment is enough to cause a vibration. Also, if you install the tube part of the drive shaft in a vise during installation, be careful not to damage the tube.

One problem caused during installation in a vise is that drive shaft yoke ears commonly become sprung inward. This results in brinelling. Brinelling is when small indentations wear into the bearing surface (**Figure 21.27**). Brinelling is often the result of a faulty U-joint installation. The joint should always feel loose and relaxed (not binding up) after a correct installation.

A problem with drive shaft angles can cause **galling**, which often happens on the ends of the trunnions. Usually galling will be found on a trunnion 180 degrees to brinelling (**Figure 21.28**).

■ UNIVERSAL JOINT REASSEMBLY

Replacement U-joints usually have a zerk fitting. If there is a zerk fitting it will probably be slanted in one direction (**Figure 21.29**).

The grease used in U-joints is an NLGI #1, with an extreme pressure (EP) additive. If the U-joint does not have a zerk fitting for lubrication, be sure to check to see that it has been packed with the lubricant before you install it in the drive shaft. Too much lubricant can bind it up, making installation difficult.

Figure 21.29 The zerk fitting is slanted toward the drive shaft.

Figure 21.30 Put a bearing cap partially into the yoke.

Figure 21.31 (A) Put the U-joint into the cap and compress the cap into the yoke. (B) Move the U-joint to see that it is not binding up. (C) Compress the remaining cap into the yoke.

Put a bearing cap partially into the yoke (**Figure 21.30**). Put the U-joint into the cap and compress the cap into the yoke (**Figure 21.31A**). Carefully install the other cap in the opposite side of the yoke. Be sure that you do not accidentally knock one of the needle bearings out of position. A very common occurrence is for one bearing to be knocked into the bottom of the cup. When the cups are pressed together all the way, the joint locks up. When you disassemble it to fix the problem, you might find a damaged needle bearing. This will require the purchase of a new U-joint.

Move the joint back and forth to see that it is free and nonbinding in both cups (**Figure 21.31B**). Place the yoke between the jaws of the vise and compress the remaining cap into the yoke (**Figure 21.31C**). While you are pressing it in, move the joint back and forth and watch for binding. Do not force anything.

NOTE: *If the bearing cap turns in the hole in the yoke, the yoke is defective. The cap is hardened and the ear of the yoke is not. A new yoke will need to be installed on the drive shaft by a machine shop.*

> **SHOP TIP** When installing the cross into the yoke be sure that the zerk fitting is angled in toward the drive shaft. Installing the cross with the zerk fitting backward will make it difficult, if not impossible, to lube the U-joint after the drive shaft is installed on the vehicle. It can also cause a bind when the shaft angle increases as a vehicle is raised on a wheels free lift.

Install one of the snaprings before completing the pressing procedure in the vise. Use a socket that is smaller than the hole that the bearing cup goes into and press the cup until it comes up against the snapring. Then, install the other cup until it is deep enough to install the remaining snapring. Move the joint in each direction to see that it moves freely without binding. If

Figure 21.32 Striking the yoke realigns the needle bearings in the bearing cup.

Figure 21.33 Install the other half of the cross onto the drive shaft.

it is slightly tight, strike the drive shaft yoke with a punch to free it up (**Figure 21.32**). A tight joint will cause severe vibration.

NOTE: *Sometimes original equipment U-joint cups are held in place with plastic material that is injected into a hole in the yoke. No clips are used to retain these joints. Disassembly requires similar techniques as other U-joints and replacement joints will be retained with clips. Follow the manufacturer's instructions.*

Some replacement U-joints use internal snaprings when replacing injection-molded plastic retainers (see **Figure 21.23**).

After the two bearing caps are installed in the slip yoke, install the other half of the cross onto the drive shaft (**Figure 21.33**).

■ DRIVE SHAFT INSTALLATION

When installing a drive shaft that bolts to a yoked flange, be sure that the U-joints fit exactly between the tabs on the flange. If the drive shaft is not installed exactly on center, there will be serious vibration. Be sure that all of the contact surfaces are clean before installing the drive shaft. Dirt or a burr on the companion flange where the U-joint cup will fit can cause the shaft to vibrate after installation. With the transmission in neutral, slide the slip yoke into the transmission. Align the marks that you made during removal and install the rear U-joint onto the companion flange. Install the retaining bolts to complete the job.

NOTE: *If a vehicle experiences vibration after the installation of new U-joints, the vibration can sometimes be corrected by removing the drive shaft and reinstalling it, turning it 180 degrees in the companion flange.*

■ TWO-PIECE DRIVE SHAFT SERVICE

On a two-piece drive shaft the center support bearing sometimes fails. The seized bearing can tear away the rubber mount that supports the outside of it, allowing the bearing to rotate with the drive shaft. This allows the drive shaft to wobble up and down, causing a vibration. The bearing is pressed off and a new one is pressed on. Be sure to press on the inside bearing race so you do not damage the bearing. Some bearings are designed to be installed in one direction only. Installing them backwards will result in damage to the rubber ring on the bearing.

If the two pieces of the drive shaft are separated, be sure that the shaft is reassembled in phase. Assembling the front and rear halves out of phase can cause extreme vibration. When the shafts were not marked before disassembly, use a tape measure or a steel rod to check the alignment of the halves.

■ DRIVE AXLES AND BEARINGS

Drive axles usually support the weight of the vehicle. Bearings are covered in Chapter 11. Refer to that chapter for more information about particular types of bearings. The coverage here is limited to specific rear-wheel bearing designs.

Semi-Floating Axle Types

Semi-floating axles (see **Figure 11.17**) are found on many vehicles. There are two semi-floating designs in common use.

Bearing Retainer Axle. On one design, called a **bearing-retained axle**, a bearing with an inner race is pressed onto the axle shaft. A bearing retainer ring (**Figure 21.34**) is pressed onto the axle shaft after the bearing is installed. The outside of the bearing fits tightly into the axle housing so oil cannot leak out of the differential. The bearing used with this axle is usually packed with grease and sealed.

A bearing retainer that is installed on the axle before the bearing is pressed on provides a means of bolting the assembly to the brake backing plate. The bolts holding the bearing retainer also clamp the brake backing plate to the axle housing.

C-Lock Axle. The other style of axle is called a **C-lock** or **C-clip axle**. On most axles of this type, the bearing rides on a hardened area of the axle, rather than having

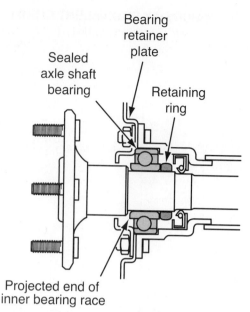

Figure 21.34 A retaining ring is pressed onto the shaft after the bearing.

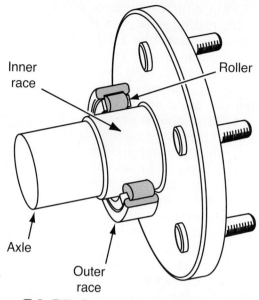

Figure 21.35 On a C-lock axle, the bearing rides on a hardened area of the axle.

an inner bearing race (**Figure 21.35**). The outer bearing race and its bearing rollers fit tightly into the end of the axle housing. The axle bearing is lubricated by oil from the differential. A lip seal keeps the oil from leaking out onto the brake shoes.

This retainer type is found only on Salisbury axles. Salisbury axles are the ones with a cover on the rear that do not have a removable center differential section (called a *third member*). A C-lock at the inside end of the axle holds the axle into the differential (see **Figure 21.43**) The outside of the C-lock fits into a recess in the differential side gear (the gear that is splined to the rear axle and drives it). When the differential pinion shaft is installed, the C-locks are trapped within the side gears.

NOTE: *With either of the semi-floating axle designs, a broken axle can result in the wheel moving out and away from the axle housing. With the C-lock axle, there will be nothing to retain it. With the bearing-retained axle, the bearing can wear through the retainer if the vehicle is driven in this condition. That is why light-duty trucks and heavier trucks use full-floating axle bearings (see Figure 11.18). The full-floating axle can break and the wheel and hub will still be supported by the bearings.*

■ INDEPENDENT REAR SUSPENSION

When a vehicle has independent rear suspension, the axles must be able to pivot independently as each wheel goes over a bump. Instead of solid axles, there are two **swing axles** (**Figure 21.36**). Each axle has a universal joint at one or both ends. These are similar to the ones used for front axles in front-wheel drive vehicles.

■ GEAR OILS

Gear oils are special heavy liquid lubricants used to lubricate gears and bearings in manual transmissions and differentials. A good oil prevents high temperatures and scoring of the parts.

A hypoid axle set is the most difficult gear application to lubricate. The gear teeth can be subjected to pressures as high as 400,000 psi. Because of their

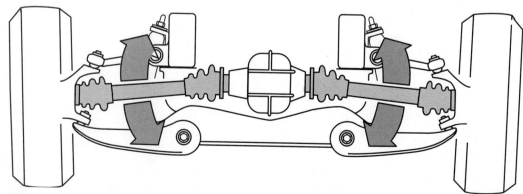

Figure 21.36 An independent rear suspension has two swing axles.

shape, there is a high amount of sliding that occurs between the mating surfaces of the gear teeth. Gears must remain separated by a film of oil, even when under heavy loads. As one gear tooth moves against another, oil is forced from between them. This oil must continuously be replenished.

When the lubricant film fails, wear occurs. Gear oils have additives similar to the ones found in engine oils. Additives prevent oxidation and corrosion, reduce friction, limit wear, and prevent foaming. Extreme pressure additives, or **EP additives**, are chemicals that cause the formation of compounds on the gear teeth that prevent welding of the gear teeth under extreme load conditions.

There are many varieties of gear oils. Some gear oils are thin, like those used in automatic transmissions. Other are thick, like those for use in differentials. Oil used in manual transmissions must remain fluid enough to allow for easy shifting in cold weather. Automatic transmission fluids must remain fluid to as low as −40°F and remain at almost the same viscosity at over 300°F.

Viscosities for gear oil range from 75 to 250. There are also multigrades like 80W-140. The W is for winter grade, like in engine oil.

NOTE: *Gear oil viscosity is not the same as that of engine oil. It is rated in centiStokes. An SAE 90 gear lube is comparable in viscosity to an SAE 40 or 50 engine oil.*

The chart in **Figure 21.37** compares viscosities of SAE gear and engine lubricants.

API Gear Oil Ratings

The **American Petroleum Institute (API)** and Coordinating Research Council (CRC) have established a system for classifying gear lubricants. The **Society of Automotive Engineers (SAE)** lists these classifications as shown in **Figure 21.38**. The GL in the number stands for gear lube. GL1 is the lowest grade and is only applicable for light-duty and low speed use. Automotive gear lubes are GL4, GL5, and GL6. GL6 is a synthetic classification.

Limited Slip Gear Oils

Limited slip gear oils require a special friction modifier additive. The fill plug is usually marked to indicate this. When there is not enough friction modifier, the clutch discs stick together and grab, rather than slip. If this friction modifier is not present, as the vehicle makes a turn a chirping or clunking sound will be heard. Be sure the lubricant you install in a limited slip axle is suited for limited slip uses. There are friction modifier additives that can be added to gear lubricants to make them compatible with limited slip axles.

NOTE: *Limited slip gear oil will work in standard axles.*

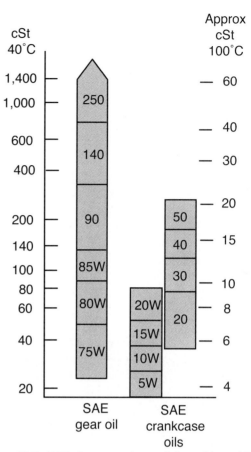

Figure 21.37 A comparison of viscosities of SAE gear and engine lubricants.

API Classification	Amount of EP Additives	Type of Service
GL-1	None	Light-duty only. Spiral bevel and worm gear axles. Some manual transmissions.
GL-2	Small amount	Worm gear axles under more severe conditions than GL-1.
GL-3	Small amount	Spiral bevel axles and manual transmissions under more severe conditions than GL-1.
GL-4	Medium amount	Hypoid gears—light service. Spur and bevel gears—medium service.
GL-5	High amount	Hypoid gears—medium service. Spur and bevel gears—heavy-duty service.
GL-6	Highest amount	Most gears—extremely heavy-duty service.

Figure 21.38 Gear lubricant ratings.

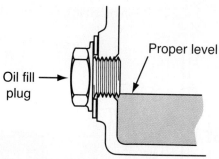

Figure 21.39 The fluid level should be at the bottom of the fill plug hole.

SHOP TIP To tell the difference between a brake cylinder leak and an axle seal leak, spray the area with water. Brake fluid will wash off and gear oil will require solvent to remove.

Lubricant Leaks

Gear oil can leak out of the differential at the pinion seals, from the cover, from the banjo housing, or from the axle seals. When axle seals leak, a rear brake could become covered with oil and not work properly. Because rear wheels do not normally steer, this is not usually an obvious problem to the driver. Because the oil is thrown by the spinning brake drum, the entire brake backing plate might be stained.

Oil should fill the axle housing as shown in **Figure 21.39**.

■ AXLE BEARING DIAGNOSIS

A worn or damaged rear axle bearing can result in a fluid leak or bearing noise. If a groan is heard that could be due to a wheel bearing, driving the vehicle can sometimes help pinpoint the problem. First, check the tires for damage and be sure they are properly inflated. Find an empty parking lot or deserted road and make slow left and right turns. The inside tire always turns at a lower rpm than the outside tire. The noise from a bad wheel bearing will change pitch as the wheel speeds up and slows down when turning from one side to the other. When the outside wheel has the bad bearing, the noise will become worse because that wheel is turning faster and is loaded more heavily due to weight shift of the vehicle to the outside. Applying the brakes can also cause the noise level from a bad bearing to lessen as they contact the drum or rotor that has the bad bearing.

After the road test, raise the vehicle on a hoist and try to move the tires and wheels by hand. The only kind of bearing that has any normal movement is the one used with a C-lock axle, which sometimes has end play that can be felt. The other kinds of wheel and axle bearings should have less than 0.005-inch end play. Up and down movement is not normal. When pushing in and out on the top and bottom of a tire on a wheel with a tapered roller bearing, a small amount of movement is normal. Testing an axle bearing in this manner should result in no movement.

CAUTION Be careful of spinning wheels and drive shafts during this test. Be especially careful that tires do not accidentally touch parts of the lift, which could result in a very serious accident!

Listening for Noises

When a noise is related to tires, it will go away when the vehicle is in the air. An axle bearing noise will also be diminished because the load has been removed. Sometimes the vibration from a bad bearing can be felt by placing your hand on a part close to it. Listen for the noise with a stethoscope. Put the end of the tool against the brake backing plates to locate bad axle bearings. Listen below the differential pinion gear for a bad pinion bearing.

■ AXLE BEARING SERVICE

Nonserviceable rear axle bearings have no service interval. When they fail, replacement is required. The axle must be removed from the housing to replace the bearings. Axle removal is commonly done by service establishments, but the presswork required to install the new bearing on the axle is usually done by parts stores, many of which have machine shops. Axle seals are sometimes part of the axle bearing; at other times they are separate from the axle bearing. Axles are supported on the inside end by the side gears in the differential. Repairing an axle seal requires removal of the axle from the differential.

Removing a Bearing-Retained Axle

Recall that axle bearings on vehicles with removable third member differentials are called *bearing-retained axles*.

NOTE: *These differentials are often called "pumpkins" because the outer ribbed structure of their casting resembles a pumpkin.*

The bearing is pressed fit on the axle shaft and fits tightly in the axle housing. There is a retaining plate that must be removed before pulling the axle. Its bolts are accessed through holes in the axle flange. A socket and extension can be used (usually with an air impact

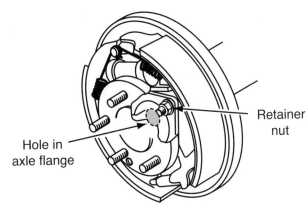

Figure 21.40 A hole in the axle flange allows for removal of the retainer nuts.

Figure 21.41 Removing an axle with a slide hammer.

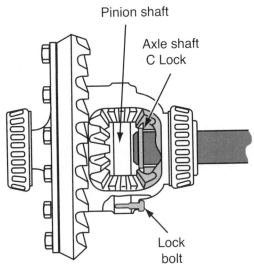

Figure 21.42 Remove the lock bolt and slide out the pinion shaft.

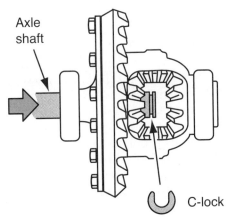

Figure 21.43 When the axle is pushed inward, the C-lock becomes free of the inside of the side gear.

wrench) to remove the bolts and nuts (**Figure 21.40**). After each nut and bolt is removed, the axle is turned to allow access to the next bolt. Some axles have four holes, but most only have one. The outside of the bearing is often tight in the hole in the axle housing. A slide hammer is often required to remove it (**Figure 21.41**).

Removing a C-Lock Axle

The cover on the back of the differential must be removed prior to the removal of a C-lock axle. The end of a C-lock axle has a machined groove that a C-lock fits into and the differential pinions must be removed to allow access to the C-locks. The pinion shaft has a lock bolt (**Figure 21.42**). After it is removed, the shaft is pulled out. After the pinion lock bolt, shaft, and pinions are removed, the axles can be pushed inward, allowing the C-locks to be removed (**Figure 21.43**).

On most C-lock axles, the bearing rides on a hardened area of the axle, rather than having an inner bearing race (see **Figure 21.35**). The outer race and bearing rollers are held in the axle housing. The hardened area of the axle can be damaged where the seal rides on it. Some replacement seals are designed to allow the lip to ride in a different area of the axle so a new sealing surface is provided.

Axle Seal Service

After the axle is removed, pry out the old seal (**Figure 21.44**). The seal can be pried out with a prybar, a seal puller, or the end of the removed axle. Check the surface of the axle shaft to see that it is clean and undamaged. Apply lubricant to the lip of the new seal before installing it. Use a seal installing tool to avoid cocking it to the side and damaging it. During reinstallation, support the axle and do not let it drag on the surface of the new seal.

> **SHOP TIP**
> - If you slide the pinion shaft out and do not turn the axle, the gears will stay in place.
> - Occasionally a pinion shaft lock bolt will break off. There are special tools available for removing these.

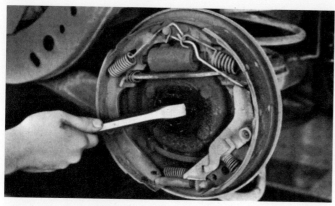

Figure 21.44 Removing the old axle seal. *(Courtesy of Chicago Rawhide)*

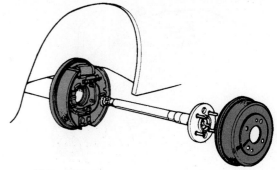

Figure 21.47 Some axles are removed without the backing plate.

■ AXLE BEARING REPLACEMENT

Some older imported cars had bearings that were pressed onto the axle after the brake backing plate. When removing the axle on these types, the brake line is first disconnected. The backing plate is unbolted and the still assembled brakes are removed along with the axle (**Figure 21.45**). A press or a special puller (**Figure 21.46**) can be used to remove the bearing. Other

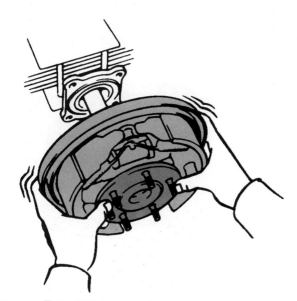

Figure 21.48 Removing an axle bearing with a slide hammer.

types of axles with press-fit bearings come out independent of the backing plate (**Figure 21.47**).

On C-lock type axles, the bearing remains in the housing after the axle is removed. The bearing is pulled with a slide hammer and special attachment (**Figure 21.48**).

Press-Fit Bearing Replacement

Do not try to press the bearing and retaining ring off at the same time. Grind a notch on the retaining ring and strike it with a chisel to weaken it prior to pressing it off (**Figure 21.49**). Presses usually have a safe cage made of large-diameter steel pipe that can be installed around the bearing when pressing it off. It is not unusual for a bearing to require in excess of 20 tons of pressure to remove it.

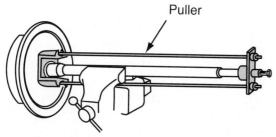

Figure 21.45 Some axles must be pulled together with the brake backing plate.

 A bearing that explodes during removal, which is not uncommon, can be compared to a grenade exploding!

Figure 21.46 A puller setup for removing an axle bearing.

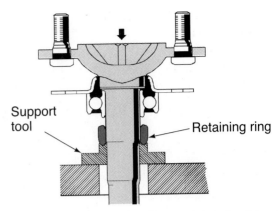

Figure 21.51 Pressing the retaining ring into place against the axle bearing.

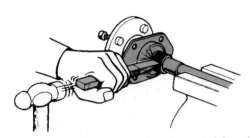

Figure 21.49 Removing a bearing retaining ring.

If the gauge on the press reads over 15 tons, something could be wrong. Be certain that you are pressing parts apart and not pressing against something solid.

■ AXLE BEARING INSTALLATION

When pressing a bearing onto the axle it is supported by the inner race (**Figure 21.50**). The cage and rollers should always be able to be turned during the installation process. This ensures that there is no load on them. First the bearing is installed on the axle. The retaining ring is often heated (150°C/300°F) for easier assembly. Then, the retaining ring is pressed into place against the axle bearing (**Figure 21.51**).

 **CAUTION** Do not attempt to press the bearing and the retaining ring onto the axle at the same time.

■ REINSTALL THE AXLE

Reinstall the axle in the housing, and bolt on the retainer. Be sure that the oil return slot in the retainer lines up with the return hole in the axle housing (**Figure 21.52**). This is similar to the oil return on the clutch bearing retainer on a manual transmission.

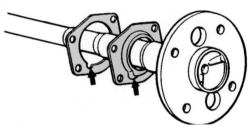

Figure 21.52 Be sure that the oil return slot in the retainer lines up with the return hole in the axle housing.

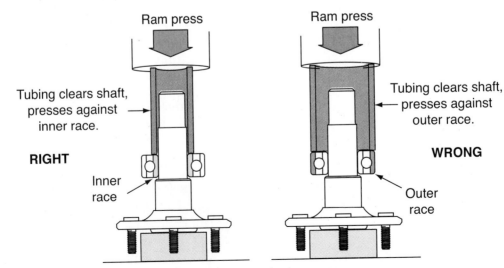

Figure 21.50 Press bearings on by putting pressure on the inner race.

FULL-FLOATING AXLE SERVICE

Full-floating axles are found on trucks that are ¾ ton and larger. Axle bearings are located in the hub and brake drum assembly (**Figure 21.53**). To remove the drum on a full floating system, the axle must be removed first. Often, there are soft metal beveled washers that surround the studs that hold the axle to the hub. Remove them by striking the outside of the axle flange with a brass hammer. Do not force a chisel in between the axle and hub or you could create a future seal leak. Some axles have threaded holes to assist in axle removal (**Figure 21.54**).

NOTE: *Place a drain pan beneath the hub assembly to catch oil that escapes.*

After the axle is removed, the locknut that holds the brake drum can be removed. There are different types of locknut arrangements. A typical locking arrangement has two nuts, the outer nut being a locknut. Lift off the brake drum. Catch the outer bearing so it does not fall on the ground.

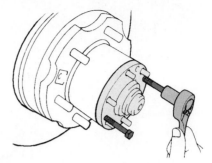

Figure 21.54 Removing a full floating axle.

Install a new inner bearing seal. During reassembly, the hub and drum are reinstalled on the outside of the axle housing. The retaining nut is tightened until it has little or no clearance.

After the hub is correctly reinstalled, install the axle and beveled washers. Pound the beveled washers in until the whole axle is seated (**Figure 21.55**).

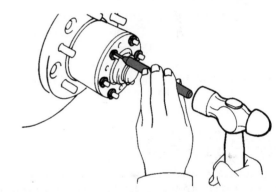

Figure 21.55 Installing the beveled washers.

CASE HISTORY

An apprentice did a brake job on a ¾ ton pickup truck. When he reinstalled the rear drum and hub, he did not pay attention to how the locking nut was supposed to work and assembled it improperly. The owner was driving his truck later that day and the axle, hub, and wheel came off during a turn. The result was that expensive repairs were required on the truck. Luckily, there was no accident and no one was hurt.

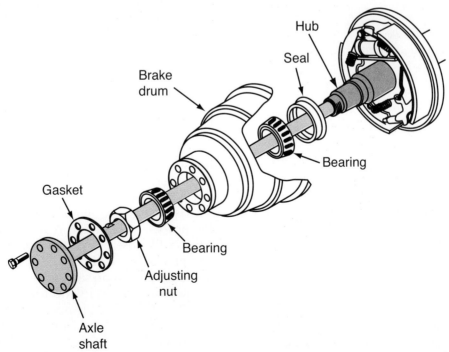

Figure 21.53 Typical full-floating axle and hub assembly. The bearings ride on the hub, not on the axle.

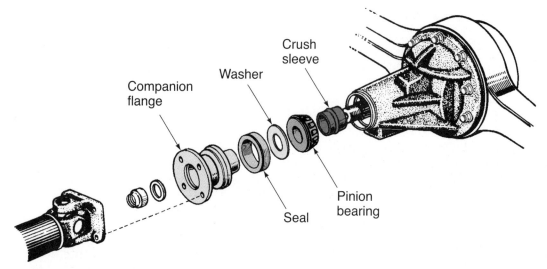

Figure 21.56 Parts involved in a pinion seal replacement.

■ DIFFERENTIAL PINION SEAL REPLACEMENT

A common repair to a differential is the replacement of the pinion seal. Parts involved in a typical pinion seal replacement are shown in **Figure 21.56**. A very important part of this job is to maintain the tension on the pinion bearing crush sleeve. One way of maintaining crush sleeve tension is to mark the nut and pinion shaft before removal (**Figure 21.57**). After reassembly, tighten an additional 1/8 turn past the mark.

Another method of maintaining pinion bearing crush sleeve tension is to use a dial indicator torque wrench when removing the nut. Note the torque required to loosen it. During reinstallation of the nut, add 5 pounds to the former torque to recompress the old crush sleeve.

NOTE: *When a bearing fails because it was too tight, the small end of the bearing will experience wear first.*

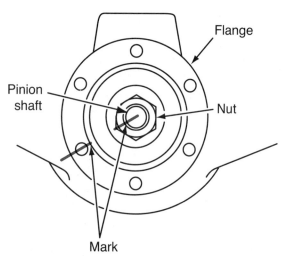

Figure 21.57 Mark the flange, nut, and pinion shaft before disassembly.

Figure 21.58 This pinion bearing failed because it was too tight. The small ends of the rollers failed first. (Reprinted with permission of Quaker State Corporation)

CASE HISTORY *A student's mother's car had a leaking pinion seal. The student removed the drive shaft and pinion flange. He used a hammer to remove the flange. Then he replaced the seal and reassembled all of the parts. He tightened the pinion nut securely without a torque wrench. Within 3,000 miles the pinion bearing failed because the bearing adjustment was too tight (**Figure 21.58**).*

Always use a long bar to hold the yoke from turning while loosening the pinion nut (**Figure 21.59**). Using an impact wrench can damage the ring and pinion or the pinion bearings.

Use a puller to remove the flange (**Figure 21.60**). Do not use a hammer. A flange can be easily bent, which will cause a drive line vibration. The student's car in a previous case history always had a drive line vibration after the pinion seal job because he bent the flange.

Remove the seal with a hammer and chisel (**Figure 21.61**). Install the new seal with a seal installer to avoid cocking or damaging it (**Figure 21.62**). Lube the lip of the new seal and the mating surface on the companion flange before reinstalling it.

Figure 21.59 Use a bar to restrain the flange while loosening the pinion nut.

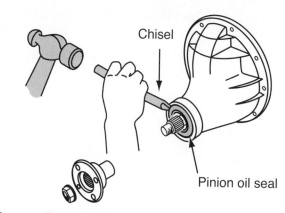

Figure 21.61 Removing a pinion seal.

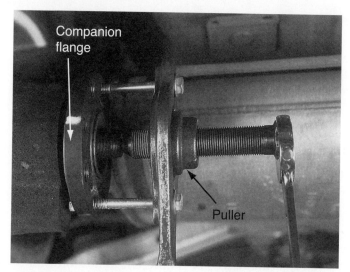

Figure 21.60 Use a puller to remove the flange.

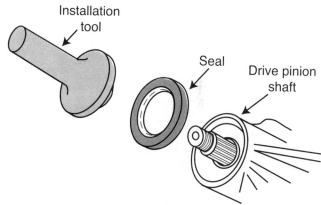

Figure 21.62 To avoid distorting the seal, install it using a seal installer.

DIFFERENTIAL BEARING PROBLEMS

A differential can experience bad carrier bearings or a bad pinion bearing. Carrier bearings and a pinion bearing will make a constant sound because the carrier is always driven at the same speed as the ring and pinion. The sound will usually become louder with an increase in speed.

REVIEW QUESTIONS

1. What is the name of the open drive shaft system?
2. What is the name of the part that slides in and out of the transmission as the distance between the transmission and differential changes?
3. How many times in one revolution does speed increase and decrease when a U-joint is operated at an angle?
4. Angles of greater than _____ degrees are not used with single U-joints because they cause excessive vibration.
5. At 30 degrees of drive shaft angle, driven yoke speed will change the maximum amount within _____ degrees of drive shaft revolution.
6. What is the name of a CV joint that uses two U-joints next to each other?
7. An SAE 90 gear lube is comparable in viscosity to an SAE 40 or 50 _____ oil.
8. What does the GL stand for in the gear lube rating?

9. What is the term for an axle that turns with the wheel?

10. What can you wrap around the U-joint cups to keep them from falling off?

11. When a drive shaft is assembled in a position where the front and rear U-joints are not parallel, this is called out-of-_____.

12. What is it called when small indentations wear into the bearing surface?

13. Which part is hardened, the U-joint cap or the ear of the drive shaft yoke?

14. When a Cardan universal joint is operated in a straight line, is the speed of the output member different than or the same as the speed of the input member?

15. _____ joints are angled to cancel each other out.

PRACTICE TEST QUESTIONS

1. Constant velocity joints are found on smaller rear-wheel drive cars.
   A. True
   B. False

2. The speed of the output yoke on a propeller shaft is constantly changing as it revolves.
   A. True
   B. False

3. A dent in the drive shaft can weaken it.
   A. True
   B. False

4. Technician A says to install a U-joint with the zerk fitting facing toward the drive shaft. Technician B says that original equipment U-joints do not usually have zerk fittings. Who is right?
   A. Technician A   B. Technician B
   C. Both A and B   D. Neither A nor B

5. Technician A says that during a turn when the outside wheel has a bad bearing, the noise can become worse. Technician B says applying the brakes can cause the noise level from a bad bearing to become higher as they contact the drum or rotor that has the bad bearing. Who is right?
   A. Technician A   B. Technician B
   C. Both A and B   D. Neither A nor B

6. Technician A says axle seals are sometimes part of the axle bearing. Technician B says axle seals are sometimes separate from the axle bearing. Who is right?
   A. Technician A   B. Technician B
   C. Both A and B   D. Neither A nor B

7. Technician A says that a limited slip axle that chatters during turns could have the wrong lubricant. Technician B says that an axle with C-locks must have the cover removed from the back of the differential housing in order to remove the axles. Who is right?
   A. Technician A   B. Technician B
   C. Both A and B   D. Neither A nor B

8. Gear lube viscosity is different than motor oil viscosity.
   A. True
   B. False

9. The most common reason a U-joint fails is because its grease dries out.
   A. True
   B. False

10. A drive shaft that is even a very small amount off-center can cause vibration.
    A. True
    B. False

CHAPTER 22

Drive Line Vibration Service

■ OBJECTIVES

Upon completion of this chapter, you should be able to:
- ✔ Understand terms related to vibration.
- ✔ Describe the different types of vibration.
- ✔ Test for vibration using test instruments.
- ✔ Check drive shaft runout.
- ✔ Balance a drive shaft.
- ✔ Check drive shaft angle.

■ KEY TERMS

amplitude
beat/boom vibration
boom
buzz
cycles
drone
electronic vibration analyzer (EVA)
first order vibration
frequency
hertz
howl
launch shudder
match mount
moan
natural frequency
reed tachometer
roughness
second order vibration
shake
snapshot
tingling
transducer
velocity
vibration
whine
working angle

■ INTRODUCTION

Tire and wheel imbalance is the most common cause of vibration complaints. Broken engine and transmission mounts are another frequent cause. After tires, wheels, and mounts have been eliminated as the cause of vibration, the items dealt with in this chapter come into play.

Vibration is something that does not usually prevent the vehicle from operating. However, it does annoy drivers and can result in their unhappiness with a repair job or the car in general.

■ VIBRATION ANALYSIS

A process of elimination is used to analyze vibration problems. The following terms are used in analyzing vibrations:

- **Vibration** describes a part that is in motion in waves, or **cycles**. **Figure 22.1** is an example of vibration as it cycles through a material.

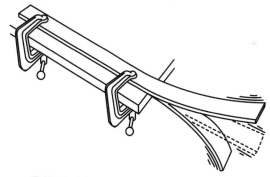

Figure 22.1 Vibration happens in cycles.

- **Frequency** is how many cycles take place in a period of time. The number of cycles in one second is measured in **hertz**, or Hz (**Figure 22.2**).
- **Amplitude** measures how intense a vibration is (**Figure 22.3**). An intense vibration has more amplitude.

481

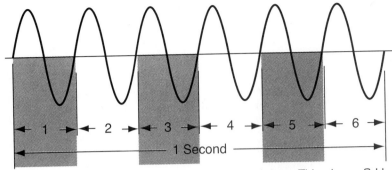

Figure 22.2 Frequency is how many cycles take place in a period of time. This shows 6 Hz.

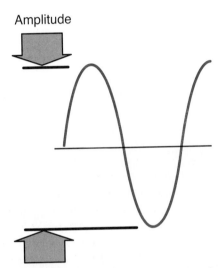

Figure 22.3 Amplitude is a measurement of the intensity of a vibration.

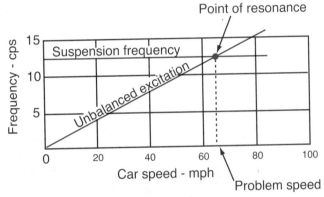

Figure 22.5 A drive shaft that is out of balance will have a frequency that can become the same as the suspension frequency at a particular speed.

- **Velocity** is a combination of amplitude and frequency (**Figure 22.4**). Velocity is not constant. It is an average measured in inches per second. If amplitude is 4 inches and frequency is 10 Hz, this is 40 in./sec. velocity.

Natural frequency, or resonant frequency, is the frequency at which a body vibrates. A tuning fork can be used to illustrate this. Unibody vehicles often have inherent vibration characteristics that cause them to vibrate at a predictable road speed. This was a problem especially with older unibody vehicles. More recently, engineers have designed vehicles so that their vibration range is outside of the normal driving range.

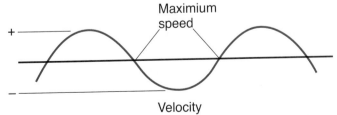

Figure 22.4 Velocity is a combination of amplitude and frequency. Maximum speed (velocity) is reached when the vibrating wave is at half travel.

Engines, suspension systems, and drive shafts are some of the things that vibrate at different frequencies. Stiffer materials tend to have higher natural vibrating frequencies. When you tighten a string on a guitar, it becomes stiffer and the sound from its vibration occurs at a higher frequency.

The severity (amplitude) of a vibration will be greatest at its **point of resonance**. An automotive suspension system has a natural frequency in the range of 10–15 Hz. The suspension's frequency is always the same, but a drive shaft that is out of balance will have a frequency that can become equal to the suspension frequency at a particular speed (**Figure 22.5**). This is the point of resonance, the speed at which the vibration will be greatest. Front and rear suspension systems are designed with frequencies that are different by 3 Hz to avoid a combination of their vibrations.

TYPES OF VIBRATIONS

First order vibration is anything that vibrates only once every revolution.

Second order vibration is two vibrations per revolution and can be caused by a U-joint. Remember, as the drive shaft rotates, the U-joints speed up and slow down twice during each revolution. This vibration is always related to road speed and is usually worse under torque.

A **beat/boom vibration** occurs when one vibration interacts with another. Canceling one of the vibrations can sometimes solve the problem.

■ VIBRATION TEST INSTRUMENTS

Because every vibration has its own frequency, vibrations can be analyzed using test instruments. One type of mechanical vibration analyzer that was invented in the 1940s senses the frequency of vibrations from 10 to 80 Hz. This tool is called a **reed tachometer** (**Figure 22.6**). It is a tachometer because it measures in hertz (cycles per second).

The reed tachometer has two rows of reeds of different lengths. The reeds vibrate at different frequencies, lower or higher as you move from left to right or vice versa. The vibration shown in the illustration is strongest at 16 Hz. The tool should be placed as near the source of the vibration as possible and will read any vibration that can be felt.

A newer and better version of the reed tachometer is the **electronic vibration analyzer (EVA)** (**Figure 22.7**). It has a 20-foot cord and an electronic pickup. The pickup is held in place with putty, Velcro, or a magnet, allowing the car to be driven while checking for vibrations. The EVA measures acceleration of a part. The bigger the reading the worse the vibration is. One side of the gauge shows the frequency of a vibration in rpm or hertz. The other side of the gauge shows the amplitude of the vibration in Gs (forces of gravity).

Like other electronic instruments, the EVA has the capability of freezing an event. This is handy when a vibration occurs quickly and then goes away. A recorded event is called a **snapshot**. A snapshot has ten frames. During playback the technician can freeze and unfreeze the display to isolate the event.

The EVA is also handy for balancing drive shafts. It has a coil of wire that the inductive pickup on an ignition timing light can hook to.

■ VIBRATION AND FREQUENCY TESTING

Many manufacturers have done extensive vibration testing. Earlier front-wheel drive cars had unibodies with resonant frequencies that came into play in the normal driving range. This has been controlled by stiffening the floor pans until their frequency is beyond the normal range of use. Different types of vibrations and their frequencies are shown in **Figure 22.8**.

Some vibrations can be felt and others can be heard. The ones that are heard go into the higher frequencies, above 100 Hz. With vibrations that cannot be heard, work on the problem with the highest amplitude and ignore the rest until the worst are fixed. The following definitions can be used to isolate vibrations that can be felt even if you do not have access to a vibration analyzer:

- **Shake** is a low-frequency vibration of 5–20 Hz. It is commonly the feeling an unbalanced tire would give, felt through the steering wheel or the seat. Shake is also called shimmy, wobble, shudder, or hop. When the vibration is road-speed related, it can be caused by tires, wheels, or brake drums. When it is engine-rpm related, check the engine and accessories.
- **Roughness** is a higher frequency vibration than shake (20–50 Hz). It is usually related to drive line parts and is similar to the feeling one gets when cutting wood with a jigsaw.
- **Buzz** comes in at a frequency of 50–100 Hz, about the frequency of an electric razor. It is often caused by a vibration in the exhaust system or an engine accessory like the air-conditioning compressor. If the buzz is road-speed sensitive, a drive line problem can be the cause.
- **Tingling** is vibration above 100 Hz. It feels like pins and needles in the steering wheel or through the floor. The reed tachometer does not measure above 100 Hz. The EVA can pick up tingling

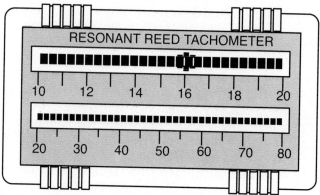

Figure 22.6 A reed tachometer.

Figure 22.7 An electronic vibration analyzer (EVA). *(Courtesy of Kent-Moore Division, SPX Corp.)*

ENGINE SPEEDS AND FREQUENCIES

	First Order Any Engine		Firing Frequency		
	RPM	Hz	4-Cylinder Second Cylinder	6-Cylinder Third Order	8-Cylinder Fourth Order
Shake	500	8.3	16.8	24.9	33.2
	750	12.5	25	37.5	50
	1000	16.5	33.3	49.8	66.6
	1500	25	50	75	100
Roughness	2000	33.3	66.6	99.9	133.2
	2500	41.6	83.2	124.8	166.4
Buzz	3000	50	100	150	200
	3500	55.3	110.6	174.9	233.2
	4000	66.6	132.4	199.8	266.4

Figure 22.8 Different types of vibrations and frequencies.

problems. Again, road speed–related is drivetrain, and engine rpm–related is engine.

The following vibrations can be heard:

- **Boom** is a low-frequency noise in the 20–60 Hz range. It is described as humming, and pressure can be felt in the ears. There can be a feeling of roughness with it. Look at the chart. Boom is normally caused by the drive line, like shake.
- **Moan** or **drone** is a noise that is compared to the hum of a bumble bee. Its range is 60–120 Hz. Powertrain mounts or exhaust system problems are possible causes.
- **Howl** is a noise like the wind across the top of a soda bottle. Its range is 120–300 Hz.
- **Whine** sounds like a turbine or mosquitos. The range is 300–500 Hz. It is usually related to gear noise.

Road Test

During the road test, use a process of elimination. Drive line vibration is slightly different than the feeling that results from tire imbalance. It is not normally identified as a possible problem until wheel balance has been done first and a vibration still remains. The vibration usually occurs at a specific rpm and goes away at other speeds, just like wheel balance.

When test-driving for drive shaft vibration, drive in high gear at the engine rpm where the vibration is worst. Then, shift into a different gear or put the transmission into neutral to see if the vibration changes or goes away entirely. When there is no change in the vibration, the drive line is a possible source of the problem. If the problem does change, one of the components whose rpm changed with the gear change is responsible. These include the engine, the clutch or torque converter, or the transmission.

Be sure you understand the difference between amplitude and frequency when attempting to diagnose vibration problems. When road-testing, note the speed at which a vibration occurs. A vibration should always occur at the same speed or frequency. If the vibration loses intensity during a second road test but occurs at the same speed, the amplitude has changed. Drive the vehicle and note the speed at which the amplitude of the vibration is the worst. Then, repair the suspected problem and retest at the same speed.

Modern test instruments for vibration are so sensitive that you could end up trying to fix a problem that cannot be felt. If you cannot feel the vibration, do not try to fix the "problem."

Testing in the Service Bay

To test for a first order vibration, raise the tires in the air and run the engine in gear to check the drive line. Move the electronic pickup of the vibration analyzer around and try to locate where the energy from the vibration is located. If there is no vibration without a load on the tires, there is no balance problem. The problem is probably related to misalignment of parts. Angles are covered later in this chapter.

While the drive shaft is spinning, check for vibrations at the front and rear of the shaft. A torque-sensitive problem is usually in the differential. This is often caused by a bearing cup that is not seated squarely in its bore.

Drive Shaft Runout

After checking for vibration check the drive shaft for runout. When there is excessive runout in a drive

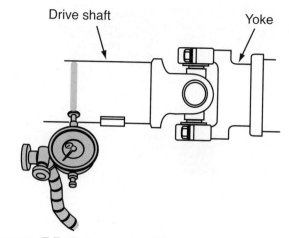

Figure 22.9 Measuring drive shaft runout.

shaft, suspect the drive shaft or the pinion flange. First measure the runout of the drive shaft (**Figure 22.9**). Mark the high spot on the shaft. Then remove it and remount it at a spot 180 degrees away on the flange. If the runout remains the same, the shaft is at fault. Otherwise the flange is the problem.

One part of a two-piece shaft is called a stub shaft (**Figure 22.10**). These are allowed 0.003 inch runout. Excessive runout here is usually the cause of vibrations felt at the center support bearing.

Some manufacturers **match mount** instead of balancing to correct runout. This means that runout of one part and runout of the other are considered during assembly. This is especially prevalent on front-wheel drive axles, which are often marked for reassembly. If a drive shaft was off by 010 inch and a flange was also off by 0.010 inch you could have either a perfectly straight combination or one that is 0.020 inch off.

NOTE: *A part with runout cannot be balanced if it causes parts to bind up once per revolution. Consider that drive shaft parts that are square can be balanced. This is not true with tires because they touch against the road surface, so they must not have runout. You cannot balance a box and expect it to run smoothly on the road.*

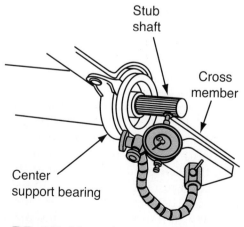

Figure 22.10 Measuring stub shaft runout.

Repairing or Replacing a Drive Shaft

Drive shafts can be aluminum or steel. Most steel drive shafts are shorter than 60 to 65 inches. A 3–3.5-inch diameter passenger car drive shaft should be shorter than 60 to 70 inches. Aluminum drive shafts can be 85 to 90 inches because aluminum has a higher "critical speed." Longer steel shafts use a center support bearing. Do not swap an aluminum drive shaft for a steel one unless the length is less than 60 to 65 inches.

Drive shafts can be straightened by a specialty machine shop. At the shop, the shaft is heated and quenched until it is straight within 0.010 inch. Longer shafts are allowed less runout in the center.

■ OTHER CAUSES OF VIBRATION

Other things besides the drive line can cause vibrations. Engine accessories can also cause vibrations. Test to eliminate outside causes of vibrations in the repair bay. If a vibration is still evident when the car is not moving, how could it be drive line related? Some engines are especially prone to vibrations. Engine accessory brackets have been designed not to react to vibration, or excite, until 400 Hz or higher. This is equivalent to 6,000 engine rpm on a V8, a driving condition that the average driver should not encounter.

The engine, torque converter, and drive shaft change in frequency as rpm changes. A rebuilt torque converter is often the cause of a vibration that is rpm related.

Check to see that a muffler is not vibrating against the frame. All muffler hangers should be in place and undamaged.

■ DRIVE SHAFT BALANCE

Drive shafts are balanced on the ends, not in the middle. The drive shaft can be compared to a very wide, low-profile tire. At the factory, weights are welded onto the ends of the drive shaft when the shaft is balanced. The balance weights should be at least 1 inch down the shaft from the weld so that the integrity of the weld is not disturbed. Power take-offs (PTOs) and shafts that spin at under 1,000 rpm do not need to be balanced. Specialty shops in the aftermarket shorten or lengthen drive shafts and do balancing.

When the drive shaft is out of balance, a technician can correct it using an on-the-car strobe-type wheel balancer. Heavy spots are counterbalanced by installing hose clamps with the screw positioned on the opposite side of the heavy spot (**Figure 22.11**). The pinion nose end of the drive shaft is where most of the first order drive line vibrations come from. Drive shaft flanges are sometimes balanced. This is to compensate for runout created during manufacturing. Weight will either be on the front of a flange or the outside depending on whether the correction was for

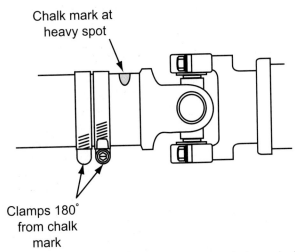

Figure 22.11 A heavy spot is counterbalanced with hose clamps.

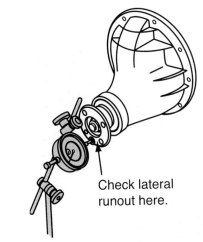

Figure 22.13 Checking flange runout.

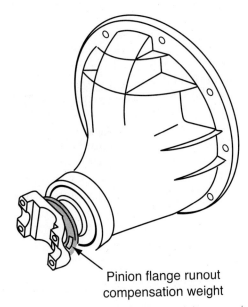

Figure 22.12 When there is weight on a pinion flange, it was put there to correct for runout.

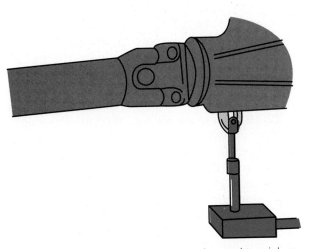

Figure 22.14 A transducer is used to pick up vibrations.

flange runout or differential balance (**Figure 22.12**). Runout of the flange is checked with a dial indicator (**Figure 22.13**). It is limited to 0.006 inch when there is no weight on the flange.

Checking Drive Shaft Balance

One end of the drive shaft is balanced first. After the other end is balanced, the first end is rechecked. A **transducer** is a good tool for testing for drive shaft imbalance. This is part of an on-the-car wheel balancer. It is mounted on the differential case under the pinion gear (**Figure 22.14**).

A transducer is a magnet that is spring loaded on both sides. It moves back and forth in a coil of wire, sending a signal to a meter in response to vibration. The drive shaft is marked anywhere with a piece of chalk. When the drive shaft is spun at the speed where the vibration occurs, the transducer triggers a strobe light. When the light illuminates the mark on the drive shaft, make a mental note of its position. When the shaft is stopped, position the mark at the same place (six o'clock, for instance). At this position, the heavy spot on the drive shaft is on the bottom of the shaft.

Large hose clamps are positioned on the shaft opposite the heavy spot. Then, the shaft is spun again to see whether more clamps are needed to correct the problem. You can move the hose clamp to one side to see if the vibration improves. If it gets worse, try moving it in the other direction. The electronic vibration analyzer has a pickup wire that can be used with an induction timing light that will determine the heavy spot.

Drive shaft balance can also be checked without special instruments. First put four numbers around the circumference of the drive shaft (**Figure 22.15**). Because imbalance can be related to runout, put the #1 mark at a point opposite to the amount of greatest runout. Put the clamp with the weight at the number one position and run the shaft. Check to see if it feels better or worse. Then, move it to the #2 position and try it again. Try all four positions to find out where to put the weight.

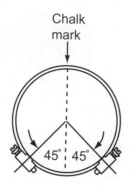

Figure 22.15 Two hose clamps can be separated or brought closer together to change the amount of weight.

When you find the best position, move the hose clamp halfway between the numbers to one side to see if the vibration improves. If it gets worse, try moving it in the other direction. When you find the best position, add another hose clamp and repeat the procedure. The clamps can be separated to fine-tune the amount of weight needed (see Figure 22.15).

Another procedure is to run the engine to turn the shaft while holding a piece of chalk near the shaft. The heavy spot will tend to be marked by the chalk. Place the hose clamp opposite to the chalk mark.

■ CHECKING DRIVE SHAFT ANGLE

To extend U-joint service life, the angle of the transmission output shaft and the front of the differential should be within ½ degree of each other. If not, either the transmission or differential should be shimmed with tapered wheel alignment shims. A ¼-inch shim will result in a change of about ¾ degree. Leaf springs can be shimmed on the rear axle housing or the transmission can be shimmed on the cross member mount. A faster shaft rpm requires a lower angle.

NOTE: *Check the condition of engine and transmission mounts. They are a frequent cause of drive shaft vibration.*

The **working angle** is the difference between the drive shaft angle and the angle of the transmission or differential (**Figure 22.16**). For passenger cars, the working angle should not be more than 4 degrees. There must always be at least a small angle (at least ½ degree) to keep the needle bearings rolling, or brinelling will occur.

The trim height (ride height) measurement (**Figure 22.17**) is important to U-joint angle and also to front-

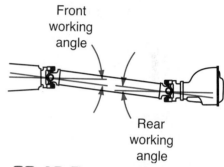

Figure 22.16 The working angle is the difference between the angle of the drive shaft and the angles of the transmission or differential.

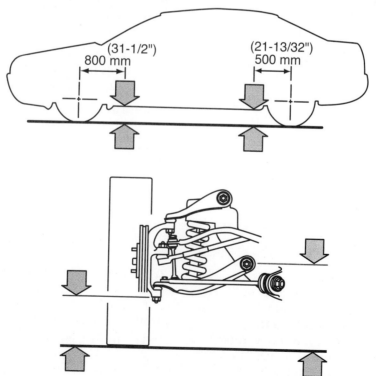

Figure 22.17 Typical ride height measurement. The lower sketch shows the typical measurement locations on an SLA system.

wheel drive tripod joints. Four-wheel drives have a front shaft that is short and has a sharp angle. Riser kits that make room for tall tires cause the angle to be even greater. Slip yokes and splines tend to wear out in these drive shafts.

Launch shudder is a common second order vibration complaint in raised trucks with a high drive shaft angle. It occurs on acceleration from a standing start to about 25 mph before disappearing.

Several methods can be used to measure the drive shaft angle. These include a digital measuring tool (**Figure 22.18**), a protractor, or a magnetically attached inclinometer. The angle is measured in both the front and the rear.

Suggested maximum drive shaft angles are as follows:

5,000 rpm—3¼ degrees
4,500 rpm—3⅔ degrees
4,000 rpm—4¼ degrees
3,500 rpm—5 degrees
3,000 rpm—5⅚ degrees

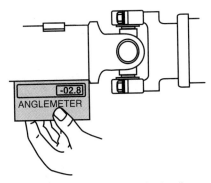

Figure 22.18 A digital drive shaft alignment tool.

2,500 rpm—7 degrees
2,000 rpm—8⅔ degrees
1,500 rpm—11½ degrees

You can see from the chart that a four-wheel drive truck with high drive shaft angles is better operated at low shaft rpm. If the transmission and differential angles are not the same, a double Cardan will be needed.

REVIEW QUESTIONS

1. The number of cycles in 1 second is measured in _____.
2. A _____ fork can be used to illustrate resonant frequency.
3. A vibration that interacts with another is called _____ a boom vibration.
4. What is the name for the vibration that is about the same frequency as an electric razor?
5. When road-testing, note the _____ at which a vibration occurs.
6. A differential vibration is often caused by a _____ cup that is not seated squarely in its bore.
7. Excessive runout of the stub shaft is usually the cause of vibrations felt at the _____ support bearing.
8. When the runout of one part and the runout of the other are considered during assembly, this is called _____ mounting.
9. The angle of the transmission output shaft and the front of the differential should be within _____ degrees of each other.
10. The difference between the drive shaft angle and the angle of the transmission or differential is called the _____ angle.

PRACTICE TEST QUESTIONS

1. Technician A says that front- and rear-suspension systems are designed to have different resonant frequencies to avoid wheel shimmy. Technician B says that a vibration that occurs twice every drive shaft revolution can be related to a universal joint. Who is right?

   A. Technician A   B. Technician B
   C. Both A and B   D. Neither A nor B

2. Technician A says that an electronic vibration analyzer can measure acceleration of a part. Technician B says that an electronic vibration analyzer has a memory function. Who is right?

   A. Technician A   B. Technician B
   C. Both A and B   D. Neither A nor B

3. Technician A says that an electronic vibration analyzer can be used with a timing light to

balance a drive shaft. Technician B says that vibrations above 100 Hz are usually felt rather than heard. Who is right?
   A. Technician A   B. Technician B
   C. Both A and B   D. Neither A nor B

4. Technician A says that a vibration should always occur at the same speed or frequency. Technician B says that if you cannot feel or hear the vibration, do not try to fix the problem. Who is right?
   A. Technician A   B. Technician B
   C. Both A and B   D. Neither A nor B

5. Technician A says that excessive runout in a drive shaft could be because of the pinion flange. Technician B says that excessive runout in a drive shaft could be because of a bent drive shaft. Who is right?
   A. Technician A   B. Technician B
   C. Both A and B   D. Neither A nor B

6. Technician A says that a square drive shaft could theoretically be balanced. Technician B says that a square tire could theoretically be balanced. Who is right?
   A. Technician A   B. Technician B
   C. Both A and B   D. Neither A nor B

7. Technician A says that a rebuilt torque converter is often the cause of drive line vibrations. Technician B says that drive shafts are balanced in the middle. Who is right?
   A. Technician A   B. Technician B
   C. Both A and B   D. Neither A nor B

8. Technician A says that heavy spots on a drive shaft can be counterbalanced by installing hose clamps with the screw positioned even with the heavy spot. Technician B says that the transmission end of the drive shaft is where most of the first order drive line vibrations come from. Who is right?
   A. Technician A   B. Technician B
   C. Both A and B   D. Neither A nor B

9. Technician A says that four-wheel drives with tall tires are more likely to have vibration problems from the drive shaft. Technician B says that shuddering in a raised truck will occur at speeds above 25 mph. Who is right?
   A. Technician A   B. Technician B
   C. Both A and B   D. Neither A nor B

10. Technician A says that the ride height measurement does not affect universal joint angle. Technician B says that shafts that spin at under 1,000 rpm do not need to be balanced. Who is right?
    A. Technician A   B. Technician B
    C. Both A and B   D. Neither A nor B

APPENDIX A

Practice ASE Certification Exam for Brakes (A5)

Use your intuition and common sense when answering ASE test questions. If you have studied and know the basics but a test question does not seem right, mark it wrong. Depending on the ASE test you take, you can miss 3 or 4 out of every 10 questions and still pass.

1. A brake pedal pulsates during braking. Technician A says this could be caused by a loose or excessively worn wheel bearing. Technician B says that a pulsating pedal could be caused by an out-of-round brake drum. Who is right?

    A. Technician A only   B. Technician B only
    C. Both A and B   D. Neither A nor B

2. Technician A says that a spongy pedal can be due to air in the hydraulic system. Technician B says that a spongy pedal can be due to a faulty master cylinder reservoir cover. Who is right?

    A. Technician A only   B. Technician B only
    C. Both A and B   D. Neither A nor B

3. A car skids too easily when stopping on wet or icy pavement. Technician A says this could be the result of a bad metering valve. Technician B says this could be caused by loose disc brake caliper mountings. Who is right?

    A. Technician A only   B. Technician B only
    C. Both A and B   D. Neither A nor B

4. A car has brakes that "drag" after the pedal is released. Technician A says this could be the result of an incorrect stoplight switch adjustment. Technician B says this could be the result of an incorrectly adjusted emergency brake. Who is right?

    A. Technician A only   B. Technician B only
    C. Both A and B   D. Neither A nor B

5. Two technicians are discussing how they check for a plugged vent (compensating) port. Technician A says to try to put a small wire or drill through it. Technician B says to have a helper depress the brake pedal while you look for fluid flow. Who is right?

    A. Technician A only   B. Technician B only
    C. Both A and B   D. Neither A nor B

6. A car has excessive pedal travel after a recent brake job. Technician A says this could be caused by self-adjusters that are installed on the wrong side of the vehicle. Technician B says this could be caused by an incorrectly adjusted parking brake. Who is right?

    A. Technician A only   B. Technician B only
    C. Both A and B   D. Neither A nor B

7. Technician A says a hole in the power brake booster diaphragm can cause engine idle to change when the brakes are applied. Technician B says that a leaking primary cup in the master cylinder will cause the pedal to slowly drop toward the floor. Who is right?

    A. Technician A only   B. Technician B only
    C. Both A and B   D. Neither A nor B

8. Technician A says the master cylinder shown in the figure operates half of the brake system independently from the other if one half fails. Technician B says a brake fluid leak into the driver's compartment is caused by a leaking

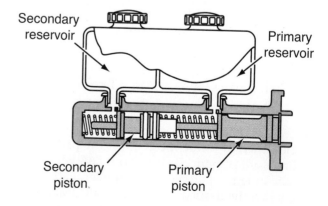

secondary cup on the primary piston. Who is right?

A. Technician A only B. Technician B only
C. Both A and B D. Neither A nor B

9. If a drum brake has a leaking wheel cylinder you should:

A. Repair the cylinder and replace the shoes on both sides of the car
B. Repair the cylinder and clean the shoes with alcohol and blow them dry
C. Do a complete brake job on all four wheels
D. Repair only the leaking wheel cylinder

10. A repeatedly worn outer left front wheel bearing could be caused by a/an:

A. Bent wheel
B. Bad grease retainer
C. Bent spindle
D. Incorrect caster adjustment

11. A vehicle creeps at a stop sign with the brakes lightly applied. Technician A says that the disc brake pads could be worn. Technician B says the proportioning valve could be defective. Who is right?

A. Technician A only B. Technician B only
C. Both A and B D. Neither A nor B

12. Two technicians are discussing vacuum power booster operation. Technician A says a steady hissing sound from under the dash when the brakes are applied with the engine idling is normal. Technician B says to test available vacuum by installing a vacuum gauge between the check valve and the brake booster. Who is right?

A. Technician A only B. Technician B only
C. Both A and B D. Neither A nor B

13. A driver complains that the brake warning light is on and there is less braking. Technician A says this could be caused by worn disc brake pads. Technician B says this could be due to a leaking brake line. Who is right?

A. Technician A only B. Technician B only
C. Both A and B D. Neither A nor B

14. Technician A says changing the rear drum brake clearance adjustment can affect the parking brake adjustment. Technician B says when adjusting the parking brake, the lever should be applied the desired amount and all clearance removed from both cables at the equalizer bar. Who is right?

A. Technician A only B. Technician B only
C. Both A and B D. Neither A nor B

15. The vacuum line leading into a power brake unit has collapsed while the engine is idling. Technician A says this could be because the wrong type of hose was installed. Technician B says the cause could be a defective check valve. Who is right?

A. Technician A only B. Technician B only
C. Both A and B D. Neither A nor B

16. Technician A says the technician in the picture is checking rotor runout. Technician B says the technician in the picture is checking thickness variations. Who is right?

A. Technician A only B. Technician B only
C. Both A and B D. Neither A nor B

17. A floating caliper disc brake has only one pad that is worn. Technician A says this is probably due to incorrect adjustment of the caliper. Technician B says this is due to a stuck sliding caliper mount. Who is right?

A. Technician A only B. Technician B only
C. Both A and B D. Neither A nor B

18. Rear brake pads are being installed on a vehicle with an integral parking brake. Technician A says a typical caliper piston must be rotated to move it deeper into its bore. Technician B says the function of the parking brake must be verified after the pads are installed. Who is right?

A. Technician A only B. Technician B only
C. Both A and B D. Neither A nor B

19. A customer complains of a clunk when stopping the car. Technician A says this could be due to a weak brake shoe return spring. Technician B says this could be due to a worn control arm bushing. Who is right?

A. Technician A only   B. Technician B only
C. Both A and B   D. Neither A nor B

20. With the ignition switch on, the brake light in the illustration is illuminated continuously. Which of the following could be the cause?

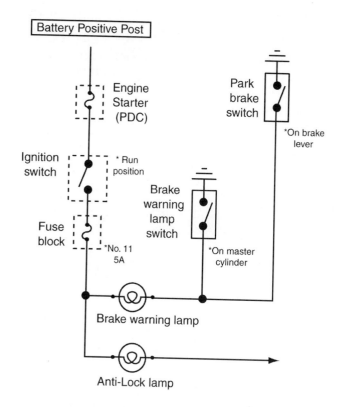

A. A parking brake switch that is continually closed (on)
B. Low fluid in one half of the master cylinder
C. A grounded wire to the parking brake switch
D. Any of the above

21. All of the following could cause a vehicle to pull to one side when stopping *except*:

A. A restricted brake hose
B. A stuck brake caliper piston
C. A loose caliper mount
D. Air in the brake line

22. Disc brake pads are being replaced. Technician A says fluid should be removed through the caliper bleed screw. Technician B says some fluid might need to be removed from the master cylinder. Who is right?

A. Technician A only   B. Technician B only
C. Both A and B   D. Neither A nor B

23. Technician A says that the tool shown in the figure can be used to check brake drum

bellmouth. Technician B says that the tool shown above can be used to check brake drum out-of-round. Who is right?

A. Technician A only   B. Technician B only
C. Both A and B   D. Neither A nor B

24. A technician is using compressed air to remove a piston from a floating caliper. Technician A says to protect the piston with a block of wood or a folded shop rag. Technician B says to drain out fluid through the bleed screw first so that you do not get fluid in your face when blowing out the piston. Who is right?

A. Technician A only   B. Technician B only
C. Both A and B   D. Neither A nor B

25. Technician A says that installing calipers on the wrong side of the vehicle can make it impossible to bleed the brakes. Technician B says drum-to-lining clearance must be adjusted correctly before adjusting the parking brake. Who is right?

A. Technician A only   B. Technician B only
C. Both A and B   D. Neither A nor B

26. A car with front disc brakes has had a recent brake job during which the service technician used an impact wrench to tighten the lug nuts. Now the car has pedal pulsation when stopping. Technician A says the car could have a warped rotor. Technician B says the cause could be a ruined wheel rim. Who is right?

A. Technician A only   B. Technician B only
C. Both A and B   D. Neither A nor B

27. A car owner complains of reduced stopping ability and a very hard brake pedal. Technician A says this could be due to stuck caliper pistons. Technician B says this could be caused by a bad power brake check valve. Who is right?

A. Technician A only   B. Technician B only
C. Both A and B   D. Neither A nor B

28. The part shown in the figure is the:

A. Proportioning valve
B. Metering valve
C. Pressure differential valve
D. Combination valve

29. The amber ABS light comes on when the engine is started, but then goes out. Technician A says an intermittent ABS problem could be the cause. Technician B says to locate an appropriate scan tool and check for diagnostic trouble codes. Who is right?

A. Technician A only B. Technician B only
C. Both A and B D. Neither A nor B

30. A technician is bleeding brakes using a pressure bleeder with 15 psi. No fluid comes out the front caliper bleed screws. Technician A says there is air in the brake line. Technician B says the metering valve pin must be activated. Who is right?

A. Technician A only B. Technician B only
C. Both A and B D. Neither A nor B

31. Technician A says a moving vehicle will send a varying AC voltage to the ABS controller. Technician B says the wheel speed sensor completes a circuit to ground when the wheel locks up. Who is right?

A. Technician A only B. Technician B only
C. Both A and B D. Neither A nor B

32. A RWD front wheel bearing has frozen to the bearing race. Technician A says hub replacement will likely be needed. Technician B says spindle replacement will likely be needed. Who is right?

A. Technician A only B. Technician B only
C. Both A and B D. Neither A nor B

33. A DTC retrieved by a scan tool lists a problem with the right rear wheel speed sensor. Technician A says to replace the wheel speed sensor, clear the code and road test the vehicle. Technician B says to use a DSO or a DMM to test the sensor. Who is right?

A. Technician A only B. Technician B only
C. Both A and B D. Neither A nor B

34. Technician A says rotor runout can be caused by using an incorrect lug nut torque sequence. Technician B says an impact wrench can cause rotor runout. Who is right?

A. Technician A only B. Technician B only
C. Both A and B D. Neither A nor B

35. Technician A says the picture shows how to bend disc brake pad tabs to limit the up-and-down movement of the pads in the caliper. Technician B says this procedure is done to prevent brake noise. Who is right?

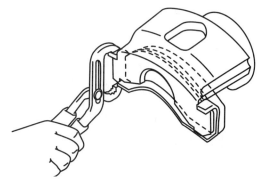

A. Technician A only B. Technician B only
C. Both A and B D. Neither A nor B

36. A car pulls to one side when braking. Technician A says this is likely due to a tire with low air pressure. Technician B says this could be caused by a stuck or lazy disc brake piston. Who is right?

A. Technician A only B. Technician B only
C. Both A and B D. Neither A nor B

37. What is the purpose of the strut bar in the brake assembly shown in the figure?

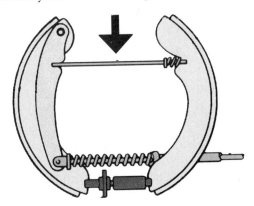

A. To equalize the braking force between the shoes during normal braking
B. To center the shoes after each brake application

C. To force the shoes into the drum when the parking brakes are applied
D. To reduce the distance between the shoes and the drum

38. With the brake pedal applied, it moves down slightly when the engine is started. Technician A says this could be due to a leaking power brake booster diaphragm. Technician B says that the cause could be a power brake booster check valve that is stuck in the closed position. Who is right?

A. Technician A   B. Technician B
C. Both A and B   D. Neither A nor B

39. Technician A says the tool shown in the figure is best for brake drum oversize measurement.

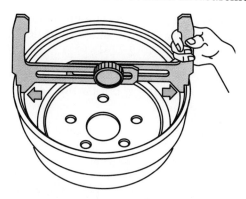

Technician B says the tool is used to adjust brake shoes. Who is right?

A. Technician A   B. Technician B
C. Both A and B   D. Neither A nor B

40. The wiring to an ABS wheel speed sensor is damaged. Technician A says to cut out the damaged section of wire and replace it using new connectors. Technician B says the entire harness section must be replaced. Who is right?

A. Technician A only   B. Technician B only
C. Both A and B   D. Neither A nor B

PRACTICE EXAM ANSWERS

1. The correct answer is C. A loose wheel bearing or an out-of-round brake drum can cause the brake pedal to pulsate.

2. The correct answer is C. A soft pedal is the result of air in the brake fluid. A faulty reservoir vent can allow air to be drawn in through the back of the master cylinder, causing a spongy brake pedal.

3. The correct answer is C. ABS tries to prevent skidding on wet pavement, but things that are out of the ordinary can also result in a skid. A defective metering valve can allow the front brakes to apply too soon, resulting in a skid. Loose parts that allow the geometric braking angles to change can also result in a skid.

4. The correct answer is C. Either of these conditions could cause the brakes to remain in the applied state.

5. The correct answer is B. Putting something through the vent port could cause a metal burr to cut the rubber cup seal. Watching for fluid movement while someone depresses the brake pedal is the correct procedure.

6. The correct answer is A. Self-adjusters installed on the wrong side of the vehicle will turn in the wrong direction and loosen the brake adjustment. An incorrectly adjusted parking brake can only cause a high pedal.

7. The correct answer is C. A leak in the diaphragm will allow air from the passenger compartment to leak into the intake manifold when the brakes are applied. A leaking primary cup in the master cylinder can allow the pedal to slip toward the floor.

8. The correct answer is C. Both technicians are correct.

9. The correct answer is A. Brake shoes that have been contaminated are replaced after the repair is made.

10. The correct answer is C. A bent spindle can put an extra load on the outer wheel bearing.

11. The correct answer is B. Worn pads will produce noise. A defective proportioning valve that blocks fluid flow to the rear brakes could allow the car to creep. Front disc brakes require more

pedal pressure to make the pads grip the smooth rotor surface.

12. The correct answer is D. A steady hiss from the back of the booster indicates a leaking diaphragm. Vacuum is tested between the manifold and booster, not between the check valve and booster. Most check valves are attached to the booster with a grommet, making testing here impossible.

13. The correct answer is B. The warning light comes on when there is a loss of pressure in one half of the hydraulic system. Some systems have brake warning lights that illuminate when the lining is worn to metal, but the most sensible answer to a complaint associated with worn disc pads would be noise.

14. The correct answer is C. Both technicians are correct.

15. The correct answer is A. A reinforced hose is necessary for the power brake vacuum supply. A defective check valve would not affect the hose, which should not collapse under any circumstances.

16. The correct answer is A. Checking for thickness variations would be done with a micrometer.

17. The correct answer is B. Calipers are not adjustable. A stuck caliper mount would allow the piston to push only one pad against the rotor.

18. The correct answer is C. Both technicians are correct.

19. The correct answer is C. A stretched brake spring can allow the shoe to slam against the drum. A worn bushing can allow a control arm to shift against its shaft during a stop.

20. The correct answer is D. All of the answers are possible causes.

21. The correct answer is D.

22. The correct answer is C. Fluid is removed through a caliper bleed screw to prevent forcing it back through the ABS system.

23. The correct answer is B. The tool shown has ledges that would prevent it from taking a bellmouth comparison reading from deep in the drum.

24. The correct answer is C. Both technicians are correct.

25. The correct answer is C. A caliper on the wrong side will have its bleed screw on the bottom, instead of the top, where trapped air will collect. Changing lining-to-drum clearance will change the parking brake adjustment.

26. The correct answer is A. A warped rotor from incorrect wheel tightening can cause pedal pulsation. A damaged rim would be unlikely to cause pedal pulsation only during braking.

27. The correct answer is C. Both technicians could be correct.

28. The correct answer is D, the combination valve.

29. The correct answer is D. Neither are correct. Technician A describes a normal situation. Checking for codes would not be necessary.

30. The correct answer is B. The problem is that the metering valve will not allow fluid flow between approximately 10–125 psi.

31. The correct answer is A. The sensors create a varying AC voltage, but do not complete a circuit to ground.

32. The correct answer is C. A frozen bearing can spin in the hub and on the spindle shaft, causing likely damage to either or both.

33. The correct answer is B. Tests must be performed before replacing parts.

34. The correct answer is C. Wheel lug nuts are tightened by hand, using a torque wrench as a final check.

35. The correct answer is C. Both technicians are correct. Noise is vibration, which is limited if the pads cannot move.

36. The correct answer is B. A low tire could cause a pull, but it would be noticeable all the time. A stuck disc brake piston would cause a pull and is the correct answer.

37. The correct answer is C. The strut is part of the parking brake actuating mechanism.

38. The correct answer is D. This is a normal condition.

39. The correct answer is B. This tool is used to adjust brake shoes. A drum mike is best for drum measurement.

40. The correct answer is B. ABS wiring is sensitive to radio interference and should not be repaired using connectors.

APPENDIX B

Practice ASE Certification Exam for Suspension and Steering (A4)

Use your intuition and common sense when answering ASE test questions. If you have studied and know the basics but a test question does not seem right, mark it wrong. Depending on the ASE test you take, you can miss 3 or 4 out of every 10 questions and still pass.

1. Which of the following could cause excessive steering effort in both directions on a vehicle with manual rack-and-pinion steering?

   A. Worn lower ball joints
   B. Rack bearing adjustment that is too tight
   C. High point adjustment that is too tight
   D. Worn upper ball joints
   E. All of the above

2. A car has worn ball joints. Technician A says the ball joints should be replaced before adjusting caster or camber. Technician B says alignment will not change when ball joints are replaced. Who is right?

   A. Technician A only   B. Technician B only
   C. Both A and B   D. Neither A nor B

3. Technician A says installing rear spring shackles that are 3 inches longer will change front wheel caster. Technician B says longer spring shackles will change toe-out on turns. Who is right?

   A. Technician A only   B. Technician B only
   C. Both A and B   D. Neither A nor B

4. A technician is adjusting the part shown in the figure. Technician A says the toe setting is being adjusted at the finish of a wheel alignment job. Technician B says this adjustment is used to center the steering wheel. Who is right?

   A. Technician A only   B. Technician B only
   C. Both A and B   D. Neither A nor B

5. A ride height check shows a sag in the right rear of a car. Technician A says this can be caused by an improperly seated left front coil spring. Technician B says this is caused by a missing sway bar link on the left front. Who is right?

   A. Technician A only   B. Technician B only
   C. Both A and B   D. Neither A nor B

6. The alignment settings shown here would result in which of these conditions?

	Readings		Specs
	Left	Right	Left or Right
Camber	+½° or +30 min.	−1½° or −1° 30 min.	0 to +½° or 0 to +30 min.
Caster	0°	0°	0° to +1°
Toe-in	1/16″ or 16 mm		1/16″ to 1/16″ or .16mm to .48 mm

   A. Left tire wear on inside; car does not pull to either side
   B. Right tire wear on inside; car pulls to left
   C. Right tire wear on outside; car pulls to left
   D. Right tire wear on outside, left tire wear on inside; car pulls to left

7. A car has excess body roll on turns. Technician A says this is caused by weak springs. Technician B says this is caused by bad shock absorbers. Who is right?

   A. Technician A only   B. Technician B only
   C. Both A and B   D. Neither A nor B

8. A car has hard steering and the wheel moves quickly back to center position. This can be caused by:

   A. Not enough caster
   B. Not enough camber
   C. Too much caster
   D. Too much camber
   E. None of the above

497

9. Technician A says bump steer can result when a steering gear needs to be adjusted. Technician B says bump steer can result from a bent idler arm. Who is right?

   A. Technician A only
   B. Technician B only
   C. Both A and B
   D. Neither A nor B

10. A car wanders and weaves. This is most likely caused by:

    A. Excessive (+) caster
    B. Excessive (−) caster
    C. Excessive (+) camber
    D. Excessive (−) camber

11. Referring to the figure, to increase caster without changing camber:

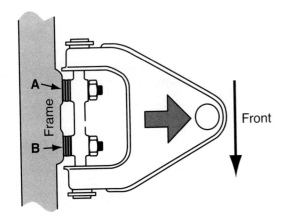

    A. Remove a shim from A and install it at B
    B. Remove a shim from B and install it at A
    C. Install a shim at A
    D. Install a shim at B

12. A power steering–equipped car is hard to steer in both directions. Technician A says this could be caused by a stuck flow-control valve. Technician B says this could be caused by a bent torsion bar in the steering gear. Who is right?

    A. Technician A only
    B. Technician B only
    C. Both A and B
    D. Neither A nor B

13. Technician A says the purpose of toe-in on a RWD card is to compensate for toe-out caused by rolling resistance of the tires. Technician B says that toe-in is needed to compensate for changes in camber. Who is right?

    A. Technician A only
    B. Technician B only
    C. Both A and B
    D. Neither A nor B

14. When checking turning radius, a technician finds a reading of 20 degrees on both wheels. Technician A says this could be caused by a bent steering arm. Technician B says that the car tires will squeal during turns. Who is right?

    A. Technician A only
    B. Technician B only
    C. Both A and B
    D. Neither A nor B

15. Which of the following, if any, helps to return the car to straight ahead?

    A. Camber
    B. Toe-in
    C. SAI
    D. None of the above

16. The front tire shown in the figure is most likely worn as a result of:

    A. Toe-in
    B. Toe-out
    C. Negative camber
    D. Positive camber
    E. None of the above

17. A manual steering car turns hard through the straight-ahead position. This is most likely caused by:

    A. Incorrectly adjusted worm bearings
    B. An incorrectly adjusted sector shaft
    C. Lack of gear lubricant
    D. None of the above

18. A car is hard to steer. Technician A says this could be because of a bent tie-rod. Technician B says that this could be caused by underinflated tires. Who is right?

    A. Technician A only
    B. Technician B only
    C. Both A and B
    D. Neither A nor B

19. A driver feels chattering in the steering wheel during cornering with a car that has a MacPherson strut suspension. With the vehicle on the shop floor, the technician can feel the left front spring bind and release as an assistant

turns the steering wheel. Technician A says this could be the result of a worn lower ball joint. Technician B says this could be due to a defective upper strut mount. Who is right?

A. Technician A only B. Technician B only
C. Both A and B D. Neither A nor B

20. Technician A says that the figure shows a pitman arm being removed. Technician B says that the tool shown is used prior to replacing a sector shaft seal. Who is right?

A. Technician A only B. Technician B only
C. Both A and B D. Neither A nor B

21. A driver hears and feels a clunk when the brakes are applied. Technician A says that upper control arm bushings could be worn. Technician B says that a ball joint could be worn. Who is right?

A. Technician A only B. Technician B only
C. Both A and B D. Neither A nor B

22. A car has a front wheel shimmy. Technician A says the car could have too much positive caster. Technician B says the tires might need to be dynamically balanced. Who is right?

A. Technician A only B. Technician B only
C. Both A and B D. Neither A nor B

23. An integral power steering gear has been overhauled. Technician A says to adjust it before installing it on the car. Technician B says that adjustment can be done just as easily on the car. Who is right?

A. Technician A only B. Technician B only
C. Both A and B D. Neither A nor B

24. Technician A says the control arms shown in the figure regulate the fore and aft positions of the axle housing. Technician B says that the control arms are used to transfer braking and acceleration torque to the frame. Who is right?

A. Technician A only B. Technician B only
C. Both A and B D. Neither A nor B

25. Which of these statements is true about rod X in the suspension shown in the figure?

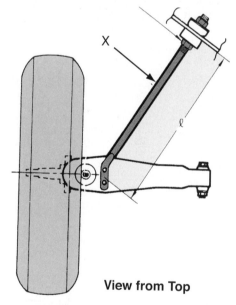

View from Top

A. Decreasing the rod length changes steering axis inclination
B. The length of the rod will not affect caster or camber
C. The rod acts as the stabilizer bar
D. Increasing the rod length changes caster

26. If SAI checks out bad, what part most likely will require replacement?

A. Upper control arm B. Ball joint
C. Spindle D. Lower control arm
E. None of the above

27. A tire is being statically balanced and 4 ounces of weight is being used to correct it. Where should the weight be put?

   A. On the outside, 180 degrees from the heavy spot
   B. On the inside, 180 degrees from the heavy spot
   C. 2 ounces each side, 180 degrees from the heavy spot
   D. 2 ounces each side, 90 degrees from the heavy spot

28. One side of a MacPherson strut front end has a bad camber setting. To adjust it:

   A. Bend it to the correct setting
   B. Insert a shim under the ball joint
   C. Replace misaligned parts
   D. None of the above

29. A vehicle with a Macpherson strut suspension has a difference of 2 degrees in camber readings from side to side. Which of the following is the most likely cause?

   A. Sagged springs
   B. Worn ball joints
   C. An off-center cradle
   D. Worn rack bushings

30. A rigid rear axle is offset in the direction shown in the figure. Which of the following is the most likely result?

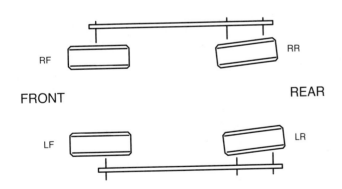

   A. Steering wheel will pull to the left under acceleration
   B. Steering wheel will not return from left or right turns
   C. Steering wheel will be off-center when driving straight
   D. Vehicle will wander when driving on a straight highway

31. A car leans to the right. Technician A says a rear coil spring could be sagged. Technician B says a right front torsion bar could be broken. Who is right?

   A. Technician A only   B. Technician B only
   C. Both A and B   D. Neither A nor B

32. After balancing the front wheels, a car vibrates when driven above 40 mph. The next thing to do is:

   A. Check driveshaft alignment
   B. Balance the rear wheels
   C. Check rear wheel bearings
   D. Align the car

33. A car has excessive negative camber that cannot be adjusted to specification. Technician A says the car likely needs new springs. Technician B says the front cross member may be sagged. Who is right?

   A. Technician A only   B. Technician B only
   C. Both A and B   D. Neither A nor B

34. A new rack-and-pinion steering gear is being installed on a car. Technician A says tapered holes in the steering linkage should be inspected. Technician B says the height of the rack must be equal from side to side. Who is right?

   A. Technician A only   B. Technician B only
   C. Both A and B   D. Neither A nor B

35. A vehicle with rack-and-pinion steering has excessive play in the steering gear. The first thing to check is the:

   A. Inner tie-rod ends
   B. Outer tie-rod ends
   C. Idler arm
   D. Pitman arm

36. A ball joint check is being performed on the front end shown in the figure. Technician A says the ball joint should be unloaded by jacking up the car under the lower control arm. Technician B says to jack the car at the frame. Who is right?

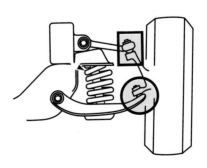

A. Technician A only   B. Technician B only
C. Both A and B   D. Neither A nor B

37. A wheel and tire assembly has too much radial runout. Technician A says to remove the wheel and put it back on the car in a different location. Technician B says to change the position of the tire on the rim. Who is right?

A. Technician A only   B. Technician B only
C. Both A and B   D. Neither A nor B

38. Which of the tires shown here resulted from worn parts?

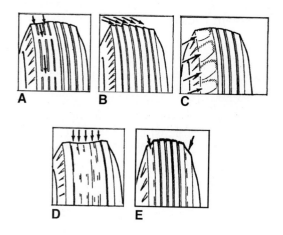

A. Tire A
B. Tire B
C. Tire C
D. Tire D
E. Tire E

39. A truck with rear leaf springs pulls to the left. This could be caused by all of the following except:

A. More positive caster on the right front wheel than the left
B. A left tire with low pressure
C. A broken center bolt in the right rear spring
D. Excessive front-wheel toe-in on the right front wheel

40. Technician A says #2 is used to adjust sector gear end play. Technician B says that adjustment of #1 is performed before adjusting sector gear lash. Who is right?

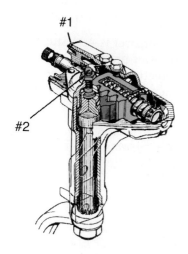

A. Technician A only   B. Technician B only
C. Both A and B   D. Neither A nor B

PRACTICE EXAM ANSWERS

1. The correct answer is B. Ball joints that are worn can cause looseness. A high point adjustment is a procedure for recirculating ball nut steering, not rack and pinion.

2. The correct answer is A. Loose or worn ball joints can change alignment angles.

3. The correct answer is A. Raising the rear of the vehicle will move caster to negative. Longer shackles will not change toe-out on turns. This angle is dictated by the angle of the steering arms attached to the steering knuckle.

4. The correct answer is C. Both technicians are correct. The tie-rod ends are connected to a threaded turnbuckle that lengthens or shortens, depending on which direction it is turned. Toe is adjusted after caster and camber. The steering wheel is brought into the center position during this adjustment.

5. The correct answer is A. An incorrectly seated coil spring will put more pressure on the opposite corner. The sway bar will not affect ride height, but will cause noise when one side's height deflects.

6. The correct answer is B. The left setting is within spec but the right is far off, leaning in at the top. This causes a pull to the left and wear on the inside of the tire.

7. The correct answer is C. Weak springs can cause the car to lean excessively. The shock cannot control the extra weight. Bad shocks can allow the weight of the vehicle to shift too quickly.

8. The correct answer is C. Too much caster can make steering difficult. It can also keep the vehicle steering in one direction.

9. The correct answer is B. Bump steer results from toe change. A bent idler arm or pitman arm on a parallelogram steering can cause this. A steering gear out of adjustment will not be a cause.

10. The correct answer is B. The lead point is behind the tire contact patch. This can cause the car to wander.

11. The correct answer is A. This will move the upper ball joint toward the rear of the car.

12. The correct answer is A. A stuck flow control valve can lower pressure, causing hard steering in both directions. A bent steering gear torsion bar can cause hard steering or self-steering in one direction only.

13. The correct answer is A. On a RWD car, tires tend to bow outward when rolling due to the resistance of the tires.

14. The correct answer is C. Both technicians are correct.

15. The correct answer is C. Steering axis inclination, the angle of the spindle, causes the vehicle to lift during a turn. The result is returnability.

16. The correct answer is C. Negative camber causes this tire wear.

17. The correct answer is B. The high point adjustment is one where the sector shaft is moved up and down. If the high point adjustment is too tight, the steering gear will bind as it is turned through center position. If it is too loose, steering will have excessive play.

18. The correct answer is C. Both technicians are correct.

19. The correct answer is B. A worn ball joint can cause looseness. A defective upper strut mount can cause binding and releasing.

20. The correct answer is C. Both technicians are correct.

21. The correct answer is C. Loose parts can shift during braking, causing a clunk.

22. The correct answer is C. Positive caster can cause a shimmy. Tires with weight off to one side of the tread can also cause a shimmy.

23. The correct answer is A. An integral power steering gear is adjusted before installing it on the vehicle.

24. The correct answer is C. Both technicians are correct.

25. The correct answer is D. Changing the length of the strut rod changes caster. The strut rod is not a stabilizer bar, which is connected to both lower control arms to reduce sway.

26. The correct answer is C. A spindle bent from an accident will affect SAI.

27. The correct answer is C. During a static balance, weight is split equally on both sides of the tire to prevent causing a dynamic imbalance.

28. The only correct answer is C. Aftermarket kits are available to adjust camber on some vehicles. Always check SAI to ensure that the spindle is not bent.

29. The correct answer is C. An incorrectly positioned cradle will tilt the struts off to one side, resulting in more negative camber on one side and more positive camber on the other.

30. The correct answer is C. Correcting the thrust angle aligns the steering wheel.

31. The correct answer is C. Both problems could cause a car to lean to that direction.

32. The correct answer is B. Driveshaft balance would also be an answer, but this is a relatively rare problem and would be done only after balancing all four wheels.

33. The correct answer is C. Negative camber can result from springs or a cross member that have sagged.

34. The correct answer is C. Tapered steering linkage connections should always be inspected when disassembled. The height of the rack is important because if incorrect, toe change and bump steer can result.

35. The correct answer is B. The first thing to check is the condition of the outer tie-rod ends because this is easiest. If the outer tie-rods pass inspection, the inner tie-rods can be felt for movement through the rubber boots at the ends of the rack.

36. The correct answer is A. This will unload the bottom ball joint, which is the load carrier.

37. The correct answer is B. Moving the position of the tire on the rim can correct this problem, while moving the wheel should have no effect on runout.

38. The correct answer is C. Worn parts can cause scalloped tire wear. A feathered edge on the tread would result from bent steering linkage. The other tires are worn as a result of incorrect tire pressure or wheel alignment.

39. The correct answer is D. Toe-in will not cause a pull because it evens out between both front tires.

40. The correct answer is C. Both technicians are correct. The worm bearing adjustment is performed prior to adjusting the sector shaft high point.

APPENDIX C

NATEF Task Correlation Grid for Brakes

Task	Chapter and Pages
A. **General Brake Systems Diagnosis**	
A.1. Identify and interpret brake system concern; determine necessary action.	Chapter 1, p. 3 Chapter 3, p. 61 Chapter 5, p. 94 Chapter 7, p. 142 Chapter 8, p. 164 Chapter 9, p. 185 Chapter 10, p. 213
A.2. Research applicable vehicle and service information, such as brake system operation, vehicle service history, service precautions, and technical service bulletins.	Chapter 1, p. 9
A.3. Locate and interpret vehicle and major component identification numbers (VIN, vehicle certification labels, calibration decals).	Chapter 1, p. 10
B. **Hydraulic System Diagnosis and Repair**	
B.1. Diagnose pressure concerns in the brake system using hydraulic principles (Pascal's law).	Chapter 4, p. 73 Chapter 5, p. 110
B.2. Measure brake pedal height; determine necessary action.	Chapter 5, p. 95
B.3. Check master cylinder for internal and external leaks and proper operation; determine necessary action.	Chapter 5, p. 96
B.4. Remove, bench bleed, and reinstall master cylinder.	Chapter 5, p. 97
B.5. Diagnose poor stopping, pulling, or dragging concerns caused by malfunctions in the hydraulic system; determine necessary action.	Chapter 5, pp. 96, 110, 112
B.6. Inspect brake lines, flexible hoses, and fittings for leaks, dents, kinks, rust, cracks, bulging, or wear; tighten loose fittings and supports; determine necessary action.	Chapter 5, p. 111
B.7. Fabricate and/or install brake lines (double flare and ISO types); replace hoses, fittings, and supports as needed.	Chapter 5, p. 112

Task	Chapter and Pages
B.8. Select, handle, store, and fill brake fluids to proper level.	Chapter 3, p. 68
B.9. Inspect, test, and/or replace metering (hold-off), proportioning (balance), pressure differential, and combination valves.	Chapter 4, p. 82
B.10. Inspect, test, and adjust height-load-sensing proportioning valve.	Chapter 4, p. 82
B.11. Inspect, test, and/or replace components of brake warning light system.	Chapter 5, p. 111
B.12. Bleed (manual, pressure, vacuum, or surge) brake system.	Chapter 5, p. 104
B.13. Flush hydraulic system.	Chapter 5, p. 104
C. **Drum Brake Diagnosis and Repair**	
C.1. Diagnose poor stopping, noise, pulling, grabbing, dragging, or pedal pulsation concerns; determine necessary action.	Chapter 6, p. 137 Chapter 8, p. 171
C.2. Remove, clean (using proper safety procedures), inspect, and measure brake drums; determine necessary action.	Chapter 1, p. 21 Chapter 8, p. 168
C.3. Refinish brake drum.	Chapter 8, p. 168
C.4. Remove, clean, and inspect brake shoes, springs, pins, clips, levers, adjusters/self-adjusters, other related brake hardware, and backing support plates; lubricate and reassemble.	Chapter 6, p. 135
C.5. Remove, inspect, and install wheel cylinders.	Chapter 6, p. 131
C.6. Preadjust brake shoes and parking brake before installing brake drums or drum/hub assemblies and wheel bearings.	Chapter 6, p. 138
C.7. Install wheel, torque lug nuts, and make final checks and adjustments.	Chapter 13, p. 294
D. **Disc Brake Diagnosis and Repair**	
D.1. Diagnose poor stopping, noise, pulling, grabbing, dragging, or pedal pulsation concerns; determine necessary action.	Chapter 7, p. 160
D.2. Remove caliper assembly from mountings; clean and inspect for leaks and damage to caliper housing; determine necessary action.	Chapter 7, p. 150
D.3. Clean and inspect caliper mounting and slides for wear and damage; determine necessary action.	Chapter 7, p. 160
D.4. Remove, clean, and inspect pads and retaining hardware; determine necessary action.	Chapter 7, p. 148
D.5. Disassemble and clean caliper assembly; inspect parts for wear, rust, scoring, and damage; replace seal, boot, and damaged or worn parts.	Chapter 7, p. 153

Task	Chapter and Pages
D.6. Reassemble, lubricate, and reinstall caliper, pads, and related hardware; seat pads and inspect for leaks.	Chapter 7, p. 155
D.7. Clean, inspect, and measure rotor with a dial indicator and a micrometer; follow manufacturer's recommendations in determining need to machine or replace.	Chapter 7, p. 149 Chapter 8, p. 171
D.8. Remove and reinstall rotor.	Chapter 8, p. 174
D.9. Refinish rotor according to manufacturer's recommendations.	Chapter 8, p. 173
D.10. Adjust calipers equipped with an integrated parking brake system.	Chapter 9, p. 193
D.11. Install wheel, torque lug nuts, and make final checks and adjustments.	Chapter 13, p. 287
E. **Power Assist Units Diagnosis and Repair**	
E.1. Test pedal free travel with and without engine running; check power assist operation.	Chapter 5, p. 96 Chapter 9, p. 194
E.2. Check vacuum supply (manifold or auxiliary pump) to vacuum-type power booster.	Chapter 9, p. 202
E.3. Inspect the vacuum-type power booster unit for vacuum leaks; inspect the check valve for proper operation; determine necessary action.	Chapter 9, p. 204
E.4. Inspect and test hydro-boost system and accumulator for leaks and proper operation; determine necessary action.	Chapter 9, p. 207
F. **Miscellaneous (Wheel Bearings, Parking Brakes, Electrical, etc.) Diagnosis and Repair**	
F.1. Diagnose wheel bearing noises, wheel shimmy, and vibration concerns; determine necessary action.	Chapter 11, p. 249
F.2. Remove, clean, inspect, repack, and install wheel bearings and replace seals; install hub and adjust wheel bearings.	Chapter 11, p. 250
F.3. Check parking brake cables and components for wear, rusting, binding, and corrosion; clean, lubricate, or replace as needed.	Chapter 9, p. 190
F.4. Check parking brake operation; determine necessary action.	Chapter 9, p. 191
F.5. Check operation of parking brake indicator light system.	Chapter 9, p. 193
F.6. Check operation of brake stop light system; determine necessary action.	Chapter 5, p. 111
F.7. Replace wheel bearing and race.	Chapter 11, p. 250
F.8. Inspect and replace wheel studs.	Chapter 13, p. 290
F.9. Remove and reinstall sealed wheel bearing assembly.	Chapter 11, p. 251

Task	Chapter and Pages
G. **Antilock Brake and Traction Control Systems**	
G.1. Identify and inspect antilock brake system (ABS) components; determine necessary action.	Chapter 10, p. 214
G.2. Diagnose poor stopping, wheel lockup, abnormal pedal feel or pulsation, and noise concerns caused by the ABS; determine necessary action.	Chapter 10, p. 225
G.3. Diagnose ABS electronic control(s) and components using self-diagnosis and/or recommended test equipment; determine necessary action.	Chapter 10, p. 225
G.4. Depressurize high-pressure components of the ABS.	Chapter 10, p. 227
G.5. Bleed ABS front and rear hydraulic circuits.	Chapter 10, p. 227
G.6. Remove and install ABS electrical/electronic and hydraulic components.	Chapter 10, p. 230

APPENDIX D

NATEF Task Correlation Grid for Suspension and Steering

Task		Chapter and Pages
A.	**General Suspension and Steering Systems Diagnosis**	
A.1.	Identify and interpret suspension and steering concern; determine necessary action.	Chapter 11, p. 235 Chapter 13, p. 281 Chapter 15, p. 343 Chapter 17, p. 382 Chapter 19, p. 416 Chapter 20, p. 443
A.2.	Research applicable vehicle and service information, such as suspension and steering system operation, vehicle service history, service precautions, and technical service bulletins.	Chapter 1, p. 9
A.3.	Locate and interpret vehicle and major component identification numbers (VIN, vehicle certification labels, calibration decals).	Chapter 1, p. 10
B.	**Steering Systems Diagnosis and Repair**	
B.1.	Disable and enable supplemental restraint system (SRS).	Chapter 17, p. 387
B.2.	Remove and replace steering wheel; center/time supplemental restraint system (SRS) coil (clock spring).	Chapter 17, p. 388
B.3.	Diagnose steering column noises, looseness, and binding concerns (including tilt mechanisms); determine necessary action.	Chapter 16, p. 370 Chapter 17, p. 382
B.4.	Diagnose power steering gear (non-rack-and-pinion) binding, uneven turning effort, looseness, hard steering, and fluid leakage concerns; determine necessary action.	Chapter 17, p. 390
B.5.	Diagnose power steering gear (rack and pinion) binding, uneven turning effort, looseness, hard steering, and fluid leakage concerns; determine necessary action.	Chapter 16, p. 370 Chapter 17, p. 391
B.6.	Inspect steering shaft universal joint(s), flexible coupling(s), collapsible column, lock cylinder mechanism, and steering wheel; perform necessary action.	Chapter 16, p. 370 Chapter 17, p. 389

Task	Chapter and Pages
B.7. Adjust manual or power non-rack-and-pinion worm bearing preload and sector lash.	Chapter 17, p. 390
B.8. Remove and replace manual or power rack-and-pinion steering gear; inspect mounting bushings and brackets.	Chapter 17, pp. 391, 403
B.9. Inspect and replace manual or power rack-and-pinion steering gear inner tie rods ends (sockets) and bellows boots.	Chapter 16, p. 369 Chapter 17, p. 386
B.10. Inspect power steering fluid levels and condition.	Chapter 17, p. 382
B.11. Flush, fill, and bleed power steering system.	Chapter 17, p. 391
B.12. Diagnose power steering fluid leakage; determine necessary action.	Chapter 17, p. 393
B.13. Remove, inspect, replace, and adjust power steering pump belt.	Chapter 17, p. 397
B.14. Remove and reinstall power steering pump.	Chapter 17, p. 393
B.15. Remove and reinstall power steering pump pulley; check pulley and belt alignment.	Chapter 17, p. 393
B.16. Inspect and replace power steering hoses and fittings.	Chapter 17, p. 402
B.17. Inspect and replace pitman arm, relay (centerlink/intermediate) rod, idler arm and mountings, and steering linkage damper.	Chapter 17, p. 384
B.18. Inspect, replace, and adjust tie-rod ends (sockets), tie-rod sleeves, and clamps.	Chapter 17, p. 386
B.19. Test and diagnose components of electronically controlled steering systems using a scan tool; determine necessary action.	Chapter 16, p. 375 Chapter 17, p. 404
C. **Suspension Systems Diagnosis and Repair**	
1. **Front Suspension**	
C.1. Diagnose short/long arm suspension system noises, body sway, and uneven riding height concerns; determine necessary action.	Chapter 15, p. 343
C.2. Diagnose strut suspension system noises, body sway, and uneven riding height concerns; determine necessary action.	Chapter 14, p. 338
C.3. Remove, inspect, and install upper and lower control arms, bushings, shafts, and rebound bumpers.	Chapter 15, p. 350
C.4. Remove, inspect and install strut rods (compression/tension) and bushings.	Chapter 15, p. 352
C.5. Remove, inspect, and install upper and/or lower ball joints.	Chapter 15, p. 352
C.6. Remove, inspect, and install steering knuckle assemblies.	Chapter 15, p. 347

Task	Chapter and Pages
C.7. Remove, inspect, and install short and long arm suspension system coil springs and spring insulators.	Chapter 15, p. 356
C.8. Remove, inspect, install, and adjust suspension system torsion bars; inspect mounts.	Chapter 15, p. 359
C.9. Remove, inspect, and install stabilizer bar bushings, brackets, and links.	Chapter 15, p. 352
C.10. Remove, inspect, and install strut cartridge or assembly, strut coil spring, insulators (silencers), and upper strut bearing mount.	Chapter 15, p. 347
C.11. Lubricate suspension and steering systems.	Chapter 11, p. 242 Chapter 15, p. 352
2. Rear Suspension	
C.1. Remove, inspect, and install coil springs and spring insulators.	Chapter 15, p. 345
C.2. Remove, inspect, and install transverse links, control arms, bushings, and mounts.	Chapter 14, p. 322 Chapter 15, p. 350
C.3. Remove, inspect, and install leaf springs, leaf spring insulators (silencers), shackles, brackets, bushings, and mounts.	Chapter 15, p. 360
C.4. Remove, inspect, and install strut cartridge or assembly, strut coil spring, and insulators (silencers).	Chapter 15, p. 347
3. Miscellaneous Service	
C.1. Inspect, remove, and replace shock absorbers.	Chapter 15, p. 343
C.2. Remove, inspect, and service or replace front and rear wheel bearings.	Chapter 11, p. 244 Chapter 21, p. 473
C.3. Test and diagnose components of electronically controlled suspension systems using a scan tool; determine necessary action.	Chapter 15, p. 360
D. Wheel Alignment Diagnosis, Adjustment, and Repair	
D.1. Differentiate between steering and suspension concerns using principles steering geometry (caster, camber, toe, etc).	Chapter 19, p. 416
D.2. Diagnose vehicle wander, drift, pull, hard steering, bump steer, memory steer, torque steer, and steering return concerns; determine necessary action.	Chapter 19, p. 416
D.3. Perform prealignment inspection; perform necessary action.	Chapter 19, p. 416
D.4. Measure vehicle riding height; determine necessary action.	Chapter 19, p. 418
D.5. Check and adjust front and rear wheel camber; perform necessary action.	Chapter 19, p. 423
D.6. Check and adjust caster; perform necessary action.	Chapter 19, p. 423

Task	Chapter and Pages
D.7. Check and adjust front wheel toe; adjust as needed.	Chapter 19, p. 428
D.8. Center steering wheel.	Chapter 19, p. 431
D.9. Check toe-out-on-turns (turning radius); determine necessary action.	Chapter 19, p. 432
D.10. Check SAI (steering axis inclination) and included angle; determine necessary action.	Chapter 19, p. 426
D.11. Check and adjust rear wheel toe.	Chapter 19, p. 437
D.12. Check rear wheel thrust angle; determine necessary action.	Chapter 19, p. 435
D.13. Check for front wheel setback; determine necessary action.	Chapter 18, p. 413 Chapter 19, p. 436
D.14. Check front cradle (subframe) alignment; determine necessary action.	Chapter 19, p. 427
E. **Wheel and Tire Diagnosis and Repair**	
E.1. Diagnose tire wear patterns; determine necessary action.	Chapter 13, pp. 282, 284 Chapter 18, p. 407 Chapter 19, p. 417
E.2. Inspect tires; check and adjust air pressure.	Chapter 13, p. 282
E.3. Diagnose wheel/tire vibration, shimmy, and noise; determine necessary action.	Chapter 13, p. 304
E.4. Rotate tires according to manufacturer's recommendations.	Chapter 13, p. 286
E.5. Measure wheel, tire, axle, and hub runout; determine necessary action.	Chapter 13, p. 298
E.6. Diagnose tire pull (lead) problem; determine necessary action.	Chapter 13, p. 287
E.7. Balance wheel and tire assembly (static and dynamic).	Chapter 13, p. 304
E.8. Dismount, inspect, repair, and remount tire on wheel.	Chapter 13, p. 290
E.9. Reinstall wheel; torque lug nuts.	Chapter 13, p. 287
E.10. Inspect and repair tire.	Chapter 13, p. 300

APPENDIX E

DOT Tire Codes

A1 Michelin (France)
A2 Loe (Brazil)
A3 General (IL)
A4 H. A. Pusan (Korea)
A5 Ciech-Stomil (Poland)
A6 Apollo (India)
A7 Bridgestone (Thailand)
A8 Bridgestone (Indonesia)
A9 General (OH)
AA General (OH)
AB General (TX)
AC General (NC)
AD General (KY)
AE General (Spain)
AF General (Portugal)
AH General (Mexico)
AJ Uniroyal (MI)
AK Uniroyal (MA)
AL Uniroyal (WI)
AM Uniroyal (CA)
AN Uniroyal (AL)
AP Uniroyal (OK)
AT Avon (England)
AU Uniroyal (Ontario)
AV Seiberling (OH)
AW Sampson (Israel)
AX Phoenix (Germany)
AY Phoenix (Germany)

B1 Michelin (France)
B2 Dunlop (Malaysia)
B3 Michelin (Nova Scotia)
B4 Taurus (Hungary)
B5 Ciech-Stomil (Poland)
B6 Michelin (SC)
B7 Michelin (AL)
B8 Michelin (Brazil)
B9 Michelin (SC)
BA Goodrich (OH)
BB Goodrich (OK)
BC Goodrich (PA)
BD Goodrich (OK)
BE Goodrich (AL)
BF Goodrich (IN)
BJ Goodrich (Germany)
BK Goodrich (Brazil)

BL Goodrich (Colombia)
BM Goodrich (Austria)
BN Goodrich (Philippines)
BP Goodrich (Iran)
BT Semperit (Austria)
BU Semperit (Ireland)
BV IRI (KY)
BW Gates (CO)
BX Gates (TH)
BY Gates (CO)

C1 Michelin (Higeria)
C2 Kelly (Brazil)
C3 McCreary (MD)
C4 Armstrong (TH)
C5 Ciech-Stomil (Poland)
C6 Mitas (Czechoslovakia)
C7 Ironsides (KY)
C8 Bridgestone (Taiwan)
C9 Seven Star (Taiwan)
CA Mohawk (OH)
CB Mohawk (AR)
CC Mohawk (VA)
CD Alliance (Israel)
CE Armstrong (CT)
CF Armstrong (IA)
CH Armstrong (CA)
CJ Inque (Japan)
CK Armstrong (TH)
CL Continental (Italy)
CM Continental (Germany)
CP Continental (Germany)
CT Continental (Germany)
CU Continental (Germany)
CV Armstrong (MS)
CW Toyo (Japan)
CX Toyo (Japan)
CY McCreary (PA)

D1 Viking-Askim (Norway)
D2 Dayton (TN)
D3 United (Ontario)
D4 Dunlop (India)
D5 Dunlop (India)
D6 Borovo (Yugoslavia)
D7 Dunlop (S. Africa)

D8 Dunlop (S. Africa)
D9 United Tire (Ontario)
DA Dunlop (NY)
DB Dunlop (AL)
DC Dunlop (Ontario)
DD Dunlop (England)
DE Dunlop (England)
DF Dunlop (England)
DH Dunlop (Scotland)
DJ Dunlop (Ireland)
DK Dunlop (France)
DL Dunlop (France)
DM Dunlop (Germany)
DN Dunlop (Germany)
DP Dunlop (Engalnd)
DT Dunlop (Australia)
DU Dunlop (Australia)
DV Verdestein (Netherlands)
DW Verdestein (Netherlands)
DX Verdestein (Netherlands)
DY Denman (OH)

E1 Chung Hsia (Taiwan)
E2 Firestone (Brazil)
E3 Seiberling (TN)
E4 Firestone (New Zealand)
E5 Firestone (S. Africa)
E6 Firestone (Tunisia)
E7 Firestone (Kenya)
E8 Firestone (Ghana)
E9 Firestone (S. Africa)
EA Metzeler (Germany)
EB Metzeler (Germany)
EC Metzeler (Germany)
ED Okamoto (Japan)
EE Nitto (Japan)
EF Hang Ah (Korea)
EH Bridgestone (Japan)
EJ Bridgestone (Japan)
EK Bridgestone (Japan)
EL Bridgestone (Japan)
EM Bridgestone (Japan)
EN Bridgestone (Japan)
EP Bridgestone (Japan)
ET Sumitoma (Japan)
EU Sumitoma (Japan)
EV Kleber (France)
EW Kleber (France)
EX Kleber (France)
EY Kleber (Germany)

F1 Michelin (Scotland)
F2 Firestone (Venezuela)
F3 Michelin (France)
F4 Fapobol (Portugal)
F5 FATE (Argentina)
F6 Firestone (Spain)

F7 Firestone (Spain)
F8 Vikrant (India)
F9 Dunlop (New Zealand)
FA Yokohama (Japan)
FB Yokohama (Japan)
FC Yokohama (Japan)
FD Yokohama (Japan)
FE Yokohama (Japan)
FF Michelin (France)
FH Michelin (France)
FJ Michelin (France)
FK Michelin (France)
FL Michelin (France)
FM Michelin (France)
FN Michelin (France)
FP Michelin (Algeria)
FT Michelin (Germany)
FU Michelin (Germany)
FV Michelin (Germany)
FW Michelin (Germany)
FX Michelin (Belgium)
FY Michelin (France)

H1 Michelin (Spain)
H2 Kumbo (Korea)
H3 Sava (Yugoslavia)
H4 Bridgestone (Japan)
H5 Hutchinson (France)
H6 Shin Hung (Korea)
H7 Li Nsia (Taiwan)
H8 Firestone (OK)
H9 Reifenberg (Germany)
HA Michelin (Spain)
HB Michelin (Spain)
HC Michelin (Spain)
HD Michelin (Italy)
HE Michelin (Italy)
HF Michelin (Italy)
HH Michelin (Italy)
HJ Michelin (Ireland)
HK Michelin (Ireland)
HL Michelin (England)
HM Michelin (England)
HN New Glasgow (Nova Scotia)
HP Michelin (Vietnam)
HT Ceat (Italy)
HU Ceat (Italy)
HV Ceat (Italy)
HW Gurmarne (Czechoslovakia)
HX Dayton (OH)
HY Dayton (OC)

J1 Philips (OK)
J2 Bridgestone (Sint)
J3 Gumarne (Czechoslovakia)
J4 Rubena (Czechoslovakia)
J5 Lee (OH)

6 Jaroslav (Russia)
7 R & J (IN)
8 Da Sing (Shanghai)
9 P.T. Latirub (Indonesia)
A Lee (ON)
B Lee (OP)
C Lee (PA)
D Lee (MO)
E Lee (VA)
F Lee (NC)
H Lee (IL)
JJ Lee (AL)
JK Lee (AN)
JL Lee (CA)
JM Lee (MA)
JN Lee (KS)
JP Lee (TX)
JT Lee (TN)
JU Lee (Ontario)
JV Lee (Ontario)
JW Lee (Ontario)
JX Lee (Quebec)
JY Lee (Argentina)

K1 Phillips (OH)
K2 Lee (KY)
K3 Kenda (Taiwan)
K4 Uniroyal (Mexico)
K5 Premant (Germany)
K6 Lee (OK)
K7 Lee (Chile)
K8 Kelley (Malaysia)
K9 Natier Tire (Taiwan)
KA Lee (Australia)
KB Lee (Australia)
KC Lee (Brazil)
KD Lee (Columbia)
KE Lee (Zaire)
KF Lee (France)
KH Lee (Germany)
KJ Lee (Germany)
KL Lee (Guatamala)
KM Lee (Luxemburg)
KN Lee (India)
KP Lee (Indonesia)
KT Lee (Italy)
KU Lee (Jamaica)
KV Lee (Mexico)
KW Lee (Peru)
KX Lee (Phillipines)
KY Lee (Scotland)

L1 Goodyear (Taiwan)
L2 Wuon Poong (Korea)
L3 Tong Shin (Korea)
L4 Cipcmp (Romania)
L5 Lassa Lastik (Turkey)
L6 Dunlop (Zimbabwe)
L7 Cipcmp (Romania)
L8 Dunlop (Zimbabwe)
LA Lee (S. Africa)
LB Lee (Turkey)
LC Lee (Thailand)
LD Lee (Turkey)
LE Lee (Venezuela)
LF Lee (England)
LH Bridgestone (Australia)
LJ Uniroyal (Belgium)
LK Llantas (Colombia)
LL Uniroyal (France)
LM Uniroyal (Germany)
LN Uniroyal (Mexico)
LP Uniroyal (Scotland)
LT Uniroyal (Turkey)
LU Uniroyal (Venezuela)
LV General (Ontario)
LW Treileborg (Sweden)
LX Mitsuboshi (Japan)
LY Mitsuboshi (Japan)

M1 Goodyear (Mor.)
M2 Goodyear (KY)
M3 Michelin (SC)
M4 Goodyear (OH)
M5 Michelin (Nova Scotia)
M6 Goodyear (OK)
M7 Goodyear (Chile)
M8 Pramier (India)
M9 Uniroyal (CT)
MA Goodyear (OH)
MB Goodyear (OH)
MC Goodyear (VA)
MD Goodyear (AL)
ME Goodyear (MI)
MF Goodyear (CA)
MH Goodyear (ME)
MJ Goodyear (KS)
ML Goodyear (MD)
MM Goodyear (NC)
MN Goodyear (IL)
MP Goodyear (TX)
MT Goodyear (PA)
MU Goodyear (Argentina)
MV Goodyear (Australia)
MW Goodyear (Australia)
MX Goodyear (Brazil)
MY Goodyear (Colombia)

N1 Maloja (Switzerland)
N2 Hurtubise (NY)
N3 Nitto (Japan)
N4 Cipcmp (Romania)
N5 Paeumant (Germany)
N6 Paeumant (Germany)

N7 Cipcmp (Romania)
N8 Lee (Malasyia)
N9 C.P. Tropical (Brazil)
NA Goodyear (Zaire)
NB Goodyear (England)
NC Goodyear (France)
ND Goodyear (Germany)
NE Goodyear (Germany)
NF Goodyear (Greece)
NH Goodyear (Guatemala)
NJ Goodyear (Luxemburg)
NK Goodyear (India)
NL Goodyear (Indonesia)
NM Goodyear (Italy)
NN Goodyear (Jamaica)
NP Goodyear (Mexico)
NT Goodyear (PA)
NU Goodyear (Philippines)
NV Goodyear (Scotland)
NW Goodyear (S. Africa)
NX Goodyear (Sweden)
NY Goodyear (Thailand)

P1 Gislaved (Sweden)
P2 Kelly (KY)
P3 Skepplanda (Sweden)
P4 Kelly (OH)
P5 Genl Popos (Mexico)
P6 Kelly (OK)
P7 Kelly (Chile)
P8 Da Sing (China)
P9 MRF (India)
PA Goodyear (Turkey)
PB Goodyear (Venezuela)
PC Goodyear (Alberta)
PD Goodyear (Quebec)
PE Seiberling (Ontario)
PF Goodyear (Ontario)
PH Kelly (MD)
PJ Kelly (HC)
PK Kelly (IL)
PL Kelly (TX)
PM Kelly (PA)
PN Kelly (OH)
PP Kelly (ON)
PT Kelly (VA)
PU Kelly (AL)
PV Kelly (MI)
PW Kelly (CA)
PX Kelly (MA)
PY Kelly (KS)

T1 Hankook (Korea)
T2 Ozos (Poland)
T3 Debickie (Poland)
T4 Carideng (Belgium)
T5 Tigar-Pirot (Hungary)

T6 Hulera Tornel (Mexico)
T7 Hankook (Korea)
T8 Goodyear (Malaysia)
T9 MRF (India)
TA Kelly (TN)
TB Kelly (Argentina)
TC Kelly (Australia)
TD Kelly (Australia)
TE Kelly (Brazil)
TF Kelly (Colombia)
TH Kelly (Zaire)
TJ Kelly (England)
TK Kelly (France)
TL Kelly (Germany)
TM Kelly (Germany)
TN Kelly (Greece)
TP Kelly (Guatemala)
TT Kelly (Luxemburg)
TU Kelly (India)
TV Kelly (Indonesia)
TW Kelly (Italy)
TX Kelly (Jamaica)
TY Kelly (Mexico)

U1 Lien Shin (Taiwan)
U2 Sumitomo (Japan)
U3 Miloje Zakie (Yugoslavia)
U4 Byers & Sons (OH)
U5 Farbenfabrik (Germany)
U6 Pneumant (Germany)
U7 Pneument (Germany)
U8 Hankang (Taiwan)
U9 Cooper (MS)
UA Kelly (Peru)
UB Kelly (Philippines)
UC Kelly (Scotland)
UD Kelly (S. Africa)
UE Kelly (Sweden)
UF Kelly (Thailand)
UH Kelly (Turkey)
UJ Kelly (Venezuela)
UK Kelly (Alberta)
UL Kelly (Quebec)
UM Kelly (Ontario)
UN Kelly (Ontario)
UP Cooper (OH)
UT Cooper (TX)
UU Cartisle (PA)
UV Kyowa (Japan)
UW Okada (Japan)
UX Tay Feng (Taiwan)
UY Cheng Shin (Taiwan)

V1 Livingston's (OH)
V2 Vsesojuznoe (Russia)
V3 Ta Msia (Taiwan)
V4 Ohtsu (Japan)

V5 Firestone (Mexico)
V6 Firestone (Mexico)
V7 Vsesojuznoe (Russia)
V8 Boras (Sweden)
V9 M & R Tire (MA)
VA Firestone (OH)
VB Firestone (OH)
VC Firestone (GA)
VD Firestone (IL)
VE Firestone (IA)
VF Firestone (CA)
VH Firestone (TN)
VJ Firestone (PA)
VK Firestone (CA)
VL Firestone (Ontario)
VM Firestone (Alberta)
VN Firestone (Quebec)
VP Firestone (Italy)
VT Firestone (Spain)
VU Universal (PA)
VV Viskators (Sweden)
VW Ohtsu (Japan)
VX Firestone (England)
VY Firestone (England)

W1 Firestone (TH)
W2 Firestone (HC)
W3 Vredestein (Nethlands)
W4 Dunlop-Oly (Australia)
W5 Firestone (Argentina)
W6 Firestone (Philippines)
W7 Firestone (Portugal)
W8 Firestone (Thailand)
W9 Firestone (Brazil)
WA Firestone (France)
WB Firestone (C.R.)
WC Firestone (Austria)
WD Firestone (Switzerland)
WE Hankang (Taiwan)
WF Firestone (Spain)
WH Firestone (Sweden)
WK Mansfield (MS)
WL Mansfield (OH)
WM Dunlop-Oly (Australia)
WN Olypmpic (Australia)
WP Schenuit (MD)
WT Madras (India)
WU Ceat (India)
WV General (Taiwan)
WW Euzkadi (Mexico)
WX Euzkadi (Mexico)
WY Euzkadi (Mexico)

X1 Tong Shin (Korea)
X2 Hwa Fong (Taiwan)
X3 Vsesojuznoe (Russia)

X4 Pars (Iran)
X5 JK Industries (India)
X6 Vsesojunzoe (Russia)
X7 Vsesojunzoe (Russia)
X8 Vsesojunzoe (Russia)
X9 Vsesojunzoe (Russia)
X10 Vsesojunzoe (Russia)
XA Pirelli (Italy)
XB Pirelli (Italy)
XC Pirelli (Italy)
XD Pirelli (Italy)
XE Pirelli (Italy)
XF Pirelli (Spain)
XH Pirelli (Greece)
XJ Pirelli (Turkey)
XK Pirelli (Brazil)
XL Pirelli (Brazil)
XM Pirelli (Argentina)
XN Pirelli (England)
XP Pirelli (England)
XT Veith-Pirelli (Germany)
XU Sam Yang (Korea)
XV Dayton (Ontario)
XW Dayton (Alberta)
XX Bandag (LA)

Y1 Goodyear (Brazil)
Y2 Dayton (NC)
Y3 Seiberling (NC)
Y4 Dayton (OH)
Y5 Tsen Tai (China)
Y6 Dunlop (Malaysia)
Y7 Bridgestone (TN)
Y8 Bombay (India)
YA Dayton (OH)
YB Dayton (OH)
YC Dayton (GA)
YD Dayton (IL)
YE Dayton (IA)
YF Dayton (CA)
YH Dayton (TN)
YJ Dayton (PA)
YK Dayton (CA)
YL Oy Nokia (Finland)
YM Seiberling (OH)
YN Seiberling (OH)
YP Seiberling (GA)
YT Seiberling (IL)
YU Seiberling (IA)
YV Seiberling (CA)
YW Seiberling (TN)
YX Seiberling (PA)
YY Seiberling (CA)

Length and Angle Conversions Used in Wheel Alignments

Inch (Fractional)	Inch (Decimal)	Metric (mm)	Degrees (Decimal)	Degrees (Fractional)	Degrees (Minutes)
1/32	0.0312	0.793	0.0625	1/16	3.75
1/16	0.0625	1.587	0.125	1/8	7.5
3/32	0.0937	2.381	0.1875	3/16	11.25
1/8	0.125	3.175	0.25	1/4	15
5/32	0.1562	3.968	0.3125	5/16	18.75
3/16	0.1875	4.762	0.375	3/8	22.5
7/32	0.2187	5.556	0.4375	7/16	26.25
1/4	0.250	6.35	0.5	1/2	30
9/32	0.2812	7.143	0.56625	9/16	33.75
5/16	0.3125	7.937	0.625	5/8	37.5
11/32	0.343	8.7317	0.6875	11/16	41.25
3/8	0.375	9.525	0.75	3/4	45
13/32	0.4062	10.318	0.8125	13/16	48.75
7/16	0.4375	11.112	0.875	7/8	52.5
15/32	0.4687	11.906	0.9375	15/16	56.25
1/2	0.500	12.7	1.0	1	60

Bilingual Glossary

A

ABS—see antilock brake system.

ABS—ver sistema de frenos antibloqueo.

Acceleration slip regulation—an electronic system that limits the amount the wheels can spin during acceleration, preventing one drive wheel from spinning faster than the opposite one.

Regulación de deslizamiento al acelerar—un sistema electrónico que limita la rotación de las ruedas durante la aceleración, evitando que una rueda de la transmisión gire más rápido que la opuesta.

Ackerman angle—also called turning radius or toe-out on turns. It causes tires to toe out during a turn because the steering arms are bent at an angle.

Ángulo Ackerman—también denominado radio de giro o divergencia al girar. Provoca la divergencia de los neumáticos al girar debido a que los brazos de la dirección se encuentran doblados en un ángulo.

Active coils—the center coils of a coil spring.

Bobinas activas—las bobinas centrales de un resorte helicoidal.

Active suspension system—an automatic suspension in which each wheel has a hydraulic cylinder to keep the car body level during all driving conditions.

Sistema activo de suspensión—una suspensión automática en la que cada rueda tiene un cilindro hidráulico que mantiene el nivel del chasis en cualquier condición de manejo.

Actuator—an electronically operated solenoid, as found in a strut or shock absorber, that controls ride firmness.

Accionador—un solenoide de operación electrónica, como el que se encuentra en un puntal o amortiguador, y que controla la firmeza de la marcha.

Adaptive suspension system—a suspension leveling system that keeps the vehicle at the same height when weight is added.

Sistema de suspensión adaptable—un sistema de nivelación de suspensión que mantiene el vehículo a la misma altura al agregarse peso.

Adjustable strut—a strut that can be adjusted manually to control ride firmness.

Puntal ajustable—un puntal que puede ajustarse manualmente para controlar la firmeza de la marcha.

Aftermarket—the distribution system from parts manufacturers to customers; *not* original equipment (OE).

Mercado secundario—el sistema de distribución de piezas de los fabricantes a los clientes; *no* son equipos originales (OE, según sus siglas en inglés).

Air spring—an air-filled cell in an air suspension system replacing conventional coil springs.

Resorte hidráulico—una célula llena de aire en un sistema de suspensión hidráulico que reemplaza los resortes helicoidales convencionales.

Alloy—when two or more metals are combined to make one

Aleación—cuando se combina dos o más metales para hacer uno solo.

All-season tires—tires with a tread design that improves wet traction.

Neumáticos para toda estación—neumáticos con un diseño de banda que optimiza la tracción en condiciones mojadas.

Aluminum space frame (ASF)—an expensive chassis developed by Alcoa Aluminum used in mass production; said to be 40 percent stiffer and 40 percent lighter than steel mono.

Armazón de espacio de aluminio (ASF, en inglés)—un chasis costoso desarrollado por Alcoa Aluminum y que se utiliza en la producción masiva. Se alega que es un 40 por ciento más rígido y un 40 por ciento más leve que el monocasco de acero.

Amplitude—the intensity of a vibration.

Amplitud—la intensidad de la vibración.

Angular contact bearing—a ball bearing assembly in which two radial thrust control bearings face each other; also called radial thrust bearing.

Rodamiento de contacto angular—un conjunto de rodamiento esférico con dos rodamientos de control de tope radial enfrentados; también denominado rodamiento de tope radial.

Anodized aluminum—aluminum with a protective plating

Aluminio anodizado—aluminio con revestimiento protector

Antilock brake controller—the computer that controls the ABS operation, also used in some systems for traction control.

Controlador de frenos antibloqueo—la computadora que controla la operación de los frenos ABS; también utilizada en algunos sistemas para control de la tracción.

Antilock brake system (ABS)—a brake system that modulates hydraulic pressure to one or more wheels to prevent wheel lockup.

Sistema de frenos antibloqueo (ABS)—un sistema de frenos que modula la presión hidráulica a una o más ruedas para evitar el bloqueo de las ruedas.

API—American Petroleum Institute

API—American Petroleum Institute (Instituto Americano del Petróleo).

Arcing—grinding brake linings to conform to the diameter of the drum, providing clearance between the ends of the shoes and the drum.

Desbastado de frenos—el desbastado de los revestimientos del freno de acuerdo con el diámetro del tambor, proporcionando espacio entre las extremidades de las zapatas y el tambor.

Arc welding—welding using electricity; stick welding.

Soldadura de arco—soldadura que utiliza electricidad.

ASE—National Institute for Automotive Service Excellence.

ASE—National Institute for Automotive Service Excellence (Instituto Nacional de Excelencia en el Servicio Automotor).

Aspect ratio—a measurement of the height-to-width ratio of a tire.

Bilingual Glossary

relación de aspecto—medida de la relación altura-ancho de un neumático.

Asymmetrical—an irregular shape, size, or position.

Asimétrico—una forma, tamaño o posición irregular.

Atmospheric pressure—the weight of the air exerted on the earth by the atmosphere; 14.7 psi at sea level.

Presión atmosférica—el peso del aire emitido sobre la tierra por la atmósfera; 14,7 psi al nivel del mar.

Atmospheric suspended power brake—an obsolete vacuum booster with atmospheric pressure on both sides of the diaphragm when the brakes are released.

Freno con cilindro maestro con suspensión atmosférica—un booster hidráulico obsoleto con presión atmosférica en ambos lados del diafragma al soltarse los frenos.

Automotive parts wholesaler—jobber.

Mayorista de piezas automotoras—proveedor de piezas automotoras al por mayor.

Auxiliary lip—a lip on a seal that controls dust.

Borde auxiliar—un borde que controla el polvo en el sello.

Axle bearing—a bearing on a live axle.

Cojinete de eje—un cojinete en un eje vivo.

B

Backbone chassis—a chassis found in some lightweight sports cars with low production numbers. It uses a long tube backbone to connect the front and rear axles.

Chasis backbone—chasis encontrado en ciertos automóviles deportivos livianos con números de producción bajos. Utiliza una columna con tubo largo para conexión con los ejes delantero y trasero.

Backing plate—the mounting surface for parts of a drum brake.

Contrafuerte—la superficie de montaje para las piezas del freno de tambor.

Ball socket—a part that connects steering linkage parts.

Toma esférica—pieza que conecta las piezas de conexión de la dirección.

Ball joint—a part used to attach the control arm to the spindle.

Junta esférica—pieza utilizada para conectar el brazo de control al eje.

Banjo fitting—a fitting resembling a banjo that allows brake fluid to make a 90-degree turn in a tight space.

Conector banjo—un conector que parece un banjo y que permite que el líquido de frenos realice un giro de 90 grados en un espacio limitado.

Barrier cream—a cream rubbed on skin prior to working in a greasy environment to aid in cleanup.

Crema barrera—crema que se frota sobre la piel antes de trabajar en un ambiente grasoso para facilitar la limpieza posterior.

Beads—coiled rings of wire at the side edges of a tire that give the tire strength so it stays firmly attached to the wheel.

Talones—bobinas de alambre en los bordes laterales del neumático que proporcionan fuerza al neumático de manera que se mantenga firmemente sujeto a la rueda.

Bead seat—the raised section on either side of the drop center area of a wheel rim.

Asiento de talón—la sección elevada en cualquier lado del área central del aro de la rueda.

Bead wire—circular wire strands molded into the bead on the inner circumference of the tire.

Alambre de talón—hilos circulares de alambre incrustados en el talón de la circunferencia interna del neumático.

Bearing cage—the part of a tapered roller bearing that holds the rollers together.

Jaula de cojinete—la parte del cojinete cónico de rodillo que une a las esferas.

Bearing cup—the outer race of a tapered roller bearing, usually pressed into the wheel hub.

Taza de cojinete—el aro externo de un cojinete cónico de rodillo, generalmente presionado dentro del cubo de la rueda.

Bearing cone—the inner race of a tapered roller bearing.

Cono de cojinete—el aro interno de un cojinete cónico de rodillo.

Bearing-retained axle—a press-fit axle with a bearing retainer ring.

Eje con cojinete de retención—eje con un aro de retención de cojinete colocado a presión.

Bearing seal—a circular metal ring with a sealing lip on the inner edge that prevents dirt and moisture from entering the bearing, while sealing in lubricant.

Sello de cojinete—un aro metálico circular un borde de sellado en el borde interior que evita la entrada de tierra y humedad en el cojinete y la pérdida de lubricante.

Beat/boom vibration—one vibration that interacts with another.

Vibración golpe/estruendo—una vibración que interactúa con otra.

Belt—a tire's cord structure, made up of plies.

Correa—la estructura de cuerda de un neumático, formada por capas.

Belted bias tire—tire construction using belts on older bias-ply tires.

Neumáticos polarizados con correa—neumáticos polarizados más antiguos que utilizan correas.

Bias-ply—a tire structure with casing plies crossing each other at an angle of 35–45 degrees, instead of 90 degrees (like a radial tire); also called diagonal or crossply.

Polarizado—estructura de neumático con telas cruzadas en un ángulo de 35–45 grados, en lugar de 90 grados (como el neumático radial); también denominado diagonal.

Bimetallic drum—a composite brake drum made of cast iron and aluminum.

Tambor bimetálico—un tambor de frenos compuesto fabricado con hierro fundido y aluminio.

Binder—adhesive, or glue, used in brake linings to bond all the other materials together.

Adhesivo—goma adhesiva utilizada en los revestimientos de los frenos para unir todos los materiales.

Bleed screw—a screw that opens and closes a bleeding port in a caliper or a wheel cylinder.

Tornillo de purga—tornillo que abre y cierra el orificio de purga de una mordaza o un cilindro de rueda.

Body-over-frame—a separate frame with all of the powertrain parts mounted in it.

Carrocería-sobre-armazón—un armazón separado en el que se encuentran montadas todas las piezas del tren de potencia.

Bonded lining—a brake lining attached to the pad or shoe by adhesive.

Revestimiento adherido—revestimiento de freno adherido con adhesivo a la almohadilla o zapata.

Boom—a low frequency noise in the 20–60 Hz range.

Estruendo—ruido de baja frecuencia dentro de los 20–60 Hz.

Booster—a vacuum-assist brake device.

Booster—dispositivo de freno asistido por vacío.

Bounce—up and down motion.

Oscilación vertical—movimiento vertical.

Brake bleeding—a procedure in which fresh brake fluid forces out air bubbles and old aerated fluid through bleeding ports in calipers and wheel cylinders.

Purga de frenos—procedimiento en el que el líquido de freno nuevo elimina las burbujas de aire y el líquido aireado viejo a través de los orificios de purga de las mordazas y los cilindros de la rueda.

Brake caliper—the part of a disc brake system mounted on the suspension or axle housing that contains a hydraulic piston and the brake pads.

BILINGUAL GLOSSARY

Mordaza del freno—la pieza de un sistema de frenos de disco montada en la suspensión o la caja del eje y que contiene un pistón hidráulico y las almohadillas de freno.

Brake fade—a loss of braking caused by excessive heat, which reduces friction between the brake linings and the rotors or drums.

Desvanecimiento de freno—pérdida de poder de frenado ocasionada por calor excesivo, y que reduce la fricción entre los revestimientos del freno y los rotores o tambores.

Brake linings—the friction material attached to a drum brake shoe or disc brake pad.

Revestimientos de freno—el material de fricción adherido a la zapata de freno de tambor o almohadilla de freno de disco.

Brake pad—a disc brake part that supports the lining friction material.

Almohadilla de freno—pieza del freno de disco para soporte del material de fricción del revestimiento.

Brake pad hardware—small parts that prevent brake vibration and noise, such as anti-rattle clips and shims.

Equipo de instalación de almohadilla de freno—piezas pequeñas que evitan la vibración y ruido en los frenos, tales como pinzas y cuñas antivibración.

Brake pad wear indicator—a warning system for excessive disc brake lining wear, either mechanical (audible) or electrical.

Indicador de desgaste de almohadilla de freno—sistema de advertencia de desgaste excesivo del revestimiento del freno de disco, ya sea mecánico (audible) o eléctrico.

Brake pressure modulator valve (BPMV)—a solenoid valve body that controls fluid pressure during antilock brake, traction control, and stability control.

Válvula del modulador de presión del freno (BPMV, en inglés)—el cuerpo de una válvula solenoide que controla la presión del líquido durante el frenado antibloqueo, control de tracción y control de estabilidad.

Brake pressure switch—an electrical switch operated by brake fluid pressure to control warning lights and to send on and off signals to a computer.

Interruptor de presión de frenos—interruptor eléctrico impulsado por la presión del líquido de freno y que controla las luces de advertencia y envía señales de apagar/encender a la computadora.

Brake shoe—on drum brakes, a metal part to which the frictional lining is attached.

Zapata de freno—en los frenos de tambor, una pieza metálica a la que está adherido el revestimiento de fricción.

Bridge-type clamp—a type of clamp used to fasten CV joint boots to the axle.

Pinza tipo puente—tipo de pinza utilizada para sujetar los fuelles de junta de la flecha homocinética al eje.

Brinelling—indentations in a bearing or race from shock loads.

Hendiduras—hendiduras en un cojinete o aro debido a cargas de choque.

Bulkhead—the metal wall between the engine and passenger compartments; also called the firewall.

Mampara—la pared metálica entre el motor y el compartimiento de pasajeros; también denominada "firewall".

Bump steer—when a wheel with tie-rods at unequal heights goes over a bump, the car momentarily steers in the direction that the wheel turns as the toe on that one wheel changes; also called orbital steer.

Dirección de baches—cuando una rueda con varillas de conexión con alturas desparejas pasa por un bache, el vehículo se mueve temporalmente en la dirección en que gira la rueda mientras cambia la convergencia en esa rueda; también denominada dirección orbital.

Bushing—a cylindrical bearing made of steel, brass, bronze, nylon, or plastic.

Buje—un cojinete cilíndrico de aluminio, bronce, nylon o plástico.

Buzz—a frequency of 50–100 Hz; about the frequency of an electric razor.

Zumbido—una frecuencia de 50–100 Hz; aproximadamente la frecuencia de una máquina de afeitar eléctrica.

C

CAFÉ—see corporate average fuel economy.

CAFÉ—ver media economía de combustible corporativa.

Camber—the inward or outward tilt of a tire at the top.

Comba—la inclinación hacia dentro o hacia fuera de un neumático en su parte superior.

Camber roll—when a cambered tire rolls in a circle, as if it were the large end of a cone.

Rodado de comba—cuando un neumático con comba gira en círculos, como si fuera la extremidad más grande de un cono.

Capillary action—the phenomenon of a liquid seeping up the surface of a tube due to surface tension; a force that allows a porous material to soak up a liquid.

Acción capilar—fenómeno en el que un líquido sube a la superficie de un tubo debido a la tensión de la superficie; fuerza que permite que un material poroso absorba un líquido.

Carbon fiber monocoque—an expensive racing chassis developed for aircraft, used on some exotic cars such as Porsche and Ferrari.

Monocasco de fibra de carbono—un chasis de carrera caro desarrollado para naves aéreas y que se utiliza en ciertos automóviles exóticos como Porsche y Ferrari.

Carcass—beads and layers of cords or plies bonded together to make the internal structure of a tire; also called the casing.

Cuerpo de la llanta—aros y capas de cuerdas o placas adheridas entre sí para fabricar la estructura interna de un neumático; también denominada alojamiento.

Cardan—the most popular universal joint design; also called cross and yoke.

Cardán—el diseño de junta universal más popular.

Caster—the forward or rearward tilt of the spindle support arm away from vertical when viewed from the side.

Ángulo de avance—la inclinación hacia delante o hacia tras del brazo de apoyo del eje, con relación a la posición vertical, visto de costado.

Center bolt—a bolt that extends through a hole in the center of all the leaves to maintain their position in a leaf spring.

Perno central—perno que pasa por un orificio en el centro de todas las láminas para que mantengan su posición en una ballesta.

Center link—a long rod that connects the pitman arm to the idler arm in a parallelogram steering linkage.

Conexión central—una varilla larga que conecta el brazo de la varilla de conexión al brazo del ralentí en una conexión de dirección de paralelograma.

Center of gravity—the point between the front and the rear wheels where the weight will be distributed evenly.

Centro de gravedad—el punto entre las ruedas delanteras y traseras donde se distribuirá el peso de forma pareja.

Center support bearing—a bearing on a two-piece drive shaft that supports the center section where the two shafts attach to each other.

Rodamiento de apoyo central—rodamiento en un eje de dirección de dos piezas que sostiene la sección central en la que los dos ejes se unen.

Chafing strips—hard strips of rubber to protect a tire's beads from damage that could result from chafing against the rim.

Tiras de rozamiento—tiras duras de hule que protegen los bordes del neumático contra daños que podrían ocurrir a raíz del rozamiento contra el aro.

Chassis—the group of parts that supports the engine and the car body, including the frame, shocks and springs, steering parts, tires, brakes, and wheels.

Chasis—grupo de piezas que sostiene el motor y la carrocería del automóvil, incluyendo el armazón, amortiguadores y resortes, piezas de la dirección, neumáticos, frenos y ruedas.

Check valve—a valve that allows fluid to flow in one direction only.

Válvula de verificación—válvula que permite el flujo del líquido en apenas una dirección.

Class A fire—a fire involving ordinary combustibles such as wood, paper, cloth, or rubbish.

Incendio de la Clase A—incendio provocado por combustibles comunes, tales como madera, papel, tela o basura.

Class B fire—a fire in which flammable liquids such as grease, oil, gasoline, or paint are involved.

Incendio de la Clase B—incendio provocado por líquidos inflamables, tales como grasa, aceite, gasolina o pintura.

Class C fire—an electrical fire.

Incendio de la Clase C—incendio eléctrico.

Class D fire—a fire involving a flammable metal such as magnesium or potassium.

Incendio de la Clase D—incendio provocado por un metal inflamable, tal como el magnesio o el potasio.

Click-type tension gauge—a small belt tension gauge that clicks when pushed against the belt and the belt's tension is read.

Medidor de tensión del tipo clic—medidor de tensión de correa pequeño que hace clic cuando se lo presiona contra la correa y mide la presión de la misma.

C-lock or C-clip axle—an axle with a groove on the inside; a retaining clip fits into the groove to keep it in place.

Eje C-lock o C-clip—eje con una ranura en su interior; un gancho de retención se encaja en la ranura para sujetarlo en su lugar.

Clock spring—a spiral cable that electrically connects the steering column to the air bag in the steering wheel.

Resorte-reloj—cable en espiral que une eléctricamente la columna de la dirección a la bolsa de aire en el volante.

Coefficient of friction—a numerical value expressing the amount of friction between two objects.

Coeficiente de fricción—valor numérico que expresa la cantidad de fricción entre dos objetos.

Coil spring—a spring steel rod wound into a coil.

Resorte de bobina—varilla de resorte de acero bobinada.

Collapsible steering column—a steering column designed to collapse in a collision to reduce driver injury.

Columna de la dirección desarmable—columna de la dirección que se desarma en caso de colisión para reducir las lesiones que pueda sufrir el conductor.

Combination valve—a brake valve with two or three functions, including the hydraulic safety switch.

Valor de combinación—válvula de freno con dos o tres funciones, incluyendo el interruptor hidráulico de seguridad.

Compact spare tire—a smaller spare tire and wheel designed for short distance driving at low speed.

Neumático de auxilio compacto—neumático de auxilio de menor tamaño diseñado para manejar distancias cortas en baja velocidad.

Composite drum or rotor—a brake drum made by combining cast iron and steel or aluminum, to reduce weight.

Tambor o rotor compuesto—tambor de freno que se fabrica combinando hierro fundido y acero o aluminio, para reducir el peso.

Composite spring—a spring made of reinforced fiberglass or graphite reinforced plastic.

Resorte compuesto—resorte fabricado con fibra de vidrio reforzada o plástico reforzado de grafito.

Compression—when the wheel moves up as the spring compresses; also called jounce.

Compresión—cuando la rueda se mueve hacia arriba a medida que se comprime el resorte.

Compression-loaded—a ball joint that relies on vehicle weight to force the ball into the joint.

Carga de compresión—una junta esférica que depende del peso del vehículo para forzar la esfera hacia dentro de la junta.

Conduit—a flexible metal housing that protects parking brake cables from dirt, rust, and damage.

Conduit—alojamiento metálico flexible que protege a los cables del freno de estacionamiento contra la tierra, el óxido y daños.

Constant velocity (CV) joint—a universal joint whose output speed is constant, like its input speed.

Junta de velocidad constante—junta universal cuya velocidad de salida es constante, así como su velocidad de entrada.

Control arms—used on independent suspensions to allow the springs to deflect (move up or down).

Brazos de control—usados en suspensiones independientes para permitir la oscilación vertical de los resortes.

Controller antilock brake (CAB)—the computer that controls the ABS.

Control de freno antibloqueo (CAB, en inglés)—la computadora que controla el ABS.

Corporate average fuel economy (CAFÉ)—a fuel economy standard that manufacturers must meet or pay a "gas-guzzler" penalty to the government.

Media de economía de combustible corporativa (CAFÉ, en inglés)—norma de economía de combustible que los fabricantes deben respetar o, de lo contrario, pagar una multa al gobierno.

Cost leaders—service items, such as oil, that are priced by most establishments with little or no markup; also called loss leaders.

Líderes de costos—artículos de servicio, tal como el aceite, cuyo precio es fijado por la mayoría de los establecimientos con poco o ningún margen. También llamados líderes de pérdidas.

Crocus cloth—a very fine polishing cloth.

Trapo pulidor—paño muy fino para pulido.

Cross and yoke—the most popular universal joint design; also called Cardan.

Junta cardán—el diseño de junta universal más popular; también denominado Cardán.

Cross groove plunge joint—a constant velocity joint with balls and grooves in the bearing races that would cross each other if they extended far enough; also called a pancake joint.

Junta de émbolo con ranura transversal—junta de velocidad constante con esferas y ranuras en los aros del rodamiento que se cruzarían entre sí si tuvieran largo suficiente; también denominada junta-panqueque.

Crosshatch—the crisscross pattern left in a bore by a hone.

Nido de abejas—el diseño en cruz que un esmeril deja en el centro.

Crowned—when a road is higher at the center than at the outside.

Abombamiento—cuando un camino es más alto en el centro que en su parte externa.

Cup expander—a metal disc on the inside of a wheel cylinder seal that holds the seal lip against the cylinder when pressure is released.

Dispositivo de expansión de taza—disco metálico en el interior de un sello de cilindro de rueda que sujeta el borde del sello contra el cilindro al soltarse presión.

Cup seal—a round rubber seal with a depressed center and raised sealing lip often used on the front of hydraulic cylinder pistons; it seals pressure when pushing forward but leaks in reverse.

Sello de taza—sello de hule redondo con depresión central y borde de sello elevado usado con frecuencia en la parte delantera de pistones de cilindro hidráulicos; sella la presión cuando se avanza, pero pierde presión en marcha atrás.

Cycle—a series of regular events.

Ciclo—una serie de eventos regulares.

D

Data link connector (DLC)—a vehicle's electrical connection point for a scan tool.

Conector de conexión de datos (DLC, en inglés)—punto de conexión eléctrica en un vehículo para conectar un dispositivo de exploración.

Deflect—move up or down.

Desviación—movimiento hacia arriba o hacia abajo

Deploy—to trigger and expand an air bag.

Abertura—activar y abrir una bolsa de aire.

Diagnostic trouble code (DTC)—a numerical code created by an electronic control system that indicates a problem in a circuit or system.

Código de diagnóstico de problemas (DTC, en inglés)—código numérico creado por un sistema de control electrónico y que indica la existencia de un problema en un circuito o sistema.

Diagonally split—a brake system used on most front-wheel drive vehicles that operates the brakes on opposite corners of the vehicle.

Partición diagonal—sistema de frenos utilizado en la mayoría de los vehículos con tracción delantera y que opera los frenos en esquinas opuestas del vehículo.

Dial caliper—a precision measuring instrument; it can make inside, outside, and depth measurements.

Calibre de disco—instrumento de medición de precisión; puede tomar medidas internas, externas y de profundidad.

Dial indicator—a precision measuring tool that measures movement and displays it on a dial.

Indicador de disco—herramienta de precisión de medición que mide el movimiento y exhibe el resultado en un disco.

Diaphragm—a flexible membrane that separates two areas from each other.

Diafragma—membrana flexible que divide dos áreas.

Directional control—the ability to steer the vehicle.

Control de la dirección—la capacidad de maniobrar el vehículo.

Directional stability—the ability to maintain straight-ahead driving or stopping.

Estabilidad de la dirección—la capacidad de conducir o detenerse en posición recta hacia delante.

Discard diameter or dimension—the allowable wear dimension of a rotor or drum; different than the allowable machining dimension.

Diámetro o dimensión descartable—dimensión de desgaste permisible de un rotor o tambor; no es igual a la dimensión de mecanizado permitida.

Disc brake—a brake that generates friction as its linings rub against the sides of a brake disc or rotor.

Freno de disco—freno que genera fricción a medida que sus revestimientos rozan contra las laterales de un disco de freno o rotor.

Dive—during hard braking, when the front of the car is pushed down and the rear of the car raises up.

Vehículo empinado—al frenar de golpe, cuando la delantera del auto baja y la trasera del auto se eleva.

DIY—do-it-yourself.

DIY—hágalo usted mismo, en inglés

DMM—Digital multimeter.

MMD—Medidor múltiple digital

Dog tracking—when the rear wheels are positioned off to one side or the other rather than directly behind the front wheels.

"Dog tracking"—cuando las ruedas traseras miran hacia un costado en lugar de estar posicionadas directamente detrás de las ruedas delanteras.

DOT—U.S. Department of Transportation.

DOT—Departamento de Transporte de los Estados Unidos.

Double flare—a type of tubing connection in which the end of the tubing is flared out and then is formed back onto itself.

Ensanchamiento doble—un tipo de conexión de cañerías en la que la extremidad del caño se ensancha y luego se vuelve a formar sobre sí misma.

Double offset plunge joint—a CV joint with six balls evenly spaced by a cage; the races have straight grooves, and the joint can plunge farther than a cross groove joint.

Junta de purga desplazada doble—una junta de velocidad constante con seis esferas espaciadas de manera pareja por una jaula; los aros tienen ranuras derechas, y la junta puede estirarse más que una junta con ranuras entrecruzadas.

Double row bearing—a ball bearing found in air-conditioning clutches and front-wheel drive bearing hubs that provides thrust handling capability and resistance to misalignment.

Rodamiento de fila doble—rodamiento de esferas encontrado en embragues de aire acondicionado y cubos de rodamiento de tracción delantera y que proporciona capacidad de manejo de tope y resistencia contra la falta de alineación.

Double wishbone suspension—a multilink suspension system in which the shape described by the links resembles a wishbone.

Suspensión de horquilla doble—sistema de suspensión multi-conexión en el que la forma que describen los conectores parece un hueso del deseo.

Drive line—the parts that transfer power from the transmission to the rear wheels.

Línea de transmisión—las piezas que transfieren potencia de la transmisión a las ruedas traseras.

Drive shaft—a shaft that transfers power from the transmission to the differential, allowing for changes in driveline length and angle when springs deflect as a vehicle goes over bumps; also called a propeller shaft.

Eje de la transmisión—eje que transfiere potencia de la transmisión al diferencial, permitiendo cambios en el largo y el ángulo de la línea de transmisión cuando los resortes oscilan al pasar el vehículo por baches; también llamado eje propulsor.

Drone—a noise compared to the hum of a bumble bee; also called moan.

Zumbido—ruido parecido al zumbido de una abeja.

Drop center—the well between the bead seats of a wheel rim that provides a means for removal and installation of a tire; also called the rim well.

Base honda—el espacio libre entre los asientos de talón del aro de la rueda que proporciona un medio de remoción e instalación de un neumático.

Drum brake—a brake that generates friction as brake shoes rub against the inside surface of a brake drum.

Freno de tambor—freno que genera fricción cuando las zapatas de freno rozan la superficie interior de un tambor de freno.

Drum-in-hat—an auxiliary internal drum-type parking brake, housed within a section of the rear disc rotor.

Freno auxiliar tipo tambor—freno de estacionamiento auxiliar interno del tipo tambor, alojado dentro de una sección del rotor del disco trasero.

Drum mike—a dial indicator tool for measuring brake drums.

Medidor de tambor—herramienta de disco indicadora para medir tambores de freno.

Dual-servo brake—a self-energizing drum brake where the primary shoe applies servo action to the secondary shoe to increase its application force; also called a duo-servo or a full-servo brake.

Freno de servo dual—freno de tambor autodinámico propia donde la zapata principal aplica acción de servo a la zapata secundaria para aumentar su fuerza de aplicación.

Dump—an ABS command to release hydraulic pressure to an individual wheel brake.

Descargar—comando del ABS para liberar la presión hidráulica hacia un freno de rueda individual.

Dynamic balance—balance of a tire and wheel in motion.

Equilibrio dinámico—el equilibrio de un neumático y una rueda en movimiento.

E

Earless clamp—a type of clamp used to fasten CV joint boots to the axle; also called a low profile or locking clamp.

Abrazadera sin oreja—tipo de abrazadera utilizada para sujetar fuelles de junta de velocidad constante al eje; también llamada abrazadera de bajo perfil o de traba.

Eccentric cam adjustment—an out-of-round adjuster used in some drum brakes.

Ajuste de árbol de levas excéntrico—dispositivo de ajuste utilizado en ciertos frenos de tambor.

Edge code—numbers and letters printed on the side of disc or drum friction material that identify the manufacturer, the lining material, and the coefficient of friction.

Código de borde—números y letras grabados en la lateral de material de fricción de disco o tambor que identifica al fabricante, el material de revestimiento y el coeficiente de fricción.

Electromechanical hydraulic (EH) unit—an ABS part with the computer and hydraulics combined.

Unidad hidráulica electromecánica—pieza del ABS con computadora e hidráulica combinada.

Electronic brake and traction control module (EBTCM)—a computer that controls antilock brake, traction control, and vehicle stability control.

Módulo de control electrónico de freno y tracción (EBTCM, en inglés)—computadora que controla el freno antibloqueo, el control de tracción y el control de estabilidad del vehículo.

Electronic control module (ECM)—the computer that controls the ABS.

Módulo de control electrónico (ECM, en inglés)—la computadora que controla el ABS.

Electronic vibration analyzer (EVA)—a tester that measures the amount of a vibration.

Analizador electrónico de vibración (EVA, en inglés)—dispositivo de prueba que mide la cantidad de vibración.

End play—the designed looseness in a bearing assembly.

Juego final—el juego diseñado en un conjunto de rodamiento.

Energy—the ability to do work.

Energía—la capacidad de realizar trabajos.

Environmental Protection Agency (EPA)—the U.S. government agency in charge of environmental protection.

Agencia de Protección Ambiental (EPA, en EE.UU.)—la entidad gubernamental de los Estados Unidos a cargo de la protección del medio ambiente.

EP additive—added to lubricants to prevent welding between metal surfaces under extreme pressures.

Aditivo EP—lubricantes agregados para evitar que la soldadura entre superficies metálicas bajo presiones extremas.

Equalizer—a piece of the parking brake linkage, usually with a provision for adjustment, that balances the force applied to each rear wheel.

Ecualizador—pieza de una conexión de freno de estacionamiento, generalmente con un dispositivo de ajuste, que equilibra la fuerza aplicada a cada rueda trasera.

F

Fade—a loss of brake lining coefficient of friction due to excessive heat, resulting in poor stopping ability.

Desvanecimiento—pérdida de coeficiente de fricción del revestimiento del freno debido a calor excesivo, y que provoca el deterioro de la capacidad de frenado.

Fast steering ratio—when a steering gear moves the front wheels from lock to lock with less rotation of the steering wheel.

Relación de maniobra rápida—cuando un engranaje de la dirección mueve las ruedas traseras de traba a traba con menor giro del volante.

Feeler gauge—a thin strip of metal used to measure clearances.

Calibrador tipo alambre—una tira fina de metal utilizada para medir espacios libres.

Fillet—the smooth curve where the shank flows into the bolt head.

Filete—la curva suave donde el mango fluye hacia dentro de la cabeza del perno.

Finish cut—a fine cut that results from a relatively fast rotation speed with a slow feed rate.

Corte de acabado—corte fino, consecuencia de una velocidad de rotación relativamente rápida con una velocidad de alimentación lenta.

Firewall—the metal wall between the engine and passenger compartments; also called the bulkhead.

Mampara—la pared metálica entre el motor y el compartimiento de pasajeros.

First order vibration—something that vibrates only once every revolution.

Vibración de primera orden—algo que vibra apenas una vez por revolución.

Fixed caliper—a disc brake caliper that does not move when the brakes are applied like a sliding caliper does; it has pistons on both the inboard and outboard sides.

Mordaza fija—una mordaza de freno de disco que no se mueve al aplicarse los frenos como lo haría una mordaza deslizante; tiene pistones en ambos lados, interno y externo.

Fixed joint—the outboard CV joint that allows for a change in the angle of the axle in response to bumps and for steering.

Junta fija—la junta de VC externa que permite un cambio en el ángulo del eje como respuesta a baches y al maniobrar.

Flange—a part bolted to the splined shaft that comes out of the front of the differential or transmission and enables a shaft to be fastened to it.

Brida—una pieza atornillada al eje con ranuras y que sale de la parte delantera del diferencial o transmisión y permite que se sujete el eje al mismo.

Flared line—tubing that is tapered outward at its end.

Caño dilatado—caño que se ensancha hacia fuera en su extremidad.

Flare nut—a hollow fitting on a fuel, brake, or hydraulic line.

Tuerca cónica—un conector hueco en una línea de combustible, frenos o hidráulica.

Flaring tool—a tool used to flare the ends of the tubing with either an SAE double flare or an ISO (metric double) flare.

Mandril—herramienta utilizada para ensanchar las extremidades del caño con ensanchamiento doble SAE o ensanchamiento ISO (doble métrico).

Flat rate—a system of pay where each job has a flat-rate time and technicians are paid for the amount of work accomplished.

Tasa fija—sistema de pago en el que cada trabajo cuenta con un tiempo de tasa fija y se les paga a los técnicos por la cantidad de trabajo realizada.

Flat-rate manual—a manual that lists the length of time repairs take to complete and includes a parts list with prices; used to estimate the cost of a repair.

Manual de tasa fija—manual en el que figura el tiempo que llevan las reparaciones para ser efectuadas y que incluye una lista de piezas con precios; se lo utiliza para estimar el costo de las reparaciones.

Floating axle—an axle design in which the bearings do not touch the axle but are located on the outside of the axle housing.

Eje flotante—diseño de eje en el que los rodamientos no tocan el eje, sino que están ubicados en la parte exterior del alojamiento del eje.

Floating caliper—a caliper that slides on a pin or in a groove. The piston, or pistons, will only be on one side of the caliper.

Mordaza flotante—mordaza que se desliza sobre un perno o dentro de una ranura. El pistón, o pistones, estará apenas en un lado de la mordaza.

Flow control valve—a power steering valve that controls fluid flow relative to demand.

Válvula de control de flujo—válvula de dirección hidráulica que controla el flujo del líquido con relación a la demanda.

Fluted lip seal—a seal with a lip that has grooves to direct the lubricant back, rather than leaking past the seal lip.

Sello con reborde con estrías—un sello con un reborde con ranuras para encaminar el lubricante de regreso, en lugar de ocasionar pérdidas de lubricante después de pasar por el reborde del sello.

Follower ball joint—a ball joint that does not support a load but maintains correct position.

Junta esférica de seguidor—junta esférica que no soporta una carga, pero que mantiene la posición correcta.

Footprint—the area of the tire tread that contacts the road.

Huella—área del neumático que toca el camino.

Force—power working against resistance to cause motion.

Fuerza—potencia que trabaja contra la resistencia para provocar movimiento.

Four-wheel alignment—a method of wheel alignment done on many newer vehicles equipped with rear-wheel adjustment capability

Alineación de cuatro ruedas—método de alineación de ruedas realizado en muchos vehículos de fabricación reciente equipados con capacidad de ajuste de ruedas traseras.

Frame-contact lift—a lift design where the wheels hang free.

Elevador de contacto con armazón—diseño de elevador en el que las ruedas quedan colgando.

Frequency—a measurement of the rate of occurrence of cycles of electric current or sound waves; indicates the number of times that a signal occurs in cycles per second (hertz).

Frecuencia—medida de la tasa de ocurrencia de ciclos de corriente eléctrica u ondas sonoras; indica la cantidad de veces que ocurre una señal en ciclos por segundo (hertz).

Friction—resistance to motion when the surface of one object moves against the surface of another object.

Fricción—resistencia al movimiento cuando la superficie de un objeto roza la superficie de otro objeto.

Front-wheel drive—vehicles made with a subframe in the front that includes the engine, transaxle, and steering/suspension system.

Tracción delantera—vehículos fabricados con un armazón secundario delantero que incluye el motor, el transeje y el sistema de dirección/suspensión.

Fulcrum—a lever's pivot point.

Punto de apoyo—punto de pivote de una palanca.

Full-floating axle—an axle design in which the bearings do not touch the axle but are located on the outside of the axle housing.

Eje flotante pleno—diseño de eje en el que los rodamientos no tocan el eje, sino que están ubicados en la parte exterior del alojamiento del eje.

Fuse—a device to protect an electrical circuit against an overload.

Fusible—dispositivo para proteger un circuito eléctrico contra sobrecargas.

G

Galling—wear caused by metal to metal contact in the absence of adequate lubrication. Metal is transferred from one surface to another.

Gripado—desgaste provocado por el contacto entre metales sin lubricación adecuada. El metal se transfiere de una superficie a la otra.

Garter spring—the spring behind a circular seal lip.

Resortes de liga—resorte detrás de un borde de sello circular.

Gas fade—brake fade that occurs very rarely when linings are extremely overheated and the organic binding agent releases a thin layer of hot gas, which causes the pad to hydroplane.

Desvanecimiento de gas—desvanecimiento de freno que ocurre raramente cuando los revestimientos se sobrecalientan demasiados y el agente de adhesión orgánico suelta una capa fina de gas caliente que hace que la almohadilla se eleve.

Gas-filled shock absorber—a shock that contains a gas charge in addition to its hydraulic fluid.

Amortiguador con carga de gas—amortiguador que contiene una carga de gas, además de fluido hidráulico.

Generic manuals—printed information on all makes of foreign and domestic vehicles.

Manuales genéricos—información impresa sobre todas las marcas de vehículos importados y domésticos.

Geometric centerline—a line drawn between the center of the front axle and the center of the rear axle.

Línea de centro geométrico—línea dibujada entre el centro del eje delantero y el centro del eje trasero.

Gravity bleeding—a brake bleeding process where old brake fluid and air drain through an open bleed screw.

Purga de gravedad—proceso de purga de frenos en el que el líquido de freno viejo y aire drenan a través de un tornillo de purga abierto.

Grease—a combination of oil and a thickening agent.

Grasa—combinación de aceite y un agente de espesamiento.

Gross vehicle weight rating (GVWR)—the total weight of a vehicle plus its maximum rated payload.

Índice de peso bruto de vehículo (GVWR, en inglés)—el peso total de un vehículo, más su carga nominal máxima.

H

Half shaft—a drive axle on a front-wheel drive vehicle; also called the axle shaft.

Medio eje—eje de la transmisión en un vehículo de tracción delantera.

Hazard Communication Standard—a document published by the Occupational Safety and Health Administration (OSHA), the basis of the Right-to-Know laws.

Norma de Comunicación de Peligros—documento publicado por la Administración de la Seguridad y Salud Ocupacional (OSHA, Occupational Safety and Health Administration), base de las leyes "Derecho de Saber".

Hazardous waste—waste with one or more of the following characteristics: ignitability, corrosivity, reactivity, and toxicity.

Residuos peligrosos—residuos con una o más de las siguientes características: inflamabilidad, corrosividad, reactividad y toxicidad.

Heat checks—small cracks on the friction surfaces of a drum or rotor.

Cuarteado por choque térmico—pequeñas fisuras en las superficies de fricción de un tambor o rotor.

Height sensor—a sensor that provides input to the computer relative to chassis height.

Sensor de altura—sensor que envía información sobre la altura del chasis a la computadora.

Height-sensing proportioning valve—a proportioning valve that adjusts brake hydraulic pressure in relation to the rear axle-to-frame height during braking.

Válvula de proporción de indicación de altura—válvula de proporción que ajusta la presión hidráulica del freno con respecto a la altura del eje trasero al armazón durante el frenado.

HEPA filter—a high-efficiency particulate air filter.

Filtro HEPA—filtro de aire de alta eficiencia.

Hertz—frequency unity of one cycle per second

Hertz—unidad de frecuencia de un ciclo por segundo.

High cordline belt—a higher quality accessory V belt with the tensile cord positioned above center.

Correa de cuerda alta—Correa en V y accesorio de alta calidad, con la cuerda tensora posicionada en su parte superior.

High point—the point where the teeth of a conventional steering gear come closer together when the vehicle is traveling straight ahead.

Punto alto—el punto en que los dientes de un engranaje de dirección convencional se acercan entre sí cuando el vehículo se mueve derecho hacia delante.

Hold valve—an ABS valve that prevents fluid from returning to the master cylinder during ABS traction control or vehicle stability control.

Válvula reguladora—válvula del ABS que evita que el líquido regrese al cilindro maestro durante el control de tracción o control de estabilidad del vehículo del ABS.

Holddown spring—a spring that holds a drum brake shoe against the backing plate.

Resorte de sujeción—resorte que sujeta la zapata del freno de tambor contra el contrafuerte.

Honing—a machining process that uses abrasive stones, rotated inside a cylinder, to restore a uniform finish.

Esmerilado—proceso de mecanizado que usa piedras abrasivas que giran dentro de un cilindro para restaurar un acabado uniforme.

Hotchkiss design—an open drive shaft.

Diseño hotchkiss—un eje abierto de la transmisión.

Howl—a noise like the wind across the top of a soda bottle; its range is 120–300 Hz.

Aullido—ruido parecido al que ocurre al soplar sobre la parte superior de una botella de refrigerante; su rango es de entre 120 y 300 Hz.

Hub—a wheel bearing and hub assembly to which the wheel rim is bolted.

Cubo—un conjunto de rodamiento y cubo de rueda al que se atornilla el aro de la rueda.

Hub-centric—when the center of the wheel has a machined counterbore that pilots on a machined area of the hub.

Cubo central—cuando el centro de la rueda tiene un refundido mecanizado que pilota en un área mecanizada del cubo.

Hydraulic—moved or affected by liquid.

Hidráulico—movido o afectado por líquido.

Hydro-boost brake—a power brake system powered off the power steering pump; developed by Bendix.

Freno hydro-boost—sistema de freno de potencia que recibe su potencia de la bomba de dirección hidráulica; creado por Bendix.

Hydroplaning—When water forms a wedge between a tire and the road, taking the vehicle out of contact with the pavement; also called aquaplaning.

Elevación sobre el agua—cuando el agua forma una cuña entre una rueda y el camino, y el vehículo pierde el contacto con el pavimento.

Hygroscopic—the tendency to attract water and absorb moisture from air.

Higroscópico—tendencia a atraer el agua y absorber la humedad del aire.

Hysteresis—a term used by chemical engineers to describe a rubber's energy absorption characteristics.

Histéresis—término usado por ingenieros químicos para describir las características de absorción de energía del hule.

I

Idler arm—a short, pivoting steering linkage connecting the vehicle frame to the center link.

Brazo de engranaje—conexión corta de la dirección de pivote que une el armazón del vehículo al conector central.

Impact socket—a thick-walled socket designed for use with an impact wrench.

Toma de impacto—toma de paredes gruesas diseñada para su uso con una llave de apriete de tornillos.

Inactive coils—coils located at the top and bottom ends of a coil spring.

Bobinas inactivas—bobinas ubicadas en las extremidades superiores e inferiores de un resorte de bobina.

Inboard side—the inside end of the axle shaft.

Lado interno—extremidad interna del eje de la rueda.

Included angle—the combination of the camber and steering axis inclination angles.

Ángulo incluido—combinación de los ángulos de inclinación de la comba y el eje de la dirección.

Independent suspension—a front or rear suspension system in which one wheel can move upward or downward without affecting the opposite wheel.

Suspensión independiente—sistema de suspensión delantera o trasera en el que una rueda puede moverse hacia arriba o hacia abajo sin afectar la rueda opuesta.

Individual toe—toe measured at each wheel in reference to the geometric centerline of the vehicle.

Convergencia individual—convergencia medida en cada rueda con respecto a la línea de centro geométrica del vehículo.

Inertia—the tendency of an object in motion to remain in motion or an object at rest to remain at rest.

Inercia—la tendencia de un objeto en movimiento de permanecer en movimiento o de un objeto en reposo de permanecer inmóvil.

Inflation pressure—the amount of air pressure required to inflate a tire, measured in pounds per square inch (psi) or kilopascals (kPa).

Presión de inflado—cantidad de presión de aire requerida para inflar un neumático, medida en libras por pulgada cuadrada (psi) o kilopascals (kPa).

Inner liner—the liner bonded to the inside of a tire that seals air into the tire.

Revestimiento interno—el revestimiento adherido a la parte interior del neumático que sella el aire dentro del neumático.

Inner race—the center of a ball or tapered roller bearing that supports the balls or rollers; also called a bearing cone.

Pista de rodadura interna—centro de un rodamiento esférico o cónico de rodillo que sostiene las esferas o rodillos; también denominada cono de rodamiento.

In phase—when the trunnions of the front and rear universal joints are parallel, or in the same plane.

En fase—cuando los muñones de las juntas universales delanteras y traseras son paralelos o se encuentran en el mismo plano.

Integral ABS—a type of antilock brake design in which the master cylinder, power booster, and ABS are combined as a complete unit.

ABS Integral—tipo de diseño de freno antibloqueo en el que el cilindro maestro, el booster de potencia y el ABS están combinados formando una unidad completa.

Integral power steering—when power steering parts are integral with the steering gear.

Dirección hidráulica integral—cuando las piezas de la dirección hidráulica están integradas al engranaje de la dirección.

Integral reservoir—a reservoir that is part of the power steering pump.

Tanque de reserva integral—tanque de reserva que forma parte de la bomba de la dirección hidráulica.

Interference fit—a precise fit between two components that provides a tight fit.

Calce de interferencia—calce preciso entre dos componentes; provee un calce justo.

ISO flare—a type of tubing flare connection with a bubble-shaped tubing end, also called a bubble flare.

Ensanchamiento ISO—tipo de conexión de caño ensanchado con una extremidad del caño con forma de burbuja; también denominado ensanchamiento de burbuja..

Isoprene—the main substance of natural rubber.

Isopreno—sustancia principal del hule natural.

J

Jam nut—a locknut.

Contratuerca—una tuerca de trabamiento.

Jobbers—a wholesale automotive parts establishment.
Mayorista—establecimiento mayorista de piezas automotoras.
Jounce—upward movement of the wheel and suspension.
Sacudida—movimiento hacia arriba de la rueda y la suspensión.

K

Kinetic energy—the energy of mechanical work or motion.
Energía quinética—energía de trabajo mecánico o movimiento.
King pin inclination (KPI)—the angle of a line through the center of the king pin in relation to the true vertical centerline of the tire, viewed from the front of the vehicle.
Inclinación del pivote de dirección (KPI, en inglés)—el ángulo de una línea que atraviesa el centro del pivote de dirección con respecto a la línea de centro vertical real del neumático, vista desde la parte delantera del vehículo.
Knockoff-type CV joint—an internal circlip CV joint retaining design.
Junta de velocidad constante eyectora—diseño de retención de junta de velocidad constante con clip interno.
kPa—kilopascal, the metric equivalent of psi.
kPa—kilopascal, el equivalente métrico de psi.

L

Ladder frame design—a frame with cross members between the side rails that resemble steps on a ladder.
Diseño de armazón tipo escalera—armazón con varillas horizontales entre los rieles laterales que parecen los escalones de una escalera.
Lands—high areas on a spool valve.
Pistas—áreas altas de una válvula de carrete.
Lateral runout—side-to-side wobble as the tire and wheel are rotated.
Descentramiento lateral—movimiento de lado a lado cuando la llanta y rueda giran.
Lateral acceleration sensor—a sensor that measures the force encountered while turning.
Sensor de aceleración lateral—sensor que mide la fuerza existente al girar.
Lathe-cut seal—a square cut seal for a caliper piston that has a square or irregular cross section; not round like an O-ring.
Sello cortado con torno—sello con corte cuadrado para pistón de mordaza, que tiene una sección transversal cuadrada o irregular y no redonda como un anillo en 'O'.
Launch shudder—a second order vibration that is a common complaint in raised trucks with a high drive shaft angle.
Vibración de lanzamiento—vibración de segunda orden que es una queja común en camiones elevados con ángulo de eje de transmisión alto.
Leading shoe—the first shoe in the direction of drum rotation in a non-servo brake.
Zapata principal—la primera zapata en la dirección de rotación del tambor en un freno que no sea servo.
Lead point—a point in front of true vertical; also called point of load.
Punto-guía—punto delante de vertical real; también llamado punto de carga.
Leaf spring—a spring made of a long, flat strip of spring steel rolled at both ends to accept a press-fit rubber insulating bushing.
Resorte de lámina—resorte hecho con una tira larga y plana de acero laminado en ambas extremidades para aceptar un buje aislador de hule a presión.
Leverage—using a lever and fulcrum, such as the brake pedal, to create a mechanical advantage.
Apalancamiento—uso de una palanca y punto de apoyo, tal como el pedal del freno, para crear una ventaja mecánica.

Lift—when the front of the vehicle lifts upward during hard acceleration.
Elevación—cuando el frente del vehículo se eleva durante al acerar de golpe.
Linear rate coil spring—a coil spring with equal spacing between the coils and a constant deflection rate regardless of load.
Resorte de bobina de tasa linear—resorte de bobina con espacios de igual tamaño entre las bobinas y una tasa de deflexión constante, independiente de la carga.
Lining fade—results when the temperature of the linings increases, resulting in more pedal effort to stop the car.
Desvanecimiento del revestimiento—ocurre cuando aumenta la temperatura del revestimiento, haciendo necesario un mayor esfuerzo de frenado para detener el vehículo.
Linkage-type power steering—a power steering system with the control valve mounted on the pitman arm end of the center link; the power cylinder connects the frame to the center link.
Dirección hidráulica del tipo eslabón—un sistema de dirección hidráulica con la válvula de control montada sobre la extremidad del brazo de la varilla de conexión; el cilindro de fuerza une el armazón al conector central.
List price—the suggested price of an item, or what the customer paying for car repair pays; also called the retail price.
Precio de lista—el precio sugerido de un artículo, o lo que el cliente paga por la reparación del automóvil; también llamado precio al por menor o precio minorista.
Live axle—an axle that turns with the wheels.
Eje vivo—eje que gira con las ruedas.
Load index—the maximum load at the designated speed rating.
Índice de carga—la carga máxima a la velocidad designada.
Load-carrying ball joint—the ball joint on the control arm that has the spring mounted on it.
Junta esférica con carga—junta esférica en el brazo de control sobre la cual está montado el resorte.
Loaded caliper—a complete rebuilt caliper assembly ready for installation.
Mordaza cargada—conjunto reacondicionado completo de mordaza, listo para instalar.
Load-leveling shock—a shock absorber in which air pressure is added to increase vehicle height.
Amortiguador nivelador de carga—amortiguador al que se agrega presión de aire para aumentar la altura del vehículo.
Load rating—a tire rating that indicates its load-carrying capability.
Índice de carga—clasificación de llanta que indica su capacidad de carga.
Lock to lock—a complete sweep of the front wheels all the way from one direction to the other.
Traba a traba—un barrido completo de las ruedas delanteras de una dirección a la otra.
Longitudinally split—a brake system used on most rear-wheel drive vehicles in which the front and rear brakes are operated as separate hydraulic systems.
División longitudinal—sistema de frenos usado en la mayoría de los vehículos con tracción trasera y en el que los frenos delanteros y traseros funcionan como sistemas hidráulicos separados.
Low pedal—when the brake pedal moves closer to the floor before the brakes apply.
Pedal bajo—cuando el pedal del freno se acerca al piso antes de que los frenos funcionen.
Lubrication service manual—a manual used during lubrication/safety checks.
Manual de servicio de lubricación—manual usado durante los chequeos de lubricación/seguridad.
Lug-centric—wheels that are centered using the wheel studs; also called stud-centric.

rueda de agarradera céntrica—ruedas centralizadas a través del uso de prisioneros de rueda.

MacPherson strut—a suspension design that incorporates the shock absorber into the front suspension.

Puntal MacPherson—diseño de suspensión que incorpora el amortiguador en la suspensión delantera.

Magnasteer—an electronic power steering system that uses magnetic force.

Magnasteer—sistema de dirección hidráulica electrónico que usa fuerza magnética.

Master cylinder—the cylinder connected to the brake pedal that provides hydraulic pressure to the brake system.

Cilindro maestro—cilindro conectado al pedal del freno y que provee presión hidráulica al sistema de frenos.

Match mount—when runout of the tire and runout of the wheel are considered during assembly so they can cancel each other out during balancing.

Combinación óptima de montaje—cuando se tiene en cuenta el descentramiento de la llanta y de la rueda durante el montaje para que se cancelen entre sí al balancear las ruedas.

Material safety data sheets (MSDS)—information sheets that provide data about potentially hazardous materials.

Hojas de datos de seguridad de materiales (MSDS)—hojas informativas con datos sobre materiales potencialmente peligrosos.

Mechanical fade—fade that occurs in drum brakes when the drum expands as temperature increases, resulting in lower pedal height.

Desvanecimiento mecánico—desvanecimiento que ocurre en los frenos de tambor cuando el tambor se expande al aumentar la temperatura y provoca una altura de freno más baja.

Memory steer—when the steering wheel does not return to the straight-ahead position after a turn.

Dirección de memoria—cuando el volante no regresa a la posición derecha después de un giro.

Metallic lining—brake friction material made from powdered metal formed by heat and pressure.

Revestimiento metálico—material de fricción de frenos hecho de metal en polvo formado por calor y presión.

Metering valve—a control valve that delays pressure to front disc brakes until the rear brakes begin to operate.

Válvula de medición—válvula de control que demora la presión a los frenos de disco delanteros hasta que los frenos traseros comiencen a funcionar.

Microfiche—a small plastic film card magnified by a microfiche reader.

Microficha—tarjeta plástica pequeña magnificada por un lector de microfichas.

Micrometer—a precision measuring device used to measure small bores, diameters, and thicknesses; also called a mike.

Micromedidor—dispositivo de medición de precisión usado para medir interiores, diámetros y espesores pequeños.

Microprocessor—a computer built on a single integrated circuit chip that can perform arithmetic and control logic functions.

Microprocesador—una computadora construida sobre un chip único de circuito integrado y que puede realizar funciones aritméticas y de lógica de control.

Moan—a noise compared to the hum of a bumble bee; also called drone.

Quejido—ruido parecido al zumbido de una abeja; también llamado zumbido.

Momentum—the force of continuing motion of a moving object equal to mass times speed.

Ímpetu—la fuerza del movimiento continuo de un objeto en movimiento igual a masa por velocidad.

Monoleaf spring—a spring made of a single strip of steel, thicker in the center and gradually tapering thinner toward the outside ends.

Resorte de una lámina—resorte hecho con una única tira de acero, más gruesa en el centro y que se va afinando gradualmente hacia las extremidades externas.

Morning sickness—a term applying to power steering or automatic transmissions that describes a lack of hydraulic pressure buildup when seals are cold and hard.

Náuseas mañaneras—término usado para direcciones hidráulicas o transmisiones automáticas y que describe la falta de acumulación de presión hidráulica cuando los sellos están fríos y duros.

Motorist Assurance Program (MAP)—a program that establishes uniform inspection guidelines to improve customer satisfaction.

Programa de Garantía al Conductor (MAP, Motorist Assurance Program)—programa que establece directrices uniformes de inspección para aumentar la satisfacción del cliente.

Multilink suspension—a suspension system with multiple links in place of A-arms.

Suspensión multiconector—sistema de suspensión con conectores múltiples en lugar de brazos en A.

Multimeter—a meter that includes a voltmeter, ohmmeter, and ammeter.

Multímetro—medidor que incluye un voltímetro, un medidor de ohms y un medidor de ams.

N

Natural frequency—the frequency at which a body vibrates; also called resonant frequency.

Frecuencia natural—la frecuencia a la que el chasis vibra; también llamada frecuencia de resonancia.

Needle bearing—a bearing with small rollers.

Rodamiento de aguja—rodamiento con rodillos pequeños.

Negative camber—when the camber line tire is tilted inward in relation to true vertical when viewed from the front of the vehicle.

Mordaza negativa—cuando la llanta de la línea de la mordaza está inclinada hacia dentro en relación con la vertical real cuando se la ve desde el frente del vehículo.

Negative caster—when the caster line tire is tilted forward when viewed from the side.

Ángulo de avance negativo—cuando la llanta de la línea de cáster (ángulo de avance) está inclinada hacia delante, vista lateralmente.

Negative scrub radius—when the steering axis inclination line meets the road surface outside the true vertical centerline of the tire.

Radio de pivotamiento negativo—cuando la línea de inclinación del eje de la dirección se encuentra con la superficie del camino fuera de la línea de centro de la vertical real de la llanta.

Neoprene—a type of oil-resistant artificial rubber.

Neopreno—tipo de hule artificial resistente al aceite.

Net price—the price that a wholesale auto repair customer pays.

Precio líquido—el precio que paga el cliente mayorista de reparación de vehículos.

NLGI—National Lubricating Grease Institute.

NLGI—National Lubricating Grease Institute (Instituto Nacional de la Grasa Lubricante).

Non-load-carrying ball joint—a ball joint that holds parts in position but does not support chassis weight; also called a follower.

Junta esférica sin carga—junta esférica que mantiene piezas en posición, pero no soporta peso del chasis; también llamada seguidor.

Noncompressible—a liquid whose volume is not decreased by an increase in pressure.

No comprimible—un líquido cuyo volumen no disminuye por el aumento de presión.
Nondirectional finish—a superior finish on the surface of a rotor.
Acabado no direccional—acabado superior en la superficie de un rotor.
Nonintegral ABS—also called an add-on or remote ABS unit; it is separate from its master cylinder and power booster.
ABS no integral—también llamado agregado o unidad de ABS remota; está separado del cilindro maestro y booster de potencia.
Non-servo brake—a drum brake with self-energizing action on the leading shoe only.
Frenos no servo—frenos de tambor con acción autodinámica apenas en la zapata principal.

O

OEM—see original equipment manufacturer.
OEM—ver fabricante de equipos originales.
Orbital steer—occurs when a wheel with tie-rods at unequal heights goes over a bump and the car momentarily steers in the direction that the wheel turns as the toe on that one wheel changes; also called bump steer.
Dirección orbital—cuando una rueda con varillas de conexión con alturas desparejas pasa por un bache, el vehículo se mueve temporalmente en la dirección en que gira la rueda mientras cambia la convergencia en esa rueda; también denominada dirección de baches.
Organic lining—brake friction material made from nonmetallic fibers.
Revestimiento orgánico—material de fricción de frenos hecha de fibras no metálicas.
Original equipment manufacturer (OEM)—the vehicles manufacturer, i.e., DaimlerChrysler, Ford, Honda.
Fabricante de equipos originales (OEM)—el fabricante de vehículos como, por ejemplo, DaimlerChrysler, Ford, Honda.
OSHA—see U.S. Occupational Safety and Health Administration.
OSHA—vea Administración de la Seguridad y la Salud Ocupacionales de los Estados Unidos.
Outboard side—the outside end nearest the wheel.
Lado externo—la extremidad externa más próxima a la rueda.
Outer race—the outer cup of a ball or roller bearing.
Taza exterior—la taza externa de una esfera o cojinete de rodillo.
Overload leaves—extra parts of a leaf spring that come into play only when the vehicle is heavily loaded.
Láminas de sobrecarga—piezas adicionales de un resorte de lámina que funcionan únicamente cuando el vehículo lleva una carga muy pesada.
Oversteer—when a vehicle turns too far in response to steering wheel movement.
Maniobra excesiva—cuando el vehículo gira demasiado en respuesta a un movimiento del volante.
oxyacetylene welding—gas welding or torch welding; can also be used to cut or heat metal.
Soldadura con oxiacetileno—soldadura con gas o soplete; también puede usarse para cortar o calentar metal.

P

Panhard rod—a rod from the chassis to the rear suspension that reduces body sway; also called the Watts rod or track bar.
Varilla Panhard—varilla del chasis a la suspensión trasera que reduce la oscilación del armazón; también denominada varilla Watts.
Parallelism—a measurement of disc brake rotor thickness uniformity; both surfaces must be within 0.001-inch thickness of each other at several places.
Paralelismo—medida de la uniformidad del espesor del rotor de freno de disco; ambas superficies deben tener una diferencia de espesor no superior a 0,001 pulgadas en varios lugares.
Parallelogram steering—a steering linkage that makes the shape of a parallelogram during a turn and has its tie-rods paralel to the lower control arms.
Dirección de paralelograma—Conector de dirección que hace la forma de un paralelograma al girar y cuyas varillas de conexión son paralelas a los brazos de control inferiores.
Parking brake—a brake that holds the vehicle stationary; independent of the service brakes.
Freno de estacionamiento—freno que mantiene al vehículo estacionario; es independiente de los frenos de servicio.
Passive suspension system—a normal suspension whose height varies according to mechanical forces on the suspension and does not adjust to these changes.
Sistema de suspensión pasivo—suspensión normal cuya altura varía de acuerdo con fuerzas mecánicas sobre la suspensión y que no se ajusta a dichos cambios.
Pedal free play—the clearance between the pushrod from the brake pedal or power booster and the primary piston in the master cylinder.
Juego libre de pedal—el espacio libre entre la varilla de empuje del pedal del freno o booster de potencia y el pistón principal del cilindro maestro.
Perimeter-type frame—a vehicle frame with the side rails near the side of the body.
Armazón del tipo perimetral—armazón de vehículo con rieles laterales próximos a la lateral de la carrocería.
Phase control unit—an electromechanical control unit that supplies hydraulic pressure to the rear wheel actuator in an electronically controlled, hydraulically actuated four-wheel steering system.
Unidad de control de fase—una unidad de control electromecánico que provee presión hidráulica al accionador de la rueda trasera en un sistema de dirección de cuatro ruedas accionado hidráulicamente y con control electrónico.
Phenolic—a thermoset plastic.
Fenólico—plástico termoestable.
Pickup coil—a sensor that generates a voltage signal as a conductor moves through a permanent magnetic field.
Bobina captadora—sensor que genera una señal de voltaje a medida que el conductor se movimenta a través de un campo magnético permanente.
Pinch bolt—a type of ball joint connection.
Perno de apriete—tipo de conexión de junta esférica.
Pinion—the drive gear in rack-and-pinion steering gear.
Piñón—el engranaje de la transmisión en el engranaje de transmisión de cremallera y piñón.
Piston stop—a raised metal part of the brake backing plate that prevents a wheel cylinder piston from coming all the way out of the cylinder bore.
Tope de pistón—una pieza metálica del contrafuerte del freno que evita que el pistón del cilindro de la rueda se salga totalmente del interior del cilindro.
Pitman arm—a short metal arm that connects the steering gear to the center link.
Brazo de varilla de conexión—un brazo metálico corto que une el engranaje de la dirección al conector central.
Pitman shaft sector shaft—part of a recirculating ball steering gear with gear teeth on a shaft that meshes with the recirculating ball gear.
Eje del sector de eje de la varilla de conexión—pieza de un engranaje de dirección esférico recirculante con dientes de engranaje en un eje que engrana con el engranaje esférico recirculante.
Placard—a label that includes maximum vehicle load, tire size, and cold inflation pressure.

Placard—rótulo que incluye la carga máxima del vehículo, el tamaño de llanta y la presión de inflación fría.

Plies—plies surround both beads and extend around the carcass of the tire to increase its load-carrying abilities.

Telas de la cubierta—Las telas de la cubierta rodean a ambos aros y se extienden alrededor del cuerpo de la llanta a fin de incrementar sus capacidades de carga.

Plunge joint—the inboard CV joint that also provides for the changing length of the axle as the spring deflects.

Junta de émbolo—la junta interna de VC que también proporciona el cambio de largo del eje durante la deflexión del resorte.

Plus sizing—when a tire with a lower profile is being installed, a wider tire and a wheel with a larger diameter will be used to make up the difference in overall tire assembly height.

Tamaño plus—Al instalar una llanta con un perfil más, se usa una llanta más ancha y una rueda con un diámetro mayor para compensar la diferencia en la altura total del conjunto del neumático.

Ply—a metal or fabric cord that is rubberized (covered with a layer of rubber) to provide tire strength to support the load of the vehicle.

Tela—una cuerda metálica o de tela cubierta por una capa de hule para proporcionar fuerza a la llanta para que pueda soportar la carga del vehículo.

P-metric radial tire size—the most common size identification system for passenger car tires.

Tamaño de llanta radial P-métrico—el sistema de identificación de tamaño más común para llantas de automóviles de pasajeros.

Polyglycol—the material from which DOT 3 and DOT 4 brake fluid are made.

Poliglicol—el material del cual se fabrican los líquidos de freno DOT 3 y DOT 4.

Positive camber—when the tire is tilted outward in relation to true vertical when viewed from the front of the vehicle.

Mordaza positiva—cuando la llanta de la línea de la mordaza está inclinada hacia fuera en relación con la vertical real, vista desde el frente del vehículo.

Positive caster—when the caster line is tilted rearward when viewed from the side.

Ángulo de avance positivo—cuando la llanta de la línea de cáster (ángulo de avance) está inclinada hacia tras, vista lateralmente.

Positive-phase steering—in a four-wheel steering system, when rear wheels steer in the same direction as the front wheels.

Maniobra de fase positiva—en un sistema de dirección de cuatro ruedas, cuando las ruedas traseras están orientadas hacia la misma dirección que las delanteras.

Positive scrub radius—when the SAI line meets the road surface inside the true vertical centerline of the tire when viewed from the front.

Radio de pivotamiento positivo—cuando la línea SAI se encuentra con la superficie del camino dentro de la línea de centro de la vertical real de la llanta, vista desde el frente.

Positive wheel slip—when the speed of the wheel is greater than vehicle speed during accelerating.

Deslizamiento positivo de rueda—cuando la velocidad de la rueda es superior a la velocidad del vehículo durante la aceleración.

Power cylinder—the cylinder and piston assembly in a linkage power steering system.

Cilindro de fuerza—el conjunto de cilindro y pistón en un sistema de dirección hidráulica con conectores.

PowerMaster—a General Motors power brake system with a hydraulic reservoir and electric pump; used in place of vacuum-assisted power brakes.

PowerMaster—un sistema de frenos de potencia con un depósito hidráulico y bomba eléctrica; se lo usa en lugar de frenos de potencia al vacío.

Preload adjustment—an adjustment where a bearing has an additional load placed against it after it is snug (zero lash).

Ajuste previo a la carga—un ajuste en el que se le coloca una carga adicional contra un cojinete, una vez bien ceñido (holgura cero).

Press-fit—when two parts with an interference fit are forced together.

Calce a presión—cuando dos piezas con calce de interferencia son unidas a la fuerza.

Pressure—force divided by area, measured in pounds per square inch (psi) or kilopascals (kPa).

Presión—Fuerza dividida por área, medida en libras por pulgada cuadrada (psi) o kilopascals (kPa).

Pressure bearing packer—a tool used to pack wheel bearings using grease under pressure.

Compactador de cojinetes a presión—herramienta utilizada para compactar cojinetes de rueda utilizando grasa bajo presión.

Pressure bleeding—a brake bleeding process that uses a tank filled with pressurized brake fluid.

Purga a presión—proceso de purga de frenos que utiliza un tanque lleno de líquido de frenos presurizado.

Pressure chamber—an area in front of the master cylinder primary cup.

Cámara de presión—área delante de la taza principal del cilindro maestro.

Pressure differential—the difference between pressures on surfaces in separate, isolated areas.

Diferencial de presión—la diferencia entre presiones sobre superficies en áreas aisladas y separadas.

Pressure differential valve—a hydraulic valve that turns on the brake warning lamp when a hydraulic failure occurs in half of the brake system.

Válvula de diferencial de presión—válvula hidráulica que enciende la luz de advertencia de frenos cuando ocurre una falla hidráulica en la mitad del sistema de frenos.

Pressure modulation—a rapid apply and release of hydraulic pressure.

Modulación de presión—aplicar y soltar rápidamente la presión hidráulica.

Pressure relief valve—a spring-loaded ball or spool valve that opens at a specified pressure to limit pressure in a hydraulic system.

Válvula de alivio de presión—una válvula esférica de resortes o válvula de carrete que se abre a una presión específica para limitar la presión en un sistema hidráulico.

Primary cup—the master cylinder's primary cup compresses fluid when the pedal is depressed.

Taza principal—la tasa principal del cilindro maestro comprime el líquido al oprimirse el pedal.

Primary shoe—the leading shoe in a duo-servo brake.

Zapata principal—zapata primaria en el freno de servo dual.

Profile—a tire's sidewall height.

Perfil—la altura lateral de una llanta.

Propeller shaft—a shaft that transfers power from the transmission to the differential, allowing for changes in drive line length and angle when springs deflect as a vehicle goes over bumps; also called a drive shaft.

Eje propulsor—eje que transfiere potencia de la transmisión al diferencial, permitiendo cambios en el largo y el ángulo de la línea de transmisión cuando los resortes oscilan al pasar el vehículo por baches; también llamado eje de dirección.

Proportioning valve—a hydraulic system valve that balances braking force between the front and rear brakes by controlling the pressure applied to rear drum brakes.

Válvula de proporcion—válvula del sistema hidráulico que equilibra la fuerza de frenado entre los frenos delanteros y traseros al controlar la presión aplicada a los frenos de tambor traseros.

Psi—pounds per square inch; the English equivalent of kilopascal.

Psi—libras por pulgada cuadrada; el equivalente inglés del kilopascal.

Puddle—in welding, when metal is melted into a puddle that is moved by the operator to complete the weld.

Charco—en la soldadura, cuando se derrite un metal hasta obtener un charco movido por el operador para completar la soldadura.

Pulse width—the amount of time a solenoid is on, measured in milliseconds.

Ancho de pulso—la cantidad de tiempo en que un solenoide está encendido, medido en milisegundos.

Pulse width modulation (PWM)—describes when the computer cycles an output on and off with an "on time" that changes (is modulated).

Modulación de ancho de pulso (PWM, en inglés)—describe cuando la computadora realiza el ciclo de apagar/encender con una "hora de encendido" que cambia (es modulada).

Pyrotechnics—a term that describes fireworks.

Pirotecnia—término que significa fuegos artificiales.

Q

Quadrasteer—a four-wheel steering system used on some SUVs.

Quadrasteer—sistema de dirección de cuatro ruedas usado en algunos vehículos utilitarios deportivos.

Quick-takeup master cylinder—a dual master cylinder that initially supplies a larger amount of brake fluid to lessen brake pedal travel.

Cilindro maestro de suministro rápido—un cilindro maestro que inicialmente provee una mayor cantidad de líquido de freno para reducir el recorrido del pedal de freno.

Quick-takeup valve—a valve in a quick-takeup master cylinder that manages the flow of brake fluid between the reservoir and the primary low-pressure chamber.

Válvula de suministro rápido—válvula en un cilindro maestro de suministro rápido que administra el flujo del líquido de frenos entre el depósito y la cámara principal de baja presión.

R

Racing slick—a tire without tread grooves.

"Racing slick"—llanta sin ranuras.

Rack—the driven shaft in a rack-and-pinion steering gear that mesh with the pinion.

Cremallera—el eje impulsor de un engranaje de dirección cremallera y piñón que engrana con el piñón.

Rack-and-pinion steering—a type of steering where the end of the steering shaft has a pinion gear that meshes with a long, horizontal rack gear.

Dirección de cremallera y piñón—tipo de dirección en el que la extremidad del eje de dirección tiene un engranaje de piñón que engrana con un engranaje de cremallera horizontal largo.

Rack piston—a hydraulic power piston on the rack of a power rack and pinion steering gear that reacts to fluid pressure to provide steering assistance.

Pistón de cremallera—pistón de potencia hidráulica en la cremallera de una cremallera de potencia y engranaje de dirección de piñón que reacciona a la presión del líquido para evitar asistencia de la dirección.

Rack seals—rack-and-pinion steering seals positioned between the rack and the housing.

Sellos de cremallera—sellos de dirección de cremallera y piñón ubicados entre la cremallera y el alojamiento.

Radial load—a vertical load.

Carga radial—la carga vertical.

Radial-ply—a tire structure with casing plies that run across the tire from bead seat to bead seat in the "radial" direction of the wheel.

Tela radial—una estructura de neumático en el que las telas corren a lo largo de la llanta de asiento de aro a asiento de aro en la dirección 'radial' de la rueda.

Radial thrust bearing—a ball wheel bearing assembly in which two radial thrust control bearings face each other; also called a angular contact bearing.

Rodamiento de tope radial—un conjunto de rodamiento esférico con dos rodamientos de control de tope radial enfrentados; también denominado rodamiento de contacto angular.

Radius plate—an alignment gauge that fits under the front wheels and measures in degrees how far a wheel is turned to the right or left.

Placa de radio—un medidor de alineación que cabe debajo de la ruedas delanteras y mide en grados la orientación de la rueda hacia la derecha o la izquierda.

Reaction disc—a part of some vacuum power brake boosters that provides pedal feel to the driver.

Disco de reacción—pieza de algunos boosters de freno de potencia al vacío que hace que el conductor "sienta" el freno.

Reaction plate and levers—a part of some vacuum power brake boosters that provides pedal feel.

Placa y palancas de reacción—pieza de algunos boosters de freno de potencia al vacío que permite que el conductor 'sienta' el freno.

Rear antilock brake system (RABS)—a two-wheel ABS used on the rear wheels of Ford light-duty trucks and SUVs.

Sistema de frenos antibloqueo trasero (RABS, en inglés)—un ABS de dos ruedas utilizado en las ruedas traseras de camiones leves y vehículos utilitarios deportivos de Ford.

Rear axle bearing—the bearing that supports the rear axle on a RWD vehicle.

Cojinete de eje trasero—El cojinete que soporta el eje trasero en un vehículo de tracción trasera.

Rear axle offset—when the rear axle assembly has one rear wheel positioned forward and the opposite rear wheel positioned rearward.

Desvío del eje trasero—cuando el conjunto del eje trasero tiene una rueda trasera posicionada hacia delante y la rueda trasera opuesta posicionada hacia tras.

Rear-wheel antilock (RWAL)—a two-wheel ABS used on the rear wheels of General Motors and Chrysler light-duty trucks and SUVs.

Antibloqueo de rueda trasera (RWAL, en inglés)—un ABS de dos ruedas utilizado en las ruedas traseras de camiones leves y vehículos utilitarios deportivos de General Motors y Chrysler.

Rear wheel toe—the distance between the fronts of the rear tires' treads compared to the rear tread distance.

Convergencia de la rueda trasera—la distancia entre los frentes de las bandas de las llantas traseras comparado con la distancia de las bandas traseras.

Rebound—downward movement of the wheel and suspension.

Rebote—movimiento hacia abajo de la rueda y la suspensión.

Recirculating ball and nut steering gear—a low friction steering gearbox in which a sector gear meshes with a ball nut that rides on ball bearings on the wormshaft to provide a smooth steering feel.

Engranaje de dirección de esfera y tuerca recirculante—una caja de transmisión de dirección de baja fricción en la que un engranaje de sector engrana con una tuerca esférica que se mueve sobre cojinetes esféricos sobre el eje helicoidal para proporcionar una sensación de maniobra suave.

Reed tachometer—an older type of mechanical vibration analyzer that senses the frequency of vibrations from 10 to 80 Hz.

Tacómetro de Reed—un tipo antiguo de analizador de vibración mecánica que detecta la frecuencia de vibraciones de 10 a 80 Hz.

Remove and replace (R&R)—the estimated time it should take a professional technician to perform a specified job.

Remover y reemplazar (RyR)—el tiempo estimado que debería llevar a un técnico profesional para realizar una tarea especificada.

Repair order (RO)—a service record used for legal, tax, and general record keeping purposes; also called a work order.

orden de reparación—un registro de servicio utilizado con fines legales, impositivos y de contabilidad general; también llamado orden de trabajo.

Replenishing port—the rear port in a master cylinder chamber; also called the inlet port.

Puerta de reposición—la puerta trasera de una cámara de cilindro maestro; también llamada puerta de entrada.

Residual check valve—a check valve that keeps a small amount of residual pressure (6–25 psi) in the brake system when the brakes are not applied.

Válvula de verificación residual—una válvula de verificación que mantiene una pequeña cantidad de presión residual (6–25 psi) en el sistema de frenos cuando los frenos no están accionados.

Residual pressure—a small amount of pressure in the brake system when the brakes are not applied that holds seals against their bores to prevent air from entering the system.

Presión residual—una pequeña cantidad de presión en el sistema de frenos cuando los frenos no están accionados que mantiene a los sellos contra sus diámetros interiores para evitar que entre aire en el sistema.

Retread—When a new tread area is vulcanized (bonded) to the tire carcass of a worn tire that has had its entire tread removed.

Banda nueva—Cuando una nueva área de banda es vulcanizada (adherida) al cuerpo de una llanta desgastada a la que se haya removido su banda completa.

Reverse fluid injection (RFI)—a type of brake bleeding that injects fluid from the wheel toward the master cylinder.

Inyección reversa de fluido—tipo de purga de frenos que inyecta fluido desde la rueda en dirección al cilindro maestro.

Ribs—part of the tire tread designed to pump water from the road through the grooves to the back of the tire, where it is thrown out onto the road.

Nervios—parte de la banda de la llanta diseñada para bombear agua del camino a través de las ranuras y hacia la parte trasera de la llanta, de donde es lanzada al camino.

Ride height—an alignment specification based on the assumption that the riding height of the vehicle is correct.

Altura de desplazamiento—especificación de alineación que se basa en la suposición de que la altura de desplazamiento del vehículo es correcta.

Right-to-Know laws—laws that require employees to be informed when the materials they handle at work are hazardous.

Leyes de Derecho de Saber—leyes que exigen que los empleados sean informados cuando los materiales que manipulen al trabajar sean peligrosos.

Rigid axle—a nonindependent axle, found on some rear ends and some heavy truck front ends; also called a straight axle or solid axle.

Eje rígido—eje no independiente encontrado en ciertas extremidades traseras y algunas extremidades delanteras de camiones pesados; también denominado eje derecho o eje sólido.

Rim—see wheel rim.

Aro—ver aro de rueda.

Rim well—the well between the bead seats of a wheel rim that provides a means for removal and installation of a tire; also called drop center.

Pozo de aro—el espacio libre entre los asientos de talón del aro de la rueda que proporciona un medio de remoción e instalación de un neumático; también llamado base honda.

Rim width—the measurement of a wheel from bead seat to bead seat.

Ancho del aro—la medida de una rueda de asiento de talón a asiento de talón.

Riveted lining—a disc or drum brake lining attached by rivets.

Revestimiento remachado—revestimiento de freno de disco o tambor sujeto por remaches.

Road crown—when the center of a road surface is higher than the sides.

Camino abombado—cuando el centro de la superficie del camino es más alto que los costados.

Road feel—the response that a driver feels during a turn when the front wheels follow the direction of the steering wheel.

Sensación de camino—la respuesta que siente el conductor al girar, cuando las ruedas traseras siguen la dirección del volante.

Roller bearing—used instead of ball bearings, they provide more surface area and carry a greater load.

Cojinete de rodillos—utilizado en lugar de cojinetes esféricos, proporciona una superficie de área mayor y tiene mayor capacidad de carga.

Rotor—the rotating part of a disc brake; also called a disc.

Rotor—pieza giratoria de un freno de disco; también llamada disco.

Rough cut—a faster machining cut.

Corte basto—corte mecanizado más rápido.

Roughness—a higher frequency vibration than shake (20–50 Hz)

Rugosidad—una vibración con mayor frecuencia que una sacudida (20-50 Hz)

Rubber Manufacturers Association (RMA)—a professional association made up of various manufacturers of rubber.

Asociación de Fabricantes de Hule (RMA–Rubber Manufacturers Association)—asociación profesional formada por diversos fabricantes de hule.

Run-flat tires—tires designed to operate safely with no air pressure for a specified distance.

Llantas "run-flat"—llantas diseñadas para manejar de manera segura sin presión de aire por una distancia especificada.

Runout—the amount that a part wobbles up and down or side to side when rotated.

Descentramiento—la cantidad que una pieza oscila horizontal o verticalmente al girar.

RWD—rear wheel drive.

RWD—Siglas en inglés para tracción en las ruedas traseras.

Rzeppa CV joint—a popular ball and socket type joint used in most outboard fixed CV joints.

Junta de VC Rzeppa—junta popular del tipo articulación cotiloidea usada en la mayoría de las juntas fijas externas de VC.

S

SAE—see Society of Automotive Engineers (SAE).

SAE—Ver Sociedad de Ingenieros de Automóviles (SAE).

Safety beads—raised metal sections on the inside edges of the bead seats.

Talones de seguridad—secciones metálicas elevadas en los bordes interiores de los asientos de talón.

Scratch cut—a cut used to verify that a rotor or drum has been correctly mounted on a brake lathe.

Corte de raspado—Corte usado para verificar que el rotor o tambor haya sido montado correctamente sobre el torno para frenos.

Scrub radius—the distance from the centerline of the tire contact patch to the point where the steering axis intersects the road when viewed from the front of the vehicle.

Radio de pivotamiento—la distancia desde la línea de centro del parche de contacto de la llanta al punto en el que el eje de la dirección intersecta el camino, visto desde el frente del vehículo.

Scuff—when the tires move sideways for a certain distance of every mile traveled due to incorrect toe.

Abrasión—cuando las llantas se mueven hacia los lados durante una cierta distancia de cada milla recorrida debido a una convergencia incorrecta.

Sealing lip—the lip of a seal that retains grease or fluid.

Reborde de sello—el reborde de un sello que retiene grasa o líquido.

BILINGUAL GLOSSARY

Secondary cup—a seal at the back of a master cylinder piston that keeps fluid from leaking out the back of the master cylinder bore.

Taza secundaria—el sello al dorso del pistón del cilindro maestro que evita la pérdida de líquido a través de la parte trasera del diámetro interior del cilindro maestro.

Secondary shoe—the trailing shoe in a duo-servo brake.

Zapata secundaria—la zapata que sigue a la primaria en el freno de servo dual.

Second order vibration—a type of vibration consisting of two vibrations per revolution, often related to a defective universal joint.

Vibración de segunda orden—tipo de vibración que consiste en dos vibraciones por revolución; con frecuencia relacionada con una junta universal defectuosa.

Section width—the widest point across a tire's cross section, usually measured in millimeters.

Ancho de sección—la punta más ancha a través de la sección transversal de la llanta, generalmente medida en milímetros.

Self-adjuster—part of a brake that provides for automatic clearance adjustment as a brake lining wears.

Dispositivo de autoajuste—pieza del freno que proporciona el ajuste automático de la holgura a medida que se desgasta el revestimiento del freno.

Self-energizing—when the leading shoe on a drum brake is forced into the brake drum.

Autodinámica—cuando la zapata principal del freno de tambor es forzada hacia dentro del tambor del freno.

SEMA—Specialty Equipment Manufacturers Association.

SEMA—Specialty Equipment Manufacturers Association (Asociación de Fabricantes de Equipos Especializados).

Semi-elliptical spring—a leaf spring.

Resorte semielíptico—resorte de láminas.

Semi-floating axle—an axle with a bearing that rides on the axle.

Eje semiflotante—eje con un cojinete montado sobre el eje.

Semi-independent rear suspension—a rear suspension system where one rear wheel can move a limited amount without affecting the opposite rear wheel.

Suspensión trasera semiindependiente—sistema de suspensión trasera en el que una rueda trasera puede moverse una cantidad limitada sin que afecte a la rueda trasera opuesta.

Semi-metallic lining—brake friction material made of a combination of organic or synthetic fibers and metal.

Revestimiento semimetálico—material de fricción de frenos hecho de una combinación de fibras orgánicas o sintéticas y metal.

Sensor—a device that sends an input to a computer.

Sensor—dispositivo que envía información a una computadora.

Series circuit—an electrical circuit with only one path for current flow.

Circuito de serie—un circuito eléctrico con sólo un camino para el flujo de corriente.

Serpentine belt—a V-ribbed belt system with all belt-driven accessories driven by a single belt.

Correa serpentina—un sistema de correa con nervios en V y con todos los accesorios a correa accionados por una única correa.

Service brakes—disc and drum brakes used to stop the vehicle; not the parking brake.

Frenos de servicio—frenos de disco y tambor usados para detener el vehículo; no es el freno de estacionamiento.

Service record—the written record of the service performed on a vehicle; also called a repair order (RO) or work order (WO).

Registro de servicio—el registro escrito del servicio realizado en el vehículo; también denominado orden de reparación u orden de trabajo.

Servo-action—when the leading shoe on a drum brake is forced into the brake drum.

Servo-acción—cuando la zapata principal del freno de tambor es forzada hacia dentro del tambor del freno.

Setback—a difference in wheelbase from one side of a vehicle to the other.

Receso—la diferencia en la base de la rueda de un lado del vehículo al otro.

Shackle—a lever that connects a leaf spring to the frame of the vehicle.

Grillete—palanca que une al resorte de láminas al armazón del vehículo.

Shake—a low-frequency vibration of 5–20 Hz.

Temblor—vibración de baja frecuencia de 5–10 Hz.

Shimmy—rapid side-to-side shaking of the front wheels.

Trepidación—temblor lateral rápido de las ruedas delanteras.

Shims—thin pieces of material used to align adjacent parts.

Cuñas—pequeñas piezas de material utilizadas para alinear piezas adyacentes.

Shock absorber—a device connecting the frame to the spring that dampens spring oscillations by converting energy from spring movement into heat energy.

Amortiguador—dispositivo que conecta el armazón al resorte que amortece las oscilaciones de resorte al convertir la energía del movimiento del resorte en energía de calor.

Shock absorber ratio—a comparison of extension control to compression control.

Relación de amortiguación—comparación de control de extensión con control de compresión.

Shoe anchor—a large pin or post that a drum brake shoe leverages against.

Ancla de zapata—un pino o perno grande contra el que la zapata de freno de tambor hace palanca.

Short-long arm (SLA) suspension—a front suspension system in which the upper control arm is shorter than the lower control arm.

Suspensión de brazo largo-corto—sistema de suspensión delantera en el que el brazo de control superior es más corto que el brazo de control inferior.

Side slip—when the rear wheels slip sideways when a vehicle is cornering at high speed.

Deslizamiento lateral—cuando las ruedas traseras se deslizan lateralmente cuando el vehículo realiza curvas a alta velocidad.

Sidewall deflection—bending of the sidewall of a tire that allows more of the tread to contact the road surface.

Deflexión de pared lateral—doblado de la pared lateral de la llanta que permite que una mayor área de la banda toque la superficie del camino.

Sipes—small grooves in the tire tread that look like knife cuts.

Surcos—pequeñas ranuras en la banda de la llanta que parecen cortes de cuchillo.

Size equivalents—when different tire sizes have the same diameter and load capacity.

Equivalentes de tamaño—cuando los distintos tamaños de llanta tienen el mismo diámetro y la misma capacidad de carga.

SLA—see short-long arm suspension.

SBC (SLA, en inglés)—ver suspensión de brazo corto-largo.

Sliding caliper—a caliper that is mounted to its support on two fixed sliding surfaces, or ways; the caliper slides on the rigid ways and does not have the flexibility of a floating caliper.

Mordaza deslizante—mordaza montada en su soporte sobre dos superficies, o caminos, deslizantes fijas; la mordaza se desliza sobre los caminos rígidos y no tiene la flexibilidad de una mordaza flotante.

Slip angle—the actual angle of the front wheels during a turn compared with the turning angle of the front wheels with the vehicle at rest.

Ángulo de deslizamiento—el ángulo real de las ruedas delanteras al girar, comparado con el ángulo de giro de las ruedas delanteras con el vehículo en reposo.

Slip plates—also called radius plates, they are part of the wheel alignment rack.

Placas de deslizamiento—también denominadas placas de radio, son parte del riel de alineación de la rueda.

Slip yoke—a part attached to a universal joint on the front end of the drive shaft.

Horqueta de deslizamiento—pieza sujeta a una junta universal en la extremidad delantera del eje de la dirección.

Slope—the numerical ratio of rear drum brake pressure to full system pressure through a proportioning valve.

Pendiente—la relación numérica de la presión de freno de disco trasero con la presión del sistema completo a través de una válvula de proporción.

Snapshot—a recorded event on a scan tool.

Reseña—hecho registrado en una herramienta de exploración.

Society of Automotive Engineers (SAE)—a professional organization made up of automotive engineers.

Sociedad de Ingenieros Automotores (Society of Automotive Engineers–SAE)—organización profesional formada por ingenieros automotores.

Solid axle—an I-beam axle type of front suspension used on some trucks.

Eje sólido—tipo de eje de cuerpo fijo con forma de "I" de la suspensión delantera utilizado en algunos camiones.

Space frame—a frame that looks like a bird cage, made up of many round and square tubes welded together.

Armazón de espacio—armazón que parece una jaula de pájaros, compuesto por muchos tubos redondos y cuadrados soldados entre sí.

Spalling—when pieces break off the bearing metal.

Astillamiento—cuando se le caen pedazos al metal del cojinete.

Specific gravity—the ratio of the weight of any liquid to the weight of water, which has a specific gravity of 1.0.

Gravedad específica—relación de peso de cualquier líquido con el peso del agua, el que tiene una gravedad específica de 1.0.

Speed rating—a tire rating that indicates the maximum safe vehicle speed that a tire can withstand.

Clasificación de velocidad—clasificación de llanta que indica la velocidad máxima segura del vehículo que una llanta puede soportar.

Speed sensitive steering (SSS)—a computer-controlled steering system that varies steering effort with vehicle speed.

Speed sensitive steering (SSS)–Dirección sensible a la velocidad—sistema de dirección controlado por computadora que hace variar el esfuerzo de maniobra según la velocidad del vehículo.

Spider—the cross part of a universal joint.

Cruceta—la parte en cruz de una junta universal.

Spiral cable—a conductive ribbon in a plastic case that electrically connects the steering column to the air bag in the steering wheel; also called a clock spring.

Cable en espiral—cinta conductora en una caja plástica que conecta eléctricamente a la columna de la dirección con la bolsa de aire del volante; también llamado resorte de reloj.

Split point—the pressure at which a proportioning valve closes, reducing further pressure to rear drum brakes.

Punto de división—la presión a la que se cierra una válvula de proporción, reduciendo la cantidad de presión a los frenos de tambor traseros.

Spread—the difference between alignment settings from side to side; also called cross caster or cross camber.

Diferencial—la diferencia entre los ajustes de alineación de lado a lado.

Spring insulator—located at the ends of a coil spring to reduce the noise transfer to the chassis.

Aislador de resorte—ubicado en las extremidades de un resorte de bobina para reducir la transferencia de ruido al chasis.

Spring rate—the load required to deflect a spring by 1 inch.

Índice de resorte—la carga requerida para que un resorte se flexione 1 pulgada.

Sprung weight—weight supported by the springs.

Peso de resorte—el peso soportado por los resortes.

Square-cut seal—a caliper piston seal with a square cross section.

Sello con corte cuadrado—sello de pistón de mordaza con una sección transversal cuadrada.

Squat—when the front of the vehicle lifts and the rear lowers during hard acceleration.

En cuclillas—cuando el frente del vehículo se eleva durante al acelerar de golpe.

Starwheel—part of a drum brake adjuster.

Starwheel—pieza de un dispositivo de ajuste de frenos de tambor.

Static—at rest.

Estático—en reposo.

Static balance—the balance or imbalance of a wheel that is at rest.

Equilibrio estático—el equilibrio o falta de equilibrio de una rueda en reposo.

Steering angle—the angle of the inside front wheel during a turn compared to the angle of the outside wheel.

Ángulo de maniobra—el ángulo de la rueda delantera interna durante un giro, comparado con el ángulo de la rueda externa.

Steering axis inclination (SAI)—the amount that the spindle support arm leans in at the top; also known as ball joint inclination (BJI) or king pin inclination (KPI).

Inclinación de eje de la dirección (SAI, en inglés)—la cantidad que el brazo de soporte del husillo se inclina en su parte superior; también conocido como inclinación de junta esférica (BJI, en inglés) o inclinación del pivote de dirección (KPI, en inglés).

Steering free play—how far the steering wheel can move before the front wheels turn.

Juego libre de dirección—la distancia que el volante puede moverse antes que giren las ruedas delanteras.

Steering knuckle—a front steering and suspension part that pivots on the ball joints and allows the wheels to steer.

Charnela de dirección—pieza de la dirección delantera y suspensión que rota sobre las juntas esféricas y permite la maniobra de las ruedas.

Steering pull—when the steering pulls to one side or the other when driving the vehicle straight ahead on a smooth, level road surface.

Tiro de dirección—cuando la dirección tira para un lado u otro al manejar el vehículo derecho hacia delante sobre un camino con superficie nivelada.

Steering radius—the angle of the inside front wheel during a turn compared to the angle of the outside wheel.

Radio de dirección—el ángulo de la rueda delantera interna durante un giro, comparado con el ángulo de la rueda externa.

Steering ratio—the relationship between degrees of steering wheel rotation and degrees of front wheel turning.

Relación de desmultiplicación de dirección—la relación entre los grados de rotación del volante y los grados del giro de las ruedas delanteras.

Steering system—allows the driver to steer the car while providing a comfortable amount of steering effort.

Sistema de dirección—permite al conductor maniobrar el automóvil proporcionando, al mismo tiempo, un esfuerzo de maniobra cómodo.

Steering wander—when the steering drifts to one side and then the other as the vehicle is driven straight ahead.

Migración de la dirección—cuando la dirección tira para un lado y luego para el otro cuando se maneja derecho hacia delante.

Steering wheel position sensor (SWPS)—the sensor that generates a voltage signal in response to the speed and amount of steering wheel rotation.

Sensor de posición del volante (SWPS, en inglés)—sensor que genera una señal de voltaje en respuesta a la velocidad y la cantidad de rotación del volante.

Stock—original equipment.

Inventario—equipos originales.

Stop to stop—steering wheel rotation from full left to full right.

Parada a parada—rotación del volante desde la extrema izquierda a la extrema derecha.

Strut actuator—an actuator that controls strut firmness electronically.

Accionador de puntal—accionador que controla electrónicamente la firmeza del puntal.

Stub shaft—the live axle end at the outside of the outboard CV joint on a front-wheel drive axle shaft.

Falso árbol—la extremidad del eje vivo en el exterior de la junta de VC exterior en un eje de dirección de tracción delantera.

Stud-centric—wheels that are centered using the wheel studs; also called lug-centric.

Rueda de agarradera céntrica—ruedas centralizadas a través del uso de prisioneros de rueda.

Subframe—a group of parts in the front of a front-wheel drive vehicle that includes the engine, transaxle, and steering/suspension system.

Armazón secundario—un grupo de piezas en la parte delantera de un vehículo con tracción delantera, que incluye el motor, el transeje y el sistema de dirección/suspensión.

Suspension linkage—provides a means of attaching the tires and wheels to the rest of the chassis.

Conexión de suspensión—proporciona un medio de sujetar las llantas y ruedas al resto del chasis.

Suspension system—a group of parts that supports the vehicle, cushioning the ride while holding the tire and wheel correctly positioned in relation to the road; parts of the suspension system include the springs, shock absorbers, control arms, ball joints, steering knuckle and spindle, or axle.

Sistema de suspensión—un grupo de piezas que soporta el vehículo, amortiguando el movimiento mientras mantiene la llanta y la rueda en posición correcta con respecto al camino; las piezas del sistema de suspensión incluyen resortes, amortiguadores, brazos de control, juntas esféricas, charnela de dirección y husillo, o eje.

Swing axle—an axle with a universal joint at one or both ends.

Eje de oscilación—un eje con una junta universal en una extremidad o ambas.

Symmetrical—a regular and balanced tread pattern.

Simétrico—un padrón de banda de rodamiento equilibrado y regular.

T

Table—the surface of a brake shoe where the lining attaches.

Mesa—la superficie de una zapata de freno donde se adhiere el revestimiento.

Tandem booster—a power brake booster with two diaphragms.

Booster en tándem—booster de freno de potencia con dos diafragmas.

Tandem master cylinder—a master cylinder with two independent hydraulic systems, required since 1967, either longitudinally split (front/rear) or diagonally split.

Cilindro maestro en tándem—cilindro maestro con dos sistemas hidráulicos independientes, exigidos desde 1967, con separación longitudinal (delantera/trasera) o diagonal.

Tap—a tool used to cut or clean up (chase) the threads in a bore.

Herramienta limpiadora—utilizada para cortar o limpiar las bandas de rodamiento en un diámetro interno.

Tapered roller bearing—a bearing with tapered rollers held in cage between an inside and outside race (cup).

Cojinete de rodillos cónicos—cojinete con rodillos cónicos alojados en una jaula entre una pista de rodadura interna y una externa (taza).

Technical service bulletins (TSB)—electronic or print literature contributed by manufacturers or member associations to inform technicians of important service changes to newer vehicles.

Boletines de servicio técnico (TSB, en inglés)—impresos o textos electrónicos de asociaciones de fabricantes o miembros para informar a los técnicos de importantes cambios de servicio para vehículos nuevos.

Temperature grade—a properly inflated tire's resistance to generating heat and its ability to dissipate heat at highway speeds.

Grado de temperatura—la resistencia de una llanta inflada adecuadamente a la generación de calor y su capacidad de disipar el calor en velocidades de carretera.

Tensile cords—tire belt cords.

Cuerdas tensoras—cuerdas de la correa de la llanta.

Tension-loaded ball joint—vehicle weight tends to pull the ball out of this type of ball joint.

Junta esférica con carga de tensión—el peso del vehículo tiende a hacer que se salga la esfera de este tipo de junta esférica.

Thrust angle—the angle formed by the thrustline and the geometric centerline.

Ángulo de tope—el ángulo formado por la línea de tope y la línea de centro geométrico.

Thrustline—a line positioned at a 90-degree angle to the rear axle and projected to the front of the vehicle.

Línea de tope—línea posicionada en un ángulo de 90 grados del eje trasero y proyectada hacia el frente del vehículo.

Tie-rod—a linkage rod from the steering arm to the steering rack or center link.

Varilla de conexión—varilla de enlace que va del brazo de la dirección a la cremallera de la dirección o conexión central.

Tingling—vibration above 100 Hz.

Hormigueo—vibración superior a 100 Hz.

Tire chains—chains temporarily installed on tires to improve traction on snow.

Cadenas para llantas—cadenas instaladas temporalmente en las llantas para lograr una mejor tracción sobre la nieve.

Tire rotation—when tires are moved to different locations on the vehicle

Rotación de llantas—cuando se cambian las llantas a distintas ubicaciones del vehículo.

Tires, batteries and accessories (TBA)—parts purchased by franchised repair shops, through the company whose branded products they sell, at very good discounts.

Llantas, baterías y accesorios (TBA, en inglés)—piezas compradas por talleres de reparación bajo franquicia, a través de la compañía cuyos productos de marca venden con descuentos excelentes.

Toe angle—the angle between the two front tires.

Ángulo de convergencia—el ángulo entre las dos llantas delanteras.

Toe change—when the toe angle changes in response to changes in ride height.

Cambio de convergencia—cuando el ángulo de convergencia cambia en respuesta a cambios en la altura de desplazamiento.

Toe-in—when the distance between the fronts of the tire treads is more when compared to the rear tread distance.

Convergencia hacia dentro—cuando la distancia entre las partes delanteras de las bandas de rodamiento de las llantas es superior comparada con la distancia de las bandas traseras.

Toe-out—when the distance between the front edges of the fronts of the tire treads is more when compared to the rear tread distance.

nvergencia hacia fuera—cuando la distancia entre los bordes delanteros de las partes delanteras de las bandas de rodamiento de las llantas es superior comparada con la distancia de las bandas traseras.

e out on turns—the steering angle of the wheel on the inside of a turn compared with the steering angle of the wheel on the outside of the turn; also known as the Ackerman angle.

nvergencia hacia fuera al girar—el ángulo de dirección de la rueda en el interior de un giro, comparado con el ángulo de dirección de la rueda en la parte exterior del giro; también conocido como ángulo de Ackerman.

orque—twisting force.

orsión—fuerza de torsión.

orque steer—when there are axle shafts of different lengths axle shafts and under hard acceleration the long one wraps up and there is a slight lag before it puts its torque to the wheel, causing the car to pull to one side.

Dirección de torsión—cuando los ejes de rueda tienen distintos largos y al acelerar con fuerza el más largo se enrolla y ocurre una demora leve para dar torsión a la rueda, haciendo que el automóvil tire hacia un lado.

Torque stick—a torsion bar extension used with an impact wrench that limits torque to a specified amount.

Varilla de torsión—Prolongación de barra de torsión utilizada con una llave de impacto que limita la torsión a una cantidad específica.

Torque tube—an enclosed drive shaft used in some vintage vehicles.

Tubo de torsión—eje de rueda encerrado utilizado en ciertos vehículos antiguos.

Torque wrench—a tool to measure clamping force on a nut or bolt.

Llave de torsión—herramienta utilizada para medir la fuerza de grampeado de una tuerca o tornillo.

Torsion bar—a straight rod that twists when working as a spring.

Barra de torsión—una varilla derecha que se contornea al funcionar como resorte.

Track bar—a bar from the suspension to the chassis that prevents sideways movement of the chassis.

Barra de enclavamiento—barra de la suspensión al chasis que evita el movimiento lateral del chasis.

Tracking—a term that refers to the relationship between the average direction that the rear tires point, when compared to the front tires.

Rastreo—término que ser refiere a la relación entre la dirección media a la que miran las llantas traseras, comparada con la de las llantas delanteras.

Traction—how well a tire grips the road.

Tracción—capacidad de la llanta de asirse al camino.

Traction control system (TCS)—a computer-controlled system that prevents wheel spin during acceleration.

Sistema de control de tracción (TCS, en inglés)—sistema controlado por computadora que evita que las ruedas giren al acelerar.

Traction grade—a UTQG rating of a tire's stopping ability on wet asphalt pavement and concrete.

Grado de tracción—clasificación de UTQG de la capacidad de frenado de una llanta sobre pavimento de asfalto y concreto mojados.

Trailing shoe—the second shoe in the direction of drum rotation from the top of a non-servo brake.

Zapata de arrastre—la primera zapata en la dirección de rotación del tambor en un freno que no sea servo.

Transaxle—a unit that combines a transmission and differential.

Transeje—unidad que combina la transmisión y el diferencial.

Transducer—a magnet that is spring-loaded on both sides and moves back and forth in a coil of wire to send a signal to a meter in response to vibration to convert energy from one form to another.

Transductor—imán con resortes en ambos lados que se mueve hacia delante y hacia tras en una bobina de alambre para enviar una señal a un medidor, en respuesta a una vibración, para convertir energía de una forma en otra.

Transitional coils—coils between the inactive and active coils of a variable rate coil spring.

Bobinas transicionales—bobinas entre las bobinas inactivas y activas de un resorte de bobina de velocidad variable.

Tread—the section of the tire that rides on the road.

Banda de rodamiento—la sección de la llanta que toca el camino.

Tread contact patch—the area of the tire tread that contacts the road.

Parche de contacto de banda de rodamiento—área de la banda de rodamiento que toca el camino.

Tread wear indicator—continuous bars that appear across a tire tread when the tread is worn to 2/32 (1/6) inch.

Indicador de desgaste de banda—barras continuas que aparecen a lo ancho de la banda de rodamiento cuando la banda presenta un desgaste de 2/32 (1/6) pulgadas.

Tread wear rating—a UTQG tire rating of the life expectancy of a tire.

Clasificación de desgaste de banda de rodamiento—clasificación de llanta del UTQG relativa a la expectativa de vida de la llanta.

Trim height—specified suspension height.

Altura de suspensión—altura de suspensión especificada.

Tripod tulip plunge joint—a CV joint with a spider assembly with three spherical rollers that can roll in and out as the axle length changes.

Junta de purga trípode del tipo tulipán—junta de VC con un conjunto de cruceta con tres rodillos esféricos que giran hacia dentro y hacia fuera, según cambia el largo del eje.

Trunnions—bearing areas ground on the ends of the cross or spider where the needle bearings ride.

Muñones—áreas de cojinete adheridas a las extremidades de la cruz o cruceta donde se desplazan los cojinetes de aguja.

Tubeless tires—a tire with a rubber valve stem that seals against a wheel rim at the bead seats.

Llanta sin cámara—llanta con un vástago de válvula de hule que se sella contra un aro de rueda en los asientos de talón.

Tube-type tire—a vintage tire with an inner tube.

Llanta con cámara—llanta antigua con una cámara interna.

Turnbuckle—a part between two rods that when turned one way shortens the tie-rod assembly and when turned the other way lengthens it.

Tensor—pieza entre dos varillas que, al ser girada en una dirección, acorta el conjunto de la varilla de conexión, y cuando se la gira en la otra dirección, lo alarga.

Turning radius—the angle of the inside front wheel during a turn compared to the angle of the outside wheel.

Radio de giro—el ángulo de la rueda delantera interna durante un giro, comparado con el ángulo de la rueda externa.

U

U-joints—located at both ends of the driveshaft, they transmit power at an angle.

Juntas en U—ubicadas en ambas extremidades del eje de la transmisión, transmiten potencia a un ángulo.

Understeer—when a car does not seem to respond to movement of the steering wheel during a hard turn.

Maniobra deficiente—cuando un automóvil parece no responder al movimiento del volante durante un giro abrupto.

Undulation—a fast rise or dip in the surface of the road.

Ondulación—una elevación o depresión repentina en la superficie del camino.

Unibody—a chassis design that has a floorpan and a small subframe section in the front and rear; also called monocoque.

BILINGUAL GLOSSARY

Diseño de una pieza—diseño de chasis con una bandeja y una pequeña sección de armazón secundario en la parte delantera y trasera; también llamado monocasco.

Unidirectional rotor—a disc brake rotor with curved cooling fins to increase airflow; unidirectional rotors are not interchangeable from side to side.

Rotor unidireccional—rotor de freno de disco con aletas de refrigeración curvas para aumentar el flujo de aire; los rotores unidireccionales no son intercambiables de un lado a otro.

Uniform tire quality grading (UTQG) system—tread wear, traction, and temperature resistance ratings to be molded into the tire sidewall, required by the U.S. Department of Transportation (DOT).

Sistema uniforme de calificación de llantas (UTQG, en inglés)—el desgaste de banda de rodamiento, tracción y resistencia a temperaturas a ser indicados en la pared de la llanta, según exigencia del Departamento de Transporte de los Estados Unidos (DOT).

Unitized body—a body design without a frame that uses all body components tied together for strength.

Carrocería unificada—diseño de carrocería sin armazón que utiliza todos los componentes de la carrocería unidos para mayor resistencia.

Unsprung weight—the weight of parts of the suspension that are *not* supported by springs.

Peso no suspendido—el peso de piezas de la suspensión que *no* cuentan con el soporte de resortes.

Upper strut bearing—a bearing for steering that allows strut rotation at the top of front MacPherson struts.

Cojinete de puntal superior—cojinete para la dirección que permite la rotación del puntal en la parte superior de puntales delanteros MacPherson.

U.S. Occupational Safety and Health Administration (OSHA)—a governmental agency charged with the enforcement of health and safety laws.

Administración de la Seguridad y Salud Ocupacionales de EE.UU. (OSHA)—una entidad gubernamental a cargo del cumplimiento de leyes de salud y seguridad.

UTQG—see uniform tire quality grading system.

UTQG—ver sistema uniforme de calificación de calidad de llantas.

V

Vacuum—in vehicle service work, a term that means air pressure lower than atmospheric pressure.

Vacío—en trabajos de servicio de vehículos, el término significa presión de aire más baja que la presión atmosférica.

Vacuum bleeding—a brake bleeding process using a vacuum pump filled with brake fluid and attached to a wheel bleeder screw to draw old brake fluid and air from the system; also known as Vacula™ or suction bleeding.

Purga al vacío—proceso de purga de frenos en el que se usa una bomba de vacío llena de líquido de frenos adjunta a un tornillo de purga de rueda para retirar líquido de freno viejo y aire del sistema; también conocida como purga Vacula' o de succión.

Vacuum suspended—the most common type of vacuum power brake booster with vacuum on both sides of the diaphragm when the brakes are released.

Suspendido al vacó—el booster de potencia al vacío más común, con vacío en ambos lados del diafragma al soltarse los frenos.

Valleys—the recessed areas between the lands of a spool valve.

Valles—áreas deprimidas entre las pistas de una válvula carrete.

Variable effort steering (VES)—a computer-controlled steering system that provides reduced effort at low speeds and increased effort at higher speeds.

Dirección de esfuerzo variable (VES, en inglés)—sistema de dirección controlado por computadora que proporciona esfuerzo reducido a velocidades bajas y mayor esfuerzo a velocidades altas.

Variable-rate coil springs—springs with an average rate based on load when compressed a predetermined amount.

Resortes de bobina de velocidad variable—resortes con una tasa media basada en la carga cuando se la comprime una cantidad predeterminada.

Variably priced items—slow-moving items that are not readily available to the customer.

Artículos de precio variable—artículos de movimiento lento que no están inmediatamente disponibles para el cliente.

Vehicle identification number (VIN)—a series of numbers and letters that identifies a vehicle.

Número de identificación del vehículo (VIN, en inglés)—una serie de números y letras que identifica el vehículo.

Vehicle roll—when a vehicle body tends to roll to one side while cornering.

Balanceo del vehículo—cuando la carrocería del vehículo tiende a balancearse hacia un lado al girar.

Vehicle speed sensor (VSS)—a sensor that produces a voltage signal used by the computer to monitor vehicle speed.

Sensor de velocidad del vehículo (VSS, en inglés)—sensor que produce una señal de voltaje usada por la computadora para monitorear la velocidad del vehículo.

Vehicle stability control system—an electronic system that prevents vehicle swerving.

Sistema de control de estabilidad del vehículo—sistema electrónico que evita la desviación del vehículo.

Velocity—a combination of amplitude and frequency.

Velocidad—una combinación de amplitud y frecuencia.

Vent port—the front port in a master cylinder; also called the compensating port.

Puerta de ventilación—la puerta delantera de un cilindro maestro; también llamada puerta de compensación.

Ventilated rotor—a rotor with cast-in cooling fins.

Rotor ventilado—rotor con aletas de refrigeración fundidas.

Vibration—a part that is in motion in waves, or cycles.

Vibración—pieza en movimiento en ondas, o ciclos.

Volume—the length, width, and height of space occupied by an object.

Volumen—el largo, ancho y altura de espacio ocupado por un objeto.

W

Wander—see steering wander.

Migración—ver migración de la dirección.

Warehouse distributors (WD)—large distribution centers that sell to auto parts wholesalers.

Distribuidores de depósito (WD, en inglés)—grandes centros de distribución que venden a mayoristas de piezas de automóvil.

Watts rod—another name for a tracking bar.

Varilla de watt—otro nombre para la barra de rastreo.

Wave seal—a type of seal that does not ride on the shaft in a straight line like a conventional seal.

Sello ondulado—tipo de sello que no va montado sobre el eje en línea recta como un sello convencional.

Ways—machined surfaces upon which another part slides.

Guía deslizante—superficies mecanizadas sobre las que otra pieza se desliza.

Wear bars or indicators—raised areas cast into the bottom of the tire tread that indicate when the tread has become worn to its safe limit.

Barras o indicadores de desgaste—áreas elevadas en la base de la banda de rodamiento que indican el momento en que la banda ha pasado del límite seguro de desgaste.

Web—the inner part of a brake shoe where the springs and linkage parts attach.

Red—parte interior de la zapata de freno donde se unen los resortes y las piezas de conexión.

Wheel alignment—an adjustment of suspension parts to original specifications to provide straight steering and minimize tire wear.

Alineación de ruedas—ajuste de piezas de la suspensión a las especificaciones originales para proporcionar dirección recta y minimizar el desgaste de las llantas.

Wheel base—the distance between the front and rear tires

Distancia entre ejes—la distancia entre las llantas delanteras y traseras.

Wheel bearing—the term for nondrive front and rear wheel bearings.

Cojinete de rueda—término usado para cojinetes de rueda delanteros y traseros sin tracción.

Wheel-contact lift—a lift that supports the vehicle by lifting it by the wheels.

Elevador de ruedas—ascensor que sostiene el vehículo elevándolo por las ruedas.

Wheel cylinder—a hydraulic cylinder of a drum brake assembly.

Cilindro de rueda—cilindro hidráulico de un conjunto de freno de tambor.

Wheel cylinder kit—a parts kit for rebuilding brake wheel cylinders.

Conjunto de cilindro de rueda—conjunto de piezas para reacondicionar cilindros de rueda de freno.

Wheel offset—the difference between the rim centerline and the mounting surface of the wheel.

Desvío de rueda—la diferencia entre la línea de centro del aro y la superficie de montaje de la rueda.

Wheel rate—a term used to describe the spring's effect at the wheel rather than at the base of the spring.

Tasa de rueda—término usado para describir el efecto del resorte en la rueda, en lugar de en la base del resorte.

Wheel rim—a strip of steel rolled and butt welded at the ends that forms the outside of a wheel, which is spot welded to the center flange.

Aro de rueda—tira de acero laminado soldado a tope en sus extremidades y que forma la parte exterior de la rueda, la que es soldada por electropuntos a la brida central.

Wheel speed sensor—a sensor that creates a voltage signal for the antilock brake system computer to monitor wheel speed.

Sensor de velocidad de la rueda—sensor que crea una señal de voltaje para que la computadora del sistema de frenos antibloqueo pueda controlar la velocidad de las ruedas.

Whine—a sound like a turbine or mosquitos.

Gemido—sonido como una turbina o mosquitos.

Wishbone—a control arm shaped like the letter A, or a wishbone.

Horquilla—brazo de control con la forma de una letra A.

Working angel—the difference between the drive shaft angle and the angle of the transmission or differential.

Ángulo de trabajo—la diferencia entre el ángulo del eje de la dirección y el ángulo de la transmisión o diferencial.

Work order (WO)—a service record used for legal, tax, and general record-keeping purposes; also called a repair order.

Orden de trabajo—un registro de servicio utilizado con fines legales, impositivos y de contabilidad general; también llamado orden de reparación.

Wormshaft—the gear that meshes with the selector shaft in a recirculating ball and nut steering gear.

Eje helicoidal—el engranaje que engrana con el eje selector en un engranaje recirculante de esferas y tuercas.

Wormshaft bearing preload—tension on the wormshaft bearing during adjustment.

Carga previa del cojinete del eje helicoidal—tensión sobre el eje helicoidal durante el ajuste.

Y

Yaw—the motion to the left or right or rotation around the vertical centerline.

Derrape—el movimiento hacia la derecha o la izquierda o rotación alrededor de la línea de centro vertical.

Yaw rate sensor—generates a voltage signal in response to sideways chassis movement for action by the vehicle stability control computer.

Sensor de tasa de derrape—genera una señal de voltaje en respuesta al movimiento lateral del chasis para que la computadora de control de estabilidad del vehículo actúe.

Z

Zero lash—the point of no clearance.

Holgura cero—el punto sin espacio libre.

Index

Note: Page numbers followed by the letter "f" denote figures.

A

Above-ground lifts, 31, 36
ABS. *See* Antilock brake system (ABS)
Acceleration slip regulation (ASR), 223
Accumulator, integral antilock brake system, 218–219
Ackerman angle, 412–413, 433, *434f*
Active suspension system, 340–341
Adapters, frame-contact lift, 33
Adaptive suspension system, 9, 337
Adjusting sleeves, 429, 430
Aftermarket, 18, 21
Aftermarket friction material certifications, 166–167
Air bags, 369–370, 387–389
Air chisels, 49–50
Air compressors, 38–39
  tools using, 38, *40f*, 48–50, *49f*
Air drills, 50
Air filter, vacuum brake booster, 198
Airgap, 215
Air lines, 38
Air pressure, tire, 281–284
Air pressure gauge, 282, *283f*
Air shocks, 335–336
  automatic suspension leveling systems, 338
  service, 346
Air springs, 325, 338, *339f*
Air tools, 38, *40f*
Alignment
  four-wheel, 423
  machines, 423
  wheel. *See* Wheel alignment
Alignment heads, compensating, 436
Alloys, 235
All-season tires, 271
Aluminum space frame (ASF) suspension system, 321
Amber brake warning light, antilock brake system (ABS), 226
American Petroleum Institute (API), 472
Amplitude, 481, *482f*
Anchor, brake shoe, 120, 121
Angular contact, 237
Anodized aluminum cylinders, 134–135
Antilock brake controller, 214
Antilock brake systems (ABS), 5, 6, 63–64, 214–232
  amber brake warning light, 226
  bleeding, 109, 227–228
  brake performance during stop, 223
  common electrical problems, 231–232
  components, 214–217
  connectors, 226
  controller, 214
  depressurizing pressure, 227
  diagnosing problems, 226–227
  dump, 220
  dynamic proportioning, 86
  electrical problems, 231–232
  electromagnetic. *See* Electromagnetic antilock brake systems (ABS)
  electromechanical hydraulic unit (EHCU), 217
  electronic brake and traction control module (EBTCM), 215
  electronic control module (ECM), 215
  electronic stability control, 224, *225f*
  false modulation, 226
  fluid service, 227–228
  flushing and bleeding, 227–228
  four-wheel, 220
  integral system. *See* Integral antilock brake systems (ABS)
  lateral acceleration sensor, 216
  nonintegral system. *See* Nonintegral antilock brake systems (ABS)
  operation of, 220–224
  pedal feel, 214
  precautions, 230–231
  pressure control, 86
  pressure dump, 220
  rear-wheel antilock (RWAL) system, 220, 231
  red brake warning light, 225–226
  safeguards, 222
  scan tool, 227, *227f*
  service, 225–232
  single channel, 220–221
  solenoid valves, 220, *221f*
  three- and four-channel, 221–222
  tone ring. *See* Tone ring
  traction control, 223–224
  two-wheel, 219–220
  types of, 217–220
  voltage induction, 215, *216f*
  wheel speed sensors. *See* Wheel speed sensors
Anti-rattle clips, 157
Antitheft lug nuts, 289
Anvil side, 22
API. *See* American Petroleum Institute (API)
Aquaplaning, 260
Arbor press, 46
Arbor speed, 176
Arc welding, 51
Armor, 88
Asbestos disposal regulations, 54
Asbestos hazards, 53–54
Asbestos linings, 164
ASE certification exams
  brake systems, 491–497
  steering system, 497–503
  suspension system, 497–503
Aspect ratio, tire, 266–267
ASR. *See* Acceleration slip regulation (ASR)
Asymmetric tread pattern, 262
Audible wear sensor, 144
Automatic suspension leveling systems, 338–339
Automotive parts wholesalers, 18
Auto parts, wholesale and retail distribution of, 18–20
Auxiliary drum parking brake, 188–189
Auxiliary lip, 242
Auxiliary transmission parking brake, 190
Auxiliary vacuum pump, 203
Auxiliary vacuum reservoir, 202
Axle and wheel seals, inspection of, 128
Axle bearings, 239
  diagnosing problems, 473
  installation, 476
  noises, 473
  replacement, 475–476
  seal service, 474
  service, 473–476
Axle nut, loosening, 448–449
Axles
  bearing-retained, 470, 473–474
  bearings. *See* Axle bearings
  C-lock/C-clip, 470–471
  drive. *See* Drive axles
  front-wheel drive. *See* Front-wheel drive axles
  full-floating, 239
  I beam, 325–326
  inspection, 128
  live. *See* Live axles
  semi-floating, 239, 470–471
  stub, 444
  swing, 471
  transaxle, 443–444
Axle sets, 137

INDEX

e shaft, 444–445
amper weights, 446–447
esign, 446–447
emoval, 448–452

ckbone chassis, 321
cking plate, brake shoe, 120, *121f*
ck protection safety, 41
lancing tires and wheels, 304–311
ll and ramp parking brake caliper, 190
ll and trunnion joint, 465
ll bearings, 235–237
ll joint inclination (BJI). *See* Steering axis inclination (SAI)
ll joints, 327, 337, *338f*
 measuring wear on, 353–354
 replacement, 355–356
 service, 352–356
 tapers, 354–355
 wear indicator, 354
all nut, 365
all sockets, 367
anjo fittings, 91, 112, *113f*
arrier creams, 55
atteries, electrical fires and vehicle, 44, *45f*
ead roller tire changer, 297–298
eads, 259
ead seater, 295, *296f*
ead seats, 277
 checking, 292–293
 seating, 294–297
 unseating, 291–292
Bead wire, 259
Bearing cage, 235, *236f*
Bearing repack, 70
Bearing-retained axle, 470
 removal of, 473–474
Bearings
 axle. *See* Axle bearings
 ball, 235–237
 drive axle, 239
 frictionless, 235, *236f*
 front-wheel drive. *See* Front-wheel drive bearings
 needle. *See* Needle bearings
 pivot, 337, *338f*
 plain, 235
 press-fit bearing replacement, 475–476
 roller, 237–238
 seals. *See* Wheel bearing seals
 wheel. *See* Wheel bearings
Bearing separator plate, 46, *47f*
Beat/boom vibration, 483
Bellmouth wear, 170, *171f*
Belt cords, 372–373
Belts, 260
 drive. *See* Drive belts, power steering
 high cordline, 372–373
 power steering drive belts. *See* Drive belts, power steering
 serpentine, 399, *400f*
Belt safety, 52
Belt tension gauge, 401
Bench bleeding of master cylinder, 99–101
Bench grinders, 47, *48f*
Bending spring, 115

Bendix brakes, 124
Bias-ply tires, 263, 264
BJI. *See* Ball joint inclination (BJI)
Bleeding
 air compressor, 39
 antilock brake system (ABS), 109, 227–228
 brakes, 104–110
 hydro-boost brakes, 209
Bleeding/purging shock absorbers, 346
Bleed screws, 104–110
Block valves, 220
Blow down, 39
Body-over-frame suspension system, 318–319
Body roll, 421
Boiling points, 69
Bolt-on type front-wheel drive bearings, 252
Bonded/riveted brake linings, 67, *68f*, 121
Boom, 484
Booster, power brake, 5, 63, *64f*, 194
 hydro-boost brakes. *See* Hydro-boost brakes
 powermaster. *See* Powermaster brakes
 service, 203–211
 vacuum brake booster. *See* Vacuum brake booster
Boot clamps, CV joint. *See* CV joint boot clamps
Boot kits, 447, *448f*
Boots, 42
Boots, CV joint. *See* CV joint boots
Boss, wheel cylinder, 123
Bottle jack, 29, *30f*
Bounce, 321
Brake adjusting gauge, 139
Brake cylinder
 clamp, 135
 hones, 133–134
Brake dimension booklets, 22
Brake drag, 96, 204
Brake drums, 168
 bellmouth wear, *171f*
 finish cut, 179
 hard spots, 170, *171f*
 hub removal, 182, *183f*
 lathes, 173–177
 lug removal, 182, *183f*
 machining, 172–177, 177–179
 measuring, 21–23
 mounting, 174–175
 out-of-round wear, 170, *171f*
 precutting inner and outer ridges, 177
 removal of, 128–130
 rough cut, 179
 service, 170
 silencing band, 175–176
 turning, 178
 wear inspection, 170
 wear limits, 171
Brake dust, 53
Brake fade, 66
Brake fluid. *See also* Reservoir, master cylinder
 adding, 101
 change interval, 102
 contamination of, 101–102

Department of Transportation (DOT)
 classifications, 68–69, 104
 hydraulic. *See* Hydraulic brake fluid
 mineral oil, 69
 routine replacement of, 102
 selecting correct, 104
 service, 101–102
 synthetic, 69
 testing methods, 102–104
Brake hoses, 86–88
 copper washers, 112, *113f*
 fittings. *See* Fittings, brake hose
 installation of, 112, *113f*
 restricted, 111–112
 retaining clip, 112, *113f*
 service, 111–112
Brake lines/tubings, 88–89
 bending, 114–115
 cutting, 113–114
 flare fittings for, 89–90, 113
 installing flared, 116
 ISO flared. *See* ISO flared brake lines/tubings
 SAE flared lines. *See* SAE flared lines
 service, 112–117
 union repairs, 117
 unrolling, 113
Brake linings, 67–68, 121
 bonded/riveted brake linings. *See* Bonded/riveted brake linings
 initial clearing adjustment, 139
 inspection, 131–132
 ordering replacement, 137
Brake Manufacturers Council (BMC) friction material certification, 167
Brake micrometer, 21
Brake pads, 67–68
Brake pad wear indicators, 86, 144
Brake parts
 inspecting and cleaning, 135–136
 washer, 53, 130–131
Brake pedal, travel and feel of, 94, *95f*, 96, 101
Brake pull, 161
Brake shoes, 67, 120
 anchors, 120, 121
 arc, 138
 clearance, 138
 removal, 132
 replacement of, 138–139
Brake spoon, 140
Brake systems, 3–5, 73–74
 antilock brake system (ABS), 63–64
 bleeding, 104–110
 diagnosing problems, 88
 disc brakes. *See* Disc brakes
 drum brakes, *63f*. *See* Drum brakes
 electrical switches, 86
 ethics in service of, 70
 Federal Motor Vehicle Safety Standards (FMVSS) regulation of, 66–67
 fluid check, 95
 fluid leaks, 95
 flushing of, 102
 friction, 65
 function in stopping, 64–65
 hybrid. *See* Hybrid brake system
 hydro-boost brakes. *See* Hydro-boost brakes

master cylinder. See Master cylinder
NATEF task correlation grid, 504–507
operation, 61
parking brake. See Parking brake
power brakes. See Power brakes
practice ASE certification exam, 491–497
pressure testing, 110–111
proportioning valves. See Proportioning valves, hybrid brake system
requirements of, 67
service brake, 61
service of, 94–117
warning lights. See Brake warning light
weight transfer, 64–66
Brake tubings. See Brake lines/tubings
Brake warning light, 84–85
 amber brake warning light, antilock brake system (ABS), 226–227
 diagnosis, 111
 red brake warning light, antilock brake system (ABS), 225–226
Breathing safety, 45
Bridge-type clamps, 454
Brinelling, 245–246, 468
Bulkhead, 194
Bump steer, 419
Bushings, 326–327, 336
 control arm, 350–351
 service, 350–352
 strut rod, 352
Buzz, 483
Bypassing, 96

C

Cab-forward suspension system design, 319, 320f
CAFE. See Corporate average fuel economy (CAFE)
Caliper, disc brake, 5f, 142
 ball and ramp parking brake, 190
 bolts, 160
 cleaning and inspecting parts, 154
 disassembly, 153–154
 dust boot, 154
 electronic, 25
 fixed, 144
 floating, 144–147
 guide pins, 160
 inspection, 150
 installation, 160
 installing disc pads in, 156–157
 reassembly, 155–156
 rebuilding, 153–157
 rebuilt, 154
 removal of, 150–152
 screw and nut parking brake caliper, 189–190
 self-adjustment reset, 193
Caliper pistons, 147, 152
 inspection of, 154–155
 removal of, 154
Caliper rubber parts kit, 153
Camber, 407–409
 adjusting, 424–426
 adjusting rear camber, 437
 four-wheel alignment, 436–437
 measuring, 423
 rules for, 433–434
 tire wear from, 417
Camber roll, 407
Capillary action, 248
Carbon fiber brake linings, 165–166
Carbon fiber monocoque chassis, 321
Carcass, 259
Cardan, 461
 double cardan U-joints, 464, 465f
Casing, 259
Caster, 409–410
 adjusting, 424–426
 four-wheel alignment, 436–437
 measuring, 423–424
 rules for, 433–434
 spread, 434
Center abutment caliper, 146
Center bolt, 324
Center link, 367
Center of gravity (CG), 37
Center support bearing, 462, 463f
Ceramic brake linings, 164, 165
Ceramic brake rotors, 169
Certifications
 aftermarket friction material certifications, 166–167
 ASE exams. See ASE certification exams
 Brake Manufacturers Council (BMC) friction material certification, 167
 D3EA friction material certification, 167
CG. See Center of gravity (CG)
Chafing strips, 260
Chain stores, retail, 20
Chain synchronizer, 36
Charcoal filter, 196
Chassis, 3
 backbone, 321
 lubricants, 241
 problems affecting disc brakes, 148
 vintage, 5–6
Check valve
 master cylinder, 78, 79f
 vacuum brake booster, 195–196, 205
Chisels
 air, 49–50
 mushroomed, 50
Chuck key, 46, 48f
Circlips, 452
Clamps
 bridge-type, 454
 CV joint boot. See CV joint boot clamps
 earless, 455
 universal, 454
Class A fire, 42, 43f
Class B fire, 42, 43f
Class C fire, 42, 43f
Class D fire, 42, 43f
Cleanliness, ensuring vehicle, 17
Click-type tension gauge, 401
Clips, 451–452
C-lock/C-clip axles, 470–471
 removal of, 474
Clock spring, air bag, 388
Clothing
 safety precautions for, 41–42
 uniforms, 18
Coefficient of friction, 65–66, 260f
Coiled tubing, 88, 89f
Coil springs, 321–322
 height adjustment, 356
 installing, 350
 linear-rate, 323
 replacement, 356–359
 service, 356–360
 variable-rate, 323
Coil spring shocks, 336
Collapsible steering column, 369–370
Collet, 116
Combination digital mikes, 25
Combination tire plug-patches, 304
Combination valve, 85–86
Compact spare tires, 273
Compensating alignment heads, 436
Compensating port. See Vent (compensating) port
Complete brake job, 69–70
Composite rotors, 169–170
Composite springs, 325
Compressed air tools, 48–50
 examples of, 49f
Compression, 7, 321
Compression fittings, high pressure, 90–91
Compression type ball joints, 337, 338f
Computer balancers, 308–309
Computerized service information, 14
Computer records, 17
Cone adapters, 174
Connectors, brake tubing, 89
Constant velocity (CV) joint. See CV joints
Control arms, 7, 326
 bushings, 350–351
Copper washers, 112, 113f
Cord, tire, 263
Corporate average fuel economy (CAFE), 265
Cost leaders, 20
Cotter pins
 removal of, 243
 selecting and installing, 249
Couple imbalance, 307, 308f
Crash protection, 320
Cross and yoke, 461
Cross groove plunge joints, 446
 service, 456
Crown, tire, 300
Crowned roads, 423–424
Crumple zone, 320
Curb riding height, 418
Curb weight, 270
Customer relations, 15–16
Custom wheels, 277–278
CV joint boot clamps
 installation of, 454–456
 small-end clamp installation, 455–456
 types of, 454–455
CV joint boots, 447–448
 clamps. See CV joint boot clamps
 installation of, 454
 removal of, 451–452
CV joints, 443, 444–445
 boots. See CV joint boots
 classifications of, 444
 construction, 445–446
 diagnosis, 448
 fixed. See Fixed joints
 knockoff-type, 452

joints (continued)
plunge. See Plunge joints
replacement of, 452–453
Rzeppa, 445. See also Fixed joints
types of, 445
universal, 463–465
U universal joints, 463–465
cles, 481

BEA friction material certification, 167
amp out, 8
ealership parts departments, 20
eceleration factor, 220
egreasers, 52–53
epartment of Transportation (DOT)
  brake fluid classifications, 68–69, 104
  tire codes, 270–271, 512–517
eploys, 388
ermatitis, 54–55
iagnostic trouble codes (DTCs), 360
iagonal split tandem master cylinder system, 79–80
ial indicator, 26, 149–150
ie grinders, 50
ifferential bearing problems, 479
ifferential pinion seal replacement, 478
igital drive shaft alignment tool, 488
igital mikes, combination, 25
igital steering angle sensor, 377
irect-acting hydraulic shock absorbers, 331
Directional control angle, 407
Directional rotors, 168–169
Directional stability, 411
Directional tires, 294
Discard diameter dimension, 23
Discard dimension, rotor, 171–172
Disc brakes, 4, 5, 63, 142–162
  advantages of, 142–144
  bleed screw, 110f
  brake pull, 161
  caliper. See Caliper, disc brake
  caliper self-adjustment reset, 193
  chassis problems, 148
  complaints, 147
  diagnosing problems, 160–162
  disadvantages of, 144
  four-wheel, 147
  friction material inspection, 148–149
  glazed linings, 160–161
  inspection, 148
  linings. See Linings, brake
  lubricants/shims, 158–159
  mounting hardware, 158
  noise prevention, 158–160
  noise suppression compound, 158, 159f
  operation of, 142–144
  pads, 144
  parking brake, 188–189
  pedal pulsation, 161
  rear. See Rear disc brakes
  repacking wheel bearings, 248
  replacing linings, 150–152
  road test, 147–148
  rotors. See Rotors
  service, 147–157
  uneven wear, 161–162

Disc brake wheel bearings, repacking, 248
Dive, 340
Dog tracking, 413
DOT codes (tires), 270–271, 512–517
Double cardan U-joints, 464, 465f
Double insulated, 45
Double offset plunge joints, 446, 456, 457f
Double-row ball bearings, 237
Double wishbone suspension system, 328–329
Drill bits, 46, 48f
Drill motors, 46, 48f
Drill press, 46, 47f
Drills, 46–47
  air, 50
Drive axle bearings, 239
Drive axles, 470–471
  full-floating axles, 477
  reinstallation of, 476
  semi-floating axles, 470–471
Drive belts, power steering, 372–373
  alignment, 398
  replacement of, 398–399
  service, 397–401
  tension, 400–401
Drive cornering scrub, 286
Drive line, 459
  as cause of vibration, 484
  vibration, 484
Driveover lifts, 34, 35f
Drive shaft, 460–470
  angle, 462–463
  angle check, 487–488
  balance, 466, 485–487
  checking angle, 487–488
  checking balance, 486–487
  diagnosing problems, 465–466
  flange, 461
  heavy spot, 485, 486f
  inspection, 466
  installation, 470
  open, 460
  out of balance, 482
  repairing/replacing, 485
  service, 466–470
  slip yoke, 460–461
  two-piece, 462, 470
  universal joints. See U-joints
  working angle, 487
Drive shaft runout, 484–485
Drive-through lifts, 34
Drone, 484
Drop center, 277
Drum brakes, 4, 5, 63, 63f, 120–140
  adjusting gauge, 192
  adjustments, 125–128
  bleed screw, 109f
  cleaning, 130–131
  design variations, 124–125
  disassembly and inspection, 128
  drums. See Brake drums
  dual-servo, 124–125
  hardware kit, 136, 137f
  leading-trailing, 124–125
  micrometer, 21
  ordering replacement parts, 137–138
  parking brake, 188

parts, 120–124
reassembly, 138–140
removal of brake drum, 128–130
residual check valve, 78, 99
self-adjusters. See Self-adjusters, drum brake
service, 128–135
servo-action, 63, 124
springs, 123–124
Drum-in-hat brake, 188–189
Drum mike, 21
  combination digital, 25
Dry boiling points, 69
DTCs. See Diagnostic trouble codes (DTCs)
Dual-diameter bore master cylinder, 80
Dual proportioning valves, hybrid brake system, 84
Dual-servo drum brake, 124–125
  self-adjustment, 125–127
Duck bill valve, 78
Dump, 220
Duplex brake, 125
Dust boot, disc brake caliper, 154, 156–157
Dust boots, 122
Dynamic balance, 307–309
Dynamic proportioning, 86
Dynamic seals, 241

E
Earless clamps, 455
Ear protection, 41, 42f
EBTCM. See Electronic brake and traction control module (EBTCM)
Eccentric cam adjustment, 425
ECM. See Electronic control module (ECM)
EHCU. See Electromechanical hydraulic unit (EHCU)
Elbow, 89
Electrical fires, vehicle batteries and, 44, 45f
Electrical safety, 45
Electrical switches, brake system, 86
Electric lights, portable, 44
Electric shock, 45
Electric welding, 51
Electromagnetic antilock brake systems (ABS), 222, 223f
  service, 231
Electromechanical hydraulic unit (EHCU), 217
Electronically controlled electric steering systems, 380
  service, 404
Electronically controlled shock absorbers, 339–340
  service, 361–362
Electronically controlled variable effort power steering, 375–377
Electronic brake and traction control module (EBTCM), 215
Electronic caliper, 25
Electronic control module (ECM), 215
Electronic four-wheel steering systems, 378–380
Electronic limited slip, 224

Electronic stability control, 224, *225f*
Electronic suspension system, 9, 360–362
Electronic vibration analyzer (EVA), 483
Emergency brake. *See* Parking brake
EP additives, 472
EP lubricants, 240–241
Equalizer, parking brake, 186–187
Ethics in brake work, 70
EVA. *See* Electronic vibration analyzer (EVA)
Exchange machine, 392
Exciter ring, 215
Exhaust ventilation system, 45
Expanding strut, 127
Extenders, frame-contact lift adapter, 33
External band brake, 5
External snaprings, 452
Extreme pressure (EP) additives, 472
Extreme pressure (EP) lubricants, 240–241
Eye protection, 40–41
Eyewash fountain, 41

F

Face shield, 40–41
False modulation, 226
FASCAR strips, hydraulic brake fluid, 103–104
Fast-fill master cylinder, 80
Fast steering ratio, 365
Federal Motor Vehicle Safety Standards (FMVSS), 66–67
Feeler gauge, 134, *135f*
Fender covers, 17
Finish cut, 179
Fins, brake drum, 168
Fire extinguishers, 42–43
Fire safety, 42–45
Firewall, 194
  vacuum brake booster, 195
First-aid kit, 40
First order vibration, 482
Fittings, brake hose, 87
  banjo fittings. *See* Banjo fittings
  compression, 90–91
  flare, 89–90
  high pressure compression, 90–91
  step-up/step-down adapter, 89
  tubing, 88–89
Fittings, zerk, 468–469
Fixed caliper, 144
Fixed joints, 445
  disassembly and inspection, 453–454
  problems, 448
Flammable materials, 43–44
Flammable storage cabinet, 43, *44f*
Flange, 461
Flapper valve, 78
Flared connections, 89–90
Flared line, 90
Flare fittings, 89–90, 113
Flare nut, 90
Flare nut wrench, 97, 113, *114f*
Flaring tool, 115, *117f*
Flash point, 44
Flat-rate manuals, 15
Flex-hone, 180, *181f*
Flexible coupling, 370
  replacement, 390

Floating caliper, 144–147
Float switch, master cylinder, 85
Floor creepers, 52
Flow control valve, power steering pump, 371
  service, 395, *396f*
Fluid
  antilock brake systems (ABS) service, 227–228
  brake. *See* Brake fluid
  lifetime, 69
  magneto-rheological (MR) shock absorber, 340
  power steering. *See* Power steering
Fluid level switch, master cylinder, 85
  service, 111
Flushing of hydraulic brake system, 102
FMVSS. *See* Federal Motor Vehicle Safety Standards (FMVSS)
Foot pads, 33
Footprint, 264
Force, 61, *62f*
Force variation, 311–312
Four-wheel alignment, 423, 434–440
  adjusting rear-wheel alignment, 437–440
  camber check, 436–437
  compensating alignment heads, 436
  measuring caster, 436–437
  performing, 436–440
  rear-wheel alignment adjustment, 437–440
Four-wheel antilock brake systems (ABS), 220
Four-wheel disc brakes, 147
Four-wheel steering systems, 377–380
  electronic, 378–380
  hydraulic, 378
Frame-contact lifts, 31–33
Frequency, 481
  natural, 323, 482
  resonant, 482
  testing, 483–485
  of vibration, 483–485
Friction, 3, 65, 260
Frictionless bearings, 235, *235f*
Friction material. *See also* Linings, brake
  aftermarket certifications, 166–167
  disc brake inspection, 148–149
  evaluation of, 166–168
Front-wheel drive axles, 444
  CV joint boots. *See* CV joint boots
  CV joints. *See* CV joints
  half shaft, 444, 457
  inspection and diagnosis, 447–448
  installation of, 457
  removal of, 450–451
  service, 447–457
  stub shaft. *See* Stub shaft
Front-wheel drive bearings, 239
  bolt-on type, 252
  front bearing inspection, 251
  front bearing replacement, 251–253
  press-on type, 252–253
  pullers for removal of, 256
  service, 250–257
Front-wheel drive vehicles
  advantages of, 443

axles. *See* Front-wheel drive axles
Fuel fires, 43–44
Full-floating axles, 239
Full-floating axles drive axles, 477
Fully hydraulic lifts, 35
Fully metallic brake linings, 165
Furnace brazing, 88

G

Galling, 468
Garter spring, 241, *242f*
Gas fade, 66
Gas-filled shock absorbers, 346, *347f*
Gasoline precautions, 44
Gas shock absorbers, 334–335
Gas welding, 51
Gearbox, steering system, 364
Gear oils, 471–473
  limited slip, 472
  ratings, 472
Gears
  oils. *See* Gear oils
  pinion, 365
  pivot lever power steering, 374–375
  rack, 365
  rack-and-pinion steering, 365–366, 402–404
  rotary valve type power steering, 375
  sector, 365
  steering systems, 364–366, 390–391
  torsion bar type power steering, 375
Generic service manuals, 12–13
Geometric centerline, 434, *435f*
Glazed linings, 160–161
Gloves, 42
Glycol, 102
Goggles, 40–41
Gravity brake bleeding, 109
Greases, 240–243
  performance classifications, 240
  types of, 241
Grease sweep, 43, 52–53
Grinders, 47–48
Grinding wheel, 47, *48f*
Grooved and drilled rotors, 170
Gross vehicle weight (GVW), 270, 284
Guide plate, *121f*
GVW. *See* Gross vehicle weight (GVW)

H

Hair, safety precautions for, 41–42
Half shaft, 444
  rebuilt, 457
Handling characteristics, 413–414
  oversteer, 414
  slip angle, 413, *414f*
  understeer, 414
Hand protection, 42
Hand tool safety, 50–51
Hard spots, 170, *171f*
Hard steering, 383
Hazardous materials, 53–56
Heat dissipation, 66
Heel, 121
Height-sensing proportioning valves, hybrid brake system, 84
Height sensors, 338
HEPA filter, 53

PA vacuum, 53–54
ptane, 56
rtz, 481
xane, 56
gh cordline belts, 372–373
gh-efficiency particulate air filter (HEPA filter), 53
gh-performance brake linings, 166
gh-performance suspension system, 329–331
gh-performance tires, 297
gh pressure compression fittings, 90–91
oists. *See* Vehicle lifts
olddown pins, 130
olddown springs, 123, *124f*, 132, *133f*
old valves, 220
ollander interchange manuals, 15
oning wheel cylinder, 133–134
ose couplings, 38
oses
 brake. *See* Brake hoses
 power steering, 371–372, 401–402
otchkiss design, 460
otline services, 15
owl, 484
HSMO. *See* Hydraulic system mineral oil (HSMO)
Hub
 greasing inside of, 246
 installation, 255
 removal of stub shaft from, 449–450
Hub-centric wheels, 277, 309
Hub removal, 182, *183f*, 244, 253–254, *256f*
Hybrid brake system, 63
 combination valve, 85–86
 metering valves, 82–83, 85–86
 proportioning valves. *See* Proportioning valves, hybrid brake system
Hydraulic brake fluid, 68–69
 FASCAR strips, 103–104
 fluid level switch, 85, 111
Hydraulic brake system. *See* Brake systems
Hydraulic control valve assembly, 217
Hydraulic equipment, 29
Hydraulic four-wheel steering systems, 378
Hydraulic jack, 30, *30f*
Hydraulic power brake boosters, 207
Hydraulic presses, 46, *47f*
Hydraulic pressure, 29, *30f*
Hydraulic safety switch, 85
 service, 111
Hydraulic service jacks. *See* Service jacks
Hydraulic shock absorbers, 331–334
 bump stops and limiters, 334
 compression and rebound resistance, 332–334
 fluid aeration, 334
Hydraulic system mineral oil (HSMO), 69
Hydro-boost brakes, 207–209
 bleeding, 209
 inspection, 208
 repair and replacement, 208–209
Hydroplaning, 260–261
Hydrovac brake, 200
Hygroscopic, 68
Hysteresis, 263

I

IATN. *See* International Automotive Technicians Network (IATN)
I beam axle, 325–326
Idler arm, 367
 inspection, 383, *384f*
 replacement, 385
Impact socket, 49
Impact wrenches, 49
Inboard side, 445
Included angle, 411
 measuring, 428
Incorrect scrub radius, 412
Incremental adjuster, 127
Independent rear suspension, 471
Independent suspension system, 325–326
 independent rear suspension, 471
Individual toe, 435
Inertia, 65
Inflation pressure, tire, 281
In-ground lifts, 31, 34–36
 maintenance, 35–36
Inner liner, 260
Inner race, 235, *236f*
In phase, 463
Integral antilock brake systems (ABS), 217, 218–219
 service, 231
Integral disc brake parking brake, 189
Integrally molded brake pads, 68
Integral power steering, 374
Integral reservoir, 391
Integral sensors, 274
Integral-type parking brake, 188
Interchange manuals, *Hollander*, 15
Internal snaprings, 452
International Automotive Technicians Network (IATN), 15
Internet service information, 15
Inverted flare, 90
ISO flared brake lines/tubings, 89, 90, 116–117
Isolation valves, 220
Isoprene, 263

J

Jacket, brake hose, 87
Jack stands, 31, *31f*
Jamb nut, 386, *387f*
Jobbers, 18–19
Joints
 ball. *See* Ball joints
 ball and trunnion, 465
 compression type ball joints, 337, *338f*
 CV. *See* CV joints
 double cardan u-joints, 464, *465f*
 load carrying ball joints, 337, 353
 universal. *See* U-joints
 wear indicator ball joints, 354
Jounce, 7, 321

K

Kinetic balance, 307
Kinetic energy, 3, 65
King pin inclination (KPI). *See* Steering axis inclination (SAI)
Knockoff-type CV joints, 452
Knuckle, 254–255

kPa (kiloPascal), 274
KPI. *See* King pin inclination (KPI)

L

Ladder frame suspension system, 318, *319f*
Lateral acceleration sensor, 216
Lathes, brake drum and rotor, 173–177
 accuracy of, 181
Lathes, on-the-car, 181–182
Launch shudder, 488
Leading-leading drum brakes, 125
Leading-trailing drum brakes, 125
 self-adjustment, 127–128
Lead point, 409
Leaf springs, 324–325
 service, 360
Lemon law, 70
Leveling devices, 335–336
Leveling wheel head, 436–437
Lifetime fluid, 69
Lifting heavy items, personal safety for, 41
Lifting safety, 37–38
Lift points, 30, *31f*, 37
Lifts. *See* Vehicle lifts
Limited slip gear oils, 472
Linear rate, 322
Linear rate coil spring, 323
Line clamp, 88
Linen service, 18
Lines, air compressor, 38
Lining fade, 66
Linings, brake. *See also* Pads, brake; Shoes
 aftermarket friction material certifications, 166–167
 asbestos, 164
 bonded or riveted. *See* Bonded/riveted brake linings
 breaking in new linings, 168
 carbon fiber, 165–166
 ceramic, 164, 165
 disc brake, 144, 150–152, 160–161
 drum brake, 121
 evaluation of friction materials, 166–168
 fully metallic, 165
 glazed, 160–161
 high performance, 166
 materials, 164–168
 noise control, 166
 nonasbestos organic (NAO), 164, 165
 organic/inorganic, 164
 performance factors, 166
 replacing disc brake, 150–152
 SAE edge code, 167–168
 selection of, 166
 semimetallic, 164, 165
 wear factors, 166
Linkages
 mechanical, 73
 parallelogram, 364, 367–368
 rack-and-pinion steering. *See* Rack-and-pinion steering
 steering linkage. *See* Steering linkage
Linkage-type power steering, 375
Lips
 auxiliary, 242
 sealing, 241–243

Liquid tire puncture sealants, 304
List price, 20
Live axles, 239, 459
Load capacities, *269f*
Load carrying ball joints, 337
    service, 353
Load index, 268
Load rating, 270
Locking clamps, 455
Locking device, 38
Lock rings, 452
Lock to lock, 365
Longitudinal split tandem master cylinder system, 79–80
Loss leaders, 20
Low pedal master cylinder action, 75–77
Low pressure warning system, 273–274
Low profile clamps, 455
Low profile tires, 297
Lubricants. *See also* Greases
    chassis, 241
    extreme pressure (EP), 240–241
    leaks, 473
Lubricants/shims, disc brake, 158–159
Lubrication service manuals, 13, *14f*
Lug-centric wheels, 309–310
Lug nuts, 279
    antitheft, 289
    tightening, 288–289
Lug removal, 182, *183f*, 287–290
Lug studs, 279
    repairing broken, 290
    swaged, 182, *183f*

M
Machine and tool safety, 46–51
    drills, 46–47
    grinders, 47–48
    hydraulic presses, 46, *47f*
Machine shop, 19
MacPherson strut suspension system, 7, *9f*, 329
    adding oil to strut body, 349–350
    coil spring installation, 350
    compressor, 348, *349f*
    removal of strut, 347–348
    service, 347–350
    strut assembly reinstallation, 350
    strut cartridge replacement, 349
    upper strut bearing inspection, 350
Magnasteer, 376
Magneto-rheological (MR) fluid shock absorbers, 340
Magnuson-Moss Warranty Act, 70
Manual brake bleeding, 105–106
    with hose, 106
Manual brakes clearance adjustment, 139–140
Manuals
    flat-rate, 15
    Hollander interchange, 15
    parts and labor estimating, 15
    service manuals. *See* Service manuals shop, 9–10
Manufacturer service manuals, 10
Mass merchandisers, 20
Master cylinder, 3, *4f*, 61, *62f*
    bail, 96
    bench bleeding, 99–101

booster. *See* Booster, power brake
check valve, 78, *79f*
disassembly of, 97–98
fluid level switch, 85, 111
inspection, 96–99
installation of reservoir, 98–*99f*
location, 79, *80f*
low pedal action, 75–77
metering valves, 82–83
operation of, 74–75
proportioning valves. *See* Proportioning valves, hybrid brake system
quick take-up, 80–82
reinstallation of, 101
removal of, 97–98
reservoir. *See* Reservoir, master cylinder
reservoir caps, 95
single piston, 75
tandem. *See* Tandem master cylinder
testing, 97
vent (compensating) port. *See* Vent (compensating) port
Match mount, 311, 485
Material safety data sheet (MSDS), 55
Measuring, 21–25
    brake drums, 21–23
    micrometer. *See* Micrometers
Mechanical drum brakes, 63
Mechanical fade, 66
Mechanical linkages, 73
Metering valves, hybrid brake system, 82–83, 107
    combination valve, 85–86
Metric micrometers, 24–25
Microfiche, 14
Micrometers, 23–25
    drum brake, 21
    metric, 24–25
    reading, 23–24
Mineral oil brake fluid, 69
Moan, 484
Moisture content testers, hydraulic brake fluid, 102–103
Monocoque suspension system, 318–320
Monoleaf springs, 325
Monotube shock absorbers, 332, *333f*, 335
Morning sickness, 391
Motor pack, 222
MSDS. *See* Material safety data sheet (MSDS)
Multilink suspension system, 329–331
Multipurpose grease, 241
Multiuse brake lathes, 173
Mushroomed chisels, 50

N
NATEF task correlation grid
    brake systems, 504–507
    steering systems, 508–511
    suspension systems, 508–511
National Automotive Technicians Education Foundation (NATEF) task correlation grid. *See* NATEF task correlation grid
National Institute for Automotive Service Excellence (ASE) certification exams. *See* ASE certification exams

National Lubricating Grease Institute (NLGI) performance classifications, 240
Natural frequency, 323, 482
Needle bearings, 238, *239f*, 462
Negative camber, 407–408
Negative caster, 409, 410
Negative scrub radius, 411–412
Neoprene, 372
Net price, 20
New high-speed rating, 270
NLGI performance classifications, 240
Noise
    axle bearing, 473
    disc brakes, 158–160
    steering systems, 383
    wheel bearings, 249–250
Noise suppression compound, 158, *159f*
Nominal diameter, 22
Nonasbestos organic (NAO) brake lining, 164, 165
Nonintegral antilock brake systems (ABS), 217, *218f*, 219

O
OEMs. *See* Original equipment manufacturers (OEMs)
On-the-car balancers, 307
On-vehicle rotor machining, 181–182
Open drive shaft, 460
Orbital sander, 180, *181f*
Orbital steer, 419
Organic/inorganic brake linings, 164
Original equipment manufacturers (OEMs), 18
Oscillations, 331
Outboard side, 445
Out-of-round wear, 23, 170, *171f*
Overall tire diameter, 276–277
Overload leaves, 325
Oversteer, 224, *225f*, 414
Oxyacetylene welding, 51

P
Pad lifts, 33
Pads, brake, 67–68, 164
    disc brakes, 144, 148–149
    friction material inspection, 148–149
    rear disc brake installation, 159–160
Parallelogram linkage, 364, 367–368
Parallelogram steering system, 6, 367
    inspection, 383–384
Parking brake, 61, 185–190
    adjustment, 191–193
    auxiliary drum, 188–189
    auxiliary transmission, 190
    ball and ramp caliper, 190
    cables, 185–187
    cable service, 191–193
    conduit, 187
    disc brake, 188–189
    drum brake, 188
    drum-in-hat brake, 188–189
    equalizer, 186–187
    incorrect adjustments, 194
    integral disc brake, 189
    integral-type, 188
    lever, 185, *186f*
    lever installation, 136

king brake (*continued*)
  screw and nut caliper, 189–190
  self-adjuster, 142, *144f*
  service, 190–194
  strut, 123, 127, 188
  types of, 188–190
  vacuum-release mechanism, 185, *186f*
  warning lamp, 187
  warning lamp service, 193–194
  king brake switch, 86
rts and labor estimating manuals, 15
rts and time guide, 15
rts pricing, 20–21
scal's Law, 73–74
ssive suspension system, 9, 337
tches, tire, 303
dal feel, 64, 94
antilock brake system (ABS), 214
vacuum-suspended power brake booster, 198
dal pulsation, 161
dal reserve, 94
rimeter frame suspension system, 318, *319f*
ripheral neuropathy, 56
rmanent magnet (PM) generators, 215
rsonal protection equipment (PPE), 40–42
rsonal safety, 40–42
hoenix Injection brake bleeding, 108–109
inch bolt, 347
inion gear, 365
inion seal replacement, differential, 478
in slider caliper, 146
istons
  caliper. *See* Caliper pistons
  single piston master cylinder, 75
itman arm, 367
  inspection, 383
  replacement, 385
itman shaft, 365
Pivot bearings, 337, *338f*
Pivot lever power steering gear, 374–375
Pivot pin, 136
Placard
  gross vehicle weight, 270
  tire size information, 265–266
Plain bearings, 235
Planetary torque multiplier, 457
Plate and lever vacuum-suspended power brake booster, 199
Plies, 259
  design, 263–265
Plugs, tire, 302–303
Plunge joints, 445
  double offset, 456
  problems, 448
  service, 456
  types of, 446
Plus sizing, 276
P-metric radial tire, 266–267
Point of load, 409
Point of resonance, 482
Polyglycol base, 68
Polyneuropathy, 56
Portable alignment gauge, *423f*
Portable electric lights, 44
Positive camber, 407–408

Positive caster, 409, 410
Positive scrub radius, 411
Power brakes, 63, *64f*, 194–211
  booster. *See* Booster, power brake
  electric hydraulic. *See* Powermaster brakes
  hydraulic boosters, 207
  hydro-boost brakes. *See* Hydro-boost brakes
  mechanical hydraulic. *See* Hydro-boost brakes
  powermaster brakes. *See* Powermaster brakes
  service, 203–211
  types of, 194–195
  vacuum brake booster. *See* Vacuum brake booster
  vacuum-suspended. *See* Vacuum-suspended power brake booster
Powermaster brakes, 209–211
  advantages, 209
  cylinder, 210
  replacement of booster, 211
  service, 210–211
Power piston, 374
Power steering, 370–377
  bleeding system of air, 392–393
  check of, 420–421
  drive belts. *See* Drive belts, power steering
  electronically controlled variable effort, 375–377
  flow control valve. *See* Flow control valve, power steering pump
  fluid, 382, 391–392, 402
  flushing system, 391–392
  hoses, 371–372, 401–402
  integral, 374
  morning sickness, 391
  pressure diagnosis, 394–395
  pressure relief valve, 371
  pulley replacement, 393–394
  pulley shaft seal service, 395, 397
  pump, 370–371
  pump replacement, 393
  pump service, 395–397
  pump shaft seal replacement, 393–394
  refilling fluid, 402
  repairing pump leaks, 393
  reservoir service, 391
  service, 391–397
  system flush, 391–392
  test drive check of, 420–421
  types of, 374–377
  variable, 404
PPE. *See* Personal protection equipment (PPE)
Precutting inner and outer ridges, 177
Preload adjustment, 390
Press-fit bearing replacement, 475–476
Press-fit races, 250
Press-fit ring, 455
Press-on type front-wheel drive bearings, 252–253
Pressure bearing packer, 246
Pressure bleeder adapters, 107
Pressure brake bleeding, 106–108
Pressure chamber, 76
Pressure dump, 220

Pressure multiplier vacuum-suspended power brake booster, 200
Pressure relief valve, power steering pump, 371
Pricing parts, 20–21
Primary cup, 75
Primary lining, 121
Profile, tire, 266
Propeller shaft, 460. *See also* Drive shaft
Proportioning valves, hybrid brake system, 82, 83–84
  combination valve, 85–86
  diagnosing problems, 110–111
psi (pounds per square inch), 274
Puddle, 51
Pullers
  for removal of front-wheel drive bearings, 256
  slide hammer, 252
Pump control switches, integral antilock brake system, 219
Pumps
  auxiliary vacuum, 203
  power steering. *See* Power steering
Purging shock absorbers, 346
Push-fit races, 250
Pyrotechnics, 388

Q

Quick take-up master cylinder, 80–82
Quick take-up master cylinder, bench bleeding, 100–101

R

RABS. *See* Rear antilock brake system (RABS)
Races
  inner, 235, *236f*
  press-fit, 250
  push-fit, 250
  replacing, 250
Racing slick, 260
Rack, 366
Rack-and-pinion steering, 6, *7f*, 364, 365–366
  damper, 366
  electronically controlled electric, 380
  gears, 365–366, 402–404
  linkage, *7f*, *364f*, 369, 384
  looseness, 391
  replacing gear units, 403–404
  service, 391
Rack gear, 365
Racks. *See* Vehicle lifts
Radial load, 235, *236f*
Radial-ply tires, 263–264
  P-metric, 266–267
Radial thrust, 237
Radius plates, 422, *422f*
Radius rods, 326
Reaction disc vacuum-suspended power brake booster, 198–199
Rear antilock brake system (RABS), 220
Rear camber, adjusting, 437
Rear disc brakes, 142
  cautions, 152–153
  pad installation, 159–160
Rear shock mounts, 335
Rear toe, adjusting, 437–440

Rear-wheel alignment adjustment, 437–440
Rear-wheel antilock (RWAL) systems, 220
　service, 231
Rear-wheel drive vehicles
　drive line, 459
　drive shaft. *See* Drive shaft
Rebound, 7, 321
Rebuilt half shafts, 457
Receiver tank, 38, *39f*
Recirculating ball and nut steering gear, 364, 365, 367
　preload adjustment, 390
　service, 390
Records, service, 16–17
Red brake warning light, antilock brake system (ABS), 225–226
Reed tachometer, 483
Refractometer testers, hydraulic brake fluid, 103
Remove and replace (R&R) job, 15
Repair order (RO), 16–17
Replenishing port, 75, 76
Reservoir
　auxiliary vacuum, 202
　integral, 391
　integral antilock brake system, 219
　master cylinder. *See* Reservoir, master cylinder
　power steering, 391
Reservoir, master cylinder, 78–79. *See also* Brake fluid
　fluid check, 95
　gasket, 101, *101f*
　installation of, 98–*99f*
Reservoir caps, 95
Residual check valve, 78
　service, 99
Residual pressure, 78
Resonant frequency, 482
Respirator, *42f*
　asbestos protection, 53
Restricted brake hoses, 111–112
Retail chain stores, 20
Retail price, 20
Retaining screws, 128, *129f*
Retracting springs, 123
Retreads, 273
Return springs, 123
　results of weak, 137
　visual check of condition of, 136–137
Reverse fluid injection (RFI), brake bleeding, 108–109
Reverse fluid injector, bench bleeding of master cylinder with, 100–101
Ribs, 262
Ride height check, 418
Rigid axle suspension system, 325–326
Rim, brake shoe, 120
Rims, wheel, 277
　removing and mounting on tires on, 290–292
Rim well, 277
Rim width, 278
Riveted brake linings. *See* Bonded/riveted brake linings
Rivet heads, 131, *132f*
RO. *See* Repair order (RO)
Road crown and pull, 423–424

Road test. *See also* Test drive
　disc brake, 147–148
　vibrations, 484
Roller bearings, 237–238
Roller power steering pump, 370, *371f*
Rotary valve type power steering gear, 375
Rotation of tires, 286–287
Rotors, *5f*, 142
　ceramic brake, 169
　composite, 169–170
　damage, 148
　directional, 168–169
　discard dimension, 171–172
　grooved and drilled, 170
　hub removal, 182, *183f*
　inspection, 149, 171–172
　lathes, 173–177
　lug removal. *See* Lug removal
　machining, 150, 172–177, 179–182
　mounting, 174–175
　nondirectional finish, 180
　on-vehicle machining, 181–182
　precutting inner and outer ridges, 177
　problems, 161
　runout, 26, 149–150, 161, 171–172
　service, 170
　silencing band, 175–176
　solid, 168
　surface finish, 180
　thickness, 171–172
　types of, 168–170
　ventilated, 168
　washing, 180–181
Rough cut, 179
Roughness, 483
Rubber, 263
Rubber bellows/boots, 369
Rubber cups, 75, *76f*
Rubber diaphragm, reservoir, 79
Rubber lube, 293–294
Run-flat tires, 272–273
Runout, 26
　drive shaft, 484–485
　rotor. *See* Rotors
　stub shaft, 485
　tire, 298–299
　wheel, 299–300
RWAL. *See* Rear-wheel antilock (RWAL) systems
Rzeppa CV joints, 445. *See also* Fixed joints

S
SAE
　brake tests, 166
　edge code, 167–168
　gear oil ratings, 472
SAE flared lines, 89, 90, 115–116
Safety
　breathing, 45
　electrical, 45
　fire, 42–45
　general precautions, 52
　hand tool, 50–51
　machine and tool. *See* Machine and tool safety
　personal, 40–42
　shop equipment, 39–40

　solvent, 56
　vehicle lifting, 37–38
Safety beads, 277
Safety containers, *43f*
Safety glasses, 40–41
SAI. *See* Steering axis inclination (SAI)
Sandwich body suspension system design, 319
Sawtooth pattern, 417
Scan tool, 227, *227f*, 360
Scratch cut, 176–177
Screw and nut parking brake caliper, 189–190
Scrub radius, 411–412
Scuff, 406
Sealing lip, 241–243
Seals
　removal of, 244–245
　tolerance of, 243
　wave, 242–243
　wheel bearing. *See* Wheel bearing seals
Seamless tubing, 88
Secondary cup, 75
Secondary lining, 121
Second order vibration, 482, 488
Sector gear, 365
Sector shaft, 365
Self-adjusters, disc brake, 142, *144f*
　ball and ramp parking brake caliper, 190
　screw and nut parking brake caliper, 189–190
Self-adjusters, drum brake, 75
　clearance adjustment with drum installed, 140
　driver habits and, 128
　installation of, 139
　parking brake strut, 127
　service, 136
Self-energizing, 63
SEMA. *See* Specialty Equipment Manufacturers Association (SEMA)
Semi-elliptical springs, 324–325
Semi-floating axles, 239, 470–471
Semihydraulic lifts, 35
Semimetallic brake linings, 164, 165
Sensors
　audible wear, 144
　automatic suspension leveling, 338–339
　digital steering angle, 377
　height, 338
　integral, 274
　lateral acceleration, 216
　tactile wear, 144
　tire, 274
　visual wear, 144, *145f*
　wheel speed. *See* Wheel speed sensors
Serpentine belt, 399, *400f*
Serpentine belt drive, 373
Service brake, 61
Service information sources, 9–15
Service jacks, 30, *30f*
　safety, 52
Service manuals
　generic, 12–13
　lubrication, 13, *14f*
　manufacturer, 10
Service records, 16–17

vice stations, 20
vice writers/advisors, 16
vo-action, 63, 124
back, 413, *414f*, 436
ckle, 324
ake, 483
immy, 307
ims, 424–425
disc brake, 158–159
apered, 439
ock absorbers, 8–9, 331–335
  air shocks. *See* Air shocks
  bleeding/purging, 346
  coil spring shocks, 336
  electronically controlled, 339–340, 361–362
  gas, 334–335
  gas-filled, 346, *347f*
  hydraulic. *See* Hydraulic shock absorbers
  magneto-rheological (MR) fluid, 340
  monotube. *See* Monotube shock absorbers
  rear shock mounts, 335
  service, 343–346
  testing, 344–345
  twin-tube. *See* Twin-tube shock absorbers
hock mounts, 346
hoes, 42, 164
hop equipment safety, 39–40
hop habits, 52–53
hop manuals, 9–10
Short/long arm (SLA)
  coil spring replacement, 357–359
  double wishbone suspension system, 328–329
  suspension system, 7, 327–329
Side slip, 377
Sidewall checks, 285–286
Sidewall deflection, 264
Sidewall markings, tire, 265–270
  aspect ratio, 266–267
  DOT codes, 270–271, 512–517
  load capacities, *269f*
  load index, 268
  load rating, 270
  new high-speed rating, 270
  ratings, 266–267
  speed rating, 267–270
  temperature grade, 275
  traction grade, 275
  tread wear ratings, 275
  uniform tire quality grading (UTQG), 274–275
Silencing band, 175–176
Simplex brake, 125
Single-box suspension system design, 319
Single channel antilock brake systems (ABS), 219–220, 220–221
Single flare, 115
Single piston master cylinder, 75
Single-plane balance, 307
Single-row ball bearings, 236, 237
Sintering, 165
Sipes, 261
SIR. *See* Supplemental inflatable restraint (SIR)

Size equivalents, tire, 275–276
Skin protection, 54
SLA. *See* Short/long arm (SLA)
Slide hammer, 129, *130f*, *474f*, *475f*
Slide hammer puller, 252
Sliding caliper, 144–147
Slip angle, 413, *414f*
Slipper power steering pump, 370, *371f*
Slip plates, 422
Slip rate, 213
Slip yokes, 460–461, 462
Slope, 84
Snap rings, 252, 451–452
Snapshot, 483
Snow chains, 272
Snow tires, 271–272
Society of Automotive Engineers (SAE). *See* SAE
Solenoid valves, 220, *221f*
Solid axle suspension system, 325–326
Solid lubricant greases, 241
Solvents, 55–56
Space frame suspension system, 320–321
Spalling, 245
Specialty Equipment Manufacturers Association (SEMA), 277
Specification charts, *12f*
Speed rating, 267–270
  new high-speed rating, 270
Spider, 461
Spindle chuck, 46, *48f*
Spindle inspection, 247–248
Spindle nut, 244, 248
Spindles, 352, 368
  wheel alignment. *See* Wheel alignment
Spindle support arm, 337
Split point, 83
Spontaneous combustion, 43
Spread, caster, 434
Spring dampers. *See* Shock absorbers
Spring rate, 322
Springs, 7, 321–325
  air, 325
  coil, 321–322
  composite, 325
  leaf, 324–325, 360
  linear-rate coil, 323
  monoleaf, 325
  semi-elliptical, 324–325
  variable-rate coil, 323
  wheel rate, 322–323
Springs, drum brake, 123–124
Spring scale, 394, *395f*
Sprung weight, 317–318
Squat, 340
SRS. *See* Supplemental restraint system (SRS)
Stabilizer bars, 336–337
  service, 352
Stabilizer links, 337
Starwheel, 125, 126
Static balance, 305–307
Static seals, 241
Steel tubing, 88
  bending, 115
  cutting, 113–114
Steering arms, 366, 368
Steering axis inclination (SAI), 410–412
  functions, 411

  measuring, 426–428
  scrub radius, 411–412
Steering column, 369–370
  service, 389–390
Steering damper, 369
Steering free play, 383
Steering knuckle, 9, 337, 347, 352, 366
Steering linkage, 366–369
  rack-and-pinion. *See* Rack-and-pinion steering
  repairs, 384–387
  tapers, 384–385
Steering pull, 420
Steering ratio, 365
Steering systems, 6–8
  diagnosing problems, 382–383
  electronically controlled electric, 380
  four-wheel. *See* Four-wheel steering systems
  gearbox, 364
  gear looseness, 383
  gears, 364–366
  hard steering, 383
  linkage. *See* Steering linkage
  lock to lock, 365
  NATEF task correlation grid, 508–511
  noise, 383
  parallelogram, 367
  parallelogram linkage, 364
  parts inspection, 383–384
  parts of, 364
  power steering. *See* Power steering
  practice ASE certification exam, 497–503
  rack-and-pinion. *See* Rack-and-pinion steering
  recirculating ball and nut steering gear. *See* Recirculating ball and nut steering gear
  service, 382–404
  tire wear, 383
  wormshaft, 365
Steering tapers, 354
Steering wheel, 389
  centering, 431–433
Steering wheel clamp, 429, *430f*
Step-bore master cylinder, 80
Step-up/step-down adapter fittings, brake tubing, 89
Stick welding, 51
Stock, 21
Stoddard solvent, 44
Stoplight switches, 86
  service, 111
Strut, parking brake, 123, 188
  expanding self-adjuster, 127
Strut rod bushings, 352
Strut rods, 326
Strut tower, 412
Stub axle, 444
Stub shaft, 253, 444
  reassembly, 256–257
  removal from hub, 449–450
Stub shaft runout, 485
Stud-centric wheels, 277
Subframe, 7
Suction brake bleeding, 108
Supplemental inflatable restraint (SIR), 370, 387

Supplemental restraint system (SRS), 387
Support rail, 35
Surface mount lifts, 36
Suspension-contact lifts, 34–35
Suspension leveling systems, 337–341
Suspension linkage, 7–8
Suspension system, 6–9
   active, 340–341
   adaptive, 9, 337
   ball joints. *See* Ball joints
   body-over-frame, 318–319
   bushings. *See* Bushings
   cab-forward design, 319, *320f*
   construction, 325–327
   designs, 318–321
   diagnosing problems, 343
   double wishbone, 328–329
   electronic, 9, 360–362
   frame types, 318–321
   high-performance, 329–331
   independent, 325–326
   ladder frame, 318, *319f*
   leveling systems, 337–341
   load carrying ball joints, 337, 353
   looseness, 420
   MacPherson strut. *See* MacPherson strut suspension system
   monocoque, 318–320
   multilink, 329–331
   NATEF task correlation grid, 508–511
   parts, 317
   passive, 9, 337
   perimeter frame, 318, *319f*
   practice ASE certification exam, 497–503
   rigid axle, 325–326
   service, 343–362
   shock absorbers. *See* Shock absorbers
   short/long arm (SLA), 327–329
   short/long arm (SLA) double wishbone, 328–329
   solid axle, 325–326
   space frame, 320–321
   spindle support arm, 337
   stabilizer bars, 336–337, 352
   torsion bar, 323–324, 359–360
   types of, 327–331
   unibody, 318–320
Swaged lug studs, 182, *183f*
Swing axles, 471
Swing caliper, 146–147
Symmetric tread pattern, 262
Synthetic brake fluid, 69
Syringe, bench bleeding of master cylinder using, 100
Systematic exposure, 56

T

Tactile wear sensor, 144
Tandem master cylinder, 77–78
   brake warning light, 84–85
   diagonal split system, 79–80
   longitudinal split system, 79–80
Tandem vacuum-suspended power brake booster, 199–200
Tapered roller bearings, 238
   construction, *239f*
Tapered shim, 439
Tapers, ball joint, 354–355

Tapers, steering linkage, 384–385
TBA. *See* Tires, batteries, and accessories (TBA)
TCS. *See* Traction control system (TCS)
Technical service bulletins (TSBs), 15
Telephone service, 16
Temperature grade, 275
Tensile cords, 372
Tension gauge, 401
Tension type ball joints, 337, *338f*
Test drive. *See also* Road test
   after brake service, 162
   disc brake, 147–148
   power steering, 420–421
   safety, 52
   wheel alignment, 420–422
Test strips, hydraulic brake fluid, 103–104
Three- and four-channel antilock brake systems (ABS), 221–222
Three-box suspension system design, 319–320
Thrust angle, 435
Thrustline, 435
Thrust load, 235
Tie-rods, 366, 367–368
   centering steering wheel, 431
   ends replacement, 385–386
   inspection, 384
   pivoting, 418–419
   rack-and-pinion, 386–387
Tilt steering column, 369–370
Tingling, 483
Tinnerman nut, 128, *129f*
Tire chains, 272
   service, 312–313
Tire changers, 291, 294–298
Tire conicity, 287
Tire cord, 263
Tire iron, 292
Tire rotation, 286–287
Tire runout gauge, 299
Tires. *See also* Wheels
   air pressure, 281–284
   all-season, 271
   aspect ratio, 266–267
   balancing, 304–305
   bead seats. *See* Bead seats
   bias-ply, 263, 264
   caster and tire wear, 410
   checking air pressure, 281–284
   check of, 420
   combination plug-patches, 304
   compact spare, 273
   construction of, 259–260
   crown, 300
   directional, 294
   footprint, 264
   force variation, 311–312
   high-performance, 297
   inflating, 294
   inflation pressure, 281
   information sticker, 265
   inspection of, 292–294
   installation of, 294–298
   kPa, 274
   liquid puncture sealants, 304
   low pressure warning system, 273–274
   low-profile, 297
   overall diameter, 276–277

   oversizing, 276
   patching, 303
   plies. *See* Plies
   plugging holes in, 302–303
   plus sizing, 276
   P-metric radial, 266–267
   pressure monitors, 273–274
   profile, 266
   psi, 274
   quality grading, 274–275
   radial-ply, 263–264
   removing and mounting on rims, 290–292
   repairing, 300–304
   retreads, 273
   rotation of, 286–287
   run-flat, 272–273
   runout, 298–299
   sensors, 274
   sidewall checks, 285–286
   sidewall deflection, 264
   sidewall markings. *See* Sidewall markings, tire
   size information placard, 265–266
   sizes, 275–277
   snow, 271–272
   temperature grade, 275
   traction. *See* Traction
   traction grade, 275
   tread. *See* Tread
   tread wear, 275
   tread wear ratings, 275
   tubeless, 260
   tube-type, 260
   types of, 271–274
   types of damage, 301
   undersizing, 277
   uniform tire quality grading (UTQG), 274–275
   valve stems, 279, 293
   wear, 284–285, 383, 410, 416–418
Tires, batteries, and accessories (TBA), 20
Tire sizes, 275–277
Tire valve stems, 279
   service, 293
Toe, 406–407
   adjusting, 429–431
   adjusting rear toe, 437–440
   excessive, 417
   in front- and rear-wheel drive vehicles, 429
   individual, 435
   measuring, 428–429
   rules for, 434
   tire wear from, 417–418
Toe change, 418–419
   check, 432–433
Toe end of brake linings, 121
Toe-in, 406, 417
Toe-out, 406, 417
Toe-out on turns, 368
Tolerance, seal, 243
Tone ring, 215, *217f*
   inspection, 230
Tool bits, 176
Tool holder, 178
Torch welding, 51
Torque arm, 322
Torque steer, 419

INDEX

que tube, 460
que wrench, 390
sion bar, 323–324
emoval and replacement, 359–360
sion bar type power steering gear, 375
ic materials, 53–56
cking, 413
ck width difference, 436
ction, 213, 223–224, 260
ction bar, 322
ction control system (TCS), 223
ction grade, 275
de magazines, 15
ammel bar, 428, *429f*
ansaxle, 443–444
ansducer, 486
ansitional coils, 323
ansmission parking brake, auxiliary, 190
ead, 260–263
  materials, 262–263
  pattern designs, 262
  ribs, 262
  sipes, 261
  wear, 275
  wear ratings, 275
read Act, 273
read depth gauge, 284, *285f*
read wear, 275
read wear ratings, 275
ripod tulip plunge joints, 446
  service, 456
roubleshooting charts, *12f*
runnions, 462
  brinelling and galling on, 468
TSBs. *See* Technical service bulletins (TSBs)
Tubeless tires, 260
Tube-type tires, 260
Tubing, brake, 88–89
Tubing bender, 114–115
Tubing cutter, 113–114
Tubing fittings, 88–89
Tubular space frame suspension system, 320–321
Tulip clamp tire changer, 298
Turnbuckle, 368
Turning radius, 412–413
  measuring, 433
Twin-tube shock absorbers, 332, *333f*, 335
Two-box suspension system design, 319
Two-piece drive shaft, 462
  service, 470
Two-wheel antilock brake systems (ABS), 219–220

U

U-joints, 370, 461–462
  Cardan, 461
  cross and yoke, 461
  diagnosing problems, 466
  disassembly, 467–468
  double cardan, 464, *465f*
  grease, 241
  needle bearings, 462
  reassembly, 468–470
  slip yokes. *See* Slip yokes
  spider, 461
  tool, 467, *468f*
  two-piece drive shaft, 462–463
  yokes. *See* Slip yokes
Ultraviolet rays, 52
Undercar repair. *See* Chassis
Understeer, 224, *225f*, 414
Unibody suspension system, 318–320
Uniforms, 18
Uniform tire quality grading (UTQG), 274–275
Union repairs of brake lines/tubings, 117
Unitized body and frame, 7
Universal clamps, 454
Universal joint grease, 241
Universal joints, 370. *See also* U-joints
Unseating bead seats, 291–292
Unsprung weight, 318
UTQG. *See* Uniform tire quality grading (UTQG)

V

Vacula™, 108
Vacuum brake bleeding, 108
Vacuum brake booster, *5f*, 195–203
  air filter, 198
  brake drag, 204
  charcoal filter, 196
  check valve, 195–196, 205
  leaking front seal, 206
  leak test, 204
  minimum vacuum, 205–206
  operation of, 196–198
  operation test, 203–204
  pushrod clearance adjustment, 207
  replacement, 206–207
  service, 203–211
  vacuum supply checks, 204–206
  vacuum-suspended power brake booster. *See* Vacuum-suspended power brake booster
Vacuum-suspended power brake booster, 196, *198f*
  auxiliary vacuum pump, 203
  auxiliary vacuum reservoir, 202
  function of, 200–202
  mounting location, 200
  pedal feel, 198
  plate and lever, 199
  pressure multiplier, 200
  reaction disc, 198–199
  tandem, 199–200
  types of, 198–200
Valve core
  installation, 298
  removal, 290
Valve stems, tire, 279
  service, 293
Vane power steering pump, 370, *371f*
Vapors, 43
Variable power steering, 404
Variable-rate coil spring, 323
Variably priced items, 20
V belts, 372
  inspection, 397–398
  tension, 401
Vehicle cleanliness, ensuring, 17
Vehicle frames, 6–7
Vehicle identification number (VIN), 10, *11f*
Vehicle lifts, 31–36
  above-ground lifts, 31, 36
  frame-contact lifts, 31–33
  in-ground lifts. *See* In-ground lifts
  locking device, 38
  safety considerations, 37–38
  surface mount lifts, 36
  types of, 31–34
  wheel-contact lifts. *See* Wheel-contact lifts
Vehicle support stands, 31
Velocity, 482
Vent (compensating) port, 75, 76
  inspection, 96
  plugged, 97
  use of to check for air in system, 97
Vibrations, 481–485
  additional causes of, 485
  beat/boom, 483
  causes of, 484–485
  drive line as cause of, 484
  electronic vibration analyzer (EVA), 483
  first order, 482
  frequencies of, 483–485
  reed tachometer, 483
  road test, 484
  second order, 482
  service bay testing, 484
  testing, 483–485
  test instruments, 483
  types of, 482–483
VIN. *See* Vehicle identification number (VIN)
Vintage chassis, 5–6
Visual wear sensor, 144, *145f*
Voltage induction, 215, *216f*
Voltometer testers, hydraulic brake fluid, 103
V-ribbed belts, 373
  alignment, 398
  replacement, 399–400
  tension, 401

W

Warehouse distributors (WDs), 18
Warning lights
  brake system. *See* Brake warning light
  parking brake, 187, 193–194
Wave seals, 242–243
WDs. *See* Warehouse distributors (WDs)
Wear bars, 262
Wear indicator ball joints, 354
Wear indicators, brake pad, 86, 144
Web, brake shoe, 120
Weight transfer, 64–66
Welding equipment, 51–52
Wet boiling points, 69
Wheel alignment, 406–414
  Ackerman angle. *See* Ackerman angle
  adjusting camber, 424–426
  adjusting caster, 424–426
  adjusting toe, 429–431
  after suspension service, 360
  angles, 406
  camber. *See* Camber
  caster. *See* Caster
  centering steering wheel, 431–433
  eccentric cam adjustment, 425

four-wheel alignment. *See* Four-wheel alignment
handling characteristics, 413–414
inspection prior to, 416
machines, 423
measuring, 421–433
measuring caster, 423–424
measuring steering axis inclination, 426–428
measuring toe, 428–429
measuring turning radius, 433
portable alignment gauge, *423f*
prealignment inspection, 416
prealignment inspection checklist, 421, *421f*
procedures, 422
rack, 421, *422f*
ride height check, 418
rules, 433–434
service, 416–440
setback. *See* Setback
steering axis inclination (SAI). *See* Steering axis inclination (SAI)
suspension looseness, 420
test drive, 420–421
tire wear inspection, 416–418
toe. *See* Toe
toe change, 418–419
toe change check, 432–433
torque steer, 419
tracking, 413
turning radius, 412–413
Wheelbase, 413
Wheelbase difference, 436
Wheel bearing greases, 241
adding, 246

Wheel bearings, 238–239
adding grease to, 246
adjustment, 248–249
disc brake, 248
front-wheel drive. *See* Front-wheel drive bearings
inspection and diagnosis, 245–248
noise, 249–250
races. *See* Races
repacking, 243
repacking disc brake, 248
retainers, 249
seals. *See* Wheel bearing seals
service, 243–257
Wheel bearing seals, 241–243
garter spring, 241, *242f*
lips, 241–243
Wheel-contact lifts, 31, *32f*, 33–34
Wheel cylinder, 121–122
honing, 133–134
inspection, 131
measuring after honing, 134
reassembly of, 135
rebuilding, 133
removing, 135
replacement of, 131
Wheel cylinder housing, 122–123
Wheel-free jacks, 34
Wheel head, leveling, 436–437
Wheel offset, 278
Wheel rate, 322–323
Wheels, 277–279. *See also* Tires
alignment. *See* Wheel alignment
balancing, 304–311
bearings. *See* Bearings
centering, 277

custom, 277–278
lug nuts. *See* Lug nuts
lug studs. *See* Lug studs
mounting, 312
rims, 277, 290–292
runout, 299–300
Wheel seals and axle, inspection of, 12
Wheel speed sensors, 215–216
air gap, 230
metal shavings, 229
replacement of, 230
service, 228–230
testing, 228–229
tone ring, 230
Wheel tramp, 284, *285f*, 306
Wheel weights, 305
Whine, 484
Wholesale and retail distribution of auto parts, 18–20
Wire wheel, 48
Wishbones, 326
WO. *See* Work order (WO)
Working angle, 487
Work order (WO), 16
Wormshaft, 365
Wrenches
flare nut wrench. *See* Flare nut wrench
impact, 49
torque, 390

Y
Yokes, slip, 460–461, 462

Z
Zerk fitting, 468–469